OMICS

Applications in Biomedical, Agricultural, and Environmental Sciences

OMICS

Applications in Biomedical, Agricultural, and Environmental Sciences

Edited by

Debmalya Barh
Vasudeo Zambare
Vasco Azevedo

CRC Press
Taylor & Francis Group
Boca Raton London New York

CRC Press is an imprint of the
Taylor & Francis Group, an **informa** business

CRC Press
Taylor & Francis Group
6000 Broken Sound Parkway NW, Suite 300
Boca Raton, FL 33487-2742

First issued in paperback 2017

© 2013 by Taylor & Francis Group, LLC
CRC Press is an imprint of Taylor & Francis Group, an Informa business

No claim to original U.S. Government works

Version Date: 20130123

ISBN 13: 978-1-4665-6281-3 (hbk)
ISBN 13: 978-1-138-07475-0 (pbk)

Library of Congress Cataloging-in-Publication Data

Omics : applications in biomedical, agricultural, and environmental sciences / editors, Debmalya
 Barh, Vasudeo Zambare, Vasco Azevedo.
   p. ; cm.
 Includes bibliographical references and index.
 ISBN 978-1-4665-6281-3 (hardcover : alk. paper)
 I. Barh, Debmalya. II. Zambare, Vasudeo. III. Azevedo, Vasco.
 [DNLM: 1. Genomics. 2. Agriculture. 3. Biomedical Research. 4. Ecology. QU 58.5]

572.8'6--dc23   2012039614

Visit the Taylor & Francis Web site at
http://www.taylorandfrancis.com

and the CRC Press Web site at
http://www.crcpress.com

CV 01.11.2019 1625

This book is dedicated to Ms. Mamata Barh (W/O Purnendu Bhusan Barh and M/O Debmalya Barh), Vice President of Institute of Integrative Omics and Applied Biotechnology (IIOAB), India.

Contents

Contents

Foreword

It is a great pleasure to write the foreword for this book, *Omics: Applications in Biomedical, Agricultural, and Environmental Sciences,* edited by Debmalya Barh, Vasudeo Zambare, and Vasco Azevedo. The book provides exciting insights into practical applications of modern "omics" technologies to real-world problems in the field of life sciences. The word *omics* generally refers to globally utilized high-throughput methods in molecular biology and their applications. Although genomics usually refers to sequencing technologies dedicated to deciphering the DNA code of living organisms, transcriptomics, proteomics, and metabolomics provide us deeper insights into the behavior of key biochemical molecules (genes, proteins, and metabolites) in response to changing environmental conditions. Nowadays, we may add the word *omics* to other expressions. Glycomics for instance, if we refer to a range of research-assisting technologies that aid with studying the dynamic effects of perturbations to a certain biological entity, glycans in this example.

One major challenge is to combine the emerging high-throughput omics data in such a way that we may use them (1) for finding so-called biomarkers and (2) for modeling the interactions between genes, proteins, and metabolites. Biomarker research concentrates on classification and feature selection methods that help with distinguishing between different tissues, diseases, bacterial colonizations, etc., for earlier, cheaper, or more accurate diagnosis in clinical practice, to name just a few. In modeling research, we aim to unravel the complex regulatory networks that control the molecular decision process of living cells. This allows for explaining and simulating cellular responses to altering internal and external conditions.

Nowadays, we utilize computational methods to draw efficient and accurate conclusions from wet laboratory data. So-called bioinformatics approaches help us understand how cells survive, reproduce, grow, differentiate, and fine-tune their behavior in accordance with changing surroundings. To form a comprehensive picture, diverse integrated omics data of different types must be analyzed appropriately; a difficult and resource-intensive task.

A diverse selection of articles discussing various applications and open challenges that arise from modern, state-of-the-art omics-driven research has been presented in this book. I certainly believe the described topics to be interesting, informative, and educating to readers from different fields of this exciting scientific area. The book is structured into three parts with chapters about current applications of omics research: biomedical sciences, agricultural sciences, and environmental sciences.

The book starts with an introduction to proteomics research with a special focus on multidimensional protein identification, presented by Dr. Elangovan. Then, in the chapters from Dr. Munshi and Dr. Kaur, we will learn how epigenetics methods can be utilized for studying genomewide methylation changes. The authors describe how these modifications alter biological programs, affect tissue development, and how they can be targeted therapeutically. The next two chapters are dedicated to pharmacogenomics. Dr. Yiannakopoulou and Dr. Overby describe how we may utilize omics technology in personalized medicine for medical decision-making and therapy optimization. Later, Dr. Arulselvi explains how proteomics, metabolomics, and genomics techniques can be coupled with classical toxicology methods to identify environmentally toxic substances and to study their effect on humans. The next chapter by Dr. Thomas introduces noncoding RNAs and their effect on human diseases followed by an article from Dr. Rao about omics techniques for the analysis if RNA splicing events. The subsequent study, presented by Dr. Chaitankar, describes theoretic information approaches for inferring gene regulatory interactions from omics data sets. With the section of Dr. Suresh, we will enjoy reading an introduction to the field of glycomics, mainly structure–function relationships of complex glycans for drug (re)design. Afterward, we will learn from Dr. Gaur and Dr. Kumar about complementing histology and imaging data with genomics and proteomics data for biomarker identification of breast cancer.

Now we are leaving the field of biomedicine to focus on applications in plant and agricultural sciences. The first publication in this section is an integrative study from Dr. Rahman covering the three traditional omics techniques (genomics, proteomics, and metabolomics) for studying crop plant development in a globally changing climate. This work is followed by an application report from Dr. Mondal about utilizing so-called next-generation sequencing technology for investigating abiotic stress response in plants. Then, Dr. Silva teaches us about computational tools for genome assembly of next-generation sequencing data of bacterial genomes, a tricky problem in agricultural genome sequencing as well. This work is followed by an article about the engineering of dark-operative chlorophyll synthesis in transgenic plastids provided by Dr. Khan. Dr. Rivera-Dominguez studies omics methods for various aspects in transgenic plant research and development. In the next section, Dr. Hemaiswarya explains how we can study and manipulate cellular pathways of microalgae for increased biosubstance productions. In the next section, Dr. Ahmad exemplifies how using the metabolic profiles of plants may aid with biomarker identification and validation. Dr. Vadivel's article follows with an introduction to nutrigenomics and a description of ways for analyzing the gene–nutrient interactions to elucidate the health-beneficial role of various bioactive molecules with reference to cancer prevention. With the last chapter in this section, Dr. Sarika provides an overview about genomic resources of insects and pests with importance in agriculture.

Now we enter the world of environmental research with the first chapter from Dr. Cozzolino about combining omics and spectroscopy techniques. Subsequently, Dr. Vakhlu provides us with an overview on metagenomics, the science of biodiversity and its utility for detecting novel enzymes, drugs, and antibiotics. Afterward, Dr. Crotti and Dr. Espinoza demonstrate how transcriptomics profiling aids in mechanistic and predictive toxicology. The following chapter by Dr. Patade describes omics approaches in biofuel production. In the last section of this book, Dr. Khurana reviews applications of omics technology in the environmental sciences, a field he calls "environomics".

In conclusion, this book covers a wide range of exciting omics application areas and highlights working solutions as well as open problems and future challenges. Thus, I sincerely believe that you will enjoy reading it.

Jan Baumbach, PhD
Formerly Head of the Computational Systems Biology Group
Max Planck Institute for Informatics
Cluster of Excellence on Multimodal Computing and Interaction
Saarland University, Saarbrücken, Germany

Head of the Computational Biology Group
University of Southern Denmark
Odense, Denmark

Preface

The Sanskrit word *om* describes completeness, and the term *omics* probably has derived from the Sanskrit *om*. Therefore, omics implies fullness. Currently, the suffix -omics is used to describe various biological fields such as genomics, proteomics, metabolomics, etc., to describe studies involving the sum total of genes, proteins, and metabolites, respectively, within an organism or a cell. With the advent of new technologies and acquired knowledge, the number of fields in omics and their applications in diverse fields is rapidly increasing in the postgenomics era. Such emerging fields, for example, pharmacogenomics, epigenomics, regulomics, splisomics, metagenomics, environomics, are budding methods to combat the global challenges associated with biomedical, agricultural, and environmental issues. So far, various "omics" have been introduced to describe related fields, and we propose the term *unknown-omics/unknownomics* to illustrate those omics fields that are not yet introduced, and we believe that under this terminology, many "omics" will be incorporated in the near future.

It is evident from the current scenario that there are various omics and also application diversities. However, so far, available books have only focused on specific omics or applications of various omics to a particular area. Therefore, there is a big gap in terms of getting an overview of various omics and their applications in various fields in biology. Keeping in mind this fact, in this book, we have taken the initiative to cover most of the conventional and recently introduced omics and their various applications in the most important fields in biology, i.e., biomedical, agricultural, environmental aspects to give an overview of real diversity of omics and their applications to our readers.

In this book, we have included 26 chapters. Each chapter describes one particular omics and their associated technologies and applications and is written by highly experienced active researchers in the corresponding field. Three parts cover these 26 chapters. Part I, The World of Omics and Applications in Biomedical Sciences, includes Proteomics: Techniques, Applications, and Challenges (Chapter 1); Epigenome: Genomic Response to Environmental Eccentricities (Chapter 2); Epigenomics: Basics and Applications (Chapter 3); Pharmacogenomics for Individualized Therapy (Chapter 4); Pharmacogenomic Knowledge to Support Personalized Medicine: The Current State (Chapter 5); Toxicogenomics: Techniques, Applications, and Challenges (Chapter 6); The World of Noncoding RNAs (Chapter 7); Spliceomics: The OMICS of RNA Splicing (Chapter 8); Computational Regulomics: Information Theoretic Approaches toward Regulatory Network Inference (Chapter 9); and Glycomics: Techniques, Applications, and Challenges (Chapter 10). Breast cancer is the one of the leading causes of death in women worldwide, and the identification of early diagnostic and prognostic molecular markers are highly desirable for designing effective treatment strategies. Keeping this in mind, in this book, in this part, we have introduced two chapters—Breast Cancer Biomarkers (Chapter 11) and Use of Protein Biomarkers for the Early Detection of Breast Cancer (Chapter 12)—to describe various markers in breast cancer, their uses, and how proteomics can be used to develop such markers.

Part II, The World of Omics and Applications in Plant and Agricultural Sciences, discusses the various cutting-edge omics technologies and their applications in plant and agriculture. This section includes Genomics, Proteomics and Metabolomics: Tools for Crop Improvement under Changing Climatic Scenarios (Chapter 13); Application of

Next-Generation Sequencing for Abiotic Stress Tolerance (Chapter 14); Next-Generation Sequencing and Assembly of Bacterial Genomes (Chapter 15); Towards Engineering Dark-Operative Chlorophyll Synthesis Pathways in Transgenic Plastids (Chapter 16); Utilization of Omics Technology to Analyze Transgenic Plants (Chapter 17); Microalgal Omics and Their Applications (Chapter 18); Plant Metabolomics: Techniques, Applications, Trends, and Challenges (Chapter 19); Nutraceuticals: Applications in Biomedicine (Chapter 20); and Genomic Resources of Agriculturally Important Animals, Insects, and Pests (Chapter 21).

Part II, The World of Omics and Applications in Environmental Sciences, discusses five topics under this budding area to give an overview on how omics approaches are utilized in environmental sciences. Chapters included are Applications of Molecular Spectroscopy in Environmental and Agricultural Omics (Chapter 22); Metagenomics: Techniques, Applications, and Challenges (Chapter 23); Toxicogenomics in the Assessment of Environmental Pollutants (Chapter 24); and Omics Approaches in Biofuel Production for a Green Environment (Chapter 25). The last chapter (Chapter 26), Environomics: Omics for the Environment, describes an overview of omics strategies to combat environmental issues and maintain environmental sustainability.

We believe that our readers will have a comprehensive outlook about various omics introduced so far in the biology field through this book, which will keep them up-to-date with cutting-edge knowledge and the recent trends in omics.

Debmalya Barh
Vasudeo Zambare
Vasco Azevedo

MATLAB® is a registered trademark of The MathWorks, Inc. For product information, please contact:

The MathWorks, Inc.
3 Apple Hill Drive
Natick, MA 01760-2098 USA
Tel: 508 647 7000
Fax: 508-647-7001
E-mail: info@mathworks.com
Web: www.mathworks.com

Editors

Debmalya Barh, MSC, MTech, MPhil, PhD, PGDM, is the founder and president of the Institute of Integrative Omics and Applied Biotechnology (IIOAB, India), a global platform for multidisciplinary research and advocacy. He is a consultant biotechnologist and has more than 12 years of rich experience in interdisciplinary omics fields. He works toward integrative omics–based translational researches, target and targeted drug discovery, early diagnostic biomarkers, pharmacogenomics, neutrigenomics, cancer, neuroscience, CVDs, infectious, and metabolic diseases, and plant, animal, and environmental biotechnology. As an active scientist in the field, he works with nearly 400 researchers from more than 80 institutes across 25 countries. He is associated with many top-ranked international publishers and has edited a number of research reference books in the areas of omics, biomarkers, and other advanced domains in biology. He also serves as an editorial and review board member for several professional international research journals of global repute.

Vasudeo Zambare, MSc, PhD, FIIOAB, is a biochemist who started his research career at Agharkar Research Institute, Pune, India, and worked in the fields of toxicogenomics, proteomics, metabolomics, and microbial mutagenomics for strain improvement of various extracellular enzymes and performed their commercial trial. He then moved to the Center for Bioprocessing Research and Development at South Dakota School of Mines and Technology (Rapid City, SD) and significantly contributed in the areas of omics approaches in lignocellulosic biomaterial for bioprocessing and enhanced production of biofuel. His current research concentrates on the use of omics technologies in agriculture, functional food, and environmental sciences. He has published more than 150 peer-reviewed articles including research, reviews, book chapters, etc., and at present, is honored to be Editor-in-Chief, technical editor, associate editor, editorial board member, and advisory member of more than 80 international research journals. He is a fellow member of several international societies including the Institute of Integrative Omics and Applied Biotechnology.

Vasco Azevedo, DVM, MSc, PhD, FESC, is a graduate from veterinary school at the Federal University of Bahia, Brazil in 1986 and obtained his masters (1989) and PhD (1993) degrees in molecular genetics, respectively, from Institut National Agronomique Paris-Grignon and Institut National de la Recherche Agronomique, France. Since 1995, he has been a full professor in Molecular Biology and Genetics at the Federal University of Minas Gerais, Brazil. Prof. Vasco is a pioneer scientist in lactic acid bacteria and *Corynebacterium pseudotuberculosis,* and is a well-known scientist in bacterial genomics, transcriptomics, proteomics, metabolomics, etc.

Apart from these areas, his researches also involve various omics approaches in livestock and agricultural biotechnology in the development of omics-based new vaccines and diagnostics against infectious diseases. He has published 160 research articles and 15 book chapters. He is an associate editor of *Genetics and Molecular Research, Microbial Cell Factories, The IIOAB Journal*, etc., and is an editorial board member of a number of international journals.

Contributors

Shoaib Ahmad, MPharm, PhD
Department of Pharmacognosy
Rayat and Bahra Institute of Pharmacy
Sahauran, India

Zakwan Ahmed, PhD
Biotechnology Division
Defence Institute of Bio-Energy Research
Goraparao, Arjunpur, Haldwani
Uttarakhand, India

Yog Raj Ahuja, PhD
Department of Genetics and Molecular
  Medicine
Vasavi Medical and Research Center
Khairathabad, Hyderabad, India

Sintia Almeida, MSc, (PhD)
Universidade Federal de Minas Gerais
Belo Horizonte-Minas Gerais, Brazil

Chinnathambi Anbazhagan, PhD
Department of Botany
Annamalai University
Chidambaram, Tamil Nadu, India

Mohommad Arif, PhD
Biotechnology Division
Defence Institute of Bio-Energy Research
Goraparao, Arjunpur, Haldwani
Uttarakhand, India

Vasco Azevedo, DVM, PhD, FESC
Instituto de Ciências Biológicas
Universida de Federal de Minas Gerais
Belo Horizonte-MG, Brazil

Dakshanamurthy Balasubramanyam, PhD
Livestock Research Station
Kattupakkam, Tamil Nadu, India

Eudes Barbosa, MSc, (PhD)
Universidade Federal de Minas Gerais
Belo Horizonte-Minas Gerais, Brazil

Debmalya Barh, PhD
Centre for Genomics and Applied Gene
  Technology
Institute of Integrative Omics and Applied
  Biotechnology
Nonakuri, Purba Medinipur
West Bengal, India

Jan Baumach, PhD
Max Planck Institute for Informatics
Cluster of Excellence on Multimodal
  Computing and Interaction
Saarland University
Saarbrücken, Germany

Adriana Carneiro, MSc, (PhD)
Universidade Federal do Pará
Belém-Pará, Brazil

Isabel S. Carvalho, PhD
IBB/CGB, Food Science Laboratory
University of Algarve
Campus de Gambelas, Faro, Portugal

Vijender Chaitankar, PhD
Department of Computer Science
Virginia Commonwealth University
Richmond, Virginia

Poonam Chilana, PhD
Centre for Agricultural Bioinformatics
Indian Agricultural Statistics Research
  Institute
New Delhi, India

Daniel Cozzolino, PhD
School of Agriculture, Food and Wine
The University of Adelaide
Glen Osmond, South Australia, Australia

Luciana B. Crotti, PhD
Department of Biochemistry
Uniformed Services University of the
    Health Sciences
Bethesda, Maryland

Nancy Daniel, MSc, MPhil
Department of Biotechnology
Periyar University
Salem, India

Vinicius De Abreu, MSc, (PhD)
Universidade Federal de Minas Gerais
Belo Horizonte-Minas Gerais, Brazil

Luis A. Espinoza, PhD
Department of Biochemistry and
    Molecular & Cell Biology
Georgetown University
Washington, District of Columbia

Rajneesh K. Gaur, PhD
Department of Biotechnology
Ministry of Science and Technology
New Delhi, India

Preetam Ghosh, PhD
Department of Computer Science
Virginia Commonwealth University
Richmond, Virginia

Atul Grover, PhD
Biotechnology Division
Defence Institute of Bio-Energy Research
Goraparao, Arjunpur, Haldwani
Uttarakhand, India

Luis Guimarães, MSc, (PhD)
Universidade Federal de Minas Gerais
Belo Horizonte-Minas Gerais, Brazil

Puja Gupta, MSc, MPhil
School of Biotechnology
University of Jammu
Jammu, India

Sanjay Mohan Gupta, PhD
Biotechnology Division
Defence Institute of Bio-Energy Research
Goraparao, Arjunpur, Haldwani
Uttarakhand, India

Houda Hachad, PharmD, M.Res
Department of Pharmaceutics
University of Washington
Seattle, Washington

Qurratulain Hasan, PhD
Department of Genetics and Molecular
    Medicine
Vasavi Medical and Research Center
Kamineni Hospital
L. B. Nagar, Hyderabad, India

Shanmugam Hemaiswarya, MPhil, PhD
IBB/CGB, Food Science Laboratory
University of Algarve
Campus de Gambelas, Faro, Portugal

Mir Asif Iquebal, PhD
Division of Biometrics and Statistical
    Modelling
Indian Agricultural Statistics Research
    Institute
New Delhi, India

Jaspreet Kaur, PhD
Department of Biotechnology
University Institute of Engineering and
    Technology
Punjab University
Chandigarh, India

Jyotdeep Kaur, PhD
Department of Biochemistry
Post Graduate Institute of Medical
    Education and Research
Chandigarh, India

Muhammad Sarwar Khan, PhD
Centre for Agricultural Biochemistry and
    Biotechnology
University of Agriculture
Faisalabad, Pakistan

Satyendra Mohan Paul Khurana, PhD
Amity Institute of Biotechnology
Amity University-Haryana
Gurgaon (Manesar), India

Srirama Krupanidhi, PhD
Department of Biosciences
Sri Satya Sai University
Prasanthi Nilayam, Andhra Pradesh, India

Annamalai Yogesh Kumar, ME
Centre for Medical Electronics
Anna University
Chennai, India

Birendra Kumar, PhD
Department of Zoology
DAV College, Muzaffarnagar
CCS University
Meerut, India
and
Department of Pharmacology
All India Institute of Medical Sciences
New Delhi, India

Maya Kumari, PhD
Biotechnology Division
Defence Institute of Bio-Energy Research
Goraparao, Arjunpur, Haldwani
Uttarakhand, India

Anderson Miyoshi, PhD
Universidade Federal de Minas Gerais
Belo Horizonte-Minas Gerais, Brazil

Tapan Kumar Mondal, PhD
National Research Centre on DNA
    Fingerprinting
National Bureau of Plant Genetic Resource
New Delhi, India

Anjana Munshi, PhD
Institute of Genetics and Hospital for
    Genetic Diseases
Osmania University
Begumpet, Hyderabad, India

Raveendran Muthurajan, PhD
Genomics and Proteomics Laboratory
Department of Plant Biotechnology
Centre for Plant Molecular Biology and
    Biotechnology
Tamil Nadu Agricultural University
Coimbatore, Tamil Nadu, India

Elangovan Namasivayam, PhD
Department of Biotechnology
Periyar University
Salem, Tamil Nadu, India

Casey Lynnette Overby, PhD
Department of Medical Education and
    Biomedical Informatics
University of Washington
Seattle, Washington

Indra Arulselvi Padikasan, PhD
Department of Biotechnology
Periyar University
Salem, India

Anne Pinto, MSc, (PhD)
Universidade Federal de Minas Gerais
Belo Horizonte-Minas Gerais, Brazil

Hifzur Rahman, MSc, (PhD)
Genomics and Proteomics Laboratory
Department of Plant Biotechnology
Centre for Plant Molecular Biology and
    Biotechnology
Tamil Nadu Agricultural University
Coimbatore, Tamil Nadu, India

Anil Rai, PhD
Centre for Agricultural Bioinformatics
Indian Agricultural Statistics Research
    Institute
New Delhi, India

Rathinam Raja, PhD
R&D, Aquatic Energy LLC
Lake Charles, Louisiana

and

IBB/CGB, Food Science Laboratory
University of Algarve
Campus de Gambelas
Faro, Portugal

Rommel Ramos, PhD
Universidade Federal do Pará
Belém-Pará, Brazil

A. R. Rao, PhD
Centre for Agricultural Bioinformatics
Indian Agricultural Statistics Research
  Institute
New Delhi, India

Ramanujam Ravikumar, PhD, MBA
Aquatic Energy LLC
Lake Charles, Louisiana

Marisela Rivera-Domínguez, PhD
Coordinación de Ciencias de los Alimentos
Centro de Investigación en Alimentación y
  Desarrollo
Hermosillo, Sonora, Mexico

Tanmaya Kumar Sahu, MSc
Centre for Agricultural Bioinformatics
Indian Agricultural Statistics Research
  Institute
New Delhi, India

Anderson Santos, PhD
Universidade Federal de Minas Gerais
Belo Horizonte-Minas Gerais, Brazil

Sarika, PhD
Centre for Agricultural Bioinformatics
Indian Agricultural Statistics Research
  Institute
New Delhi, India

Paula Schneider, PhD
Universidade Federal do Pará
Belém-Pará, Brazil

Anu Sharma, MCA
Centre for Agricultural Bioinformatics
Indian Agricultural Statistics Research
  Institute
New Delhi, India

Artur Silva, PhD
Universidade Federal do Pará
Belém-Pará, Brazil

Nishtha Singh, MSc
Centre for Agricultural Bioinformatics
Indian Agricultural Statistics Research
  Institute
New Delhi, India

Surender Singh, PhD
Department of Pharmacology
All India Institute of Medical Sciences
New Delhi, India

Siomar Soares, MSc, (PhD)
Universidade Federal de Minas Gerais
Belo Horizonte-Minas Gerais, Brazil

D. V. N. Sudheer Pamidimarri, PhD
Department of Chemical and Biochemical
  Engineering
Dongguk University
Seoul, South Korea

Veeraperumal Suresh, PhD
Department of Botany
Annamalai University
Chidambaram, Tamil Nadu, India

Keita Sutoh, PhD
Life Science Institute Co. Ltd.
Tokyo, Japan

Winnie Thomas, MPhil
Department of Genetics and Molecular
    Medicine
Vasavi Medical and Research Center
Khairathabad, Hyderabad, India

Martín-Ernesto Tiznado-Hernández, PhD
Coordinación de Tecnología de Alimentos
    de Origen Vegetal
Centro de Investigación en Alimentación y
    Desarrollo
Hermosillo, Sonora, Mexico

Vellingiri Vadivel, PhD
Institute for Biological Chemistry and
    Nutrition
University of Hohenheim
Stuttgart, Germany

Konagurtu Vaidyanath, PhD
Department of Biotechnology
Post-Graduate College of Science
Osmania University
Saifabad, Hyderabad, India

Jyoti Vakhlu, PhD
School of Biotechnology
University of Jammu
Jammu, India

Ramanathan Valarmathi, PhD
Department of Plant Biotechnology
Tamil Nadu Agricultural University
Coimbatore, Tamil Nadu, India

Dinesh K. Yadav, PhD
Amity Institute of Biotechnology
Amity University-Haryana
Gurgaon (Manesar), India

Neelam Yadav, PhD
Amity Institute of Biotechnology
Amity University-Haryana
Gurgaon (Manesar), India

Patade Vikas Yadav, PhD
Biotechnology Division
Defence Institute of Bio-Energy Research
Goraparao, Arjunpur, Haldwani
Uttarakhand, India

Purnmasi Ram Yadav, PhD
Department of Zoology
DAV College, Muzaffarnagar
CCS University
Meerut, India

Eugenia Ch. Yiannakopoulou, MD, PhD
Department of Basic Medical Lessons
Technological Educational Institute of
    Athens
Athens, Greece

Vasudeo Zambare, PhD
Centre for Genomics and Applied Gene
    Technology
Institute of Integrative Omics and Applied
    Biotechnology
Nonakuri, Purba Medinipur
West Bengal, India

Part I

The World of Omics and Applications in Biomedical Sciences

1

Proteomics: Techniques, Applications, and Challenges

Elangovan Namasivayam

CONTENTS

1.1 Introduction

The term proteome refers to the totality of the proteins in a cell, tissue, or organism. Proteomics is the study of these proteins—their identity, their biochemical properties and functional roles, and how their quantities, modifications, and structures change during development and in response to internal and external stimuli. The field of proteomics has been propelled by advances in mass spectrometry (MS) and other techniques that have made it possible to analyze proteins in large numbers of biological samples rapidly and at low cost.

In 2001, the Human Proteome Organization postulated the aims of proteomics in a more precise manner as the identification of all proteins coded by human genomes (and other genomes, especially those of model organisms) with the subsequent assessment of (a) their (protein) expression in different cell types of the organism (expression proteomics), (b) protein distribution in subcellular compartments of the organelles, (c) posttranslational modifications of the proteins, (d) protein–protein interactions (b–d constitute structural proteomics), and (e) the relation between protein structure and function (functional proteomics). The proteome also changes in response to cancer and other diseases, making the proteome of great interest to medical researchers. For example, cancer cells often secrete specific proteins or fragments of proteins into the bloodstream and other body fluids such as urine and saliva. Researchers hope to discover groups or patterns of proteins—called protein signatures—in these easily accessible fluids that would provide information about the risk, presence, and progression of disease. This knowledge could ultimately help doctors better detect cancer before the symptoms manifest and customize treatment to the individual patient.

The proteome of an organism is much larger and more complex than its genome. Many genes have the potential to produce more than one version of the protein they encode. In addition, proteins are frequently modified by cells after they are made. The protein composition of an organism or a tissue changes constantly as new proteins are made, existing proteins are eliminated and proteins become modified/demodified in response to internal and external stimuli. In contrast, a person's genome remains relatively unchanged over the course of his or her lifetime.

1.1.1 Branches of Proteomics

1.1.1.1 Protein Separation

All proteomic technologies rely on the ability to separate a complex mixture so that individual proteins are more easily processed with other techniques (Figure 1.1).

1.1.1.2 Protein Identification

Well-known methods include low-throughput sequencing through Edman degradation. High-throughput proteomic techniques are based on MS, usually peptide mass fingerprinting on simpler instruments or *de novo* repeat detection sequencing on instruments capable of more than one round of MS. Antibody-based assays can also be used, but are unique to one sequence motif.

1.1.1.3 Protein Quantification

Gel-based methods are used, including differential staining of gels with fluorescent dyes (difference gel electrophoresis). Gel-free methods include various tagging or chemical modification methods, such as isotope-coded affinity tags (ICATs), metal-coded affinity tags, or combined fractional diagonal chromatography.

1.1.1.4 Protein Sequence Analysis

This is one of the more bioinformatic branches, dedicated to searching databases for possible protein or peptide matches as well as the functional assignment of domains, prediction of function from sequence, and evolutionary relationships of proteins.

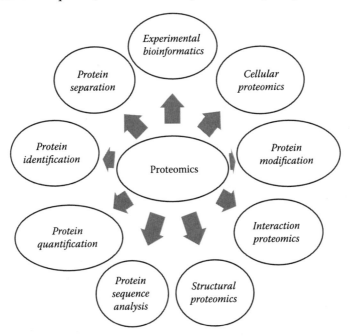

FIGURE 1.1
Branches of proteomics.

1.1.1.5 Structural Proteomics

This concerns the high-throughput determination of protein structures in three-dimensional space. Common methods include x-ray crystallography and nuclear magnetic resonance spectroscopy.

1.1.1.6 Interaction Proteomics

This concerns the investigation of protein interactions on the atomic, molecular, and cellular levels (protein–protein interaction).

1.1.1.7 Protein Modification

Almost all proteins are modified from their pure translated amino acid sequence, so-called posttranslational modification. Specialized methods have been developed to study phosphorylation (phosphoproteomics) and glycosylation (glycoproteomics).

1.1.1.8 Cellular Proteomics

This is a new branch of proteomics whose goal is to map the location of proteins and protein–protein interactions in whole cells during key cell events. It focuses on the use of techniques such as x-ray tomography and optical fluorescence microscopy.

1.1.1.9 Experimental Bioinformatics

A branch of bioinformatics, as it is applied in proteomics, coined by Mathias Mann. It involves the mutual design of experimental and bioinformatics methods to create (extract) new types of information from proteomics experiments.

1.1.2 Technologies Used in Proteomics

One- and two-dimensional gel electrophoresis are used to identify the relative mass of a protein and its isoelectric point. X-ray crystallography and nuclear magnetic resonance are used to characterize the three-dimensional structures of peptides and proteins. However, low-resolution techniques such as circular dichroism, Fourier transform infrared spectroscopy, and small angle x-ray scattering can be used to study the secondary structure of proteins. Tandem MS combined with reversed phase chromatography or 2D electrophoresis (2DE) is used to identify (by *de novo* peptide sequencing) and quantify all the levels of proteins found in cells.

Matrix-assisted laser desorption ionization (MALDI) and time of-flight (TOF) mass spectrometry (nontandem), is often used to identify proteins by peptide mass fingerprinting. Less commonly, this approach is used with chromatography or high-resolution MS (or both). This technique is becoming less frequently used and the scientific world no longer accepts absolute identification of a protein based solely on peptide mass fingerprint data. Affinity chromatography, yeast two-hybrid techniques, fluorescence resonance energy transfer, and surface plasmon resonance are used to identify protein–protein and protein–DNA binding reactions.

X-ray tomography is used to determine the location of labeled proteins or protein complexes in an intact cell. Software-based image analysis is utilized to automate the quantification and detection of spots within and among gel samples. Although this technology is widely used, the intelligence has not yet been perfected. For example, the leading software tools in this area tend to agree on the analysis of well-defined well-separated protein spots, but they deliver different results and tendencies with less-defined, less-separated spots—thus necessitating manual verification of results [1,2].

1.2 Protein Digestion Techniques

1.2.1 Introduction

Modern MS instruments are capable of measuring the molecular weights of intact proteins with a fairly high degree of accuracy. So why not perform proteomics simply by measuring the masses of intact proteins? [3]. MS instruments are now well-suited to the analysis of peptides. Moreover, the data obtained from MS analysis of peptides can be taken directly for comparison to protein sequences derived from protein and nucleotide sequence databases. A key element of the search algorithms that assign protein identity from comparisons of peptide MS data to database information is the knowledge that certain proteolytic enzymes cleave the proteins to peptides at specific sites.

1.2.2 Proteases

Protease refers to a group of enzymes whose catalytic function is to hydrolyze (break down) the peptide bonds of proteins. They are also called proteolytic enzymes or proteinases. Proteases differ in their ability to hydrolyze various peptide bonds. Each type of protease has a specific kind of peptide bonds it breaks. Examples of proteases include fungal protease, pepsin, trypsin, chymotrypsin, papain, bromelain, and subtilisin. The activity of proteases is inhibited by protease inhibitors. One example of protease inhibitors is the serpin superfamily, which includes α1-antitrypsin, C1 inhibitor, antithrombin, α1-antichymotrypsin, plasminogen activator inhibitor-1, and neuroserpin.

1.2.2.1 Trypsin

Trypsin and chymotrypsin, like most proteolytic enzymes, are synthesized as inactive zymogen precursors (trypsinogen and chymotrypsinogen) to prevent unwanted destruction of cellular proteins, and to regulate when and where enzyme activity occurs Trypsin is by far the most widely used protease in proteomic analysis. Trypsin cleaves proteins at lysine and arginine residues, unless either of these is followed by a proline residue in the C-terminal direction. The spacing of lysine and arginine residues in many proteins is such that many of the resulting peptides are of a length well-suited to MS analysis. This "dual-specificity" means that trypsin will cut proteins more frequently than will a protease that cuts at only one amino acid residue. As a general rule, a 50 kDa protein will yield approximately 30 tryptic peptides. An advantage of trypsin for proteomics work is that the enzyme displays good activity both in solution and in "in-gel" digestion.

1.2.2.2 Glu-C (V8 Protease)

Staphylococcus aureus V8 protease, also known as endoproteinase Glu-C, is specific for cleavage at the carboxy-terminus of glutamic acid and aspartic acid residues. There are two major problems with any proteolytic digestion. The first problem is that the reaction must be quenched to stop the digestion, thus diluting the sample. The second problem is that the sample is contaminated with the protease and possibly autodigested protease fragments. These problems are surmounted with immobilized V8 protease. With immobilized proteases, no quenching is necessary and there are no problems with protease contamination of the sample or autodigestion of the protease. Simply separate the sample solution from the agarose matrix and these problems are eliminated.

1.2.2.3 Other Proteases

Several other enzymes are used for proteomic analysis. These include Lys-C, chymotrypsin, Asp-N, and several "nonspecific" proteases such as subtilysin, pepsin, proteinase K, and pronase. The cleavage specificity of these enzymes generally is not ideal for most proteomic analyses. Those enzymes that cleave at only one amino acid residue tend to yield fewer, larger fragments that do not provide useful sequence information in tandem MS analyses.

1.2.2.4 Cyanogen Bromide

Proteins can also be cleaved with some chemicals like cyanogen bromide. Cyanogen bromide is a pseudohalogen compound with the formula CNBr. CNBr cleaves proteins at methionine residues. Working with cyanogen bromide is very difficult because is volatile, and readily absorbed through the skin or gastrointestinal tract. Therefore, toxic exposure may occur by inhalation, physical contact, or ingestion. It is acutely toxic, causing a variety of nonspecific symptoms.

1.2.3 In-Gel Digestions

A commonly used approach to digestion of proteins separated by one-dimensional or two-dimensional sodium dodecyl sulfate (SDS)–polyacrylamide gel electrophoresis (PAGE) is

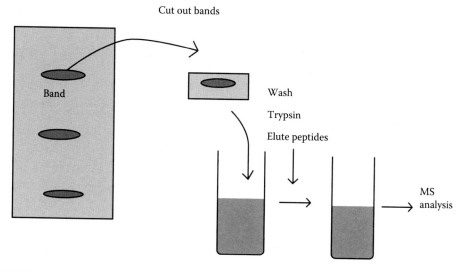

FIGURE 1.2
Schematic representation of in-gel digestion.

referred to as "in-gel" digestion (Figure 1.2). The band or spot of interest is cut from the gel, destained, and then treated with a protease (most commonly trypsin). The enzyme penetrates the gel matrix and digests the protein into peptides, which then are eluted from the gel by washing and then analyzed by MS [4].

1.3 Protein Separation and Processing

Proteins are very diverse. They differ by size, shape, charge, hydrophobicity, and their affinity for other molecules. Several approaches are being employed for the separation of complex mixtures. To study proteins, it is crucial to obtain a sample which contains only the interested molecule. All these properties can be exploited to separate them from one another so that they can be studied individually. Popular protein separation methods include one-dimensional electrophoresis, 2D gel electrophoresis, and various chromatography methods (ion exchange chromatography, gel-filtration chromatography, affinity chromatography, high-pressure liquid chromatography); among these, 2DE is a prominent method.

1.3.1 Chromatography Methods

The most powerful methods for fractionating proteins make use of column chromatography, which takes advantage of differences in protein charge, size, binding affinity, and other properties. A porous solid material with appropriate chemical properties (the stationary phase) is held in a column and a buffered solution (the mobile phase) percolates through it. The protein-containing solution, layered on the top of the column, percolates through the solid matrix as an ever-expanding band within the larger mobile phase. Individual proteins migrate faster or more slowly through the column depending on their properties.

1.3.1.1 Ion Exchange Chromatography

Ion exchange may be defined as a process concerned with the reversible exchange of ions in solution with ions electrostatically bound to an inert support medium. Ion exchange separations are usually carried out in columns packed with an ion exchanger. An ion exchanger is a solid material that has charged groups to which ions are electrostatically bound. Ion exchangers can exchange these ions for ions in aqueous solutions. It is used in column chromatography to separate molecules according to the charge. There are two types: anions and cations. Protein purification using ion exchange chromatography usually employs positively charged anion exchangers because the majority of proteins are negatively charged at neutral pH.

1.3.1.2 Gel Filtration Chromatography

Gel filtration chromatography is also known as gel permeation chromatography or molecular sieve chromatography. Molecules that are small enough to penetrate into the matrix are delayed and travel more slowly through the column. It is a special type of partition chromatography in which the separation of macromolecules is based only on molecular size. A column of tiny particles is prepared from an inert substance that contains small pores. If a solution containing different molecular weight substances is passed through the column, molecules larger than the pores move only in the space between the particles and hence are not retarded by the column material. However, molecules smaller than the pores diffuse in and out of particles. In this way, they are slowed in their movement down the column. Hence, molecules are eluted from the column in the order of decreasing size or decreasing molecular weight.

1.3.1.3 Affinity Chromatography

This technique is useful for the separation of proteins, polysaccharides, nucleic acids and other classes of naturally occurring compounds. It exploits the functional specificities of biological systems. It can rapidly achieve separations that are time-consuming, difficult, or even impossible using other techniques. A complex mixture of substances in solution can be applied to an affinity chromatography column and purification of the substances of interest can be accomplished in a single step.

1.3.1.4 High-Performance Liquid Chromatography

A modern refinement in chromatographic methods—high-performance liquid chromatography (HPLC) makes use of high-pressure pumps that speeds the movement of the protein molecules down the column, as well as higher-quality chromatographic materials that can withstand the crushing force of the pressurized flow. By reducing the transit time on the column, HPLC can limit diffusional spreading of protein bands and thus greatly improve resolution. In most cases, several different methods must be used sequentially to purify a protein completely, and before the most effective one can be found. Trial and error can often be minimized by basing the procedure on purification techniques developed for similar proteins [5].

1.3.2 One- and Two-Dimensional Gel Electrophoresis

For many proteomics applications, 1DE is the method of choice to resolve protein mixtures. In 1DE, proteins are separated on the basis of molecular mass. Because proteins are solubilized in SDS, protein solubility is rarely a problem. Moreover, 1DE is simple to perform, is reproducible, and can be used to resolve proteins with molecular masses of 10 to 300 kDa. The most common application of 1DE is the characterization of proteins after some form of protein purification. This is because of the limited resolving power of a 1DE. If a more complex protein mixture, such as a crude cell lysate, is encountered then 2DE can be used. In 2DE, proteins are separated by two distinct properties. They are resolved according to their net charge in the first dimension and according to their molecular mass in the second dimension. The combination of these two techniques produces resolution far exceeding that obtained in 1DE.

One of the greatest strengths of 2DE is the ability to resolve proteins that have undergone some form of posttranslational modification. This resolution is possible in 2DE because many types of protein modifications confer a difference in charge as well as a change in mass on the protein. One such example is protein phosphorylation. Frequently, the phosphorylated form of a protein can be resolved from the nonphosphorylated form by 2DE. In this case, a single phosphoprotein will appear as multiple spots on a 2D gel. In addition, 2DE can detect different forms of proteins that arise from alternative mRNA splicing or proteolytic processing.

The primary application of 2DE continues to be protein expression profiling. In this approach, the protein expression of any two samples can be qualitatively and quantitatively compared. The appearance or disappearance of spots can provide information about differential protein expression, whereas the intensity of those spots provides quantitative information about protein expression levels. Protein expression profiling can be used for samples from whole organisms, cell lines, tissues, or bodily fluids. Examples of this technique include the comparison of normal and diseased tissues or of cells treated with various drugs or stimuli. An example of 2DE used in protein profiling and another application of 2DE is in cell map proteomics. 2DE is used to map proteins from microorganisms, cellular organelles, and protein complexes. It can also be used to resolve and characterize proteins in subproteomes that have been created by some form of purification of a proteome. Because a single 2DE gel can resolve thousands of proteins, it remains a powerful tool for the cataloging of proteins. Many 2DE databases have been constructed and are available on the World Wide Web.

A number of improvements have been made in 2DE over the years. One of the biggest improvements was the introduction of immobilized pH gradients, which greatly improved the reproducibility of 2DE. The use of fluorescent dyes has improved the sensitivity of protein detection and specialized pH gradients are able to resolve more proteins. The speed of running 2DE has been improved and 2D gels can now be run in the mini gel format. In addition, there have been efforts to automate 2DE. Hochstrasser's group has automated the process of 2DE from gel-running to image-analysis and spot-picking. The use of computers has aided the analysis of complex 2D gel images. This is a critical aspect of 2DE because a high degree of accuracy is required in spot detection and annotation if artifacts are to be avoided. Recently, a molecular scanner was developed to record 2DE images. Software programs such as Melanie compare computer images of 2D gels and facilitate both the identification and quantification of protein spots between samples.

Despite efforts to automate protein analysis by 2DE, it is still a labor-intensive and time-consuming process. A typical 2DE experiment can take 2 days, and only a single sample can be analyzed per gel. In addition, 2DE is limited by both the number and type of proteins that can be resolved. For example, the protein mixture obtained from a eukaryotic cell lysate is too complex to be completely resolved on a single 2D gel [6].

1.4 Protein–Protein Interactions

Protein–protein interactions are essential to cellular mechanisms at all levels in biologically responsive systems. These interactions occur extracellularly and include ligand–receptor interactions, cell adhesion, antigen recognition and virus–host recognition. Intracellular protein–protein interactions occur in the formation of multiprotein complexes, during the assembly of cytoskeletal elements and between receptor–effector as well as effector–effector, molecules of signal transduction pathways. Finally, assembly of transcriptional machinery involves protein interactions. Several *in vitro* and *in vivo* approaches have been introduced for the purpose of physical identification of protein–protein partners in a protein complex. Most utilized approaches depend on copurification coupled with MS including immunoprecipitation, biotinylation, protein affinity chromatography, yeast two-hybrid system, and phage display.

1.4.1 Immunoprecipitation and Coimmunoprecipitation

If the antibody against a protein or its fusion is available, immunoprecipitation becomes a method of choice for *in vitro* coimmunoprecipitation proteins interacting with the target protein. Despite coimmunoprecipitation being a key in studying protein–protein interactions, if the protein(s) comigrate on the SDS-PAGE gel with antibody subunits, the resolution and separation of coimmunoprecipitation proteins becomes problematic. The cross-linking step of the IgG to protein A or G matrix has reduced this problem significantly, resulting in a coimmunoprecipitation profile without the interference of antibody subunits. However, the success of an immunoprecipitation or coimmunoprecipitation experiment depends on the quality of the antibody (i.e., purity and specificity), and sample preparation. To maintain the integrity of protein–protein interactions, it is always favorable to utilize a sample prepared under nondenaturing conditions and binding conditions should be kept as close as possible to physiological pH. To overcome this problem, there is now a wide selection of commercially available immunoprecipitation kits with sample preparation reagents included [e.g., Sieze IP (Thermo Fisher Scientific Inc. Rockford, IL USA) kits from Pierce].

1.4.2 Biotinylation

Recently, biotinylation of tags has been utilized in the recovery and characterization of ribonucleoprotein complexes. This approach utilizes the cotransfection of the cell with two plasmids. One expresses an RNA-binding protein tagged with a biotin acceptor peptide, a protease recognition site is placed between RNA-binding protein and biotin acceptor peptide, the other plasmid expresses BirA enzyme. The biotinylation of the biotin acceptor peptide tag only occurs in the presence of BirA upon the addition of exogenous biotin. RNA-binding protein complexes with their corresponding RNA can be recovered by incubation with streptavidin-sepharose beads, followed by protease treatment. Although this approach has been used for the recovery and quantification of relevant mRNAs, with a slight twist, it can be used to study protein–protein interactions. The coexpression of BirA with the bait protein tagged with biotin acceptor peptide in cells deficient of the bait gene will cure for this deficiency. When coupled with streptavidin-sepharose beads, interacting proteins can be purified. Utilizing this approach for studying protein–protein interactions seems lucrative; however, the generation of a null mutant of the gene expressing for the bait protein will be a prerequisite and thus DNA manipulation is required for the study of nonessential genes.

1.4.3 Yeast Two-Hybrid System

The yeast two-hybrid system requires DNA manipulation to study protein–protein interactions (Figure 1.3). In the first step of the yeast two-hybrid system, the target protein is cloned into the "bait" vector.

In this way, the gene encoding the bait protein is placed into a plasmid next to the gene encoding a DNA-binding domain from some transcription factor (e.g., LexA), thus generating the DNA-binding domain–bait fusion. If a bait protein has no ability to activate the reporter genes in yeast per se, the yeast cell will not be able to survive on plates lacking histidine and will not turn blue (lacZ–). Separately, a second gene or a library of cDNAs encoding potential interactors, collectively called "prey," is cloned in the frame adjacent to an activation domain of the Gal4 yeast transcription factor.

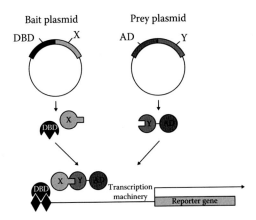

FIGURE 1.3
Principle of the yeast two-hybrid system.

Because a prey protein has no ability to bind DNA-binding domain–responsive elements, the yeast cell will again not be able to survive on plates lacking histidine and will not turn blue (lacZ–). If the two proteins fused to the DNA-binding domain and activation domain (AD) physically interact, they will bring the DNA-binding domain and the AD close enough together to restore a functional transcription factor. The reporter genes, for example, a nutritional selection marker (HIS3) or lacZ, are transcribed resulting in histidine prototrophy and blue coloration of yeast cells. Thus, the interaction of two proteins is measured by the reconstitution of a hybrid transcription factor and the consequent activation of a set of specific reporter genes (Figure 1.4). As mentioned previously, common reporter genes include an auxotrophic gene for growth selection and a secondary reporter gene for color selection. The power of the yeast two-hybrid system lies in this combined approach because growth selection enables the sampling of highly complex cDNA libraries encoding millions of potential binding partners: only those clones which encode an interacting protein survive growth selection and are analyzed further using a convenient color assay (e.g., BlueTech). The complexity of such an analysis is time-consuming and requires expertise, for this reason, the yeast two-hybrid custom services market is expanding.

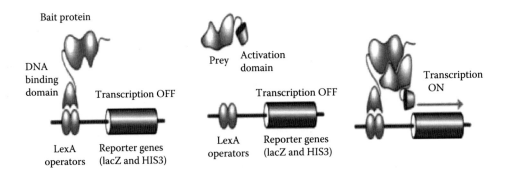

FIGURE 1.4
Yeast two-hybrid system.

1.4.4 Protein Affinity Chromatography

A protein can be covalently coupled to a matrix such as sepharose under controlled conditions and used to select ligand proteins that bind and are retained from an appropriate extract. Most proteins pass through such columns or are readily washed off under low-salt conditions; proteins that are retained can then be eluted by high-salt solutions, cofactors, chaotropic solvents, or SDS. If the extract is labeled *in vivo* before the experiment, there are two distinct advantages: labeled proteins can be detected with high sensitivity, and unlabeled polypeptides derived from the covalently bound protein can be ignored (these might be either proteolytic fragments of the covalently bound protein or subunits of the protein which are not themselves covalently bound). This method was first used 20 years ago to detect phage and host proteins that interacted with different forms of *Escherichia coli* RNA polymerase. Proteins were retained by an RNA polymerase-agarose column (which was shown to be enzymatically active; Figure 1.5).

1.4.5 Phage Display

Phage display technology was first introduced in 1985 by George Smith. It was used as an expression vector, capable of presenting a foreign amino acid sequence accessible to binding an antibody. Since then, a large number of phage displayed peptide and protein libraries have been constructed, leading to various techniques for screening such libraries. Phages are viruses that infect bacterial cells and many of the vectors used in recombinant DNA research are phages that infect the standard recombinant DNA host: the bacterium *E. coli*. The key feature of recombinant DNA vectors, including phages, is that they accommodate segments of "foreign" DNA species of human DNA, for instance, or even stretches of chemically synthesized DNA. As vector DNA replicates in its *E. coli* host, then, the foreign "insert" replicates along with it as a sort of passenger (Figure 1.6).

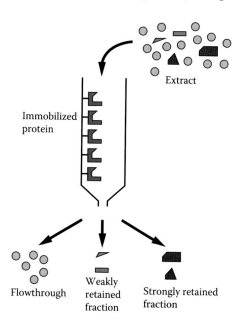

FIGURE 1.5
Protein affinity chromatography.

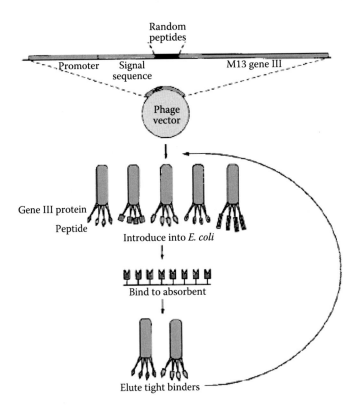

FIGURE 1.6
(See color insert.) A peptide library in a filamentous phage vector.

Phage display allows the presentation of large peptide and protein libraries on the surface of a filamentous phage, which leads to the selection of peptides and proteins, including antibodies, with high affinity and specificity to almost any target. The technology involves the introduction of exogenous peptide sequences into a location in the genome of the phage capsid proteins. The encoded peptides are expressed or "displayed" on the phage surface as a fusion product with one of the phage coat proteins.

This way, instead of having to genetically engineer different proteins or peptides one at a time and then express, purify, and analyze each variant, phage display libraries containing most variants can be constructed simultaneously. Phage particles withstand very harsh conditions, such as low pH and low temperatures, without losing bacterial infectivity. Thus, protocols using low pH and high-concentration urea have been used to dissociate bound phage from a target. In addition, bound phage does not need to be eluted from a microtiter well or animal tissue before bacterial infection. Instead, infection can proceed after the addition of bacteria directly into the well or to the homogenized organ or tissue.

The strength of phage technology is its ability to identify interactive regions of proteins and other molecules without preexisting notions about the nature of the interaction. The past decade has seen considerable progress in the applications of phage display technology. Different screening methods have allowed the isolation and characterization of peptides binding to several molecules *in vitro*, and in the context of living cells, in animals and in humans. It is a rapid procedure and should be widely applicable [7].

1.5 Protein Identification Methods

There are numerous methods available for protein separation. Regardless of the choice of a given proteomic separation technique, gel-based or gel-free, a mass spectrometer is always the primary tool for protein identification. Before the advent of MS, protein sequences were determined by Edman degradation. However, during the last decade, significant improvements have been made in the application of MS for the determination of protein sequences.

1.5.1 Edman Degradation

In the early 1960s, a technique called the protein sequence was presented by Edman and Begg (1967) for the N-terminal sequencing of proteins. This chemical degradation technique allowed the extraction of individual amino acids in a cycle-dependent manner from the N-termini of proteins. The retention time observed by HPLC was then used to identify the individual amino acid sequences. By repeating the chemical degradation cycle, it was possible to obtain the amino acid sequence at the N-terminus of typically up to 20 amino acids. Using this approach, it is possible to painfully *de novo* sequence a protein or to sequence a sufficient length of the protein to be able to clone the gene. The availability of protein and DNA sequence databases has now facilitated the work and allowed protein identification using limited N-terminal amino acid sequencing.

1.5.2 Protein Identification by MALDI-TOF and Accurate Peptide Masses

1.5.2.1 MALDI-TOF MS

The development of the MALDI-TOF technique in the field of MS has allowed rapid and accurate mass measurements of analytes. This technique allows the transfer of peptides from a solid state to the gas phase, whereas the TOF mass spectrometer rapidly separates peptides according to their m/z ratio. Proteins isolated from 2D gel electrophoresis are digested, desalted, and then spotted on a MALDI plate for cocrystallization with a saturated matrix solution. Alternatively, protein digests can be separated by HPLC and eluting peptides deposited on a MALDI plate. The target plate is then introduced into the vacuum chamber of the mass spectrometer.

The identification of proteins by MALDI-TOF MS is generally achieved by measuring the m/z ratio of the peptides predominantly of charge +1. The combination of accurate peptide mass measurement with the availability of protein sequence databases forms the basis of protein identification by MALDI-TOF. This is often called peptide mass fingerprinting. The measured masses present in tryptic digests are tabulated, and the known contaminants are deleted from the list. The reduced mass list is then used to search the protein databases. Different software packages have been developed for the identification of proteins based on the accurate measurement of peptide masses. The simplest method for scoring is to add the number of peptide masses that match with the predicted masses for each entry in a protein database. The database entries are then ranked according to the number of hits. This forms the basis behind software such as Pep Sea and MS-Fit (http://prospector.ucsf.edu). Typically, these software packages work well for quality experimental data.

Generally, all these software packages perform well when good-quality spectra are available. The ones that provide more advanced scoring schemes perform better when less information is available or when the quality of the MS spectra is reduced. Regardless of the software, the identification of the protein depends on the number of peptides observed, the accuracy of the measurement, and the size of the genome of the particular species. For smaller genomes, such as yeast and *E. coli*, protein identification using the MALDI-TOF mass spectrometer is generally successful. For larger genomes, the rate of success drops significantly using MALDI-TOF.

1.5.2.2 Postsource Decay

MALDI-TOF MS also offers the possibility of recording the fragmentation patterns obtained from a peptide. This is achieved using a technique called postsource decay. Postsource decay is achieved by increasing the laser power beyond the value needed to generate ions. The precursor ions are transferred from the MALDI plate to the gas phase. The excess energy induces the precursor ions (peptides) to fragment along their backbone. Generally, these ions are not seen in conventional MALDI-TOF analysis because of their lower kinetic energy. Fortunately, on the MALDI-TOF equipped with a reflectron, the lower kinetic energy of peptide fragments can be compensated for by changing the settings on the reflectron. Postsource decay is typically achieved by acquiring spectra for specific mass ranges with different settings on the reflectron. All the spectra are stitched together to make a full postsource decay spectrum.

1.5.2.3 The Surface-Enhanced Laser Desorption/Ionization–TOF MS Approach

The discovery, identification, and validation of proteins associated with a particular disease state is a difficult and laborious task and often requires hundreds, if not thousands, of samples to be analyzed. Presently, disease biomarker discovery is generally carried out using a 2D-PAGE comparison to identify differences in protein expression, and the patterns between diseased and normal samples. After 2D-PAGE fractionation and staining, the protein(s) of interest are removed, proteolytically or chemically digested, and identified by MS. Although 2D-PAGE separation provides excellent resolution, the need for protein staining and the subsequent sample-handling limits the sensitivity of the overall approach. Surface-enhanced laser desorption/ionization (SELDI)–TOF MS is a novel approach, which was introduced by Hutchens and Yip. The Protein Chip Biology System uses SELDI-TOF MS to retain proteins on a solid-phase chromatographic surface that is later ionized and detected by TOF MS. This versatile instrumentation is presently being used for projects ranging from the identification of disease biomarkers to the study of biomolecular interactions.

The Protein Chip Biology System allows for protein profiling from a variety of complex biological materials such as serum, blood, plasma, intestinal fluid, urine, cell lysates, and cellular secretion products with limited sample preparation and for protein–protein and protein–DNA interactions. The system is most effective at profiling low-molecular weight proteins, providing a complementary visualization technique to 2D-PAGE. SELDI-TOF MS, however, is more sensitive, and requires smaller amounts of sample than 2D-PAGE. Protein profiles have been generated from as few as 25 to 50 cells. The Protein Chip platform has been used to discover biomarkers for diseases such as cancer, neurological disorder, and pathogenic organisms. It has been proven to be useful in the discovery of potential diagnostic markers for prostate, bladder, breast and ovarian cancers, Alzheimer disease, and profiling of urinary proteins to assess renal function.

SELDI-TOF MS to profile low-molecular weight peptides and proteins secreted by different cancer cell lines grown in serum-free medium. The chromatographic surfaces that make up the various Protein Chip Arrays are uniquely designed to retain proteins from a complex sample mixture according to specific properties such as hydrophobicity, charge, etc. The molecular weight of the retained proteins can then be measured by TOF MS. This mini review will discuss the application, advantages and limitations of the SELDI-TOF MS approach to protein profiling and protein biomarker detection and identification [8].

1.5.3 Hybrid Instruments

1.5.3.1 MALDI-Pulsar System

Recently, a set of novel MS/MS-capable mass spectrometers have been developed based on MALDI ionization. These instruments combine the MALDI ionization technique with the fragmentation of ions by collision-induced dissociation [9] [an example of the MALDI-Pulsar from Sciex (AB SCIEX Framingham, MA 01701, USA) was recently introduced]. This instrument, as its name alludes to, includes a MALDI ionization interface followed by a set of quadruples and a collision cell. The first quadruple is used as an ion guide. It is followed by a second quadruple, which is either used as an ion guide, a precursor scan, or for a parent ion selection. The first quadruple is followed by a collision cell that can be used to fragment ions by collision-induced dissociation. A pulsing grid set at a 45° angle is positioned after the collision cell and deflects the ions upward into a TOF mass analyzer equipped with a reflectron.

Although a pulsing laser is still used to generate ions off the MALDI plate, it is not used for the timing and correction of the masses of the observed ions. In this design, the ionization of the sample is decoupled from the acceleration of the ions into the TOF tubes. Furthermore, significant collisional cooling is present in the first quadruple, reducing the distribution of the kinetic energy of the ions. Therefore, the geometry of the plates does not affect the mass accuracy of the instrument. This is a significant difference over conventional MALDI-TOF in which great care must be taken regarding the plate geometry. This means that internal or close external calibrants are not necessary to maintain the mass accuracy of the MALDI-Pulsar system.

1.5.3.2 Electrospray Ionization MS

Electrospray ionization (ESI) is also a very popular approach to introduce protein and peptide mixtures to mass spectrometers. Typically, ESI is used in conjunction with a triple quadrupole, ion trap, or hybrid quadruple TOF mass spectrometer. ESI is mainly popular because it provides a direct interface between the mass spectrometer and the atmospheric pressure. It can also be readily coupled to separation techniques such as HPLC, liquid chromatography, and capillary zone electrophoresis or it can be used for the continuous infusion of samples. Until the introduction of the MALDI-Pulsar, ESI-based mass spectrometers were the only viable approach for the generation of MS/MS spectra. Furthermore, ESI produces more multiply charged ions that provide richer MS/MS spectra.

1.5.3.3 HPLC–MS/MS

HPLC has been the technique of choice for the separation of analytes online with ESI mass spectrometers [10]. In reversed phase mode, HPLC concentrates and separate analytes according to their hydrophobicity. Different approaches have been developed to introduce peptides into a mass spectrometer by HPLC. What they all have in common are peptide solutions that are loaded and separated on an HPLC column made of C18-like material. They differ, however, in terms of their flow rates, sample paths, robustness, and sensitivity.

1.5.3.4 Coupling to the Electrospray Mass Spectrometers

In recent years, three approaches have emerged for the coupling of m-HPLC columns to ESI-MS. The first approach consists of establishing an electrical contact directly at the end of the electrospray needle. This is typically achieved by gold-coating a part of the tip. This allows the direct application of a high-voltage potential to the tip of the column, which is necessary for the generation of the ESI process. The direct application of the high voltage generates, by far, the most stable electrospray process; however, the thin gold layer applied to the tip will be removed over time, and the electrical connection can be lost.

Alternatively, high voltages can be indirectly applied to the electrospray needle by using a liquid junction before the needle. This is typically achieved using an ultralow dead-volume connector placed between the transfer line and the electrospray needle. This approach has the advantage of being more robust in terms of maintaining the electrical connection; however, it has the disadvantage of generating a less stable, although still adequate, electrospray process. The last approach is a combination of the previous two approaches, and it consists of using a fully external gold-coated needle coupled to an ultralow dead-volume connector to which a high voltage is applied. In this fashion, the electric connection is always maintained while an optimum electrospray process is obtained.

The electrical liquid junction in commercial connections or homemade connections can cause debilitating problems when used improperly. It is important to understand that electrolysis does occur in the liquid junction, generating a small amount of gas. Generally, the linear flow is strong enough to carry the gas forward with no formation of bubbles; however, in improperly fitted connections, the linear flow could be reduced when entering the liquid junction to a level that allows the formation of gas bubbles.

1.5.4 Bioinformatics for the Identification of Protein Based on MS/MS

All the ESI-based tandem mass spectrometers are generally used to perform MS and MS/MS spectra. Although the mass accuracy, resolution, sensitivity, and quality of the fragmentation patterns are different, they contain information that can be used to identify a peptide provenance by protein/DNA database searching. The MS/MS spectra contain the fragmentation patterns related to the amino acid sequence of specific peptides. The analysis of MS/MS spectra is more intensive than the interpretation of MS data. The approaches that are used for the interpretation of these spectra can be classified into three subgroups according to the level of user intervention required.

1.5.4.1 No Interpretation

Clearly, one would like to be able to identify proteins without having to do any interpretation of the MS/MS spectra. For high-throughput analysis, this becomes essential. A few algorithms have been developed to search protein/DNA databases with uninterpreted MS/MS spectra. They all require that the partial or full sequence of the protein should be known and included in a database. All of the software packages provide a list of possible matches between individual MS/MS spectrum and peptide sequences obtained from the database; however, the scoring algorithms used to determine the validity of the matches are very different. These algorithms are explained in more detail below.

Mascot by Matrix Sciences (MATRIX SCIENCE- Boston, MA 02110 USA) (http://www.matrixscience.com) can be freely accessed online (for noncommercial entities) [11]. It was built on the basis of the Molecular Weight Search (MOWSE) scoring algorithm from

Papin. The MOWSE algorithm computes a matrix of probability in which each row represents an interval of 100 Da in peptide mass, and each column represents an interval of 10 kDa in intact protein mass. As each sequence entry is processed, the appropriate matrix elements are increased to accumulate statistics on the size distribution of peptide masses as a function of the protein mass. Therefore, this matrix represents the peptide distribution by protein molecular weight present in the database. The matrix will change as the database increases. From this matrix, a score is ascribed based on the measured masses. The same algorithm is used for peptide MS/MS. In Mascot, they have further incorporated the probability that a match is a random event. This allows the probabilistic evaluation and ordering of all the potential matches, as well as a probabilistic evaluation of the best match. Protein Prospector from University of California, San Francisco (http://prospector.ucsf.edu/) is also an example of a web-based MS/MS search engine. The identification of the protein is typically unambiguous, and is achieved by the number of peptides that matches to the same protein.

Furthermore, although protein identification has been performed with as little as one peptide using this algorithm, unambiguous identification of the provenance of a protein is often achieved by the multitude of peptides that matches with the same entry in a database. The SEQUEST software is computing intensive, and for high-throughput demand, it can rapidly paralyze the best dual-CPU server. In reality, the slowness of SEQUEST is due to the recurrent scans of the selected database to find the top 500 isobaric peptides. The larger the database, the longer it takes to scan the databases. An improved version of the software, called Turbo-SEQUEST, predigests and orders the databases and has greatly improved the search speed.

1.5.4.2 *Partial Interpretation*

Although the fully automated approaches are favorable for automation; historically, the computing power was not available for the rapid identification of proteins. Another subgroup of database search engines, based on the partial interpretation of the MS/MS spectra, was developed to perform faster searches of databases, although it requires human intervention. The most popular partial interpretation of MS/MS spectra is the "sequence-tag" approach [12]. It consists of reading the mass spacing between specific fragments of an MS/MS spectrum. This allows the generation of a short peptide sequence (tag). The tag is then used to pinpoint the possible peptide sequences from isobaric peptides in databases, whereas the residual mass information before and after the tag confirms one or a few peptide matches. Every MS/MS spectrum requires the generation of a tag followed by database searching. Unambiguous identification of the protein is established by the multitude of peptides that match to the same protein.

1.5.4.3 De Novo *Sequencing*

Finally, the last option is the full interpretation of the MS/MS spectra, often called *de novo* sequencing. Obviously, the other automated and semiautomated approaches are used before this approach. Although many genomes have been recently sequenced, the genomic data only represents a fraction of the world's genomic pool, and often no databases are available for specific organisms. The requirements for *de novo* sequencing are more stringent. The MS/MS spectra must be of good quality in terms of intensity and coverage of the peptide sequence. The MS/MS spectra of peptides contain ladder-type information, which, in principle, indicates their amino acid sequence. Experienced mass spectrometrists can manually extract the peptide sequence from the MS/MS spectra.

In large-scale proteomic studies, the throughput of analysis is a critical factor. Therefore, once enough MS/MS spectra have been generated to unambiguously identify a protein, generating MS/MS spectra on the residual peptides can be a waste of time. In some cases, however, such as when dealing with expressed sequence tags or protein mixtures, it is important to increase the number of MS/MS generated.

1.6 Stable Isotope Methods

Stable isotope tagging methods provide a useful means of determining with high precision the relative expression level of individual proteins between samples in a mass spectrometer. Because two or more samples tagged with different numbers of stable isotopes can be mixed before any processing steps, sample-to-sample recovery differences are eliminated.

1.6.1 Isobaric Tags for Relative and Absolute Quantification

A common complaint about global stable isotope profiling strategies is the time required for the separation of the peptides to prevent confounding overlaps in the mass spectrum. ABI (http://www.appliedbiosystems.com) recently announced isobaric tags for relative and absolute quantification (iTRAQ), which is potentially more cost-effective than other methods because up to four samples can be screened simultaneously with four isotopically distinguishable reagents. The iTRAQ methodology utilizes isobaric tags containing both reporter and balancer groups.

The reporter is quantitatively cleaved during collision-induced dissociation to yield an isotope series representing the quantity of a single peptide of known mass from each of up to four different samples. This quantification group (the reporter) is "balanced" by a second group (the balancer) depleted of the same stable isotopes, which maintains each isotopic tag at exactly the same mass. Because the peptide remains attached to the isobaric tags until collision-induced dissociation is performed, the peptide is simultaneously fragmented for sequence identification. The current generation of iTRAQ reagents labels lysine residues and the N termini of peptides, meaning that most peptides are multiply labeled. Therefore, iTRAQ suffers the same peptide overabundance problem and must be coupled with one or more dimensions of chromatographic or electrophoretic separation before MS analysis to limit the number of isobaric-tagged peptides in the first MS dimension. The advantage of iTRAQ over these methods is that the label is cleaved in the tandem MS before quantification.

Because differences in peptide levels can only be determined after tandem MS, the first MS dimension cannot be used to prescreen peptides for differential expression before tandem MS identity determination. Therefore, each and every peptide must be subjected to tandem MS analysis, making iTRAQ both time-consuming and sample-intensive for biomarker discovery applications. Furthermore, any untagged isobaric chemical noise may confound tandem MS sequencing of the iTRAQ-labeled peptides. Another issue with this method is the problem of protein variants. Any variant of the peptide of interest will not be isobaric with the same tagged peptides from control samples. Such nonisobaric peptides can be detected by their absence, but may be falsely interpreted as downregulation of the parent protein. Furthermore, such peptides may be isobaric with other peptides, confounding separately.

All cysteines in the healthy sample are modified with one isotopic version of the tag and the cysteines in the perturbed sample are tagged with the opposite isotopic reagent. The protein samples are then mixed and proteolyzed. Cysteine-tagged peptides are enriched by affinity chromatography of the iminobiotin tag through an avidin column and are later chromatographed by reversed phase HPLC, alone or in combination with ion exchange liquid chromatography such as strong cation exchange, and introduced into a mass spectrometer for quantification of expression differences. Further analysis of differentially expressed peptides by tandem MS allows the sequencing and identification of any differentially expressed peptides [13].

1.6.2 Isotope-Coded Affinity Tag

ICAT has been successfully used in several applications; these include the determination of microsomal protein levels in human myeloid leukemia cells undergoing pharmacologically induced differentiation and the identification of changes in the protein composition of lipid rafts isolated from control and stimulated Jurkat T-cells, and cleavable ICAT (cICAT) reagents have been used to identify protein expression differences in cystic fibrosis. The original [1H]- and [2H]-isotopic ICAT tags led to variable retention times in reversed phase HPLC for the same peptide. This issue has been rectified in second-generation ICAT reagents, which incorporate [12C] and [13C] isotope pairing. The iminobiotin affinity tag can also be problematic. Excess label or any endogenous biotin in the sample matrix (e.g., serum) may reduce the affinity column capacity because of competition for binding sites. Furthermore, ICAT peptides are sometimes isobaric with nontagged peptides eluting from the affinity chromatography step, leading to false-positive and false-negative detections. The ICAT tag itself can lead to collisional energy losses during fragmentation, resulting in sequencing difficulties. The third-generation ICAT (cICAT) reagents contain an acid-cleavable group between the biotin and the isotopically labeled linker so that a smaller attachment group is left on the cysteine residue after acidification, allowing more robust fragmentation and sequencing analysis.

An even greater problem resulting from peptide-selective ICAT strategies is a loss of peptide redundancy for a given protein species. Even with the low relative abundances of cysteine and histidine, the number of peptides from a tissue sample must still be further reduced by one or two dimensions of HPLC before MS analysis [14].

1.7 Proteomics in Data Analysis

Proteomics has undergone tremendous advances over the past few years, and technologies have noticeably matured. Despite these developments, biomarker discovery remains a very challenging task due to the complexity of the samples (e.g., serum, other bodily fluids, or tissues) and the wide dynamic range of protein concentrations. To overcome these issues, effective sample preparation (to reduce complexity and to enrich for lower abundance components whereas depleting the most abundant ones), state-of-the-art MS instrumentation, extensive data processing, and data analysis are required. Most of the serum biomarker studies performed to date seem to have converged on a set of proteins that are repeatedly identified in many studies and that represent only a small fraction of the entire blood proteome.

1.7.1 Data Processing

An important, frequently underestimated element of an integrated data analysis system is the data acquisition and signal preprocessing. A related issue is the format used to capture and store the data.

1.7.1.1 Signal Processing

Often, instruments are operated as a black box and are not always used to the maximum of their performance, whereas data preprocessing is often performed in default mode. For instance, data quality increases significantly (at the expense of data volume) by acquiring data in profile mode and by subsequently (postacquisition) using more elaborate algorithms to determine signal and noise, and derive more accurate measurements. Peak detection (peak picking) is a key element, often neglected, in the data analysis. It is usually part of the instrument software, and the users have limited control over it. This step is often performed automatically during the data acquisition, and the critical parameters are not always explicitly documented. However, high-quality data combined with effective preprocessing tools (i.e., algorithms for noise reduction, peak detection, and monoisotopic peak determination) are the basis of a reliable data analysis.

1.7.1.2 Data Format

A large variety of instrument platforms (ion traps, quadruple/TOF, ion cyclotron resonance, TOF/TOF, etc.) from various manufacturers are available for proteomics studies. Each instrument type will generate spectra with its own characteristics (signal-to-noise ratio, resolution, accuracy, etc.) usually in a proprietary data format. Data processing algorithms are not fully documented and are usually restricted to one instrument platform, thus limiting portability to other data processing tools and comparison of results. The definition of a generic mass spectrometric data format such as mzXML1 and the Human Proteome Organization's Proteomics Standards Initiative have been the first steps to overcoming this problem. The use of a standardized file format allows analyzing data within a pipeline that is independent of the instrument platform. Although conversion into the mzXML format requires additional computing resources and may increase the file size, a generic format broadly accepted by the community, including the manufacturers, will foster sharing and exchanging data in the future. In this context, the concrete plan to merge the mzXML and mzDATA formats into a single unified file format is encouraging.

1.7.2 Peptide Identification and Validation

Peptide identification via database searches is very computationally intensive and time-demanding. High-quality data allow more effective searches due to tighter constraints, that is, tolerance on precursor ion mass and charge state assignment, which will drastically reduce the search time in case of an indexed database. In addition, accurate mass measurements of fragment ions further simplify the database searches and add confidence to the results. Once the initial output of the database search engine has been obtained, it is essential that the reliability of the assignments of spectra to peptide sequences is statistically validated. Such analyses generate reliable estimates of the false positive and false negative error rates, values that are critical to meaningfully compare results from multiple experiments or platforms. The Peptide Prophet algorithm has been designed to achieve this goal. The error rates in a data set can also be estimated by performing a search using a "reversed database" (i.e., a database in which the sequences were scrambled to produce only false positive identifications and thus ascertain the false positive error rate). In contrast to more specialized tools, reversed database search results do not estimate the false negative error rate of a data set.

1.7.3 Protein Identification and Validation

The association of identified peptides with their precursor proteins is a very critical and difficult step in shotgun proteomics strategies because many peptides are common to several proteins, thus leading to ambiguous protein assignments. Therefore, it becomes critical to have an appropriate tool that is able to assess the validity of the protein inference and associate a probability to it. Protein prophet combines probabilities assigned to peptides identified by MS/MS to compute accurate probabilities for the proteins present. Protein prophet weights peptides that have a reliable score/probability and from all corresponding proteins derives the simplest list of proteins that explains the observed peptides. Obviously, proteins with multiple peptide matches have a much greater confidence in their assignment than proteins identified by one single peptide. In fact, there is a massive amplification of the false positive error rate at the protein level compared with the peptide level. This emphasizes the importance of this step and the need for a tool that is able to predict peptide and protein association.

1.7.4 Quantification

Quantification is a further critical step in biomarker studies because the primary focus is on peptides (proteins) that show differences in expression between two sets of samples; peptides that are invariant present much less interest. Systematic quantification of all peptides across multiple data sets is actually a very demanding task that has not yet been fully resolved. Strategies emphasizing the quantitative aspect tend to decouple identification and quantification and perform two independent experiments. Basically, two main approaches have been applied. The first is based on stable isotope labeling and requires derivatization of the peptides from the various sample sets with different reagents that have different isotopic compositions. The resulting products are then pooled together and analyzed in one single liquid chromatography-tandem MS experiment. The relative quantity of a specific analyte is then determined from the relative signal intensity of the signal in the full spectrum. Subsequently, the analyte in question is identified by database searching of the corresponding MS/MS signal.

The second approach, which is more relevant to larger biomarker studies [i.e., analysis of a larger set of samples from normal (control) and disease (or treated) patients], analyzes each sample individually and then compares the multiple liquid chromatography/MS runs subsequently.

1.7.5 Data Repositories

The large volume of such data sets has made the publication of detailed results using classic mechanisms very challenging. Sharing and exchange of data and results requires the definition of standard formats for the data at all levels (including raw mass spectrometric data, processed data, and search results) as well as a better definition (and/or standardization) of the parameters used for the data processing or the database searches. In this way, a broad range of data and results generated by a variety of tools can be captured into widely accessible repositories, including results of database search engines, *de novo* sequencing algorithms, and statistical validation tools (e.g., PeptideProphet and ProteinProphet). In all cases, metadata describing the parameters used for the analysis are essential and have to be included.

Additional data, including annotation and clinical information about the samples, ought to be incorporated into the database as well. Data and results repositories, such as PeptideAtlas (Institute for Systems Biology), Global Proteome Machine Database (Beavis Informatics), or Proteomics Identifications (PRIDE) Database (European Bioinformatics Institute), facilitate the exchange of results and information. Initially limited to peptide sequences and proteins with a minimum of parameters (i.e., retention/elution time, mass, charge state, and signal intensity) these databases are rapidly expanding to include actual spectra, links to the associated proteins, and genomics data.

One can envision expanding the repertoire of parameters captured as long as robust normalization methods are defined. For instance, elution times have limited value unless they are translated into normalized values that can be generalized across the entire database [15].

1.8 Multidimensional Protein Identification Technology

1.8.1 Introduction

A critical challenge in the postgenomic era is the development of protein analytical methods to carry out large-scale qualitative and quantitative analyses of proteins in tissues, cells, organelles, and complexes. Although 2D-PAGE remains a widely implemented separation method in proteomics, there is growing interest in alternative approaches to 2D-PAGE. Several biases in the use of 2D-PAGE result in the technology's limited capability to detect and identify low-abundance proteins with extremes in isoelectric point and molecular weight and very hydrophobic proteins [16,17]. Generally speaking, the fact that 2D-PAGE provides limited coverage of a proteome has driven the need to develop alternate separation strategies for proteomics.

1.8.2 Development of the Multidimensional Protein Identification Technology

A variety of multidimensional chromatographic approaches for separating proteins and peptides have been described as alternatives to 2D-PAGE; initial descriptions of two-dimensional chromatography coupled to MS systems focused on protein separation and failed to identify large numbers of proteins in a sample. Efforts to overcome this problem resulted in the development of a method called multidimensional protein identification technology (MudPIT). Multidimensional MudPIT couples two-dimensional chromatography of peptides in MS-compatible solutions directly to tandem MS, allowing for the identification of proteins from highly complex mixtures. Since the initial descriptions of MudPIT, this approach has been implemented in the analysis of whole proteomes, organelles, and protein complexes. Key aspects of many of the analyses are the validation of MudPIT data sets with alternate strategies and the integration of MudPIT data sets with other biochemical, cell biology, or molecular biology approaches. In MudPIT, a biphasic microcapillary column packed with reversed phase and strong cation exchange HPLC grade materials is loaded with a complex peptide mixture generated from a biological sample. The biphasic microcapillary column is interfaced with a quaternary HPLC pump coupled to a tandem mass spectrometer and acts as the ion source for the tandem mass spectrometer. Peptides are eluted off of the biphasic microcapillary column and directly ionized and analyzed in the tandem mass spectrometer. The resulting tandem MS data are searched using the SEQUEST algorithm, which interprets the tandem mass spectra (MS/MS) generated and identifies the peptide sequence from which it was generated, resulting in the determination of the protein content of the original sample.

1.8.3 Validation of Large-Scale Database

An important usage of large-scale, proteome-wide data sets revolves around the validation of the information generated, in which a large-scale data set is defined as a data set generated from the analysis of an organelle, cell, or organism. For example, initial analysis of the *Saccharomyces cerevisiae* proteome via MudPIT resulted in the detection and identification of 1484 proteins including proteins with extremes in molecular weight, p*I*, hydrophobicity (integral membrane proteins), and low predicted abundance. Subsequent large-scale qualitative analyses of proteomes via MudPIT have included the detection and identification of more than 2400 proteins from *Oryza sativa* (rice) leaves, roots, and seeds, and more than 2400 proteins from several life cycle stages of the primary causative agent of human malaria, *Plasmodium falciparum*. MudPIT data sets generally could contain more than 1000 proteins; to attempt to decipher the biological relevance of this number of proteins can be a little intimidating. Validation of these data sets first requires the selection of which of the identified proteins might provide novel insight into the biology of an organism. Although there has recently been extensive discussion regarding false positives in proteomics data sets, researchers generally agree that proteins detected and identified by at least two high-quality spectral matches are likely to be correct. Because MudPIT data sets generally contain more information than can be validated, appropriate choices for the selection of proteins to follow-up on are proteins of novel biological interest with the most high-quality unique peptide identifications. Upon the selection of appropriate proteins to validate with follow-up strategies, the validation strategy itself must be chosen. In most cases, for large-scale data sets, this requires imaging approaches to validate the detection and identification of proteins from organelles or whole proteomes.

1.8.4 Validation of Protein Complex Data Sets

Currently, many researchers aware of MudPIT and similar proteomic strategies think of these approaches as methods for the analysis of whole proteomes rather than protein complexes; however, MudPIT is proving to be an excellent tool for the comprehensive analysis of multiprotein complexes. In this strategy, a protein complex of interest is affinity-purified from a given cell type and approximately one-quarter of the material after final purification is analyzed via SDS-PAGE or Western blotting (or both). Some fraction of the remaining material is also directly analyzed via MudPIT strategies—that is, bands from SDS-PAGE gels are not cut out and are analyzed via MS and the solution from the purification is directly analyzed via MudPIT. The advantage is that more material can be loaded onto a MudPIT column than can be loaded into a given SDS-PAGE lane, in which there is, moreover, a significant amount of sample loss in extracting proteins from gel matrices. Applying the same data analysis strategies described in the previous section, the proteins in a multiprotein complex may be determined.

1.8.5 Application

MudPIT has been used in a wide range of proteomics experiments, including large-scale catalogues of proteins in cells and organisms, profiling of organelle and membrane proteins, identification of protein complexes, determination of posttranslational modifications, and quantitative analysis of protein expression. The technology was also used to identify novel proteins released by pancreatic cancer cells involved in extracellular matrix remodeling.

1.8.6 Future Directions

Although the application of MudPIT and MudPIT-like technologies is having an effect on biology, the analysis of proteomes using these approaches is, in many ways, in its infancy. Software strategies for interrogating the quantitative aspects of MS for MudPIT and other approaches [18] are beginning to be made available, which should enable more researchers to implement quantitative proteomics strategies in a similar fashion. It is important to keep in mind that data from these emerging technologies needs to be validated using alternate strategies. Regardless of whether one is carrying out a qualitative or quantitative proteomic analysis of protein complexes, organelles, cells, tissues, or organisms, proteomics data should be validated via alternate approaches. Because many proteomics technologies mature to the point of implementation in biological experimentation, it is increasingly important to think of approaches like MudPIT as tools to be integrated with other methods rather than as a standalone technology.

1.9 Applications of Proteomics

1.9.1 Application of Proteomics in Heart Disease

Diseases resulting in heart failure (e.g., hypertensive heart disease, ischemic heart disease) or specific heart muscle disease (e.g., dilated cardiomyopathy) are among the leading causes of morbidity and mortality in developed countries. The underlying molecular causes of cardiac dysfunction in most heart diseases are still largely unknown, but are likely to result from underlying alterations in gene and protein expression. Proteomics now allows us to examine global alterations in protein expression in the diseased heart and will provide new insights into the cellular mechanisms involved in cardiac dysfunction and should also result in the generation of new diagnostic and therapeutic markers.

1.9.2 Proteomics of Heart Disease

The causes of cardiac dysfunction in most heart diseases are still largely unknown, but are likely to result from underlying alterations in gene and protein expression. Proteomic studies are therefore likely to provide new insights into the cellular mechanisms involved in cardiac dysfunction and may also provide new diagnostic and therapeutic markers. Proteomic studies of human heart tissue are complicated by factors such as disease state, tissue heterogeneity, genetic variability, medical history, and therapeutic interventions. The current battery of proteomic technologies make it possible to characterize global alterations in protein expression associated with processes of human disease. Although the application of proteomics to human heart disease is in its infancy, it is already clear from studies of dilated cardiomyopathy, both in human patients and appropriate models, that it complements traditional candidate gene/protein studies and global genomic approaches and promises to provide new insights into the cellular mechanisms involved in cardiac dysfunction. An additional benefit of these proteomic studies should be the discovery of new diagnostic or prognostic biomarkers and the identification of potential drug targets for the development of new therapeutic approaches for combating heart disease.

1.9.3 Application of Plant Proteomics

In this postgenomic era, a more thorough understanding of gene expression and function can be achieved through the characterization of the products of expression—the proteins—which are essential biological determinants of plant phenotypes. The analysis of plant proteomes has drastically expanded in the last few years. MS technology, stains, software and progress in bioinformatics have made the identification of proteins relatively easy. The assignment of proteins to particular organelles and the development of better algorithms to predict subcellular localization are examples of how proteomic studies are contributing to plant biology.

In plants, studies with pathogens could have considerable biological variation and this can be a problem for the proteomics studies as well. Furthermore, proteomics studies of plant–pathogen interactions face intrinsic difficulties because by definition the plant–pathogen interaction is a complex one, involving the protein complements of two organisms. Because of intimate physical contact, it can be hard to distinguish proteins that are differentially expressed by the plant in response to the pathogen from those of the pathogen itself. Despite these problems, there is currently no substitute for *in planta* studies with pathogens. To minimize biological variability, particular attention should be paid to the experimental design. Inoculation in a block design, pooling tissue samples, and independent replication of the experiment are all important measures [19].

A simplified model of the actual plant–pathogen interaction has been used whereas others devised interesting experimental setups in which the partners of the interaction can be separated. Furthermore, there are studies where the proteins were identified but a host or pathogen origin could not be assigned with any degree of certainty. In addition, there are studies of plant defense that use mutant plants, such as lesion-mimic mutants, which do not directly involve a pathogen at all [20,21].

1.9.3.1 Application of Proteomics in Plant Pathology

The importance of agriculture and maintaining sustainable crop production around the world cannot be understated, particularly considering the projected significant increase in the global population in the future. Reliable and accurate disease diagnosis and pathogen detection are crucial to optimize crop yield and quality. Due to the myriad of biological organisms, including viruses, bacteria, nematodes, and fungi that can potentially cause devastating consequences to crop production worldwide, any technology that is utilized for plant disease detection should ideally be able to identify both prokaryotic and eukaryotic species.

1.9.3.2 Traditional Methods for Detection of Plant Pathogens

Traditionally, detection of phytopathogenic organisms may have involved the use of time-consuming culturing and cultivation on specific media with a subsequent morphological or biochemical analysis of the growth [22]. Pathogen identification procedures that involve the analysis of disease symptoms in or on infected plant material may lead to the implementation of improper disease management practices, especially in cases where different organisms that are not inhibited by the same antimicrobial agents are able to cause diseases that produce similar symptoms in plant hosts. There are also limitations to molecular biology–based and antibody-based techniques, because many of these protocols require reagents that are highly specific for individual pathogens. The recent interest in proteomics-based studies has led some researchers to examine the potential application of proteome-level investigations for phytopathogen detection.

Use of proteomics-based technology to study phytopathogens may eventually become the focus of more research programs and facilities; it is currently being used for the analysis of proteome-level changes in plants upon pathogen infection [23]. Proteomics can be used to investigate those processes that occur during plant microbe interactions such as those during symbiotic relationships at a molecular level. Such studies can lead to the identification of proteins (and consequently genes) that may be used to genetically engineer crops for enhanced disease tolerance and other beneficial traits.

Plant–microbe interactions are perhaps much broader than originally thought. From the initial objectives of identification of individual proteins to the development of high-throughput proteomics-based technologies, proteomics research has developed remarkably in the last few years. As strategies for high-throughput proteome analysis continue to develop, the capability of proteomics to identify novel targets for the development of plants with increased tolerance or resistance to various phytopathogens will likely continue to improve. The integration of proteome-based techniques along with information generated from genomic sequencing, traditional plant pathology, and genetic engineering will lead to a better understanding of various host–pathogen interactions, which may ultimately contribute to the development of novel disease-resistant varieties of agriculturally significant crops.

1.9.4 Proteomics and Diabetic Research

Both type 1 and type 2 diabetes mellitus are heterogeneous diseases with alterations in many genes and their products. Not all transcriptional alterations lead to protein changes, which makes it very important to, in conjunction with mRNA expression studies, also address changes in cellular protein levels. Various proteomic techniques are available for measuring many protein changes simultaneously.

1.9.4.1 The Application of Proteomics to Diabetes Research

There are key organs and tissues known to contribute to the development of diabetes mellitus and its complications. To define the underlying pathophysiological processes which contribute to disease development and to identify key pathways to target in the development of novel therapeutics will require a variety of approaches. These will include, for example, analyzing tissue-specific and cell-specific protein profiles under conditions that mimic diabetes mellitus and different environmental stimuli thought to contribute to the development of diabetes mellitus, and analyzing samples from patients with diabetes mellitus, insulin resistance, or impaired glucose tolerance and comparing them with healthy subjects. Diabetes is a disease in which many organs are affected and because all tissues are in contact with blood, proteins secreted or leaking from the different tissues are reflected in the circulation.

1.9.5 Application of Proteomics in Neurology

Proteomics is a burgeoning field that may provide a valuable approach to evaluate the posttraumatic central nervous system. Proteomics will likely be very useful for developing diagnostic predictors after central nervous system injury and for mapping changes in proteins after injury to identify new therapeutic targets. Neurotrauma results in complex alterations to the biological systems within the nervous system, and these changes evolve over time. Exploration of the "new nervous system" that develops after injury will require methods that can both fully assess and simplify this complexity.

Neurodegenerative diseases are central nervous system disorders characterized by progressive neuronal cell loss and intraneuronal accumulation of fibrillary materials. Abnormal protein–protein interactions may allow the precipitation of these proteins, forming intracellular and extracellular aggregates. These abnormal interactions could play a role in the dysfunction and neuronal death that characterizes several common neurodegenerative diseases, such as Alzheimer disease and Parkinson disease. A number of mechanisms seem to be involved in the neurodegenerative process, including alterations in calcium homeostasis in the endoplasmic reticulum, which contribute to neuronal apoptosis, excitotoxicity, and unregulated proteolysis. Mitochondrial dysfunction may also be linked to neurodegenerative disease through free radical generation, impaired calcium buffering, and mitochondrial permeability transition. Both apoptosis and necrotic cell death are observed in neurodegenerative diseases. Another possible mechanism is the disorganization of the cytoskeleton leading to neuronal degeneration.

1.9.5.1 Proteomic Aspects of Brain Tissue in Neurodegenerative Diseases

Proteomic analysis data has become an important resource in the investigation of neurodegenerative diseases. Proteomic profiling, in particular, has enabled researchers to investigate a vast number of proteins at once. Such principles have been utilized to detect specific alterations in the protein expression levels of various regions of a brain with Alzheimer disease compared with a control brain. This may, in turn, facilitate the construction of hypotheses of the mechanisms by which the disease progresses.

1.9.6 Proteomics and Bioremediation

Bioremediation is the recently developed technology which explores the microbial potential for biodegradation of xenobiotics compounds. Microorganisms display a remarkable range of contaminant degradation ability that can efficiently and effectively restore natural environmental conditions. Attempts have been made to interpret some areas of genomics and proteomics which have been used in bioremediation studies. Bioinformatics requires the study of microbial genomics, proteomics, systems biology, computational biology, phylogenetic trees, data mining and application of major bioinformatics tools for determining the structures, and biodegradative pathways of xenobiotic compounds. Naturally occurring microbial consortia have been utilized in a variety of bioremediation processes. Recent developments in molecular microbial ecology offer new tools that facilitate molecular analyses of microbial populations at contaminated and bioremediated sites. Information provided by such analyses aid in the evaluation of the effectiveness of bioremediation and the formulation of strategies that might accelerate bioremediation.

1.9.6.1 Application of Proteomics in Bioremediation

Environmental pollutants have become a major global concern, given their undesirable recalcitrant and xenobiotic compounds. A variety of polynuclear aromatic hydrocarbons, xenobiotics, and chlorinated and nitroaromatic compounds have been depicted to be highly toxic, mutagenic, and carcinogenic for living organisms [24]. Bioremediation is defined as a process by which microorganisms are stimulated to rapidly degrade hazardous organic pollutants to environmentally safe levels in soils, sediments, substances, materials, and ground water. Recently, the biological remediation process has also been devised to either precipitate effectively or immobilize inorganic pollutants such as heavy metals. Stimulation of microorganisms is achieved by the addition of growth substances such as nutrients, terminal electron acceptors/donors, or some combination, thereby resulting in an increase in organic pollutant degradation and biotransformation. Energy and carbon are obtained through the metabolism of organic compounds by the microbes involved in the bioremediation processes [25]. Biodegradation is nature's way of recycling wastes, or breaking down organic matter into nutrients that can be used by other organisms. The degradation is carried out by the microorganisms: bacteria, fungi, insects, worms, etc., by taking nutrients such as C, N, and P from the contaminant which upon long-term acclimatization convert the toxic compound into an environmentally friendly compound. By harnessing these natural forces of biodegradation, people can reduce wastes and clean up some types of environmental contaminants.

Bioremediation of a contaminated site typically works in one of two ways:

1. To enhance the growth of whatever pollution-eating microbes might already be living at the contaminated site
2. Specialized microbes are added to degrade the contaminants (less common)

The fields of biodegradation and bioremediation offer many interesting and unexplored possibilities from the bioinformatics point of view. They need to integrate a huge amount of data from different sources: chemical structure and reactivity of the organic compounds; sequence, structure, and function of proteins (enzymes); comparative genomics; environmental biology, etc. Bioinformatics provides the database for microarrays, gene identification, and microbial degradation pathways of compounds [26].

1.9.6.2 Bioremediation Using Proteomics

The cellular expression of proteins in an organism varies with environmental conditions. The changes in physiological response may occur due to the organism's adaptive responses to different external stimuli, such as the presence of toxic chemicals in the environment. The advent of proteomics has allowed an extensive examination of global changes in the composition or abundance of proteins, as well as the identification of key proteins involved in the response of microorganisms in a given physiological state. Using a proteomics approach, the physiological changes in an organism during bioremediation provide further insight into bioremediation-related genes and their regulation. The recent combined approaches of transcriptomics and proteomics have revealed new pathways for aerobic and anaerobic biodegradation of toxic wastes that will certainly pave the way for further identification of new signature proteins [27].

1.9.6.3 Tracking the Insights of Bioremediation Using Proteomics

The cellular expression of proteins in an organism varies with environmental conditions. The changes in physiological response may occur due to the organism's adaptive responses to different external stimuli, such as the presence of toxic chemicals in the environment. The advent of proteomics has allowed an extensive examination of global changes in the composition or abundance of proteins, as well as the identification of key proteins involved in the response of microorganisms in a given physiological state [28].

The recent combined approaches of transcriptomics and proteomics have revealed new pathways for aerobic and anaerobic biodegradation of toxic wastes that will certainly pave the way for further identification of new signature proteins. The growing demands of genomics and proteomics for the analysis of gene and protein function from a global bioremediation perspective are enhancing the need for microarray-based assays enormously. In the past, protein microarray technology has been successfully implicated for the identification, quantification, and functional analysis of proteins in basic and applied proteome research [29]. Other than the DNA chip, a large variety of protein microarray–based approaches have already verified that this technology is capable of filling the gap between transcriptomics and proteomics [30]. However, in bioremediation, microarray-based protein–protein interaction studies still need to progress to understand the chemotaxis phenomenon of any site-specific bacterium toward the environmental contaminant.

1.9.7 Proteomics in Molecular Marker Discovery

The development of toxicological and clinical biomarkers for disease diagnosis, quantification of toxicant/drug responses, and rapid patient care are major concerns in modern biology. Even after human genome sequencing, identification of specific molecular signatures for unambiguous correlation with toxicity and clinical interventions is a challenging task. Differential protein expression patterns and protein–protein interaction studies have started unraveling rigorous molecular explanations of multifactorial and toxicant-borne diseases. Proteome profiling is extensively used to investigate the etiology of diseases, develop predictive biomarkers for toxicity and therapeutic interventions, and to develop potential strategies for the treatment of complex and toxicant-mediated diseases.

Current molecular tools that include proteomics have erased recent distinctions among fields of pathology, toxicology, and molecular genetics [31]. Following human genome sequencing, biologists initiated multidimensional approaches to identify complete protein sets to understand differential expression, posttranslational modifications, protein–protein interactions, and structural aspects.

1.9.7.1 Biomarkers for Toxicity Assessment

The identification of toxicant-responsive proteins provides better means of toxicity assessment, toxicant classification, and exposure monitoring than current indicators [32]. Proteomics-based techniques are used for the development and validation of toxicity biomarkers for heavy metals, polycyclic aromatic hydrocarbons, and other environmental pollutants and toxicants. Proteomics-based approaches developed new families of biomarkers that permitted the characterization and efficient monitoring of cellular perturbations, provided an increased understanding of the influence of genetic variation on toxicological outcomes, and allowed the definition of environmental causes of genetic alterations and their relationship to human disease [31]. Assessment of the toxicity of drugs and environmental chemicals in target tissue are being done using several tools ranging from 2D-PAGE to protein microarrays. Proteome profiling predicts the toxicity response and resistance of several anticancer, cardiovascular, and anti-inflammatory drugs. Proteome profiling is also used to develop biomarkers for chemical toxicity [32].

1.9.7.2 Clinical Biomarkers

Two-dimensional gel electrophoresis in combination with MS allowed the identification of cancer, cardiovascular, neurodegenerative, and other complex disease-specific plasma/serum and cerebrospinal fluid biomarkers. Protein microarrays also represent a powerful tool to identify and characterize groups of proteins involved in tumor progression. Proteomics also offer potential for tracing mechanisms of complex diseases, differential diagnosis, classification of diseases, and therapeutic monitoring. Proteomics in combination with functional imaging, biosensors, and computational biology generated an unprecedented effect on the discovery and development of RNA-based gene expression profiles for the prediction of response, resistance, and toxicity of both new and existing anticancer, cardiovascular, and anti-inflammatory drugs [33].

1.9.7.3 Biomarkers for Cancer

The diagnosis of cancer at an early stage is difficult due to the complex structure and frequent modifications in participating proteins in response to environmental and other stresses, and proteomics has helped a lot in this direction. Marker proteins are released into the blood and other body fluids from cancerous tissues. Model systems are extensively used to find out novel proteins in blood and other body fluids associated with target tissues, for example, ovarian cancer [34]. Biomarkers are developed in blood, body fluids, tissues, and target organs for cancer diagnosis, classification, and therapeutic interventions. Proteomic tools are used to identify tumors at an early stage and, therefore, patients can safely avoid surgery or radiation therapy [35].

1.9.7.4 Biomarkers for Diagnosis

Proteomics provided enormous data-gathering capabilities by developing several clinical biomarkers in serum, plasma, urine, tissues, and other biological samples. Serum proteome profile is used to monitor the development and progression of ovarian and colon cancers [34]. In addition to proteome profiles, proteolytic degradation patterns in serum peptidome detect cancer and distinguish indolent and aggressive tumor. A reproducible MS approach to identify tumor-specific peptidome patterns in serum samples was developed along with an automated procedure for simultaneous measurement of peptides in serum [36].

1.9.7.5 Cardiovascular Biomarkers

Increasing incidence of cardiovascular diseases, specifically atherosclerosis and heart failure, prioritized the search for novel biomarkers. Biomarkers are needed for the diagnosis, prognosis, therapeutic monitoring, and risk stratification of acute myocardial infarction and heart failure, leading causes of mortality and morbidity [37]. Proteomics allowed protein identification, characterization, expression, and posttranslational modification assessment involved in cardiovascular diseases and the development of suitable biomarkers [38]. Genetic markers have enormous potential for the identification of cardiovascular disease at an early stage of onset [39]. Functional proteomics of a protein complex–based portrait in cardiac cell signaling examined and briefly evaluated and analyzed multiprotein complex that showed several advantages over classic methods.

1.9.7.6 Reproductive Toxicity Biomarkers

Surrogate tissue analyses incorporating contemporary genomic and proteomic tools are used for the determination of toxicant exposure and effect, disease state, and identification of target tissues at an early clinical stage [31].

1.9.7.7 Skeletal Toxicity Biomarkers

Proteomics allows the development of biomarkers for skeletal toxicity assessments. Compound-induced changes in urine samples using SELDI-TOF demonstrated urinary parvalbumin-α as a specific, potential, noninvasive, and easily detectable biomarker for skeletal muscle toxicity in rats [40]. Several oxidative stress-related and metabolism-related protein biomarkers have been developed; however, comprehensive analysis of the proteins defined is necessary to develop more sensitive toxicity biomarkers.

1.9.7.8 Limitations of Proteomics in Biomarker Development

Proteomics has opened a path toward understanding the underlying mechanism of toxicity and clinical interventions and being used to increase the speed and sensitivity of toxicological screening by identifying toxicity biomarkers [41]. Proteomics provides insights into the mechanism of action of a wide range of substances, from metals to peroxisome proliferators, increased predictability of early drug development, and identified noninvasive biomarkers of toxicity or efficacy. The utilization of safety biomarkers for the prediction of compound-related toxicity provided several advantages in drug discovery and development, particularly at an early stage [42]. Proteomics is being used for understanding and predicting clinical and toxicity responses and therefore has been used to sort out several major problems in medical sciences and toxicology; however, many obstacles remain to be resolved before biomarkers get widespread practical applications. The major challenge is the discrimination of changes due to interindividual variation, experimental background noise in protein profiling, and posttranslational modifications. Despite intensive research, a very limited number of plasma proteins have been validated as biomarkers for disease [35]. Although proteome approaches have provided opportunities to define the molecular mechanism of toxicity and clinical interventions, reproducibility in expression depends on experimental conditions across different laboratories; therefore, it is still a challenge for proteomics. Despite extensive efforts in the development of new technologies in proteomics, it is not possible to properly detect hydrophobic proteins and proteins present in very high or minute quantities using proteomics-based approaches such as 2D-PAGE; however, fractionating proteins into organelle components increases the probability of detecting low-copy number proteins. Most of the drug targets are hydrophobic proteins, therefore, application of diagnostic protein expression profiling in a predictive context is still a challenge and needs further investigation. Similarly, increases in automation and the development of new techniques in proteomics could control high-throughput speeds, another major drawback of proteomics.

1.9.8 Tumor Metastasis and Proteomics

1.9.8.1 Application of Proteomics in Studying Tumor Metastasis

There are two main expected outcomes from proteomic analysis of tumor metastasis. The first is to discover new molecular markers from the profiling of metastatic tumors. The second is to decipher the intracellular signaling pathways that lead cancer cells to be metastatic. Expression and functional proteomics are, respectively, suited for the purposes of the metastasis studies. The resulting data would provide knowledge bases for the early detection and prediction of metastasis and for the identification of novel targets for drug development and therapeutic intervention. Although vast progress has been made in diagnostic technology and therapeutic treatment during the last decades, cancer mortality remains high, accounting for approximately 25% of deaths in the developed world [43]. Metastasis is by far the leading cause of death in patients with cancer, and is responsible for more than 90% of all cancer mortalities [44]. Tumor metastasis is the spread of cancer cells from the original site to other parts of the body. It is a very complex and multistep process and is often referred to as a cascade. The process of metastasis formation begins with some tumor cells breaking adhesions with neighboring cells and detaching from the primary tumor. Those cells then dissolve the extracellular matrix, migrate, and invade surrounding tissues, or travel via the circulatory system, invade, survive, and proliferate at distant new sites [45].

1.9.9 Proteomic Drug Discovery

Proteomics is highly complementary to genomic approaches in the drug discovery process and, for the first time, offers scientists the ability to integrate information from the genome, expressed mRNAs, their respective proteins, and subcellular localization. It is expected that this will lead to important new insights into disease mechanisms and improved drug discovery strategies to produce novel therapeutics. Drug discovery is a prolonged process that uses a variety of tools from diverse fields. To accelerate the process, a number of biotechnologies, including genomics, proteomics, and a number of cellular and organismic methodologies, have been developed. Proteomics development faces interdisciplinary challenges, including both the traditional (biology and chemistry) and the emerging (high-throughput automation and bioinformatics). Emergent technologies include two-dimensional gel electrophoresis, MS, protein arrays, isotope-encoding, two-hybrid systems, information technology, and activity-based assays. These technologies, as part of the arsenal of proteomics techniques, are advancing the utility of proteomics in the drug-discovery process.

The drug discovery process in the pharmaceutical industry includes the definition and validation of new drug targets. This involves many phases including target identification, lead identification, small-molecule optimization, and preclinical/clinical development. Efficiency in this process relies on timely knowledge of biological cause-and-effect in the course of disease and treatment, which ultimately rests on the knowledge of protein function and regulation. One of the key steps, target identification, has been fostered through applied genomics, primarily because both high-throughput methods and tools that allow nucleic acid amplification have enabled large-scale profiling of expressed genes [46]. However, analysis of the information produced by genomics, when measured against comparable information regarding protein expression, has led to the conclusion that message abundance fails to correlate with protein quantity [47]. Furthermore, posttranslational processes such as protein modifications or protein degradation remain unaccounted for in genomic analysis [48]. Because both cell function and its biochemical regulation depend on protein activity, and because the correlation between message level and protein activity is low, the measurement shifts from genomics to proteomics. This shift has occurred not only in target discovery but also in many other areas of the process, including patient treatment and care [49].

For drug discovery, the ideal proteomics method would be one that is:

1. Sensitive enough to detect low-abundance proteins
2. Able to detect activity over in addition to abundance
3. Able to detect protein–protein and protein–small molecule interactions
4. Easily implemented and performed quickly

Research in proteomics seeks to satisfy all, or some, of these conditions by developing new methods to understand protein function and interactions in a biological context. Proteomics and two-hybrid screening technologies are then used to study the interactions of these targets, which provide immediate validation of the target in the context of a disease-specific cellular response. The proprietary technology of the Anadys Pharmaceuticals and Pharmacia is called GATE (genetics-assisted target evaluation), in addition to technologies for high-throughput screening of challenging targets for drug development that include ATLAS (any target ligand affinity screen) for protein targets, SCAN (screen for compounds with affinity for nucleic acids) for RNA targets, and riboproteomics (systematic annotation of RNA–protein interactions that affect RNA metabolism: transport, splicing, translation, and decay). Analysts are using these technologies to translate human and microbial genomic information into small-molecule therapeutics.

1.9.10 Proteomics and Autoimmune Disease

Array-based autoantibody profiling may provide a long-awaited platform for technology to determine the specificity of autoimmune responses in individuals and cohorts of patients. This technology will drive the development and selection of antigen-specific therapies for use in the clinic. It is possible, in a manner parallel to the use of skin testing in the allergy clinic to select desensitization therapy, that antigen arrays could be applied to select customized antigen-specific therapy for individual patients with autoimmune disease. An antigen array–based strategy will be initially utilized to develop therapies that could be used to treat patients with a specific autoimmune disease and to select patients to receive such therapies. Once antigen-specific therapies prove safe and effective, and if inclusion of the exact set of targeted autoantigens in the therapy proves crucial for efficacy in individual patients, then customized therapies based on the selection of agents including the targeted autoantigens in individual patients could be developed in the coming decades.

One potential application of protein array technology is in clinical diagnostic testing. These novel protesomics technologies will enable complex analysis of protein levels within and between cells (e.g., comparing transcriptional and protein profiles of distinct lymphocyte populations), as well as parallel analysis of cytokine and chemokine levels in biological fluids. Such studies could identify and further define critical signaling pathways involved in disease pathogenesis, cell surface receptors that may serve as novel targets of drug discovery, and proteomic characteristics of lymphocytes that are anergic, autoreactive, or regulatory. The powerful combination of genomics and proteomics has the potential to forever change how we study the basic biology of autoimmunity. Antigen-specific therapies involve the administration of targeted autoantigens in a manner that induces immune tolerance to treat autoimmunity. Such therapies include

1. Oral administration of antigen to induce "oral tolerance" [50]
2. Administration of native peptides via intravenous or other routes [51]
3. Administration of altered peptide ligands [52]
4. Administration of whole protein antigens [53]
5. Administration of other biomolecules such as DNA, or proteins and peptides with posttranslational modifications [54]
6. Administration of DNA-tolerizing vaccines encoding the targeted self proteins [55,56].

Future antigen-specific therapies based on this strategy will deliver multiple targeted epitopes or protein antigens as tolerizing agents. Such therapies could deliver a consensus dominant targeted epitope or antigen, and could also deliver a cocktail of 20 or more of the consensus targeted epitopes or antigens to treat patients with a specific disease or subset of that disease. Using antigen arrays to guide the development and selection of antigen-specific DNA-tolerizing vaccine therapy, a strategy termed "reverse genomics." Applying a reverse genomics strategy to develop and select antigen-specific DNA-tolerizing vaccines to treat experimental autoimmune encephalomyelitis (EAE) [57] and collagen-induced arthritis, a model for rheumatoid arthritis (RA).

1.10 Challenges in Proteomics

1.10.1 Introduction

The complete human genome project (HUGO) was officially completed in the first quarter of 2003. The major challenge in proteomics is that in any given tissue, there are approximately 10,000 different proteins being expressed at levels that vary by as much as six orders of magnitude. Chemical biologist Stuart Schreiber advocates building a "perturbogen" library of small molecules that could specifically activate or deactivate every protein in the human body. Schreiber estimates that such a project would require at least a decade. One purpose of the library would be to enable biomedical engineers to develop therapies more efficiently. A number of scientists have developed ways to break the proteome down into meaningful pieces that can be studied more easily. Notably, activity-based proteomics developed by Benjamin Cravatt III uses specially designed chemical probes to analyze classes of active enzymes within a tissue.

1.10.2 The Challenge of Reproducible MS Analysis

MS is one of the most versatile technologies used in biology. However, not all is well in the discipline of proteomics, and much fuzzy thinking and bad data have unfortunately found their way into the literature. Determining significant differences between MS data sets from biological samples is one of the major challenges for proteome informatics. Accurate and reproducible protein quantization in complex samples in the face of biological and technical variability has long been a desired goal for proteomics. MS is well known for

TABLE 1.1

The Four-Step Approach to Fixing Proteomics

Step	Experiment Phase	Have I Reached This Step?
1. Check your system	Basic technical functionality	Are you confident you get the same result every time you run a standard on your system?
2. Run pilot experiments	Exploratory	Do you know everything you need to know to design and run a conclusive experiment?
3. Confirm your results	Within-laboratory confirmatory	Can you reliably reproduce and confirm the results you got from your pilot using enough replicates?
4. Reproduce across laboratories	Cross-laboratory confirmatory	Are you confident these results are biological in nature, reproducible by the community as a whole, and are not influenced by specific protocols within your laboratory?

the potential it offers for discovery and validation within proteomics research. However, it has also been recognized that MS is a technically challenging technique; not a black box solution to delivering reproducible quantitative proteomics results each time you run the same experiment. This was highlighted in the results of a 3-year Human Proteome Organization quality control study using test samples to assess issues of reproducibility and sampling in proteomics. Only seven of the twenty-seven participating laboratories were 100% successful in the first part of the study, which dealt with protein identification. The laboratories were also subjected to a subsequent sampling and reporting exercise with peptides of very similar molecular weights. Only one of the twenty-seven laboratories was successful in this attempt.

The four-step approach to fixing proteomics applies to any technique so it helps deliver cross-laboratory reproducibility for MS. It allows you to create experiments that incrementally increase the level of reliability and reproducibility of your proteomics data, controlling the technical variation with rigorous, standardized approaches in each case. Table 1.1 summarizes the four steps, where they can fit into your experimental approach and the questions you can use to check if you have achieved each step.

1.10.3 Conclusion

There are many challenges in protein array, liquid chromatography MS, and data analysis. The challenge to be met is not an easy task and will, in many aspects, be determined by the speed and success by which the analytical field can make milestone progress. Proteomics can be expected to have a wide-ranging effect on the future of biological research.

References

1. Abhilash, M. 2009. Application of proteomics. *The Internet Journal of Genomics and Proteomics* 4:1.
2. Thongboonkerd, V., and J.B. Klein. 2004. Overview of proteomics. In *Proteomics in Nephrology.* Contrib Nephrol. Basel: Karger.
3. Walker, J.M. 1996. *Protein Protocols Handbook*. Totowa: Humana Press.
4. Jensen, O.N., M. Wilm, A. Shevchenko, and M. Mann. 1999. Sample preparation methods for mass spectrometric peptide mapping directly from 2-DE gels. *Methods Molecular Biology* 112:513–530.

5. Veerakumari, L. 2006. Bioinstrumentation. Chennai, India: MJP Publishers.
6. Daniel, F. 2005. *Industrial Proteomics: Applications for Biotechnology and Pharmaceuticals*. New York: John Wiley & Sons, Inc.
7. Macro, A. 2005. Phage display technology—application and innovations. *Genetics and Molecular Biology* 28:1–9.
8. Haleem, J.I. 2002. The SELDI-TOF mass spectrometry approach to proteomics: Protein profiling and biomarker identification. *Biochemical and Biophysical Research Communications* 292:587–590.
9. Shevchenko, A., A. Loboda, W. Ens, and K.G. Standing. 2000. MALDI quadruple time-of-flight mass spectrometry: A powerful tool for proteomic research. *Analytical Chemistry* 72:2132–2141.
10. LaCourse, W.R. 2002. Column liquid chromatography: Equipment and instrumentation. *Analytical Chemistry* 72:37R–51R.
11. Pappin, D.J.C., P. Hojrup, and A.J. Bleasby. 1993. Rapid identification of proteins by peptide mass fingerprinting. *Current Biology* 36:327–332.
12. Wilkins, M.R., D.F. Hochstrasser, J.C. Sanchez, A. Bairoch, and R.D. Appel. 1996. Integrating two dimensional gel databases using the Melanie II software. *Trends Biochemical Sciences* 21:496–497.
13. Luke, V.S., and P.H. Michael. 2005. Stable isotope methods for high precision proteomics. *Drug Discovery Today: Targets* 10:353–363.
14. Von Heller, P.D. 2003. The application of new software tools to quantitative protein profiling via isotope-coded affinity tag (ICAT) and tandem mass spectrometry: II. Evaluation of tandem mass spectrometry methodologies for large-scale protein analysis and the application of statistical tools for data analysis and interpretation. *Molecular and Cellular Proteomics* 2:426–427.
15. Bruno, D., and R. Aebersold. 2006. Challenges and opportunities in proteomics data analysis. *Molecular and Cellular Proteomics* 5:1921–1926.
16. Jones, A.M., M.H. Bennett, J.W. Mansfield, and M. Grant. 2006. Analysis of the defense phosphoproteome of *Arabidopsis thaliana* using differential mass tagging. *Proteomics* 6:4155–4165.
17. Jung, Y.H., R. Rakwal, and G.K. Agrawal. 2006. Differential expression of defense/stress-related marker proteins in leaves of a unique rice blast lesion mimic mutant (blm). *Journal of Proteome Research* 5:2586–2598.
18. Kim, S.T., S.G. Kim, and Y.H. Kang. 2008. Proteomics analysis of rice lesion mimic mutant (spl1) reveals tightly localized probenazole-induced protein (PBZ1) in cells undergoing programmed cell death. *Journal of Proteome Research* 7:1750–1760.
19. Lopez, M.M., E. Bertolini, and A. Olmos. 2003. Innovative tools for the detection of plant pathogenic viruses and bacteria. *International Microbiology* 6:233–243.
20. Gygi, S.P., G. Corthals, and Y.I. Zhang. 2004. Evaluation of two-dimensional gel electrophoresis–based proteome analysis molecules. *Analytical and Bioanalytical Chemistry* 378:1952–1961.
21. Wu, C.C., and J.R. Yates, III. 2003. The application of mass spectrometry to membrane proteomics. *Nature Biotechnology* 21:262–267.
22. Li, X.J., H. Zhang, J.A. Ranish, and R. Aebersold. 2004. Automated statistical analysis of protein abundance ratios from data multidimensional protein identification technology. *Molecular Cell* 14:685–691.
23. Padliya, N.D., and B. Cooper. 2006. Mass spectrometry–based proteomics for the detection of plant pathogens. *Proteomics* 6:4069–4075.
24. Zhang, C., and G.N. Bennett. 2005. Biodegradation of xenobiotics by anaerobic bacteria. *Applied Microbiology and Biotechnology* 67:600–618.
25. Fulekar, M.H. 2005. *Environmental Biotechnology*. New Delhi: Oxford & IBH Publication.
26. Ellis, L.B., C.D. Hershberger, and L.P. Wackett. 2000. The University of Minnesota Biocatalysis/Biodegradation Database: Microorganisms, genomics and prediction. *Nucleic Acids Research* 28:377–379.
27. Chen, X., P.H. Shao, F.S. Chao, M.D. Chang, Y.S. Ji, and Y. Chen. 2009. Interaction of *Pseudomonas putida* CZ1 with clays and ability of the composite to immobilize copper and zinc from solution. *Bioresource Technology* 100:330–337.
28. Vasseur, C., J. Labadie, and M. Hébraud. 1999. Differential protein expression by *Pseudomonas fragi* submitted to various stresses. *Electrophoresis* 20:2204–2213.

29. Labaer, J., and N. Ramachandran. 2005. Protein microarrays as tools for functional proteomics. *Current Opinion in Chemical Biology* 9:14–19.

30. Link, A.J., J. Eng, and D.M. Schieltz. 1999. Direct analysis of protein complexes using mass spectrometry. *Nature Biotechnology* 17:676–682.

31. Aardema, M.J., and J.T. MacGregor. 2002. Toxicology and genetic toxicology in the new era of toxic genomics: Impact of proteomics technologies. *Mutation Research* 499:13–25.

32. Merrick, B.A., and M.E. Bruno. 2004. Genomic and proteomic profiling for biomarkers and signature profiles of toxicity. *Current Opinion in Molecular Therapeutics* 6:600–607.

33. Ross, J.S., W.F. Symmans, L. Pusztai, and G.N. Hortobagyi. 2005. Pharmacogenomics and clinical biomarkers in drug discovery and development. *American Journal of Clinical Pathology* 124:S29–S41.

34. Rockett, J.C., M.E. Burczynski, A.J. Fornace, P.C. Herrmann, S.A. Krawetz, and D.J. Dix. 2004. Surrogate tissue analysis: Monitoring toxicant exposure and health status of inaccessible tissues through the analysis of accessible tissues and cells. *Toxicology and Applied Pharmacology* 194:189–199.

35. Coombes, K.R., J.S. Morris, J. Hu, S.R. Edmonson, and K.A. Baggerly. 2005. Serum proteomics profiling—A young technology begins to mature. *Nature Biotechnology* 23:291–292.

36. Villanueva, J., D.R. Shaffer, J. Philip, and C.A. Chaparro. 2004. Differential exoprotease activities confer tumour-specific serum peptidome patterns. *The Journal of Clinical Investigation* 116:271–284.

37. Stanley, B.A., R.L. Gundry, R.J. Cotter, and E.J.E. Van. 2004. Heart disease, clinical proteomics and mass spectrometry. *Disease Markers* 20:167–178.

38. Gallego-Delgado, J., A. Lazaro, J.I. Osende, M.G. Barderas, L.M. Blanco-Colio, M.C. Duran et al. 2005. Proteomic approach in the search of new cardiovascular biomarkers. *Kidney International* 99:S103–S107.

39. Gibbons, G.H., C.C. Liew, and M.O. Goodarzi. 2004. Genetic markers: Progress and potential for cardiovascular disease. *Circulation* 109:IV47–IV58.

40. Dare, T.O., H.A. Davies, J.A. Turton, L. Lomas, T.C. Williams, and M.J. York. 2002. Application of surface-enhanced laser desorption/ionization technology to the detection and identification of urinary parvalbumin-alpha: A biomarker of compound-induced skeletal muscle toxicity in the rat. *Electrophoresis* 23:3241–3251.

41. Kennedy, S. 2002. The role of proteomics in toxicology: Identification of biomarkers of toxicity by protein expression analysis. *Biomarkers* 7:269–290.

42. Yamamoto, T., T. Fukushima, R. Kikkawa, H. Yamada, and I. Horii. 2005. Protein expression analysis of rat testes induced testicular toxicity with several reproductive toxicants. *Journal of Toxicological Sciences* 30:111–126.

43. Brooks, S.A., and U. Schumacher. 2001. *Metastasis Research Protocols*. Totowa: Humana Press.

44. Hanahan, D., and R.A. Weinberg. 2000. The hallmarks of cancer. *Cell* 100:57–70.

45. Meyer, T., and I.R. Hart. 1998. Mechanisms of tumour metastasis. *European Journal of Cancer* 34:214–221.

46. Nutall, M.E. 2001. Drug discovery and target validation. *Cells, Tissues, Organs* 169:265–271.

47. Pandey, A., and M. Mann. 2000. Proteomics to study genes and genomes. *Nature* 405:837–846.

48. Mann, M. 1999. Quantitative proteomics. *Nature Biotechnology* 17:954–955.

49. Simpson, R.J., and D.S. Dorow. 2001. Cancer proteomics: from signaling networks to tumor markers. *Trends Biotechnology* 19:S40–S48.

50. Weiner, H.L., A. Friedman, A. Miller, S.J. Khoury, A. Al-Sabbagh, and L. Santos. 1994. Oral tolerance: Immunologic mechanisms and treatment of animal and human organ-specific autoimmune diseases by oral administration of autoantigens. *Annual Reviews in Immunology* 12:809–837.

51. Gaur, A., B. Wiers, A. Liu, J. Rothbard, and C.G. Fathman. 1992. Amelioration of autoimmune encephalomyelitis by myelin basic protein synthetic peptide-induced energy. *Science* 258:1491–1494.

52. Brocke, S., K. Gijbels, M. Allegretta, I. Ferber, C. Piercy, and T. Blankenstein. 1996. Treatment of experimental encephalomyelitis with a peptide analogue of myelin basic protein. *Nature* 379:343–346.

53. Critchfield, J.M., M.K. Racke, and P. Zuniga. 1994. T cell deletion in high antigen dose therapy of autoimmune encephalomyelitis. *Science* 263:1139–1143.
54. Furie, R.A., J.M. Cash, M.E. Cronin, R.S. Katz, M.H. Weisman, and C. Aranow. 2001. Treatment of systemic lupus erythematosus with LJP 394. *Journal of Rheumatology* 28:257–265.
55. Ruiz, P., H. Garren, I. Urbanek-Ruiz, D. Hirschberg, L. Nguyen, and M. Karpuj. 1999. Suppressive immunization with DNA encoding a self-protein prevents autoimmune disease: Modulation of T cell costimulation. *Journal of Immunology* 162:3336–3341.
56. Garren, H., and L. Steinman. 2000. DNA vaccination in the treatment of autoimmune disease. *Current Directions in Autoimmunity* 2:203–216.
57. Garren, H., P.J. Ruiz, T.A. Watkins, P. Fontoura, L.T. Nguyen, and E.R. Estline. 2001. Combination of gene delivery and DNA vaccination to protect from and reverse Th1 autoimmune disease via deviation to the Th2 pathway. *Immunity* 15:15–22.

2

Epigenome: Genomic Response to Environmental Eccentricities

Anjana Munshi, Srirama Krupanidhi, and Yog Raj Ahuja

CONTENTS

2.1 Introduction

In the pre-Mendelian era, inheritance was believed to be a merging of characteristics of both parents. Mendel's experiments indicated that the reciprocal crosses between parents with contrasting characters produced the same results. But there was existing evidence showing that reciprocal crosses may not always be alike. A classic example is a cross between a female horse and a male donkey, which produces a mule, whereas the reciprocal cross produces a hinny. No suitable explanation was available until lately when the significance of epigenetics started to be realized. In the Greek language, the prefix "epi" means features that are "above" or in addition to something. Therefore, epigenetic traits exist in addition to the traditional molecular basis for inheritance.

Historically, the word epigenetics was coined by Waddington (1905–1975), who proposed that "phenotypes arise from genotypes through programmed changes." He defined epigenetics as a branch of biology that studies the causal interaction between genes and their products which bring the phenotype into being [1]. Originally, the term was defined in a development context to depict permanent changes in gene activation and deactivation required for cellular differentiation. Subsequently, epigenetics was defined as the study of mitotically or meiotically (or both) heritable changes in gene function that cannot be explained by changes in DNA sequence [2]. With the information now available, the term epigenetics has broadened to encompass changes that are both heritable and transient in nature [3]. In fact, epigenetic mechanisms provide an extra layer of transcriptional/translational/posttranslational controls that regulate how genes are expressed. Because gene expression is involved in all life processes, DNA replication, cell division, transcription, translation, and posttranslation patterns are all under epigenetic control. Because the word "epigenetics" is similar to genetics, many parallel words have originated. "Epigenome" (which refers to the epigenetic state of the genome) is parallel to genome and "epigenetic code" (a set of epigenetic features that create different phenotypes in different cells) is parallel to "genetic code."

There are about 25,000 genes in the human genome. More than 600 genes have been reported to be modulated by epigenetic modifications [4]. Despite their identical DNA sequences, monozygotic twins or cloned animals can differ in their susceptibility to disease [5,6]. Prenatal and postnatal environmental factors have the potential to modify epigenetic programming and bring about subsequent changes which may have relevance in health and disease. The epigenetic modifications are most likely the result of altered DNA methylation or posttranslational chemical modifications of the chromatin, especially histone modifications [7–10]. Furthermore, it is now known that the epigenome can be modulated by a variety of environmental factors [11–13] (Figure 2.1). Emerging evidence is supportive of the importance played by noncoding RNAs (ncRNA)/micro-RNAs and protein interactions in the management of the epigenome. In this review, however, we intend to focus on methylation and histone modifications with a sprinkle of ncRNAs and protein interactions vis-a-vis the environmental modulations of the epigenome (Figure 2.1).

2.1.1 DNA Methylation

The most common epigenetic modification of mammalian DNA involves the enzymatic transfer of a methyl group to the fifth position of cytosine residues in CpG dinucleotides [14]. After 6 years of Waddington's narration on epigenetics, DNA methylation was found to be an epigenetic marker [15]. DNA methylation is known to be involved in tissue-specific

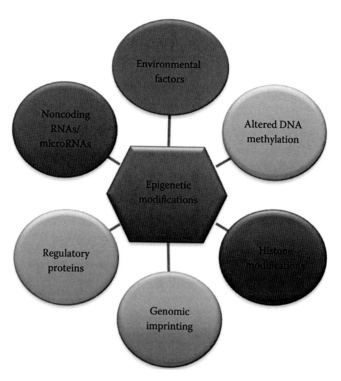

FIGURE 2.1
Different kinds of epigenetic modifications.

patterns of gene expression, cellular differentiation, genomic imprinting, and X chromosome inactivation [16,17]. In the human genome, a majority of the promoters have a high CpG content (high-CpG-density promoters), whereas the remainder have a low CpG content (low-CpG-density promoters). High-CpG-density promoters are associated with two categories of genes: housekeeping genes and tightly regulated key developmental genes, for example, genes involved in neural precursor cell differentiation. The low-CpG-density promoters are associated mainly with tissue-specific genes [18]. Methylated CpG sites do not occur as frequently as expected in mammalian DNA (~1/100 bases) but these are clustered at a high frequency (~1/10 bases) in CpG islands, which are typically unmethylated in the 5′ upstream sequences of nearly 50% to 60% of the active genes in normal adult tissues [19].

The regulatory influences of CpG sequences are most important in the 5′ promoter regions of active genes. However, there are reports of some gene intron and 3′ methylation influences [20]. Usually, there is an inverse relationship between the methylation status of a gene and its expression. However, this relationship may not be universal, for example O^6–alkylguanine-DNA alkyl transferase has been shown to be hypomethylated when it is inactivated [21]. This inverse relationship has also been shown in imprinted growth-regulating genes such as IGF2R [22]. In a majority of the genes in the genome, in addition to promoter, first exons are strongly enriched in unmethylated domains and depleted in methylated domains [23]. The methylated domains are found predominantly in interspersed and tandem repetitive sequences and exons other than the first exons [24].

DNA methylation is catalyzed by DNA methyltransferases [25]. The methylation of DNA is initiated and established with the DNA methyltransferases DNMT3 family

(DNMT3A and DNMT3B) [25,26]. The expression of DNMT3A and DNMT3B is coordinated by DNMT3L [27], lymphoid-specific helicase [28] micro-RNAs [29], and Pi RNAs [30]. DNA methylation has been divided into two types: *de novo* and maintenance methylation. DNMT 3A and DNMT 3B catalyze *de novo* methylation, which is important for the establishment of methylation patterns in early embryos during development and carcinogenesis [26]. DNMT1 maintains these methylation patterns set by *de novo* methylation. DNMT1 is localized to the replication fork during cellular division and conducts maintenance methylation [31,32]. However, DNMT1 is not able to maintain the methylation of many CpG dense regions [33]. Therefore, *de novo* activities of DNMT3A and DNMT3B are important to reestablish the methylation patterns so that they are not lost because of the inefficient activity of DNMT1.

The genetic elements can be successfully silenced by histone modifications and chromatin structure. However, these modifications are easily reversible and are poor gatekeepers for long-term silencing [34]. Therefore, mammalian cells require an additional mechanism for prolonged silencing of these sequences. DNA methylation is an important component of this process and is a stable modification which is inherited throughout cellular divisions. DNA methylation in promoters prevents the reactivation of silent genes, even though the repressive histone marks are reversed [35]. This allows the daughter cells to retain the same expression pattern as their precursor cells. This is important for many cellular processes including the silencing of repetitive elements, X-inactivation, genomic imprinting, and development.

Without doubt, DNA methylation has been correlated with transcriptional silencing for more than 20 years, the mechanisms by which DNA methylation inhibits transcription have begun to be uncovered recently. Many models have been suggested. One model suggests that DNA methylation directly impedes the binding of transcriptional factors to their target sites and thereby prohibits transcription. Other proposed mechanisms are based on the idea that methylation of CpG sequences can alter chromatin structure by affecting histone modifications and nucleosome occupancy in the promoter regions. Many transcription factors are targeted to GC-rich sequences, and methylation of CpG sites within these sequences prevents the binding of these proteins to these sites [36,37].

2.1.2 Histone Modifications

Histone modifications have emerged as vital components of the epigenetic mechanisms that regulate the genetic information encoded in DNA. The basic unit of chromatin is the nucleosome, which consists of 147 bp of DNA wrapped nearly twice around the octamer, containing two copies of each of the core histones H2A, H2B, H3, and H4. Each nucleosome is separated by a 10- to 16-bp linker DNA and this bead-on-a-string arrangement constitutes a chromatin fiber of approximately 10 nm in diameter. Each core histone within the nucleosome contains a globular domain which mediates histone–histone interactions and also bears a highly dynamic amino acid terminal tail around 20 to 35 residues in length that is rich in basic amino acids. These tails extend from the surface of the nucleosome. Histone H2A has an additional approximately 37 amino acid, carboxy-terminal domain protruding from the nucleosome. All histone proteins are modified inside the nucleus of the cell, but only few modifications have been studied thus far. Nature uses several modifications in proteins to complement and expand its chemical repertoire. In fact, more than 200 distinct posttranslational modifications (PTMs) have been identified to date, and the number and variety of modifications continue to grow. Histones are subject to an enormous number of PTMs including acetylation and methylation of lysines (k) and arginines (R), phosphorylation of serine (S) and threonines (T), ubiquitination, sumoylation, and biotinylation of lysines as well as ADP ribosylation (Figures 2.2 and 2.3). These modifications are recognized by specific protein–protein modules and can regulate each other as well [38].

Recent findings have revealed that these histone tails do not contribute to the structure of individual nucleosomes nor to their stability but play an important role in the folding of nucleosomal arrays into higher-order chromatin structures [39]. Eukaryotes have developed many histone-based strategies to introduce variation into the chromatin fiber because the chromatin is the physiological template for all the DNA-mediated processes and these strategies are likely to control the structure or function (or both) of the chromatin fiber. Two of the most common strategies employed are the PTMs of histones and replacement of major histone species by the variant isoforms [40]. Many studies have shown that the site-specific combinations of histone modifications correlate well with particular biological functions such as transcription, silencing, heterochromatinization, DNA repair, and replication. These observations have led to the idea of a "histone code," although the degree of specificity of these codes may vary as particular combinations of histone marks do not always dictate the same biological function. Moreover, there is a clear indication that mistakes in PTMs may be involved in many human diseases, especially cancer [41]. More than three decades ago, it was noticed that transcriptionally active histones were acetylated *in vivo*, and this led to the speculation that covalent modifications of histones might have a role in determining the states of gene activity [42]. The histone N-terminal

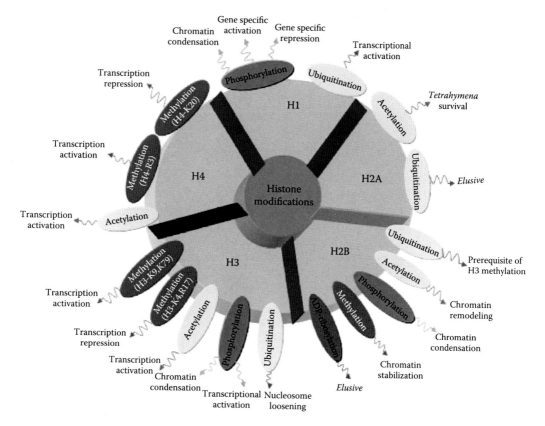

FIGURE 2.2
Pictorial representation of the histone modifications and their biological roles.

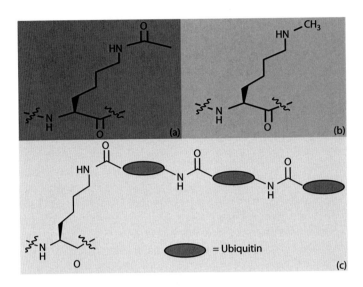

FIGURE 2.3

A typical example of histone acetylation (a), methylation (b), and ubiquitination (c). (Reprinted from *Journal of Genetics and Genomics*, 36, A. Munshi, G. Shafi, N. Aliya, and A. Jyothy, Histone modifications dictate specific biological readouts; Copyright 2009, with permission from Elsevier.)

tails, exposed on the nucleosome surface, are subject to a variety of enzyme-catalyzed PTMs. As previously mentioned, a repertoire of modifications has been documented to occur on histones. Moreover, each lysine residue can accept one, two, or even three methyl groups and, similarly, an arginine can be either monomethylated or dimethylated, which adds to the complexity. Hence, it is quite evident that histone modifications are implicated in diverse biological processes. Lately, the precise mechanistic roles have started to come into focus but the picture is not clear yet.

Multiple signaling pathways converge on histones to mediate covalent modifications of specific amino acids [43]. Although Allis and Strahl proposed the histone code in 2000, up until now it is not clear how the code is established and maintained. Many of the modification sites are close enough to each other and it seems that the modification of histone tails by one enzyme might influence the rate and efficiency at which other enzymes use the newly modified tails as a substrate [44]. The model proposed by Strahl and Allis [45] has set the ground for researchers to discover posttranslational writers (i.e., the enzymes that modify the histones posttranslationally) and their readers (i.e., the modules or motifs that recognize the histones based on their modified state) [45]. Although the basic composition of the nucleosome is the same over long stretches of chromatin, specific patterns of modification on the nucleosome create local structural and functional diversity, delimiting chromatin subdomains. Therefore, the challenge of histone PTMs has shifted from simply identifying sites of modification to identifying combinations of histone modification patterns that dictate specific biological readouts. Currently, the histone code is under a lot of investigation and is gaining experimental support as well.

Methylation of DNA at cytosine residues as well as PTMs of histones, including phosphorylation, acetylation, methylation, and ubiquitylation, contributes to the epigenetic information carried by the chromatin. These changes play an important role in the regulation of gene expression by modulating the access of regulatory factors to the DNA. The use of a combination of biochemical, genetic, and structural approaches has allowed the demonstration of the role of chromatin structure in transcriptional control. The structure of nucleosomes has been elucidated and enzymes involved in DNA or histone modifications have been extensively characterized. Although it has been known for many years that histones in eukaryotes are modified by acetylation, it is only in the past decade that the role of histone acetylation in transcription regulation has come under focus [46,47]. Of all the histone modifications, acetylation is the most studied one. In the beginning, many of the enzymes responsible for the acetylation of histones were known as transcriptional coactivators and later as enzymes.

During specific biological processes, selected lysines such as lysine 9 and 14 are acetylated. Acetylation of lysine residues at the core histone N-terminal is achieved through enzymes called histone acetyltransferases. However, the steady state balance of this modification is achieved through the orchestrated action of histone acetyltransferases and one more species of enzymes, namely, histone deacetylases [47,48]. Histone methylation was first reported more than 40 years ago, but this is among the least understood PTMs. In contrast to acetylation and phosphorylation, lysine methylation seems to be a relatively stable histone modification. The most heavily methylated histone is H3 followed by H4. Histone K methylation occurs on lysine residues [4,9,27], and on position 36 on H3 and on position 20 on H4. The H1 amino terminus also seems to possess K methylation. Therefore, lysine methylation might be providing an ideal epigenetic mark for more long-term maintenance of chromatin states [49]. Histone methyltransferases are the enzymes that direct the site-specific methylation of lysine residues, for example, lysine 9 and lysine 4 in the amino terminus of histone H3. It has been reported that the standard histone methyltransferases contain the evolutionarily conserved 130 amino acid SET domain and thereby stimulate some of the biological processes such as gene activation or repression [50].

The different histone modifications either act in a sequential manner or in combination to form a "histone code" or "epigenetic code," which is read by other proteins to bring about specific biological events. The epigenetic code might operate alongside the genetic code to interpret the DNA sequence. Workman [51] is of the opinion that exploitation of the histone code is likely to be the new "omics." This could lead to therapeutic exploitation of histone modifications and Workman has termed this as "histonomics." The histone-modifying enzymes might turn out to be a rich source of potential targets. Any diseases which are modulated by modification of histones can be targeted by interfering with the enzymes responsible for this. Therefore, drugs that can target these enzymes and thereby treat diseases such as cancer, neurological conditions, and other disorders might be the future treatment modalities. A detailed account of histone modifications has been published elsewhere [52].

2.1.3 ncRNAs

There is increasing evidence that most epigenetic mechanisms of gene expression control include regulation by ncRNAs, such as micro-RNAs, small RNAs, and long or large RNAs. ncRNAs with different regulatory functions have become a common feature of mammalian transcriptomes, especially after the discovery that most of the eukaryotic genome is transcribed into RNAs that do not code for any protein [53–55]. Cytosine methylation and histone modification, which are related to gene expression regulation in complex organisms, are directed by ncRNAs [53]. These ncRNAs are very important in various epigenetic modification mechanisms including transposon activity and silencing, position effect variegation, X chromosome inactivation, and paramutation. Increasing evidence is showing that ncRNAs are always transcribed from an imprinted region of the mammalian genome. This is very similar to the mechanisms used by the large ncRNA XIST during X chromosome inactivation in female cells. This suggests that ncRNAs are also involved in this type of silencing process in autosome chromosomes [56]. The role of ncRNAs in epigenetics has been reviewed by Costa [57].

Micro-RNAs constitute one of the largest families of gene regulators that are found in plants and animals. Most mature micro-RNAs are the product of RNA polymerase II–transcribed transcripts that have been processed by two RNase III enzymes, Drosha and Dicer. Mostly, they negatively regulate posttranscriptionally their targets in one of two ways depending on the degree of complementarity between micro-RNA and the target. First, micro-RNAs that bind with perfect or nearly perfect complementarity to protein-coding mRNA sequences induce the RNA-mediated interference pathway; mRNA transcripts are cleaved and degraded by multiprotein micro-RNA–induced silencing complex. This mechanism of micro-RNA–mediated gene silencing is commonly found in plants [58]. Secondly, micro-RNAs exert their regulatory effects by binding to imperfect complementarity sites within the 3′ UTRs of their mRNA targets, repressing target gene expression. Reduction of protein levels in this way is often modest; however, many such RNAs probably collectively fine-tune gene expression [59]. Although the protein levels of these genes are dampened, their mRNA levels are barely affected. This mechanism of micro-RNA–mediated gene expression control is mostly seen in animals [60].

Micro-RNAs and their targets seem to perform complex regulatory networks. For example, a simple micro-RNA can bind to and regulate many different mRNA targets and, conversely, several different micro-RNAs can bind to and cooperatively control a single mRNA target [61]. Micro-RNAs have key roles in diverse regulatory pathways, including control of metabolism, immunity, developmental timing, cell proliferation, cell differentiation, organ development, senescence, and apoptosis.

2.1.4 Regulatory Proteins

In addition to DNA methylation, histone modifications, and ncRNAs, certain proteins are also involved in the control of gene expression. For example, the polycomb group of proteins are involved in transcriptional repression, whereas the trithorax group of proteins act antagonistically to polycomb group of proteins to maintain gene transcriptional activation [62]. Both the polycomb and trithorax groups of proteins act as mediators of epigenetic memory [63]. Any abnormality in the expression of such proteins is likely to result in a diseased condition. For example, abnormalities in the expression of polycomb group of genes have been involved in cancer progression in mammals.

It has been observed that methylation per se does not prevent transcription, but instead transcriptional silencing of methylated loci is due to methyl CpG binding domain proteins that alter chromatin structure [64]. The methyl CpG binding domain proteins interact with and recruit histone deacetylases, which deacetylate the associated chromatin to render it transcriptionally incompetent [65]. There are several other roles that proteins play in gene expression such as in signal transduction, in ribosome function, in controlling the rate of transcription as well as translation, in insulating chromatin by creating regulatory boundaries, and in intercellular communication through gap junctions.

2.1.5 Genomic Imprinting

Genomic imprinting is an epigenetic mechanism that produces functional differences between the paternal and maternal genomes and plays an essential role in mammalian development and growth. There are approximately 156 genes in our genome that are subject to genomic imprinting in which one parent copy of the gene is expressed whereas the other is silent [66]. There is a rapid and possibly active genomewide demethylation in both male and female primordial germ cells resulting in the erasure of imprints [67]. This is accompanied by reactivation of the inactive X chromosome in the female germ cells [68]. The same mechanism also erases any aberrant epigenetic modification, preventing the inheritance of epimutations [69]. The precise mechanisms responsible for epigenetic erasure and demethylation in primordial germ cells are not yet clear. The initiation of new imprints occurs subsequently during gametogenesis and primarily during oogenesis [70] (Figure 2.4). Imprinting is controlled by the methylation of stretches of regulatory DNA, the imprinting control regions [71]. Interestingly, to get ready for the establishment of imprints, methyl marks on histones need to be removed [72].

The imprinted genes are particularly susceptible to environmental factors because their regulation is tightly linked to epigenetic mechanisms. Because imprinted genes are functionally haploid, the health consequences of genomic imprinting may sometimes be potentially disastrous. Monoallelic expression eliminates the protection that diploidy normally affords against deleterious effects of recessive mutations. The imprinted traits are passed on to the progeny via the sperm and the egg through sexual reproduction. Once the imprints are set, they are maintained in the somatic cells through subsequent cell divisions [73]. For normal development, genomic imprints of both the parents are necessary. Dysregulation of imprinted genes can occur in somatic cells, either by epigenetic or genetic mutations. If one of the imprints is missing, it leads to diseases. Prader–Willi syndrome and Angelman syndrome, two dysmorphic developmental disorders associated with learning, are caused due to missing paternal and maternal imprints, respectively, of the same chromosome region 15q11–13.

There is a relative degree of specificity of expression of imprinted genes, such that some genes are only monoallelic in expression in a particular tissue, whereas they have biallelic expression in other tissues. Angelman syndrome UBE3A gene is maternally imprinted in brain but is nonimprinted in other tissues [74]. The imprinted genes are expressed in a parent of origin-specific manner. However, for many genes reported to be imprinted, repression of the silent allele is partial. Recent studies in humans, mice, and plants have shown that different alleles of nonimprinted genes are also not expressed equally at the mRNA level [75]. Quantitative variability in the expression of imprinted alleles and differential expression of nonimprinted alleles allows organisms to achieve adaptation to a changing environment [76].

The expression of imprinted genes is species-, tissue-, and developmental stage–dependent, and there are indications that imprinting may have played an important role in mammalian evolution [77]. For example, the epigenetic mechanisms may thus be associated with the evolution of intrauterine development in mammals that requires the formation of a placenta. This is reflected by the fact that 80% of imprinted genes that have been identified to date are expressed in the placenta [78]. Different theories have been proposed to explain the evolutionary significance of imprinting, and of these, the parent–offspring conflict theory proposed by Trivers (1974) has a general appeal [79]. According to this theory, genes in parents and genes in offspring can have conflicting interests at the level of organization because they are all interested only in their own propagation. One such conflict is over maternal nutritional resources provided through the placenta to the fetus, where maternal genes in the fetus are predicted to be more conservative in extracting resources from the mother (to promote maternal survival to sustain multiple pregnancies), whereas paternal genes are greedier and want more nutrition (to promote fetal survival necessary for the survival of a species).

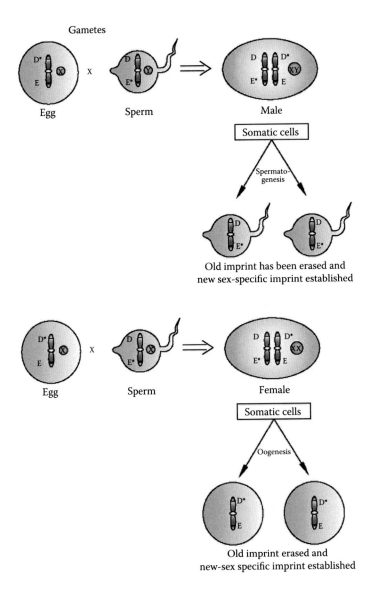

FIGURE 2.4
The diagram shows the chromosome carrying two imprinted genes, D and E. D is subjected to imprinting in the female germ line and E in the male germ line. D is imprinted when present on a maternally inherited chromosome and E is imprinted when present on a paternally inherited chromosome. The chromosome may pass through the male and female germ line in successive generations: a man may hand over a chromosome inherited from his mother and a woman can transmit a chromosome, which she received from her father. Therefore, a mechanism must be present which erases the old imprint from the germ line before establishing a new sex-specific imprint. (Reprinted from *Journal of Genetics and Genomics*, 34, A. Munshi and D. Shanti, Genomic imprinting—the story of other half and the conflicts of silencing, Copyright 2007, with permission from Elsevier.)

2.1.6 Differentiation

For the development of any organism, specific genes are switched on at the proper time in a stepwise order, and when their job is done, they are switched off. This orchestra is played in a coordinated fashion by direct/indirect interactions among DNMTs, histone-modifying enzymes, histone subtypes, chromatin remodeling factors, and ncRNAs. Quantitative and qualitative epigenetic changes must be maintained through multiple cell divisions to achieve normal development. During this process, about 250 different cell types, distributed in various organs and tissues, are produced in the human body [80]. Each cell within a multicellular organism has distinguishable characteristics established by its unique pattern of gene expression [81]. After fertilization, the zygote undergoes a series of divisions leading to blastocyst stage of the embryonic development. The blastocyst has a coat and an inner mass of cells which are totipotent embryonic stem cells. The number of genes that are highly expressed in undifferentiated embryonic stem cells is much higher than in adult tissues [82], and many genes are downregulated as the embryonic stem cells differentiate to a particular cell lineage [84]. During early embryonic development, genomic packaging receives a variety of signals to eventually set up cell type–specific expression pattern of genes. This process of regulated chromatinization leads to cell type–specific epigenomes. The expression state attained during the differentiation process needs to be maintained subsequently.

DNA methylation has been suggested to play an important role in the development. Some germ line and tissue-specific promoters have been shown to undergo DNA methylation during cellular differentiation. However, there are many developmental genes which are not silenced by DNA methylation [84–86]. Therefore, the question arises if DNA methylation really plays a substantial role in the development process. A study carried out by Weber et al. [87] has provided new evidence implicating DNA methylation in the silencing of germ line–specific regions. This study has found a subset of genes that were methylated in fibroblasts but not in sperm. Quite a good number of these differentially methylated genes were found to be germ line–specific. This shows that somatic gene DNA methylation does contribute to differentiation by repressing key genes in the germ line and irreversibly forcing the cell on a path to differentiation [87].

2.2 Environmental Modulations of the Epigenome

The transcriptomes of an organism are continually changing in response to developmental and environmental clues. A given genotype can give rise to different phenotypes, depending on environmental conditions. The potential noncorrespondence between genetic and epigenetic information allows a single genotype to adopt multiple epigenotypes conferring different phenotypes. The epigenome is particularly susceptible to deregulation during gestation, neonatal development, puberty, and old age. It is most vulnerable to environmental factors during embryogenesis because the DNA synthesis rate is high, and the elaborate DNA methylation patterning required for normal tissue development is established during early development. Three potential epigenetic susceptible targets for environmentally induced effects are (1) transposable elements, (2) the promoter regions of housekeeping genes, and (3) *cis*-acting regulatory elements of imprinted genes.

These genome targets contain CpG islands that are normally methylated or differentially methylated, respectively [77].

Experiments with model systems have shown that tobacco smoke, lead, arsenic, cadmium, nickel, ionizing, nonionizing, and UV radiation are epigenetic mutagens or carcinogens [88]. Other environmental agents including viruses, such as human papilloma virus, Epstein–Barr virus, human hepatitis virus, and bacteria such as *Helicobacter pylori* may also alter the expression of host genes via epigenetic strategy. Methylation of viral genomes, binding of viral proteins to host gene promoters, hypermethylation of host genes, changes in chromatin modification patterns, and recruitment of histone deacetylases are some of the mechanisms suggested for virally and bacterially mediated epigenetic changes responsible for the carcinogenesis of hepatocellular, gastric, nasopharyngeal, and cervical epithelia [4]. Too little methylation across the genome or too much methylation in the CpG islands can cause cancer by activating nearby oncogenes and by silencing tumor suppressor genes, respectively. In short, various aspects of the environment, including exposure to mutagens or other agents such as diet, pathogens, and behavior patterns (e.g., maternal nursing) can also induce epigenetic variation [14]. Moreover, evidence that epigenetic alterations may be inherited transgenerationally is accumulating, thereby affecting the health of future generations. Often, these effects are not fully penetrant and are inherited in a complex non-Mendelian manner. Some of the endogenous or exogenous environmental factors (Figure 2.5) that need special attention will be discussed in the subsequent sections.

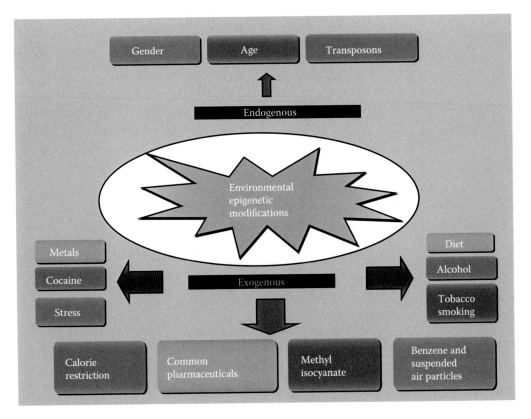

FIGURE 2.5
Endogenous and exogenous environmental factors modulating the epigenome.

2.2.1 Endogenous Environmental Factors

2.2.1.1 Transposons

Transposable elements (including retrotransposons and DNA transposons) are parasitic elements. Transposons are groups of mobile genetic elements that replicate, increase in copy number, and persist in nature by moving and inserting into genes. When activated, they may cause genetic mutations and transcriptional noise [89]. Transposable elements–related sequences are numerous and scattered throughout the mouse and human genome. Approximately 40% of the human genome constitutes methyl-rich transposable elements, of which 9% are retrotransposons [90]. The Alu family alone consists of several hundred thousand elements which are shown to be transcriptionally silent in somatic cells. ncRNAs, histone modifications, and DNA methylation are mechanisms that might act in synergy in transposon silencing as well as in affecting transposition [30,91]. Mutations in the components of epigenetic silencing, the DNMT1 or the RNA interference pathway, lead to increased transcription and mobility of transposable elements [92]. Transposable elements have also been shown to be activated by different kinds of cellular stress [93], and stress induced by environmental or dietary agents.

Most transposable elements in the mammalian genome are normally silenced by CpG methylation. The epigenetic state of a subset of transposable elements, however, is metastable and can affect regions encompassing neighboring genes. The epigenetic metastability of such regions renders them susceptible to nutritional influences during early development. Hence, epigenetic alterations among metastable epialleles are a likely mechanistic link between early nutrition and adult disease susceptibility [94]. Activation of a transposable element under dietary stress is illustrated by the following example. In wild-type (A/A) mice, there is a subapical yellow band on each black hair shaft causing the brown (agouti) coat color, which is controlled by a metastable gene. The A^{vy} (variable yellow agouti) allele is the result of an insertion of a murine IAP (intracisternal A particles, an endogenous retrovirus) transposable element [95]. The degree of 1AP methylation varies dramatically among A^{vy}/a isogenic individual mice, causing a wide distribution in coat color from yellow (unmethylated) to agouti (methylated). Feeding pregnant female mice a diet supplement with high doses of folic acid, choline, and vitamin B_{12} shifts the coat color of their offspring by hypermethylation of the A^{vy} IAP promoter leading to the agouti phenotype [96].

2.2.1.2 Age

Cellular senescence seems to be accompanied by a redistribution of heterochromatin from constitutive heterochromatin to other normally euchromatic sites, especially to specialized domains of facultative heterochromatin, called senescence-associated heterochromatin foci [97]. The age-associated divergence of genetically identical MZ is consistent with a stochastic component of the change. A stochastic and presumably unregulated component of the age-associated change in chromatin structure, which seems inherent, given the underlying biochemistry of chromatin (global hypomethylation, CpG island hypermethylation, and senescence-associated heterochromatin foci), may lead to gradual deterioration in cell and tissue function with age [98]. Mostly common disease incidence starts to rise during middle age and increases decade after decade thereafter. Random changes in chromatin or methylation patterns from one generation to the next, or within the life span of a single generation, could contribute greatly to the total variation in epigenotypes [99].

Phenotypic variation of a common disease may also be influenced by age and environmental factors. Epigenetic variation, even at a single locus, might also help explain the quantitative nature of common disease phenotypes despite its being a polygenic trait. Although it is likely that common diseases are multigenic, even a single locus could show a gaussian distribution of phenotype based on the quantitative nature of epigenetic variation. The causes of age-related epigenetic changes are largely unknown, although some studies have suggested the role of environmental exposures or modifier genes [100]. The environmental factors promoting these changes in methylation patterns may include cadmium, nickel, arsenic, chronic UV exposure, or infectious agents (e.g., *H. pylori*) and dietary factors such as excessive alcohol intake, folate deficiency, aging, and epigenetic drift (loss of imprinting; reviewed by Herceg [4]). Also, DNA hypomethylation increases with age due to an age-dependent decrease in the proficiency of DNA repair [101].

2.2.1.3 Gender

Gender can be considered an environmental variable that includes the cellular, metabolic, physiological, anatomical, and even behavioral differences between males and females. Therefore, sex might interact with the genotype in a manner similar to other environmental factors. Nearly all human diseases have sex-specific differences in prevalence, age of onset, and severity. For example, cardiovascular disease, which is predominant in men throughout adulthood, has a higher rate of occurrence in postmenopausal women compared with men of the same age. Again, in comparison with men, cardiovascular disease is extremely uncommon in premenopausal women, and the protective role of estrogen has been demonstrated [102]. It has been shown that estrogen can inhibit cardiomyocyte apoptosis, which otherwise could have contributed significantly to the loss of cardiomyocytes [103]. Estrogen decreases Reactive Oxygen Species (ROS) production and increases intracellular antioxidants such as superoxide dismutase [104]. In infancy or childhood, neural tube defects, congenital dislocation of the hip, and scoliosis are more common among girls, whereas autism, stuttering, and pyloric stenosis are more common among boys.

In adulthood, major depression and Alzheimer disease are more common in women, whereas schizophrenia, Parkinson disease, and colorectal cancer are more common in men. Both the immune and endocrine systems probably contribute to sexual dimorphism in the incidence of many common diseases. However, some of the differences between males and females may also be due to differences in genetic and epigenetic architecture [105]. To examine the effects of chronic maternal stress on cognitive function in the offspring during young age, pregnant Wistar rats were subjected to restraint stress. Prenatal maternal restraint stress impaired memory retention during young age in both sexes; however, the performances of the females were better than that of the males. This difference in performance could be due to activation of estrogenic effects [106].

2.2.2 Exogenous or Environmental Factors

2.2.2.1 Tobacco Smoking

Epigenetic disturbances due to cigarette smoking have not only been seen in benign conditions like chronic obstructive pulmonary disease, but also during the development of malignancy. Aberrant promoter methylation of several genes, including p16 and MGMT (O^6-methylguanine-methyltransferase), has been detected in DNA from sputum in 100% of patients (with a history of smoking) with lung squamous cell carcinoma even up to 3 years before clinical manifestation of cancer [107].

2.2.2.2 Alcohol

Chronic alcohol administration causes chromatin changes [108,109]. Alcohol consumption leads to a state of folate deficiency which, in turn may affect gene expression by disturbing DNA methylation patterns or by inducing base substitution, DNA breaks, and gene amplification. Alcohol may inhibit the activity of DNA methylase, the important determinant of epigenetic regulation [110]. The epigenetic changes associated with alcohol-metabolizing genes may also enhance the other epigenetic toxic effects of alcohol on different organs. For example, fetal-alcohol syndrome in the newborn is the result of alcohol abuse by pregnant women [111]. Maternal alcohol consumption has also been seen to be associated with lasting changes in DNA methylation in the fetal heart, which may contribute to ischemic injury later in life [112]. Chronic alcohol consumption may cause liver cirrhosis, which in turn may lead to cancer due to aberrant genetic and epigenetic changes.

2.2.2.3 Cocaine

Dopamine controls many brain functions, including roles in behavior and cognition, motor activity, motivation, and reward. Drug-induced changes in gene expression, largely driven by dopaminergic signaling, occur in the nucleus accumbens and other relevant regions of the brain [113]. Repeated cocaine exposure has been seen to induce a distinct set of genes (e.g., cdk5 and bdnf) in the nucleus accumbens, a brain reward region. Some of these genes remain elevated for days to weeks even after withdrawal of cocaine [114]. Associated with the changes in gene expression, increased histone H3 acetylation was observed on the promoters of both cdk5 and bdnf [115].

2.2.2.4 Metals

Heavy metals, which are environmental toxins, disrupt DNA methylation and chromatin structure, ending up in some disease [116]. For example, mercury has been implicated in neurodevelopmental disorders. Usually, mercury is found to be the main culprit in Minamata disease, in which motor neurons are badly affected. Mercury has also been seen to be implicated in the causation of autism, another neurodegenerative disorder [117]. Experiments in tissue derived from old monkeys that were exposed to lead (Pb) as infants, demonstrated that the APP (amyloid precursor protein) mRNA, APP and Aβ are elevated [118]. It is plausible that developmental exposure to Pb could induce the demethylation process of the APP promoter in old age, thus elevating APP expression. Developmental Pb exposure also increases 8-oxo dG production due to the elevation of oxidative stress in old age [119]. Elevation of APP, Aβ, and 8-oxodG are indicators of Alzheimer disease.

A study on sperm chromatin structure in monkeys found that alterations at environmentally relevant blood Pb levels caused a decrease in levels of protamine–DNA interactions that may have altered sperm chromatin condensation [120,121]. Alterations in sperm chromatin structure point to the potential of Pb toxicity in decreasing fertility. To identify the genes that are altered owing to infantile exposure to Pb in the frontal association cortex of 23-year-old *Cyanomolgus* monkeys, microarray analysis of about 588 neurobiology-related human genes was conducted. The expression profile of 22 genes, belonging to the neurotransmitter, growth receptor, and signal pathways, changed owing to early-life exposure to Pb; 20 of the altered genes were abundant in CpG dinucleotides in their regulatory regions. About one-third of genes elevated and the remaining one-third of genes were repressed [122]. This study showcased the multisystem epitoxicity of Pb.

Arsenic has been recognized to be multisite carcinogen [123]. Long-term exposure to arsenic is associated with an inhibitory effect of expression and activity of DNA methyltransferases, which in turn may result in the depletion of cellular pools of S-adenosyl methionine leading to global DNA hypomethylation, a phenomenon believed to promote the activation of oncogenes [124]. Many studies also support the role of arsenic in altering one or more DNA repair processes [125]. *In vitro* studies showed that the addition of cadmium (Cd) to hepatic nuclear extracts inhibited DNA methyltransferase [126]. The Cd effect on DNA-methyltransferase indicated that the action of Cd via decreasing DNA methylation may be responsible for its carcinogenic properties [127].

2.2.2.5 Benzene and Suspended Air Particles

Low-level occupational exposure of gas station attendants to benzene, and traffic police to traffic-derived particles has resulted in significant epigenetic alterations, as characterized by significant reduction of long interspersed nucleotide elements-1 and melanoma antigen-1 gene methylation in blood DNA samples, as compared with control subjects [128].

2.2.2.6 Methyl Isocyanate

The Bhopal gas tragedy took place in 1984 and has thus far been the worst industrial holocaust in India. There was a leakage of 30 to 40 tons of methyl isocyanate, a toxic gas, spreading over approximately 45 km². In addition to a large number of animals, the human death toll reached 6000, with approximately 200,000 injured. In a 2009 study of the long-term effects of methyl isocyanate, Mishra et al. [129] showed that *in utero*, in the first trimester of pregnancy, methyl isocyanate exposure from more than two decades ago caused a persistently hyperresponsive cellular and humoral immune state in affected individuals. This modulation of the immune system is possibly at the level of the epigenome.

2.2.2.7 Common Pharmaceuticals

Without doubt, it is now well established that various environmental agents cause epigenetic changes, it has been recently hypothesized that even commonly used pharmaceutical drugs can cause persistent epigenetic changes, which can be manifested in the persistence of drug-induced adverse events (side effects). The drugs may alter epigenetic homeostasis by direct or indirect mechanisms. The drugs which affect chromatin architecture on DNA methylation can cause direct effects, for example, the antihypertensive drugs hydralazine inhibits DNA methylation. Drugs interfering with gene expression mechanisms cause indirect effects, for example, isotretinoin has transcription factor activity [10].

2.2.2.8 Diet

Folates are members of the B complex vitamins, which are required for the synthesis of nucleotide precursors and the methylation of a wide variety of essential biological substances including phospholipids, proteins, DNA, and neurotransmitters [130]. Folates are required for the maintenance of intracellular homeostasis. Folates cannot be synthesized *de novo* by mammals; hence, an efficient intestinal absorption process is required. Folates play an essential role in one-carbon transfer involving remethylation of homocysteine to methionine, which is a precursor of S-adenosyl methionine, the primary methyl group donor for most biological methylations. Dietary deficiency in folic acid and methionine, which are necessary for normal biosynthesis of *S*-adenosyl methionine, leads to aberrant imprinting of insulin-like growth factor 2 [12], which in turn affects the growth and development of an individual. Neural tube defects are classic abnormalities due to folic acid deficiency. There is increasing evidence for epigenetic modulation in response to an abnormal intrauterine environment associated with epigenetic downregulation of genes involved in pancreatic β-cell function. Studies of individuals with compromised folate pathways have found that genomic DNA methylation correlates positively with folate status and negatively with homocysteine status [131]. Cancer, cardiovascular diseases, schizophrenia, and autism are complex diseases that have been linked to levels of folate and homocysteine. Vitamin intake (or depletion), calorie restriction (CR), and diet variables contribute to and affect epigenetic signatures in ways that have yet to be fully determined.

2.2.2.9 Calorie Restriction

McCay et al. [132] demonstrated that rats fed with CR lived longer as compared with rats fed with a standard diet. The effects of CR on life span are very similar among diverse eukaryotes (e.g., yeast, insects, fish, and mammals) [133]. CR induces a decrease in protein oxidation, lipid peroxidation, and ROS production and this translates into a decrease in mitochondrial metabolism [134]. The delaying effect of CR on aging is based on increased genomic stability through increased expression of DNA repair machinery [135], including increased fidelity of DNA replication [136]. Consequently, the effect of CR on life span could be elicited through a long-term activation of a detoxifying machinery, cell cycle progression, and inhibition of apoptosis and senescence [137]. Also implicit to the CR response is gene inactivation, which seems to occur via two different mechanisms: regulation of individual genes to modulation of specific transcription factors and heterochromatinization of large parts of a genome that might include many genes [138].

2.2.2.10 Stress

Individual cells respond to epigenetically stressful conditions, including but not limited to heat, by producing heat shock proteins (HSPs), which help them to cope by keeping cellular processes working smoothly in the face of adversity. The HSPs have two functions: to inhibit undesirable happenings and to promote desirable ones, so that a stable and productive bond forms between protein partners. For example, when HSP90 function was suppressed in *Drosophila melanogaster*, a large number of pre-existing genetic mutations were unmasked, indicating that potentially deleterious effects were being buffered by HSP90 [139]. During exercise, the body is under oxidative stress, and in response to stress, HSPs are produced. These HSPs in turn stimulate the immune response. Consequently, exercise has an important role in the enhancement of an individual's immunity [140]. Individuals who regularly work out (nonstrenuous) in gymnasia may experience beneficial effects such as enhanced antioxidant enzyme systems [141], upregulated DNA repair enzymes [142], or increased clearance of cells damaged by apoptosis [143].

Regular exercise has multiple benefits, but strenuous exercise generates an excessive amount of free radicals (ROS), which cause oxidative stress. In case antioxidant defense of the body is not adequate to take care of the excessive ROS, they react with membranes, proteins, nucleic acids, and cellular components to initiate cellular damage and degeneration [144]. In the exercising skeletal muscle, a large number of genes are activated after exercise. These include stress response genes (for HSPs), metabolic priority genes (for management of blood glucose and muscle glycogen), and metabolic/mitochondrial enzyme genes (for producing proteins which convert food into energy).

Physical activity lowers the glycemic index of carbohydrates. Exercise, through certain signaling pathways, promotes the level of glucose transporter protein, GLUT4. During a sedentary lifestyle, GLUT4 levels decrease, leading to atherosclerosis, obesity, type 2 diabetes, cancer, weakness in old age, and many other chronic diseases [145]. Cancer cells, because they are abnormal, are under constant stress and, as a result, they are generating higher levels of HSPs. HSP90 in particular is believed to help cancer cells survive stressful conditions just as it does for normal cells. Therefore, inhibiting HSP90 could make malignant cells more vulnerable to toxic therapies. Hence, pharmacological inhibitors with an increasing specificity for HSPs are being tested in patients with cancer in combination with chemotherapy [140].

Depression and mood disorders are signs of physiological stress. Microarray analysis of specimens from the nucleus accumbens revealed hundreds of epigenetically deregulated gene promoters implicated in the pathophysiology of depression and related mood disorders. Antidepressant treatment resulted in mood restoration associated with the normalization of methylation status of many of the deregulated promoters [146]. Besides humans, the effects of stress have also been studied in other organisms. Due to thermal stress from global warming, coral health has suffered a great deal around the world. To understand the molecular and cellular basis of gene expression changes associated with thermal stress, bleaching, and declining health in Caribbean coral, DeSalvo et al. [147] used a complementary DNA microarray containing 1310 genes. On the basis of differentially expressed genes (due to epigenetic modulations), in their transcriptomic study, they postulated that oxidative stress in thermal stressed corals causes a disruption of Ca^{2+} homeostasis, which in turn leads to cytoskeleton and cell adhesion changes, decreased calcification, and initiation of cell death via apoptosis and necrosis [147].

2.2.2.11 Early-Life Environmental Experiences

The reduced fearfulness of high maternally groomed rats is due to an increase in the number of glucocorticoid receptors in the hippocampus. High-grooming by mothers is associated with DNA hypomethylation and histone acetylation of the glucocorticoid receptor promoter resulting in its expression. This increase in the expression of glucocorticoid receptor in turn results in more glucocorticoid receptors in the hippocampus. The epigenetic marks seem to maintain the glucocorticoid receptor expression state for the rest of the rat's life [148,149].

Epidemiological studies have shown that intrauterine growth retardation affecting fetal development predisposes the offspring to obesity or insulin resistance/diabetes in later life [150]. For example, African-Americans have a higher rate of low birth weight followed by a higher rate of cardiovascular diseases in adults, relative to whites [151]. This phenomenon has been explained by thrifty phenotype hypothesis. According to this hypothesis, there is an increased level of insulin resistance in the offspring of starved mothers. Increased insulin resistance causes energy conservation and reduced somatic growth to allow the offspring a better chance of survival in an environment in which nutrition is poor. However, with insulin resistance, the higher levels of plasma insulin, glucose, and fatty acids become a problem if food becomes abundantly available [152]. Lately, it has been established that changes in DNA methylation and histone modifications occur from nutritional restriction during gestation [153], and these changes persist to adulthood [154]. Aside from starvation, maternal vitamin/mineral/trace element deficiencies may also predispose the offspring to insulin resistance and associated diseases in their later life. This animal model study reiterates the importance of maternal micronutrient status in determining the offspring's body, which predisposes the offspring to adult onset diseases [155].

2.2.2.12 Low-Dose Radiation

Low-dose radiation sources have become nearly ubiquitous in our environment as a result of nuclear tests, radiation accidents, and diagnostic, therapeutic, and occupational exposures. Exposure to low-dose ionizing radiation causes a number of long-term effects. One of them is genomic instability in terms of chromosomal aberrations and gene mutations. It has a transgenerational nature and thus may be a precursor to tumorigenesis and genetic defects. The transgenerational nature of genomic instability indicates the possible involvement of epigenetic mechanism [156,157]. Kovalchuk et al. [158] observed that chronic low-dose radiation exposure was a more potent inducer of epigenetic effects than acute exposure. They also noted sex-specific and tissue-specific differences in p16$^{INK\alpha}$ (a tumor suppressor) as well as global genomic DNA methylation patterns in the liver and muscle of exposed male and female mice upon low-dose radiation exposure.

2.2.2.13 Electromagnetic Fields

Technological developments have added a large number of gadgets in our daily life, and most of these gadgets run with the help of electromagnetic fields (EMFs). At home or at the workplace, we have televisions, computers, air-conditioners, and other gadgets for entertainment and to facilitate work. On top of that, mobile phones have become personal companions. Exposure to these gadgets adds an extra load of EMFs over and above the low level of natural EMFs from the cosmos necessary for our survival. No doubt, our lifestyle would be incomplete and progress insufficiently without these gadgets. Nevertheless, an overexposure to EMFs may have harmful health effects.

Assisted reproduction technologies (ARTs) have been linked to low success rates as well as to increased risk of intrauterine growth retardation, premature birth, low birth weight, and prenatal death [159]. There have been several reports of an increased risk of epigenetic and genomic imprinting disorders associated with ART [160]. For example, the prevalence of ART use was approximately four to nine times greater among children with Beckwith–Wiedemann syndrome (BWS) than among children in the general population. When molecular analysis of patients with BWS conceived through *in vitro* fertilization or intracytoplasmic sperm injection was carried out, it revealed a loss of maternal-specific DNA methylation at imprinting centers [161]. Thirteen out of fourteen reported cases of BWS associated with ART involved hypomethylation of the LIT1 gene [159], compared with about one in three control patients with BWS. In addition to BWS, increased incidences of Angelman syndrome and retinoblastoma, also due to epigenetic disturbances, have been reported in children born through ART [162]. One of the reasons for an increase in these disorders may be the exposure of the stem cells in early embryonic development to EMFs during incubation before implantation [163]. Gene expression and methylation status in animal embryos can get affected in *in vitro* culture conditions [164]. *In vitro* embryo culture in animal models (sheep) has also been found to lead to reduced viability and growth, developmental abnormalities, behavioral changes, and loss of imprinting at the IGF2R locus [165]. A wide variety of cell types, including mouse embryonic stem cells had altered methylation levels during *in vitro* culture [166]. As a matter of fact, there is ample evidence to support the contention that EMFs affect the gene expression and differentiation process through epigenetic mechanisms [167]. The low success rate seen in ARTs may also be due to the epigenetic disruption caused by EMFs during incubation of early embryos [168].

Cloning involves the transplantation of a somatic nucleus into an enucleated oocyte. During reprogramming of somatic nuclei to totipotency, there are possibilities in which some imprints may be erased leading to fetal and growth anomalies as well as postnatal health problems [169,170], consequently affecting the success rate of cloning, which has been seen to be rather low. There may be some disturbances in imprints during cloning, which also involves ART.

2.3 Transgenerational Epigenetic Inheritance

Certain eating patterns, or smoking at critical periods in life, cause epigenetic changes. These changes may persist for more than one generation. Epidemiological studies have highlighted that these factors including lifestyle can cause semipermanent changes in the germ line during critical periods and may be responsible for health problems in subsequent generations. Male F1 offspring of gestating rats exposed to vincolozolin showed decreased spermatogenic capacity. Their phenotype was transmitted to subsequent generations through the male germ line, affecting even the F4 generation; it was found that the endocrine disruptors reprogrammed the epigenetic state of the male germ line during development and induced the presence of new imprinted–like genes/DNA sequences that were stably transmitted through the male germ line [171].

Feeding low-protein or unpalatable diet to rats for 12 generations progressively reduced birth weight, which did not return to control values until three generations after reinstating a balanced diet [172]. Transgenerational diabetes was observed in F1 and F2 generations of a grand maternal (P1) rat which was infused with glucose during the last week of pregnancy (third trimester). A similar situation is seen in humans, in which maternal F1 and grand maternal (P1) type 2 diabetes is associated with gestational diabetes in F2 [173]. During the winter famine of 1944 to 1945 in the Netherlands, previously adequately nourished women were subjected to low calorie intake and associated environmental stress. Pregnant women exposed to famine in late pregnancy gave birth to smaller babies who later developed an increased risk of insulin resistance. F2 offspring of females exposed *in utero* in the first trimester also had reduced birth size independent of any effect on F1 maternal birth weight [174]. Individuals who were prenatally exposed to this famine had abundant food later on but had, six decades later, less DNA methylation of the imprinted insulin-like growth factor 2 gene compared with their unexposed same-sex siblings [175]. Although the molecular nature of transgenerational epigenetic memory is not properly understood, a complex network of ncRNAs seems to play a significant role in this memory.

2.4 Conclusions

Like many scientific disciplines, the study of epigenetics began with a collection of unexpected and unusual phenomenon in both plants and animals, for example, position–effect variegation in *Drosophila*, X chromosome inactivation in mammals, and RNA interference in heterochromatin formation in fission from yeasts and plants. No doubt, not all of the mysteries have been solved, a considerable headway has been made in understanding the molecular mechanisms involved in epigenetic regulation. Although in general, focus has been on CpG islands for methylation and silencing of a gene, it has been found that about 40 human genes do not contain bona fide CpG islands [176]. Eckhardt et al. [177] carried out DNA methylation profiling of 873 genes located on chromosomes 6, 20, and 22, and showed that 17% of these genes were differentially methylated and that only one-third of these showed inverse correlation between methylation and transcription. In other words, there is more to gene silencing than CpG island methylation. Various studies have shown that nucleosomal remodeling is a key component to the epigenetic silencing of genes. Nucleosomal remodeling in turn is linked to DNA cytosine methylation, histone modifications, and ncRNAs. A lot of research is converging on the study of covalent and noncovalent modifications of DNA and histones as well as on the mechanisms by which the overall chromatin structure gets altered by such modifications. Research into epigenetics certainly holds promise for revealing the inner complexity of biological systems and also yields the hope that new therapeutic possibilities and strategies would be found, targeting the various nuances involved in the regulation of chromatin architecture. Esteller and coworkers studied 80 pairs of identical twins, ranging between 3 and 74 years old and found that epigenetic differences were hardly detectable in the youngest twins, but increased markedly with age. These epigenetic changes had a striking effect on gene activity; 50-year-old twins had more than three times the genes that differ in activity in comparison with pairs aged 3 years. "So we are more than our genes." Cells must read another code, "the epigenetic code," which overlays the "genetic code." The future of epigenetic therapy holds tremendous potential not only for individualized health care but also for disease diagnostic and preventive measures.

Acknowledgments

We are indebted to P. Diwakar and Saleha Bhanu Vayyuri for help in literature search, and Preetha Shetty, Ch. Rosy, and P. Nagarjuna for secretarial assistance.

References

1. Waddington, C.H. 1942. Genes, genetics and epigenetics: The epigenotype. *Endeavour* 1:18–20.
2. Russo, V.E.A., R.A. Martienssen, and A.D. Riggs. 1996. *Epigenetic Mechanisms of Gene Regulation.* Woodbury: Cold Spring Harbor Laboratory Press.
3. Bird, A. 2007. Perceptions of epigenetics. *Nature* 447:396–398.
4. Herceg, Z. 2007. Epigenetics and cancer: Towards an evaluation of the impact of environmental and dietary factors. *Mutagenesis* 22:91–103.
5. Esteller, M. 2008. Epigenetics in cancer. *New England Journal of Medicine* 358:1148–1159.
6. Fraga, M.F., M. Herranz, J. Espada, E. Ballestar, M.F. Paz, S. Ropero et al. 2004. A mouse skin multi stage carcinogenesis model reflex aberrant DNA methylation patterns of human tumors. *Cancer Research* 64:5527–5534.

7. Holliday, R., and J.E. Pugh. 1975. DNA modification mechanisms and gene activity during development. *Science* 187:226–232.

8. Riggs, A.D. 1975. X inactivation, differentiation and DNA methylation. *Cytogenetics and Cell Genetics* 14:9–25.

9. Scarno, E. 1971. The control of gene function in cell differentiation and in embryogenesis. *Advanced Cytopharmacology* 1:13–24.

10. Csoka, A.B., and M. Szyf. 2009. Epigenetic side-effects of common pharmaceuticals: A potential new field in medicine and pharmacology. *Medical Hypotheses* 73:770–780.

11. Anway, M.D., A.S. Cupp, M. Uzumcu, and M.K. Skinner. 2005. Epigenetic transgenerational actions of endocrine disruptors and male fertility. *Science* 308:1466–1469.

12. Waterland, R.A., D.C. Dolinoy, J.R. Lin, C.A. Smith, X. Shi, and K.G. Tahiliani. 2006. Maternal methyl supplements increase offspring DNA methylation at axin fused. *Genesis* 44:401–406.

13. Roth, T.L., F.D. Lubin, A.J. Funk, and J.D. Sweatt. 2009. Lasting epigenetic influence of early-life adversely on the BDNF gene. *Biological Psychiatry* 65:762–769.

14. Richards, E.J. 2006. Inherited epigenetic variation—revisiting soft inheritance. *Nature Reviews Genetics* 7:305–401.

15. Hotchkiss, R.D. 1948. The quantitative separation of purines, pyrimidines and nucleosides by paper chromatography. *Journal of Biological Chemistry* 175:315–332.

16. Cedar, H., and A. Razin. 1990. DNA methylation and development. *Biochimica et Biophysica Acta* 1049:1–8.

17. Lyon, M. 1992. Some milestones in the history of X chromosome inactivation. *Annual Reviews Genetics* 26:17–28.

18. Mikkelsen, T.S., M. Ku, D.B. Jaffe, B. Issac, E. Lieberman, G. Giannoukos et al. 2007. Genome-wide maps of chromatin state in pluripotent lineage committed cells. *Nature* 448:553–560.

19. Cross, S.H., and A.P. Bird. 1995. CpG islands and genes. *Current Opinion Genetics Development* 5:309–314.

20. Bird, A. 1992. The essentials of DNA methylation. *Cell* 70:5–8.

21. Cairns-Smith, S., and P. Karran. 1992. Epigenetics silencing of DNA repair enzyme O^6 methyl guanine-DNA methyl transferase in Mex-human cells. *Cancer Research* 52:5257–5263.

22. Stöger, R., P. Kubicka, C.G. Liu, T. Kafri, A. Razin, H. Cedar et al. 1993. Maternal IGF-2R locus identifies the expressed locus as carrying the imprinted signal. *Cell* 73:61–71.

23. Rollins, R.A., F. Haghighi, J.R. Edwards, R. Das, M.Q. Zhang, J. Ju et al. 2006. Large scale structure of genomic methylation patterns. *Genome Research* 16:157–163.

24. Suzuki, M.M., and A. Bird. 2008. DNA methylation landscapes: Provocative insights from epigenomics. *Nature Reviews Genetics* 9:465–476.

25. Goll, M.G., and T.H. Bestor. 2005. Eukaryotic cytocine methyl transferases. *Annual Reviews Biochemistry* 74:481–514.

26. Okano, M., D.W. Bell, D.A. Haber, and E. Li. 1999. DNA methyl transferases DNMT 3a and DNMT3b are essential for *de novo* methylation and mammalian development. *Cell* 99:247–257.

27. Gowher, H., K. Liebert, A. Hermann, G.X., and A. Jeltsch. 2005. Mechanism of stimulation of catalytic activity of Dmmt3A and Dmmt3D DNA. (Cytocine-C′5)-methyl transferases by DNMT 3L. *Journal of Biological Chemistry* 280:13341–13348.

28. Zhu, H., T.M. Geiman, S. Xi, Q. Jiang, S. Anja, T. Chen et al. 2006. Lsh is involved in *de novo* methylation of DNA. *EMBO Journal* 25:335–345.

29. Fabbri, M., R. Garzon, A. Cimmino, Z. Liu, N. Zanesi, E. Calegari et al. 2007. Micro RNA 29 family reverts aberrant methylation in lung cancer by targeting DNA methyltransferases 3A and 3B. *Proceedings of the National Academy of Sciences of the United States of America* 104:15805–15810.

30. Kato, M., A. Miura, J. Bender, E.J. Steven, and K. Tetsuji. 2003. Role of CG and non-CG methylation in immobilization of transposons in arabidopsis. *Current Biology* 13:421–426.

31. Leonhardt, H., A.W. Page, H.U. Weier, and T.H. Bester. 1992. A targeting sequence directs DNA methyl transferase to sites of DNA replication in mammalian nuclei. *Cell* 71:865–873.

32. Liu, Y., E.J. Oakeley, L. Sun, and J.P. Jost. 1998. Multiple domains are involved in the targeting of the mouse DNA methyltransferase to the DNA replication foci. *Nucleic Acids Research* 26:1038–1045.
33. Liang, G., M.F. Chan, Y. Tomigahara, Y.C. Tsai, F.A. Gonzales, E. Li et al. 2002. Cooperatively between DNA methyltransferases in the maintenance methylation of respective elements. *Molecular Cell Biology* 22:480–491.
34. Takeuchi, T., Y. Watanabe, T. Takano-Shimizu, and S. Kondo. 2006. Roles of jumonji and jumonji family genes in chromatin regulation and development. *Developmental Dynamics* 35:2449–2459.
35. McGarvey, K.M., A.F. Jill, G. Eriko, M. Joost, J. Thomas, and B.B. Stephen. 2006. Silenced tumor suppressor genes reactivated by DNA demethylation do not return to a fully euchromatic state. *Cancer Research* 66:3541–3549.
36. Comb, M., and H.M. Goodman. 1990. CpG methylation inhibits proenkephalin gene expression and binding of transcription factor Ap-2. *Nucleic Acid Research* 18:3975–3982.
37. Prendergast, G.C., D. Lawe, and E.B. Ziff. 1991. Association of Myn, the murine homolog of max, with c-Myc stimulates methylation-sensitive DNA binding and ras cotransformation. *Cell* 65:395–407.
38. Yap, K.L., and M.M. Zhou. 2006. Structure and function of protein modules in chromatin biology. *Results and Problems in Cell Differentiation* 41:1–23.
39. Peterson, C.L., and M.A. Laniel. 2004. Histones and histone modifications. *Current Biology* 14:R546–R551.
40. Taverna, S.D., D.C. Allis, and S.B. Hake. 2007. Hunting for post-translational modifications that underline the histone code. *International Journal of Mass Spectrometry* 259:40–45.
41. Somech, R., S. Izrael, and A.J. Simon. 2004. Histone deacetylase inhibitors a new tool to treat cancer. *Cancer Treatment Reviews* 30:461–472.
42. Allfrey, V.G., R. Faulkner, and A.E. Mirsky. 1964. Acetylation and methylation of histones and their possible role in the regulation of RNA synthesis. *Proceedings of the National Academy of Sciences of the United States of America* 51:786–794.
43. Cheung, P., C.D. Allis, and P. Sassone-Corsi. 2000. Signaling to chromatin through histone modifications. *Cell* 103:263–271.
44. Chung, D. 2002. Histone modification: The 'next wave' in cancer therapeutics. *Trends Molecular Medicine* 8:S10–S11.
45. Strahl, B.D., and C.D. Allis. 2000. The language of covalent histone modifications. *Nature* 403:41–45.
46. Grunstein, M. 1997. Histone acetylation in chromatin structure and transcription. *Nature* 389:349–352.
47. Kou, M.H., and C.D. Allis. 1998. Roles of histone acetyltransferases and deacetylases in gene regulation. *Bioessays* 20:615–626.
48. Brownell, J.E., and C.D. Allis. 1996. Special HATs for special occasions: Linking histone acetylation to chromatin assembly and gene activation. *Current Opinion in Genetics Development* 6:176–184.
49. Jenuwein, T., and C.D. Allis. 2001. Translating the histone code. *Science* 293:1074–1080.
50. Jenuwein, T. 2001. Re-SET-ting heterochromatin by histone methyltransferases. *Trends in Cell Biology* 11:266–273.
51. Workman, P. 2001. Scoring a bull's-eye against cancer genome targets. *Current Opinion in Pharmacology* 1:342–352.
52. Munshi, A., G. Shafi, N. Aliyaa, and A. Jyothy. 2009. Histone modifications dictate specific biological readouts. *Journal of Genetics and Genomics* 36:75–88.
53. Costa, F.F. 2005. Non-coding RNAs: New players in eukaryotic biology. *Gene* 357:83–94.
54. Willingham, A.T., and T.R. Gingeras. 2006. TUF love for 'junk' DNA. *Cell* 125:1215–1220.
55. Mattick, J.S., and I.V. Makunin. 2006. Non coding RNA. *Human Molecular Genetics* 11:R17–R29.
56. Pauler, F.M., M.V. Koerner, and D.P. Barlow. 2007. Silencing by imprinted non coding RNAs: Is transcription the answer? *Trends in Genetics* 6:284–292.
57. Costa, F.F. 2008. Non coding RNAs, epigenetics and complexity. *Gene* 410:9–17.

58. Tang, G., B.J. Reinhart, D.P. Bartel, and P.D. Zamora. 2003. A biochemical framework for RNA silencing in plants. *Genes and Development* 17:49–63.

59. Wu, W., M. Sun, G.M. Zou, and J. Chen. 2007. MicroRNA and cancer: Current status and prospects. *International Journal of Cancer* 120:953–960.

60. Lim, L.P., N.C. Lau, P. Garrett-Engele, A. Grimson, J.M. Schelter, J. Castle et al. 2005. Microarray analysis shows that some microRNAs down regulate large numbers of target mRNAs. *Nature* 433:769–773.

61. Lewis, B.P., C.B. Burge, and D.P. Bartel. 2005. Conserved seed pairing, often flanked by adenosines indicates that thousands of human genes are microRNA targets. *Cell* 120:15–20.

62. Schwartz, Y.B., and V. Pirrotta. 2007. Polyclonal silencing mechanisms and the management of genomic programmes. *Nature Reviews Genetics* 8:9–22.

63. Ringrose, L., and R. Paro. 2007. Polycomb/trithorax response elements and epigenetic memory of cell identity. *Development* 134:223–232.

64. Jaenisch, R., and A. Bird. 2003. Epigenetic regulation of gene expression: How the genome interacts intrinsic and environmental signals. *Nature Genetics* 33:245–254.

65. Wade, P.A., A. Gegonne, P.L. Jones, E, Ballestar, F. Aubry, and A.P. Wolffe. 1999. Mi-2 complex couples DNA methylation to chromatin remodelling and histone deacetylation. *Nature Genetics* 23:62–66.

66. Luedi, P.P., F.S. Dietrich, J.R. Weidman, J.M. Bosko, R.L. Jirtle, and A.J. Hartemink. 2007. Computational and experimental identification of novel human imprinted genes. *Genome Research* 17:1723–1730.

67. Mann, J.R. 2001. Imprinting in the germline. *Stem Cells* 19:289–294.

68. Tam, P.L., S.X. Zhou, and S.S. Tan. 1994. X-chromosome activity of the mouse primordial germ cells revealed by the expression of an X-linked lac Z transgene. *Development* 120:2925–2932.

69. Morgan, H.D., H.G. Sutherland, D.I. Martin, and E. Whitelaw. 1999. Epigenetic inheritance at the agouti locus in the mouse. *Nature Genetics* 23:314–318.

70. Murphy, S.K., and R.L. Jirtle. 2003. Imprinting evolution and the price of silence. *Bioassays* 25:577–588.

71. Feil, R., and F. Berger Convergent evolution of genomic imprinting in plants and mammals. *Trends in Genetics* 23:192–199.

72. Ciccone, D.N., H. Su, S. Hevi, F. Gay, H. Lei, J. Bajko et al. 2009. KDM1B is a histone H3K4 demethylase required to establish maternal genomic imprints. *Nature* 461:415–418.

73. Gosden, R., and A.P. Feinberg. 2007. Genetics and epigenetics nature's pen and pencil set. *New England Journal of Medicine* 356:731–733.

74. Fung, D.C., B. Yu, K.F. Cheong, A. Smith, and R.J. Trent. 1998. UBE3A "mutations" in two unrelated and phenotypically different Angelman syndrome patients. *Human Genetics* 102:487–492.

75. Khatib, H. 2007. Is it genomic imprinting or preferential expression? *Bioessays* 29:1022–1028.

76. Beaudet, A.L., and Y.A. Jiang. 2002. Rheostat model for a rapid and reversible form of imprinting-dependent evolution. *American Journal of Human Genetics* 70:1389–1397.

77. Dolinoy, D.C., and R.L. Jirtle. 2008. Environmental epigenetic in human health. *Environmental and Molecular Mutagenesis* 49:4–8.

78. Hemberger, M. 2007. Epigenetic landscape required for placental development. *Cellular and Molecular Life Sciences* 64:2422–2436.

79. Munshi, A., and S. Duvvuri. 2007. Genomic imprinting—The story of the other half and the conflicts of silencing. *Journal of Genetics and Genomics* 34:93–103.

80. Krupanidhi, S., S.S. Madukar, and Y.R. Ahuja. 2009. Epigenetics in cellular differentiation. In *Emerging Trends in Modern Biology 2009*, edited by K.R.S.S. Rao and C.V. Ramnadevi. Guntur, Andra Pradesh, Nagajunasagar: Centre for Biotechnology, Acharya Nagarjuna University.

81. Minard, M.E., A.K. Jain, and M.C. Barton. 2009. Analysis of epigenetic alterations to chromatin during development. *Genetics* 47:559–572.

82. Ramalho-Santos, M., S. Yoon, Y. Matsuzaki, R.C. Mulligan, and D.A. Melton. 2002. "Stemness" transcriptional profiling of embryonic and adult stem cells. *Science* 298:597–600.

83. Bhattacharya, B., T. Miura, R. Brandenberger, J. Mejido, Y. Luo, A.X. Yang et al. 2004. Gene expression in human embryonic stem cell lines: Unique molecular signature. *Blood* 103:2956–2964.
84. Bird, A. 2002. DNA methylation patterns and epigenetic memory. *Genes and Development* 16:6–21.
85. Strichman-Almashanu, L.Z., R.S. Lee, P.O. Onyango, E. Perlman, F. Flam, M.B. Frieman, and A.P. Feinberg. 2002. A genome wide screen for normally methylated human CpG islands that can identify novel imprinted genes. *Genome Research* 12:543–554.
86. Fazzari, M.J., and J.M. Greally. 2004. Epigenomics: Beyond CpG islands. *Nature Review Genetics* 5:446–455.
87. Weber, M., I. Hellmann, M.B. Stadler, L. Ramos, S. Pääbo, M. Rebhan, and D. Schübeler. 2007. Distribution, silencing potential and evolutionary impact of promoter DNA methylation in the human genome. *Nature Genetics* 39:457–466.
88. Ahuja, Y.R., A. Munshi, and P. Jehan. 2009. Environmental modulation of epigenome and its consequences. *Proceedings of the Andra Pradesh Academy of Sciences India* 13:87–92.
89. Smit, F.A. 1999. Interspersed repeats and other mementos of transposable elements in mammalian genomes. *Current Opinion in Genetics Development* 9:657–663.
90. International Human Genome Sequencing Consortium. 2004. Finishing the euchromatic sequence of the human genome. *Nature* 431(7011):931–945.
91. Aravin, A.A., R. Sachidanandam, A. Girard, K. Fejes-Toth, and G.J. Hannon. 2007. Developmentally regulated piRNA clusters implicate M1L1 in transposon control. *Science* 316: 744–747.
92. Slotkin, R.K., and R. Martienssen. 2007. Transposable elements and the epigenetic regulation of the genome. *Nature Reviews Genetics* 8:272–285.
93. Chu, W.M., R. Ballard, B.W. Carpick, B.R. Williams, and C.W. Schmid. 1998. Potential Alu function: Regulation of the activity of double-stranded RNA-activated kinase PKR. *Molecular and Cellular Biology* 18:58–68.
94. Waterland, R.A., J.R. Lin, C.A. Smith, and R.L. Jirtle. 2003. Post-weaning diet affects genomic imprinting at the insulin-like growth factor 2 (IGF 2). *Human Molecular Genetics* 15:705–716.
95. Duhl, D.M., H. Vrieling, K.A. Miller, G.L. Wolff, and G.S. Barsh. 1994. Neomorphic agouti mutations in obese yellow mice. *Nature Genetics* 8:59–65.
96. Cooney, C.A., A.A. Dave, and G.L. Wolff. 2002. Maternal methyl supplements in mice affect epigenetic variation and DNA methylation of offspring. *The Journal of Nutrition* 132:2393S–2400S.
97. Narita, M., S. Núñez, E. Heard, M. Narita, A.W. Lin, S.A. Hearn et al. 2003. Rb mediated heterochromatin formation and silencing of E2F target gene during cellular senescence. *Cell* 113:703–716.
98. Sedivy, J.M., G. Banumathy, and P.D. Adams. 2008. Aging by epigenetics: A consequence of chromosome damage? *Experimental Cell Research* 314:1909–1917.
99. Bjornsson, H.T., M.F. Daniele, and P.F. Andrew. 2004. An integrated epigenetic and genetic approach to common human disease. *Trends in Genetics* 20:350–358.
100. Ahuja, N., Q. Li, A.L. Mohan, S.B. Baylin, and J.P. Issa. 1998. Aging and DNA methylation in colorectal mucosa and cancer. *Cancer Research* 58:5489–5494.
101. Gorbunova, V., A. Seluanov, Z. Mao, and C. Hine. 2007. Changes in DNA repair during aging. *Nucleic Acids Research* 35:7466–7474.
102. Lund, L.H., and D. Manciani. 2004. Heart failure in women. *The Medical Clinics of North America* 88:1321–1345.
103. Narula, J., F.D. Kolodgie, and R. Virmani. 2000. Apoptosis and cardiomyopathy. *Current Opinion in Cardiology* 15:183–188.
104. Strehlow, K., S. Rotter, S. Wassmann, O. Adam, C. Grohé, K. Laufs et al. 2003. Modulation of antioxidant enzyme expression and function of estrogen. *Circulation Research* 93:170–177.
105. Ober, C., D.A. Loisel, and Y. Gilad. 2008. Sex-specific genetic architecture of human disease. *Nature Reviews Genetics* 9:911–922.

106. Cherian, S.B., L. Bairy, and M.S. Rao. 2009. Chronic prenatal restraint stress induced memory impairment in passive avoidance task in post weaned males and female wistar rats. *Indian Journal of Experimental Biology* 47:893–899.

107. Belinsky, S.A. 2004. Gene promoter hypermethylation as a biomarker in lung cancer. *Nature Reviews Cancer* 4:707–717.

108. Mahadev, K., and M.C. Vemuri. 1998. Effect of ethanol on chromatin and nonhistone nuclear proteins in rat brain. *Neurochemistry Research* 23:1179–1184.

109. Kim, J.S., and S.D. Shukla. 2006. Acute *in vivo* effect of ethanol (binge drinking) on histone H3 modification in rat tissues. *Alcohol* 41:126–132.

110. Hamid, A., N.A. Wani, and J. Kaur. 2009. New perspectives on folate transport in relation to alcoholism induced folate malabsorption-association with epigenome stability and cancer development. *FEBS Journal* 276:2175–2191.

111. Nelson, J.L. 1992. Genetic narratives: Biology, stories, and the definition of the family. *Health Matrix* 2:71–83.

112. Zhang, H., A. Darwanto, T.A. Linkhart, L.C. Sowers, and L. Zhang. 2007. Maternal cocaine administration causes an epigenetic modification of protein kinase gene expression in fetal rat heart. *Molecular Pharmacology* 71:1319–1328.

113. Tsankova, N., W. Renthal, A. Kumar, and E.J. Nestler. 2007. Epigenetic regulation in psychiatric disorders. *Nature Reviews Neuroscience* 8:355–367.

114. Bibb, J.A., J. Chen, J.R. Taylor, P. Svenningsson, A. Nishi, G.L. Snyder et al. 2001. Effects of chronic exposure to cocaine are regulated by the neuronal protein Cdk5. *Nature* 410:376–380.

115. Kumar, A., K.H. Choi, W. Renthal, N.M. Tsankova, D.E. Theobald, H.T. Truong et al. 2005. Chromatin remodeling is a key mechanism underlying cocaine-induced plasticity in striatum. *Neuron* 48:303–314.

116. Sutherland, J.E., and M. Costa. 2003. Epigenetics and the environment. *Annals of the New York Academy of Sciences* 983:151–160.

117. Adams, J.B., C.E. Holloway, F. George, and D. Quig. 2006. Analysis of toxic metals and essential minerals in the hair of Arizona children with autism and associated conditions and their mothers. *Biological Trace Element Research* 110:193–209.

118. Wu, J., M.R. Basha, B. Brock, D.P. Cox, F. Cardozo-Pelaez, C.A. McPherson et al. 2008. Alzheimer's disease (AD)–like pathology in aged monkeys after infantile exposure to environmental metal lead (Pb): Evidence for a developmental origin and environmental link for AD. *The Journal of Neuroscience* 28:3–9.

119. Bolin, C.M., R. Basha, D. Cox, N.H. Zawia, B. Maloney, D.K. Lahiri, and F. Cardozo-Pelaez. 2006. Exposure to lead and the developmental origin of oxidative DNA damage in the aging brain. *FASEB Journal* 20:788–790.

120. Foster, W.G., McMahon, and D.C. Rice. 1996. Sperm chromatin structure is altered in Cytomolgus monkeys with environmentally relevant blood lead levels. *Toxicology and Industrial Health* 12:723–735.

121. Quintanilla-Vega, B., D. Hoover, W. Bal, E.K. Silbergeld, M.P. Waalkes, and L.D. Anderson. 2000. Lead effects on protamine–DNA binding. *American Journal of Industrial Medicine* 38:324–329.

122. Zawia, N.H., D.K. Lahiri, and F. Cordozo-Pelaez. 2009. Epigenetics, oxidative stress and Alzheimer disease. *Free Radical Biology and Medicine* 46:1241–1249.

123. Breslow N.E. and N.E. Day. 1980. Statistical Methods in Cancer Research. International Agency for Research on Cancer 1980.

124. Feinberg, A.P., and B. Tycko. 2004. The history of cancer epigenetics. *Nature Reviews Cancer* 4:143–153.

125. Andrew, A.S., J.L. Burgess, M.M. Meza, E. Demidenko, M.G. Waugh, J.W. Hamilton et al. 2006. Arsenic exposure is associated with decreased DNA repair *in vitro* and in individuals exposed to drinking water arsenic. *Environmental Health Perspectives* 114:1193–1198.

126. Poirier, L.A., and T.I. Vlasova. 2002. The prospective role of abnormal methyl metabolism in cadmium toxicity. *Environmental Health Perspectives* 110:793–795.

127. Takiguchi, M., W.E. Achanzar, W. Qu, G. Li, and M.P. Waalkes. 2003. Effects of cadmium on DNA (cytosine-5) methyltransferase activity and DNA methylation status during cadmium-induced cellular transformation. *Experimental Cell Research* 286:355–365.

128. Pogribny, I.P., and F.A. Beland. 2009. DNA hypomethylation in the origin and pathogenesis of human disease. *Cellular and Molecular Life Sciences* 66:2249–2261.

129. Mishra, P.K., S. Dabadghao, G.K. Modi, P. Desikan, A. Jain, I. Mittra et al. 2009. *In utero* exposure to methyl isocyanate in the Bhopal gas disaster: Evidence of persisting hyperactivation of immune system two decades later. *Occupational and Environmental Medicine* 66:279.

130. Balamurugan, K., and H.M. Said. 2006. Role of reduced folate carrier in intestinal folate uptake. *American Journal of Physiology Cell Physiology* 291:189–193.

131. Friso, S, S.W. Choi, D. Girelli, J.B. Mason, G.G. Dolnikowski, P.J. Bagley et al. 2002. A common mutation in the 5,10-methylenetetrahydrofolate reductase gene affects genomic DNA methylation through an interaction with folate status. *Proceedings of the National Academy of Sciences of the United States of America* 99:5606–5611.

132. McCay, C.M., M.F. Crowell, and L.A. Maynard. 1935. The effect of retarded growth upon the length of life span and upon the ultimate body size. *Journal of Nutrition* 10:63–79.

133. Kennedy, B.K., K.K. Steffen, and M. Kaeberlain. 2007. Ruminations on dietary restrictions and aging. *Cellular and Molecular Life Sciences* 64:1323–1328.

134. Sohal, R.S., H.H. Ku, S. Agarwal, M.J. Forster, and H. Lal. 1994. Oxidative damage, mitochondrial oxidant generation and antioxidant defences during aging and in response to food restriction in the mouse. *Mechanism of Aging and Development* 74:121–133.

135. Heydari, A.R., A. Unnikrishna, L.V. Lucente, and A. Richardson. 2007. Calorie restriction and genome stability. *Nucleic Acids Research* 35:7485–7496.

136. Srivastava, V.K., S. Miller, M.D. Schroeder, R.W. Hart, and D. Busbee. 1993. Age-related changes in expression and activity of DNA polymerase: Some effects of dietary restriction. *Mutation Research* 295:265–280.

137. Masoro, E.J. 2005. Overview of calorie restriction and aging. *Mechanism of Aging and Development* 126:913–922.

138. Vaquero, A., and D. Reinberg. 2009. Calorie restriction and the exercise of chromatin. *Genes and Development* 23:1849–1869.

139. Whitesell, L., and S.L. Lindquist. 2005. HSP 90 and the chaperoning of cancer. *Nature Reviews Cancer* 5:761–772.

140. Srivastava, P.K. 2008. New jobs for ancient chaperones. *Scientific American* 299:50–55.

141. Shern-Brewer, R., N. Santanam, and C. Wetztein. 2000. The paradoxical relationship of aerobic exercise and the oxidative theory of atherosclerosis. In *Hand Book of Oxidants and Antioxidants in Exercise*, edited by Sen, C., L. Packer, and O. Hanninen, 1053–1069. Amsterdam: Elsevier.

142. Radák, Z., P. Apor, J. Pucsok et al. 2003. Marathon running alters the DNA base excision repair in human skeletal muscle. *Life Sciences* 72:1627–1633.

143. Kaiser, J. 2003. Sipping from a poisoned chalice. *Science* 302:879–881.

144. Gandhi, G. 2009. Exercise-induced genetic damage: A review. *International Journal of Human Genetics* 9:69–96.

145. Booth, F.W., and P.D. Neufer. 2005. Exercise controls gene expression. *American Scientist* 93:28–35.

146. Wilkinson, M.B., G. Xiao, A. Kumar, Q. LaPlant, W. Renthal, D. Sikder et al. 2009. Imipramine treatment and resiliency exhibit similar chromatin regulation in the mouse nucleus accumbens in depression models. *The Journal of Neuroscience* 29:7820–7832.

147. DeSalvo, M.K., C.R. Voolstra, S. Sunagawa, J.A. Schwarz, J.H. Stillman, M.A. Coffroth et al. 2008. Differential gene expression during thermal stress and bleaching in the Caribbean coral *Montastraea faveolata*. *Molecular Ecology* 17:3952–3971.

148. Weaver, I.C., N. Cervoni, F.A. Champagne, A.C. D'Alessio, S. Sharma, J.R. Seckl et al. 2004. Epigenetic programming by maternal behaviour. *Nature Neuroscience* 7:847–854.

149. Weaver, I.C., A.C. D'Alessio, S.F. Brown, I.C. Hellstrom, S. Dymov, S. Sharma et al. 2007. The transcription factor nerve-growth factor–induced protein which mediates epigenetic programming: Altering epigenetic marks by immediate early genes. *The Journal of Neuroscience* 27:1756–1768.

150. Barker, D.J. 1994. Programming the body. In *Methods, Babies and Diseases in the Later Life,* edited by Barker, D.J., 14–36. London: BMJ Publishing Group.

151. Kuzawa, C.W., and E. Sweet. 2009. Epigenetics and the embodiment of race: Developmental origins of US racial disparities in cardiovascular health. *American Journal of Human Biology* 21:2–15.

152. Armitage, J.A., P.D. Taylor, and L. Poston. 2005. Experimental models of developmental programming: Consequences of exposure to an energy rich diet during development. *The Journal of Physiology* 565:3–8.

153. Ke, X., Q. Lei, S.J. James, S.L. Kelleher, S. Melnyk, S. Jernigan et al. 2006. Uteroplacental insufficiency affects epigenetic determinants of chromatin structure in brains of neonatal and juvenile IUGR rats. *Physiological Genomics* 25:16–28.

154. Szyf, M. 2009. The early life environment and the epigenome. *Biochimica et Biophysica Acta* 1790:878–885.

155. Raghunath, M., L. Venu, I. Padmavathi, Y.D. Kishore, M. Ganeshan, K. Anand Kumar et al. 2009. Modulation of macronutrient metabolism in the offspring by maternal micronutrient deficiency micronutrient deficiency in experimental animals. *Indian Journal of Medical Research* 130:655–665.

156. Barber, R., M.A. Plumb, E. Boulton, I. Roux, and Y.E. Dubrova. 2002. Elevated mutation rates in the germline of first and second generation offspring of irradiated male mice. *Proceedings of the National Academy of Sciences of the United States of America* 99:6877–6882.

157. Nagar, S., L.E. Smith, and W.F. Morgan. 2003. Characterization of a novel epigenetic effect of ionizing radiation: The death inducing effect. *Cancer Research* 63:324–328.

158. Kovalchuk, O., P. Burke, J. Besplug, M. Slovack, J. Filkowski, and I. Pogribny. 2004. Methylation changes in muscle and liver tissues of male and female mice exposed to acute and chronic low-dose X-ray-irradiation. *Mutation Research* 548:75–84.

159. Niemitz, E.L., and A.P. Feinberg. 2004. Epigenetics and assisted reproductive technology: A call for investigation. *American Journal of Human Genetics* 74:599–609.

160. Laprise, S.L. 2009. Implications of epigenetics and genomic imprinting in assisted reproductive technologies. *Molecular Reproduction and Development* 76:1006–1018.

161. Gicquel, C., V. Gaston, J. Mandelbaum, J.P. Siffroi, A. Flahault, and Y. Le Bouc. 2003. *In vitro* fertilization may increase the risk of Beckwith–Wiedemann syndrome related to the abnormal imprinting of the KCN1OT gene. *American Journal of Human Genetics* 72:1338–1341.

162. Jacob, S.M., and H. Kelle. 2005. Gametes and embryo epigenetic reprogramming affect developmental outcome: Implication for ART. *Pediatric Research* 58:437–444.

163. Malgoli, D., F. Gobba, and E. Ottaviani. 2003. Effects of 50 Hz magnetic fields on signalling pathways of MLP-induced shape changes in invertebrate immunocytes: The activation of alternative stem cells pathway. *Biochimica et Biophysica Acta* 1620:185–190.

164. Doherty, A.S., M.R. Mann, K.D. Tremblay, M.S. Bartolomei, and R.M. Schultz. 2000. Differential effects of culture on imprinted H19 expression in the preimplantation mouse embryo. *Biology of Reproduction* 62:1526–1535.

165. Young, L.E., K. Fernandes, T.G. McEvoy, S.C. Butterwith, C.G. Gutierrez, C. Carolan et al. 2001. Epigenetic changes in IGF 2 is associated with fetal overgrowth after sheep embryo culture. *Nature Genetics* 27:153–154.

166. Dean, W., L. Bowden, A. Aitchison, J. Klose, T. Moore, J.J. Meneses et al. 1998. Altered imprinted gene methylation and expression in completely ES cells derived mouse fetuses. *Development* 125:2273–2282.

167. Ahuja, Y.R., S.C. Bhargavac, and K.S. Ratnakarde. 2005. Electric and magnetic fields in stem cell research. *Electromagnetic Biology and Medicine* 24:121–134.

168. Ahuja, Y.R., V. Vijayalakshmi, and K. Polasa. 2007. Stem cell test: A practical tool in toxicogenomics. *Toxicology* 231:1–10.

169. Tamashiro, K.L., T. Wakayama, R.J. Blanchard, D.C. Blanchard, and R. Yanagimachi. 2000. Postnatal growth and behavioral development of mice cloned from adult cumulus cells. *Biology of Reproduction* 63:328–334.

170. Solter, D. 2000. Mammalian cloning: Advances and limitations. *Nature Reviews Genetics* 1:199–207.

171. Chang, H.S., M.D. Anway, S.S. Rekow, and M.K. Skinner. 2006. Transgenerational epigenetic imprinting of the male germ-line by endocrine disruptor exposure during gonadal sex determination. *Endocrinology* 147:5524–5541.

172. Stewart, R.J.C., R.F. Preece, and H.G. Shepppard. 1975. Twelve generations of marginal protein deficiency. *The British Journal of Nutrition* 33:233–253.

173. Cooney, C.A. 2006. Germ cells carry the epigenetic benefits of grandmother's diet. *Proceedings of the National Academy of Sciences of the United States of America* 103:17071–17072.

174. Painter, R.C., T.J. Roseboom, and O.P. Bleker. 2005. Prenatal exposure to the Dutch famine and disease in later life: An overview. *Reproduction Toxicology* 20:345–352.

175. Heijmans, B.T., E.W. Tobi, A.D. Stein, H. Putter, G.J. Blauw, E.S. Susser et al. 2008. Persistent epigenetic differences associated with prenatal exposure to famine in humans. *Proceedings of the National Academy of Sciences of the United States of America* 105:17046–17049.

176. Takai, D, and P.A. Jones. 2002. Comprehensive analysis of CpG islands in human chromosomes 21 and 22. *Proceedings of the National Academy of Sciences of the United States of America* 99:3740–3745.

177. Eckhardt, F., J. Lewin, R. Cortese, V.K. Rakyan, J. Attwood, M. Burger et al. 2006. DNA methylation profiling of human chromosomes 6, 20 and 22. *Nature Genetics* 38:1378–1385.

3

Epigenomics: Basics and Applications

Jaspreet Kaur and Jyotdeep Kaur

CONTENTS

3.1 Introduction

Significant advances have been made in the field of genetics in the last 50 years and completion of the Human Genome Project provides opportunities in the field of medicine to develop better diagnostics and targeted gene therapies. A new field that will have an effect on medicine is epigenetics. It is the study of heritable changes in gene function that do not involve the DNA sequence but provide control of gene expression. In broad terms, it encompasses all inherited cellular material other than DNA, for example, proteins, RNAs, and various transcription factors present in oocytes that persist beyond fertilization through embryonic development and have a role in establishing patterns of gene expression. More commonly, however, the term refers to an increasing number of reversible covalent modifications of DNA and proteins that control gene expression and are stably transmitted to descendent cells without alteration in the DNA sequence. In other words, all the cell types in an organism have the same DNA sequence, yet they have high variability in their phenotypes because of differential expression of genes in these cells as regulated by epigenetics.

Epigenome simply means above the genome. The Greek prefix *epi* implies "on top of" or "in addition to" and it pertains to chemical modification or marking system including DNA methylation and chromatin remodeling, histone acetylation/deacetylation (post-translational modifications of histones), and noncoding RNAs that regulate gene expression without any change in the genetic code itself. Epigenetic mechanisms are essential for the normal development and maintenance of tissue-specific gene expression patterns in mammals. The ability to modify epigenetic programming is crucial for cell differentiation and the development of complex organisms having the same set of genetic instructions. Epigenetic mechanisms are essential for the maintenance of tissue-specific gene expression patterns in mammals.

The regulation of gene expression in eukaryotes is carried out by a large number of independent mechanisms operating at different levels. One of the most important and extensively studied mechanisms that play a role in gene regulation is DNA methylation. This modification, which is present almost exclusively at the 5′ position of cytosine residues in CpG-containing islets, is found in the DNA of all mammalian cells. DNA methylation is the earliest genetic mark that plays a critical role in gene silencing. In addition to DNA methylation, modifications of histone proteins orchestrate DNA organization and gene expression.

It is now well documented that the active regions of chromatin have unmethylated DNA and have high levels of acetylated histones, whereas the inactive regions of chromatin contain methylated DNA and deacetylated histones [1,2]. Thus, an epigenetic "tag" is placed on targeted DNA, marking it with a special status that specifically activates or silences genes. However, unraveling the relationship between these components has led to rapidly evolving new concepts on how they interact and stabilize each other (Figure 3.1).

Disruption of one or another of these interacting systems can lead to overexpression or silencing of genes resulting in "epigenetic diseases." In this chapter, the principles of epigenetic mechanisms (the roles of DNA methylation, histone modification, and RNA

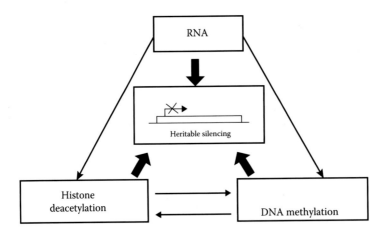

FIGURE 3.1
Interaction between the miRNA, histone modification, and DNA methylation in heritable silencing. Involvement of RNA interference in gene silencing has been demonstrated clearly in experiments with yeasts and plants.

silencing), the epigenomics of X chromosome inactivation, genomic imprinting, and metastable epialleles have been discussed. Also, deregulations in the normal epigenetic patterns leading to the initiation and progression of various diseases (with special reference to cancer) and their futuristic therapies have been addressed.

3.2 Basic Mechanisms of Epigenetic Gene Regulation

The eukaryotic genome is organized into a highly compact and dynamic nucleoprotein structure composed of DNA, proteins, and RNA. All DNA-templated processes, such as replication, transcription, and repair, require that the chromatin structure be accessible to their enzymatic machineries, meaning that upon requirement, the compact nucleoprotein structure should fold and unfold. Coiling of DNA around a histone octamer in the nucleosome is now recognized as a cornerstone in the control of gene expression (Figure 3.2).

Nucleosomes repress all genes except those whose transcription is brought about by specific positive regulatory mechanisms. Repression of transcription by nucleosomes is mediated in different ways by interfering with the interaction of activator and repressor proteins, polymerases, transcription factors, etc. [3], but the structural basis of repression in nucleosomes depends on histone configuration and its acetylation status, and there is a correlation between DNA methylation and histone deacetylation. A protein that binds to methylated DNA recruits a multiprotein histone deacetylase (HDAC) complex [4]. Thus, a general repression mechanism is the basis for cell type–specific regulation of the large eukaryotic genome [5,6]. The gene expression in eukaryotes is regulated by a coordinated interplay of chromatin remodeling complexes, general and gene-specific transcription factors, RNA polymerases, and enzymatic complexes that control DNA and histone covalent modifications. Although chromatin-remodeling enzymes in combination with

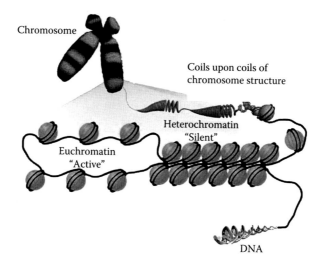

FIGURE 3.2
Organization of eukaryotic DNA into a dynamic nucleoprotein structure formed by coiling of DNA onto histone octamer in the nucleosome. Interaction of activator and repressor proteins, polymerases, transcription factors, etc., influence this organization along with histone configuration and DNA methylation status.

transcription factors can induce gene expression, the propagation of silenced genes during mitosis and meiosis is performed by DNA and histone-modifying proteins [7]. These protein complexes directly or indirectly influence the activity of basal transcriptional machinery and thus influence gene expression.

3.2.1 DNA Methylation and the Enzymes Responsible for Methylation Status

DNA methylation is the earliest proposed genetic mark and has long been recognized as an epigenetic silencing mechanism of fundamental importance [8,9]. The sites of DNA methylation are CpG-rich palindromic sequences, and methylation is mainly on the 5′ position of cytosine (5mC) residues. In mammalian genomes, CpG dinucleotides are somewhat depleted on average, but are found densely clustered within sequences known as CpG islands. These CpG islands are typically in the range of 0.5 to 2 kb in size and are located within 1 kb of transcription start sites called promoters. Under normal circumstances, most, although not all, CpG sites of promoters are unmethylated. In contrast to these CpG islands, the remainder of the mammalian genome contains a high percentage of CpG dinucleotides, both in unique sequences and repetitive elements, which are heavily methylated. Methylation of CpGs in the promoter region has been shown to repress transcription, but some reports suggest that CpG methylation in the promoter region does not always inhibit gene expression and also transcriptionally active and inactive genes have been found to be unmethylated [10].

The methylation of these CpG sites in the genome is maintained by a number of conserved set of proteins called DNA methyltransferases (DNMTs) and has multifaceted roles for the silencing of transposable elements, for defense against viral sequences, and for the transcriptional repression of certain genes. Epigenetic patterns are largely maintained during somatic cell proliferation, and are important in diverse physiological and pathophysiological phenomena. Among the various types of epigenetic modifications of the genome, DNA methylation is certainly one of the most stable. To study the degree of methylation in DNA, methylation-specific restriction enzymes can be used. Some restriction enzymes that have CG as part of their restriction site are inhibited if the C is methylated, others are not. Two of the most commonly used restriction enzymes are *Hpa*II (*Hemophilus parainfluenzae*) and *Msp*I (from *Moraxella* species). *Hpa*II cleaves the unmethylated sequence CCGG, whereas *Msp*I cleaves similar sequences irrespective of the state of methylation of the internal C. Thus, comparison of the restriction maps produced by the two enzymes can detect the levels of methylation in the DNA. In addition, various other techniques used to study the methylation of cytosines in DNA include bisulfite sequencing and cloning, methylation-specific polymerase chain reaction, and high-resolution melting analysis.

Approximately 70% of the cytosine residues in the sequence 5′-CpG-3′ have been found to be methylated at the 5′ position in the DNA of higher animals, where CpGs are present in abundance in transposons, which are interspersed repetitive sequences constituting more than 45% of the human genome. The methylated CpGs in transposons are thought to be a mechanism to suppress transposon reactivation and thereby prevent genomic instability. With the exception of methyl cytosines at the 5′ ends of a majority of genes, CpGs are less frequently found elsewhere in the mammalian genome possibly because of the deamination of methylated cytosine to thymine. DNMTs involved in maintaining cytosines in the methylated state through repeated rounds of cell division of eukaryotes have been classified into two groups: *de novo* and maintenance methyltransferases. Maintenance DNMTs maintain the preexisting methylation patterns by methylating the complementary strand of hemimethylated DNA after DNA replication, *de novo* DNMTs establish those patterns.

Five DNMTs have been identified in mice and have been studied for their methylating activity under *in vitro* and *in vivo* conditions [11].

Mammals have three methyltransferases from the DNMT 3 family: 3A, 3B, and 3L. DNMT 3A and DNMT 3B are essential for *de novo* methylation and mammalian development [12], and are found to methylate both hemimethylated and unmethylated DNA, whereas DNMT 3L does not possess any methylating activity. DNMT 3L is thought to have a role in the upregulation of DNMT 3A activity, as reviewed by Goll and Bestor [13]. Promoter silencing by DNA methylation need not be seen exclusively as a gene inactivation mechanism because the inactivation of one promoter can in fact activate another nearby promoter. Two mechanisms have been proposed to explain the role of DNA methylation in gene silencing. In the first case, binding of transcriptional factors is inhibited by CpG methylation, leading to gene silencing, whereas in the other, proteins that bind to methylated DNA, called methylated DNA-binding proteins (MBDs) recruit proteins that modify histones and thus inhibit transcription by chromatin remodeling, as reviewed by Bird [14]. To date, six methyl-CpG-binding proteins, including MECP2, MBD1, MBD2, MBD3, MBD4, and Kaiso, have been identified in mammals. MECP2 contains a methyl-CpG-binding domain at its amino terminus and a transcription repression domain in the middle. MBDs1–4 were cloned on the basis of their sequence homology to MECP2 in the MBD. All of these proteins, except MBD3, bind preferentially to the methylated CpG islands. Moreover, MECP2, MBD1, and MBD2 function as transcription repressors, whereas MBD4 is a DNA glycosylase and is involved in DNA mismatch repair. Kaiso, which lacks an MBD domain, binds methylated CGCG through its zinc-finger domain, as reviewed by Clouaire and Stancheva [15].

3.2.2 Histone Modifications

Each histone of the nucleosome is organized into two domains, a histone fold (central core particle) and an unstructured N-terminal tail. Although histone modifications occur throughout the entire sequence, the tail extending outside the core particle is highly modified. Thus, the structural basis of repression/derepression by nucleosomes depends on histone configuration. Specific amino acids of histones have been found to have modifications such as acetylation, methylation, phosphorylation, polyribosylation, sumoylation, and ubiquitination [16]. Such posttranscriptional modifications of histones on conserved lysine residues of amino terminal tail domains have been studied over the past few years. Although the role of many of these modifications is only partly understood, certain modifications have been shown to regulate gene expression.

3.2.2.1 Histone Acetylation/Deacetylation

Of the various modifications, acetylation and the enzyme histone acetyltransferase have been extensively studied. Acetylation converts the positively charged amino group of lysine into uncharged amide linkage causing the chromatin structure to open up. Hypoacetylation of lysine residues is associated with the formation of heterochromatin and gene silencing, whereas euchromatic domains are associated with hyperacetylated histones. Histone acetylation occurs at a low level ubiquitously throughout most of the genome, resulting from a balance between several histone acetyl transferase and HDAC activities [17]. Histone acetylation by acetyltransferase activity thus relieves the repression of chromatin, but this is not sufficient for transcriptional activation because it fails to disrupt the core particle of the nucleosome. There are, however, alternative "multiprotein complexes" that remodel the structure of chromatins in an ATP-dependent manner. Two families of such complexes, called SWI/SNF, perturb the core structure, as shown by increased susceptibility of the DNA to nuclease attack, as reviewed by Kornberg [18].

3.2.2.2 Histone Methylation and Other Modifications

Histone methylation can be a marker for both active and inactive regions of chromatin. Histone H3 methylation of lysine 9 at the N-terminus (H3-K9) is a hallmark of inactive DNA and is distributed in heterochromatic regions such as centromeres and telomeres [19]. In contrast, methylation of lysine 4 of histone H3 (H3-K4) is found predominantly on regions of active genes. Furthermore, the lysine methylation can be monomeric, dimeric, or trimeric and other posttranslational modifications such as phosphorylations and ribosylations lead to multiple possible combinations of different modifications, constituting a "histone code" which can be read by different cellular factors [20,21]. Links between histone modification and DNA methylation have been found as interactions between HDACs, histone methyltransferases, and methyl cytosine binding proteins lead to the activation of DNMTs at their specific target sequences [22,23].

3.2.3 Noncoding RNAs

The role of RNA in posttranscriptional silencing has been reported recently. RNA (in the form of antisense RNAs, noncoding RNAs, or interfering RNA) can inhibit transcription by the formation of heterochromatin. In *Arabidopsis thaliana*, iRNA has been found to affect DNA methylation and histone methylation [24]. The role of iRNA has been elucidated in *Schizosaccharomyces pombe* after the deletion of different components of the iRNA machinery, and a loss of the characteristic H3-K9 methylation has been reported within the same region [25,26]. Although similar effects have not been reported in mammals, silencing of some imprinted genes has been found to be mediated by antisense RNA [27]. RNA thus influences DNA methylation to specific loci and histone H3-K9 methylation, thereby causing heritable silencing.

Micro-RNAs (miRNAs) are expressed in a tissue-specific manner and control a wide array of biological processes including cell proliferation, apoptosis, and differentiation. Like normal genes, the expression of miRNAs can be regulated by epigenetic mechanisms [28]. In addition, miRNAs can also modulate epigenetic regulatory mechanisms inside a cell by targeting the enzymes responsible for DNA methylation and histone modifications [29].

3.3 Epigenomics in Health

3.3.1 Genomic Imprinting

Genomic imprinting is a form of gene regulation in which epigenetic chromosomal modifications result in differential gene expression in a parent of origin manner. The majority of autosomal genes are expressed from both parental alleles but approximately 1% of autosomal genes are imprinted, with expression from only one parent allele; for a review, see Lewis and Reik [30]. There are reports that in mice, the paternal genome gets demethylated at a majority of CG sites a few hours after entering the egg. Methylation of the maternal genome, however, persists and is called genome imprinting. It is the selective inactivation of genes derived from one or the other parent, which results in the transcription of either the maternal or paternal allele but not both. This monoallelic expression is due to gamete-specific epigenetic marks that silence one allele and at the same time increase the susceptibility to diseases because a single mutation or an epigenetic event can alter the function of haploid loci of maternal or paternal origin ([31,32] and reviewed by Jirtle [33]). It was not until 1991 that the first imprinted genes were identified.

After the reports that insulin-like growth factor (IGF2), a potent growth factor [34], and IGF2 receptor (IGF2R) [35] are imprinted, approximately 80 imprinted genes have been identified in mice and humans, and about one-third being imprinted in both species [36]. Imprinted genes and metastable epialleles are regulated by epigenetic mechanisms and thus are susceptible to epigenetic modifications, particularly by environmental factors, as reviewed by Dolinoy and Jirtle [37]. Because imprinted genes are functionally haploid (one allele is already inactive because of imprinting), the consequences of genomic imprinting are disastrous. Epigenetic alterations in the imprinted genes during development result in a number of pediatric disorders such as Prader–Willi syndrome, Angelman syndrome, and Beckwith-Wiedemann syndrome, as reviewed by Murphy and Jirtle [31]. Several diseases resulting from genetic and epigenetic modifications at imprinted loci have been reported and will be discussed in Section 3.5.1.1. There have been recent reports of imprinted noncoding RNAs that may play a central role in the imprinting mechanism.

3.3.2 X Chromosome Inactivation

In mammals, one of the two X chromosomes has been found to be transcriptionally silenced during early female development. Either the paternally or the maternally inherited X chromosome is inactivated randomly due to the expression of large noncoding Xist RNA. Xist expression is controlled by a mechanism that ensures that one X chromosome remains active in a diploid cell. Xist RNA has been identified in mouse and human [38]. By blocking the epigenetic pathways in somatic cells, including histone deacetylation and DNA methylation, partial reactivation of the Xi has been obtained [39]. This has led to the conclusion that gene silencing on the Xi is mediated by a number of epigenetic factors that act together to maintain the inactive state of the chromosome. X chromosome inactivation and imprinted genes are classic examples of CpG island methylation during development [40].

3.3.3 Metastable Epialleles

Metastable epialleles are defined as loci that are epigenetically modified in a variable and reversible manner, such that a distribution of phenotypes occurs from genetically identical cells. A few genes such as murine A^{vy} (viable yellow agouti), $Axin^{Fu}$ (axin fused), and $Capb^{IAP}$ (CDK5 activator-binding protein intracisternal A particle gene) with metastable epialleles have been identified. Metastable epialleles are variably expressed due to epigenetic modifications established in early development [41]. They are often associated with retroelements and transgenesis. The extent of DNA methylation depends on maternal nutrition and environmental exposures during early development [42–45]. The murine yellow agouti (A^{vy}) allele is the most extensively studied metastable epiallele and allelic expression is correlated with the epigenetic status of the transposon associated with the promoter region of the gene [42]. Epigenetic status of a subset of transposable elements is metastable and varies from hypomethylated to hypermethylated. This variable epigenetic status can affect the expression of neighboring genes, causing variability in genetically identical individuals [46].

3.4 Epigenomics and Environment

Gene–environment interactions play an important role in the susceptibility of an individual to developing chronic diseases and to the impairment of cognitive functions of that individual. There are reports suggesting that such interactions result in epigenetic changes which may play a causative role for the development of diseases. Environmental exposure in the form of various xenobiotics, chemical, nutritional, and physical factors can alter gene expression by mutations and epigenetic changes at critical epigenetically labile genomic regions [47]. DNA methylation patterns have been extensively studied and reported to fluctuate in response to changes in diet, inherited genetic polymorphism, and environmental chemicals, as reviewed by Sutherland and Costa [48]. Environmental agents such as aromatic hydrocarbons and metals destabilize the genome. Common polymorphisms in the methyl THF reductase gene lead to altered DNA methylation patterns and diseases in response to these agents. In the viable yellow agouti (Avy) mouse, maternal diet affects the coat color distribution of offspring by perturbing the establishment of methylation at the Avy metastable epiallele.

Dietary supplementation to A^{vy} mice during pregnancy with methyl donors such as choline, betaine, folic acid, and vitamin B_{12}, shifts the coat color distribution in offspring and decreases the yellow coat color. It is also associated with reduced risk of developing obesity, diabetes, and cancer, and the changes were shown to be caused by increased DNA methylation of a transposable element upstream of the agouti gene [42,49]. Hence, the A^{vy} mouse can be employed as a sensitive epigenetic biosensor to assess the effects of dietary methionine supplementation on locus-specific DNA methylation because methyl groups transferred for DNA methylation are derived from methionine. Recent developments in epigenomic approaches that survey locus-specific DNA methylation on a genomewide scale offer broader opportunities to assess the effects of high methionine intake on mammalian epigenomes. Methionine excess may actually impair DNA methylation by inhibiting remethylation of homocysteine. Although little is known regarding the effect of dietary methionine supplementation on mammalian DNA methylation, the available data suggest that methionine supplementation can induce hypermethylation of DNA in specific genomic regions. Because locus-specific DNA hypomethylation is implicated in the etiology of various cancers and developmental syndromes, clinical trials of "promethylation" dietary supplements are already underway. However, aberrant hypermethylation of DNA could be deleterious. Determination of the effects of methionine supplementation on therapeutic maintenance of DNA methylation (without causing excessive and potentially adverse locus-specific hypermethylation) is a very important area for cancer prevention.

3.5 Epigenomics in Cancer

3.5.1 DNA Methylation

The methylation of CpG islands in the promoter region silences gene expression and is a normal event that occurs in cells to regulate gene expression. However, when aberrant DNA methylation of tumor suppressor genes occurs in tumors, it is implicated in neoplastic transformation. The low level of DNA methylation in tumors as compared with that in their normal-tissue counterparts was one of the first epigenetic alterations to be observed in human cancer. This initial observation of global DNA hypomethylation in human tumors was followed by the identification of hypermethylated tumor suppressor genes, and later, the discovery of inactivation of miRNA genes by DNA methylation [50].

3.5.1.1 DNA Hypomethylation

There is widespread documentation of an overall level of genomic hypomethylation coupled with gene-specific hypermethylation in cancer cells. Hypomethylation events are more generalized, and could lead to the activation of endogenous retroviral elements and dormant protooncogenes. However, gene-specific hypomethylation is unlikely to play a major role in oncogenesis because promoter CpG islands are normally unmethylated, with the notable exceptions of imprinted alleles and genes on the inactive X chromosome. Decreased cytosine methylation was basically found to affect satellite DNA, repetitive sequences, and CpG sites located in the coding and intron regions of DNA, which allow the transcription of alternative messenger RNAs that are transcribed from a gene [51]. A recent large-scale study of DNA methylation with the use of genomic microarrays has detected extensive hypomethylated genomic regions in gene-poor areas. During the development of a neoplasm, the degree of hypomethylation of genomic DNA increases as the lesion progresses from a benign proliferation of cells to an invasive cancer [52].

Three mechanisms have been proposed to explain the contribution of DNA hypomethylation to the development of a cancer cell: generation of chromosomal instability, reactivation of transposable elements, and loss of imprinting. Undermethylation of DNA can favor mitotic recombination, leading to deletions and translocations [53] and it can also promote chromosomal rearrangements. This mechanism was seen in experiments in which the depletion of DNA methylation by the disruption of DNMTs caused aneuploidy. Global hypomethylation occurs mainly at DNA-repetitive elements and might contribute to the genome instability frequently seen in cancers. Hypomethylation might also contribute to the inappropriate activation of oncogenes and transposable elements. Hypomethylation of DNA in malignant cells can reactivate intragenomic endoparasitic DNA, such as L1 (long interspersed nuclear elements), and Alu (recombinogenic sequence) repeats. These undermethylated transposons can be transcribed or translocated to other genomic regions, thereby further disrupting the genome. The hypomethylation of DNA can have unpredictable effects. The progeny of a mouse deficient in DNA methylation and a Min mouse, which has a genetic defect in the adenomatous polyposis coli (*APC*) gene and is prone to colon adenoma, have been observed to develop fewer tumors than expected. On the other hand, another DNMT-defective mouse strain has shown an increased risk of lymphoma. Moreover, hypomethylation was shown to suppress the later stages of intestinal tumorigenesis while promoting early precancerous lesions in the colon and liver through genomic deletions, as reviewed by Esteller [54].

Gene promoters that are physiologically methylated after embryogenesis and development might become reactivated during genomic hypomethylation and result in aberrant expression. The cancer/testis antigen (*CAGE*) is an example of such an expression [55]. Other genes that are overexpressed in hypomethylation include *NAT1* in breast cancer [56], *CD30* in anaplastic large cell lymphoma [57], and *c-Myc* in viral-related hepatocellular carcinoma [58]. In addition, DNA hypomethylation can lead to the activation of growth-promoting genes, such as *R-Ras* and *MAPSIN* in gastric cancer, S-100 in colon cancer, and MAGE (melanoma-associated antigen) in melanoma [59]. Hypomethylation has also been shown to be associated with loss of imprinting of the IGF gene that is expressed only from paternally inherited alleles. However, aberrant methylation of the IGF2 imprinting control region on the maternal allele leads to additional expression of the maternal copy. Thus, the levels of IGF2 increase approximately twofold and this elevated level is associated with cancer susceptibility [60]. Loss of imprinting of *IGF2* is also a risk factor for colorectal cancer, and disrupted genomic imprinting contributes to the development of Wilms' tumor. In animal models, mice with a loss of imprinting of *IGF2* or overall defects in imprinting have an increased risk of cancer [61].

A causal relationship between genomic instability and hypomethylation is supported by studying the mechanisms of the developmental disorder, ICF syndrome (immunodeficiency, chromosome instability, and facial abnormalities), which is caused by loss-of-function mutations in the DNA methyltransferase *DNMT3B* [62]. However, global hypomethylation is not the only epigenetic abnormality that may contribute to genomic instability evident from the example; CENPA, the H3-like histone found in centromeric nucleosomes, becomes overproduced in colorectal cancer and causes aneuploidy [63]. The question now is, how might global hypomethylation lead to tumor formation? In cancer cells, hypomethylation is particularly evident in pericentromeric regions, like those found on chromosomes 1 and 16 [64]; thus, predisposing them to recombination events. Recurrent unbalanced chromosomal translocations with breakpoints in the pericentromeric DNA sequences have been studied in breast cancers, ovarian cancers, and Wilms' tumors [65]. Alternatively, hypomethylation of latent viral sequences might allow them to re-express and contribute to tumor progression. For example, cervical cancer latency has been linked to hypermethylation of the HPV16 genome, and reactivation of the HPV genome due to progressive hypomethylation has been observed to result in cervical dysplasia [66]. Epstein–Barr virus latency follows a similar pattern in associated lymphomas.

3.5.1.2 DNA Hypermethylation

The genome regions subject to hypermethylation in cancer cells are the CpG islands associated with gene promoters. A study revealed that out of the 45,000 CpG islands in the human genome, 600 exhibit methylation patterns in tumors different from those in normal tissues [67]. These methylation changes seem to occur early in the neoplastic process, and some are even cancer-type–specific, suggesting that CpG island hypermethylation is mechanistically involved in carcinogenesis rather than being a consequence of neoplastic transformation. An increase in DNA methylation also occurs with ageing. Consequently, the increased cancer predisposition observed with ageing may be partially attributable to the age-dependent increases in genome methylation. The methylation of CpG dinucleotides results in spontaneous deamination to form a TpG dinucleotide pair. This C–T base substitution is not readily recognized by DNA repair proteins, thus contributing to inefficient repair and subsequent accumulation of the mutation in the genome. In addition to this type of susceptibility to mutations, these mutated sequences could indirectly affect the interactions between DNA and various proteins of the transcription machinery [67].

Hypermethylation of the CpG islands in the promoter regions of tumor suppressor genes is a major event in the origin of many cancers. The initial reports of hypermethylation of the CpG islands in the promoter region of the retinoblastoma tumor suppressor gene (*Rb*) [68] were followed by findings that later confirmed the observation that hypermethylation of the promoter region CpG island was a mechanism of inactivation of many tumor suppressor genes such as *VHL* (associated with von Hippel–Lindau disease), *p16^{INK4a}*, *hMLH1* (a homologue of MutL *Escherichia coli*), *BRCA1* (breast cancer susceptibility gene 1), and Rassf1a [69]. We observed promoter region CpG island methylation of six tumor suppressor genes (*CDKN2B, CDKN2A, APC, CDH1, SOCS1,* and *GSTP1*) in hepatitis virus–related hepatocellular carcinoma. Notably, the methylation of these genes started as early as in the hepatitis stage of viral infection thus confirming the role of DNA methylation of tumor suppressor genes in the pathogenesis of hepatocellular carcinoma [58].

Further research in this area suggested that hypermethylation of the CpG island promoter can affect genes involved in the cell cycle, DNA repair, metabolism of carcinogens, cell-to-cell interaction, apoptosis, and angiogenesis, all of which are already known to participate in the development of cancer [70]. Hypermethylation of these genes can occur at different stages in the development of cancer and in different cellular networks, and involve their interactions to cause the disease (Table 3.1). Such interactions can be seen when hypermethylation inactivates the CpG island of the promoter of the DNA repair genes *hMLH1, BRCA1, MGMT* (O^6-methylguanine-DNMT), and the gene associated with Werner's syndrome (*WRN*) [70].

The profiles of hypermethylation of the CpG islands in tumor suppressor genes are specific to the cancer type [67]. Each tumor type can be assigned specific, defining CpG island methylations which can be termed as "hypermethylomes." The functional significance of gene silencing by hypermethylation can be appreciated in the context of Knudson's two-hit hypothesis of tumor suppressor gene inactivation. Several studies have now shown that tumors can stably maintain mutations in one allele of a gene while the other allele is hypermethylated, thereby leading to functional inactivation [71]. Not only are tumor suppressor genes inactivated by hypermethylation, other classes of genes can be inactivated by promoter hypermethylation. These are genes belonging to the family of transcription factors, cell cycle regulation, signal transduction pathways, and tissue remodeling. Silencing of the expression of a transcription factor, and consequently, its downstream targets, is the most reasonable way of inactivating the gene. Examples of transcription factors that are hypermethylated in cancer are *OCT3/4* in testicular cancer, *NFATC1* in lymphoma, and *RUNX3* in esophageal cancer [72]. Also, the *GATA-4* and *GATA-5* transcription factors and their downstream target genes are inactivated by hypermethylation in colorectal and gastric cancers [73]. The other part that promoter hypermethylation might play in cancer is to promote gene inactivation by classic mutational events. This phenomenon is mostly observed with hypermethylation of DNA repair genes. For example, hypermethylation of

TABLE 3.1

Genes Silenced by DNA Methylation in Various Cancers

Gene	Function	Cancer Type
APC	Regulation of catenin cell adhesion	Colorectal, gastrointestinal
BRCA1	DNA damage repair	Breast, ovarian
CDH1	Homotypic epithelial cell–cell adhesion	Bladder, breast, colon
p15 (CDKN 2B)	Cyclin-dependent kinase inhibitor	Acute leukemia, Burkitt lymphoma, multiple myelomas
p16 (CDKN 2A)	Cyclin-dependent kinase	Lung, gliomas, colon, bladder
Rb	Sequesters E2F transcription factor	Retinoblastoma
VHL	Inhibits angiogenesis, regulates transcription	Renal cell carcinoma
GATA4, GATA5	Transcription factors	Endometrial, lung hepatoma, colon, breast
GSTP1	Carcinogen detoxification	Prostate, breast, renal
DAPK1	Apoptosis	Lung, prostate, hematological

Notes: BRCA1, breast cancer 1; Rb, retinoblastoma; VHL, von Hippel–Lindau 1; GATA, GATA binding protein; GSTP1, glutathione S-transferase pi 1; DAPK1, death-associated protein kinase 1.

the DNA mismatch repair gene *MLH1* is a frequent finding in sporadic colorectal tumors characterized by microsatellite instability. This *MLH1* hypermethylation is also found in the normal mucosa of the same patients, confirming that hypermethylation precedes the development of colon cancer [74]. Another example is of *MGM*, another DNA repair gene that undergoes promoter hypermethylation and epigenetic silencing in cancer [75].

The protein MGMT normally removes O^6-methylguanine adducts from DNA which, if left unrepaired, would result in G→A transition mutations in key genes. Hence, DNA hypermethylation–dependent silencing of tumor suppressor genes in cancer is now well established, but how genes are targeted for this aberrant DNA methylation is still unclear. Tumor-specific CpG island methylation can occur through a sequence-specific instructive mechanism by which DNMTs are targeted to specific genes following their association with oncogenic transcription factors. Aberrant hypermethylation and silencing of specific target gene promoters by the promyelocytic leukemia-retinoic acid receptor (PML-RAP) fusion protein in acute promyelocytic leukemia is an example of such a mechanism [76]. Large stretches of DNA can become abnormally methylated in cancer, causing some CpG islands to be hypermethylated as a result of their location inside such genomic regions that have undergone large-scale epigenetic reprogramming. Another proposed mechanism is about the role of histone marks in the tumor-specific targeting of *de novo* methylation. An interesting observation is that regions that are hypermethylated in cancer are often premarked with the H3K27me3 polycomb in embryonic stem cells [77], suggesting a link between developmental regulation and tumorigenesis. This observation also partially explains the theory of the "CpG island methylator phenotype," which hypothesizes that there is a coordinated methylation of a subset of CpG islands in tumors because many of these CpG island methylator phenotype loci are known polycomb targets [78].

3.5.2 Histone Modifications

Histones are the proteins mediating DNA compaction into chromatin. The histones are acetylated at specific amino acids by histone acetyltransferases helping the chromatin to form a relaxed or open conformation, the euchromatin, allowing the transcription factors access to the DNA, at which point gene expression takes place. On the other hand, deacetylation of histones by the enzymes known as HDACs closes the chromatin, the heterochromatin, blocking transcription factor access and hence gene expression. DNA methylation and histone modifications are interrelated phenomena that together determine gene expression [77]. A number of proteins involved in DNA methylation (e.g., DNMTs, MBDs) interact directly with histone-modifying enzymes such as Histone Methyltransferasc (HMT) and HDACs. Therefore, most of the factors affecting DNA methylation levels will also affect histone modifications. DNMTs recruit HDACs, leading to histone deacetylation and transcriptional repression. Methylated DNA is bound by a family of methyl-binding proteins that also recruit HDACs as well as other ATP-dependent chromatin-remodeling proteins, which together cause chromatin condensation and gene inactivation. Additionally, CpG island promoter methylation can block transcription by interfering with the binding of transcription factors to their target binding sites [79].

Studies have shown that there is a global loss of acetylated H4K16 and trimethylated H4K20 as a part of chromatin changes during tumorigenesis. Such loss of histone acetylation is mediated by the overexpression of HDACs in various cancers. Histone acetyltransferases, which are also involved in the maintenance of histone acetylation, can also be altered in cancer [80]. In addition to the changes in histone acetylation in cancer, cancer cells also exhibit alterations in histone methylation patterns. Aberrant methylation of

H3K9 and H3K27 has been shown to result in gene silencing in various cancers [81,82]. These altered histone marks have been attributed to the dysregulation of HMTs, which might be responsible for the silencing of various tumor suppressor genes. A H3K27 HMT, EZH2, was found to be overexpressed in breast and prostate cancer, whereas another H3K9 HMT had a role in the progression of leukemia. In addition to HMTs, histone demethylases, which are involved in the maintenance of histone methylation, have been implicated in tumor progression. For example, lysine demethylases LSD1 and Jumonji C are upregulated in various cancers, as reviewed by Sharma et al. [83].

3.5.3 Nucleosome Positioning

Although the roles of DNA methylation and histone modifications in cancer initiation and progression are well established, the changes in chromatin structure that accompany such modifications are not yet well understood. Recent studies have revealed that nucleosome remodeling in concert with DNA methylation and histone modifications also play a central role in tumor-specific gene silencing. DNA methylation–induced silencing of tumor suppressor genes in cancer involves changes in nucleosome positioning. Reactivation of such silenced genes using DNMT inhibitors is associated with the loss of nucleosomes from the promoter region [84]. Gene silencing due to nucleosome remodeling can occur by the transmission of repressive epigenetic marks to tumor suppressor gene promoters. Recently, it has been demonstrated that the NuRD (nucleosome remodeling and deacetylase) corepressor complex plays a central role in aberrant gene silencing. The NuRD complex facilitates the recruitment of the polycomb repressive complex 2 and DNMT3A to the promoters of target genes, resulting in their permanent silencing. NuRD also interacts with the methyl-binding domain 2 protein and the later directs the NuRD complex to methylate DNA, thus maintaining the DNA methylation marks and hence silencing of promoters, refer to the review by Kanwal and Gupta [85].

3.5.4 miRNAs

Because miRNAs regulate the genes involved in transcriptional regulation, cell proliferation, and apoptosis (the most common processes deregulated in cancer), alterations in their expression can promote tumorigenesis. miRNAs can function as either tumor suppressors or oncogenes depending on their target genes. Two miRNAs, miR-15a and miR-16-1, clustered in 13q14, the most commonly deleted region in B-cell chronic lymphocytic leukemias have been reported. Subsequently, a global reduction of miRNA expression has been found in tumors; however, the upregulation of some miRNAs has also been reported. Many tumor suppressor miRNAs targeting growth-promoting genes are repressed in cancer. For example, miR-15 and miR-16, which target *BCL2*, are downregulated in chronic lymphocytic leukemia, whereas let-7, which targets the *RAS* oncogene, is downregulated in lung cancer [86]. Furthermore, miR-127, which targets an antiapoptotic gene, *BCL6*, is significantly repressed in prostate and bladder tumors [87], whereas miR-101, against polycomb group protein EZH2, shows reduced expression in bladder transitional cell carcinoma [83]. There are miRNA which are upregulated in cancer and are called oncogenic miRNA. These miRNAs generally target growth-inhibitory pathways. For example, human glioblastoma show upregulation of miR-21 [88], whereas miRNA-155 is upregulated in breast, lung, and several hematopoietic malignancies [89]. Similarly, the oncogenic miR-17–miR-92 cluster, which targets the proapoptotic gene Bim, is found overexpressed in various malignancies [90] (Table 3.2).

TABLE 3.2

miRNAs Altered in Various Human Cancers

miRNAs	Target Gene(s)	Cancer Type
miR-127	Bcl-6	Bladder cancer
miR-124	CDK6	Colon cancer
miR-17, miR-92	c-MYC	Lung cancer
miR-155	RHOA	Burkitt's lymphoma, breast, colon
miR-146	NF-κB	Breast, pancreas and prostate cancers
miR-520	CD44	Breast cancer
miR-127, miR-199a	BCL6, E2F1	Cervical cancer
miR-190, miR-196	HGF	Pancreatic cancer
miR-340, miR-421, miR-658	MYC, RB, PTEN	Lymph node metastasis and gastric
let-7a-3	RAS, IGF-II	Lung and ovarian cancer
miR-9	NF-κB	Ovarian and lung cancer
miR-181	VGFR	Lung cancer
miR-145	ER	Colon and breast cancer

Notes: Bcl-6, B cell lymphoma protein 6; CDK6, cyclin D kinase 6; RHOA, Ras homologue gene family member; NF-κB, nuclear factor κB; CD44, cluster differentiation 44; HGF, hepatocyte growth factor; VGFR, vascular endothelial growth factor; ER, estrogen receptor.

Changes in miRNA expression can be achieved through various mechanisms including chromosomal abnormalities, transcription factor binding, and epigenetic alterations. DNA hypermethylation in the miRNA 5′ regulatory region is a mechanism that can account for the downregulation of miRNA in tumors. For example, in colon cancer cells with disrupted DNMTs, hypermethylation of the CpG island of miRNAs does not occur. The methylation silencing of miR-124a is a common epigenetic lesion in tumors [91]. Considering their well-documented role in tumorigenesis, p53 and Myc are among the most relevant examples of transcription factors involved in miRNA regulation. For instance, miR-34 family and the miR-17–92 cluster are direct transcriptional targets of p53 [92] and Myc [93], respectively. Impairment of miRNA processing can influence miRNA expression patterns. The expression levels of Dicer or Drosha are altered in several tumors, which can downregulate the miRNA processing machinery [94]. Importantly, Merritt et al. [95] have found that the association of low levels of Dicer and Drosha are associated with a poor clinical outcome in ovarian cancer.

3.6 Epigenomics in Other Diseases

Each cell has its own epigenetic pattern that is carefully maintained to regulate gene expression. Any perturbation in the pattern can lead to a diseased state or predisposition to acquired diseases such as cancers (discussed previously) and neurodegenerative disorders. In humans, DNMT 3B is nonfunctional (an essential enzyme required for the establishment of DNA methylation patterns as already explained previously) in a rare autosomal disorder called ICF syndrome [96]. There is loss of methylation of satellite DNA at pericentromeric regions of chromosomes 1, 9, and 16, leading to the cytogenetic abnormalities and clinical features of ICF. Mutations in the proteins binding to methylated DNA can cause phenotypic abnormalities such as mutated MECP2 genes causing abnormal gene expression patterns within the first year of life, leading to a neurological disorder called Rett syndrome in humans and a related disorder in mice [97].

Abnormal expression of imprinted genes during development results in a number of severe pediatric syndromes such as Prader–Willi syndrome, Angelman syndrome, and Beckwith–Wiedemann syndrome. In these imprinting disorders, abnormality is due to the absence of maternal or paternal copy of the imprinted gene, or loss of methylation at imprinting control regions or biallelic expression (uniparental disomy), as reviewed by Murphy and Jirtle [31]. Silencing of imprinted genes such as *CDKNIC* is found in most sporadic cases of Beckwith–Wiedemann syndrome. Mutations in nonimprinted regions of the gene can also affect the regulation of imprinted genes and result in disorders such as mutations in the single locus *Callipyge* (CLPG1) gene in the telomeric region of chromosome 18, which results in fast twitch hypertrophy [98]. All these studies indicate that embryogenesis is the most vulnerable stage for epigenetic alterations as most epigenetic marks are generally stable in somatic cells. During developmental periods of gametogenesis and early preimplantation embryos, the epigenome undergoes extensive reprogramming as reported by several researchers [42,43,45,99] and as reviewed by Dolinoy and Jirtle [37].

Fragile X chromosome results from gene silencing, when a CGG in the FMR1 region expands and gets methylated *de novo*, causing a visible "fragile" site on the X chromosome. Age-dependent epigenetic changes have been found to be involved in the development of neurological disorders, autoimmunity, and cancers (as discussed in detail in the previous section). These disorders have age-related alterations in DNA methylation patterns [100]. Immunity and related disorders such as autoimmune diseases are associated with the loss of epigenetic control over complex processes in immune cells to mount a specific immune response. Shifts in both acetylation and methylation are required to mount an immune response against an antigen. Abnormal DNA methylation has been found in patients with lupus erythematosus in which T cells exhibit decreased methyltransferase activity and hypomethylated DNA [101,102].

3.7 Epigenetic Therapy

Because of their dynamic nature and potential reversibility, epigenetic modifications are becoming appealing therapeutic targets in cancer. Two main classes of epigenetic drugs, that is, DNMT inhibitors and HDAC inhibitors, are currently used in clinical trials for the treatment of cancers. The mechanism of action of these drugs is that these are the nucleoside analogs and get incorporated into the DNA of rapidly growing tumor cells during replication and inhibit DNA methylation by trapping DNMTs onto the DNA [103]. This drug-induced reduction of DNA methylation causes the activation of tumor suppressor genes that are aberrantly silenced in cancer, thus inhibiting growth in cancer cells [104]. 5-Aza-CR (azacitidine) and 5-aza-CdR (decitabine) have now been approved by the Food and Drug Administration for use in the treatment of myelodysplastic syndromes, and promising results have also emerged from the treatment of other hematological malignancies such as acute myeloid leukemia and chronic myeloid leukemia using these drugs [105]. The possible clinical use of other improved DNA methylation inhibitors such as zebularine, which can be orally administered, is currently under investigation. Because these drugs are incorporated into the DNA, concerns have been raised regarding their potential toxic effects on normal cells. However, these drugs mainly target rapidly dividing tumor cells as they act on dividing cells. Nevertheless, nonnucleoside compounds, which can effectively inhibit DNA methylation without being incorporated into the DNA are being developed for small-molecule inhibitors such as SGI-1027, RG108, and MG98, as reviewed by Datta et al. [106].

The second class of agents is the HDAC inhibitors, which seem to be more promising targets for epigenetic anticancer therapy. The HDAC inhibitors butyrate, trichostatin A, depsipeptide, and suberolyl anilide hydroxamic acid cause the induction of expression of several cell cycle regulatory molecules that act as inhibitors of cell cycle progression, cyclin-dependent kinases and result in cell cycle arrest [107]. The cross-talk between DNA methylation and histone deacetylation has led scientists to explore the effect of combinatorial therapy for cancer treatment, which involves the use of both DNA methylation and HDAC inhibitors together. Such combination treatment strategies have been found to be more effective than individual treatment approaches. For example, the derepression of certain putative tumor suppressor genes was only seen when 5-Aza-CdR and trichostatin A were combined. The antitumorigenic effects of depsipeptide were enhanced when leukemic cells were simultaneously treated with 5-Aza-CdR [108].

Apart from DNA methylation and HDAC inhibitors, HMT inhibitors have recently been explored as another approach toward epigenetic therapy. DZNep, one such HMT inhibitor, was shown to induce apoptosis in cancer cells by targeting polycomb-repressive complex 2 proteins [109]. The use of miRNAs as potential therapeutic targets has also been determined by several workers. Studies have suggested that the introduction of synthetic miRNAs mimicking tumor suppressor miRNAs can be used to selectively repress oncogenes in tumors. MiR-101, which targets EZH2 [104], can be used to regulate the aberrant epigenetic machinery in cancer, which may assist in restoring the normal epigenome. However, efficient delivery methods need to be developed to effectively use this strategy.

3.8 Conclusion

After the identification of potentially labile regions in humans, epigenetic approaches for screening and diagnosis, which will be highly useful in enabling the clinicians to identify individuals at high risk before the onset of the disease, should be developed. Because epigenetic profiles are potentially reversible, prevention and treatment in the form of nutritional supplementation and pharmaceutical therapies may be developed to counteract abnormal epigenetic profiles.

References

1. Peterson, C.L., and M.A. Laniel. 2004. Histones and histone modifications. *Current Biology* 14:R546–551.
2. Espino, P.S., B. Drobic, K.I. Dunn, and J.R. Davie. 2005. The Ras-MAPK signal transduction pathway, cancer and chromatin remodeling. *Journal of Cellular Biochemistry* 94:1088–1102.

3. Workman, J.L., and R.E. Kingston. 1998. Alterations of nucleosome structure as a mechanism of transcriptional regulation. *Annual Reviews in Biochemistry* 67:545–579.
4. Nig, H.H., and A. Bird. 1999. DNA methylation and chromatin modification. *Current Opinion in Genetic Development* 12:599–606.
5. Ramakrishnan, V. 1997. Histone H1 and chromatin higher order structure. *Critical Reviews in Eukaryotic Gene Expression* 7:215–230.
6. Grunstein, M. 1998. Yeast heterochromatin: Regulation of its assembly and inheritance by histones. *Cell* 93:325–328.
7. Craig, J.M. 2005. Heterochromatin—many flavors, common themes. *Bioassays* 27:17–28.
8. Holiday, R., and J.E. Pugh. 1975. DNA modification mechanisms and gene activity during development. *Science* 187:226–232.
9. Riggs, A.D. 1975. X-inactivation, differentiation and DNA methylation. *Cytogenetic and Cell Genetics* 14:9–25.
10. Walsh, C.P., and T.H. Bestor. 1999. Cytosine methylation and mammalian development. *Genes Development* 13:26–34.
11. Lei, H., S.P. Oh, M. Okano, R. Jüttermann, K.A. Goss, R. Jaenisch, and E. Li. 1996. *De novo* DNA cytosine methyl transferase activities in mouse embryonic stem cells. *Development* 122:3195–3205.
12. Okano, M., D.W. Bell, D.A. Haber, and E. Li. 1999. DNA methyltransferases DNMT 3a and 3b are essential for *de novo* methylation and mammalian development. *Cell* 99:247–257.
13. Goll, M.G., and T.H. Bestor. 2005. Eukaryotic cytosine methyltransferases. *Annual Reviews in Biochemistry* 74:481–514.
14. Bird, A. 2002. DNA methylation patterns and epigenetic memory. *Genes Development* 16:6–21.
15. Clouaire, T., and I. Stancheva. 2008. Methyl-CpG binding proteins: Specialized transcriptional repressors or structural components of chromatin? *Cellular and Molecular Life Science* 65:1509–1522.
16. Nightingale, K.P., L.P. O'Neill, and B.M. Turner. 2006. Histone modifications: Signaling receptors and potential elements of a heritable epigenetic code. *Current Opinion in Genetic Development* 16:125–136.
17. Vogelauer, M., L. Rubbi, I. Lucas, B.J. Brewere, and M. Grunstein. 2002. Histone acetylation regulates the time of replication origin firing. *Molecular Cell* 10:1223–1233.
18. Kornberg, R.D. 1999. Eukaryotic transcriptional control. *Trends in Cell Biology* 9:46–49.
19. Lechner, M., and T. Jenuwein. 2002. The many faces of histone lysine methylation. *Current Opinion in Cellular Biology* 14:286–298.
20. Fischle, W., Y. Wang, and C.D. Aliss. 2003. Binary switches and modification cassettes in histone biology and beyond. *Nature* 425:475–479.
21. Strahl, B.D., and C.D. Allis. 2000. The language of covalent histone modifications. *Nature* 403:41–45.
22. Fuks, F., W.A. Burgers, A. Brehm, L. Hughes-Davies, and T. Kouzarides. 2000. DNA methyl transferase DNMT 1 associates with histone deacetylase activity. *Nature Genetics* 24:88–91.
23. Fuks, F., P.J. Hurd, D. Wolf, X. Nan, and A.P. Bird. 2003. Methyl CpG binding protein MeCP2 links DNA methylation to histone methylation. *Journal of Biological Chemistry* 278:4035–4040.
24. Zilberman, D., X. Cao, and S.E. Jacobsen. 2003. ARGONAUTE4 control of locus specific siRNA accumulation and DNA and histone methylation. *Science* 299:716–719.
25. Hall, I.M. 2002. Establishment and maintenance of a heterochromatin domain. *Science* 297:2232–2237.
26. Volpe, T.A., C. Kidner, I.M. Hall, G. Teng, S.I. Grewal, and R.A. Martienssen. 2002. Regulation of heterochromatin silencing and histone H3 lysine-9 methylation by RNAi. *Science* 297:1833–1837.
27. Rougeulle, C., and E. Heard. 2002. Antisense RNA in imprinting: Spreading silence through air. *Trends in Genetics* 18:434–437.
28. He, L., and G.J. Hannon. 2004. Micro RNAs: Small RNA's with a big role in gene regulation. *Nature Reviews in Genetics* 5:522–531.

29. Fabbri, M., R. Garzon, A. Cimmino, Z. Liu, N. Zanesi, E. Callegari et al. 2007. MicroRNA-29 family reverts aberrant methylation in lung cancer by targeting DNA methyltransferases 3A and 3B. *Proceedings of the National Academy of Sciences of the United States of America* 104:15805–115810.

30. Lewis, A., and W. Reik. 2006. How imprinting centres work. *Cytogenetic and Genome Research* 113:81–89.

31. Murphy, S.K., and R.L. Jirtle. 2003. Imprinting evolution and the price of silence. *Bioessays* 25:577–588.

32. Jirtle, R.L., and J.R. Weidman. 2007. Imprinted and more equal. *American Society* 95:143–149.

33. Jirtle, R.L. 2009. Epigenome: The program for human health and disease. *Epigenomics* 1:13–16.

34. DeChiara, T.M., E.J. Robertson, and A. Efstratiadis. 1991. Parental imprinting of the mouse insulin like growth factor II gene. *Cell* 64:849–859.

35. Barlow, D.P., R. Stroger, B.G. Herrmann, K. Saito, and N. Schweifer. 1991. The mouse insulin like growth factor type-2 receptor is imprinted and closely linked to the Tme locus. *Nature* 349:84–87.

36. Morison, I.M., J.P. Ramsay, and H.G. Spencer. 2005. A census of mammalian imprinting. *Trends in Genetics* 21:457–465.

37. Dolinoy, D.C., and R.L. Jirtle. 2008. Environmental epigenomics in human health and disease. *Environmental and Molecular Mutagen* 49:4–8.

38. Brown, C.J., A. Ballabio, J.L. Rupert, R.G. Lafreniere, M. Grompe, and R. Tonlorenzi. 1991. A gene from the region of the human X inactivation centre is expressed exclusively from the inactive X chromosome. *Nature* 349:38–44.

39. Csankovszki, G., A. Nagy, and R. Jaenisch. 2000. Synergism of Xist RNA, DNA methylation, and histone hypoacetylation in maintaining X chromosome inactivation. *Journal of Cellular Biology* 153:773–784.

40. Birds, A., and T. Kouzarides. 2000. Chromosomes and expression mechanisms. *Current Opinion in Genetic Development* 10:141–143.

41. Rakyan, V.K., M.E. Bleuwitt, R. Druker, and E. Whitelaw. 2002. Metastable epialleles in mammals. *Trends in Genetics* 18:348–351.

42. Waterland, R.A., and R.L. Jirtle. 2003. Transposable elements: Targets for early nutritional effects on epigenetic gene regulation. *Molecular and Cellular Biology* 23:5293–5300.

43. Dolinoy, D.C., J. Weidman, R. Waterland, and R.L. Jirtle. 2006. Maternal genistein alters coat color and protects A^vy mouse offspring from obesity by modifying the fetal epigenome. *Environmental Health Perspectives* 114:567–572.

44. Waterland, R., and D.C. Dolinoy. 2006. Maternal methyl supplements increase offspring DNA methylation at axin (fused). *Genesis* 44:401–406.

45. Dolinoy, D.C., D. Huang, and R.L. Jirtle. 2007. Maternal nutrient supplementation counteracts bisphenol-A induced DNA hypomethylation in early development. *Proceedings of the National Academy of Sciences of the United States of America* 104:13056–13061.

46. Drucker, R., and E. Whitelaw. 2004. Retroposons derived elements in the mammalian genome: A potential source of disease. *Inheritance and Metabolic Diseases* 27:319–330.

47. Waterland, R.A., and R.L. Jirtle. 2004. Early nutrition, epigenetic changes at transposon and imprinted genes, and enhanced susceptibility to adult chronic diseases. *Nutrition* 20:63–68.

48. Sutherland, J.E., and M. Costa. 2003. Epigenetics and the environment. *Annals New York Academy of Sciences* 983:151–160.

49. Jirtle, R.L., and M.K. Skinner. 2007. Environmental epigenomics and disease susceptibility. *Nature Reviews in Genetics* 8:253–262.

50. Bernstein, B.E., A. Meissner, and E.S. Lander. 2007. The mammalian epigenome. *Cell* 128:669–681.

51. Feinberg, A.P., and B. Tycko. 2004. The history of cancer epigenetics. *Nature Reviews in Genetics* 4:143–153.

52. Fraga, M.F., M. Herranz, and J. Espada. 2004. A mouse skin multistage carcinogenesis model reflects the aberrant DNA methylation patterns of human tumours. *Cancer Research* 64:5527–5534.

53. Eden, A., F. Gaudet, A. Waghmare, and R. Jaenisch. 2003. Chromosomal instability and tumors promoted by DNA hypomethylation. *Science* 300:455–455.

54. Esteller, M. 2008. Epigenetics in cancer. *New England Journal of Medicine* 358:1148–1159.
55. Lee, T.S., J.W. Kim, G.H. Kang, N.H. Park, and Y.S. Song. 2006. DNA hypomethylation of CAGE promoters in squamous cell carcinoma of uterine cervix. *Annals New York Academy of Sciences* 1091:218–224.
56. Kim, S.J., H.S. Kang, H.L. Chang, Y.C. Jung, and H.B. Sim. 2008. Promoter hypomethylation of the N-acetyltransferase 1 gene in breast cancer. *Oncology Reports* 19:663–668.
57. Watanabe, M., Y. Ogawa, K. Itoh, T. Koiwa, and M.E. Kadin. 2008. Hypomethylation of CD30 CpG islands with aberrant JunB expression drives CD30 induction. *Laboratory Investigation* 88:48–57.
58. Kiran, M., Y.K. Chawla, and J. Kaur. 2009. Methylation profiling of tumor suppressor genes and oncogenes in hepatitis virus–related hepatocellular carcinoma in northern India. *Cancer Genetics and Cytogenetics* 195:112–119.
59. Klose, R.J., and A.P. Bird. 2006. Genomic DNA methylation: The mark and its mediators. *Trends in Biochemical Sciences* 31:89–97.
60. Reik, W., M. Constancia, W. Dean, K. Davies, L. Bowden, A. Murrell, and R. Feil. 2000. IGF2 imprinting in development and disease. *International Journal of Developmental Biology* 44:145–150.
61. Kaneda, A., and A.P. Feinberg. 2005. Loss of imprinting of IGF2: A common epigenetic modifier of intestinal tumor risk. *Cancer Research* 65:11236–11240.
62. Tuck-Muller, C.M., A. Narayan, F. Tsien, D.F. Smeets, and J. Sawyer. 2000. DNA hypomethylation and unusual chromosome instability in cell lines from ICF syndrome patients. *Cytogenetics and Cellular Genetics* 89:121–128.
63. Tomonaga, T., K. Matsushita, S. Yamaguchi, T. Oohashi, and H. Shimada. 2003. Overexpression and mistargeting of centromere protein-A in human primary colorectal cancer. *Cancer Research* 63:3511–3516.
64. Qu, G.Z., P.E. Grundy, A. Narayan, and M. Ehrlich. 1999. Frequent hypomethylation in Wilms tumors of pericentromeric DNA in chromosomes 1 and 16. *Cytogenetics and Cellular Genetics* 109:34–39.
65. Qu, G., L. Dubeau, A. Narayan, M.C. Yu, and M. Ehrlich. 1999. Satellite DNA hypomethylation versus overall genomic hypomethylation in ovarian epithelial tumors of different malignant potential. *Mutation Research* 423:91–101.
66. Badal, V., L.S. Chuang, E.H. Tan, S. Badal, and L.L. Villa. 2003. CpG methylation of human papillomavirus type 16 DNA in cervical cancer cell lines and in clinical specimens: Genomic hypomethylation correlates with carcinogenic progression. *Journal of Virology* 77:6227–6234.
67. Costello, J.F., M.C. Fruhwald, D.J. Smiraglia, L.J. Rush, G.P. Robertson, X. Gao et al. 2000. Aberrant CpG island methylation has non-random and tumor-type-specific patterns. *Nature Genetics* 25:132–138.
68. Sakai, T., J. Toguchida, N. Ohtani, D.W. Yandell, J.M. Rapaport, and T.P. Dryja. 1991. Allele-specific hypermethylation of the retinoblastoma tumor-suppressor gene. *American Journal of Human Genetics* 48:880–888.
69. Herman, J.G., and S.B. Baylin. 2003. Gene silencing in cancer in association with promoter hypermethylation. *New England Journal of Medicine* 349:2042–2054.
70. Esteller, M. 2007. Cancer epigenomics: DNA methylomes and histone-modification maps. *Nature Reviews in Genetics* 8:286–298.
71. Shames, D.S., J.D. Minna, and A.F. Gazdar. 2007. DNA methylation in health, disease, and cancer. *Current and Molecular* Medicine 7:85–102.
72. Icobuzio-Donahue, C.A. 2009. Epigenetic changes in cancer. *Annual Reviews in Pathology and Mechanism of Disease* 4:229–249.
73. Akiyama, Y., N. Watkins, H. Suzuki, K.W. Jair, and M. van Engeland. 2003. GATA-4 and GATA-5 transcription factor genes and potential downstream antitumor target genes are epigenetically silenced in colorectal and gastric cancer. *Molecular and Cellular Biology* 23:8429–8439.
74. Ricciardiello, L., A. Goel, V. Mantovani, T. Fiorini, and S. Fossi. 2003. Frequent loss of hMLH1 by promoter hypermethylation leads to microsatellite instability in adenomatous polyps of patients with a single first-degree member affected by colon cancer. *Cancer Research* 63:787–792.

75. Jacinto, F.V., and M. Esteller. 2007. Mutator pathways unleashed by epigenetic silencing in human cancer. *Mutagen* 22:247–253.

76. Rhee, I., K.W. Jair, R.W. Yen, C. Lengauer, and J.G. Herman. 2000. CpG methylation is maintained in human cancer cells lacking DNMT1. *Nature* 404:1003–1007.

77. D'Alessio, A.C., and M. Szyf. 2006. Epigenetic tete-a-tete: The bilateral relationship between chromatin modifications and DNA methylation. *Biochemistry and Cell Biology* 84:463–476.

78. Babushok, D.V., E.M. Ostertag, and H.H. Kazazian, Jr. 2007. Current topics in genome evolution: Molecular mechanisms of new gene formation. *Cellular and Molecular Life Sciences* 64:542–554.

79. Perini, G., D. Diolaiti, A. Porro, and G. Della Valle. 2005. *In vivo* transcriptional regulation of N-Myc target genes is controlled by E-box methylation. *Proceedings of the National Academy of Sciences of the United States of America* 102:12117–12122.

80. Fraga, M.F., E. Ballestar, A. Villar-Garea, M. Boix-Chornet, J. Espada, G. Schotta et al. 2005. Loss of acetylating at Lys16 and trimethylation at Lys20 of histone H4 is a common hallmark of human cancer. *Nature Genetics* 37:391–400.

81. Song, J., J.H. Noh, J.W. Eun, Y.M. Ahn, S.Y. Kim, S.H. Lee et al. 2005. Increased expression of histone deacetylase 2 is found in human gastric cancer. *Acta Pathologica, Microbiologica, et Immunologica Scandinavica* 113:264–268.

82. Nguyen, C.T., D.J. Weisenberger, M. Velicescu, F.A. Gonzales, J.C. Lin, G. Liang et al. 2002. Histone H3-lysine 9 methylation is associated with aberrant gene silencing in cancer cells and is rapidly reversed by 5-aza-2-deoxycytidine. *Cancer Research* 62:6456–6461.

83. Sharma, S., K.K. Theresa, and A.J. Peter. 2010. Epigenetics in cancer. *Carcinogenesis* 31:27–36.

84. Lin, J.C., S. Jeong, G. Liang, D. Takai, M. Fatemi, Y.C. Tsai et al. 2007. Role of nucleosomal occupancy in the epigenetic silencing of the MLH1 CpG island. *Cancer Cell* 12:432–444.

85. Kanwal, R., and S. Gupta. 2010. Epigenetics and cancer. *Journal of Applied Physiology* 109:598–605.

86. Ventura, A., and T. Jacks. 2009. MicroRNAs and cancer: Short RNAs go a long way. *Cell* 136:586–591.

87. Saito, Y., G. Liang, G. Egger, J.M. Friedman, J.C. Chuang, G.A. Coetzee et al. 2006. Specific activation of microRNA-127 with downregulation of the proto-oncogene BCL6 by chromatin-modifying drugs in human cancer cells. *Cancer Cell* 9:435–443.

88. Chan, J.A., A.M. Krichevsky, and K.S. Kosik. 2005. MicroRNA-21 is an antiapoptotic factor in human glioblastoma cells. *Cancer Research* 65:6029–6033.

89. Kluiver, J., B.J. Kroesen, S. Poppema, and A. van den Berg. 2006. The role of microRNAs in normal hematopoiesis and hematopoietic malignancies. *Leukemia* 20:1931–1936.

90. Mendell, J.T. 2008. miRiad roles for the miR-17–92 cluster in development and disease. *Cell* 133:217–222.

91. Lujambio, A., S. Ropero, and E. Ballestar. 2007. Genetic unmasking of an epigenetically silenced microRNA in human cancer cells. *Cancer Research* 67:1424–1429.

92. He, L., X. He, and L.P. Lim. 2007. A microRNA component of the p53 tumour suppressor network. *Nature* 447:1130–1134.

93. O'Donnell, K.A., E.A. Wentzel, K.I. Zeller, C.V. Dang, and J.T. Mendell. 2005. c-Myc-regulated microRNAs modulate E2F1 expression. *Nature* 435:839–843.

94. Chiosea, S., E. Jelezcova, U. Chandran, J. Luo, G. Mantha, R.W. Sobol et al. 2007. Overexpression of Dicer in precursor lesions of lung adenocarcinoma. *Cancer Research* 67:2345–2350.

95. Merritt, W.M., Y.G. Lin, L.Y. Han, A.A. Kamat, W.A. Spannuth, and R. Schmandt. 2008. Dicer, Drosha, and outcomes in patients with ovarian cancer. *New England Journal of Medicine* 359:2641–2650.

96. Xu, W., D.G. Edmondson, and Y.A. Evrard. 2000. Loss of Gcn 512 leads to increased apoptosis and mesodermal defects during mouse development. *Nature Genetics* 26:229–232.

97. Kriaucionis, S., and A. Bird. 2003. DNA methylation and Rett syndrome. *Human Molecular Genetics* 12:221–227.

98. Cockett, N.E., S.P. Jackson, T.L. Shay, D. Neilson, S.S. Moore, M.R. Steele et al. 1994. Chromosomal localization of Callipyge gene in sheep using bovine DNA markers. *Proceedings of the National Academy of Sciences of the United States of America* 91:3019–3023.
99. Reik, W., W. Dean, and J. Walter. 2001. Epigenetic reprogramming in mammalian development. *Science* 293:1089–1093.
100. Richardson, B. 2003. Impact of ageing on DNA methylation. *Ageing Research and Reviews* 2:245–261.
101. Wilson, C.B., K.W. Makar, M. Shnyreva, and D.R. Fitzpatrick. 2005. DNA methylation and the expanding epigenetics of T cell lineage commitment. *Seminars in Immunology* 17:105–119.
102. Richardson, B. 2003. DNA methylation and autoimmune disease. *Clinical Immunology* 109:72–79.
103. McCabe, M.T., J.C. Brandes, and P.M. Vertino. 2009. Cancer DNA methylation: Molecular mechanisms and clinical implications. *Clinical Cancer Research* 15:3927–3937.
104. Yoo, C.B., and P.A. Jones. 2006. Epigenetic therapy of cancer: Past, present and future. *Nature Reviews in Drug Discovery* 5:37–50.
105. Plimack, E.R., H.M. Kantarjian, and J.P. Issa. 2007. Decitabine and its role in the treatment of hematopoietic malignancies. *Leukemia and Lymphoma* 48:1472–1481.
106. Datta, J., K. Ghoshal, W.A. Denny, S.A. Gamage, D.G. Brooke, P. Phiasivongsa et al. 2009. A new class of quinoline-based DNA hypomethylating agents reactivates tumor suppressor genes by blocking DNA methyltransferase 1 activity and inducing its degradation. *Cancer Research* 69:4277–4285.
107. Bots, M., and R.W. Johnstone. 2009. Rational combinations using HDAC inhibitors. *Clinical Cancer Research* 15:3970–3977.
108. Carew, J.S., F.J. Giles, and S.T. Nawrocki. 2008. Histone deacetylase inhibitors: Mechanisms of cell death and promise in combination cancer therapy. *Cancer Letters* 269:7–17.
109. Tan, J., L. Zhuang, X. Jiang, W. Chen, and P.L. Lee. 2007. Pharmacologic disruption of Polycomb-repressive complex 2-mediated gene repression selectively induces apoptosis in cancer cells. *Genes Development* 21:1050–1063.

4

Pharmacogenomics for Individualized Therapy

Eugenia Ch. Yiannakopoulou

CONTENTS

4.1 Fundamentals of Pharmacogenomics

The term "pharmacogenetics" was coined in 1950s as it became evident that there was an inherited basics for differences in disposition and effects of drugs [1]. In 1990s, the concept of pharmacogenomics was introduced in the literature. In 2001, Pirmohamed [2] reported that pharmacogenetics is defined as the study of genetic variation in drug response, whereas the term pharmacogenomics was a broader term encompassing all genes in the genome that may determine drug response. He considered this distinction arbitrary as both terms are frequently used interchangeably. According to Jose de Leon [3], pharmacogenetics is commonly used in the context of studies that investigate the role of candidate genes, whereas pharmacogenomics is generally used in the exploratory context of genomewide association studies. Roses [4] has made a distinction between the two types of pharmacogenetics: safety pharmacogenetics is aimed at avoiding adverse drug reactions (ADRs), whereas efficacy pharmacogenetics is aimed at predicting response to medications. A drug's activity is the result of interactions with proteins involved in absorption, distribution, metabolism, elimination, and cellular targets. Genetic variation in any of these proteins can have a significant effect on drug response. Pharmacogenomics affects many aspects of therapy, including disposition of the drug (pharmacokinetics), efficacy of the drug (pharmacodynamics), dose selection, and adverse events of the drug—whether dose related or not [5,6]. Thus, pharmacogenomics is expected to contribute to personalized medicine [7].

The clinical importance of genetic variations depends on the allele frequency and on the effect size of the outcome parameter. Yet, it also depends on the therapeutic range and safety of the drug because pharmacokinetic changes might not be important for each drug, but is especially important for drugs which have to be carefully monitored because of a small dose range for response and a high risk for ADRs. The importance of genotyping is larger in drug therapies of which the individual response can hardly be predicted. For example, in antidepressant drug therapy, 30% of the patients will not respond sufficiently to drug therapy and therapeutic effects are first expected after 2 weeks of treatment.

4.1.1 Types of Genetic Variations

At the nucleotide level, genetic variability is observed as (i) polymorphisms and (ii) larger insertions, deletions, and duplications (copy number variations, CNVs). CNVs are reported to affect 12% of the human genome. Up to now, limited attention has been given to CNVs, although it seems that CNVs may be important for pharmacogenomics. Less common genetic variations, such as translocations, inversions, and substitutions, may also have some relevance to pharmacogenomics.

Polymorphisms represent common variations in the DNA sequence that may lead to reduced activity of the encoded gene although, in some cases, this may lead to increased activity. Unlike somatic mutations, they are stable and heritable. Polymorphisms include single nucleotide polymorphisms (SNP), microsatellites, and minisatellites.

Microsatellites (or tandem repeats) are multiple copies of repeated DNA sequences at a length of 0.1 to 10 kb. Microsatellites consist of repeated small DNA sequences up to 4 nucleotides. Microsatellites and minisatellites are less common genetic variations than SNPs, but have some relevance in pharmacogenomics. An SNP represents a single base exchange that may or may not cause an amino acid exchange in the encoded protein. SNPs that alter drug response are often found in the coding region of DNA and they can cause amino acid substitutions in the translated protein (nonsynonymous SNP). The resulting variant protein can have altered biological behavior relative to the wild-type protein.

However, current evidence suggests that even silent SNPs, namely, those that do not result in amino acid change (synonymous SNP) can lead to the synthesis of a protein with the same amino acid sequence but can affect protein folding kinetics and create an improperly folding and functionally different protein. Noncoding SNPs in the upstream promoter region of a gene can affect DNA transcription and subsequent protein expression. The frequency of SNPs is greater than 1% in the population and accounts for more than 90% of genetic variation in the human genome. The number of SNPs in the human genome varies between 1 and 10 million. It is estimated that there is one SNP present for every 300 nucleotides in the human genome. Between 50,000 and 250,000 SNPs are distributed in and around the coding genes. Many SNPs do not confer phenotypic alterations. Alleles are alternate forms of a gene. The gametic haplotype is the sum of alleles at a genetic locus.

4.1.2 Micro-RNAs and Pharmacogenomics

Micro-RNAs (miRNAs) are evolutionarily conserved small noncoding RNAs (21–25 nucleotides in length) that comprise one of the more abundant classes of gene regulatory molecules in plants, animals, viruses, and even in unicellular organisms [8]. miRNAs regulate protein output in the cell via silencing specific gene expression, either inhibiting the translation of proteins by binding to the target transcript in the 3′ untranslated region or destabilizing the target mRNA. Thus, upregulation of a target miRNA is expected to enhance the silencing of the target miRNAs, whereas reduced activity of miRNA is expected to result in the upregulation of target mRNAs. However, it is not a one-to-one relationship. A number of miRNAs may bind one mRNA and cooperate to silence gene expression. On the other hand, a single miRNA can affect the expression of several miRNAs [9,10]. Current evidence suggests that miRNAs are involved in the regulation of multiple pathways including cell death, cell proliferation, stress resistance, and fat metabolism.

Recently, a novel class of functional polymorphisms termed miR/SNPs was reported, defined as a polymorphism present at or near miRNA binding sites of functional genes that can affect gene expression by interfering with miRNA function. It is now known that miRNA polymorphisms can alter drug response or drug resistance. A recent study demonstrated that MiRSNP 829 C-T present in the DHFR gene near a miR24 binding site was associated with methotrexate drug resistance sensitivity [9]. Thus, the pharmacogenomics of miRNA is a novel interesting field of research with clinical implications because miRNAs are promising drug targets, are differentially expressed in malignant versus normal cells, and regulate the expression of important proteins in the cell.

4.1.3 Epigenetics and Pharmacogenetics

Pharmacoepigenetics is another current field of research that is not yet well understood. Epigenetics is an aspect of genetic variation that may affect drug response. The term epigenetics refers to heritable traits in the cells and organisms that do not involve changes to the underlying DNA sequence [11]. Thus, the term epigenetics refers to changes in gene expression, caused commonly by environmental factors. These changes may persist through cell division and for the remainder of the organism's life. Epigenetic processes include methylation of DNA, acetylation, phosphorylation, ubiquitylation, and sumoylation of histones [12]. Silencing of genes (through epigenetic mechanisms) that mediate the therapeutic effects of certain drugs results in abrogating or blunting the drug's effects. An example is that of the expression of ERa gene which is recognized as an important predictor of response to endocrine therapy of breast cancer. Loss of ERa expression is often associated with promoter hypermethylation of ERa gene [12].

On the other hand, it has been hypothesized that commonly prescribed medications can cause persistent epigenetic changes through direct or indirect mechanisms. Direct effects may be caused by drugs that alter chromatin architecture or DNA methylation. For example, the antihypertensive drug hydralazine alters DNA methylation. A two-stage mechanism is suggested for indirect effects. According to the hypothesis, acute exposure to a drug affects signaling pathways that may lead to alterations of transcription factor activity at gene promoters. Altered expression of receptors and signaling molecules follows. Chronic exposure results in cell adaptation with subsequent permanent modifications to DNA methylation and chromatin structure. Possibly, some drug adverse events such as systemic lupus erythematosus are due to epigenetic changes caused by the drugs. The relevance of pharmacoepigenetics to the drug response has to be elucidated in the future.

4.1.4 Pharmacogenetic Biomarkers

Personalized medicine can be accomplished with the introduction of pharmacogenomic biomarkers. Biomarkers are defined as a characteristic that is objectively measured and evaluated as an indicator of normal biological processes, pathogenic processes, or pharmacological response to a therapeutic intervention.

Pharmacogenomic biomarkers could provide tools to avoid overdosing and subsequent adverse events, avoid underdosing and subsequent lack of efficacy, avoid drug use by hypersensitive persons, and rescue drugs previously withdrawn because of adverse events.

Pharmacogenomic biomarkers could be genetic variants of drug-metabolizing enzymes, drug transporters, and drug targets. Since 2005, according to the Food and Drug Administration (FDA), genetic variants of two metabolic enzymes, CYP2D6 and thiopurine S-methyltransferase (TMPT), are considered valid pharmacogenomic biomarkers [13].

4.2 Pharmacogenomics and Pharmacokinetics

Knowledge of the genetic factors that affect pharmacokinetics far exceeds knowledge of the genetic factors that affect pharmacodynamics. This discrepancy is attributed to a number of reasons. Cloning of most metabolizing enzymes and drug transport proteins has been accomplished in the last 20 years. Sequencing of the human genome has allowed the characterization of the variability of drug-metabolizing and transport genes. Thus, it is straightforward to understand the genetic differences in drug disposition that result in different exposures among patients that are prescribed the same dose. Knowledge of the polymorphisms of pharmacokinetic genes can be applied in all therapeutic areas and chemical entities. On the other hand, factors that affect drug efficacy are drug/drug class–specific or pathway-specific (or both). Thus, the search for genetic factors affecting efficacy must be done for each new chemical entity or at least for each new class of drugs.

4.2.1 Pharmacogenomics and Drug Metabolism

Genetically based differences in drug metabolism are reflected in differences in clearance, half-life, and maximal plasma concentrations. Differences in pharmacokinetic parameters such as oral clearances could be overcome by dose adjustments. Theoretical dose adjustments make sense for those drug therapies in which similar plasma concentration time courses also lead to similar clinical effects. However, for some drugs, plasma concentrations correlate poorly with clinical efficacy, and empirical dose findings rather than plasma concentrations are used to avoid adverse drug events.

4.2.1.1 Cytochrome P450 Polymorphisms

The CYP enzymes are encoded by the CYP genes. These genes are categorized into families and subfamilies based on the sequence similarities observed in their encoded proteins. Thus, genes encoding proteins showing similarity in their amino acid sequence greater than 40% are assigned to the same family, whereas genes encoding proteins showing similarity greater than 55% are assigned to the same subfamily. Humans have 18 CYP gene families and 44 CYP gene subfamilies. The CYP gene superfamily includes the families that are identified by a number (e.g., CYP1, CYP2) and are classified into subfamilies identified by a letter (e.g., CYP1A, CYP1B). Each subfamily includes the genes identified by a number (e.g., CYP1A1, CYP1A2), and the allele families identified by an asterisk followed by a number (e.g., CYP1A1*1), in which the single alleles are identified by a letter (e.g., CYP1A1*1A).

Cytochrome P450 is the most important group of phase I drug-metabolizing enzymes. It has been estimated that up to 80% of drugs undergo oxidation reactions catalyzed by these enzymes. CYP3A4 is the human enzyme known to be involved in the metabolism of the largest number of medications. It has been estimated that CYP3A4 is responsible for approximately 50% of all cytochrome P450–mediated reactions of prescribed drugs. CYP3A4 does not seem to be subject to polymorphisms, although it is known that there is interindividual variation in overall levels of activity [14]. CYP2D6 and CYP2C9 are responsible for 25% and 20%, respectively, of all cytochrome P450–mediated reactions of prescribed drugs. On the other hand, CYP1A2, CYP2A6, CYP2B6, CYP2C8, CYP2C19, and CYP3A5 have a more limited contribution in drug metabolism. CYP2D6, CYP2A6, CYP2C19, and CYP3A5 are subject to polymorphisms leading to the absence of enzyme activity. CYP2C9 activity is very low in some individuals due to two common polymorphisms.

The CYP2B6 human cytochrome is less well studied. Clinically used drug substrates include cytostatics (cyclophosphamide), HIV drugs, antidepressants, antimalarials, anesthetics (propofol), and synthetic opioids (methadone). Contrary to the model polymorphisms of CYP2D6 and CYP2C19, which were discovered by ADRs, a pharmacogenetic study of CYP2B6 was initiated by reverse genetics approaches and subsequent functional and clinical studies [15]. CYP2B6 is one of the most polymorphic human CYPs with more than 100 SNPs and numerous complex phenotypes.

4.2.1.2 *CYP2D6 Polymorphisms*

Common substrates of CYP2D6 are largely lipophilic bases and include some antidepressants, antipsychotics, antiarrhythmics, bera-adrenororeceptor antagonists and opioids. The CYP2D6 activity ranges considerably within a population and includes ultrarapid metabolizers, extensive metabolizers, and poor metabolizers. More than 80 allelic variants and a series of subvariants of CYP2D6 have been identified and characterized. They include fully functional alleles, alleles with reduced function, and nonfunctional alleles which convey a wide range of enzyme activities ranging from no activity to ultrarapid metabolism of substrates. Thus, adverse effects or a lack of effect may occur if standard doses are applied. Approximately 2% to 10% of the population are poor metabolizers and therefore resistant to opioid analgesic effects. Thus, there is remarkable interindividual variability in the adequacy of pain relief when uniform doses of codeine are prescribed. It is estimated that 95% of poor metabolizers from Europe have two copies of any combination of four alleles termed CYP2D6*3, CYP2D6*4, CYP2D6*5, and CYP2D6*6 with each encoding defective forms of CYP2D6.

The remaining 5% of poor metabolizers are homozygous or heterozygous for a range of relatively rare loss of function alleles including CYP2D6*7, CYP2D6*8, CYP2D6*11, CYP2D6*12, CYP2D6*13, CYP2D6*14, CYP2D6*15, CYP2D6*16, CYP2D6*18, CYP2D6*19, CYP2D6*20, CYP2D6*21, CYP2D6*38, CYP2D6*40, CYP2D6*42, CYP2D6*44, CYP2D6*56, and CYP2D6*62. Most inactivating mutations in CYP2D6 are point mutations resulting in splicing defects or deletions that result in a truncated protein or no protein at all. Many individuals are characterized as intermediate metabolizers. Intermediate metabolizers are heterozygous for one of the inactivating mutations or homozygous for alleles associated with impaired metabolism. In European populations, two alleles associated with impaired metabolism, CYP2D6*9 and CYP2D6*41, are quite common. CYP2D6*9 encodes a protein with deletion of one amino acid. CYP2D6*41 includes a number of different polymorphisms including two nonsynonymous mutations, which are also seen in the CYP2D6*2 allele, an upstream polymorphism at position −1584 and a base substitution in intron 6. The nonsynonymous mutations do not seem to alter the enzyme activity, but the intron 6 polymorphism is associated with altered RNA splicing, resulting in lower levels of protein.

On the other hand, the CYP2D6 gene is subject to CNVs that are often associated with the ultrarapid metabolizer phenotype. Marked decreases in drug concentrations have been observed in ultrarapid metabolizers with tramadole, venlafaxine, morphine, mirtazapine, and metoprolol. The ultrarapid phenotype is most commonly associated with a single gene duplication event. However, some ultrarapid metabolizers have 13 copies of CYP2D6 arranged as tandem repeats. Depending on the country of origin, 1% to 8% of Europeans have one extra copy of CYP2D6*1 or CYP2D6*2, resulting in faster than average metabolism. The number of active CYP2D6 alleles present in an individual's genome predicts CYP2D6 enzyme activity, with those heterozygous for poor metabolizer alleles showing lower levels of enzyme activity than those with two normal alleles.

Drugs that are more affected by CYP2D6 polymorphisms are commonly those in which CYP2D6 represents a substantial metabolic pathway either in the activation to form active metabolites or in the clearance of the drug. For example, encainide metabolites are more potent than the parent drug and thus QRS prolongation is more common in extensive metabolizers than in poor metabolizers. On the other hand, propaphenone is a more potent β-blocker than its metabolites and β-blocker activity of propaphenone is more prominent in poor metabolizers than in extensive metabolizers because the drug accumulates in poor metabolizers. Because the contribution of CYP2D6 is greater for metoprolol than for carvedilol, propranolol, and timolol, a stronger gene–dose effect is seen with metoprolol than with the other β-blockers.

CYP2D6 genotyping has not entered clinical practice up to now. Reasons include the general difficulty of introducing genetic tests into clinical practice, withdrawal from the market of a number of CYP2D6 key substrates due to problems experienced by poor metabolizers, and because certain classes of CYP2D6 substrates such as tricyclic depressants are less commonly used than when the polymorphism was described. However, current evidence suggests that CYP2D6 genotype may be relevant to the outcome of treatment with codeine and tamoxifen. Several CYP2C8 polymorphisms have been described. The variant alleles CYP2C8*2, CYP2C8*3, CYP2C8*5, CYP2C*7, and CYP2C*8 alter enzyme activity. The most common of these are CYP2C8*2 in African subjects, CYP2C8*3 and CYP2C8*4 in Caucasian subjects, and CYP2C8*5 in Asian subjects. Typical CYP2C8 substrates include paclitaxel, verapamil, amodiaquine, troglitazone, and ibuprofen. It has been demonstrated that CYP2C8 variant alleles are associated with diclofenac-induced hepatotoxicity and acute gastrointestinal bleeding induced by nonsteroidal anti-inflammatory drugs (NSAIDs).

Human cytochrome P450 2C9 CYP2C9 accounts for approximately 20% of total hepatic CYP and metabolizes approximately 15% of clinically used drugs including S-warfarin, tolbutamide, phenytoin, losartan diclofenac, and celecoxib. Concerning CYP2C9, a total of 34 polymorphic alleles have been described, with CYP2C9*2 and CYP2C9*3 being the most common. These two alleles differ from the wild-type CYP2C9*1 allele by a single point mutation. CYP2C9*2 is characterized by a 430C>T exchange in exon 3 resulting in an Arg144Cys amino acid substitution. CYP2C9*3 is characterized by a 1075A>T exchange in exon 7, causing an Ile359leu substitution in the catalytic site of the enzyme.

It has been demonstrated in several populations that there is a linkage between CYP2C8 and CYP2C9 polymorphisms. Of particular interest is the linkage between CYP2C8*3 and CYP2C9*2 polymorphisms. Individuals simultaneously carrying both variant alleles demonstrate the lowest functional activity in the metabolism of CYP2C8 and CYP2C9 substrates. CYP2C9 polymorphisms are particularly important to the metabolism of drug substrates with narrow therapeutic indices including warfarin and phenytoin. Evidence suggests that there is a relationship between the required warfarin dose and the CYP2C9 genotype.

CYP2C9 polymorphism plays a role in treatments using nonsteroidal anti-inflammatory agents, At least 16 different registered NSAIDs are at least partially metabolized by CYP2C9. These include acenofenac, acetylsalicylic acid, azapropazone, celecoxib, diclofenac, frurbiprofen, ibuprofen, indomethacin, lornoxicam, mefenamic acid, meloxicam, naproxen, phenylbutazone, piroxicam, and tenoxicam [16]. Significant intergenotypic differences have been reported in the pharmacokinetics of celecoxib, flurbiprofen, ibuprofen, and tenoxicam, which could be translated into dose recommendations based on CYP2C9 genotype. Celecoxib was one of the first drugs for which the manufacturer's drug information recommended caution when administering to poor metabolizers of CYP2C9 substrates, as they could have abnormally low levels [16]. However, the relative contribution of CYP2C9 on the pharmacokinetics of diclofenac has been found to be independent from CYP2C9 polymorphisms in several studies [16].

Oral hypoglycemic drug treatment is affected by CYP2C9 polymorphism. Oral clearance of glyburide, glimepiride, glipizide, and tolbutamide in carriers of the CYP2C9*1/*3 genotypes has been found to be approximately 50% of that in CYP2C9*1/*1 homozygotes. In homozygous carriers of the CYP2C9*3/*3 genotype, the clearance was only 20% of that in wild-type individuals [16]. Two relatively common alleles of CYP2C19 associated with the absence of enzyme activity have been identified. Individuals with the CYP2C19 deficiency have two variant alleles present, although heterozygotes may also show impaired metabolism. CYP2C19 is responsible for the metabolism of a limited number of commonly prescribed drugs, including the proton pump inhibitor omeprazole [17].

4.2.1.3 Pharmacogenomics of Drug-Metabolizing Phase II Enzymes

Phase II enzymes conjugate phase I metabolites, other intermediates, or the parent compound for renal or biliary excretion. Phase II enzymes include glutathione S-transferases (GSTs), UDP glycurosyltransferases (UGT), N-acetyltransferases, GSTs, NADH quinone oxidases, and others. Knowledge on pharmacogenomics of phase II enzymes is not extensive. However, it is evident that defective alleles of UGTs, TMPT, and other biotransformation enzymes can influence the outcome of therapy (Table 4.1). For example, defective alleles for N-acetyltransferase 2 are responsible for the well-described acetylation polymorphism that leads to impaired clearance and hepatotoxicity of isoniazide [18].

On the other hand, GSTs participate in the detoxification of chemotherapeutic agents and genetic variations in GST genes (GSTP1 Ile105val, copy number variants of GSTM1 and GSTT1 that lead to diminished enzyme activity) have been associated with increased chemotherapeutic treatment benefit in patients with colorectal cancer.

4.2.1.3.1 UDP Glycurosyltransferases

UGTs catalyze the glucoronidation of many lipophilic xenobiotics and endobiotics to make them more water-soluble and therefore enhance their elimination. UGTs are quite important for the metabolism of psychiatric drugs including some antipsychotics, some antidepressants, some mood stabilizers, and some benzodiazepines. The human UGT superfamily has been classified into the UGT1 and UGT2 families, and further classified into three subfamilies, UGT1A, UGT2A, and UGT2B. Overall, 30 UGT isoforms with overlapping specificities have been identified. Functional polymorphisms have been described for UGT1A1, UGT1A6, UGT1A7, UGT2B4, UGT2B7, and UGT2B15.

In general, polymorphisms of UGTs have not been investigated adequately because of the overlapping activities of the UGTs and the lack of selective probes, the complexity of the glucuronidation cycle, and the difficulty of developing analytic methods to measure glucuronides. Variations in drug metabolisms due to altered UGT activity as a consequence of polymorphisms have been described for UGT1A1 and UGT2B7. All nine functional members of the UGT1A subfamily are encoded by a single gene locus on chromosome 2q17. Several polymorphisms of the UGT1A1 gene have been reported. The best studied are the UGT1A1 promoter polymorphisms defined by a variable length "TA" tandem repeat in the regulatory TATA box of the UGT1A1 gene promoter. Individuals that are homozygous for seven promoter TA repeats (namely those carrying two copies of the gene UGT1A1*28) express a 2.5-fold lower level of UGT1A1 protein than those carrying the reference gene UGT1A1*1 with six TA repeats. Evidence suggests that UGT1A1*28 contributes only 40% of the observed variability in UGT1A1 enzyme activity. Other polymorphic variants with five to eight TA repeats have been reported.

TABLE 4.1

Pharmacogenomics of Phase II Metabolizing Enzymes

Gene	Affected Drugs	Effect
UGT1A1	Irinotecan	Myelosuppression and delayed type diarrhea
TMPT	Azathioprine	Fatal myelosuppression
TMPT	Mercaptopurine	Fatal myelosuppression
TMPT	Thioguanine	Fatal myelosuppression
NAT2	Isoniazide	Altered clearance
UGT2B17 deletion	Exemestane	Possible effect on drug efficacy and safety
GSTP1	Oxaliplatin	Chronic neuropathy

Polymorphisms resulting in absent or no UGT1A1 activity have been associated with heritable unconjugated hyperbilirubinemia syndromes: type I and II Crigler–Najjar syndromes and Gilbert syndrome. Gilbert syndrome is common among Caucasians and is associated with the presence of an extra seventh dinucleotide TA insertion (UGT1A1*28) in the TATAA box of the UGT1A1 promoter region. A recent study has suggested a potential role of UGT2B17 deletion in exemestane metabolism [19]. Exemestane is a third-generation aromatase inhibitor used in the treatment of breast cancer in postmenopausal women. Reduction to the form 17-dihydroxyexemestane and subsequent glucuronidation to exemestane-17-*O*-glucuronide is a major pathway for exemestane metabolism. Literature data suggests that the UGT2B17 deletion polymorphism might play a role in exemestane metabolism.

4.2.2 Pharmacogenomics and Drug Transporters

Transporters can be classified as influx or efflux transporters, which are located either at the basolateral or apical membrane in polarized cells. The interplay of influx and efflux transporters with phase I and phase II metabolism enzymes is required for the sequential traverse of the basolateral and apical membranes. For example, in the liver, an uptake transporter as the organic anion transporter polypeptide B1 (OATPB1) may extract its drug substrates from the portal blood into hepatocytes. After metabolism, other transporters such as MDR1 may be important for the efflux of metabolites from the hepatocytes.

4.2.2.1 ATP-Binding Cassette Transporters

Transporters concerned with outward drug transport are mainly members of the ATP-dependent ABC transporter (ATP-binding cassette transporter) family, the largest family of transmembrane proteins that use ATP-derived energy to transport various substances over cell membranes. ABC transporters are found in tissue barriers and excretory organs, in which they efflux substances from an organ or the body against a concentration gradient (e.g., excretin into bile). Due to their role in the efflux of chemotherapeutic and other drugs from their target cells, they are known as multidrug resistance transporters. There are at least 49 ABC transporter genes.

The human ABCB1 gene located on chromosome 7q21 is composed of 29 exons. More than 50 SNPs and other polymorphisms affect the function of the well-characterized ABCB1 gene encoding the transporter P-glycoprotein. P-glycoprotein is expressed in liver, kidney, and intestine. A synonymous SNP in exon 27 (C3435T) was the first variant to be associated with altered expression in the human gastrointestinal tract, although the SNP does not change the encoded amino acid. Pgp expression in the duodenum of individuals with the CC genotype was twofold higher when compared with that in individuals with the TT genotype [20] and was associated with significantly decreased plasma concentrations of the Pgp substrate digoxin after oral administration, suggesting that digoxin absorption was lower in individuals with high Pgp levels. However, later relevant studies had discordant results. Thus, the effect of ABCB1 polymorphisms on drug disposition has not been clarified. Moreover, polymorphisms have been reported for other members of the ABC family, but their clinical significance has not been elucidated.

4.2.2.2 Solute Carrier Transporters

Solute carrier transporters (SLC) are found throughout the body and play significant roles in cellular homeostasis and the distribution of nutrients. SLC transporters consist of channels, facilitated transporters, and active transporters and include 40 families. The families of solute carriers with importance in drug disposition are SLC22 [organic cation transporters (OCTs) and organic anion transporters (OATs)] and SLCO (organic anion transporting polypeptides OATPs). The drug transporters of the SLCO/OATP family are mainly concerned with the inward transport of drugs into cells such as hepatocytes and renal tubular cells. The family's natural substrates include bile salts, thyroid hormones, and prostaglandins. Some of the forms also transport benzylopenicillin and statins. A member of the SLCO1B1 family is a key mechanism in the transport of pravastatin, rosuvastatin, pitavastatin, simvastatin, fluvastatin, atorvastatin, lovastatin, and cerivastatin across the hepatic sinusoidal membrane with greater significance for hydrophilic statins. Sixteen SNPs have been described for SLCO1B1.

A nonsynonymous polymorphism in SLCO1B1 has been associated with the pharmacokinetics of several widely prescribed drugs. These include some but not all statins and this polymorphism may also be associated with ADRs of statins. SLCO1B1 polymorphisms also affect the pharmacokinetics of the antidiabetic drug repaglinide, the antihistamine fexofenadine, and the endothelin A receptor antagonist atrasentan. SLC22A1 and SLC22A2 encode the cationic transporters OCT1 and OCT2, and play important roles in the transport of cationic cells into hepatocytes and renal tubular cells, respectively. Evidence suggests that coding region polymorphisms in these genes are relevant to metformin response and cisplatin-induced nephrotoxicity.

4.3 Pharmacogenomics and Drug Efficacy (Pharmacodynamics)

Drugs react with specific target proteins to exert their pharmacological effects, such as receptors, enzymes, or proteins involved in signal transduction, cell cycle control, or other cellular events. Many of the genes encoding these drug targets exhibit polymorphisms, which in many cases alter their sensitivity to specific medications. Examples include polymorphisms in β-adrenergic receptors (βARs) and sensitivity to β-agonists in patients experiencing asthma, angiotensin converting enzyme (ACE) and sensitivity to ACE inhibitors, angiotensin II T1 receptor and vascular reactivity to phenylephrine or response to ACE inhibitors, sulfonylurea receptor and responsiveness to sulfonylurea hypoglycemic agents, and 5-hydrotryptamine receptor and responsiveness to neuroleptics such as clozapine.

The pharmacogenomics of drug targets is far more complicated than the pharmacogenomics of pharmacokinetic genes, as pharmacokinetic genes tend to exhibit high penetrance and are predominantly monogenic, whereas drug target genes tend to be more polygenic. Thus, drug targets display a genetic influence of unknown magnitude emerging from the activity of multiple genes, with each conferring only a small effect. The complexity of the pharmacogenomics of drug targets is well illustrated in the case of antihypertensive drugs. Variations in two genes encoding ACE and nitric oxide synthase influence the effect of antihypertensive treatments. Polymorphism in the sodium channel γ-subunit promoter region is significantly associated with blood pressure response to hydrochlothiazide. In addition, SNPs in angiotensinogen, apolipoprotein B, and adrenoreceptor α2A significantly predict the change in left ventricular mass during antihypertensive treatment. Although common variants can influence the blood pressure response to a given class of antihypertensive medications, the results of relevant studies of polymorphisms are conflicting [21–23].

4.3.1 Polymorphisms in Receptors

Most of the studies have focused on G protein–coupled receptors, 5HT (hydroxytryptamine) receptors, and adrenergic receptors. βARs are widely expressed in cardiovascular cells with β1 adrenoreceptors being the major adrenoreceptor class found in the heart. Pharmacological stimulation or blockade of βAR signaling is applied in the treatment of cardiogenic shock, hypertension, ischemia, arrhythmias, and heart failure. Interindividual variability in the response of βAR agonists and antagonists has motivated investigation of genetic variability in the genes encoding βAR signaling pathway members [24]. The β1AR is encoded by the ADRB1 gene. The intronless ADRB1 gene has two common nonsynonymous polymorphisms in the coding region. Although many other SNPs have also been reported, their frequencies have been lower than 0.01, have been found in only a few individuals, and have never been found in a homozygous state. All these indicate that these polymorphisms are not expected to have a significant effect on public health. The most common β1AR SNP leads to a substitution of Gly for Arg at amino acid 389 (G389A), within the receptor's fourth intracellular loop. Relevant publications report differences in β-blocker response on the basis of phenotype both in studies of hypertension and of heart failure. However, the main issues have to be resolved before the extraction of valid conclusions in the pharmacogenomics not only of β1AR but also of all the adrenergic receptors [25].

The β1AR, β2AR, α2AAR, and α2CAR genes have polymorphisms in their promoter, 5′ untranslated region (UTR), and 3′UTR regions, as well as in their coding regions. Thus, although individual analysis of a promoter SNP, for example, could reveal a phenotype, it is difficult to know the effect of this individual SNP within the context of all other variations in these intronless genes. This is somewhat different from a coding SNP, which has a marked phenotype in terms of ligand binding, or G protein coupling where these functions are clearly relevant to drug response.

Therefore, it has been proposed that promoter, 5′UTR, and 3′UTR SNPs should be studied in the context of the other SNPs as they occur in nature, namely, in the context of haplotypes (combinations of polymorphisms). The number of theoretical haplotypes is 2N where N is the number of SNPs. However, the human population is fairly young and there has not been enough time for too many recombination events for the number of haplotypes to approach the theoretical one. It has been estimated that the number of haplotypes for any gene is approximately equal to the number of SNPs *1.1. Thus, the effect of these polymorphisms or of the haplotypes on the variability on drug response should also be considered. However, the need to genotype more SNPs, and to impute haplotypes, would require more subjects for clinical trials. Moreover, additional power considerations would be needed because of multiple comparisons.

Finally, it is recognized that other adrenergic receptors and other nonreceptor components are involved in the sympathetic signaling network of the heart. Therefore, multiple polymorphisms in multiple genes should be considered in the context of β-blocker pharmacogenomics. The statistical analysis of these complex interactions will be demanding; necessitating some novel approaches. It has been suggested that because many of the β-blockers have unique properties, it could be anticipated that for each of them, a distinct group of polymorphisms predicts the response. On the other hand, it has been proposed that some primary polymorphisms may predict the response to all β-blockers.

4.3.2 Polymorphisms in Target Enzymes

Widely prescribed drugs targeting enzymes include the coumarin anticoagulant (which inhibits vitamin K epoxide reductase), ACE inhibitors, statins [which inhibit HMGCoA

(3-hydroxy-3-methyl-glutaryl-CoA) reductase], and NSAIDs (which inhibit prostaglandin H synthases I and II). HMGCR is the gene encoding HMGCoA reductase that converts HMGCoA into mevalonic acid, which is the rate limiting step in cholesterol biosynthesis. Two polymorphisms in noncoding regions of HMGCR have been associated with a decreased response to statin treatment.

4.4 Pharmacogenomics and Drug Toxicity

There are two types of ADRs: type A and type B. Type A ADRs are dose-dependent, referring to the augmentation of pharmacological action (Table 4.2). Polymorphisms of pharmacokinetic genes may be important for narrow therapeutic window drugs, with poor metabolizers having increased risk of ADRs. Polymorphisms of drug targets may also be important for type A ADRs. Type B idiosyncratic ADRs are not predicted by pharmacological action. Idiosyncratic ADRs are thought to account for up to 20% of all ADRs, although some researchers regard this an overestimate, with 5% being closer to reality [26].

Two mechanisms have been proposed for type B ADRs: idiosyncratic metabolic pathways leading to reactive metabolites and immunological responses. Moreover, genetic variations of pharmacokinetic genes have been implicated in the etiology of idiosyncratic ADRs. Moreover, some type B ADRs are associated with the Human Leucocyte Antigen (HLA) system. The genetic contribution to the occurrence of idiosyncratic adverse drug events is likely to vary with the drug, ethnic group, and the phenotype of ADR. Although a single gene locus may be responsible for the ADR, the genetic effect size will usually be small or moderate, and the clinical manifestation of the ADR will be the combined effect of genotype and environmental factors. Thus, each gene will contribute to the risk of developing the idiosyncratic ADR, but each individual gene is neither necessary nor sufficient by itself to cause the ADR. The goal of pharmacogenomics should be the prediction of the drug response before the patient takes the drug through preprescription genotyping [27].

4.4.1 Cytochrome P450 Polymorphisms and Drug Safety

Individual susceptibility to drug toxicity is, to a large extent, defined by polymorphisms in cytochrome P450. Examples of drug toxicities that can be predicted by P450 polymorphisms

TABLE 4.2

HLA Polymorphisms and Adverse Drug Events

Drug	HLA Polymorphism	Adverse Event
Gold salts	HLA-B8 and DRW3	Proteinuria and nephrotic syndrome
D-Penicillamine	HLA-B8 and DRW3	Proteinuria and nephrotic syndrome
Carbamazepine	HLA-B*1502	Stevens–Johnson syndrome/toxic epidermal necrolysis
Allopurinol	HLA-B*5801	Stevens–Johnson syndrome/toxic epidermal necrolysis
Flucloxacillin	HLA-B*5701	Hepatotoxicity
Coamoxiclav	HLA DRB1*101	Hepatotoxicity
Lumaricoxib	HLA-DQA1*0102	Hepatotoxicity
Abacavir	HLA-B *5701, HLA-DR7, and HLA-DQ3	Hypersensitivity syndrome

TABLE 4.3

CYP Genetic Variations Resulting in Adverse Drug Events

CYP Gene	Drug	Adverse Event
CYP2D6	Codeine	Opioid intoxication
CYP2D6	Tramadol	Opioid intoxication
CYP2B6	Methadone	Fatal intoxication
CYP2D6	Mitrazapine	Cardiovascular toxicity
CYP2D6	Metoclopramide	Extrapyramidal symptom
CYP2C9	NSAIDs	Gastrointestinal bleeding
CYP2C9	Phenytoin	Neurological toxicity
CYP2C19	Phenytoin	Neurological toxicity
CYP2C19	Clopidogrel	Bleeding
CYP2C8	NSAIDs	Gastrointestinal bleeding
CYP2C8	Diclofenac	Hepatotoxicity

include those exerted by codeine, tramadol, methadone, warfarin, clopidogrel, and aceno-coumarol (Table 4.3). The translation of this information into clinical practice has been slow, although an increasing number of pharmacogenomics labels are assigned, with the predictive genotyping before treatment being mandatory, recommended, or only for informational reasons.

4.4.1.1 Codeine Toxicity

CYP2D6 ultrarapid metabolizers might be at greater risk for opioid-related adverse events and might benefit from a lower dose of opioids. The analgesic drug codeine is an important CYP2D6 substrate. It is activated to morphine exclusively by CYP2D6. Excessive activation of codeine in ultrarapid metabolizers with one additional copy of CYP2D6 has been described in two case reports. The first described a patient who suffered from a life-threatening opioid intoxication after being prescribed a cough medicine containing codeine. Upon genotyping, it was shown that this patient had at least three copies of CYP2D6. The second case report concerned the death of a breastfed baby 13 days after birth. His mother was prescribed codeine as an analgesic postdelivery. Postmortem examination of stored breast milk samples showed morphine levels four times higher than expected. Upon genotyping, the mother was found to be heterozygous for a CYP2D6*2A allele and a CYP2D6*2X2 gene duplication. Thus, the mother had three functional CYP2D6 alleles and was classified as an ultrarapid metabolizer.

4.4.1.2 Tramadol Toxicity

Tramadol is an example of an analgesic drug that is metabolized by CYP2D6, to generate a pharmacologically active product, the analgesic opioid receptor agonist O-desmethyltramadol. Case reports have indicated that carriers of gene duplications, being CYP2D6 ultrarapid metabolizers, are at high risk for toxic responses to tramadol treatment. A case report described a man with renal insufficiency and CYP2D6 ultrarapid metabolizer genotype who developed postoperative respiratory insufficiency under intravenous tramadol treatment.

4.4.1.3 Methadone Toxicity

Interindividual differences in sensitivity to methadone have been observed. Fatal poisonings occur typically at concentrations between 0.4 and 1.8 mg/mL. However, in susceptible individuals, death may occur at much lower concentrations. In one study, 40 postmortem cases were reported wherein methadone had been implicated as the cause of death [28]. The CYP2B6*6 allele, that results in slow metabolism, was associated with highest postmortem methadone concentration.

4.4.1.4 NSAIDs and Gastrointestinal Bleeding

The presence of CYP2C9 variant alleles is associated with acute gastrointestinal hemorrhage due to NSAIDs. In 2003, acute gastrointestinal bleeding was reported in a 71-year-old patient undergoing long-term treatment with acenocoumarol after treatment with indomethacin. Upon genotyping, the patient was found to be homozygous for the CYP2C9*3 allele and the authors suggested that the interaction of genetically impaired metabolisms of indomethacin and acenocoumarol, as well as a putative interaction in the metabolism of both CYP2C9 substrates, was the cause of bleeding.

The first study that analyzed the effect of CYP2C9 polymorphisms in NSAID-induced gastrointestinal bleeding was published in 2004 [29]. According to the investigators, carriers of CYP2C9 variant alleles were more prone to developing acute gastrointestinal bleeding when they received NSAIDs that were CYP2C9 substrates. However, a subsequent study with small sample size and underrepresentation of the CYP2C9*3 allele concluded that the CYP2C9 genotype was not significant as a risk factor for NSAID-induced ulcers. In 2007, another study found a high frequency of carriers of CYP2C9*3 allele and, to a lesser extent, of CYP2C9*2 allele among patients who developed gastrointestinal bleeding with NSAIDs [30]. In 2008, these results were confirmed, indicating a role for CYP2C8*3 in this risk as well [31].

4.4.1.5 Antidepressants and Cardiotoxicity

Mitrazapin is a tetracyclic antidepressant prescribed for depressed people with sleep disorders. Differences have been shown in pharmacokinetics among mitrazapin enantiomers. S-mitrazapine is efficiently hydroxylated by CYP2D6. On the other hand, R-enantiomer seems to have a more pronounced effect on heart rate and blood pressure. In CYP2D6 ultrarapid metabolizers, administration of higher doses could be considered for the achievement of appropriate antidepressant and sedative effect. However, a high risk of cardiovascular adverse effects would be expected due to nonultrarapid elimination of cardiovascular active R-enantiomer.

4.4.1.6 Metoclopramide and Toxicity

A case report described two patients who experienced acute dystonic reactions after receiving the antiemetic drug metoclopramide. Upon genotyping, both patients were poor metabolizers of CYP2D6. Metoclopramide is both a substrate and an inhibitor for CYP2D6 and extrapyramidal syndrome because metoclopramide has been observed in 1 of 500 patients. Drug-induced extrapyramidal syndrome is more frequent in patients who are deficient for CYP2D6 and therefore the investigators recommend a dose reduction in these patients.

4.4.1.7 ADP P2Y12 Inhibitors

Clopidogrel is a prodrug that is converted to its active metabolite in a cytochrome 450–catalyzed reaction. CYP2C19 is the main enzyme responsible for the activation of clopidogrel. The active clopidogrel metabolite inhibits the ADP P2Y12 receptor with resultant inhibition of platelet aggregation. The CYP2C*19 allele is associated with increased enzyme activity. Thus, carriers of this allele have more pronounced protective effect of clopidogrel after myocardial infarction. However, increased response to clopidogrel in carriers of the CYP2C*19 allele is associated with an increased risk of bleeding.

4.4.1.8 Antiepileptic Drugs and Toxicity

According to a case report, one woman developed neurological signs of phenytoin intoxication after 10 days of treatment with phenytoin. Serum concentrations were measured and an elimination time of 103 h was detected, which is five times longer than normal. Phenytoin metabolism is mainly catalyzed by CYP2C9 and, to a lesser extent, by CYP2C19. Upon genotyping, it was revealed that the patient was homozygous for CYP2C9*3 and heterozygous for CYP2C19*2, implying that CYP2C9*3 was responsible for the drug overdose. According to another case report, an African American female presented to the emergency department with mental confusion, slurred speech, memory loss, and an inability to stand. All these symptoms were due to phenytoin toxicity. Clearance of phenytoin was extremely poor with an elimination half-life of 13 days, that is, approximately 13 times longer than normal. Upon genotyping, she was found to be homozygous for allele CYP2C9*6.

4.4.1.9 Antiretroviral Drugs and Toxicity

Efavirenz is a nonnucleoside reverse transcriptase inhibitor. Central nervous system adverse events have been reported with efavirenz including headache, dizziness, insomnia, and fatigue. High plasma concentrations seem to increase the risk of these adverse events. The CYP2B6*6 allele results in an enzyme with slow metabolic capacity and is associated with high plasma concentrations of efavirenz. Thus, treatment of patients homozygous for the CYP2B6*6 allele is expected to decrease the incidence of adverse events in these patients.

4.4.1.10 Drug-Induced QT Interval Prolongation

Drug-induced QT interval prolongation and its more severe adverse event, the polymorphic ventricular tachycardia torsades de pointes, has been one of the most common reasons for drug withdrawal. Many drugs are known to cause QT interval prolongation, and many studies focus on single cases or case series using candidate gene approaches. Several genetic factors that modulate the risk of drug-induced QT interval prolongation have been identified. These are genes responsible for the congenital long QT syndrome, drug metabolism genes (mainly CYP2D6 and CYP3A4), and genes in other regulatory pathways. Although QT interval prolongation is common, torsades de pointes is rare. Thus, it is not known if the genetic factors that predispose for QT interval prolongation are the same with those that predispose for torsades de pointes.

4.4.2 HLA Polymorphisms and Drug Safety

HLA class I and II alleles have been associated with ADRs (Table 4.2). The adverse events of gold salts, D-penicillamine and tiopronin have been associated with particular HLA alleles or haplotypes (or both). Gold salts have been used in the treatment of patients with rheumatoid arthritis. These drugs are associated with a variety of adverse events including skin rashes, thrombocytopenia, granulocytopenia, aplastic anemia, interstitial pneumonitis, gastrointestinal side effects, chrysiasis of the cornea and lens, with proteinuria and nephrotic syndrome being the most common. A genetic predisposition to gold toxicity has been suggested. Proteinuria and nephrotic syndrome is also a common ADR of D-penicillamine.

The possible relation between HLA antigens and toxicity of D-penicillamine and sodium aurothomalate has been investigated in patients with rheumatoid arthritis. Nineteen of 24 patients in whom proteinuria developed were positive for HLA-B8 and DRw3 antigens. Several investigators confirmed the association between gold-induced proteinuria and DR3 and B8, but others were unable to confirm it. The allele HLA-B*1502 has been strongly associated with the carbamazepine-induced Stevens–Johnson syndrome/toxic epidermal necrolysis in Han Chinese. The FDA recommends HLA-B*1502 genotyping in patients with South Asian ancestry [32]. Allopurinol causes Stevens–Johnson syndrome and toxic epidermal necrolysis in patients carrying the HLA-B*5801 allele.

Flucloxacillin is an antistaphylococcal β-lactam antibiotic. It can cause a cholestatic hepatitis, either in isolation or accompanied by a rash. The Drug-Induced Liver Injury Genetics Network has shown a strong association between the HLA-B*5701 and flucloxacillin-induced hepatotoxicity [33]. The same allele has been identified in a genomewide association study [34]. The mechanistic basis for the association between HLA-B*5701 and flucloxacillin-induced hepatotoxicity has not been identified. Moreover, flucloxacillin is a ligand for the nuclear receptor PXR, and genetic polymorphisms of this gene have been associated with the hepatotoxic effect of flucloxacillin. Coamoxiclav-induced liver injury has been associated with HLA DRB1*101 [33]. HLA-DQA1*0102 has been associated with liver-induced toxicity of the COX-II inhibitor lumaricoxib, which has been withdrawn from the market because of toxicity [35].

Abacavir, a guanosine nucleoside analogue, is a potent HIV-1 reverse transcriptase inhibitor whose main treatment-limiting toxicity is a drug hypersensitivity syndrome characterized by fever, gastrointestinal and respiratory symptoms, and rash which typically occurs in the second week of treatment [36]. This hypersensitivity reaction can be fatal and affects 5% to 8% of Caucasian patients and 3% of African Americans. Rechallenge after presumed hypersensitivity is contraindicated, as it can result in severe morbidity or even mortality. Immunological etiology was suggested by the clinical manifestations and by the finding of drug-reactive T cells.

After extensive investigation of major histocompatibility complex, the investigators found a strong association between abacavir hypersensitivity and the haplotype comprising HLA-B *5701, HLA-DR7, and HLA-DQ3. This association has been verified in two other cohorts. In Caucasians, this reaction is observed only in patients with the HLA-B *5701 genotype and 50% of patients with this genotype develop this adverse reaction during abacavir treatment. The association of HLA-B *5701 with immunologically confirmed abacavir hypersensitivity reactions has also been found in black people. In 2008, the evidence was evaluated by the regulators and both the FDA and European Medicines Evaluation Agency changed the drug label for abacavir. HLA genotyping is mandatory before starting treatment with abacavir [36].

4.4.3 Transporter-Mediated Adverse Drug Events

Although less thoroughly investigated than CYP and HLA polymorphisms associated with ADRs, transporter pharmacogenomics has been implicated in a number of ADRs, encompassing chemotherapeutic agents and statins (Table 4.4). Statins can cause myopathy. Myopathy occurs in 1 of 10,000 patients (for standard doses 20–40 mg of simvastatin), but increases for higher doses (e.g., 80 mg of simvastatin). Statin-induced muscle toxicity caused by statins most commonly manifests as an asymptomatic increase of CPK levels and, in the most severe cases, can cause rhabdomyolysis and death.

TABLE 4.4

Transporter Polymorphisms and ADRs

Gene	Drug	Effect
SLCO1B1	Simvastatin	Myopathy
ABCB1	Paclitaxel	Neutropenia, neurotoxicity
SLCO1B1	Methotrexate	Gastrointestinal toxicity
SLCO1B1	Irinotecan	Severe neutropenia
OCT2	Cisplatin	Ototoxicity and nephrotoxicity
ABCB1	Docetaxel	Neutropenia, neurotoxicity

The mechanism of muscle damage is unclear and various hypotheses have been proposed, including a recent one in which atrogin 1, a gene involved in muscle atrophy, is involved in the pathogenesis of statin-induced toxicity. Various candidate gene studies have been performed but most findings either have shown a low odds ratio, or have not been replicated. Recently, a genomewide association study of subjects [37] receiving 80 mg of simvastatin showed a strong association of simvastatin-induced myopathy with OATPB1, transporter gene SLCO1B1, and in particular with the noncoding SNP OATP1B1*5 (rs4363657). This polymorphism reduces hepatic transport of simvastatin with subsequent increase of plasma concentrations of the drug. This result was subsequently replicated in a study of 16,644 persons on lower doses of simvastatin (40 mg) [38].

Moreover, it is not known if it is a drug effect or a class effect. Thus, in a recent study investigating severe statin-associated myopathy in a cohort of patients using various statin medications, the SLCO1B1 rs4363657 genotype was not significantly associated with myopathy in patients treated with atorvastatin [39]. In addition, CYP polymorphisms have been implicated in statin-induced rhabdomyolysis. A case report of rhabdomyolysis after cerivastatin administration has been reported in a patient that was found homozygous for the 475delA variant of CYP2C8. Blood levels of cerivastatin metabolites were elevated in that patient. Variants of the CYP2D6 result in inappropriate or extensive metabolism of simvastatin. Thus, defective allele carriers show increased incidence of adverse events.

4.5 Pharmacogenomics and Drug Interactions

Drug–drug interactions show significant intergenotypic differences. Poor metabolizers of CYP2D6 (those with alleles expressing no functional enzymes) do not exhibit the drug–drug interactions predicted by *in vitro* studies because there is no functional CYP2D6 activity to be inhibited. Ultrarapid metabolizers may also not exhibit the expected drug–drug interaction. The ultrarapid metabolizers have a large functional reserve of CYP2D6 activity that would need toxic doses of an inhibitor to elicit a drug–drug interaction. Usually, individuals who display a drug interaction are intermediate metabolizers, or those who have inherited CYP2D6 alleles with reduced or altered affinity for CYP2D6 substrates.

The dependence of drug interactions from CYP2D6 genotype has been evidenced for encainide, metoprolol, mexiletine, desipramine, codeine, and propafenone. However, it should be emphasized that in poor metabolizers, drug interactions in alternative pathways of metabolism could occur, for example, between propafenone and rifampicin, an enzyme inducer. CYP2D6 that primarily metabolizes propafenone by hydroxylation is not inducible. However, coadministration of rifampicin has been reported to decrease the oral bioavailability of propafenone from 30% to 10% in extensive metabolizers and from 81% to 48% in poor metabolizers. This drug interaction was due to enhanced clearance of propafenone through N-dealkylation (mediated by CYP3A4 and CYP1A2) and glycuronidation. Relevant enzymes were induced regardless of CYP2D6 genotype. This is an example of a clinically relevant metabolic reaction of rifampicin with propafenone due to the induction of CYP2D6-independent pathways by rifampicin.

4.5.1 Drug Interactions of Clopidogrel and CYP2C19 Polymorphisms

Clopidogrel, an antithrombotic drug is a prodrug that is converted to its active metabolite by CYP2C19. Other enzymes including CYP3A4, CYP2B6, and CYP2C9 have been implicated in the metabolism of clopidogrel. The active metabolite targets P2Y12, a G protein–coupled receptor. Drugs that inhibit CYP2C19 may cause reduced or complete loss of CYP2C19 activity and similar clinical consequences as CYP2C19 mutants. Proton pump inhibitors including omeprazole, lansoprazole, esomeprazole, rabeprazole, and pantoprazole are often coadministered with clopidogrel to prevent gastrointestinal bleeding. These proton pump inhibitors are substrates and inhibitors of CYP2C19 and CYP3A4 [40]. Thus, proton pump inhibitors potentially reduce or block clopidogrel activation through inhibition of CYP2C19 and CYP3A4. The label of clopidogrel includes the information that coadministration of clopidogrel and omeprazole, an inhibitor of CYP2C19, reduces the pharmacological activity of clopidogrel if given concomitantly or 12 h apart.

4.5.2 Transporter-Mediated Drug Interactions and Gene Polymorphisms

The drug transporter OATP1B1 encoded by SLCO1B1 is involved in the hepatic uptake of statins. Genetic variants of SLCO1B1 have been recognized as risk factors for statin-induced adverse events. There are case reports of rhabdomyolysis due to concomitant administration of simvastatin and cyclosporine (which is an OATP1B1 inhibitor). In addition, gemfibrozil inhibits OATP1B1 thus reducing the hepatic uptake of statins with resultant increase of plasma concentrations of pravastatin, simvastatin, cerivastatin, and lovastatin [41]. The label of simvastatin includes a statement on the risk of myopathy upon coadministration with cyclosporine or gemfibrozil [40].

4.6 Applications of Pharmacogenomics in Patient Care

4.6.1 CYP2C9-VKORC1 Polymorphisms and Warfarin

Warfarin is a coumarin anticoagulant that is prescribed for the prevention and treatment of thromboembolism, and is metabolized mainly by CYP2C9. Vitamin K epoxide reductase complex subunit 1 (VKORC1) has been identified as the intracellular target of warfarin. Warfarin inhibits VKORC1, leading to reduced levels of vitamin K. Reduced vitamin K (vitamin K epoxide) is a necessary cofactor in the γ-carboxylation of clotting factors II, VII, IX, and XI as well as of the proteins C and S. Inhibition of VKORC1 results in decreased levels of activating clotting factors and, therefore, the ability of blood to clot is reduced but not completely eliminated. Up to 20-fold interindividual variation is observed with warfarin treatment. This variation, coupled with the narrow therapeutic index of the drug, results in an annual 1% to 5% risk of major bleeding. Polymorphisms within VKORC1 CYP2C9 genes are common, with more than two-thirds of the Caucasian population and up to 90% of East Asians manifesting at least one variant. Affected individuals require, on average, lower doses of warfarin to maintain a therapeutic International Normalized Ratio and more time to achieve stable dosing. Carriers of variant alleles are at a higher risk for bleeding complications, especially upon the induction of warfarin treatment.

Current evidence suggests that a personalized dosing regimen using the genotype data from CYP2C19 and VKORC1 genes seems far superior to the conventional strategy that used only coagulation testing for maintaining adequate efficacy in the prevention of stroke in patients with atrial fibrillation and in avoiding bleeding due to unintended overdosing [42]. In 2007, the FDA recommended genetic tests to improve the initial estimate of warfarin dose. This was the first FDA recommendation to consider genetic testing when first initiating a medication.

4.6.2 CYP2C19 Polymorphisms and Clopidogrel

According to TRITON-TIMI 38 study, individuals with genetic variants of CYP2C19 exhibited reduced conversion rates of clopidogrel to its active metabolite. Thus, these individuals had an increased risk of thrombosis with the standard "population based" dosage regimen. Carriers of the reduced function allele CYP2C19*2 had 32% lower plasma concentrations of clopidogrel in comparison with noncarriers. Moreover, they exhibited a 53% increase in the risk of death from cardiovascular disease, myocardial infarction, or stroke and three times higher risk for stent thrombosis in comparison with noncarriers of the defective allele. These data have prompted the FDA to include, in the clopidogrel label, the statement that poor metabolizers of clopidogrel may not get the full benefit of clopidogrel and that tests are available to identify genetic variants of CYP2C19.

4.6.3 CYP Polymorphisms and Tamoxifen

Tamoxifen can be considered a prodrug. The parent drug has weak affinity for the estrogen receptors but undergoes excessive biotransformation catalyzed by phase I and II enzymes into active and inactive metabolites. N-dimethyltamoxifen is the primary metabolite formed via CYP3A4/5. N-dimethyltamoxifen is a weak antiestrogen, but it is subsequently metabolized into a hydroxytamoxifen, N-didesmethyltamoxifen and 4-hydroxy-N-desmethyltamoxifen (known as endoxifen). 4-Hydroxytamoxifen is a minor primary metabolite whose production is catalyzed by multiple enzymes including CYP2D6. Endoxifen and 4-hydroxytamoxifen each have at least 10-fold higher affinity for estrogen receptors than tamoxifen, and are associated with equivalent antiestrogenic potency. In patients receiving chronic tamoxifen therapy, endoxifen is found in serum concentrations 6-fold to 12-fold higher than 4-hydroxytamoxifen. Thus, it has been suggested that endoxifen is the most important metabolite required for tamoxifen treatment [43]. For this reason, the pharmacogenetics of CYP2D6 is expected to affect endoxifen concentrations and possibly long-term tamoxifen-associated outcomes [44]. Pharmacokinetic studies have demonstrated that women who are poor metabolizers of CYP2D6, either by genotype (PM/PM) or by a CYP2D6 inhibitor like some of the serotonin reuptake inhibitors (paroxetine or fluoxetine) that are often coprescribed to alleviate hot flashes, have lower endoxifen plasma concentrations than patients with normal CYP2D6 metabolisms.

Evidence from a large number of retrospective studies implies that the CYP2D6 genotype plays a role in tamoxifen treatment outcome, indicating that patients with decreased

CYP2D6 metabolism as a result of genetic or environmental factors are less likely to derive clinical benefit from adjuvant tamoxifen therapy than patients with normal CYP2D6 metabolism. Thus, carriers of variant CYP2D6 allele such as CYP2D6*4 and CYP2D6*10 have an increased risk of breast cancer relapse and lower event-free survival rates when compared with extensive metabolizers. However, other studies suggest that CYP2D6 activity is not necessary for tamoxifen bioactivation and efficacy. Therefore, mandatory use of CYP2D6 genetic test requires additional data from randomized clinical trials. Moreover, a potential effect of CYP2C19 polymorphism on breast cancer risk prolapse during tamoxifen therapy has also been suggested.

4.6.4 Thiopurine S-Methyltransferase

Thiopurine drugs such as mercaptopurine, azathioprine, and thioguanine are prodrugs used in the treatment of acute lymphoblastic anemia, autoimmune disorders, and inflammatory bowel disease. TMPT catalyzes the S methylation of these prodrugs into inactive metabolites. An elevated risk of severe or even fatal myelosuppression is observed in TMPT-deficient patients treated with thiopurine drugs [45–48]. Polymorphisms of the TMPT gene affect enzyme activity. Population studies investigating TMPT polymorphisms associated the presence of polymorphic alleles in the open reading frame of the TMPT gene with low enzyme activity due to enhanced degradation of the variant protein. Decreased enzyme activity in 80% to 95% of patients has been attributed to three of nine variant alleles: TMPT*2, TMPT*3A prevalent in Caucasians, and TMPT*3C prevalent in Asian, African, and African American populations. Approximately 10% of the general population has one nonfunctional allele resulting in intermediate TMPT activity and approximately 0.3% has two nonfunctional alleles and no TMPT activity. Genotyping for the common TMPT alleles is recommended because it can identify patients at risk for toxicity [49,50]. However, although there is a strong phenotype–genotype association for TMPT, most of the thiopurine-induced adverse reactions and the efficacy of therapy cannot be explained by TMPT polymorphisms.

4.6.5 Uridine Glucurosyltransferase 1A1 (UGT1A1)

The combination of 5-fluorouracil with other drugs such as irinotecan has changed the survival rates of patients with colorectal cancer. Irinotecan is a captothecin derivative. It is converted to its active metabolite, SN-38, by carboxy-elastase enzymes in the liver and is subsequently conjugated primarily by UGT1A1 for elimination in the bile and urine [51]. Irinotecan and, in particular, its active metabolite SN-38, stabilizes the DNA topoisomerase I complex by binding to it, thus preventing the resealing of single-strand breaks. Irinotecan prevents replication from proceeding, which results in double-strand breaks and thus in its antitumor effect and also in adverse effects from rapidly dividing tissues such as bone marrow and intestinal mucosa.

The main dose-limiting toxicities of irinotecan are myelosuppression and delayed-type diarrhea, which are attributed to increased levels of the active metabolite SN-38 [52]. Reduced expression of UGT1A1 due to polymorphisms of the gene is associated with lower SN-38 glucuronidation and increased incidence of toxicity. In 2004, a small clinical study [53] showed that the homozygous UGT1A1*28 genotype was associated with an increased risk of severe neutropenia and severe delayed-type diarrhea. In 2005, the FDA suggested a reduction by at least one level in the starting dose of irinotecan for patients homozygous for the UGT1A1*28 allele. However, there is no clear recommendation on the necessity of screening patients for UGT1A1*28 allele before the administration of irinotecan [45,48,49,52].

TABLE 4.5

Pharmacogenomic Markers that Predict Response to Targeted Cancer Treatment

Pharmacogenomic Marker	Drug	Cancer Site
CYP2D6	Tamoxiphen	Breast
ERBB2	Trastuzumab	Breast
CYP3A4	Dazanitib	Acute lymphoblastic leukemia
KRAS	Cetuximab	Colorectal
KRAS	Panitumumab	Colorectal
BCR-ABL1	Imatinib	Chronic myelogenous leukemia
KIT	Imatinib	Chronic myelogenous leukemia–acute lymphoblastic anemia
EGFR	Erlotinib	Lung
EGFR	Gefitinib	Lung
EGFR	Cetuximab	Colorectal

4.6.6 Targeted Therapies

Advances in pharmacogenomics have led to the development of novel therapies that are collectively known as molecular targeted therapies to highlight their specificity and their interference with key molecular events responsible for the malignant phenotype (Table 4.5).

4.6.6.1 Breast Cancer

Gene amplification and thus overexpression of the human epidermal growth factor receptor number 2 (HER2) develops in 20% to 25% of all breast cancers and is associated with an aggressive tumor phenotype. The monoclonal antibody, trastuzumab, specifically targets HER2 to inhibit proliferation and induce tumor cell death. Due to the necessity of HER2 overexpression for treatment response, trastuzumab is indicated only for approximately 10% of breast cancer patients who overexpress Her2/new (determined by immunohistochemistry) or show amplification of HER2 (determined by fluorescence *in situ* hybridization) [54]. This drug is an example of a protein therapeutic for which an obligatory biomarker assay and diagnostic test has been developed to identify patients who are most likely to benefit from this drug [55].

4.6.6.2 Colon Cancer

The monoclonal antibodies panitunumab and cetuximab target epidermal growth factor receptor 1 (HER1, also known as EGFR). They are indicated in metastatic colon cancer, as tumor growth often depends on signaling via EGFR, which involves downstream V-kras2 Kirsten rat sarcoma viral oncogene homologue (KRAS) [56]. Due to HER1 dependence, these monoclonal antibodies are used only for treatment of HER1-expressing tumors. Patients carrying activating KRAS mutations are nonresponders to panitunumab and cetuximab. Thus, KRAS genotyping is essential for identifying responders.

4.7 Pharmacogenomic Studies: Biostatistical Issues

Currently, there are two primary research paradigms for population-based genetic association studies, both of which are based on genotyping SNPs: the candidate gene association

TABLE 4.6

Candidate Gene Studies versus Genomewide Association Studies

	Candidate Gene Studies	Genomewide Association Studies
Focus	One or more genes	Equal weight to all genes in the genome
Appropriate for	Low allele frequencies, small effect sizes, limited population	Difficult to identify genetic associations
Analysis	Statistical methods for type I error inflation are applied	Computationally demanding, high risk of false positive discovery
Interpretation	Easy for positive findings, difficult for negative findings	Not straightforward
Drawback	Less obvious relevant genes may not be investigated, wrong polymorphisms may be selected for investigation	High percentage of false positives

studies and the genomewide association studies (Table 4.6). Both types of studies have identified genotype associations with diseases or drugs, but replication and validation also suffer in both studies.

4.7.1 Candidate Gene Association Studies

Candidate gene approaches focus on one or more genes or pathways, are hypothesis driven, and are chosen based on the evidence that the gene product is involved in variation in pharmacokinetics or pharmacodynamics. The hypothesis is that genetic variations in genes that play a significant role in pharmacokinetics or pharmacodynamics of a drug would likely affect the drug's efficacy or safety. In this approach, initial polymorphism discovery was carried out using a reference set of genomic DNA from a collection of ethnically diverse individuals. The value of candidate gene association studies is apparent when allele frequencies are low, effect sizes are small, or the study population is limited or unique. They are also valuable for validating previous reports of genetic associations with disease in different populations. However, candidate gene association studies sometimes waste information because of an inefficient study design and suboptimal SNP selection strategies.

Two approaches are possible in candidate gene association studies: the candidate gene approach or the candidate pathway approach. Currently, knowledge of the functions of biochemical pathways is stronger than knowledge of the functions of individual genes, and new tools are available for assigning genes to functional pathways. Thus, a meaningful approach in candidate gene association studies is to hypothesize at the level of pathways and include all known genes in the pathway as candidate genes. Compared with the study of individual genes, the inferences derived from a candidate pathway are enhanced by allowing global conclusions about the association between an entire biochemical pathway and disease or drug.

The statistics assessing the associations between the SNPs and the outcome are of primary interest. The type I error inflation due to multiple comparisons is addressed by the Bonferroni correction or by determining the false discovery rates and p values. The positive finding of a candidate gene approach is easy to interpret and can provide clinically relevant information. However, a negative result can be interpreted in many different ways. The sample size may be too small to detect an effect, the causal genetic polymorphism may not have been included in the study [57], or there may be a true absence of an effect. The major disadvantage of candidate gene studies is that only obvious candidates selected on the basis of the biology of their drug or on current understanding of disease pathogenesis are likely to be included and thus other less obvious genes, which could also have relevance to the disease susceptibility, may not be examined. The other disadvantage is that the genetic polymorphisms chosen for investigation in candidate gene studies may often have been selected in a rather arbitrary way. Thus, it has been suggested that it seems inappropriate to genotype at one SNP position when there are multiple SNPs, particularly when molecular studies have shown haplotype effects.

Typically, candidate gene studies include samples of 40 European Americans, 40 African Americans, and 40 Asian Americans. This provides for a 95% probability for detecting polymorphisms with an allele frequency of 0.03 (3%) if the polymorphism is completely confined to one racial group, and 0.01 if the allele is found in the three racial groups. Subsequently, any nonsynonymous polymorphisms as well as those in the promoter or any other untranslated region are studied *in vitro* in transfected cell systems. From these results, the generation of transgenic mice is considered, thus providing the opportunity to explore variants in the cell type of interest, in the organ of interest, and under appropriate physiological conditions. Small cohorts of genotyped patients can be studies for short-term physiologic phenotypes, such as the response to the infusion of agonists or antagonists. Thus, combining these multiple approaches, a phenotype can be established at the molecular and physiological levels. Based on this information, a hypothesis-driven clinical trial can be designed to investigate the potential effect of the genetic variant in question on drug response.

4.7.2 Genomewide Association Studies

The genomewide approach gives equal weight to all genes in the genome and can be used when little is known regarding the gene–drug effect in an effort to be unbiased. Genomewide association studies are large-scale hypothesis-generating studies that investigate massive numbers of SNPs or CNVs in parallel to identify markers of signature that are correlated with the drug effect of interest. The hypothesis of genomewide association studies is that any genetic variant in the human genome can contribute to variation in drug effect.

These approaches are suitable for difficult-to-identify genetic associations that may exist for idiosyncratic ADRs [58]. The major advantage of the genomewide approach is that new genetic loci can be identified within the human genome without *a priori* knowledge regarding the relation of genes or genetic pathways with their target. The advances made in bioinformatics allowed limiting the region to be searched around a newly identified locus and also investigating many unrelated individuals in association studies, in contrast to the former family-based association studies.

Analyzing data from genomewide association studies requires large computational capacity and has a high risk of false-positive discovery [55]. Given a *p* value of .05, five false-positive findings are expected among 100 tests performed. When one million tests are performed, a large number of positive findings can be false discoveries. Thus, the threshold for significance in these studies is typically set at $p = 10^{-7}$ to 10^{-8}. Moreover, statistical methods that reduce false discovery rate (e.g., Bonferroni correction) are applied. In addition, it is standard procedure to replicate any apparently significant association in a second, apparently larger cohort. Possible associations that lie close to genomewide significance can also be re-examined in replication cohorts. However, it has been emphasized that even with increasingly large samples achieved through population-based cohorts, the challenge to identify the few true positives among the false positives using conventional statistical methods remains truly demanding.

Most genomewide association studies have involved the initial genotyping of between 500 and 2000 disease cases and up to 5000 controls for 500,000 up to 1,000,000 SNPs. Two distinct technologies for genotyping are provided by two companies, Affymetrix (Santa Clara, CA) and Illumina (San Diego, CA). It has been evidenced that the currently available chips, namely, Affymetrix 500 K, Affymetrix 6.0, Illumina 300 K, Illumina 610 K, and Illumina 1M show comparable statistical power to detect associations. However, there are differences in SNPs covered by the two manufacturers. Thus, Illumina chips mostly cover specific populations, whereas Affymetrix chips are less population specific. Moreover, Affymetrix chips include more SNPs that correlate with one another with the disadvantage of slightly decreasing power but with the advantage of fewer quality control losses and possibly stronger association signals.

Elucidation of the pathophysiology associated with a newly identified locus by genome-wide association studies can be really demanding, as illustrated by the following example. A 98 kb interval on chromosome 9p21.3 had been repeatedly found to be associated with coronary heart disease. However, no protein coding genes had been found in this locus. Three years later, investigators, applying gene knockout techniques, demonstrated that distant gene regulatory functions are located in this locus. Previous research efforts had identified a large antisense noncoding RNA gene in that locus that altered the expression of genes controlling proliferation relevant to atherosclerosis. Interpretation of genome-wide association studies is not straightforward.

References

1. Vogel, F. 1959. Moderne Probleme der human Genetic. *Ergebnisse der Human Genetics und Kinderheilkunde* 12:52–125.
2. Pirmohamed, M. 2001. Pharmacogenetics and pharmacogenomics. *Journal of Clinical Pharmacology* 52:345–347.
3. de Leon, J. 2009. The future of personalized prescription in psychiatry. *Pharmacological Research* 59:81–89.
4. Roses, A.D. 2004. Pharmacogenetics and drug development: The path to safer and more effective drugs. *Nature Reviews. Genetics* 5:645–656.
5. Evans, W. 1999. Translating functional genomics into rational therapeutics. *Science* 286:487–491.
6. Lesko, L., R. Salerno, B. Spear, D.C. Anderson, T. Anderson, C. Brazell et al. 2003. Pharmacogenetics and pharmacogenomics in drug development and regulatory decision making: Report of the first FDA-PWG-PhRMA-DruSafe Workshop. *Journal of Clinical Pharmacology* 43:342–358.
7. Ginsburg, G.S., and H.F. Willard. 2009. Genomic and personalized medicine: Foundations and applications. *Translational Research* 154:277–287.
8. Lema, C., and M. Cunningham. 2010. MicroRNAs and their implications in toxicological research. *Toxicology Letters* 198:100–105.
9. Mishra, P.J., P.J. Mishra, D. Barenjee, and J.R. Bertino. 2008. MiRNSPs or MiR polymorphisms, new players in microRNA mediated regulation of the cell: Introducing microRNA pharmacogenomics. *Cell Cycle* 7:853–858.
10. Mishra, P., and J. Bertino. 2009. MicroRNA polymorphisms: The future of pharmacogenomics, molecular epidemiology and personalized medicine. *Pharmacogenomics* 10:399–416.
11. Nebert, D., G. Zhang, and E. Vesell. 2008. From human genetics and genomics to pharmacogenetics and pharmacogenomics: Past lessons, future directions. *Drug Metabolism Reviews* 40:187–224.
12. Lo, P.K., and S. Sukumar. 2008. Epigenomics and breast cancer. *Pharmacogenomics* 9:1879–1902.
13. Freedman, A., L. Sansbury, W. Figg, A.L. Potosky, S.R. Weiss Smith, M.J. Khoury et al. 2010. Cancer pharmacogenomics and pharmacoepidemiology: Setting a research agenda to accelerate translation. *Journal of the Natural Cancer Institute* 102:1698–1705.
14. Daly, A. 2010. Pharmacogenetics and human genetic polymorphisms. *Biochemical Journal* 429:435–449.

15. Zanger, U.M., K. Klein, T. Saussele, J. Blievernicht, M.H. Hofmann, and M. Schwab. 2007. Polymorphic CYP2B6: Molecular mechanisms and emerging clinical significance. *Pharmacogenomics* 8:743–759.

16. Kirchheiner, J., and A. Seeringer. 2007. Clinical implications of pharmacogenetics of P450 drug metabolising enzymes. *Biochimica et Biophysica Acta* 1770:489–494.

17. Kruth, P., and M. Wehling. 2003. Pharmacogenomic Was ist fur den Internist relevant? *Der Internist* 44:1524–1530.

18. Daly, A., and Ch. Day. 2009. Genetic association studies in drug induced liver injury. *Seminars in Liver Disease* 29:400–411.

19. Sun, D., G. Chen, R.W. Dellinger, A.K. Sharma, and P. Lazarus. 2010. Characterization of 17-dihydroexemestane glucuronidation: potential role of the UGT2B17 deletion in exemestane pharmacogenetics. *Pharmacogenetics and Genomics* 20:575–585.

20. Evans, W. 2003. Marshalling the human genome to individualize drug treatment. *Gut* 52(Suppl II):ii10–ii18.

21. Arnett, D., St. Claas, and St. Glasser. 2006. Pharmacogenetics of antihypertensive treatment. *Vascular Pharmacology* 44:107–118.

22. Arnett, D., St. Claas, and A. Lynch. 2009. Has pharmacogenetics brought us closer to personalized medicine for initial drug treatment of hypertension? *Current Opinion in Cardiology* 24:333–339.

23. Rafic, S., S. Anand, and R. Roberts. 2010. Genome wide association studies of hypertension. *Journal of Cardiovascular and Translational Research* 3:189–196.

24. Dorn, G., and S. Liggert. 2008. Pharmacogenomics of β-adrenergic receptors and their accessory signaling proteins in heart failure. *Clinical and Translational Science* 1:255–262.

25. Dorn, G., and S. Lippert. 2009. Mechanisms of pharmacogenomic effects of genetic variation within the cardiac adrenergic network in heart failure. *Molecular Pharmacology* 76:466–480.

26. Pirmohamed, M. 2010. Pharmacogenetics of idiosyncratic adverse drug reactions. In *Adverse Drug Reactions Handbook of Experimental Pharmacology*, edited by Uetrecht, J., 478–491. New York/Heidelberg: Springer-Verlag.

27. Ryan, T., J. Stevens, and C. Thomas. 2008. Strategic applications of pharmacogenomics in early drug discovery. *Current Opinion in Pharmacology* 8:654–660.

28. Ernst, E., A. Bartu, A. Popescu, K.F. Ileutt, R. Hansson, and N. Plumley. 2002. Methadone-related deaths in Western Australia 1993-99. *Australian and New Zealand Journal of Public Health* 26:364–370.

29. Martínez, C., G. Blanco, J.M. Ladero, E. García-Martín, C. Taxonera, F.G. Gamito, M. Diaz-Rubio, and J.A. Agúndez. 2004. Genetic predisposition to acute gastrointestinal bleeding after NSAIDs use. *British Journal of Pharmacology* 141:205–208.

30. Pilotto, A., D. Seripa, M. Franceschi, C. Scarcelli, D. Colaizzo, E. Grandone, V. Niro, A. Andriulli, G. Leandro, F. Di Mario, and B. Dallapiccola. 2007. Genetic susceptibility to nonsteroidal anti-inflammatory drug-related gastroduodenal bleeding: Role of cytochrome P450 2C9 polymorphisms. *Gastroenterology* 133:465–471.

31. Blanco, G., C. Martinez, J.M. Ladero, E. Garcia-Martin, C. Taxonera, F.G. Gamito, M. Diaz-Rubio, and J.A. Agundez. 2008. Interaction of *CYP2C8* and *CYP2C9* genotypes modifies the risk for nonsteroidal anti-inflammatory drugs-related acute gastrointestinal bleeding. *Pharmacogenetics and Genomics* 18:37–43.

32. Loscher, W., U. Klotz, F. Zimprich, and D. Schmidt. 2009. The clinical impact of pharmacogenetics on the treatment of epilepsy. *Epilepsia* 50:1–23.

33. Daly, A. 2010. Drug induced liver injury: Past, present and future. *Pharmacogenomics* 11:607–611.

34. Daly, A.K., P.T. Donaldson, P. Bhatnagar, Y. Shen, I. Peter, A. Floratos et al. 2009. *HLA-B*5701* genotype is a major determinant of drug-induced liver injury due to flucloxacillin. *Nature Genetics* 41:816–819.

35. Russmann, S., A. Jetter, and J. Kullak-Ubkick. 2010. Pharmacogenetics of drug induced liver injury. *Hepatology* 52:748–761.

36. Phillips, E., and S. Mallal. 2008. Pharmacogenetics and the potential for individualization of antiretroviral therapy. *Current Opinion in Infectious Diseases* 21:16–24.

37. Link, E., S. Parish, J. Armitage, L. Bowman, S. Heath, F. Matsuda, I. Gut, M. Lathrop, R. Collins, and SEARCH Collaborative Group. 2008. SLCO1B1 variants and statin-induced myopathy–a genomewide study. *The New England Journal of Medicine* 359:789–799.

38. Holmes, M., T. Shah, C. Vickery, L. Smeeth, A.D. Hingorani, and J.P. Casas. 2009. Fulfilling the promise of personalized medicine? Systematic review and field synopsis of pharmacogenetic studies. *PLoS One* 4:e7960.

39. Brunham, L.R., P.J. Lansberg, L. Zhang, F. Miao, C. Carter, G.K. Hovingh, H. Visscher, J.W.Jukema, A.F. Stalenhoef, C.J. Ross, B.C. Carleton, J.J. Kastelein, and M.R. Hayden. 2012. Differential effect of the rs4149056 variant in SLCO1B1 on myopathy associated with simvastatin and atorvastatin. *The Pharmacogenomics Journal* 12:233–237.

40. Bai, J. 2010. Ongoing challenges in drug interaction safety: From exposure to pharmacogenomics. *Drug Metabolism and Pharmacokinetics* 25:52–71.

41. Schmitz, G., A. Schmitz Madry, and P. Ugoksai. 2007. Pharmacogenetics and pharmacogenomics of cholesterol lowering therapy. *Current Opinion in Lipidology* 18:164–173.

42. Gandara, E., and Ph. Wells. 2010. Will there be a role for genotyping in warfarin therapy? *Current Opinion in Hematology* 17:439–443.

43. Higgins, M., and V. Stearns. 2010. CYP2D6 polymorphisms and tamoxifen metabolism: Clinical relevance. *Current Oncology Reports* 12:7–15.

44. Hertz, D., H. McLeod, and J. Hoskins. 2009. Pharmacogenetics of breast cancer therapies. *Breast* 18:559–563.

45. Offit, K., and E. Mark. 2010. New pharmacogenomic paradigm in breast cancer treatment. *Journal of Clinical Oncology* 28:4665–4673.

46. Cheng, Q., and W. Evans. 2005. Cancer pharmacogenomics may require both qualitative and quantitative approaches. *Cell Cycle* 4:1506–1509.

47. Kager, L., and W. Evans. 2006. Pharmacogenomics of acute lymphoblastic leukemia. *Current Opinion in Hematology* 13:260–265.

48. Huang, S., and M. Ratain. 2009. Pharmacogenetics and pharmacogenomics of anticancer agents. *CA: A Cancer Journal for Clinicians* 59:42–55.

49. Ezzeldin, H., and R. Diasio. 2006. Genetic testing in cancer therapeutics. *Clinical Cancer Research* 12:4137–4141.

50. Lee, W., C. Lockhart, R. Kim, and M. Rothenberg. 2005. Cancer pharmacogenomics: Powerful tools in cancer chemotherapy and drug development. *The Oncologist* 10:104–111.

51. Bandrés, E., R. Zarate, N. Martinez, A. Abajo, N. Bitarte, and J. Garíia-Foncillas. 2007. Pharmacogenomics in colorectal cancer: The first step for individualized therapy. *World Journal of Gastroenterology* 13:5888–5901.

52. Palomaki, G., L. Bradley, M. Douglas, K. Kolor, and W.D. Dotson. 2009. Can UGT1A1 genotyping reduce morbidity and mortality in patients with colorectal cancer treated with irinotecan? An evidence-based review. *Medical Genetics* 11:21–34.

53. Marcuello, E., A. Altés, A. Menoyo, E. Del Rio, M. Gómez-Pardo, and M. Baiget. 2004. UGT1A1 gene variations and irinotecan treatment in patients with metastatic colorectal cancer. British *Journal of Cancer* 91:678.

54. Lesko, L. 2010. DNA, drugs and chariots: On a decade of pharmacogenomics at the US FDA. *Pharmacogenomics* 11:507–512.

55. Kirk, R., J. Hung, S. Horner, and J.T. Perez. 2008. Implications of pharmacogenomics for drug development. *Experimental Biology and Medicine* 233:1484–1497.

56. Pohl, A., J. Lurje, P.C. Manegold, and H.J. Lenz. 2008. Pharmacogenomics and genetics in colorectal cancer. *Advanced Drug Delivery Reviews* 61:375–380.

57. Franke, B., B.M. Neale, and S.V. Faraone. 2009. Genome-wide association studies in ADHD. *Human Genetics* 126:13–50.

58. Shrinivasan, B., J. Chen, Ch. Cheng, D. Conti, S. Duan, B.L. Fridley et al. 2009. Methods for analysis in pharmacogenomics: Lessons from the Pharmacogenetics Research Network Analysis Group. *Pharmacogenomics* 10:243–251.

5

Pharmacogenomic Knowledge to Support Personalized Medicine: The Current State

Casey Lynnette Overby and Houda Hachad

CONTENTS

5.1 The Vision: Pharmacogenomics and Personalized Medicine

Much attention is currently focused on the promise that the human genome project will enable the integration of genome discoveries into clinical applications that will potentially improve health care. However, the current approaches in medical practice do not generally take into account the genetic profile of individuals. A vision of personalized, predictive, preventive, and participatory (P4) health care has been proposed [1,2] that would account for each patient's individual genetic characteristics to tailor drug therapy. Pharmacogenomics (PGx)—the study of genetic influence on individual responses to medication—is key to this health care model. PGx information includes both genotypic data (e.g., from single nucleotide polymorphism arrays), and genomic knowledge (e.g., the recommended drug dosage adjustments for a particular genotype). This chapter focuses on the current state and role of pharmacogenomic knowledge in translation research. Translation research supports a P4 model of health care because it is required to effectively move PGx discoveries to evidence-based practice. This form of research has been described as four iterative phases with feedback loops to allow integration of new knowledge [3]. Phase 0 (T0) translation research is discovery research, phase I (T1) is research to develop a candidate health application, phase II (T2) is research that evaluates a candidate application and develops evidence-based recommendations, phase III (T3) is research that assesses how to integrate an evidence-based recommendation into clinical care and prevention, and phase IV (T4) is research that assesses the health outcomes and population impact (these phases are summarized in Figure 5.1).

Within the drug development process, there have already been several U.S. Food and Drug Administration (FDA)–approved PGx diagnostics (genetic tests), and drugs for which there is PGx information in their labels [4]. Moreover, there has been oversight by the regulatory authorities over this field since 2004 and in the white paper titled *Innovation or*

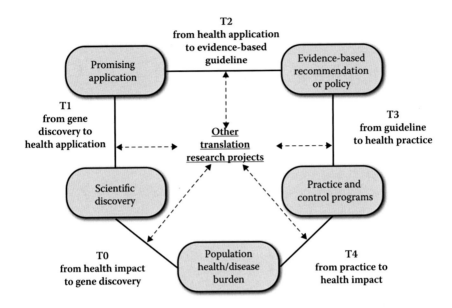

FIGURE 5.1
Summary of the iterative phases of translation research. (Adapted from Khoury, M.J. et al., *Genet. Med.* 9:665–674, 2007 and Khoury, M.J., *Clin. Pharmacol. Ther.* 87: 635–638, 2010.)

Stagnation: Challenge and Opportunity on the Critical Path to New Medical Products [5], attention was brought to how emerging PGx techniques showed promise for improving on the safety, efficacy, and quality of drug products. Subsequently, in 2005, the FDA published the white paper *Guidance for Industry on Pharmacogenomics Data Submission* [6], with the goal of promoting the use of PGx in drug development and encouraging public sharing of data and information on PGx test results.

Increasing the prevalence of PGx activities to provide a competitive advantage through product differentiation and biomarker-driven disease stratification and modification is a future vision for drug development. Drug discovery, for example, might involve identifying potential drug targets based on a biological understanding of the disease in question, and exploring the interactions between compounds and genetic variants of these potential targets (T0 research). Phases I and II of the clinical trials investigate, respectively, whether a new treatment is safe to use in a certain population, and whether it improves the patient's condition. During these early stage clinical trials, PGx evidence can help explain the unexpected variability in pharmacokinetic and pharmacodynamic responses, and can be applied prospectively as inclusion/exclusion criteria for phase III trials (T1 research). Phase III clinical trials often compare a new treatment to the current standard of care, and additional information is gathered for the purpose of evaluating the overall benefit–risk relationship of the drug. Information gathered during this phase is used to determine label information (T2 research). In cases in which drug candidates are only effective in a subpopulation, knowledge gained from PGx evaluations might explain these safety outcomes. In some instances, pairing a new drug with a required or recommended PGx test to identify responders and those less likely to experience adverse affects increases the likelihood of regulatory approval.

TABLE 5.1

Forty-Six Drugs for which the FDA Provides Information about Genetic
Testing of Drug-Metabolizing Enzymes

Drug	Drug-Metabolizing Gene/Enzyme
Diazepam	CYP2C19
Drospirenone and ethinyl estradiol	CYP2C19
Esomeprazole	CYP2C19
Nelfinavir	CYP2C19
Prasugrel	CYP2C19
Rabeprazole	CYP2C19
Voriconazole	CYP2C19
Celecoxib	CYP2C9
Warfarin	CYP2C9
Aripiprazole	CYP2D6
Atomoxetine	CYP2D6
Carvedilol	CYP2D6
Cevimeline	CYP2D6
Clozapine	CYP2D6
Codine sulfate	CYP2D6
Dextromethorphan and quinidine	CYP2D6
Doxepin	CYP2D6
Fluoxetine and olanzapine	CYP2D6
Fluoxetine HCL	CYP2D6
Metoprolol	CYP2D6
Propafenone	CYP2D6
Propranolol	CYP2D6
Protriptyline	CYP2D6
Quinidine	CYP2D6
Risperidone	CYP2D6
Terbinafine	CYP2D6
Tetrabenazine	CYP2D6
Thioridazine	CYP2D6
Timolol	CYP2D6
Tiotropium	CYP2D6
Tolterodine	CYP2D6
Tramadol and acetominophen	CYP2D6
Venlafaxine	CYP2D6
Capecitabine	DPYD
Fluorouracil	DPYD
Chloroquine	G6PD
Dapsone	G6PD
Primaquine	G6PD
Rasburicase	G6PD
Rifampin, isoniazid, and pyrazinamide	NAT
Isosorbide and hydralazine	NAT
Azathioprine	TPMT

(continued)

TABLE 5.1 (Continued)

Forty-Six Drugs for which the FDA Provides Information about Genetic
Testing of Drug-Metabolizing Enzymes

Drug	Drug-Metabolizing Gene/Enzyme
Mercaptopurine	TPMT
Thioguanine	TPMT
Irinotecan	UGT1A1
Nilotinib	UGT1A1

Source: U.S. Food and Drug Administration. 2011. Table of Pharmacogenomic
Biomarkers in Drug Labels. http://www.fda.gov/drugs/scienceresearch/
researchareas/pharmacogenetics/ucm083378.htm.

Notes: CYP2C19, Cytochrome P450-2C19; CYP2C9, Cytochrome P450-2C9; DPYD,
Dihydropyrimidine dehydogenase; G6PD, Glucose-6-phosphate dehydroge-
nase; NAT, N-acetyltransferase; TPMT, Thiopurine methyltransferase; UGT1A1,
UGP Glucuronosyltransferase 1 family, polypeptide A1.

Within clinical practice, tailoring treatment to individual characteristics such as demo-
graphics, patient history, and concurrent medications is a common practice. With the
recent advances in genomics and molecular biology, clinicians are now interested in
incorporating knowledge of an individual's genetic profile and understanding how certain profile characteristics give rise to certain "phenotypes." In a report for the President's
Council of Advisors on Science and Technology titled *Priorities for Personalized Medicine*, it
was concluded that the development and application of genomics-based molecular diagnostics is essential to expand the promise of personalized medicine [8]. Although it is
common for molecular assays to be used for diagnostics, the majority of these tests are
based on single-gene mutations. Modern genomic technologies, which are rapidly emerging, have the capability to evaluate several genes at once, which is necessary to understand
the interplay of different genes in the physiological response to therapy. Genomics-based
molecular diagnostics can be applied to assess individual risk of disease, identify patients
who will benefit from particular interventions, or tailor medication dose to accommodate
individual variations.

PGx profile data in particular can be used to make dosage adjustments to prevent drug
toxicity, therefore leading to a reduction of drug-related deaths and adverse drug effects.
In the United States, the number of deaths from drug-induced causes in 1999 (19,128 deaths)
more than doubled by 2007 (38,371 deaths) [9]. Drug-metabolizing enzymes, drug transporters, and human leukocyte antigens (HLAs) are the three main categories of genes currently associated with adverse drug reactions (ADRs) [10,11]. Type A ADRs in particular
are typically dose-related and occur relatively frequently, accounting for more than 50% of
all ADRs [12]. Variants in drug-metabolizing enzymes and drug transporters can affect the
clearance of drugs and can lead to this form of ADR if no dose adjustment is made. HLAs
have been implicated in type B ADRs, which are unpredictable and occur in susceptible
individuals. There are currently two drugs for which the FDA recommends genetic testing
for an HLA variant before drug administration [4]. These drugs include carbamazepine
(HLA-B*1502) and abacavir (HLA-B*5701). There are 46 drugs for which the FDA provides
information about genetic testing of drug-metabolizing enzymes (Table 5.1). It is possible
to find equivalent doses for different pharmacokinetic genotypes [12]; therefore, genetic
testing might facilitate preemptive genotype-guided prescribing. This practice has already
proven more efficient and safer than the traditional "population average" protocol with
warfarin (the most prescribed anticoagulation therapy worldwide) [13].

Translation research can help facilitate the adoption of genomics-based molecular diagnostics such as the relevant content that is provided as information in FDA drug labels. Research on the most appropriate testing practices and patient management based on test results (T3 research) is useful for developing clinical practice guidelines and programs. In addition, research that provides data on the potential effect of introducing genetic testing, when compared with the standard of practice (T4 research), can facilitate an understanding of population and public health impact.

In the following section, we investigate some of the possible reasons for low clinical uptake of PGx information use, and discuss ways of providing access to pharmacogenomics knowledge that might address some of these issues.

5.2 The Need for Pharmacogenomics Knowledge to Support Personalized Medicine

Lack of uptake of pharmacogenomics discoveries in clinical practice may be due in part to the lack of translation research efforts. It has been estimated that no more than 3% of genomics research focuses on T2 to T4 research that aims to validate genomic discoveries for use in practice [3]. Despite the development of several promising genomic and personalized medical applications, few have undergone the rigorous evaluations needed for regulatory approval [14]. Similar forms of evaluations are also required for professional organizations to develop and implement authoritative recommendations in support of interpretation activities of testing laboratories and clinical practitioners; and for payer organizations to make coverage decisions. The inability to perform rigorous evaluations is due in part to what is described as the "evidence dilemma," in view of the fact that there is often a lack of sufficient evidence to weigh the benefits and risks of population screening or routine use of PGx applications in clinical practice [7]. T2 clinical validity and utility evaluation research, and T3 diffusion research activities in particular are required for regulatory agencies, payer organizations, and professional organizations to, respectively, make regulatory decisions, coverage decisions, and decisions about what practices to endorse.

There are some early examples of projects implementing pharmacogenomics in clinical practice that further eliminate the need for translation research efforts. With a focus on "high-throughput" molecular diagnostics, two technologies that have PGx uses include Roche's AmpliChip® CYP450 Test (Basel, Switzerland) and Genomic Health's Oncotype DX® (Redwood City, CA). Oncotype DX is a gene expression assay primarily used for breast cancer prognosis [15]. However, the test also has PGx uses with respect to administering adjuvant chemotherapy in conjunction with Tamoxifen (used for estrogen receptor-positive breast cancer tumors). The AmpliChip CYP450 test is a microarray-based PGx test that detects variations in CYP2D6 and CYP2C19 genes. The assay aids in dosing decisions for drugs metabolized through these genes (e.g., tamoxifen) [16]. A recent study investigating the discordant rates of clinical adoption for these two tests revealed that high and low adoption were more closely correlated with endorsements of patient groups and medical professional societies than with the (a) characteristics of the genetic tests and (b) clinical validity, utility, and cost-effectiveness of those tests [17]. These results further support the need for translational research, particularly T3 diffusion research, by both professional organizations and patient groups.

A recent study found that approximately 24% of Americans already receive drugs affected by known biomarkers [18] and more than one million seek care for adverse reactions every year [19]. We are now beginning to see more cases in which PGx applications have adequate knowledge on drug response–phenotype correlations to appear in drug labels [4]. However, it is still rare for genotype data to be considered as a critical determinate in drug therapy decisions, such as setting drug dose [20]. In 2008, in a study conducted by Medco Research and the American Medical Association, among 10,303 physician survey respondents in the United States, 98% agreed that patient genetic profiles may influence drug therapy but only 13% had ordered a pharmacogenetic test in the previous 6 months [21].

Lack of access to appropriate information necessary to support clinical decision-making hinders the ability to incorporate existing pharmacogenomic test results into clinical practice. For clinicians to adopt genomics-based molecular diagnostics such as those for which relevant content is provided for information in FDA drug labels, T3 and T4 research is required. For example, much of the information that is made available in FDA drug labels about genomics-based molecular diagnostics does not provide guidance on what testing should be performed, for whom, and how test results should be interpreted [22]. In 2006, a study reviewing PGx information in the drug labels and in the literature for the top 200 prescribed drugs showed that 71.3% of the drugs had published PGx information in the literature, but only three had drug labels with information sufficient to guide dosing [23]. T3 research on the most appropriate testing practices and patient management based on test results is required to develop clinical practice guidelines and programs for incorporating genomics-based molecular diagnostics. Furthermore, more T4 research such as the warfarin study that provides data on the potential effect of introducing genetic testing when compared with standards of practice [13] are needed.

In addition to the need for more translation research efforts, there are also several other systematic barriers and challenges to incorporating P4 medicine in clinical practice [24]. For example, lack of clinical uptake may also be due in part to the challenge of dealing with rapidly changing genetic knowledge. A previous evaluation shows that providers cite this challenge as the greatest obstacle to providing information to their patients [25]. This reflects the need for accelerated translation of genomic knowledge into a form that will assist clinicians in their use of genomics-based molecular diagnostics. Basing prescribing decisions on genetic tests is also often beyond the present scope of medical training [26], if not human cognitive ability. Moreover, most practitioners are unfamiliar with the guidelines for use of genetic tests; and few are implemented, evaluated, or enforced [27]. In the study conducted by Medco Health Solutions, Inc. and the American Medical Association, they found that only 10% of physician survey respondents believed they are adequately informed about PGx testing [21].

A 2003 study on barriers to the adoption of genetic counseling, testing, and interpretation services concluded that educational programs are needed to facilitate the implementation of genetic services across a broader set of physicians [28]. This is still true and therefore a key requirement for accelerating the PGx research findings into clinical practice is to enhance education and guidance for health care professionals to support accurate use and interpretation of genetic tests. With a focus on PGx, the Secretary's Advisory Committee on Genetics, Health and Society report, *Realizing the Potential of Pharmacogenomics: Opportunities and Challenges* [29], suggests that guidance for physicians include support for understanding criteria for PGx genetic testing, and understanding use of results for patient care. Clinical decision support software that provides PGx knowledge (e.g., genetic testing protocols) at the point-of-care can aid in this process and will become especially important for multigene pharmacogenomic protocols expected in the future [30].

In another report, the Secretary's Advisory Committee on Genetics, Health and Society recommended that a Web-based registry or repository of information be made available to provide up-to-date and accurate information for available genetic tests [27]. Providing these forms of support with PGx knowledge sources, integrated at different points of need within a clinicians' workflow, have the potential to influence uptake of PGx tests in clinical practice. For example, a study in 2004 describes a conceptual framework for evaluating PGx tests and consists of the following: (1) medical need, (2) clinical validity and utility of a test, (3) ease of use of the test, and (4) choice of treatments based on the results of the test [31]. Available PGx knowledge resources may support each of these points of evaluation.

5.3 Major Sources of Pharmacogenomics Knowledge

A list of regulatory agencies, professional organizations, and payor organizations within the United States that are participating in T2 to T4 evaluations regarding the use of PGx data in clinical practice are shown in Table 5.2. In addition to the efforts of stakeholder organizations, knowledge bases are required to support translation research.

To date, there are more freely available PGx knowledge resources that are able to support T0 and T1 translation research compared with those available to support T2 to T4 translation research. Some open source resources are summarized in Table 5.3 and incorporate findings from two studies that characterize drug information sources [32,33].

Freely available knowledge bases to support T2 and T3 research include the GeneTests knowledge base [34], Genetic Test Registry [35,36], the GAPP Knowledge Base [37], and the *PLoS Currents: Evidence on Genomic Tests* publication [38]. The GeneTests knowledge base provides an online laboratory directory (laboratories offering in-house molecular genetic testing, specialized cytogenetic testing, and biochemical testing for inherited disorders), an online genetic clinic directory (providers of genetic evaluation and counseling services), and provide gene reviews documents (containing clinical guidance in areas such as testing strategy, interpretation genetic test results, and genetic counseling). The Genetic Testing Registry encourages providers of genetic tests to publicly share information about the availability and utility of their tests, provides information on the locations of laboratories that offer particular tests, and facilitates genetic and genomic data-sharing for research and new scientific discoveries. The GAPP Knowledge Base, being developed by the Genomic Applications in Practice and Prevention Network (GAPPNet), is an online resource that provides access to information on applications of genomic research for use in public health and health care. Current features include the *GAPPFinder* (a searchable database of genetic tests in transition to practice), *Evidence for Genomic Applications* (an online, open access journal that links to published evidence reviews and recommendations), *Evidence Aggregator* (an application that facilitates searching evidence reports, systematic reviews, recommendations, or guidelines in genetic tests and genomic applications), and *Project Locator* (an online database for archiving genomic translation projects). The *PLoS Currents: Evidence on Genomic Tests* is an open access publication provided by the U.S. Centers for Disease Control and Prevention and the Public Library of Science. The publication provides brief summaries of evidence for the clinical validity and clinical utility of genomic tests and is intended to complement other efforts described above.

TABLE 5.2

Regulatory Agencies, Professional Organizations, and Payer Organizations within the United States that are Participating in T2 to T4 Evaluations Regarding the Use of PGx Data in Clinical Practice

Organization	Role/Resources	Examples
Regulatory organization— Centers for Disease Control, Office of Public Health Genomics, Evaluation of Genomic Applications in Practice and Prevention [39]	Evidence reports	Can UGT1A1 genotyping reduce morbidity and mortality in patients with metastatic colorectal cancer treated with irinotecan? An evidence-based review [40]
Regulatory organization— National Institutes of Health, National Institute of General Medical Sciences, Pharmacogenetic Research Network, Clinical Pharmacogenetics Implementation Consortium [47]	Recommendations for implementing specific pharmacogenomic tests and practices	TPMT - thiopurines [41]; CYP2C19 - clopidogrel [42]; CYP2C9, VKORC1 - warfarin [43]; CYP2D6 - codine [44]; HLA-B - abacavir [45]; SLCO1B1 - simvastatin [46]
Regulatory organization— U.S. FDA, Interdisciplinary Pharmacogenomics Review Group [48]	Review voluntary exploratory data submissions Qualification of exploratory biomarkers into valid biomarkers Technical recommendations	Table of valid genomic biomarkers in the context of approved drug labels [4]
Regulatory organization— Agency for Healthcare Research and Quality	Technology assessments	Systematic reviews on selected pharmacogenetic tests for cancer treatment: CYP2D6 for tamoxifen in breast cancer, KRAS for anti-EGFR antibodies in colorectal cancer, and BCR-ABL1 for tyrosine kinase inhibitors in chronic myeloid leukemia [49]
Professional organization— American Society of Clinical Oncology (ASCO)	Clinical practice guidelines Provisional clinical opinion ASCO guideline endorsements Clinical evidence review	Testing for KRAS gene mutations in patients with metastatic colorectal carcinoma to predict response to anti-epidermal growth factor receptor monoclonal antibody therapy [50]
Professional organization— National Comprehensive Cancer Network (NCCN)	Clinical practice guidelines (NCCN guidelines) NCCN Task Force Reports	ER and/or PgR testing in breast cancer [51]
Professional organization— College of American Pathologists (CAP)	Reference resources and publications	ER/PgR guideline and resources (CAP and ASCO joint guideline) [52] HER2 testing guidelines (CAP and ASCO joint guideline) [53]
Payer organization—Medco Health Solutions	Medco Research Institute, pharmacogenomics community	Warfarin study (Medco/Mayo Clinic) [54] Physician survey (Medco/American Medical Association) [21] Physician Adoption Study [55]
Payer organization—CVS/ Caremark Pharmacy Services	Pharmacogenomic testing program (in partnership with Generation Health Inc.)	Pegasys and Copegus (treatment of hepatitis C); Gleevec, Tasigna, Sprycel, Tarceva, and Tykerb (oncology drugs) [56]

TABLE 5.3

Characteristics of Open Source PGx Resources that Support T0 and T1 Translation Research

Database Content/Feature	ChemBank [57]	DrugBank [58]	PharmGED [59]	PharmGKB [60]	SuperCYP [61]
Curated Facts from Publications					
Natural language descriptions/ comments	Yes	Yes	Yes	Yes	—
Quantitative data	Yes	—	—	—	—
Variant details	—	—	Yes	—	Yes
Genotype details	—	—	—	Yes	—
Disease/phenotype details	—	Yes	—	Yes	Yes
Population details	—	—	—	—	Yes
Study Details					
Design and drug/chemical compound administration	—	—	—	—	—
Genotyping method	Yes	—	—	—	Yes
Phenotyping method	—	—	—	—	—
Alleles tested for	—	—	—	—	Yes
Pharmacokinetic measurements	—	—	Yes	—	—
Pharmacodynamic measurements/ protocol	—	—	Yes	—	—
Side effects	—	—	Yes	—	—
Drug/Chemical Compound Information					
Pharmacology and interactions	Yes	Yes	—	Yes	—
Absorption, distribution, metabolism, elimination, toxicity	—	Yes	—	Yes	—
Biophysical properties	Yes	Yes	—	Yes	Yes
Metabolizing enzymes	—	Yes	Yes	Yes	Yes
Targets	Yes	Yes	Yes	Yes	Yes
Gene associations	—	Yes	Yes	—	—
Drug or Chemical Compound Target/Metabolizing Enzyme Information					
Related drugs/chemical compounds	Yes	Yes		Yes	Yes
Related diseases	Yes	Yes	Yes	Yes	—
Biophysical properties	Yes	Yes	—	Yes	Yes
Gene/protein sequence	—	Yes	Yes	Yes	Yes
Pathway information	—	Yes	—	Yes	—
Detailed single nucleotide polymorphism data	—	Yes	—	Yes	Yes
Other					
External database links	Yes	Yes	Yes	Yes	Yes
Data analysis capabilities	Yes	Yes	—	Yes	—
Publicly available	Yes	Yes	Yes	Yes	Yes

5.4 Bridging the Gap between the Current State of Pharmacogenomics Knowledge and Its Integration

Research in drug development, health care cost-effectiveness, and evaluations by professional organizations have become information-intensive fields with PGx studies increasing in both number and throughput. Nearly double the number of articles on PGx studies were published in 2007 when compared with 2002 [62]. In addition, in 2009 there were at least 12 PGx genomewide association studies conducted [63]. To interpret data, draw accurate conclusions, and make appropriate decisions, scientists, economic researchers, and professional organizations must be able to leverage PGx knowledge contained in the biomedical literature base. Moreover, for PGx data to be utilized by care providers, knowledge must be summarized and presented in a way that allows for easy interpretation of the data and allows clinicians to make decisions at the point-of-care.

We have previously shown the presence of PGx knowledge resources in support of personalized medicine, and the translation of PGx findings into clinical practice. However, the context under which PGx knowledge resources can be of use should be better understood before they can be of most use. In addition, there are several barriers to the diffusion of findings into clinical practice due to the need for educational programs, supportive infrastructure, and support for ethical concerns such as privacy and confidentiality.

Toward understanding the context under which P4 medicine might be implemented in clinical practice, the U.S. Department of Health and Human Services undertook its Personalized Health Care Initiative. In 2006, a personalized health care workgroup of the American Health Information Community was formed to foster a community-based approach to facilitate the incorporation of clinically useful genetic/genomic information and analytical tools to support clinical decision-making [64]. This working group completed an assessment of the informational needs of several stakeholders including patients, clinicians, organizations, and systems. This assessment resulted in a compiled list of information needs at three phases of performing a genetic/genomic test: (1) a preanalytic phase, (2) an analytic phase, and (3) a postanalytic phase [65]. The preanalytic phase includes events such as determining which genetic test is appropriate to answer a clinical question and collecting/transporting a sample to the test site; the analytic phase involves sample analysis; and the postanalytic phase includes the reporting and interpretation of results. Genomic/PGx knowledge specific to the information needs at these different phases of performing a test are required to facilitate the incorporation of PGx findings into clinical use. These needs are already being used to specify data exchange requirements for health information technology.

Health care information technology is necessary to give health care providers access to patient genotype data and supportive knowledge in real time; however, there is currently a lack of infrastructure to support these technologies. Toward achieving the necessary infrastructure to support the use of genotyping technologies on a broad scale in the United States, as part of the Health Information Technology for Economic and Clinical Health (HITECH) and Act section of the American Recovery and Reinvestment Act (ARRA), President Obama and Congress have provided over $20 billion in funding. The HITECH Act calls for the implementation of interoperable electronic health records in 10 years. The implementation of clinical decision support tools will also play a role in achieving the goal of improving safety, quality, and efficiency of care with the use of electronic health records [66]. For example, such tools might apply logic to choices about medical prescriptions given the patient data captured in health records. Currently, it is required that at least one clinical decision support tool is implemented within the electronic health record for clinicians and hospitals to receive incentive payments [67].

As mentioned previously, endorsements by professional and patient groups are correlated with clinical uptake of genomics-based diagnostics. It is clear that information about pharmacogenomic technologies and findings will need to be presented in a way that facilitates the assessment of value by these groups. The rise and fall of whole-body computed tomography screening centers is evidence that the public can effectively make these types of assessments [68]. In addition, to ease any possible ethical concerns the public may have regarding potential misuse of genetic information, in 2008, the Genetic Information Nondiscrimination Act was passed in the United States. This act ensures that genetic information will not be misused to deny health insurance or employment.

It is evident that educational programs are also needed to facilitate uptake and appropriate use of genomics-based diagnostics in clinical practice. A survey of pharmacogenomics instruction in medical schools shows that most medical schools in the United States and Canada have started incorporating pharmacogenomics material into their curriculum [69]. This is indicative of an emerging clinical workforce prepared to integrate the use of PGx testing in caring for their patients. In summary, we have provided an overview of motivating factors to incorporating PGx data in the drug development process and into clinical practice. We have also described ways in which PGx knowledge and translation research can help support the goals of personalized medicine, and have outlined organizations and resources that are currently available. Finally, we outlined current progress toward overcoming barriers to incorporating the use of PGx information in drug development and clinical practice. Overall, given the current initiatives, educational programs, and increasing number of quality pharmacogenomic resources, we believe several of the barriers to using PGx information can be overcome as part of the larger vision of a personalized, predictive, preventive, and participatory (P4) health care model.

References

1. Zerhouni, E.A. 2007. Translational research: Moving discovery to practice. *Clinical Pharmacology and Therapeutics* 81:126–128.
2. Hood, L., L. Rowen, D.J. Galas, and J.D. Aitchison. 2008. Systems biology at the Institute for Systems Biology. *Briefings in Functional Genomics and Proteomics* 7:239.
3. Khoury, M.J., M. Gwinn, P.W. Yoon, N. Dowling, C.A. Moore, and L. Bradley. 2007. The continuum of translation research in genomic medicine: How can we accelerate the appropriate integration of human genome discoveries into health care and disease prevention? *Genetics in Medicine* 9:665–674.
4. U.S. Food and Drug Administration. 2011. Table of Pharmacogenomic Biomarkers in Drug Labels. http://www.fda.gov/drugs/scienceresearch/researchareas/pharmacogenetics/ucm083378.htm. Date accessed May 2011.
5. U.S. Food and Drug Administration. 2004. Innovation or Stagnation: Challenge and Opportunity on the Critical Path to New Medical Products.
6. U.S. Food and Drug Administration. 2005. Guidance for Industry: Pharmacogenomic Data Submissions.
7. Khoury, M.J. 2010. Dealing with the evidence dilemma in genomics and personalized medicine. *Clinical Pharmacology and Therapeutics* 87:635–638.

8. President's Council of Advisor on Science and Technology. 2008. Priorities for Personalized Medicine.

9. Xu, J., K.D. Kochanek, S.L. Murphy, and B. Tejada-Vera. 2010. Deaths: Final data for 2007. *National Vital Statistics Reports* 58:19.

10. Wilke, R.A., D.W. Lin, D.M. Roden, P.B. Watkins, D. Flockhart, I. Zineh et al. 2007. Identifying genetic risk factors for serious adverse drug reactions: Current progress and challenges. *Nature Reviews in Drug Discovery* 6:904–916.

11. Nakamura, Y. 2008. Pharmacogenomics and drug toxicity. *New England Journal of Medicine* 359:856–858.

12. Brockmoller, J., and M.V. Tzvetkov. 2008. Pharmacogenetics: Data, concepts and tools to improve drug discovery and drug treatment. *European Journal of Clinical Pharmacology* 64:133–157.

13. Caraco, Y., S. Blotnick, and M. Muszkat. 2008. CYP2C9 genotype-guided warfarin prescribing enhances the efficacy and safety of anticoagulation: A prospective randomized controlled study. *Clinical Pharmacology Therapy* 83:460–470.

14. Deverka, P.A. 2009. Pharmacogenomics, evidence, and the role of payers. *Public Health Genomics* 12:149–157.

15. Dobbe, E., K. Gurney, S. Kiekow, J.S. Lafferty, and J.M. Kolesar. 2008. Gene-expression assays: New tools to individualize treatment of early-stage breast cancer. *American Journal of Health System Pharmacy* 65:23–28.

16. Jain, K.K. 2005. Applications of AmpliChip CYP450. *Molecular Diagnosis* 9:119–127.

17. Yoo, J.K. 2009. What makes personalized medicine work? An Empirical Analysis of the Role of Product Attributes. Medical Professional Societies and Patient Groups in the Diffusion of Four Breast Cancer Genetic Tests. Master's Thesis, Massachusetts Institute of Technology.

18. Frueh, F.W., S. Amur, P. Mummaneni, R.S. Epstein, R.E. Aubert, T.M. DeLuca et al. 2008. Pharmacogenomic biomarker information in drug labels approved by the United States Food and Drug Administration: Prevalence of related drug use. *Pharmacotherapy* 28:992–998.

19. Lazarou, J., B.H. Pomeranz, and P.N. Corey. 1998. Incidence of adverse drug reactions in hospitalized patients: A meta-analysis of prospective studies. *Journal of the American Medical Association* 279:1200–1205.

20. Bhathena, A., and B.B. Spear. 2008. Pharmacogenetics: Improving drug and dose selection. *Current Opinion in Pharmacology* 8:639–646.

21. Medco Health Solutions, Inc., and the American Medical Association. 2009. National Pharmacogenomics Physician Survey: Who Are the Physicians Adopting Pharmacogenomics and How Does Knowledge Impact Adoption?

22. Overby, C.L., P. Tarczy-Hornoch, J.I. Hoath, I.J. Kale, and D.L. Veenstra. 2010. Feasibility of incorporating genomic knowledge into electronic medical records for pharmacogenomic clinical decision support. *BMC Bioinformatics* 11 Suppl 9, S10.

23. Zineh, I., G.D. Pebanco, C.L. Aquilante, T. Gerhard, A.L. Beitelshees, B.N. Beasley, and A.G. Hartzema. 2006. Discordance between availability of pharmacogenetics studies and pharmacogenetics-based prescribing information for the top 200 drugs. *Annual Pharmacotherapy* 40:639–644.

24. Deverka, P.A., T. Doksum, and R.J. Carlson. 2007. Integrating molecular medicine into the US health-care system: Opportunities, barriers, and policy challenges. *Clinical Pharmacology Therapy* 82:427–434.

25. Wilkins-Haug, L., L. Hill, L. Schmidt, G.B. Holzman, and J. Schulkin. 1999. Genetics in obstetricians' offices: A survey study. *Obstetrics and Gynecology* 93:642–647.

26. Menasha, J.D., C. Schechter, and J. Willner. 2000. Genetic testing: A physician's perspective. *Mount Sinai Journal of Medicine* 67:144–151.

27. Secretary's Advisory Committee on Genetics, Health and Society (SACGHS). 2008. U.S. System of Oversight of Genetic Testing: A Response to the Charge of the Secretary of Health and Human Services.

28. Suther, S., and P. Goodson. 2003. Barriers to the provision of genetic services by primary care physicians: A systematic review of the literature. *Genetics in Medicine* 5:70.

29. HHS Secretary's Advisory Committee on Genetics, Health and Society (SACGHS). 2008. Realizing the Potential of Pharmacogenomics: Opportunities and Challenges.

30. McKinnon, R.A., M.B. Ward, and M.J. Sorich. 2007. A critical analysis of barriers to the clinical implementation of pharmacogenomics. *The Clinical Risk Management* 3:751–759.

31. Shah, J. 2004. Criteria influencing the clinical uptake of pharmacogenomic strategies. *British Medical Journal* 328:1482.

32. Mattingly, C.J. 2009. Chemical databases for environmental health and clinical research. *Toxicology Letters* 186:62–65.

33. Sharp, M., O. Bodenreider, and N. Wacholder. 2008. A framework for characterizing drug information sources. Annual Symposium proceedings/AMIA Symposium. AMIA.

34. Pagon, R.A., P. Tarczy-Hornoch, P.K. Baskin, J.E. Edwards, M.L. Covington, M. Espeseth et al. 2002. GeneTests-GeneClinics: Genetic testing information for a growing audience. *Human Mutation* 19:501–509.

35. Kuehn, B.M. 2010. NIH launching genetic test registry. *Journal of the American Medical Association* 303:1685.

36. Khoury, M.J., M. Reyes, M. Gwinn, and W.G. Feero. 2010. A genetic test registry: Bringing credible and actionable data together. *Public Health* 13:360–361.

37. GAPP Knowledge Base. 2010. [database on the Internet].

38. Gwinn, M., W.D. Dotson and M.J. Khoury. 2010. PLoS currents: Evidence on genomic tests—at the crossroads of translation. *PLoS Currents* 2.

39. Teutsch, S.M., L.A. Bradley, G.E. Palomaki, J.E. Haddow, M. Piper, N. Calonge et al. 2009. The Evaluation of Genomic Applications in Practice and Prevention (EGAPP) initiative: Methods of the EGAPP working group. *Genetics in Medicine* 11:3.

40. Palomaki, G.E., L.A. Bradley, M.P. Douglas, K. Kolor, and W.D. Dotson. 2009. Can UGT1A1 genotyping reduce morbidity and mortality in patients with metastatic colorectal cancer treated with irinotecan? An evidence-based review. *Genetics in Medicine* 11:21–34.

41. Relling, M.V., E.E. Gardner, W.J. Sandborn, K. Schmiegelow, C.H. Pui, S.W. Yee et al. 2011. Clinical pharmacogenetics implementation consortium guidelines for thiopurine methyltransferase genotype and thiopurine dosing. *Clinical Pharmacology and Therapeutics* 89:387–391.

42. Scott, S.A., K. Sangkuhl, E.E. Gardner, C.M. Stein, J.S. Hulot, J.A. Johnson et al. 2011. Clinical Pharmacogenetics Implementation Consortium guidelines for cytochrome P450-2C19 (CYP2C19) genotype and clopidogrel therapy. *Clinical Pharmacology and Therapeutics* 90:328–332.

43. Johnson, J.A., L. Gong, M. Whirl-Carrillo, B.F. Gage, S.A. Scott, C.M. Stein et al. 2011. Clinical pharmacogenetics implementation consortium guidelines for CYP2C9 and VKORC1 genotypes and warfarin dosing. *Clinical Pharmacology and Therapeutics* 90:625–629.

44. Crews, K.R., A. Gaedigk, H.M. Dunnenberger, T.E. Klein, D.D. Shen, J.T. Callaghan et al. 2012. Clinical pharmacogenetics implementation consortium (CPIC) guidelines for codeine therapy in the context of cytochrome P450 2D6 (CYP2D6) genotype. *Clinical Pharmacology and Therapeutics* 91:321–326.

45. Martin, M.A., T.E. Klein, B.J. Dong, M. Pirmohamed, D.W. Haas, and D.L. Kroetz. 2012. Clinical pharmacogenetics implementation consortium guidelines for HLA-B genotype and abacavir dosing. *Clinical Pharmacology and Therapeutics* 91:734–738.

46. Wilke, R.A., L.B. Ramsey, S.G. Johnson, W.D. Maxwell, H.L. McLeod, D. Voora et al. 2012. The clinical pharmacogenomics implementation consortium: CPIC guideline for SLCO1B1 and simvastatin-induced myopathy. *Clinical Pharmacology and Therapeutics* 92:112–117.

47. Relling, M.V., R.B. Altman, M.P. Goetz and W.E. Evans. 2010. Clinical implementation of pharmacogenomics: Overcoming genetic exceptionalism. *The Lancet Oncology* 11:507–509.

48. Goodsaid, F. and F.W. Frueh. 2007. Implementing the U.S. FDA guidance on pharmacogenomic data submissions. *Environmental and Molecular Mutagen* 48:354–358.

49. Teruhiko, T., I.J. Dahabreh, P.D. Castaldi, and T.A. Trikalinos. 2009. Systematic reviews on selected pharmacogenetic tests for cancer treatment: CYP2D6 for tamoxifen in breast cancer, KRAS for anti-EGFR antibodies in colorectal cancer, and BCR-ABL1 for tyrosine kinase inhibitors in chronic myeloid leukemia. Technology Assessment Report. Project ID: GEN609.

50. Allegra, C.J., J.M. Jessup, M.R. Somerfield, S.R. Hamilton, E.H. Hammond, D.F. Hayes et al. 2009. American Society of Clinical Oncology provisional clinical opinion: Testing for KRAS gene mutations in patients with metastatic colorectal carcinoma to predict response to anti-epidermal growth factor receptor monoclonal antibody therapy. *Journal of Clinical Oncology* 27:2091–2096.
51. Allred, D.C., R.W. Carlson, D.A. Berry, H.J. Burstein, S.B. Edge, L.J. Goldstein et al. 2009. NCCN Task Force report: Estrogen receptor and progesterone receptor testing in breast cancer by immunohistochemistry. *Journal of the National Comprehensive Cancer Network* 7, Suppl 6.
52. Hammond, M.E.H., D.F. Hayes, M. Dowsett, D.C. Allred, K.L. Hagerty, S. Badve et al. 2010. American Society of Clinical Oncology/College of American Pathologists guideline recommendations for immunohistochemical testing of estrogen and progesterone receptors in breast cancer. *Journal of Clinical Oncology* 28:2784–2795.
53. Wolff, A.C., M.E.H. Hammond, J.N. Schwartz, K.L. Hagerty, D.C. Allred, R.J. Cote et al. 2007. American Society of Clinical Oncology/College of American Pathologists guideline recommendations for human epidermal growth factor receptor 2 testing in breast cancer. *Journal of Clinical Oncology* 25:118–145.
54. Epstein, R.S., T.P. Moyer, R.E. Aubert, D.J. O'Kane, F. Xia, R.R. Verbrugge et al. 2010. Warfarin genotyping reduces hospitalization rates: Results from the MM-WES (Medco–Mayo Warfarin Effectiveness Study). *Journal of the American College of Cardiology* 55:2804–2812.
55. Medco Health Solutions. 2010. Nationwide Implementation and Adoption of Pharmacogenetic Testing through a Novel Pharmacy-Based Approach.
56. CVS Caremark. 2010. Generation Health Outline Target Medications That Will Be the Focus of New Pharmacogenomics Partnership [press release].
57. Seiler, K.P., G.A. George, M.P. Happ, N.E. Bodycombe, H.A. Carrinski, S. Norton et al. 2008. ChemBank: A small-molecule screening and cheminformatics resource database. *Nucleic Acids Research* 36:D351–359.
58. Wishart, D.S. 2008. DrugBank and its relevance to pharmacogenomics. *Pharmacogenomics* 9:1155–1162.
59. Zheng, C.J., L.Y. Han, B. Xie, C.Y. Liew, S. Ong, J. Cui et al. 2007. PharmGED: Pharmacogenetic effect database. *Nucleic Acids Research* 35:D794–799.
60. Altman, R.B. 2007. PharmGKB: A logical home for knowledge relating genotype to drug response phenotype. *Nature Genetics* 39:426.
61. Preissner, S., K. Kroll, M. Dunkel, C. Senger, G. Goldsobel, D. Kuzman et al. 2010. SuperCYP: A comprehensive database on cytochrome P450 enzymes including a tool for analysis of CYP–drug interactions. *Nucleic Acids Research* 38:D237–243.
62. Holmes, M.V., T. Shah, C. Vickery, L. Smeeth, A.D. Hingorani, and J.P. Casas. 2009. Fulfilling the promise of personalized medicine? Systematic review and field synopsis of pharmacogenetic studies. *PLoS One* 4:e7960.
63. Daly, A.K. 2010. Genome-wide association studies in pharmacogenomics. *Nature Reviews Genetics* 11:241–246.
64. Glaser, J., D.E. Henley, G. Downing, and K.M. Brinner. 2008. Advancing personalized health care through health information technology: An update from the American Health Information Community's personalized health care workgroup. *Journal of the American Medical Information Association* 15:391–396.
65. U.S. Department of Health and Human Services (DHHS). 2010. Personalized Health Care Workgroup: Genetic/genomic test priority area.
66. Blumenthal, D., and M. Tavenner. 2010. The "meaningful use" regulation for electronic health records. *New England Journal of Medicine* 363:501–504.
67. Jha, A.K. 2010. Meaningful use of electronic health records: The road ahead. *Journal of the American Medical Association* 304:1709–1710.

68. Burger, I.M., N.E. Kass, J.H. Sunshine and S.S. Siegelman. 2008. The use of CT for screening: A National survey of radiologists' activities and attitudes. *Radiology* 248:160–168.
69. Green, J.S., T.J. O'Brien, V.A. Chiappinelli, and A.F. Harralson. 2010. Pharmacogenomics instruction in US and Canadian medical schools: Implications for personalized medicine. *Pharmacogenomics* 11:1331–1340.

6

Toxicogenomics: Techniques, Applications, and Challenges

Indra Arulselvi Padikasan and Nancy Daniel

CONTENTS

6.1 Introduction

Emerging technologies have enabled scientists to get an unprecedented amount of information about basic cellular processes. This new information is already being put to use in improving diagnostic tests and therapeutic agents in clinical medicine. In the future, it promises to improve our ability to measure and predict the effects of chemicals on human health. Collectively, these technologies have been termed the "omics" because this suffix is added to roots describing the particular part of the cellular machinery being studied: "toxicogenomics" refers to the responses of genes to toxic exposures, "proteomics" indicates the responses of proteins, and "metabolomics" refers to the responses of metabolites.

To obtain information about how chemicals affect human health, scientists have traditionally focused on measuring exposures and looking for adverse health outcomes, from poor performance on neuropsychological tests to cancer and death. The scientists' two main tools are toxicologic studies, conducted on animals or cultured cells in laboratories, and epidemiologic studies, which observe differences in diseases among groups of people. Most toxicological studies intentionally expose experimental animals to controlled doses, and then look for effects such as tumors, behavioral changes, altered reproductive function, or changes in blood proteins or the microscopic appearance of tissues that indicate organ damage. Newer *"in vitro"* toxicological tests use cultured cells or tissues instead of whole animals to gain insight into biological responses to toxic exposures. Epidemiological studies typically start by observing diseases within groups of people, and then attempt to estimate past exposure. These methods have provided a great deal of information over the past decades to help protect the public from harm, but also have significant limitations. Cutting-edge techniques are being used to more thoroughly investigate not just the end-stage, externally observable effects of chemicals on rats or humans but also their more subtle and incremental effects on the inner workings of individual cells. With these new insights, medical scientists are increasingly able to detect diseases, such as cancer, in their early stages, and are now working on interventions to stop the disease at these earlier points.

In addition to providing insight into the cause and progress of diseases, including those caused by chemical exposure, these technologies are also being used to understand how individuals differ in their responses to chemicals, whether therapeutic or toxic agents. By detecting tiny differences in the genes (and thus the proteins) involved in responding to chemicals, scientists are gradually discovering why some people get sick from exposures or experience side effects from drugs, whereas others do not. This field of "pharmacogenetics" will help pharmaceutical companies and physicians tailor individualized drugs and dosages to a patient's genetic profile. For those concerned with chemical safety, "toxicogenetics" will aid our understanding of how individuals vary in their susceptibility to harm from chemical exposure. The public interest community has a critical role to play in helping guide the application of this powerful new science. As with all new technologies, alongside the societal benefits come societal risks. One future benefit is likely to be faster and less expensive toxicologic testing that relies less on the use of animals and reduces uncertainties in how chemicals affect human health. But public interests groups must engage in the science policy process to ensure that shortcuts are not taken and public health protection is not compromised.

6.1.1 Background

Astounding progress has been made in sequencing and characterizing the human genome. This progress has been so fast that a complete first draft of the human genome sequence was announced in June 2000, far sooner than expected by many experts in the field. The Human Genome Project and other genome sequencing projects have accelerated progress in many important scientific areas. In particular, vastly powerful new technologies have been developed for expression profiling of messenger RNAs (mRNAs) and proteins. As a result of these advances, important scientific questions that have long been intractable to toxicology and environmental health are now open to investigation. Toxicologists can utilize these new methods to obtain a more fundamental understanding of chemical-induced and drug-induced disease processes. cDNA microarray and proteomics technologies assess changes in gene expression on a genomewide basis, providing a "global" perspective about how an organism responds to a specific stress, drug, or toxicant. This information can define cellular networks of response genes, identify target molecules of toxicity, provide future biomarkers and alternative testing procedures, and identify individuals with increased susceptibility to environmental agents or drugs. These are but a few of the difficult issues which these new tools will help resolve. The field of toxicology is extremely diverse and its research community is crowded with competing and collaborating participants.

It is clear that advances in toxicogenomics will be more rapid and efficient if these stakeholders join forces and work together in a coordinated manner. The National Institute of Environmental Health Sciences (NIEHS) is uniquely positioned to coordinate efforts directed toward centralizing activities in the field of toxicogenomics and allowing the free distribution of information associated with investigations centered on the discovery of genome–environment interactions. Thus, the NIEHS has established the National Center for Toxicogenomics (NCT) to coordinate international research efforts to develop the field of toxicogenomics. The NCT will provide a unified strategy and a public database, and develop the informatics infrastructure to promote the development of the field of toxicogenomics. The NIEHS will pay special attention to toxicogenomics as applied to the prevention of environmentally related diseases. The NCT will work to allow all partners in this unprecedented enterprise to share equally in its benefits and achievements.

6.1.2 Evolution

The field of genetic toxicology began its development before the biochemical basis of heredity was understood. Early investigators observed that physical and chemical agents could cause heritable mutations. The role of radiation in producing heritable changes in a living organism was first reported by Muller. Auerbach was the first to report the ability of chemicals to cause mutations. These early observations of induced change in heritable traits formed a core of study that evolved into the field of genetic toxicology. It has been evolved from early gene expression studies, which described the response of a biological system to a particular toxicant or panel of reference agents, toward more mature investigations that integrate several omics domains with toxicology and pathology data. Exposure and outcome-specific patterns of gene, protein, and metabolite profiles have been used to identify molecular changes that can be used as biomarkers of toxicity and can provide insights into mechanisms of toxicity and disease causation. Crucial to this evolution were extensive genome sequencing and annotation efforts, which are still ongoing, and the ability to describe response profiles in genetically and toxicologically important species, such as mouse, rat, dog, and human. Another important contribution to toxicogenomics has been the formation of collaborative research consortia that bring together scientists from regulatory agencies, industrial laboratories, academia, and governmental organizations to identify and address important issues for the field.

6.2 Types of Toxicogenomics

6.2.1 Comparative/Predictive Toxicogenomics

There are two main applications for a toxicogenomic approach, comparative/predictive and functional. Comparative genomic, proteomic, or metabolomic studies measure the number and types of genes, protein, and metabolites, respectively, that are present in normal and toxicant-exposed cells, tissues, or biofluids. This approach is useful in defining the composition of the assayed samples in terms of genetic, proteomic, or metabolic variables. The biological sample derived from toxicant, or sham-treated animals can be regarded as an n-dimensional vector in gene expression space with genes as variables along each dimension. The same analogy can be applied for protein expression or nuclear magnetic resonance (NMR) analysis data thereby providing n-dimensional fingerprints or profiles of the biological sample under investigation. Thus, this aspect of toxicogenomics deals with automated pattern recognition analysis aimed at studying trends in data sets rather than probing individual genes for mechanistic information. The need for pattern recognition tools is mandated by the volume and complexity of data generated by genomic, proteomic, and metabolomic tools, whereas human intervention, such as that required for repetitive computation, is kept to a minimum. Automatic toxicity classification methods are very desirable and prediction models are well suited for this task.

The data profiles reflect the pharmacological or toxicological effects, such as disease outcome of the drug or toxicant being utilized. The underlying goal is that a sample from an animal exposed to an unknown chemical, or displaying a certain pathological end point, can then be compared with a database of profiles corresponding to exposure conditions with well-characterized chemicals, or to well-defined pathological effects, to glean/predict some properties regarding the studied sample. These predictions, as we view them, fall into two major categories, namely, (1) classification of samples based on the class of compound to which animals were exposed and (2) classification of samples based on the histopathology and clinical chemistry that the treated animals displayed.

Such data will allow insight into the gene, protein, or metabolite perturbations associated with pharmacologic effects of the agent or toxic end points that ensue. If array data can be "phenotypically anchored" to conventional indices of toxicity.

It will be possible to search for evidence of injury before its clinical or pathological manifestation. This approach could lead to the discovery of potential early biomarkers of toxic injury. "Supervised" predictive models [1] have been used for many years in the financial sectors for evaluating future economic prospects of companies, and in geological institutes for predicting adverse weather outcomes using past or historical knowledge. They have also been utilized to make predictions, using clinical and radiographic information, regarding the diagnosis of active pulmonary tuberculosis at the time of presentation at a health care facility that can be superior to a physicians' opinion. Predictive modeling will undoubtedly revolutionize the field of toxicology by recognizing patterns and trends in high-density data, and forecasting gene–environment, protein–environment, or metabolite–environment interactions relying on historical data from well studied compounds and their corresponding profiles. During the development of a predictive model, a number of issues must be considered. These include the representativeness of the variables to the entity being modeled and the quality of the databases consulted. The NCT at the NIEHS, is building a database to store many variables (e.g., dose, time, and biological systems) and observations (e.g., histopathology, body weight, and cell cycle data) that accompany the process of compound evaluation studies (*in vivo* or *in vitro*) [2]. Recording these parameters will greatly enhance the process of parameter selection in subsequent efforts such as predictive modeling or mechanism of action interpretation. Predictive modeling can be fragmented into a multistage process. The primary stage predictive modeling includes hypothesis development, organization, and data collection. Secondary stage modeling includes initial model development and testing. Tertiary stage modeling includes continued application of the model, ongoing refinement, and validation. Ideally, tertiary stage modeling is a perpetual process whereby lessons learned from previous model applications are incorporated into new and future applications—maintaining or increasing the predictive robustness of the model.

6.2.2 Functional Toxicogenomics

Functional toxicogenomics is the study of genes' and proteins' biological activities in the context of compound effects on an organism. Gene and protein expression profiles are analyzed for information that might provide insight into specific mechanistic pathways. Mechanistic inference is complex when the sequence of events following toxicant exposure is viewed in both dose and time space. Gene and protein expression patterns can indeed be highly dependent on the toxicant concentrations furnished at the assessed tissue and the time of exposure to the agent. Expression patterns are only a snapshot in time and dose space. Thus, a comprehensive understanding of potential mechanisms of action of a compound requires establishing patterns at various combinations of time and dose. This will minimize the misinterpretation of transient responses and allow the discernment of delayed alterations that could be related to adaptation events or which may be representative of potential biomarkers of pathophysiological end points. Studies that target the temporal expression of specific genes and protein in response to toxicant exposure will lead to a better understanding of the sequence of events in complex regulatory networks. Algorithms, such as self-organizing maps, can categorize genes or proteins based on their expression pattern across a continuum of time points [3]. These analyses might suggest relationships in the expression of some genes or proteins depending on the concerted modulation of these variables. An area of study which is of great interest to toxicologists is the mechanistic understanding of toxicant-induced pathological end points. The premise that perturbations in gene, protein, or metabolite levels are reflective of adverse phenotypic effects of toxicants offers an opportunity to phenotypically anchor these perturbations. This is quite challenging because phenotypic effects often vary in the time–dose space of the studied agent and may have regional variations in the tissue. Furthermore, very few compounds exist that result in only one phenotypic alteration at a given coordinate in dose and time. Thus, objective assignment of measured variables to multiple phenotypic events is not possible under these circumstances. However, by studying multiple structurally and pharmacologically unrelated agents that share pathological end points of interest, one could tease out gene, protein, or metabolite modulations that are common between the studied compounds. Laser capture microdissection may also be used to capture regional variations such as zonal patterns of hepatotoxicity. This concept will allow the objective assignment of measurable variables to phenotypic observations that will supplement traditional pathology. It is noteworthy to mention that stand-alone, gene and protein expression, or metabolite fluctuation analyses are not expected to produce decisive inferences on the role of genes or proteins in certain pathways or regulatory networks. However, these tools constitute powerful means to generate viable and testable hypotheses that can direct future endeavors on proving or disproving the involvement of genes, proteins, and metabolites in cellular processes. Ultimately, hypothesized mechanistic inferences have to be validated by the use of traditional molecular biology techniques that include the use of specific enzyme inhibitors and the examination of the effects of overexpression or deletion of specific genes or proteins on the toxic end point or mechanism of compound action being studied.

- Species comparisons can be used to identify regions of functional constraint or positive selection.
- Algorithms for motif detection can be used to predict regulatory elements (and possibly transcription factors).
- cDNA microarrays can be used to assess gene expression.
- Antibody arrays can be used to assess protein levels.
- Protein arrays can be used to assess enzyme function.

An area of study which is of great interest to toxicologists is the mechanistic understanding of toxicant-induced pathological end points. The premise that perturbations in gene, protein, or metabolite levels are reflective of adverse phenotypic effects of toxicants offers an opportunity to phenotypically anchor these perturbations. This is quite challenging because phenotypic effects often vary in the time–dose space of the studied agent and may have regional variations in the tissue. Furthermore, very few compounds exist that result in only one phenotypic alteration at a given coordinate in dose and time. Thus, objective assignment of measured variables to multiple phenotypic events is not possible under these circumstances. However, by studying multiple structurally and pharmacologically unrelated agents that share pathological end points of interest, one could tease out gene, protein, or metabolite modulations that are in common between the studied compounds. Laser capture microdissection may also be used to capture regional variations such as zonal patterns of hepatotoxicity. This concept will allow the objective assignment of measurable variables to phenotypic observations that will supplement traditional pathology.

6.3 Toxicogenomic Technologies

6.3.1 Genomic Technologies

Transcriptomics, proteomics, and metabolomics are genomic technologies with great potential in toxicological sciences. Toxicogenomics involves the integration of conventional toxicological examinations with gene, protein, or metabolite expression profiles. An overview, together with selected examples of the possibilities of genomics in toxicology, is given. The expectations raised with the use of toxicogenomics are earlier and more sensitive detection of toxicity. Furthermore, toxicogenomics will provide a better understanding of the mechanism of toxicity and may facilitate the prediction of toxicity of unknown compounds. Mechanism-based markers of toxicity can be discovered and improved interspecies and *in vitro–in vivo* extrapolations will drive model developments in toxicology. Toxicological assessment of chemical mixtures will benefit from new molecular biological tools.

The dramatic evolution of rapid DNA sequencing technologies during the 1980s and 1990s culminated in sequences of the human genome and the genomes of dozens of other organisms, including those used as animal models for chemical toxicity. Complete genome sequences provided catalogues of genes, their locations, and their chromosomal contexts. They serve as general reference maps but do not yet reflect individual variations, including single nucleotide polymorphisms (SNPs) and groups of SNP alleles (haplotypes) that account for the individual genetic variation underlying different responses to toxic chemical and environmental factors.

6.3.1.1 *Microarrays*

Microarray technology provides a format for the simultaneous measurement of the expression of thousands of genes in a single experimental assay and has quickly become one of most the powerful and versatile tools for genomics and biomedical research. The methods of toxicogenomics might be classified into omics studies (e.g., genomics, proteomics, and metabolomics) and population study focusing on risk assessment and gene–environment interaction. In omics studies, microarray analysis is the most popular approach. According to a prior hypothesis, up to 20,000 genes falling into several categories (e.g., xenobiotics metabolism, cell cycle control, DNA repair, etc.) can be selected for analysis. The appropriate type of samples and species should be selected in advance. Multiple doses and varied exposure durations are suggested to identify those genes clearly linked to toxic response. Microarray experiments can be affected by numerous nuisance variables including experimental designs, sample extraction, type of scanners, etc.

DNA chips, or microarrays, allow quantitative comparisons of the expression levels of potentially thousands of individual genes between different biological samples. This facilitates, for instance, comparisons of normal tissue with diseased tissue, and of control cell lines with toxicant-treated cell lines. Construction of arrays involves the immobilization of DNA sequences (either cDNA sequences or oligonucleotides), corresponding to the coding sequence of genes of interest, on a solid support such as a glass slide or nylon membrane. The mRNA prepared from cells or tissues can be labeled and hybridized, usually in the form of a reverse-transcribed cDNA copy, to a microarray and visualized using phosphor imager scanning or other appropriate methodologies. Subsequent analysis using appropriate software allows the determination of the extent of hybridization of the labeled probes to the corresponding arrayed cDNA spots, and a comparison of control samples with test samples permits quantitative assessment of changes in gene expression associated with treatment [4].

Examples of the assessment of immunotoxicity by gene expression prowling, presented and discussed here, show that microarray analysis is able to detect known and novel effects of a wide range of immunomodulating agents. Besides the elucidation of mechanisms of action, toxicogenomics is also applied to predict consequences of exposing biological systems to toxic agents. Successful attempts to classify compounds using signature gene expression prowling have been reported. It has been suggested that gene expression profiling using DNA microarrays may be used to characterize the potential mechanisms of action of environmental contaminants through the identification of gene expression networks:

- Identify modes of action for previously uncharacterized toxicants based on correlations with the molecular signatures of toxicants with well-characterized etiologies
- Assess toxicant-induced gene expression as a biomarker of chemical exposure
- As certain overlapping patterns of gene expression in animal species exposed to specific toxicants
- Use microarray information to extrapolate the effects of toxicants from one animal species to another
- Characterize the biological effects of complex chemical mixtures
- Examine the effects of chronic versus acute exposure to chemicals
- Characterize genetic polymorphisms in populations and assess the importance of modifying individual susceptibilities to a contaminant or toxicant

The application of microarray analysis toward toxicology is rapidly being embraced by the pharmaceutical industry as a useful tool to identify safer drugs in a faster, more cost-effective manner. Almost all major pharmaceutical companies have devoted significant resources to develop and apply microarray analysis toward toxicology. Microarray technology has been employed in immunology to study molecular functions associated with immune-related genes to better understand immune function and regulation, also referred to as "immunomics." Gene transcripts that are specifically expressed in immune cell subsets and participate in their maintenance and function have been identified.

6.3.1.1.1 cDNA Microarray

Complementary DNA sequences that are involved in converting mRNA into DNA can be ligated to a cloning vector. Once the cDNA strand has been synthesized, the RNA member of hybrid molecules can be partially degraded by treating with RNaseH. The remaining RNA fragments then serve as primers for DNA polymerase II, which synthesizes the second cDNA strand; resulting in a double-stranded DNA fragment that can be ligated into a vector and cloned. In a cDNA microarray, each gene of interest is represented by a long DNA fragment (200–2400 bp) typically generated by polymerase chain reaction (PCR) and spotted on glass slides using robotics (i.e., pin or inkjet method). In Affymetrix arrays, the probes are short oligonucleotides (15–25 bp) synthesized directly onto a solid support using photolabile nucleotide chemistry. Spotted oligonucleotide arrays were recently developed using synthetic oligonucleotides (30–100 bp).

In the past several years, numerous systems were developed for the construction of large-scale DNA arrays. All of these platforms are based on cDNAs or oligonucleotides immobilized to a solid support. In the cDNA approach, cDNA (or genomic) clones of interest are arrayed in a multi-well format and amplified by PCR. The products of this amplification, which are usually 500 to 2000 bp clones from three regions of the genes of interest, are then spotted onto solid support by using high-speed robotics. By using this method, microarrays of up to 10,000 clones can be generated by spotting onto a glass substrate. Sample detection for microarrays on glass involves the use of probes labeled with fluorescent or radioactive nucleotides.

Fluorescent cDNA probes are generated from control and test RNA samples in single-round reverse-transcription reactions in the presence of fluorescently tagged dUTP (e.g., Cy3-dUTP and Cy5-dUTP), which produce control and test products labeled with different fluors. The cDNAs generated from these two populations, collectively termed the "probe," are then mixed and hybridized to the array under a glass coverslip. The fluorescent signal is detected by using a custom-designed scanning confocal microscope equipped with a motorized stage and lasers for fluor excitation. The data are analyzed with custom digital image analysis software that determines the ratio of fluor 1 to fluor 2, corrected for local background, for each DNA feature. The strength of this approach lies in the ability to label RNAs from control and treated samples with different fluorescent nucleotides, allowing for the simultaneous hybridization and detection of both populations on one microarray. This method eliminates the need to control for hybridization between arrays. The research groups of Dr. Patrick Brown and Dr. Ron Davis at Stanford University spearheaded the effort to develop this approach, which has been successfully applied to studies of *Arabidopsis thaliana* RNA, yeast genomic DNA, tumorigenic versus nontumorigenic human tumor cell lines, human T-cells, yeast RNA, and human inflammatory disease–related genes. The most dramatic result of this effort was the first report published of the yeast *Saccharomyces cerevisiae*.

Advantages

- Proper for comparison of global gene expression between different environmental exposures
- The content of each microarray is determined by the researcher
- The cost per array is relatively low

Disadvantages

- Variable amount of DNA in each spot
- Specificity of the hybridization to the relatively large cDNA inserts.

6.3.1.1.2 ToxBlot II Microarray

To our knowledge, ToxBlot II is the largest custom cDNA microarray available for toxicity characterization. ToxBlot II contains immobilized probe sequences for almost 13,000 human genes representing entire gene families and pathways of relevance to a broad spectrum of toxic end points with particular emphasis on the research and investigative interests in our laboratory. In the design and construction of the ToxBlot II array, human clones from public domain or proprietary sources were classified by function and toxicological interest using a combination of in-house approaches. In addition to the use of high-density "discovery" arrays from commercial vendors, our laboratories have a program of in-house custom toxicology array design and manufacture: ToxBlot. A custom approach allows the selection of the genes to be represented on the array, permitting arrays to be focused on areas of particular interest to mechanistic or investigative toxicology research programs.

Although the major costs of custom array manufacture are capital investment in appropriate technologies, once cDNA sets have been amplified, purified, and verified, the unit array cost is relatively modest, being in the order of $20 per array. For the construction of the ToxBlot arrays, approximately 2000 sequences of human or marine origin were identified as being of potential relevance to various forms of toxicity or normal cell regulatory or signaling pathways. In-house cDNA clone sets, of both public domain and proprietary origin, were used as the source material for array construction. After PCR amplification, purification, and (where appropriate) sequence verification, the representative cDNA sequences were immobilized on nylon membranes. Both human and mouse ToxBlot arrays were constructed, in each case, comprising approximately 2400 cDNA sequences, spanning approximately 600 genes of the relevant species. Reproducibility is ensured by each gene being represented by four individual spots on each array and the inclusion of two nonoverlapping cDNAs for each gene (in duplicate) wherever possible. To aid in the interpretation of differential gene expression results, we have assembled a database on gene function, distribution, and known allelic variations for each represented gene.

6.3.1.1.3 Oligonucleotide Microarray

Oligonucleotide microarrays are constructed either by spotting prefabricated oligos on a glass support or by the more elegant method of direct *in situ* oligo synthesis on the glass surface by photolithography. The strength of this approach lies in its ability to discriminate DNA molecules based on single base pair difference. Affymetrix GeneChip/ Oligonucleotide microarrays are constructed either by spotting prefabricated oligos on a glass support or by the more elegant method of direct *in situ* oligo synthesis on the glass surface by photolithography. The strength of this approach lies in its ability to discriminate DNA molecules based on single base pair difference. This allows the application of this method to the fields of medical diagnostics, pharmacogenetics, and sequencing by hybridization as well as gene expression analysis.

The application of this method is to the fields of medical diagnostics, pharmacogenetics, and sequencing by hybridization as well as gene expression analysis. Fabrication of oligonucleotide chips by photolithography is theoretically simple but technically complex. The light from a high-intensity mercury lamp is directed through a photolithographic mask onto the silica surface, resulting in deportation of the terminal nucleotides in the illuminated regions. The entire chip is then reacted with the desired free nucleotide, resulting in selected chain elongation. This process requires only $4n$ cycles (where n = oligonucleotide length in bases) to synthesize a vast number of unique oligos, the total number of which is limited only by the complexity of the photolithographic mask and the chip size. Sample preparation involves the generation of double-stranded cDNA from cellular poly(A) + RNA followed by antisense RNA synthesis in an *in vitro* transcription reaction with biotinylated or fluor-tagged nucleotides. The RNA probe is then fragmented to facilitate hybridization. If the indirect visualization method is used, the chips are incubated with fluor-linked streptavidin (e.g., phycoerythrin) after hybridization. The signal is detected with a custom confocal scanner. This method has been applied successfully to the mapping of genomic library clones, to *de novo* sequencing by hybridization, and to evolutionary sequence comparison of the *BRCA1* gene. In addition, mutations in the cystic fibrosis and BRCA1 gene products and polymorphisms in the human immunodeficiency virus-1 clade B protease gene have been detected using this method. Oligonucleotide chips are also useful for expression monitoring, as demonstrated by the simultaneous evaluation of gene expression patterns in nearly all open reading frames of the yeast strain. More recently, oligonucleotide chips have been used to help identify single nucleotide polymorph *S. cerevisiae* organisms in the human and yeast genomes.

Advantages

- Synthesized probe is typically of known concentration, and of known sequence. Most of the process can be automated, leading to fewer sample mix-ups and less drop-out of samples.
- Proper for experiments related with SNPs.

Disadvantages

- Relatively high cost of synthesizing large numbers of large oligonucleotides.
- Selection of small sequences from the whole gene is problematic.
- Nonrenewable nature of the resource.

6.3.1.1.4 Yeast DNA Microarray

Yeast DNA microarrays have been used to monitor gene expression levels as a function of toxin exposure and as a means of determining the mechanisms of toxicity. To analyze toxicity, the concentration of the toxicant has to be determined. Concentrations that are often used are the ones that lead to strong growth inhibition but not cellular death. Lethal concentrations may result in cell death without any response, whereas low concentrations may result in only slight growth inhibition, which would cause little response. The incubation period may vary according to the goals of the study and the transcriptional response may differ accordingly. To analyze early responses, the incubation time may be short, whereas the detoxifying response might not be measured if the incubation time is too long.

Gene expression is a unique way of characterizing how cells and organisms adapt to changes in the external environment. The measurements of gene expression levels upon exposure to a chemical can be used to both provide information about the mechanism of action of the toxicant and to form a sort of "genetic signature" for the identification of toxic products. The development of high-quality, commercially available gene arrays has allowed this technology to become a standard tool in molecular toxicology. The field of toxicogenomics has progressed rapidly since the application of DNA chips to toxicology was proposed in the late 1990s. The comparison of the gene expression profile with a database of profiles of known hepatotoxins indicated that hepatic toxicity of the new chemical is mediated by the aryl hydrocarbon nuclear receptor.

6.3.1.2 SNPs

The ultimate approach for SNP discovery is individual whole-genome sequencing. Although not yet feasible, rapidly advancing technology makes this approach the logical objective for SNP discovery in the future. In the meantime, rapid advances in microarray-based SNP analysis technology have redefined the scope of SNP discovery, mapping, and genotyping. Currently, polymorphism studies are used to assess individuals that have "susceptible" alleles for the gene implicated in the disease. Genetic changes in normal individuals lead to considerable structural, functional, and behavioral deviations. There are analyses that put forward a general approach to breaking down genetic networks systematically across a biological scale from base pairs to behavior. Genetic susceptibility and toxicity is now well-established by laboratory research in toxicology and pharmacogenetics that polymorphisms (e.g., single base changes, insertions, or deletions) exist in most of the genes involved in biological processes such as the uptake, metabolism, and excretion of drugs and environmental xenobiotics, DNA repair, cell cycle control, and membrane signaling [5]. The information obtained from these polymorphic studies could be used in target validation. If a target is determined to be highly polymorphic, it could be abandoned.

Asthma and other respiratory diseases are very complex diseases influenced by genetic susceptibility, chance, and environmental influences. Genetics, particularly the study of differences within populations using SNP analysis, has the potential to highlight a number of biological mechanisms implicated in disease status and could lead to the discovery of numerous targets for therapeutic intervention applicable to different forms of the disease. New microarray-based genotyping technology enables "whole-genome association" analyses of SNPs between individuals or between strains of laboratory animals. In contrast to the candidate gene approach described above, whole-genome association identifies hundreds to thousands of SNPs in multiple genes. Arrays used for these analyses can represent a million or more polymorphic sequences mapped across a genome [6]. This approach makes it possible to identify SNPs associated with disease and susceptibility to toxic insult. As more SNPs are identified and incorporate into microarrays, the approach samples a greater fraction of the genome and becomes increasingly powerful. The strength of this technology is that it puts a massive amount of easily measurable genetic variation in the hands of researchers in a cost-effective manner ($500–$1000 per chip). Criteria for including the selected SNPs on the arrays are a critical consideration because these affect the inferences that can be drawn from the use of these platforms.

6.3.1.3 Genome Sequencing

DNA sequencing, the process of determining the exact order of the 3 billion chemical building blocks (called bases and abbreviated A, T, C, and G) that make up the DNA of the 24 different human chromosomes, was the greatest technical challenge in the Human Genome Project. Achieving this goal has helped reveal the estimated 20,000 to 25,000 human genes within our DNA as well as the regions controlling them. The resulting DNA sequence maps are being used by twenty-first century scientists to explore human biology and other complex phenomena. Meeting the Human Genome Project sequencing goals requires continual improvements in sequencing speed, reliability, and costs. Previously, standard methods were based on separating DNA fragments by gel electrophoresis, which was extremely labor-intensive and expensive. Total sequencing output in the community was approximately 200 million base pairs for 1998. In January 2003, the DOE Joint Genome Institute alone sequenced 1.5 billion bases for the month.

Gel-based sequencers use multiple tiny (capillary) tubes to run standard electrophoretic separations. These separations are much faster because the tubes dissipate heat well and allow the use of much higher electric fields to complete sequencing in shorter times. Beckman Coulter Genomics has generated whole-genome sequence for a diverse variety of organisms with an equally wide range of associated genome sizes. Beckman Coulter Genomics' years of extensive experience with Sanger genome sequencing, high-throughput genomics, and large-scale sequencing projects enabled a strong transition of many whole-genome sequencing projects to next-generation technologies. Factors considered for the best experimental design for a whole-genome sequencing project include the size of the target genome, the anticipated complexity of the genome, the desired level of finishing, or completeness, and whether the genome requires *de novo* assembly or can be mapped.

The application of these technologies to toxicology has ushered in an era in which genotypes and toxicant-induced genome expression, protein, and metabolite patterns can be used to screen compounds for hazard identification, monitor individuals' exposure to toxicants, track cellular responses to different doses, assess mechanisms of action, and predict individual variability in sensitivity to toxicants. It may also yield a number of substantial dividends, including assisting predevelopment toxicology by facilitating more rapid screens for compound toxicity, allowing compound selection decisions to be based on safety as well as efficacy, the provision of new research leads, a more detailed appreciation of molecular mechanisms of toxicity, and an enhanced ability to extrapolate accurately between experimental animals and humans in the context of risk assessment.

6.3.1.4 Southern Hybridization

Southern hybridization is a powerful procedure for studying gene structure and function. This method involves the separation of DNA according to size on an agarose or polyacrylamide gel and subsequent transfer of the DNA to suitable nitrocellulose or nylon-based membranes. DNA molecules bound to the membrane are then fixed and hybridized with a radioactive probe, specific for the gene under study. The fixing step is particularly important for quantitatively reproducible results, especially if membranes are to be rehybridized several times. The traditional fixing procedure involved baking the membrane for 2 h at 80°C.

1. More recently, it has been shown that efficient cross-linking of DNA to the membrane could be obtained by 30 s to 3 min exposure of UV irradiation.
2. Earlier studies described the use of a microwave oven to fix bacterial colonies to nylon membranes.
3. Fix the DNA within paraffin tissue blocks.
4. In Southern blots, treatment for 2 min in a microwave oven efficiently fixed DNA to the nylon membrane, which can withstand repeated reuse in hybridization reactions.

Toxicogenomics uses these new technologies to analyze genes, genetic polymorphisms, mRNA transcripts, proteins, and metabolites. The foundation of this field is the rapid sequencing of the human genome and the genomes of dozens of other organisms, including animals used as models in toxicology studies. Furthermore, toxicogenomics will provide a better understanding of the mechanism of toxicity and may facilitate the prediction of toxicity of unknown compounds. Mechanism-based markers of toxicity can be discovered and improved interspecies and *in vitro–in vivo* extrapolations will drive model developments in toxicology. Toxicological assessment of chemical mixtures will benefit from the new molecular biological tools. In our laboratory, toxicogenomics is predominantly applied for the elucidation of the mechanisms of action and discovery of novel pathway-supported, mechanism-based markers of liver toxicity.

6.3.1.5 Polymerase Chain Reaction

The PCR is a scientific technique in molecular biology to amplify a single or few copies of a piece of DNA across several orders of magnitude, generating thousands to millions of copies of a particular DNA sequence. The method relies on thermal cycling, consisting of cycles of repeated heating and cooling of the reaction for DNA melting and enzymatic replication of the DNA. Primers (short DNA fragments) containing sequences complementary to the target region, along with a DNA polymerase (after which the method is named), are key components to enable selective and repeated amplification. As PCR progresses, the DNA generated is itself used as a template for replication, setting in motion a chain reaction in which the DNA template is exponentially amplified. PCR can be extensively modified to perform a wide array of genetic manipulations. In the short time since its invention by Kary Mullis, PCR has revolutionized the approach to molecular biology. The effect of PCR on biological and medical research has been like a supercharger in an engine, dramatically speeding the rate of progress of the study of genes and genomes. We can isolate essentially any gene from any organism using PCR, and has become cornerstone of genome sequencing projects. It can establish diagnostic tests to detect mutant forms of a gene.

Compounds which induce toxicity through similar mechanisms lead to characteristic gene expression patterns. The concept that structurally similar compounds may have similar biological profiles, the so-called generalized neighborhood behavior, is harder to demonstrate. Approximately 625 compounds were screened from a fully combinatorial library for their gene expression profiles *in vitro* over a selected toxicity panel of 56 genes. Out of 12 different scaffolds, four families showed noncorrelating, uniform distribution among clusters, whereas eight families showed neighborhood behavior of varying strengths. Structurally dissimilar compounds may have highly similar biological activity, on the other hand, compounds of the same scaffold family do not all share the same biological effects based on toxicology-related gene expression fingerprints.

6.3.1.5.1 The New Frontier in Toxicology and Environmental Health Science

Microarray analysis only recently became a widely used biological tool in the late 1990s. It uses silicon or glass chips, capable of being coated with dense, discrete spots of nucleic acids, which are termed "probes." A "target," fluorescent cDNA, RNA, or cRNA, can then be bound to the array of probes to determine the gene expression pattern—which genes' RNA products were indeed present in the target sample. The target binds to the probe as a result of nucleic acid base pairing, also known as hybridization. This seemingly simple procedure allows scientists to screen tens of thousands of genes for an expression or particular sensitivity after cells have been subjected to a particular environmental stimulus. Toxicogenomics has taken advantage of this very principle. In the field of toxicogenomics, microarrays are used to make research easier, systematic, more efficient, and faster. Toxicologists have the responsibility of establishing or disproving whether exposure to molecules, including industrial and consumer by-products, has a cause–effect relationship with human disease. Consider the example of the connection between cigarette smoke and lung cancer. Toxicologists have established both physiological and statistical data, piled a mile high, proving the relationship between cigarette smoke and the development of lung cancer. Presently, given the analytical power of microarrays, toxicologists can do the following: determine whether there are discrete gene changes linking cigarette smoke exposure with lung dysfunction and disease; assess whether an uncharacterized agent in cigarettes can be harmful by matching data with that of known toxicants; and assess drug sensitivities and effectiveness by identifying lung disease biomarkers, commonly occurring genetic traits in instances of bodily dysfunction, specifically tissue or organ dysfunction.

6.3.1.5.2 Recent Applications of DNA Microarray Technology to Toxicology and Ecotoxicology

Several national and international initiatives have provided the proof-of-principle tests for the application of gene expression for the study of the toxicity of new and existing chemical compounds. In the last few years, the field has progressed from evaluating the potential of the technology to illustrating the practical use of gene expression profiling in toxicology. The application of gene expression profiling to ecotoxicology is at an early stage, mainly because of the many variables involved in analyzing the status of natural populations. Nevertheless, significant studies have been carried out on the response to environmental stressors both in model and in nonmodel organisms. It can be easily predicted that the development of stressor-specific signatures in gene expression profiling will soon have a major effect on ecotoxicology. International collaborations could play an important role in accelerating the application of genomic approaches in ecotoxicology.

6.3.2 Transcriptomic Technologies

6.3.2.1 Northern Blot

Northern blotting is an important procedure for the study of gene expression because the RNA present in a given cell or tissue type roughly represents the portion of the genome that is expressed in that cell or tissue. The analysis of RNA is central to a wide range of molecular–biology studies because it is often important to obtain information about the expression of genes with organisms. Filter hybridization of size-separated RNAs with a labeled nucleic acid probe, a technique known as Northern blot analysis, is designed to address the study of RNA sequences.

1. The procedure derived from Southern blot analysis is based on the ability of complementary single-stranded nucleic acids to form hybrid molecules.
2. In brief, RNA is separated according to size on an agarose gel under denaturing conditions, transferred (blotted), and irreversibly bound to a nylon or nitrocellulose membrane, and hybridized with a 32P-labeled nucleic acid probe. After washing off the unbound and nonspecifically bound probe, hybridizing RNA molecules are revealed as autoradiographic bands on an x-ray film. It detects the steady-state level of accumulation of a given RNA (usually mRNA) sequence in the sample, and allows one to relate mRNA levels to the physiological and morphological properties of living organisms. Several factors control mRNA abundance.

a. Gene transcription

b. mRNA processing and transport, and

c. mRNA stability

Of these, gene transcription is the most important determinant of mRNA expression. In addition to the abovementioned factors, translational and posttranslational controls play additional roles in the overall regulation of gene expression, which ultimately results in a specific level of protein being synthesized. Thus, caution should be used in interpreting Northern hybridization data because several factors, in addition to mRNA levels, concur with phenotypic expression. Although techniques that study the localization of mRNAs at the single-cell level are more sensitive, methods for the analysis of rare transcripts, that is, reverse transcriptase–mediated PCR (RT-PCR) and RNase protection assay, have recently become common. Northern hybridization remains a widely used procedure for the analysis of gene expression. For example, this technique has been extensively used for the characterization of cloned cDNAs and genes, the study of tissue/organ specificity of gene expression, the analysis of the activity of endogenous and exogenous genes in transgenics and the study of gene expression in relation to development, biotic, and abiotic factors.

Applications

1. Northern blot analysis is used as a widespread technique in molecular biology studies. Valuable information on the size and occurrence/abundance of transcripts in samples of various origins can be obtained using this method. Such information may provide important clues on gene structure and regulation.

2. It is also a valuable tool for the characterization of large numbers of DNA sequences isolated in genome random-sequencing projects.

3. The presence of an RNA band on a Northern blot when used as a probe is a good indication of the ability to encode proteomics.

4. Many biological problems have benefited from the analysis of RNA by Northern blotting, including gene characterization and regulation, analysis of development, and responsiveness to internal and external stimuli.

6.3.2.2 Serial Analysis of Gene Expression

The technology for assaying gene, protein, and metabolic expression profiles are not new inventions. Measurements of gene expression have evolved from the single measures of steady-state mRNA using Northern blot analysis to the more global analysis of thousands of genes using DNA microarrays and serial analysis of gene expression (SAGE)—the two dominant technologies. The advantage of global approaches is the ability of a single investigation to query the behavior of hundreds, thousands, or tens of thousands of biological molecules in a single assay. SAGE allows the entire collection of transcripts to be catalogued without assumptions about what is actually expressed (unlike microarrays, in which one needs to select probes from a catalogue of genes). SAGE is a technology based on sequencing strings of short expressed sequence tags (EST) representing both the identity and the frequency of occurrence of specific sequences within the transcriptome. It is costly and gives a relatively low throughput because each sample to be analyzed requires a SAGE tag library to be constructed and sequenced. Massively parallel signature sequencing speeds up the SAGE process with a bead-based approach that simultaneously sequences multiple tags, but it is costly.

6.3.2.3 Expressed Sequence Tags

ESTs are small pieces of DNA sequence (usually 200–500 nucleotides long) that are generated by sequencing either one or both ends of an expressed gene. It is a partial gene sequence and, when used in map construction, they provide a quick way of locating the genes' position, even though the identity of the gene might not be apparent from the EST sequence. The idea is to sequence bits of DNA that represent genes expressed in certain cells, tissues, or organs from different organisms and use these "tags" to fish a gene out of a portion of chromosomal DNA by matching base pairs. The challenge associated with identifying genes from genomic sequences varies among organisms and is dependent on genome size as well as the presence or absence of introns, the intervening DNA sequences interrupting the protein coding sequence of a gene.

Sequencing only the beginning portion of the cDNA produces what is called a 5′ EST. A 5′ EST is obtained from the portion of a transcript that usually codes for a protein. These regions tend to be conserved across species and do not change much within a gene family. Sequencing the ending portion of the cDNA molecule produces what is called a 3′ EST. Because these ESTs are generated from the 3′ end of a transcript, they are likely to fall within noncoding or untranslated regions, and therefore tend to exhibit less cross-species conservation than decoding sequences. Partial cDNA sequencing to generate ESTs is currently being used to quickly and efficiently obtain detailed profiles of genes expressed in various tissues, cell types, or developmental stages.

Genome projects have taken advantage of EST studies because ESTs represent a particular type of sequence-tagged sites that are useful for the physical mapping of genomes. Analysis of ESTs constitutes a useful approach for gene identification that, in the case of human pathogens, might result in the identification of new targets for chemotherapy and vaccine development. ESTs can serve the same purpose as sequence-tagged sites, with the additional bonus of pointing directly to express genes. One of the most interesting applications of the EST database is gene discovery. A significant development with important implications in this field has been the enormous growth of the EST database. Novel genes can be found by querying the EST database with a protein or DNA sequence.

6.3.2.4 RT-PCR

RT-PCR remains the gold standard in the accurate determination of gene expression changes [7] applied to studying continuous neomycin and gentamicin treatment in monkeys. Use of novel nanocapillary, quantitative real-time PCR Open Array technology combines outstanding analytical performance with the medium throughput ideal for such a sample-per-feature ratio. Applying hybrid clustering on the gene expression data, the correlation between molecular scaffold and biological fingerprint was analyzed. Structurally highly dissimilar, hepatotoxicity compounds showed similar fingerprints on our toxicity panel; however, compounds of the same scaffold and of unknown biological effect do not always share similar fingerprints.

6.3.2.4.1 Toxicogenomic Screening of Small Molecules Using High-Density, Nanocapillary RT-PCR

Preliminary tests were performed with the in-house Toxico-Screen DNA microarrays and with the traditional QRT-PCR technique, after which we shifted to the Open Array nanocapillary QRT-PCR technology that has recently appeared on the market (BioTrove Inc., Boston, MA). This later technology merges the high-throughput of DNA microarrays with the sound characteristics of QRT-PCR; therefore, it is ideal for the toxicogenomic screening of chemical libraries. After determining the gene expression pattern of each compound, we analyzed the results using hierarchical and K-means clustering methods, looking for a correlation between chemical scaffold and biological effect. The system combines high accuracy, precision, and dynamic range characteristics of QRT phase reactions, which are run in parallel in 33 nL through-holes—the size of a microscope slide in the software-controlled, completely standardized environment of a thermal cycler. The custom-selected primer pairs are immobilized in the Open Array plate, generating a custom-based screening platform for a subset of genes. These technologies are ideal for toxicogenomics screening. In comparison to microarrays, the higher analytical performance is due to the more standardized and automated loading and incubation of samples.

6.3.2.5 Micro-RNAs

Micro-RNAs (miRNAs) are noncoding regulatory RNA molecules that bind target mRNAs and suppress their translation into proteins. When an organism is exposed to a toxic compound, cells respond by altering the pattern of gene expression, including miRNAs. Altered miRNA expression affects protein translation, which in turn alters cellular physiology, causing adverse biological effects. Moreover, different types of cellular stress have been shown to affect miRNA expression as a mechanism of adaptation or tolerance to stress factors to survive. Most mammalian mRNAs are targets of miRNAs [8]. First described as having a role in developmental processes in worms and flies, miRNAs have been implicated in various aspects of vertebrates' development, cell differentiation, and disease. Moreover, due to their multiple roles as posttranscriptional regulators of gene expression, miRNAs have been shown to play a role in virtually every cellular process; and like other key regulators, aberrant expression of miRNAs is associated with cellular dysfunction and disease. Most attention has been focused on the implications of miRNAs in cancer. Accumulating evidence suggests that numerous miRNAs are aberrantly expressed in human cancers [9] and contribute to confer the hallmark traits of the cancer cell: self-sufficiency in growth signals is sensitivity to antigrowth signals and self-renewal, evasion of apoptosis, limitless replication potential, angiogenesis, invasion, and metastasis. In addition to genetic factors, epigenetic processes have a function that is equally important in neoplasia. Epigenetics describes the mechanisms that results in inheritable alterations in gene expression profiles without an accompanying change in DNA sequence. miRNA expression appears to be under epigenetic regulation. Conversely, miRNAs themselves seem to mediate epigenetic modifications.

The roles of RNA can be identified in nearly all aspects of biological processes, their roles as key mediators of cellular response to extracellular signals, as well as being regulators for appropriate control of tissue homeostasis, might be more relevant than previously suspected. Increasing evidence that the expression of miRNAs is affected by several known toxicants as well as oxidative and other forms of cellular stress certainly suggests the important role of miRNAs in toxicology, which could provide a link between environmental influences and gene expression. Cellular gene expression is altered in response to an organism being exposed to an exogenous agent, such as environmental stress factors, toxins, drugs, and chemicals. Toxicogenomics studies the relationship between toxicant exposure and alterations in genomewide gene expression patterns. Analysis of the resulting molecular signatures provide new tools to identify mechanisms of toxicity, as well as to classify compounds based on the biological response they elicit and to identify clusters of genes detective or predictive of certain types of toxic response, which are employed as biomarkers. However, to have a full understanding of environmental interactions with the genome, epigenetic factors also need to be considered. Noncoding RNAs (including miRNAs), DNA methylation, and histone modifications are the main mechanisms of epigenetic regulation of gene expression. Representative examples of miRNAs involved in response to toxicants, their respective target genes, and the biological effects elicited are described. In contrast with stress-related miRNAs, the toxicological research of miRNAs associated with specific toxicants started a few years ago.

miRNAs have generated a great deal of interest in toxicology because of the numerous envisioned applications that stem from their key role as regulators of gene expression. However, miRNA regulatory networks are complex because a single miRNA can target numerous mRNAs, often in combination with other miRNAs. Therefore, understanding miRNAs' roles in the toxicological processes requires an overall toxicogenomic approach. Chemogenomics allows advanced *in silico* screening of chemical compounds and protein target identification by the use of virtual screening databases of small molecules. It can be expected that this approach will be a productive means to generate novel leads for future drug discovery programs. Because miRNAs regulate protein expression, these RNA profiles could become useful tools to predict effective drug–target interaction. Allowing the expression profile of miRNAs involved in the regulation of the drug's protein target will favor a high expression of the protein target and subsequently an effective drug–target interaction. Conversely, a high miRNA expression profile will be associated with fewer probabilities for the drug to interact effectively with its target. Likewise, if miRNA profiling works for predicting effective drug–target interaction, it may also work to predict toxicant–target that interact and define individual susceptibility to toxicants. On the other hand, miRNA profiling data looks promising as a tool to predict the potential toxicity of unknown compounds.

Treatments explored the interaction of single-walled carbon nanotubes with human cells by profiling miRNA expression using a miRNA microarray platform. Normal human bronchial epithelial cells cultures were treated *in vitro* with single-walled carbon nanotubes and other low-micron and nanoscale materials. miRNA expression profiles were obtained at 0 and 24 h after treatment and compared with those obtained with silica and carbonyl iron as positive and negative controls, respectively. The data indicated that silica caused the highest perturbation, whereas single-walled carbon nanotubes and carbonyl iron generated similar profiles in miRNA expression when compared with untreated controls. Upregulated and downregulated miRNAs and their target mRNAs were identified. miRNAs signatures generated by known toxicants (i.e., silica) may be useful in predicting toxicity by their comparison to profiles of unknown compounds. Thus, miRNA signatures of a known toxic compound may include miRNAs related to cellular response to stress, xenobiotic metabolism, and DNA repair. These signatures, derived from supervised classification algorithms, may effectively identify potential toxic compounds (the interpretation for all these techniques is given in Table 6.1).

TABLE 6.1

Gene Expression Analysis Methods

S. No.	Methods	Comments
1	Northern blot	Standard practice; but low throughput
2	Subtractive cloning	Not always complete
3	Differential display	Full-length cloning required; potential to identify rare mRNAs
4	EST/SAGE	Costly and requires a dedicated sequencing facility
5	Gridded filters	Cannot multiplex probes derived from two different tissue samples
6	High-density arrays	Identification of differentially expressed genes dependent on arrayed elements

6.3.2.5.1 *miRNAs and Their Implications in Toxicological Research*

Their roles as key mediators of cellular response to extracellular signals as well as being regulators for appropriate control of tissue homeostasis might be more relevant than previously suspected. Increasing evidence that the expression of miRNAs is affected by several known toxicants as well as oxidative and other forms of cellular stress certainly suggest an important role for miRNAs in toxicology, which could provide a link between environmental influences and gene expression. Cellular gene expression is altered as a response of an organism being exposed to exogenous agents such as environmental stressors, toxins, drugs, and chemicals. Toxicogenomics studies the relationship between toxicant exposure and alterations in genomewide gene expression patterns. Analysis of the resulting molecular signatures provides new tools to identify mechanisms of toxicity, as well as to classify compounds based on the biological response they elicit and to identify clusters of genes detective or predictive of certain types of toxic response, which are employed as biomarkers. However, to have a full understanding of environmental interactions with the genome, epigenetic factors need also to be considered. Noncoding RNAs (including miRNAs), DNA methylation, and histone modifications are the main mechanisms of epigenetic regulation of gene expression. Representative examples of miRNAs involved in response to toxicity.

6.3.3 Proteomic Technologies

Proteomics is the study of proteomes, which are collections of proteins in living systems. Because proteins carry out most functions encoded by genes, analysis of the protein complement of the genome provides insights into biology that cannot be drawn from studies of genes and genomes. MS, gene and protein sequence databases, protein and peptide separation techniques, and novel bioinformatics tools are integrated to provide the technological platform for proteomics [10]. In contrast to "the genome," there is no single, static proteome in any organism; instead, there are dynamic collections of proteins in different cells and tissues that display moment-to-moment variations in response to diet, stress, disease processes, and chemical exposures. There is no technology analogous to PCR amplification of nucleic acids that can amplify proteins, so they must be analyzed at their native abundances, which span more than six orders of magnitude. Each protein may be present in multiple modified forms; indeed, variations in modification status may be more critical to function than absolute levels of the protein. A related problem is the formation of protein adducts by reactive chemical intermediates generated from toxic chemicals and endogenous oxidative stress [11]. Protein damage by reactive chemical intermediates may also perturb endogenous regulatory protein modifications. All these characteristics add to the challenge of proteome analysis. In contrast to the microarray technologies applied to gene expression, most analytical proteomic methods represent elaborate serial analyses rather than truly parallel technologies.

It is likely that not all of the toxicant-induced molecular changes can be identified at the mRNA level. In particular, protective detoxifying mechanisms require fast modification, protein folding, or redistribution of proteins already present in the cell. Protein analysis technologies may have a better chance to visualize processes that do not involve active biosynthesis, or at least they will be complementary to gene expression analysis. For many years, it has been possible to array complex protein mixtures by two-dimensional gel electrophoresis, which combines the separation of proteins by isoelectric focusing in the first dimension followed by sodium dodecyl sulfate–polyacrylamide gel electrophoresis (SDS-PAGE) based on molecular weight in the second dimension. The product is a rectangular pattern of protein spots that are typically revealed by Coomassie blue, silver, or fluorescent staining. However, the amount of each protein separated is at least one order of magnitude below the amount needed for chemical characterization and therefore the development of semipreparative methods for purifying proteins in parallel and more sensitive techniques for protein characterization were needed.

6.3.3.1 SDS-PAGE

In the gel-based approach, proteins are resolved by electrophoresis or another separation method and protein features of interest are selected for analysis. This approach is best represented by the use of two-dimensional SDS-PAGE to separate protein mixtures, followed by the selection of spots and identification of the proteins by digestion to peptides, MS analysis, and database searching. Gel-based analyses generate an observable "map" of the proteome analyzed, although liquid separations and software can be coupled to achieve analogous results. The reproducibility of two-dimensional gel separations has dramatically improved with the introduction of commercially available immobilized pH gradient strips and precast gel systems [12]. Comparative two-dimensional SDS-PAGE with differential fluorescent labeling (e.g., differential gel electrophoresis) offers powerful quantitative comparisons of proteomes [13]. Moreover, modified protein forms are often resolved from unmodified forms, which enable separate characterization and quantitative analysis of each. Although two-dimensional gels have usually been applied to global analyses of complex proteomes, they have great potential for comparative analyses of smaller subproteomes (e.g., multiprotein complexes). Problems with gel-based analyses stem from the poor separation characteristics of proteins with extreme physical characteristics, such as hydrophobic membrane proteins. A major problem is the limited dynamic range for protein detection by staining (200- to 500-fold), whereas protein abundances vary by more than a million-fold [14]. This means that abundant proteins tend to preclude the detection of less abundant proteins in complex mixtures. This problem is not unique to gel-based approaches.

6.3.3.2 Mass Spectrometry

Mass spectrometry (MS) measures the mass-to-charge ratios (m/z) of ions under vacuum, and various combinations of ionization sources and mass analyzers have been constructed. It has become more compatible with the analysis of biopolymers, proteins, nucleic acids, and carbohydrates using ionization methods like electrospray ionization and matrix-enhanced or surface-enhanced laser desorption ionization techniques that allow rapid characterization of proteins or protein fragments. It analyses large [≥10,000 mass units (u)] biologically derived polymers. The components involved include an ionization source, a mass analyzer, and a detector. These proteomic methods allow for the analysis of the functional and structural proteins in a sample. Methods for simultaneously monitoring small molecules involved in intermediary metabolic pathways (metabolomics) are also at hand. The ability to monitor defense responses in humans via proteomics or metabolomics at subpathological doses is of particular importance because it is now possible to conduct human studies that could not be previously carried out at overtly toxic exposures. These technologies provide complementary information to gene expression data. Clearly, posttranslational modifications of proteins, such as phosphorylation, will not be evident as changes in gene expression. Also, nucleic acids may not be available for analysis in all cases (e.g., invasive procedures would be needed to obtain samples from many human tissues), although proteins may be secreted or diffuse into accessible compartments or be more amenable to imaging techniques. The use of all of these tools will be important for obtaining a comprehensive picture of toxicological changes in cellular constituents. They have been converted into a mature, robust, sensitive, and rapid technology that has dramatically advanced the ability to identify proteins and constitute the basis of the emerging field of proteomics.

The integration of MS with protein and peptide separation technologies has enabled the characterization of complex proteomes as well as a mechanistic study of chemically induced protein modifications [10]. Because of the extraordinarily wide range of protein expression levels (>106), no proteomic approach is capable of analyzing all proteins in the cell, tissue, or organism. Global (as opposed to comprehensive) proteome analyses have been applied to derive insight into the mechanisms of toxic action. Two-dimensional gel electrophoresis and MS-based protein identification have revealed a number of proteome changes that are characteristic of chemical toxicities [15,16]. In these cases, a contextual or exploratory approach identified proteins and peptides that are correlated with injury and the adaptive biological responses that followed. Although many of the changes observed may encode mechanistic information about chemically induced injury, distinguishing proteomic changes that represent causes versus effects require more sophisticated experimental approaches.

Shotgun analyses begin with direct digestion of protein mixtures to complex mixtures of peptides, which then are analyzed using liquid chromatography-coupled MS (LC-MS). The resulting collection of peptide tandem mass spectrometry (MS-MS) spectra is searched against databases to identify corresponding peptide sequences and then the collection of sequences is reassembled using computer software to provide an inventory of the proteins in the original sample mixture. A key advantage of shotgun proteomics is its unsurpassed performance in the analysis of complex peptide mixtures [17,18]. Peptide MS-MS spectra are acquired by automated LC-MS-MS analyses in which ions corresponding to intact peptides are automatically selected for fragmentation to produce MS-MS spectra that encode peptide sequences. This approach enables automated analyses of complex mixtures without user intervention. However, selection of peptide ions for MS-MS fragmentation is based on the intensity of the peptide ion signals, which favors acquisition of MS-MS spectra from the most abundant peptides in a mixture. Thus, the detection of low-abundance peptides in complex mixtures is somewhat random. Application of multidimensional chromatographic separations (e.g., ion exchange and then reversed-phase high-performance liquid chromatography) "spreads out" the peptide mixture and greatly increases the number of peptides for which MS spectra are acquired [17]. This increases the detection of less abundant proteins and modified protein forms [19].

Proteomics is the study of proteomes, which are collections of proteins in living systems. Because proteins carry out most functions encoded by genes, analysis of the protein complement of the genome provides insights into biology that cannot be drawn from studies of genes and genomes. MS, gene and protein sequence databases, protein and peptide separation techniques, and novel bioinformatics tools are integrated to provide the technological platform for proteomics [10]. In contrast to "the genome," there is no single, static proteome in any organism; instead, there are dynamic collections of proteins in different cells and tissues that display moment-to-moment variations in response to diet, stress, disease processes, and chemical exposures. There is no technology analogous to PCR amplification of nucleic acids that can amplify proteins, so they must be analyzed at their native abundances, which span more than six orders of magnitude. Each protein may be present in multiple modified forms; indeed, variations in modification status may be more critical to function than absolute levels of the protein. A related problem is the formation of protein adducts by reactive chemical intermediates generated from toxic chemicals and endogenous oxidative stress [11]. Protein damage by reactive chemical intermediates may also perturb endogenous regulatory protein modifications. All these characteristics add to the challenge of proteome analysis. In contrast to the microarray technologies applied to gene expression, most analytical proteomic methods represent elaborate serial analyses rather than truly parallel technologies.

6.3.4 Metabolomic Technologies

Metabolomics is the study of small-molecule components of biological systems, which are the products of metabolic processes. Metabolic intermediates reflect the actions of proteins in biochemical pathways and thus represent biologic states in a way that is analogous to proteomes. As with proteomes, metabolomes are dynamic and change in response to nutrition, stress, disease states, and even diurnal variations in metabolism. Unlike genomes, transcriptomes, and proteomes, metabolomes comprise a chemically diverse collection of compounds that range from small peptides, lipids, and nucleic acid precursors and degradation products to chemical intermediates in biosynthesis and catabolism as well as metabolites of exogenous compounds derived from the diet, environmental exposures, and therapeutic interventions. A consequence of the chemical diversity of metabolome components is the difficulty of comprehensive analysis with any single analytical technology. Because metabolites reflect the activities of RNAs, proteins, and the genes that encode them, metabolomics allows for functional assessment of diseases and drug and chemical toxicity. Metabolomic technologies, employing NMR spectroscopy and MS, are directed at simultaneously measuring dozens to thousands of compounds in biofluids (e.g., urine) or in cell tissue extracts. A key strength of metabolomic approaches is that they can be used to noninvasively and repeatedly measure changes in living tissues and living animals, and that they measure changes in the actual metabolic flow. As with proteomics, the major limitation of metabolomics is the difficulty of comprehensively measuring diverse metabolites in complex biological systems. The effect of cellular processes is ultimately reflected in the metabolite levels, which can be measured in intracellular as well as in extracellular fluids. This may be advantageous to toxicologists because noninvasive samples can be obtained such as blood (plasma), urine, or other body fluids.

6.3.4.1 NMR Spectroscopy

This is an attractive approach because a wide range of metabolites can be quantified simultaneously without extensive sample preparation. An NMR spectrum is generated with resonance peaks characteristic for all small biomolecules. In this way, a metabolic fingerprint is obtained, characterizing the biological fluid under study. In the spectrum, individual signals derived from metabolites can be quantified and identified based on reference spectra whenever available. The NMR spectra of biological fluids are very complex due to the mixture of numerous metabolites present in these fluids. Variations between samples are often too small to be recognized by eye. To find significant differences, multivariate data analysis is needed to explore recurrent patterns in a number of NMR spectra. Correlation between variables in the complex and large data sets (thousands of signals in spectra) is related to a target variable such as toxicity status. The combination of mass spectrometric profiling with multivariate data analysis provides a fingerprinting methodology. For example, a factor spectrum was used to identify the metabolite NMR peaks and resulting fingerprint, which differed in the urine obtained from rats exposed to the hepatotoxicant bromobenzene as compared with controls.

Genomic and proteomic methods do not offer the information needed to gain understanding of the resulting output function in a living system. Neither approach addresses the dynamic metabolic status of the whole animal. The metabolomic approach is based on the premise that toxicant-induced pathological or physiological alterations result in changes in relative concentrations of endogenous biochemical. Metabolites in body fluids such as urine, blood, or cerebrospinal fluid are in dynamic equilibrium with those inside cells and tissues, thus toxicant-induced cellular abnormalities in tissues should be reflected in altered biofluid compositions. An advantage of measuring changes in body fluids is that these samples are much more readily available from human subjects. High-resolution NMR spectroscopy (1H NMR) has been used in a high-throughput fashion to simultaneously detect many cellular biochemicals in urine, bile, blood plasma, milk, saliva, sweat, gastric juice, seminal, amniotic, synovial, and cerebrospinal fluids [20–22]. In addition, intact tissue and cellular suspensions have also been successfully analyzed for metabolite content using magic-angle-spinning 1H NMR spectroscopy.

Although it is possible to establish the identity of many, but not all, of the peaks in NMR spectra of urine and biofluids, the value of the data has been in the analyses of collections of spectral signals. These pattern recognition approaches have been used to identify distinguishing characteristics of samples or sample sets. Unsupervised analyses of the data, such as principal components analysis, have proven useful for grouping samples based on sets of similar features [23]. These similar features frequently reflect chemical similarity in metabolite composition and thus similar courses of response to toxicants. Supervised analyses allow the use of data from biochemically or toxicologically defined samples to establish models capable of classifying samples based on multiple features in the spectra [24].

NMR-based metabolomics of urine measure global metabolic changes that have occurred throughout an organism. However, metabolite profiles in urine can also indicate tissue-specific toxicities. Principal components analyses of urinary NMR data have shown that the development and resolution of chemically induced tissue injury can be followed by plotting trajectories of principal components analysis–derived parameters. Although the patterns themselves provide a basis for analyses, some specific metabolites have also been identified (based on their resonances in the NMR spectra). Mapping these metabolites onto known metabolic pathways makes it possible to draw inferences about the biochemical and cellular consequences and mechanisms of injury [25]. An interesting and important consequence was the identification of endogenous bacterial metabolites as key elements of diagnostic metabolomic profiles [26]. Although the interplay of gut bacteria with drug and chemical metabolism had been known previously, recent NMR metabolomic studies indicate that interactions between host tissues and gut microbes have a much more pronounced effect on susceptibility to injury than had been appreciated previously A critical issue in the application of metabolomics is the standardization of methods, data analysis, and reporting across laboratories. A recent cooperative study by the Consortium for Metabolomic Toxicology indicated that NMR-based technology is robust and reproducible in laboratories that follow similar analytical protocols [27].

6.3.4.2 GC-MS

MS-based analyses offer an important alternative approach to metabolomics. This analysis records the mass-to-charge ratios and retention times of metabolites. Because most small molecules produce singly charged ions, the analyses provide molecular weights of the metabolites. Analysis of standards in the same system and the use of MS-MS analysis can establish the identity of the components of interest. However, apparent molecular weight measurement alone is often insufficient to generate candidate metabolite identifications; frequently, hundreds or thousands of molecules are being analyzed. Nevertheless, accurate information about molecular weight, where possible, is of great value in identification. For this reason, LC-MS metabolomic analyses are most commonly done with higher mass accuracy MS instruments, such as LC Time-of-Flight Mass Spectrometry (TOF), quadruple TOF, and Fourier Transform Ion Cyclotron Resonance (FT-ICR) MS instruments [28].

LC-MS analyses are done with both positive and negative chemical ionization. These four ionization methods provide complementary coverage of diverse small-molecule chemistries. The principal mode of analysis is via "full scan" LC-MS, in which the mass range of the instrument is repeatedly scanned. NMR-based and MS-based approaches provide complementary platforms for metabolomic studies and an integration of these platforms will be needed to provide capabilities that are most comprehensive. Clearly, either platform can detect metabolite profile differences sufficiently to distinguish among toxicities. What is not yet clear is the degree to which either approach can resolve subtly different phenotypes. Finally, the use of stable isotope dilution LC-MS-MS analysis provides a method for absolute quantification of individual proteins in complex samples [29]. The use of stable isotope–labeled standard peptides that uniquely correspond to proteins or protein forms of interest are spiked into proteolytic digests from complex mixtures, and the levels of the target protein are measured relative to the labeled standard. This approach holds great potential for targeted quantitative analysis of candidate biomarkers in biological fluids.

The greatest potential advantage of MS-based methods is sensitivity. MS analyses can detect molecules at levels up to 10,000-fold lower compared with NMR [28]. Both GC-MS and LC-MS approaches have been used, although the limits of volatility of many metabolites reduce the range of compounds that can be analyzed successfully with GC-MS. Metabolomics is the analysis of collections of small-molecule intermediates and products of diverse biological processes. Metabolic intermediates reflect the actions of proteins in biochemical pathways and thus represent biological states analogously to proteomes. As with proteomes, metabolome are dynamic and change in response to nutrition, stress, disease states, and even diurnal variations in metabolism. Unlike genomes, transcriptomes, and proteomes, metabolomes comprise a chemically diverse collection of compounds that range from small peptide, lipid, and nucleic acid precursors and degradation products to chemical intermediates in biosynthesis and catabolism as well as metabolites of exogenous compounds derived from the diet, environmental exposure, and therapeutic interventions. A consequence of the chemical diversity of metabolome components is the difficulty of comprehensive analysis with any single analytical technology.

6.3.5 Biomarkers

Toxicogenomic-based biomarkers will likely comprise an agglomeration of responses that allow for further stratification of the population to identify sensitive groups, which could then be treated more effectively while minimizing the risk of unacceptable toxicities. Ideally, these biomarkers will be mechanistically based and causally associated with the adverse effect, which is expected to further minimize uncertainties in the source-to-outcome continuum and extrapolations between species (rodent to human) and across models (*in vitro* to *in vivo*). Classifications based on mechanisms of action will identify biomarkers with greater predictive accuracy that could be used for exposure assessments in humans and extended to include wildlife species. Moreover, they will provide evaluations of the appropriateness of cross-species extrapolations by assessing the degree of conservation of mechanisms of toxicity, which would facilitate the implementation of mechanistically based, chemical-specific uncertainty factors that account for both within-species and across-species variability. Furthermore, toxicogenomics provides strategies for the comprehensive assessment of mixtures because all possible chemical, gene, protein, metabolite, and network interactions that may be important in eliciting mixture-specific toxicities can be considered. However, these applications, including the identification of biomarkers, will require broad acceptance and comprehensive validation procedures such as those proposed by the Interagency Coordinating Committee on the Validation of Alternative Methods and the European Centre for the Validation of Alternative Methods.

Systemic exposure load chemical analytical determination can be done by the exposure of urine with a biomarker, its metabolites or specific reaction products with macromolecules (e.g., DNA or protein adducts) can be used as exposure biomarkers. The objective of such methods is to detect toxic compounds circulating in plasma or excreted in urine, as a reflection of the systemic exposure load. Specific DNA adducts, DNA breaks, oxidative DNA damage, and micronuclei (e.g., in blood lymphocytes and selected somatic cells) may be indicative of an increased 224 G. This is based on the assumption that increased DNA damage will enhance the probability of mutations occurring in critical target genes and cells, or that increased DNA damage is the result of a higher load of genotoxic agents that will enhance the process of carcinogenesis (by inducing DNA damage as well as other molecular processes of carcinogenesis). The damage may be detected in peripheral blood lymphocytes, in any primary or cultured cell system, or by analyzing the excretion in urine or plasma of reaction products indicative of genotoxic interactions.

6.3.5.1 Susceptibility Biomarkers

In the future, it will be more and more important to analyze the effect of food hazards on the basis of genetic and other susceptibilities. Three main types of susceptibilities for cancer can be distinguished in Figure 6.1.

1. *Predetermining damage.* One possibility is to identify individuals at high risk because they carry high-penetrance genetic alterations in cancer target genes. The further development of biomarkers could include the isolation of cells from these individuals to search for additional associated parameters, which could enhance the detection of alterations at much earlier stages (e.g., specific adducts in p53 genes).

2. *Predisposing alterations.* The second possibility is to identify individuals at different degrees of risk because they carry frequent low-penetrating alterations in

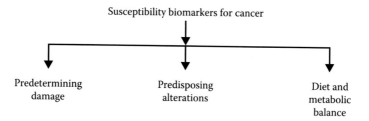

FIGURE 6.1
Types of susceptibilities for cancer.

genes (genetic polymorphisms), which occur frequently and are more indirectly related to the process of carcinogenesis. These include the genetic polymorphisms for activating enzymes (which usually catalyze oxidation reactions and phase I enzymes). These genetic variants may be associated with either enhanced or decreased rates of metabolic conversion by the specific enzyme. Depending on the type of metabolic conversion (activation or deactivation), the result will be more or less genotoxic exposure and thus risk.

3. *Diet and metabolic balance.* Future biomarker approaches will consider more and more modulating influences of the diet as a susceptibility factor and thus on biomarker responses. Some influences may lead to increased genetic damage, making the cells more vulnerable to additional toxic compounds. Some may lead to less damage, making the cells more tolerant to additional exposure-related factors. Biomarkers for detecting toxicity, in particular cancer-related end points after ingestion of hazardous food compounds, are available. Many of these, however, need to be validated for their applicability, reliability, and predictivity. Moreover, biomarker techniques that are useful for the identification of potentially causative hazards for diseases other than cancer are much less advanced.

6.3.5.2 Toxicogenomics in Allergic Response Diagnosis

In vitro models aiming to replace traditional animal tests for determining the skin irritation potential of a test substance have been developed, evaluated in prevalidation studies, and recently validated by the European Centre for the Validation of Alternative Methods. To investigate the usefulness of toxicogenomic technologies to identify novel mechanistic end points for skin irritation responses, the present work challenged the human reconstituted epidermis model validated by the European Centre for the Validation of Alternative Methods with four irritant chemicals and four nonclassified chemicals tested at subcytotoxic concentrations. Using a specifically designed low-density DNA array, approximately 50 out of 240 genes were found to be significantly and differentially expressed between tissues exposed to irritant and nonirritant chemicals for at least one test chemical when compared with the seven others. These genes are involved in cell signaling, stress response, cell cycle, protein metabolism, and cell structure. Among them, 16 are expressed in the same way whatever the irritant compound applied. Differential gene expressions might represent new or additional end points useful for the mechanistic understanding and perhaps also the hazard assessment of the skin irritation potential of chemicals and products.

6.3.5.3 Methods of In vitro Toxicology

In vitro methods are common and widely used for screening and ranking chemicals, and have also been taken into account sporadically for risk assessment purposes in the case of food additives. However, the range of food-associated compounds amenable to *in vitro* toxicology is considered much broader, comprising not only natural ingredients, including those from food preparation, but also compounds formed endogenously after exposure; permissible/authorized chemicals including additives, residues, and supplements; chemicals from processing and packaging; and contaminants. A major promise of *in vitro* systems is to obtain mechanism-derived information that is considered pivotal for adequate risk assessment. This report critically reviews the entire process of risk assessment by *in vitro* toxicology, encompassing ongoing and future developments, with major emphasis on cytotoxicity, cellular responses, toxicokinetics, modeling, metabolism, cancer-related end points, developmental toxicity, prediction of allergenicity, and finally, development and application of biomarkers. It describes in-depth the use of *in vitro* methods in strategies for characterizing and predicting hazards to humans. The major weaknesses and strengths of these assay systems are addressed, together with some key issues concerning major research priorities to improve hazard identification and characterization of food-associated chemicals.

6.3.6 Toxicoinformatics

Bioinformatics is a research area that, in general, develops algorithms to process, manage, and present a vast amount of biological data for microarray technology and, more specifically, toxicogenomic bioinformatics is responsible for creating the necessary tools for statistical analysis. Whereas in most research fields, the analytical weight of the experiment lies with the design, this is not the case for toxicogenomics. Equal weight is given to the experimental design and the statistical analysis of the data. The first critical task to be completed is processing the microarray data. In the case of nucleic acid arrays, the cDNA, mRNA, or cRNA targets are fluorescently labeled before complementary base pairing to the probe. As for whole-cell arrays of reverse transfection, the expression vector with the target gene usually contains the green fluorescent protein gene, fused downstream of the gene of interest and its promoter.

After scanning the microarrays, the toxicologist meets the critical step, matching each fluorescence intensity relative to every other one. Bioinformatics is essential for creating the computer software to handle this task. To analyze the qualitative data scanned from the microscope or camera, it must be translated into quantitative or numerical statistics. The software that toxicologists use therefore relies heavily on mathematical algorithms. The basic analysis that can be done without a computational strategy is to note the observed signal. Because the amount of the probe on the array exceeds the amount of the target, the observed signal at any given position is a good estimate of the abundance of cognate target in the sample.

6.4 Applications in Toxicogenomics

Toxicogenomics combines transcript, protein, and metabolite profiling with conventional toxicology to investigate the interaction between genes and environmental stress in disease causation. The patterns of altered molecular expression that are caused by specific exposures or disease outcomes have revealed how several toxicants act and cause disease. Despite these success stories, the field faces noteworthy challenges in discriminating the molecular basis of toxicity. We argue that toxicology is gradually evolving into a system that will eventually allow us to describe all the toxicological interactions that occur within a living system under stress, and use our knowledge of toxicogenomic responses in one species to predict the modes-of-action of similar agents in other species.

6.4.1 Application to Hazard Screening

Toxicogenomics can be used to screen for hazardous or potentially hazardous chemical compounds. A screening test can be defined as a preliminary test designed to detect a state or property more quickly and cheaply compared with more elaborate tests. In predictive toxicology, the property being detected by screening tests is generally hazardous. Screening tests may not give complete information on toxicity, such as the time course, chronic effects, or dose–response characteristics. Screening data provide an input to the hazard identification step in risk assessment but do not allow full determination of risk. The chemical and pharmaceutical industries also use screening tests to detect desirable properties, such as the ability to bind to specific target receptors. The use of genomic techniques to screen for desirable, pharmacologic properties may be analogous to hazard screening [30]. However, this is not the principal focus of this chapter. Toxicogenomic technologies may be incorporated directly into existing, more traditional hazard screening tests; they may be the basis of new tests that substitute for more traditional tests, or they may generate mechanistic insights that enable more basic tests, such as receptor binding or other physicochemical assays, to be conducted for screening compounds.

Comprehensive hazard assessment for a chemical substance generally requires a variety of *in vitro* and *in vivo* toxicologic assays as well as evaluations of physical properties. The selection of individual screening tests depends greatly on the setting and specific regulatory requirements. For example, the current practice of the U.S. Environmental Protection Agency under the Toxic Substances Control Act, in the absence of more extensive preexisting data, is to screen new chemicals based solely on physicochemical data using quantitative structure–activity relationship models. In this setting, chemical tests may be limited to properties such as boiling point, octanol–water partition coefficient, vapor pressure, and solubility. If environmental fate and transport of substances are not primary concerns, short-term *in vivo* rodent assays may be used, such as a 28-day feeding study, which examines histopathology in most critical target organs. More comprehensive screening programs have adopted batteries of tests that provide information on different types of toxicity but remain insufficient to fully assess chemical risks. As one example, the Organization of Economic Cooperation and Development has developed the Screening Information Data Set, which consists of 21 data elements.

Each toxicity test involves administering a measured amount of a compound to whole organisms or to cells in culture, and then measuring indicators of toxic outcomes. Compared with more extensive tests, screening tests tend to use higher and fewer doses of the compound being studied, fewer test subjects, a shorter period of observation, and less extensive evaluation of the toxic outcomes. To reduce the use of mammals for laboratory testing, there is a strong impetus to develop and validate screening tests that use cultured cells or lower-order animals, such as worms. The incorporation of toxicogenomics into screening tests involves measuring gene, protein, or metabolite changes in response to specific doses of an administered test compound at specific time points, with or without the parallel measurement of more traditional markers of toxicity.

The critical question about new toxicogenomic techniques is whether they can improve hazard screening by making tests faster, more comprehensive, less reliant on higher-order animals, and more predictive and accurate without being prohibitively expensive. For a screening test to be useful, it must be capable of detecting the property or state being tested when it truly exists. This is the definition of the "sensitivity" of a screening test. In many cases, screening tests are designed to be highly sensitive, sometimes at the expense of the specificity of the test or the ability of the test to return a negative result when the property or state of concern does not exist. Another way to describe this quality is that hazard screening tests often accept a higher rate of false-positive results to avoid not detecting a hazard because of a high rate of false-negative results. When the data generated by screening tests are continuous, as is the case with gene and protein expression and metabolite assays, the selection of thresholds for positive and negative results plays a dominant role in determining the sensitivity and specificity of the test. When larger values of the test are more likely to indicate the presence of a particular hazard, selection of a relatively low value as the threshold for a positive result will lead to greater sensitivity and lower specificity (i.e., fewer false-negatives and more false-positives). Conversely, a high threshold for a positive result will lead to lower sensitivity and higher specificity (i.e., more false-negatives and fewer false-positives). A critical challenge in designing and validating toxicogenomic screening tests is to identify and define a "gold standard" for hazard screening—the indication of the true state of toxicity against which the sensitivity and specificity of the screening test can be measured.

6.4.1.1 Bioreactivity of Diesel Exhaust Particles

In vitro triple cell coculture models consisting of human epithelial cells (16HBE14o-), monocyte-derived macrophages, and dendritic cells recently demonstrated that macrophages and dendritic cells create a transepithelial network between the epithelial cells to capture antigens without disrupting the epithelial tightness. The expression of the different tight junction proteins in macrophages and dendritic cells, and the formation of tight junction–like structures with epithelial cells have been demonstrated. Immunofluorescent methods combined with laser scanning microscopy and quantitative real-time PCR were used to investigate if exposure to diesel exhaust particles (0.5, 5, 50, 125 μg/mL) for 24 h could modulate the expression of the tight junction mRNA/protein of occludin in all three cell types. Only the highest dose of diesel exhaust particles (125 μg/mL) seemed to reduce the occludin mRNA in the cells of the defense system but not in epithelial cells, however, although the occludin arrangement in the latter cell type was disrupted. The transepithelial electrical resistance was reduced in epithelial cell monocultures but not in the triple cell cocultures following exposure to high concentrations of diesel exhaust particles. Cytotoxicity was not found in either epithelial monoculture or in triple cell cocultures after exposure to different concentrations of diesel exhaust particles.

6.4.1.2 Bioreactivity of Chemical Compounds

Within larger chemical screening initiatives, toxicogenomic technologies could enhance screening-level determinations of basic modes of action. Two such applications involve the High Production Volume Chemical Challenge Program and the initial evaluation of premanufacturing notices for new chemicals under the Toxic Substances Control Act. Categorization based on chemical structures could eventually be combined with transcriptome or proteome profiling. Premanufacturing notices might then contain toxicogenomic data that show different, nontoxic profiles compared with other chemicals of similar structures. Alternatively, scientists from the U.S. Environmental Protection Agency may request rapid toxicogenomic assays of new chemicals and use the profiles generated to make more rigorous determinations of safety. Additional database development and demonstration of predictive accuracy would be required before true toxicogenomic screens would be useful in this setting.

Current toxicogenomic assays are not sufficiently developed to replace more traditional toxicologic screening tests, but there are important opportunities to build the familiarity and databases needed to inform future toxicogenomic screening regimens by adding toxicogenomic tests to existing chemical programs. For example, The National Toxicology Program uses established methodologies to test relatively large numbers of animals, which is both costly and time-consuming. However, the chronic exposure experiments provide a rare source of biological material that could be used to advance toxicogenomic applications. Strategically adding appropriate toxicogenomic assays to these testing protocols will generate novel, phenotypically correlated data that would be especially useful in building databases to aid in predictive toxicology. Such data will be suited for helping to determine whether short-term *in vivo* studies using toxicogenomic technologies can substitute for chronic bioassays.

For *in vivo* studies, toxicogenomics may allow ongoing noninvasive sampling (especially for proteomic or metabolomic end points) and provide useful mechanistic information. For certain end points (as demonstrated currently with different types of hepatotoxicity), reliance on toxicogenomic data to predict potential hazards will be possible. Several companies seem to be working on using *in vitro* toxicogenomic assays to evaluate multiple cellular effects. Surrogate screens using panels of genes in *in vitro* systems may be deployed rather cheaply and produce large amounts of data quickly. The critical question for the industry is whether adding toxicogenomic technologies to more standard toxicology testing or other *in vitro* methods will result in overall cost savings. To date, the economic value of toxicogenomic testing in drug development has not been demonstrated through rigorous studies. In the absence of such data, it is likely that pharmaceutical companies will make individual judgments about incorporating toxicogenomic assays. Regardless of the degree to which pharmaceutical companies use toxicogenomic screens to identify possible hazards, for the foreseeable future, full interpretation of risk will be done in the context of additional, required testing information.

6.4.1.3 Assessment of Herbicides and Pesticides

Herbicides are agrochemicals that control the growth of undesired weeds, bringing about a significant overall increase in crop productivity. The herbicide 2,4-D (2,4-dichlorophenoxy-acetic acid) is among the most successful selective organic herbicides used in agriculture. However, the widespread and intensive use of 2,4-D might give rise to several toxicological and environmental problems and has led to the emergence of herbicide-resistant weeds. This could lead to significant economic losses and to the use of higher herbicide application rates with deleterious consequences to the environment and human health.

It is expected that the molecular mechanisms described might have an effect on the design or optimization of safe and effective weed control strategies. It is also expected that further coordinate studies, using *S. cerevisiae* and *Arabidopsis* models, will accelerate the understanding of the molecular mechanisms governing 2,4-D toxicity in addition to adaptation and resistance to the herbicide, one of the most exciting and less explored topics in plant biology. This has important implications for agriculture, environmental safety, and human health. The expansion of all this fundamental knowledge establishes the basis for the development of rapid, inexpensive, and reliable assays that can be used to screen a large number of chemicals for toxicity. There is a pressing need for such assays, as the majority of the chemicals in commercial use have not been comprehensively tested for human toxicity. Environmental genomics, in particular toxicogenomics, helps to understand gene–environment interactions.

6.4.1.3.1 Environmental Genomics: Mechanistic Insights into Toxicity and Resistance to Herbicide 2,4-D

Genomic information and tools are beginning to be used to increase our understanding of how organisms of all types interact with their environment. The study of the expression of all genes, at the genome, transcriptome, proteome, and metabolome level, in response to exposure to a toxicant, is known as toxicogenomics. Here, we show how this new field of environmental genomics has enhanced the development of fundamental knowledge on the mechanisms behind the toxicity of and resistance to the herbicide 2,4-D. Although 2,4-D is one of the most successfully and widely used herbicides, its intensive use has led to the emergence of resistant weeds and might give rise to several toxicological problems when present in concentrations that are higher than those recommended. Many current research works summarize recent mechanistic insights into 2,4-D toxicity and the corresponding adaptive responses based on studies carried out using *S. cerevisiae* and *A. thaliana* as model organisms.

6.4.1.3.2 Toxicogenomics to Study the Hepatic Effects of Food Additives and Chemicals

Toxicogenomics is predominantly applied for the elucidation of mechanisms of action and discovery of novel pathway-supported mechanism-based markers of liver toxicity. In addition, Researchers aim to integrate transcriptome, proteome, and metabolome data supported by bioinformatics to develop a systems biology approach for toxicology. Transcriptomics and proteomics studies on bromobenzene-mediated hepatotoxicity in the rat are discussed [31]. Finally, an example is shown in which gene expression profiling together with conventional biochemistry led to the discovery of novel markers for the hepatic effects of the food additives butylated hydroxytoluene, curcumin, propyl gallate, and thiabendazole.

6.4.2 Applications of Toxicogenomics in Pharmaceuticals

6.4.2.1 Drug Discovery and Drug Safety

New drugs are screened for adverse reactions using a laborious, costly process, and still some promising therapeutics are withdrawn from the marketplace because of unforeseen human toxicity. Novel higher-throughput methods in toxicology need to be developed. These new approaches should provide more insight into potential human toxicity than current methods. Toxicogenomics, the examination of changes in gene expression after exposure to a toxicant, offers the potential to identify a human toxicant earlier in drug development and to detect human-specific toxicants that cause no adverse reaction in rats.

Gene expression changes associated with toxicologic changes typically reflect a large number of complex pharmacological, physiological, and biochemical processes, most of them interacting with each other and related to a multitude of toxicological end points. Consequently, numerous gene expression changes are not necessarily involved in the toxicologic process being investigated, but are simply secondary or indirect consequences of this toxicologic change. For instance, a general, nonspecific sign of toxicity is anorexia or decreased food consumption, which can affect the homeostasis of several tissues. This has, for instance, been elegantly studied in rats using a variety of pathologic end points. One of the most promising applications involves the screening and prioritization of commercial chemicals and drug candidates that warrant further development and testing. This consists of comparing their toxicogenomic profiles to databases containing profiles of known toxicants and identifying biomarkers of exposure and toxicity that can be used in high-throughput screening programs.

The large number of compounds, tissues, corroborative toxicologic and pathologic changes, and gene expression data in these reference databases allows one to strengthen statistical inferences. The concept of databases in toxicology is not novel, and to better illustrate the requirement for toxicogenomics databases, one has only to relate to the use of databases for serum chemistry, hematology, pathology, or carcinogenesis. Without proper historical databases and experience, interpretation of these data reveals little about the mechanism of action of toxicologic changes. Ideally, toxicogenomic databases consist of many known pharmaceutical agents, toxicants, and control compounds at multiple doses and time points, with biological replicates for each condition. The reference compounds profiled in the database should reflect a variety of toxic mechanisms, and represent different structure–activity relationships. In some situations, time course data can be very useful to identify gene expression changes linked to a time-dependent toxic response and increase the chances of observing a true toxic response over that of a single time point. The number of animals required for each time point or dose is also an important consideration.

In drug discovery and development, testicular toxicity is of particular interest because testicular changes are typically subtle in the early stages without well-established correlating biomarkers (such as changes in testicular weight) or striking morphologic changes. Several elegant studies have demonstrated how gene expression profiling can elucidate the molecular basis of testicular toxicity. For instance, gene expression changes in the testis were evaluated after exposure of mice to bromochloroacetic acid, a known testicular toxicant. Using a custom nylon DNA array, numerous changes in gene expression were detected in genes with known functions in fertility, such as Hsp70-2 and SP22, as well as genes encoding proteins involved in cell communication, adhesion, and signaling, supporting that the toxicologic effect was the result of disruption of cellular interactions between Sertoli cells and spermatids [20,22]. Likewise, Lee and coworkers, using several testicular toxicants, demonstrated that upregulation of Fas is a common and critical step for initiating germ cell death; thus, if Sertoli cells were the target of the toxicant, a concomitant upregulation of Fas ligand would also occur to eliminate Fas-positive germ cells. Such gene expression studies provide rapid methods to not only understand the mechanism of testicular toxicity but also to develop counterscreens to analyze backup compounds.

6.4.2.1.1 Risk Assessment

The goal of toxicology is the assessment of possible risk to humans. An emerging technology with the potential to have a major effect on risk assessment is toxicogenomics. Toxicogenomics helps in the extrapolation of findings across species and increases predictability. Biomarkers are valuable in the evaluation of compounds at earlier development phases, improving clinical candidate selection. Caution regarding the interpretation of the results is still necessary. Nevertheless, toxicogenomics will accelerate preclinical safety assessments and improve the prediction of toxic liabilities, as well as of potential risk accumulation for drug–drug or drug–disease interactions. These applications are analogous with the development of diagnostic signatures and classification protocols for disease states which can identify more effective treatment regimens for selected populations and can also be used to monitor drug efficacy during clinical trials.

6.4.2.2 Nanomedicines

Synthetic polymers and nanomaterials display selective phenotypic effects in cells, and in the body, signal transduction mechanisms involved in inflammation, differentiation, proliferation, and apoptosis. When physically mixed or covalently conjugated with cytotoxic agents, bacterial DNA, or antigens, polymers can drastically alter specific genetically controlled responses to these agents. These effects, in part, result from cooperative interactions of polymers and nanomaterials with plasma cell membranes and trafficking of polymers and nanomaterials to intracellular organelles. Cells and whole organism responses to these materials can be phenotype- or genotype-dependent. In selected cases, polymer agents can bypass limitations to biological responses imposed by the genotype, for example, phenotypic correction of immune response by polyelectrolytes. Overall, these effects are relatively benign as they do not result in cytotoxicity or major toxicities in the body. Collectively, however, these studies support the need for assessing pharmacogenomic effects of polymer materials to maximize clinical outcomes and understand the pharmacological and toxicological effects of polymer formulations of biological agents, that is, polymer genomics.

6.4.2.2.1 Polymers and Nanomaterials

Polymer-based drugs ("polymer therapeutics") emerged from the laboratory bench in the 1990s as a promising therapeutic strategy for the treatment of certain devastating human diseases. A number of polymer therapeutics are presently on the market or undergoing clinical evaluation to treat cancer, rheumatoid arthritis, HIV/AIDS, hepatitis C, multiple sclerosis, and other diseases. Most of them represent small drug molecules or therapeutic proteins that are chemically linked to water-soluble polymers such as poly(ethylene glycol) to increase drug solubility, drug stability, or enable the targeting of drugs to tumors. Separately, poly(ethylene glycol)–coated liposomes carrying chemotherapeutic drugs have been approved for clinical use.

6.4.2.2.2 Toxicogenomics of Nanomaterials

It is increasingly evident that polymers and nanomaterials may affect cell signaling through interactions with cell plasma membranes. They can also be transported within a cell via endocytosis and interfere with normal cell function by interacting with intracellular molecular targets. Compared with the relatively simple polymer systems considered in the previous sections, nanoparticles may have more complex structures, and display multiple functional groups at the surface, which can be hydrophilic, lipophilic, charged, or even chemically reactive. Nanoparticles are small and have tremendous surface areas and thus can exhibit profound effects on cell function. The interaction of the nanoparticles with the lipid membranes and the transport of nanoparticles into cells are very poorly understood. However, there is an increasing body of evidence that the surface properties of nanoparticles can lead to considerable toxicity.

6.4.2.2.3 Toxicogenomics in Drug Response Study

Adverse drug responses are an important postmarketing public health issue, occurring many times in subsets of treatment populations. Promising new approaches to predicting physiological responses to drugs are focused on "genomic responses" or toxicogenomics.

This article provides a current perspective on toxicogenomics technologies that are aimed at (1) providing new tools and systems for more rapid, accurate, and complete toxicity assessments in advance of human exposure; (2) enhancing the thoroughness and accuracy of toxicity assessments achievable with currently available test systems; and (3) predictive assessments of individualized risk for developing adverse drug reactions.

6.4.2.2.4 *Mechanisms of Drug-Induced Hepatotoxicity during Drug Discovery and Development*

Hepatotoxicity is a common cause of failure in drug discovery and development, and is also frequently the source of adverse drug reactions. Therefore, a better prediction, characterization, and understanding of drug-induced hepatotoxicity could result in safer drugs and a more efficient drug discovery and development process. Among the "omics" technologies, toxicogenomics (or the use of gene expression profiling in toxicology) represents an attractive approach to predicting toxicity and gaining a mechanistic understanding of toxic changes. In this review, we illustrate, using selected examples, how toxicogenomics can be applied to investigate drug-induced hepatotoxicity in animal models and *in vitro* systems. In general, this technology can not only improve the discipline of toxicology and risk assessment but also represent an extremely effective, hypothesis-generating alternative to rapidly understanding the mechanisms of hepatotoxicity.

6.4.2.2.5 *Polymer Genomics—Toxicology of Nanomedicines*

Synthetic polymers and nanomaterials display selective phenotypic effects in cells and in the body signal transduction mechanisms involved in inflammation, differentiation, proliferation, and apoptosis. When physically mixed or covalently conjugated with cytotoxic agents, bacterial DNA, or antigens, polymers can drastically alter specific genetically controlled responses to these agents. These effects, in part, result from cooperative interactions of polymers and nanomaterials with plasma cell membranes and trafficking of polymers and nanomaterials to intracellular organelles. Cells and whole organism responses to these materials can be phenotype-dependent or genotype-dependent. In selected cases, polymer agents can bypass limitations to biological responses imposed by the genotype, for example, phenotypic correction of immune response by polyelectrolytes. Overall, these effects are relatively benign as they do not result in cytotoxicity or major toxicities in the body. Collectively, however, these studies support the need for assessing the pharmacogenomic effects of polymer materials to maximize clinical outcomes and understand the pharmacological and toxicological effects of polymer formulations of biological agents, that is, polymer genomics.

6.4.3 Applications in Disease Diagnosis

6.4.3.1 *Cancer Biology*

Specific genotoxic events such as gene mutations and chromosome damage are considered hallmarks of cancer. The genotoxicity testing battery enables relatively simple, rapid, and inexpensive hazard identification, namely, by assessing a chemical's ability to cause genetic damage in cells. In addition, the 2-year rodent carcinogenicity bioassay provides an assessment of a risk associated with the chemical to develop cancer in animals. Although the link between genotoxicity and carcinogenicity is well documented, this relationship is complicated due to the effect of nongenotoxic mechanisms of carcinogenesis and by character of the *in vitro* genotoxicity assays and specific end points making the interpretation of test results in light of human risk and relevance difficult. In particular, the specificity of test results has been questioned. Therefore, the development of novel scientific approaches bridging genotoxicity and carcinogenicity testing via understanding underlying mechanisms is extremely important for facilitating cancer risk assessment. In this respect, toxicogenomic approaches are considered promising as these have the potential to provide generic insights into molecular pathway responses.

The pathogenesis of neoplasia is a multistage process that consists of at least three operationally defined stages beginning with initiation, followed by a promotion, and then a progression phase [32]. Its full complexity can only be studied in an *in vivo* model representing all aspects of exposure and target organ biology. The fact that the gene expression profile analysis provides only a snapshot of molecular processes at the time that regards time points of analysis and dose. These particularly aim at developing alternative methods for animal tests based on generating genomic risk profiles by microarray technologies for carcinogenicity, immunotoxicity and reprotoxicity. Predicting the toxic properties of a substance by assessing gene expression profiles requires that a large database is first built for well-known classes of toxic substances That database consists of gene expression profiles for many substances per class. For each class, a gene expression profile can be identified with mathematical models through which that class can be distinguished from others. Next, by comparing the gene expression profile of a new substance with those of the different toxic classes, the best fit with a specific class can be determined.

6.4.3.2 Autoimmune Disease

Gene therapy vectors based on lentiviruses persist in the host and are ideally suited for long-term therapies of genetic disorders. However, recent incidences of T-cell leukemia in X-SCID children receiving gene therapy revealed discrepancies between the preclinical and clinical studies. Divergent results on the potential oncogenic property of integrating vectors obtained from different animal models raise further concerns about the relevance and sensitivity of available preclinical systems used to assess their toxicity. Evaluation of transcriptional responses to vector transduction in different *in vivo* models may provide us with early indications of the potential adverse effects of these vectors and comparative information on the sensitivity and suitability of these models. For this purpose, we used cDNA microarray to examine transcriptional changes in BALB/c and CD-1 (IRC) mice, the two common murine strains used in pharmacological and toxicological studies. Our results revealed a significant difference in the transcriptional responses to vector transduction between the two mouse strains. Modest gene changes, in terms of gene numbers and gene expression, were observed in BALB/c mice, whereas the expression of 15 oncogenes was upregulated in CD-1 (ICR) mice. We confirmed the upregulation of oncogenes, such as N-myc, A-Raf, Fli-1, Wnt-2b, and c-Jun in CD-1 (IRC) mice using RT-PCR. This study provided an insight into the function of lentiviral transduction in different mouse strains. Distinctive toxicogenomic profiles of two mouse strains should be considered in the context of future development of sensitive models for toxicity evaluation of lentiviral vector products.

A recent study using retroviral gene marking strongly suggested that retroviral induction of leukemia in C57B1/6J mice required the combined action of insertion oncogene activation and signal interference by transgene [33]. The requirement for complementary factors or a high level of oncogene expression may explain why we did not detect tumors even though oncogene expression was observed. It has also been reported that hepatic cells overexpressing N-myc display a striking propensity to undergo apoptosis [34]. It is likely that the processes regulating the expression of oncogenesis also play a role in their normal pattern of regulation during cellular growth. It may provide an insight into the function of lentiviral transduction in different mouse strains and support future development of suitable models for toxicity evaluation of gene therapy products.

6.4.3.3 Allergic Responses

Dermal irritation is defined as the production of "reversible damage of the skin following the application of a test substance." Determining the potential of chemicals to cause acute skin irritation is important for establishing procedures for the safe handling, packaging, and transport of chemicals, as well as for general safety assessment purposes. A recent study using retroviral gene marking strongly suggested that retroviral induction of leukemia in gene marking of C57B1/6J mice required the combined action of insertion oncogene activation and signal interference by transgene [33]. The requirement for complementary factors or a high level of oncogene expression may explain why we did not detect tumors even though oncogene expression was observed. It has also been reported that hepatic cells overexpressing N-myc display a striking propensity to undergo apoptosis [34]. It is likely that the processes regulating the expression of oncogenes also play a role in their normal pattern of regulation during cellular growth.

Microarray technology is finding increasing application in the characterization of immune function and immunopathological processes [35–38]. An area in which such analyses are predicted to have a particularly important effect is allergy, and in particular in providing new approaches to the identification and characterization of sensitization hazards. Exciting opportunities exist for both chemical and protein allergy, but in the context of exploring the potential application of toxicogenomic analyses, attention here is focused on the former. Researching gene regulation events during toxicological processes has, until comparatively recently, focused in some detail on the study of a small number of genes. For example, the regulation of the alcohol-inducible cytochrome P4502E1 has been characterized at high resolution, including the transcriptional and translation mechanisms through which enzymatic activity is regulated [39]. Such studies help identify the specific factors involved in modulating individual gene expression and thereby allow characterization of the interplay between discrete signaling and regulatory pathways. This in turn permits the definition of the role of transcriptional regulation in the development of toxicity [40]. The advent of microarray technology (the major technique used in transcript profiling) modifies this working model by allowing a measurement of changes in the transcriptional regulation of thousands of genes simultaneously, but without necessarily understanding the mechanisms through which expression of any of the genes takes place, or indeed what the consequences of altered expression may be.

6.4.3.4 Stem Cell Tests

During early embryonic development, at blastocyst stage, the embryo has an outer coat of cells and an inner cell mass. The inner cell mass is the reservoirs of embryonic stem (ES) cells, which are pluripotent, that is, have the potential to differentiate into all cell types of the body. Cell lines have been developed from ES cells. In addition, there are embryonic germ (EG) cell lines developed from progenitor germ cells, and embryonic carcinoma (EC) cell lines developed from teratomas. These cell lines are being used for the study of basic and applied aspects in medical therapeutics, and disease management. Another potential of these cell lines is in the field of environmental mutagenesis. In addition to ES cells, there are adult stem cells in and around different organs and tissues of the body. It is now possible to grow pure populations of specific cell types from these adult stem cells. Treating specific cell types with chemical or physical agents and measuring their response offers a shortcut to test the toxicity in various organ systems in the adult organism. For example, to evaluate the genotoxicity of a chemical (e.g., drug or pesticide) or a physical agent (e.g., ionizing radiation or nonionizing electromagnetic radiation) during embryonic development, a large number of animals are being used. As an alternative, use of stem cell lines would be a feasible proposition. Using stem cell lines, efforts are being made to standardize the protocols, which will not be only useful in testing the toxicity of a chemical or a physical agent, but also in the field of drug development, environmental mutagenesis, biomonitoring and other studies.

6.4.3.4.1 Stem Cell Test—A Practical Tool in Toxicogenomics

During early embryonic development, at blastocyst stage, the embryo has an outer coat of cells and an inner cell mass (ICM). ICM is the reservoirs of embryonic stem (ES) cells, which are pluripotent, that is, have the potential to differentiate into all cell types of the body. Cell lines have been developed from ES cells. In addition, there are embryonic germ (EG) cell lines developed from progenitor germ cells, and embryonic carcinoma (EC) cell lines developed from teratomas. These cell lines are being used for the study of basic and applied aspects in medical therapeutics, and disease management. Another potential of these cell lines is in the field of environmental mutagenesis. In addition to ES cells, there are adult stem cells in and around different organs and tissues of the body. It is now possible to grow pure populations of specific cell types from these adult stem cells. Treating specific cell types with chemical or physical agents and measuring their response offers a shortcut to test the toxicity in various organ systems in the adult organism. For example, to evaluate the genotoxicity of a chemical (e.g., drug or pesticide) or a physical agent (e.g., ionizing radiation or nonionizing electromagnetic radiation) during embryonic development, a large number of animals are being used. As an alternative, use of stem cell lines would be a feasible proposition. Using stem cell lines, efforts are being made to standardize the protocols, which will not be only useful in testing the toxicity of a chemical or a physical agent, but also in the field of drug development, environmental mutagenesis, biomonitoring and other studies.

6.4.3.4.2 Toxicogenomics Profile Comparison, Treated with HIV-1–Based Vectors for Gene Therapy

Gene therapy vectors based on lentiviruses persist in the host and ideally suited for long-term therapies of genetic disorders. However, recent incidences of T cell leukemia in X-SCID children receiving gene therapy reveal discrepancy among the preclinical and clinical studies. Divergent results on the potential oncogenic property of integrating vectors obtained from different animal models further raise concern about the relevance and sensitivity of available preclinical systems used to assess their toxicity. Evaluation of transcriptional responses to vector transduction in different *in vivo* models may provide us with early indications of the potential adverse effects of these vectors and comparative information on the sensitivity and suitability of these models. For this purpose, we used cDNA microarray to examine transcriptional changes in BALB/c and CD-1 (IRC) mice, the two common murine strains used in pharmacological and toxicological studies. Our results revealed a significant difference in the transcriptional responses to vector transduction between the two mouse strains. Modest gene changes, in terms of gene numbers and gene expression were observed in BALB/c mice, whereby expression of 15 oncogenes was upregulated in CD-1 (ICR) mice. We confirmed the upregulation of oncogenes, for example, N-myc, A-Raf, Fli-1, Wnt-2b, and *c-Jun* in CD-1 (IRC) mice using RT-PCR. This study provided an insight into the function of lentiviral transduction in different mouse strains. Distinctive toxicogenomic profiles of two mouse strains should be considered in the context of future development of sensitive models for toxicity evaluation of lentiviral vector products.

6.4.3.4.3 Toxicogenomics to Study Mechanisms of Genotoxicity and Carcinogenicity

Specific genotoxic events such as gene mutations and chromosome damage are considered hallmarks of cancer. The genotoxicity testing battery enables relatively simple, rapid and inexpensive hazard identification, namely by assessing a chemical's ability to cause genetic damage in cells. In addition, the 2-year rodent carcinogenicity bioassay provides an assessment of a risk associated with the chemical to develop cancer in animals. Although the link between genotoxicity and carcinogenicity is well documented, this relationship is complicated due to the effect of nongenotoxic mechanisms of carcinogenesis and by character of the *in vitro* genotoxicity assays and specific end points making the interpretation of test results in light of human risk and relevance difficult. In particular, the specificity of test results has been questioned. Therefore, the development of novel scientific approaches bridging genotoxicity and carcinogenicity testing via understanding underlying mechanisms is extremely important for facilitating cancer risk assessment. In this respect, toxicogenomics approaches are considered promising as these have the potential of providing generic insight in molecular pathway responses. The goal of this report is therefore to review recent progress in the development and application of toxicogenomics to the derivation of genomic biomarkers associated with mechanisms of genotoxicity and carcinogenesis. Furthermore, the potential for application of genomic approaches to hazard identification and risk assessment is explored.

6.5 Challenges

Toxicogenomics is not a promise for the future, it is a tool that is available to us now and which, if used correctly and within the guiding principles of good experimental biology, will bring huge dividends. Concern has already been voiced that a potential problem is the misinterpretation, or overinterpretation of genomic analyses, particularly in the context of determining product safety. It must be recognized that the interaction of xenobiotics with biological systems will, in many instances, result in some changes in gene expression, even under circumstances in which such interactions are benign with respect to adverse effects. The challenge again is to ensure that sound judgments and the appropriate toxicological skills and experience are brought to bear on the data generated, so that toxicologically relevant changes in gene expression are distinguished.

The greatest challenge of toxicogenomics is no longer data generation but effective collection, management, analysis, and interpretation of data. Although genome sequencing projects have managed large quantities of data, genome sequencing deals with producing a reference sequence that is relatively static in the sense that it is largely independent of the tissue type analyzed or a particular stimulation. In contrast, transcriptomes, proteomes, and metabolomes are dynamic and their analysis must be linked to the state of the biological samples under analysis.

6.5.1 Human Umbilical Cord Risk Assessment

Attempts have been made to apply toxicogenomic analysis of umbilical cords using DNA microarray for future risk assessment. Because the umbilical cord is part of the fetal tissue, it is possible to estimate the effects of chemicals on the fetus by analyzing the alteration of gene expression. This type of toxicogenomic analysis could be a powerful and effective tool for developing a new risk assessment strategy to help investigators understand and possibly prevent the long-term effects caused by fetal exposure to multiple chemicals. Endocrine Disrupters (EDs) are chemicals that disturb the function of natural hormones in humans and wildlife [41,42]. In animal experiments, potential EDs have adverse effects on the development or function of the reproductive and nervous systems, particularly when exposure occurs during fetal or neonatal periods [43–45]. Similarly, human fetuses and infants are significantly more sensitive to a variety of environmental toxicants than adults.

Microarray technology has been applied in the field of toxicology using animal experiments and will be applied to the field of risk assessment of human exposure to several environmental toxicants. A recent study using cDNA microarray reported the identification of lead-sensitive genes in immortalized human fetal astrocytes. The study showed the potential of DNA microarrays in the discovery of novel toxicant-induced gene expression alterations and in understanding the mechanisms underlying lead neurotoxicity. The use of umbilical cords for risk assessment of chemical effects on the fetus seems to be an ideal method.

- *Step 1*: global gene expression analysis of the umbilical cords by DNA microarray.
- *Step 2*: a combined analysis of the data from step 1 and exposure assessment data in each umbilical cord. In this analysis, a relationship between a global gene expression profile and chemical exposure levels will be clarified.
- *Step 3*: a DNA microarray analysis of the alteration of gene expression using an *in vitro* experiment in human umbilical cord–derived cells after exposure to chemicals. For this purpose, a human umbilical vein endothelial cell (HUVEC) is one of the candidates, as we found that HUVEC genes changed their expression after chemical exposure. Although HUVEC can be applied in this step, other cells from umbilical cords may also be applicable.
- *Step 4*: an integrated analysis of a comparison between steps 2 and 3. In this integrated analysis, biological reactions at the molecular level caused by exposure to chemicals in fetuses can be detected. To extend the toxicogenomic analysis method to develop a new risk assessment strategy, comprehensive studies are required to clarify the correlation between data from toxicogenomic analysis, data from animal experiments observing the adverse effects of chemical exposure, and data from human prospective studies. By establishing the toxicogenomic analysis with the *in silico* integrated analysis of human umbilical cords, the new risk assessment for multiple chemical exposures will be practical.

The umbilical cord is not the target organ affected by the chemicals. Others doubt if toxicogenomic analysis using microarrays is possible with umbilical cords. To answer these questions, and to prove the importance of toxicogenomic analysis using umbilical cords, we conducted the following two preliminary experiments. First, the total RNA was purified from umbilical cords ($n = 5$) and RT-PCR analysis was conducted. As a result, mRNAs of cytochrome P450 (CYP) 1A1, CYP1A2, CYP1B1, and sex hormone receptors were detected in human umbilical cords. We also found the difference in the expression levels of mRNA from CYP1A1 in each umbilical cord (unpublished data). These results show that umbilical cords can be targets for toxicogenomic analysis. Second, to conduct a toxicogenomic analysis of umbilical cords, alteration of the gene expression profile was analyzed in five umbilical cords using the human I cDNA microarray (human 15154 genes).

6.5.2 Cancer Research

The development and application of new and improved toxicogenomic and related technologies for the prediction and characterization of toxic responses will ultimately provide (1) better resolution and accuracy than current animal models used today; (2) more information from alternatives to animal models such as primary cell strains, cell lines, and *in vitro* tissue models, combined with higher throughput than current *in vivo* models; (3) *in vitro* models of human systems offering more "human relevant" data than current animals models for toxicity and safety; and (4) more and better data from human clinical studies, with renewed focus on sampling and end points aimed at the definition of mechanism(s) of action and overall physiological effect. The ultimate goal sought from the combined contributions of all of these technologies is the development of functional and applied physiomics, a biosystem-wide characterization and modeling of complex biological and biochemical responses to genetic and environmental factors. The availability of physiomics will enable comprehensive characterizations of the mechanisms and effects involving all aspects of adsorption, distribution, metabolism and excretion of drugs. In the end, these technologies will contribute to more and better medicines that will be discovered and developed in less time and for lower costs. They will also provide large contributions toward the elimination of the "trial and error" and "one drug fits all" methods that are an inherent part of drug development and prescription medicine today.

6.5.3 Future Prospects

The development and application of new and improved toxicogenomic and related technologies for the prediction and the characterization of toxic responses will ultimately provide:

- Better resolution and accuracy than current animal models used today.
- More information from alternatives to animal models such as primary cell strains, cell lines, and *in vitro* tissue models, combined with higher throughput than current *in vivo* models.
- *In vitro* models of human systems offering more "human relevant" data than current animal models for toxicity and safety.
- More and better data from human clinical studies, with renewed focus on sampling.

End points aimed at the definition of mechanism(s) of action and overall physiological effects. The ultimate goal sought from the combined contributions of all of these technologies is the development of functional and applied physiomics, a biosystem-wide characterization and modeling of complex biological and biochemical responses to genetic and environmental factors. The availability of physiomics will enable comprehensive characterizations of mechanisms and effects involving all aspects of adsorption, distribution, metabolism, and excretion of drugs. In the end, these technologies will contribute to more and better medicines that will be discovered and developed in less time and for lower costs. They will also provide large contributions toward the elimination of the "trial and error" and "one drug fits all" methods that are an inherent part of drug development and prescription medicine today. As tools that enable the simultaneous analysis of multiple targets and pathways are applied, a more comprehensive classification of cellular perturbations that result from toxicity will emerge. Already, toxicologists are starting to look at toxicity through the wider lens of toxicogenomic methods that provide a more complete view of complex networks of gene/protein changes. This is analogous to the difference between attempting to assess neurotoxicity by monitoring the activity of a single neuron at an arbitrary site versus simultaneously monitoring the function of thousands of neurons in multiple neural networks throughout the body. It is likely that these new technologies will reveal that many currently held beliefs are incomplete or incorrect. However, these new technologies, and those yet to be developed, will undoubtedly lead us to a better understanding of toxicity.

6.5.3.1 Toxicogenomics—Risk Assessment of Delayed Long-Term Effects of Multiple Chemicals in Human Fetuses

Previous studies analyzing umbilical cords show that human fetuses in Japan are exposed to multiple chemicals. Because of these findings, we believe it is necessary to establish a new strategy for examining the possible delayed long-term effects caused by prenatal exposure to multiple chemical combinations and evaluating the health risk to human fetuses. In this commentary, we describe our attempts to apply toxicogenomic analysis of umbilical cords, using DNA microarrays for future risk assessment. Because the umbilical cord is part of the fetal tissue, it is possible to estimate the effects of chemicals on the fetus by analyzing the alteration of gene expression. This type of toxicogenomic analysis could be a powerful and effective tool for developing a new risk assessment strategy to help investigators understand and possibly prevent long-term effects caused by fetal exposure to multiple chemicals. Worldwide cooperation is needed to establish a new strategy for risk assessment using toxicogenomic analysis that focuses on the human fetus.

6.5.3.2 Toxicogenomics Approaches—Improving Prediction of Chemical Carcinogenicity

Although scientific knowledge of the potential health significance of chemical exposures has grown, experimental methods for predicting the carcinogenicity of environmental agents have not been substantially updated in the last two decades. Current methodologies focus first on identifying genotoxicants under the premise that agents capable of directly damaging DNA are most likely to be carcinogenic to humans. Emphasis on the distinction between genotoxic and nongenotoxic carcinogens is also motivated by assumed implications for the dose–response curve; it is purported that genotoxicants would lack a threshold in the low-dose region, in contrast with nongenotoxic agents. However, for the vast majority of carcinogens, little if any empirical data exists to clarify the nature of the cancer dose–response relationship at low doses in the exposed human population. Recent advances in the scientific understanding of cancer biology—and increased appreciation of the multiple effects of carcinogens on this disease process—support the view that environmental chemicals can act through multiple toxicity pathways, modes, or mechanisms of action to induce cancer and other adverse health outcomes. Moreover, the relationship between dose and a particular outcome in an individual could take multiple forms depending on genetic background, target tissue, internal dose, and other factors besides mechanisms or modes of action; interindividual variability and susceptibility in response are, in turn, key determinants of the population dose–response curve. New bioanalytical approaches (e.g., transcriptomics, proteomics, and metabolomics) applied in human, animal, and *in vitro* studies could better characterize a wider array of hazard traits and improve the ability to predict the potential carcinogenicity of chemicals.

6.5.3.3 Computational Toxicology—A Tool for Early Safety Evaluation

Although inappropriate pharmacokinetic properties were a major cause of attrition in the 1990s, safety issues are recognized as today's single largest cause of drug candidate failure. It is expected that the right balance of *in vivo, in vitro,* and computational toxicology predictions, applied as early as possible in the discovery process, will help reduce the number of safety issues. This review focuses on recent developments in computational toxicology. Direct modeling of toxic end points has been deceiving and hampered the wide acceptance of computer predictions. The current trend is to make simpler predictions, closer to the mechanism of action, and to follow them up with *in vitro* or *in vivo* assays.

6.6 Conclusion

The suggestion that toxicogenomic data such as changes in gene expression, protein levels, or metabolite levels may be used in risk assessment creates considerable unease with some stakeholders. Conversely, others are actively pursuing toxicogenomic approaches to identify putative high-throughput biomarkers to rank and prioritize lead candidates that warrant further development. The anxiety on one hand, and enthusiasm on the other, has created a discord within the risk assessment community on how to proceed with toxicogenomics. The concerns are justifiable due to the potential naïve and premature use of the data, which could have dire consequences for all stakeholders including the general public. Nevertheless, there is general agreement that toxicogenomics will play an increasingly larger role in regulatory decision-making. Given the opportunity, effective and productive communication and collaboration will be a critical factor in establishing protocols for the interpretation and incorporation of toxicogenomics into quantitative risk assessment, which will iteratively evolve in the presence of existing strategies as all stakeholders gain further experience with these emerging technologies.

6.7 Limitations

For toxicogenomic technologies to be widely applied to screen compounds for toxicity, it will be essential to demonstrate that toxicogenomic-based screens are sufficiently robust and comprehensive for a wide range of compounds and toxic end points. There are several aspects of this generalizability.

- Because the screening of environmental chemicals has different requirements than the screening of candidate drugs, algorithms designed for screening candidate drugs will need to be modified and validated for environmental chemicals.

- The database of toxicogenomic and traditional toxicologic data used for training and validating screening algorithms must be extensive and comprehensive enough to allow accurate predictions for a wide range of chemicals.
- The costs of toxicogenomic assays must be compatible with implementation in screening.
- Throughput of toxicogenomic assays must meet or exceed norms for currently used screens.

References

1. Jonic, S., T. Jankovic, V. Gajic, and D. Popovic. 1999. Three machine learning techniques for automatic determination of rules to control locomotion. *IEEE Transactions on Biomedical Engineering* 46:300–310.
2. Tennant, R.W. 2002. The National Center for Toxicogenomics: Using new technologies to inform mechanistic toxicology. *Environmental Health Perspectives* 110:A8–A10.
3. Kreuzer, K.A., U. Lass, A. Bohn, O. Landt, and C.A. Schmidt. 1999. LightCycler technology for the quantitation of bcr/abl fusion transcripts. *Cancer Research* 59:3171–3174.
4. Brown, P.O., and D. Botstein. 1999. Exploring the new world of the genome with DNA microarrays. *Nature Genetics* 21:33–37.
5. Saeed, A.I., V. Sharov, J. White, J. Li, W. Liang, N. Bhagabati et al. 2003. TM4: A free, open-source system for microarray data management and analysis. *Biotechniques* 374–378.
6. Gunderson, K.L., F.J. Steemers, G. Lee, L.G. Mendoza, and M.S. Chee. 2005. A genome-wide scalable SNP genotyping assay using microarray technology. *Nature Genetics* 37(5):549–554.
7. Davis, J.W., F.M. Goodsaid, C.M. Bral, L.A. Obert, and G. Mandakas. 2004. Quantitative gene expression analysis in a nonhuman primate model of antibiotic-induced nephrotoxicity. *Toxicology and Applied Pharmacology* 200:16–26.
8. Friedman, R.C., K.K. Farh, C.B. Burge, and D.P. Bartel. 2009. Most mammalian mRNAs are conserved targets of microRNAs. *Genome Research* 19:92–105.
9. Sotiropoulou, G., G. Pampalakis, E. Lianidou, and Z. Mourelatos. 2009. Emerging roles of microRNAs as molecular switches in the integrated circuit of the cancer cell. *RNA* 15:1443–1461.
10. Aebersold, R., and M. Mann. 2003. Mass spectrometry–based proteomics. *Nature* 422(6928): 198–207.
11. Liebler, D.C., B.T. Hansen, J.A. Jones, H. Badghisi, and D.E. Mason. 2003. Mapping protein modifications with liquid chromatography-mass spectrometry and the SALSA algorithm. *Advances in Protein Chemistry* 65:195–216.
12. Righetti, P.G., and A. Bossi. 1997. Isoelectric focusing in immobilized pH gradients: Recent analytical and preparative developments. *Analytical Biochemistry* 247(1):1–10.
13. Von Eggeling, F., A. Gawriljuk, W. Fiedler, G. Ernst, U. Claussen, J. Klose et al. 2001. Fluorescent dual colour 2D-protein gel electrophoresis for rapid detection of differences in protein pattern with standard image analysis software. *International Journal of Molecular Medicine* 8(4):373–377.
14. Gygi, S.P., G. Corthals, L. Zhang, Y. Rochon, and R. Aebersold. 2000. Evaluation of two-dimensional gel electrophoresis-based proteome analysis technology. *Proceedings of the National Academy of Sciences of the United States of America* 97(17):9390–9395.
15. Ishimura, R., S. Ohsako, T. Kawakami, M. Sakaue, T. Aoki, and C. Tohyama. 2002. Altered protein profile and possible hypoxia in the placenta of 2,3,7,8-tetrachlorodibenzo-p-dioxin exposed rats. *Toxicology and Applied Pharmacology* 185(3):197–206.
16. Xu, X.Q., C.K. Leow, X. Lu, X. Zhang, and J.S. Liu. 2004. Molecular classification of liver cirrhosis in a rat model by proteomics and bioinformatics. *Proteomics* 4(10):3235–3245.

17. Wolters, D.A., M.P. Washburn, and Y.R. Yates, III. 2001. An automated multi-dimensional protein identification technology for shotgun proteomics. *Analytical Chemistry* 73(23):5683–5690.

18. Washburn, M.P., R. Ulaszek, R. Deciu, D.M. Schieltz, and J.R. Yates, III. 2002. Analysis of quantitative proteomic data generated via multidimensional protein identification technology. *Analytical Chemistry* 74(7):1650–1657.

19. MacCoss, M.J., W.H. McDonald, A. Saraf, R. Sadygov, J.M. Clark, J.J. Tasto et al. 2002. Shotgun identification of protein modifications from protein complexes and lens tissue. *Proceedings of the National Academy of Sciences of the United States of America* 99(12):7900–7905.

20. Robertson, D.G., M.D. Reily, R.E. Sigler, D.F. Wells, D.A. Paterson, and T.K. Braden. 2000. Metabonomics: Evaluation of nuclear magnetic resonance (NMR) and pattern recognition technology for rapid in vivo screening of liver and kidney toxicants. *Toxicology Science* 57:326–337.

21. Nicholls, A.W., E. Holmes, J.C. Lindon, J.P. Shockcor, R.D. Farrant, and J.N. Haselden. 2001. Metabonomic investigations into hydrazine toxicity in the rat. *Chemical Research in Toxicology* 14:975–987.

22. Waters, M.D., K. Olden, and R.W. Tennant. 2003. Toxicogenomic approach for assessing toxicant-related disease. *Mutation Research* 544:415–424.

23. Holmes, E., J.K. Nicholson, and G. Tranter. 2001. Metabonomic characterization of genetic variations in toxicological and metabolic responses using probabilistic neural networks. *Chemical Research in Toxicology* 14:182–191.

24. Stoyanova, R., J.K. Nicholson, J.C. Lindon, and T.R. Brown. 2004. Sample classification based on Bayesian spectral decomposition of metabonomic NMR data sets. *Analytical Chemistry* 76(13):3666–3674.

25. Griffin, J.L., S.A. Bonney, A. Mann, A.M. Hebbachi, and G.F Gibbons. 2001. An integrated reverse functional genomics and metabolic approach to understanding orotic acid–induced fatty liver. *Physiology Genomics* 17(2):140–149.

26. Robosky, L.C., D.F. Wells, D.A. Egnash, M.L. Manning, M.D. Reily, and D.G. Robertson. 2005. Metabonomic identification of two distinct phenotypes in Sprague-Dawley (Crl:CD[SD]) rats. *Toxicology Science* 87(1):277–284.

27. Lindon, J.C. 2003. Contemporary issues in toxicology: The role of metabonomics in toxicology and its evaluation by the COMET project. *Toxicology and Applied Pharmacology* 187:137–146.

28. Wilson, I.D., R. Plumb, J. Granger, H. Major, R. Williams, and E.M. Lenz. 2005. HPLC-MS-based methods for the study of metabonomics. *Journal of Chromatography. B, Analytical Technologies in the Biomedical and Life Sciences* 817(1):67–76.

29. Gerber, S.A., J. Rush, O. Stemman, M.W. Kirschner, and S.P. Gygi. 2003. Absolute quantification of proteins and phosphoproteins from cell lysates by tandem MS. *Proceedings of the National Academy of Sciences of the United States of America* 100(12):6940–6945.

30. Lum, P.Y., Y.D. He, J.G. Slatter, J.F. Waring, and N. Zelinsky 2007. Gene expression profiling of rat liver reveals a mechanistic basis for ritonavir-induced hyperlipidemia. *Genomics* 90:464–473.

31. Stierum, R., W. Heijne, A. Kienhuis, B. van Ommen, and J. Groten. 2005. Toxicogenomics concepts and applications to study hepatic effects of food additives and chemicals. *Toxicology and Applied Pharmacology* 207(2 Suppl):179–188.

32. Pitot, III H.C., and Y.P. Dragan. 2001. Chemical carcinogenesis. In *Casarett & Doull's Toxicology*, edited by Klaassen, C.D., vol. 6:266–280. New York: McGraw-Hill Companies Inc.

33. Li, Z., J. Dullmann, B. Schiedlmeier, M. Schmidt, C. von Kalle, J. Meyer et al. 2002. Murine leukaemia induced by retroviral gene marking. *Science* 296:497.

34. Ueda, K., and D. Ganem. 1996. Apoptosis is induced by N-myc expression in hepatocytes, a frequent event in hepadnavirus oncogenesis and is blocked by insulin-like growth factor II. *Journal of Virology* 70:1375–1383.

35. Ellisen, L.W., R.E. Palmer, R.G. Maki, V.B. Truong, and A. Tamayo. 2000. Cascades of transcriptional induction during human lymphocyte activation. *European Journal of Cell Biology* 80:321–328.

36. Hamalainen, H., H. Zhou, W. Chou, H. Hashizume, R. Heller, and R. Lahesmaa. 2001. Distinct gene expression profiles of human type 1 and type 2 T helper cells. *Genome Biology* 2(7):1–11.
37. Schaffer, A.L., A. Rosenwald, E.M. Hurt, J.M. Giltnane, L.T. Lam, and L.M. Staudt. 2001. Signatures of the immune response. *Immunity* 15:375–385.
38. Schmidt-Weber, C.B., J.G. Wohlfahrt, and K. Blaser. 2001. DNA arrays in allergy and immunology. *International Archives of Allergy and Immunology* 126:1–10.
39. Novak, R.F., and K.J. Woodcroft. 2000. The alcohol-inducible form of cytochrome P450 (CYP 2E1): Role in toxicology and regulation of expression. *Archives of Pharmaceutical Research* 23:267–282.
40. Stevens, J.L., H. Liu, M. Halleck, R.C. Bowes, Q.M. Chen, and B. van de Water. 2000. Linking gene expression to mechanisms of toxicity. *Toxicology Letters* 112–113.
41. Andersson, A.M., K.M. Grigor, E.R.D. Meyts, H. Leffers, and N.E. Skakkebaek, eds. 2001. *Hormones and Endocrine Disrupters in Food and Water: Possible Impact on Human Health.* Copenhagen: Munksgaard.
42. Safe, S.H. 2001. Endocrine disruptors and human health—is there a problem? An update. *Environmental Health Perspectives* 108:487–493.
43. Newbold, R.R. 2001. Effects of developmental exposure to diethylstilbestrol (DES) in rodents: Clues for other environmental estrogens. *APMIS: Acta Pathologica, Microbiologica, et Immunologica Scandinavica* 109:S261–S271.
44. Newbold, R.R., B.C. Bullock, and J.A. McLachlan. 1984. Müllerian remnants of male mice exposed prenatally to diethylstilbestrol. *Teratogenesis, Carcinogenesis, and Mutagenesis* 7:337–389.
45. Williams, K., C. McKinnell, P.T.K. Saunders, M. Walker, J.S. Fisher, K.J. Turner et al. 2001. Neonatal exposure to potent and environmental oestrogens and abnormalities of the male reproductive system in the rat: Evidence for importance of the androgen-oestrogen balance and assessment of the relevance to man. *Human Reproduction Update* 7:236–247.

7

The World of Noncoding RNAs

Winnie Thomas, Qurratulain Hasan, Konagurtu Vaidyanath, and Yog Raj Ahuja

CONTENTS

7.1 Introduction

Noncoding RNAs (ncRNAs) gained international attention when Fire and colleagues [1] discovered the ability of double-stranded RNA (dsRNA) to silence gene expression in *Caenorhabditis elegans.* The discovery of small interfering RNAs (siRNA) earned Andrew Z. Fire and Craig C. Mello the Nobel prize in Medicine and Physiology in 2006. ncRNAs interact *inter se* with messenger RNAs (mRNAs), DNA, and proteins to form networks that can regulate gene activity and have infinite potential [2].

7.1.1 Evolutionary Significance of ncRNAs

A large number of ncRNAs have been identified, and many are highly conserved within the animal or plant kingdom but are divergent between the two kingdoms. A comparative study showed that the complexity of an organism is inversely proportional to the protein coding fraction of its genome (i.e., 90% in prokaryotes, 25% in nematodes, and 2% in mammals). Conversely, the proportion of noncoding DNA has a linear increase with the evolutionary scale [3]. Because ncRNAs are part of the transcribed nontranslated portion of the genomes, they seem to be the molecular difference in the evolution of complex organisms [4].

7.1.2 Functions of ncRNAs

ncRNAs can regulate transcription by interacting with transcription factors, RNA polymerase, or DNA itself. The small dsRNA, neuron-restrictive silencer element (NRSE), directs the activation of neuronal genes containing NRSE sequences by triggering the RE-1-silencing transcription factor transcriptional machinery [5] components, which are also targets of micro-RNAs (miRNAs) [6].

Large numbers of ncRNAs has been identified and among these are small ncRNAs, such as siRNAs (19–21 nt), miRNAs (22–25 nt), piwi-interacting RNAs (26–31 nt), and short nucleolar (sno) RNAs (60–300 nt) involved in gene expression. In addition, several long ncRNAs, ranging from 0.5 to more than 100 kb, regulate gene expression by modifying chromatin structure.

7.2 Small ncRNAs

Small ncRNAs constitute one of the largest families of gene regulators that are found in plants and animals. These are generally produced by fragmentation of longer precursors [7]. Most mature small ncRNAs are the product of RNA polymerase II–transcribed transcripts that have been processed by two RNase III enzymes, Drosha and Dicer. They are regulatory molecules that, besides protecting cells from the intrusion of any exogenous nucleic acids (like viruses), are involved in maintaining genomic integrity by silencing transcription from undesired loci (retrotransposons and repeat sequences) [8].

A large class of small RNAs (sRNAs) is involved in the regulation of gene expression primarily at posttranscriptional level. miRNAs and siRNAs represent this class of sRNA in eukaryotes, whereas Hfq-binding sRNAs are the major base pairing sRNAs in bacteria. It was earlier assumed that Hfq-binding sRNAs do not encode proteins and execute their functions only through base pairing, referred to as riboregulation. Later, Wadler and Vanderpool [9] demonstrated that an Hfq-binding sRNA of *Escherichia coli* acts not only by base pairing but also by serving as an mRNA template for a small functional protein to deal with the same metabolic stress. Thus far, no example for eukaryotic sRNAs encoding a small protein is known.

7.2.1 miRNAs

Genes for miRNAs are located in all chromosomes and half of all miRNAs are identified in clusters [10]. miRNA genes are found in introns of noncoding or coding genes and in exons of noncoding genes [11].

7.2.1.1 Biogenesis of miRNA

miRNAs are transcribed by RNA polymerase II (pol II) as precursor molecules. These long primary transcripts of miRNA genes (pri-miRNA) are subsequently cleaved by Drosha (an RNase III endonuclease) to produce a stem-loop structured precursor (pre-miRNA) approximately 70 nt long. Precursor miRNA with a short hairpin structure is delivered by Exportin-5 (Exp5) through the nuclear pores to the cytoplasm [12]. Then, precursor miRNAs are processed by Dicer, which chops long dsRNAs into approximately 22-nt duplexes of mature miRNAs. Dicer-interacting proteins are helpful in unwinding dsRNAs and loading one stand of the ds-miRNA onto the effector complex, miRNA-induced silencing complex (micro-RISC). The other strand (passenger strand) of the ds-miRNA is destroyed [13]. Figure 7.1 represents the overall process of miRNA biogenesis.

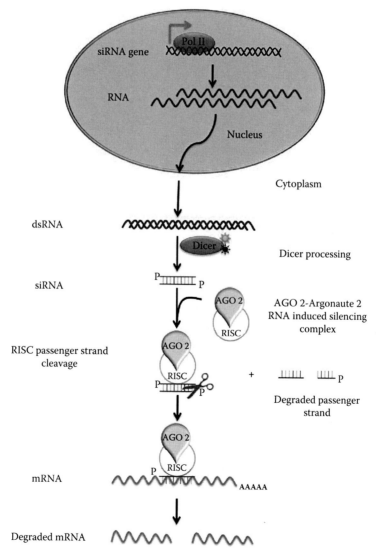

FIGURE 7.1
Biogenesis of miRNA.

7.2.1.2 Mechanism of Action of miRNA

The miRNA negatively regulates their targets posttranscriptionally in one of two ways depending on the degree of *complementarity* between ncRNA and the target. First, miRNAs that bind with perfect or nearly perfect *complementarity* to protein coding mRNA sequences induce the RNA-mediated interference (RNAi/siRNA) silencing complex (RISC). This mechanism of RNA-mediated gene silencing is commonly found in plants [14]. Second, miRNAs exert their regulatory effects by binding to imperfect complementarity sites within the 3′ untranslated regions (UTR) of their mRNA targets repressing target gene expression. Reduction of protein levels in this way is often modest, and many such RNAs probably collectively fine-tune gene expression [15]. Although the protein levels of these genes are dampened, the mRNA levels of these genes are barely affected. This mechanism of miRNA-mediated gene expression control is mostly seen in animals [16]. miRNAs and their targets seem to perform complex regulatory networks. For example, a simple miRNA can bind to and regulate many different mRNA targets and, conversely, several different miRNAs can bind to and cooperatively control a single mRNA target [17]. When a miRNA is misexpressed, it has the potential to regulate many targets that it might never encounter in its endogenous expression domain [18].

7.2.1.3 Functions of miRNA

Although most miRNAs play roles in full or partial gene silencing, Kuwabara et al. [5] reported that miRNAs could also function in gene expression in adult neural stem cells. They are involved both in the transcriptional activation and coactivation of many genes [19]. It has been estimated that there are at least 1000 known functional miRNA genes [20], and more than one-third of all known human genes are probably regulated by miRNAs [21]. Whereas some miRNAs are ubiquitously expressed, others have an expression pattern that depends on the developmental stage or cell type [22]. miRNAs in the gametes may have a direct role in the differentiation and development of the early zygote or may play a part in postfertilization epigenetic reprogramming. miRNAs, which are expressed in the mammalian brain at different levels, seem to be critical at dictating neuronal cell identity during development and play a pivotal role in neurite growth, synaptic development, neuronal plasticity, and possibly in learning and memory [23]. In short, miRNAs have key roles in diverse regulatory pathways, including control of metabolism, immunity, developmental timing, cell proliferation, cell differentiation, organ development, senescence, and apoptosis.

There is much more dynamic interplay between genome and environment, and miRNAs have an important role in this interplay [24]. miRNAs can direct both genetic and epigenetic modifications. Recent studies have shown that miRNAs have a significant role in determining chromatin structure and gene regulation [25]. miRNAs can direct the cytosine methylation and histone modifications that are implicated in chromatin states and in turn in gene expression regulation [26,27]. Also, miRNAs regulate the cell cycle in embryonic stem cells [28]. Evidence suggests that miRNAs may control epigenetic notations of specific regions of the genome, providing a mechanism for precise DNA and histone modifications [29].

miRNAs take part in regulating cell activity by generating pathways parallel to others already present. In this way, they guarantee their support, allowing either the reinforcement of signaling in normal condition, or its partial maintenance in case the parallel pathway is lost. This kind of action can be exemplified by MiR-34a/b/c activity. This family of miRNAs participates in the execution of p53 tumor suppressor activity [30,31].

Chromatin modification and associated epigenetic memory are regulated by miRNA signaling. Abnormal expression of miRNAs can disrupt signaling networks in the cells, resulting in pathological changes. Mutated or inappropriately expressed miRNAs are involved in most human cancers, obesity, diabetes, hair loss, brain disease, and skeletal muscle injury [32,33]. Depletion of Dicer in human cells leads to a significant enhancement of Ataxin-3–induced toxicity [34]. The expression of polyglutamate repeats in Ataxin-3 has been linked to neurodegeneration. Which miRNAs are involved in this human neurodegenerative disorder remains to be determined.

miRNA levels change over the life of an individual and are associated with many disorders acquired during the aging process [35]. Recent evidence suggests that miRNAs may be a contributing factor in neurodegeneration leading to Parkinson disease and Alzheimer's disease [36,37]. Postmortem brain studies of schizophrenia have revealed changes in the expression of certain proteins involved in synaptic neurotransmission and development. Changes in the expression of these proteins seem to be due to alterations in the levels of miRNAs, for example, miR181b and miR219, in the cortex [38,39]. Emerging evidence suggests that miRNAs play significant roles in the production, action, and secretion of insulin and also in diverse aspects of the glucose and lipid metabolisms. Not only are the altered levels of miRNAs critical in the development and progression of diabetes but also in diabetic complications such as nephropathy, retinopathy, and cardiac hypertension [40].

7.2.1.3.1 *miRNAs and Centromere Behavior*

The centromere is of vital importance for genetic stability. It is the DNA region which ensures the separation of chromosomes in mitosis and meiosis. Defects in chromosome segregation are associated with human disease. For example, defects in meiotic chromosome segregation may lead to the production of aneuploid embryos with too few or too many chromosomes [41], and mitotic chromosome segregation errors may contribute to tumor formation [42]. It is fascinating to know that centrosome behavior (centromere assembly and propagation) is governed by epigenetic mechanisms. A centromeric histone variant, CenH3, also known as centromere protein-A (CENP-A), and other histone modifications play key roles at centromeric chromatin in determining centromere identity and kinetochore assembly. In addition, many CENP-A–interacting proteins and factors that affect its localization have been identified [43]. Pezer and Ugarkovic [44] have proposed that miRNA derived from centromeric repeats plays an active role, mostly through the RNAi pathway, in the formation of pericentromeric and centromeric heterochromatin, both of which are important for proper centromere function.

The formation of the kinetochore and centromeric heterochromatin in fission yeast is dependent on cell cycle–regulated centromeric repeat–derived RNAs and RNAi pathways. The RNAi pathway, along with directed histone modifications, also regulates the organization of the nucleolus in *Drosophila* [45].

7.2.1.3.2 *miRNAs and Trans-Splicing Events*

Noncoding miRNAs may also be involved in trans-splicing events in other genes. For example, CDC2L2 gene located on chromosome 1 is associated with fine-tuning of the cell division related and apoptotic activities in testis [46]. Recently, our team has shown that CDC2L2 mRNA is trans-spliced by a ncRNA transcript from Y chromosome. In the long arm of Y chromosome there is a major heterochromatic block, and there is a consensus sequence of GGAAT in this block, which is transcribed specifically in the testis. A ncRNA transcript from Yq12 in the heterochromatic block trans-splices the CDC2L2 mRNA from a chromosome 1p36.3 locus to generate a testis-specific chimeric βsr13 isoform of CDC2L2.

This trans-splicing event could be involved in regulating the activity of CDC2L2 in the testis [47].

7.2.1.3.3 miRNAs and Cancer

miRNAs posttranscriptionally regulate the target mRNA transcripts. Many of these target mRNA transcripts are involved in cell proliferation, differentiation and apoptosis, processes commonly altered during tumorigenesis [48]. Several miRNAs have emerged as candidate components of oncogenes as well as tumor suppressor regulators. For example, miR372 and miR373 have been implicated as proto-oncogenes in testicular cancer [49] and let-7 has shown a potential as tumor suppressor in various cancers [50]. Loss of let-7 correlates with over-expression of Ras proteins [51]. Recent studies have shown that miR-34 family is direct transcriptional target of p53. MiR-34 activation can mimic p53 activities, including the induction of cell cycle arrest and promotion of apoptosis. Loss of miR-34 can impair p53 mediated cell death [30,31].

Reduced levels of certain miRNAs have been seen in cancer cells in comparison with normal cells [52], and this reduction could be due to the reduced expression of Drosha or Dicer, which are involved in the biogenesis of miRNAs. For example, reduced expression of Dicer is associated with a poor prognosis in lung cancer [53]. Again, through the analysis of human and mouse models of B-cell lymphoma, it was shown that a prominent consequence of oncogenic overactivation of Myc is widespread repression of miRNA expression. Much of this repression is likely to be a direct result of Myc binding to miRNA promoters. When the expression of repressed miRNA was strengthened, it diminished the tumorigenic potential of lymphoma cells [54].

Chromosomal regions harboring miRNAs are sites of frequent genomic alterations involved in cancer. The main genetic alterations related to miRNA expression deregulation are chromosomal abnormalities, although some mutations have also been described. It is remarkable that more than 50% of miRNA genes are located in fragile sites, minimal regions of amplification, or regions of loss of heterozygosity and common breakpoint regions frequently associated with cancer [55]. B-cell chronic lymphocytic leukemia, associated with the over-expression of antiapoptotic oncogenic protein Bcl-2, is the most common adult leukemia in developed countries. Chronic lymphocytic leukemia is commonly associated with the loss of chromosomal region 13q14 [56], and within this deleted area are the transcription sites of miR-15a and miR-16-1, which seem to inhibit Bcl-2 protein activity [57].

In the United States, prostate cancer affects African American males at a much higher rate than Caucasian males. No target genes have thus far been identified for this difference. A microarray study on 10 Caucasians and 8 African Americans showed significantly overexpressed as well as downregulated miRNAs in African American compared with Caucasian individuals. The list of miRNAs that were at least three times differentially expressed included miR-301, miR-26a, miR-1b-1, and miR-30c-1 [56].

There are times when the behavior of a miRNA may be erratic. miRNAs being designated as oncogenes or antioncogenes, depending on the level of their expression, may not always be consistent. Emerging evidence has shown that many of the miRNAs may have a dual role. Variability created by SNPs in the binding sites of miRNAs and their target sites on mRNA, along with other surrounding influences due to genetic and epigenetic architecture, may determine the role of the miRNA; whether it is going to behave as an oncogene or an antioncogene [58]. Leung et al. [59] found that all Argonaute family members are relocated within the cell upon stress. The association of Argonaute with other RNA-binding proteins has recently been shown to switch a normally repressive miRNA into an activator of its target when cells are under stress, for example, during the starvation of serum or amino acids [60,61]. Tumor cells experience a variety of stresses, such as hypoxia, nutrient deficiency, and of course, exposure to irradiation and chemotherapy during treatment. For example, in one context, the c13orf25 cluster miRNAs can act as an oncogene, whereas in another context, it seems to antagonize the effects of different oncogenes, acting like a classic tumor suppressor gene. In addition to stress, short miRNA expression can be altered by several other mechanisms in human cancer: chromosomal abnormalities, mutations and polymorphisms (SNPs), defects in miRNA biogenesis (altered Drosha or Dicer activity), and epigenetic changes such as altered DNA methylation and histone deacetylase inhibition [62]. Sorting out the miRNA regulatory networks will be challenging.

7.2.2 siRNAs

The siRNA are formed not only from mRNA but also from noncoding transcripts such as centromeric DNA transcripts and, in particular, from long terminal repeat (LTR) of transposons [63].

7.2.3 Biogenesis of siRNA

siRNAs are processed from dsRNA that form by base-pairing of complementary RNA with the single-stranded RNA transcript. Dicer cleaves dsRNA into shorter approximately 20-nt-long double-stranded siRNAs. One siRNA strand then assembles into an RNA-induced silencing complex (RISC). The main components of the RISC complex are proteins of the Argonaute family. Ago2 is the sole enzyme capable of endonucleolytic cleavage in mammals [64]. The siRNA in this complex then identifies the mRNA based on sequence complementarity. RISC then cleaves the mRNA in the middle and the resulting mRNA halves are degraded by other cellular enzymes. Biogenesis of siRNA is represented in Figure 7.2.

7.2.4 Functions of siRNA

Although often described as junk DNA sequences, heterochromatin is essential for normal chromosomal organization [65], and centromere and telomere function [66,67]. In addition, it functions to silence gene expression, reduce the frequency of recombination, promote long-range chromatin interactions, and ensure accurate chromosome segregation during mitosis [68,69]. Pal-Bhadra et al. [70] investigated the mechanism of heterochromatin silencing in Drosophila and noticed that, in addition to DNA methylation and histone modifications, RNAi machinery had a significant role to play in it.

siRNA has an important role in defending cells against parasitic genes such as viruses and transposons. There are various studies which report that siRNA induced interferon response [71–75]. However, this response varies between species and cell types; in one study, siRNA generated no interferon response in mice, whereas the same siRNA caused a potent response in human monocytes [76]. siRNAs can also activate dendritic cells through Toll-like receptors (TLRs), a subset of which (TLR3,TLR7, TLR8, and TLR9) recognizes foreign double-stranded nucleic acids [71,74]. Hornung et al. [71] recently identified a nine-nucleotide sequence in siRNA that activated the TLR7, and thereby caused an immune response. Immunostimulatory motifs in siRNA have also been noted in another study [73]. These studies now allow siRNA to be designed for the stimulation of immune responses.

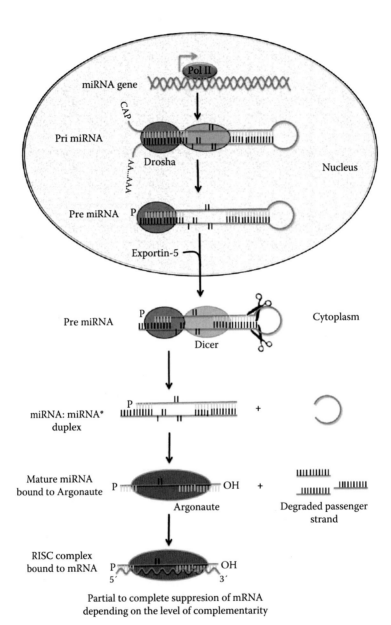

FIGURE 7.2
(See color insert.) Biogenesis of siRNA.

7.2.4.1 siRNA and HIV

HIV was the first infectious agent targeted by siRNAi, synthetic siRNAs and expressed shRNAs have been used to target several early and late HIV-encoded RNAs in cell lines and in primary hematopoietic cells including the TAR element [77], *tat, rev* [78], *gag* [79], *env* [80], *vif, nef* [77] and reverse transcriptase [81]. Despite the success of RNAi-mediated inhibition of HIV-encoded RNAs in cell culture, targeting the virus directly represents a substantial challenge for clinical applications because the high viral mutation rate will lead to mutants that can escape being targeted [82].

7.2.4.2 *siRNA and Neurodegeneration*

For neurodegenerative diseases, decreasing the levels of mutant protein and thus preventing the downstream deteriorating effects seems to be one of the best strategies. Efficient and sequence-specific siRNAs have been synthesized to target the mutant gene containing CAG repeat expansion or its adjacent gene sequences [83,84]. Targeting the CAG repeats may set an offsite suppression of normal genes or other genes containing CAG repeats that may have some important biological function. Therefore, siRNAs have been designed which can carry out allele-specific silencing of the diseased gene without affecting its normal counterpart, that is, it can target the mutant gene that differs from normal genes even by one nucleotide [85]. Such a stringent scheme has already been developed for some neurodegenerative diseases such as Machado–Joseph disease (or SCA3), in which the targeted region is the SNP linked to the mutation site. Success in RNAi research has given a new outlook to the treatment of neurodegenerative diseases, which still have no known management options.

siRNA has been used with great success to knock down the expression of selected target genes, with therapeutic value in both *in vitro* and animal models of various diseases such as cancer and neurodegenerative disease. This has led to speculation about siRNA- and RNAi-based therapy being a being a source of a major class of drugs in the future.

7.3 Piwi-Associated RNAs

Piwi-associated RNAs (piRNAs) are generated from long single-stranded transcripts in a process independent of Drosha and Dicer. These sRNAs (26–31 nt) associate with a subfamily of Argonaute proteins called Piwi proteins. Piwi is one subgroup of Argonaute proteins, which is a germ line–specific protein. Tens of thousands of piRNAs have been identified. Together with their Piwi partners, they are essential for the development of germ cells [7]. piRNAs accumulate at the onset of male meiosis, and sperm maturation arrests at different stages in Mili and Miwi knockout mice, piRNAs are thought to play an essential role during gametogenesis [86]. Although the function and targets of piRNAs are unclear, piRNAs function along with piwi to silence transposons by heterochromatin formation from flies to vertebrates [87]. In addition, piRNAs have also been observed in the regulation of euchromatin formation in the subtelomeric regions of *Drosophila* [88].

7.4 Long ncRNAs

Transcription actually occurs essentially everywhere, including both in the coding regions and noncoding regions, and often on both strands [89,90]. Recently, Bumgarner et al. [91] have shown that the interplay between the expressions of two long ncRNAs transcribed on opposite strands can exert epigenetic metastatic control on the transcription of the adjacent protein-encoding FLOW11 gene in yeast (*Saccharomyces cerevisiae*). This regulatory mechanism has a profound effect on the life cycle of yeast. When FLOW11 is on, diploid cells grow in filaments called pseudohyphae, and haploid cells invade the agar when grown in plates; when FLOW11 is off, neither of these events occurs and the cells grow in their familiar budding pattern.

Long ncRNAs exhibit cell type–specific repression [92], localized to specific subcellular compartments [93], and are associated with human diseases [94]. Recent work on Alzheimer's disease has identified an approximately 2 kb ncRNA, which is induced in response to numerous cell stresses, and which increases the stability of the BACE1 mRNA, thus leading to even more Aβ peptides [95], which are involved in neuronal damage.

Long ncRNAs can act on domains ranging in size from a single promoter to an entire chromosome, and they can function *in cis* or *in trans* to establish chromatin conformations which either activate or repress transcription [96]. They can have numerous molecular functions, including modulating transcriptional patterns, regulating protein activities, serving structural or organizational roles, altering RNA processing events, and serving as precursors to sRNAs [97]. In mammals, transcription of long ncRNAs contributes to various processes including T-cell receptor recombination [98], maintenance of telomeres [99], and inactive X chromosome perinuclear localization [100]. Long ncRNAs are also involved in genomic imprinting and X chromosome inactivation.

7.4.1 Long ncRNAs and Genomic Imprinting

RNA transcripts important for gene imprinting were among the first long ncRNAs identified in humans and mouse. Gene imprinting is an epigenetic process that selectively represses one or more genes from a parental allele [101]. It is a *cis*-acting epigenetic process by which a subset of autosomal genes is expressed in a parent of origin-specific manner. Thus far, more than 90 such genes have been identified in mammals and they are generally grouped in clusters of 2 to 12 genes. A typical imprinted cluster consists of several protein coding genes and at least one long ncRNA, which invariably shows bidirectional reciprocal expression to the coding genes [102].

A key feature of imprinted gene clusters is the presence of an imprint control element (ICE). The ICE is epigenetically modified on only one parental chromosome by a DNA methylation "imprint," which is acquired during maternal or paternal gametogenesis and is maintained on the same chromosome in the diploid embryo. The ICE carries histone modifications that are specific to the DNA-methylated allele, that is, repressive marks are associated with the DNA-methylated ICE, whereas active histone marks are associated with the unmethylated ICE [103].

7.4.2 Long ncRNAs and X Chromosome Inactivation

X chromosome inactivation in female mammals is epigenetic dosage compensation by an unknown mechanism. For the inactivation of X chromosome, the long noncoding Xist RNA (inactive X-specific transcript RNA), which operates *in cis*, is essential [104]. The X-inactivation center requires the expression and spread of the Xist ncRNA over one of the X chromosomes to induce a cascade of chromatin changes that ultimately result in transcriptional repression. These changes include the incorporation of the histone variant macroH2A, DNA methylation and recruitment of PcG (polycomb group) proteins. These chromatin changes allow the inactivated X chromosome to be stably silenced at later stages of development, even in the absence of Xist [105]. Tsix (inactive X-specific transcript), a noncoding transcript with antisense orientation of Xist, acts as a negative modulator of Xist expression by blocking Xist RNA accumulation along the future active X chromosome [106].

7.4.3 RNA Granules

In higher organisms, granules that contain protein, mRNAs, and ncRNAs are found in the cytoplasm of somatic and germ cells. Specific components of these RNA granules can alter DNA and RNA sequences and can repress transcription in a form of cytoplasmic inheritance [107]. For example, in cloning, when a somatic nucleus is transferred to an enucleated egg, the cytoplasmic RNA granules present in this egg reprogram the somatic nucleus to a state of totipotency so that embryogenesis can get started in the clone.

During mitosis, germ cell granules shrink and disappear from the part of the cell that will become the soma, and fuse or enlarge in the part of the cell that will remain in the germ line [108]. The differentiation of somatic and germ line daughter cells is determined in part by the absence or presence of germ cell granules. These are also acquired to specify the germ line in the next generation, which represents a clear example of epigenetic inheritance [107].

7.4.4 Paraspeckles

Paraspeckles are ribonucleoprotein bodies in the interchromatin space of mammalian cell nuclei. A number of RNA-binding proteins (including paraspeckle protein complex-1) together with the long RNA-NEAT-1 form paraspeckles [109]. Paraspeckles are not present in human embryonic stem cells but only appear upon differentiation [110]. These bodies are approximately 0.5 to 1.0 μm in size, and their numbers vary within both cell populations depending on the cell type. These structures play a role in regulating the expression of certain genes in differentiated cells [111].

7.4.5 Diagnostic and Therapeutic Potential of ncRNAs

The cell cycle is a tightly orchestrated process during normal development. miRNAs seem to play a central role in achieving this unique cell cycle structure. Manipulating the expression or function of this large family of cell cycle–promoting miRNAs may provide an important therapeutic avenue. Emerging evidence suggests that expression profiling of miRNAs may be used in diagnostics [112]. Different cancer types have distinct miRNA profiles. Accurately predicting miRNA targets for any known miRNA will provide a useful tool to accelerate the progress of miRNA studies in developmental or cancer biology, as well as in pathology. Analyzing miRNA levels in serum is a promising area for identifying biomarkers of cancer [113]. Recent studies have demonstrated that tumor-derived miRNAs can be detected in sputum and blood in a stable form that is protected from endogenous RNase activity. Unlike mRNAs, endogenous miRNAs can be robustly and reliably measured in sputum or plasma/serum over days of storage or freeze–thawing cycles—remaining largely intact [114]. sRNAs can easily be measured from the formalin-fixed tissue specimens used routinely in hospital pathology laboratories; thus potential miRNA-based diagnosis could fit simply into the standard hospital workflow.

Novel therapeutic interventions based on the manipulation of miRNA levels could also be put to use [115]. For example, modified antisense oligonucleotides can artificially regulate the expression of miRNAs [116]. In the same way as classic tumor suppressor genes, in some cases, epigenetic mechanisms are responsible for silencing miRNA expression in cancer, including CpG island methylation and repressive histone modification. miRNA methylation signatures could be useful in the diagnosis and prognosis of cancer [117].

Endogenous cellular miRNAs that target viral RNAs have been reported as well. In one scenario, the cell uses miRNA to impede viral replication. miR-32 restricts the replication of the retrovirus PFC-1 in cell culture [118]. RNAi-based therapies in viral diseases seem promising and are gaining interest [119]. In addition, miRNAs can be integrated into stem cells inducing them to differentiate for the repair or regulation of damaged cells, structures, or tissues [33]. siRNAs have also been used to decrease the drug resistance of cells *in vitro* by inhibiting the repression of MDR1, a multidrug transporter, with a major role in multidrug resistance [120]. miRNA-based therapy could also be useful to enhance sensitivity to conventional drugs used in cancer treatment. Delivery of the molecules to the right cells is a technical hurdle. Many individual miRNAs home in on dozens or even hundreds of genes. It is also important to identify all the genes that miRNAs influence to ensure that modifying their expression will not have untoward effects.

Acknowledgments

We thank P. Diwakar, Raja Rukmini, Rukhsana Sultana, and Saleha Bhanu for literature survey, and Khaliq Mohiuddin and Waseem Gul Lone for secretarial assistance.

References

1. Fire, A., S. Xu, M.K. Montgomery, and S.A. Kostas. 1998. Potent and specific genetic interference by double stranded RNA in *Caenorhabditis elegans*. *Nature* 391:806–881.
2. Mattick, J.S., and M.J. Gagen. 2001. The evolution of controlled multitasked gene networks: The role of introns and other noncoding RNAs in the development of complex organisms. *Molecular Biology and Evolution* 18:1611–1630.
3. Taft, R.J., M. Pheasaunt, and J.S. Mattick. 2007. The relationship between non-protein coding DNA and eukaryotic complexity. *Bioessays* 29:288–299.
4. Costa, F.F. 2008. Non-coding RNAs, epigenetics and complexity. *Gene* 410:9–17.
5. Kuwabara, T., J. Hsieh, and K. Nakashima. 2004. A small modulatory dsRNA specifies the fate of adult neural stem cells. *Cell* 116:779–793.
6. Makeyev, E.V., and T. Maniatis. 2008. Multilevel regulation of gene expression by microRNAs. *Science* 319:1789–1790.
7. Grosshans, H., and W. Filipowicz. 2008. The expanding world of small RNAs. *Nature* 451: 414–416.
8. Naqvi, A.R., M.N. Islam, N.R. Choudhury, and Q.M.R. Haq. 2009. The fascinating world of RNA interference. *International Journal of Biological Sciences* 5:97–117.
9. Wadler, C.S., and C.K. Vanderpool. 2007. A dual function for a bacterial small RNA: SgrS performs base pairing–dependent regulation and encodes a functional polypeptide. *Proceedings of the National Academy of Sciences of the United States of America* 104:20454–20459.
10. Bartel, D.P. 2004. MicroRNAs: Genomics, biogenesis, mechanism and function. *Cell* 116:281–297.
11. Rodriguez, A., S. Griffiths-Jones, J.L. Ashurst, and A. Bradley. 2004. Identification of mammalian microRNA host genes and transcription units. *Genome Research* 14:1902–1910.

12. Yi, R., Y. Qin, I.G. Macara, and B.R. Cullen. 2003. Expotin-5 mediates the nuclear export of pre-microRNAs and short hairpin RNAs. *Genes & Development* 17:13011–3016.
13. Chu, C.-Y., and T.M. Rana. 2007. Small RNAs: Regulation and guardians of the genome. *Journal of Cellular Physiology* 213:412–419.
14. Tang, G., B.J. Reinhart, D.P. Bartel, and P.D. Zamora. 2003. A biochemical framework for RNA silencing in plants. *Genes & Development* 17:49–63.
15. Wu, W., M. Sun, and G.M. Zou. 2007. MicroRNA and cancer: Current status and prospects. *International Journal of Cancer* 120:953–960.
16. Lim, L.P., L.N. Lau, G.P. Engelle, and A.I. Grimson. 2005. Microarray analysis shows that some microRNAs downregulate large numbers of target mRNAs. *Nature* 433:769–773.
17. Lewis, B.P., C.B. Burge, and D.P. Bartel. 2005. Conserved seed pairing, often flanked by adenosines, indicates that thousands of human genes are microRNA targets. *Cell* 120:15–20.
18. Farh, K.K., A. Grimson, C. Jan, B.P. Lewis, and W.K. Johnson. 2005. The wide spread impact of mammalian micro RNAs on mRNA repression and evolution. *Science* 310:1817–1821.
19. Goodrich, J.A., and Kugel, J.F. 2009. From bacteria to humans, chromatin elongation and activation to repression: The expanding roles of non-coding RNAs in regulating transcription. *Critical Reviews in Biochemistry and Molecular Biology* 44:3–15.
20. Filipowicz, W., S.N. Bhattacharyya, and N. Sonenberg. 2008. Mechanism of post-transcriptional regulation by microRNAs: Are the answers in sight? *Nature Reviews Genetics* 9:102–114.
21. Davalos, V., and M. Esteller. 2010. MicroRNAs and cancer epigenetics: A macrorevolution. *Current Opinion in Oncology* 221:35–45.
22. Landgraf, P. 2007. A mammalian microRNA expression atlas based on small RNA library sequencing. *Cell* 129:1401–1414.
23. Presutti, C., J. Rosai, S. Vincenti, and S. Nasi. 2006. Non-coding RNA and brain. *BMC Neuroscience* 7 Suppl 1:S5.
24. Mattick, J.S., P.P. Amaral, M.E. Dinger, T.R. Mercer, and M.F. Mehler. 2009. RNA regulation of epigenetic process. *Bioessays* 31:51–59.
25. Roloff, T.C., and U.A. Nubert. 2008. Chromatin, epigenetic and stem cells. *European Journal of Cell Biology* 84:123–135.
26. Bernstein, E., and C.D. Allis. 2005. RNA meets chromatin. *Genes & Development* 19:1635–1655.
27. Costa, F.F. 2005. Non-coding RNA: New players in eukaryotic biology. *Gene* 357:83–94.
28. Wang, X., and R. Blelloch. 2009. Cell cycle regulation by microRNAs in embryonic stem cells. *Cancer Research* 69:4093–4096.
29. Mattick, J.S., and I.V. Makunin. 2005. Small regulatory RNAs in mammals. *Human Molecular Genetics* 14:R121–R132.
30. He, L., X. He, L.P. Lim, S.E. De, Z. Xuan, Y. Liang et al. 2007. A microRNA component of the p53 tumor suppressor network. *Nature* 447:1130–1134.
31. He, L., X. He, S.W. Lowe, and G.J. Hannon. 2007. MicroRNAs join the p53 network: Another piece in the tumor suppression puzzle. *Nature Reviews. Cancer* 7:819–822.
32. He, L., J.M. Thomason, M.T. Hemann, and E. Hernando-Monge. 2005. A microRNA polycistron as a potential human oncogene. *Nature* 435:828–833.
33. Yang, Z., and J. Wu. 2007. MicroRNAs and regenerative medicine. *DNA and Cell Biology* 26:257–264.
34. Bilen, J., N. Liu, and B.G. Burnett. 2006. MicroRNA pathways modulate polyglutamine induced neurodegeneration. *Molecular Cell* 24:157–163.
35. Zhao, Y., J.F. Ransom, and A. Li. 2007. Dysregulation of cardiogenesis, cardiac conduction and cell cycle in mice lacking mi RNA-1-2. *Cell* 129:303–317.
36. Nelson, P.T., W.X. Wang, and B.W. Rajeev. 2008. MicroRNAs (miRNAs) in neurodegenerative diseases. *Brain Pathology* 18:130–138.
37. Wang, W.X., B.W. Rajeev, and A.J. Stromberg. 2008. The expression of micro RNA miR-107 decreases early in Alzheimer's disease and may accelerate disease progression through regulation of beta-site amyloid precursor protein-cleaving enzyme 1. *Journal of Neuroscience* 28:1213–1223.

38. Beveridge, N.J., P.A. Tooney, A.P. Carroll, and E. Gardiner. 2008. Dysregulation of miRNA-181b in the temporal cortex in schizophrenia. *Human Molecular Genetics* 17:1156–1168.
39. Kocertha, J. 2009. MicroRNA-219 modulates NMDA receptor-mediated neurobehavioral dysfunction. *Proceedings of the National Academy of Sciences of the United States of America* 106:3507–3512.
40. Pandey, A.K., P. Agarwal, K. Kaur, and M. Datta. 2009. MicroRNAs in diabetes: Tiny players in big disease. *Cellular Physiology and Biochemistry* 23:221–232.
41. Lamb, N.E., and T.J. Hassold. 2004. Non-disjunction: A view from ringside. *New England Journal of Medicine* 351:1931–1934.
42. Weaver, B.A., A.D. Silk, C. Montagna, P. Verdier-Pinard, and D.W. Cleaveland. 2007. Aneuploidy acts both oncogenically and as a tumor suppressor. *Cancer Cell* 11:25–36.
43. Allshire, R.C., and G.H. Karpen. 2008. Epigenetic regulation of centromeric chromatin: Old dogs, new tricks? *Nature Reviews Genetics* 9:923–937.
44. Pezer, Z., and C.D. Ugarkovic. 2008. Role of non-coding RNA and heterochromatin in aneuploidy and cancer. *Seminars in Cancer Biology* 18:123–130.
45. Peng, J.C., and G.H. Karpen. 2007. H3K9 methylation and RNA interference regulate nucleolar organization and repeated DNA stability. *Nature Cell Biology* 9:25–35.
46. Gururajan, R., M. Lahti, and J. Grenet. 1998. Duplication of genomic region contaning the Cdc2L1-2 and MMP21-22 genes on human chromosome 1p36.3 and their linkage to D1Z2. *Genome Research* 8:929–939.
47. Jehan, Z., S. Vallinayagam, and S. Tiwari. 2007. Novel noncoding RNA from human Y distal heterochromatic block Yq12. generates testis-specific chimeric CDC2L2. *Genome Research* 17:433–440.
48. Kumar, M.S., J. Lu, and K.L. Mercer. 2007. Impaired MicroRNA processing enhances cellular transformation and tumorigenesis. *Nature Genetics* 39:673–677.
49. Voorhoeve, P.M. 2005. A genetic screen implicates miRNA372 and miRNA373 as oncogenes in testicular germ cell tumors. *Cell* 124:1169–1181.
50. Johnson, S.M., H. Grosshans, and J. Shingara. 2005. RAS is regulated by the let-7 microRNA family. *Cell* 120:635–647.
51. Yanaihara, N., N. Caplen, E. Bowman, M. Seike, K. Kumamoto, M. Yi. et al. 2006. Unique microRNA molecular profiles in lung cancer diagnosis and prognosos. *Cancer Cell* 9:189–198.
52. Lee, Y.S., and A. Dutta. 2008. MicroRNAs in cancer. *Annual Review of Pathology* 4:199–227.
53. Karuba, Y., H. Tanaka, and H. Osada. 2005. Reduced expression of Dicer associated with poor prognosis in lung cancer patients. *Cancer Science* 96:111–115.
54. Chang, Y.C., D. Yu, and Y.S. Lee. 2008. Widespread microRNA repression by Myc contributes to tumorigenesis. *Nature Genetics* 40:43–50.
55. Calin, G.A., C. Sevignani, C.D. Dumitru et al. 2004. Human microRNA genes are frequently located at fragile sites and genomic regions involved in cancer. *Proceedings of the National Academy of Sciences of the United States of America* 101:2999–3004.
56. Calin, G.A., and C.M. Croce. 2007. Chromosome rearrangements and micro RNAs: A new cancer link with clinical implications. *Journal of Clinical Investigation* 117:2059–2066.
57. Cimmino, A., G.A. Calin, and M. Fabbri. 2005. miR-15 and miR-16 induce apoptosis by targeting BCL2. *Proceedings of the National Academy of Sciences of the United States of America* 102:13944–13949.
58. Fabbri, M., N. Valeri, and G.A. Calin. 2009. MicroRNAs and genomic variations: From Proteus tricks to Prometheus gift. *Carcinogenesis* 30:912–917.
59. Leung, A.K., J.M. Calabrease, and P.A. Sharp. 2006. Quantitative analysis of Argonaut protein reveals microRNA-dependent localization to stress granules. *Proceedings of the National Academy of Sciences of the United States of America* 103:18125–18130.
60. Bhattacharyya, S.N., R. Habermancher, and U. Martine. 2006. Relief of microRNA-mediated translational repression in human cells subjected to stress. *Cell* 125:1111–1124.
61. Vasudevan, S., and J.A. Steitz. 2007. AU-rich-element–mediated upregulation of translation by FXR1 and Argonaute 2. *Cell* 128:1105–1118.

62. Iorio, M.V., and C.M. Croce. 2009. MicroRNAs in cancer: Small molecules with a huge impact. *Journal of Clinical Oncology* 27:5848–5856.

63. Hennig, W. 2004. The revolution of the biology of the genome. *Cell Research* 14:1–7.

64. Liu, J., F.V. Rivas, C.G. Marsden, and J.M. Thomson. 2004. Argonaute2 is the catalytic engine of mammalian RNAi. *Science* 305:1437–1441.

65. Dernburg, A.F., J.W. Sedat, and R.S. Hawley. 1996. Direct evidence of a role for heterochromatin in meiotic chromosome segregation. *Cell* 86:135–146.

66. Bernard, P., J.F. Maure, and J.F. Partridge. 2001. Requirement of heterochromatin for cohesion at centromeres. *Science* 294:2539–2542.

67. DeLange, T. 2005. Sheltern: The protein complex that shapes and safeguards human telomeres. *Genes & Development* 19:2100–2110.

68. Jia, S., and T. Yamada. 2004. Heterochromatin regulates cell-type specific long-range chromatin interactions essential for directed recombination. *Cell* 119:469–448.

69. Pidoux, A.L., and R.C. Allshire. 2004. Kinetochore and heterochromatin domains of the fission yeast centromere. *Chromosome Research* 12:521–533.

70. Pal-Bhadra, M., B.A. Leibovitch, S.G. Gandhi, M. Rao, U. Bhadra, J.A. Birchler, and S.C. Elgin. 2004. Heterochromatin silencing and HP 1 localization in Drosophila are dependent on the RNAi machinery. *Science* 303:669–672.

71. Hornung, V., M. Guenthner-Biller, and C. Bourquin. 2005. Sequence-specific potent induction of IFN-alpha by short interfering RNA in plasmacytoid dendritic cells through TLR7. *Nature Medicine* 11, 263.

72. Jackson, A.L., S.R. Bartz, and J. Schelter. 2003. Expression profiling reveals off-target gene regulation by RNAi. *Nature Biotechnology* 21:635–637.

73. Judge, A.D., V. Sood, and J.R. Shaw. 2005. Sequence-dependent stimulation of the mammalian innate immune response by synthetic siRNA. *Nature Biotechnology* 23, 457.

74. Kariko, K., P. Bhuyan, and J. Capodici. 2004. Exogenous siRNA mediates sequence-independent gene suppression by signaling through Toll-like receptor 3. *Cells, Tissues, Organs* 117:132–138.

75. Sledz, C., M. Holko, M. de Veer, R. Silverman, and B. Williams. 2003. Activation of the interferon system by short-interfering RNAs. *Nature Cell Biology* 5:834–839.

76. Sioud, M., and D.R. Sorensen. 2003. Cationic liposome-mediated delivery of siRNAs in adult mice. *Biochemical and Biophysical Research Communications* 312:1220–1225.

77. Jacque, J.M., K. Triques, and M. Stevenson. 2002. Modulation of HIV-1 replication by RNA interference. *Nature* 418:435–438.

78. Coburn, G.A., and B.R. Cullen. 2002. Potent and specific inhibition of human immunodeficiency virus type 1 replication by RNA interference. *Journal of Virology* 76:9225–9231.

79. Novina, C.D., M.F. Murray, D.M. Dykxhoorn, P.J. Beresford, J. Riess, S.K. Lee et al. 2002. siRNA-directed inhibition of HIV-1 infection. *Nature Medicine* 8, 681–686.

80. Park, W.S. 2002. Prevention of HIV-1 infection in human peripheral blood mononuclear cells by specific RNA interference. *Nucleic Acids Research* 30:4830–4835.

81. Surabhi, R.M., and R.B. Gaynor. 2002. RNA interference directed against viral and cellular targets inhibits human immunodeficiency virus type 1 replication. *Journal of Virology* 76:12963–12973.

82. Boden, D., O. Pusch, and F. Lee. 2003. Human immunodeficiency virus type 1 escape from RNA interference. *Journal of Virology* 77:11531–11535.

83. Caplen, N.J., J.P. Taylor, and V.S. Statham. 2002. Rescue of polyglutamine-mediated cytotoxicity by double-stranded RNA–mediated RNA interference. *Human Molecular Genetics* 11:175–184.

84. Miller, V.M., H. Xia, G.L. Marrs, C.M. Gouvion, G. Lee, B.L. Davidson, and H.L. Paulson. 2003. Allele-specific silencing of dominant disease genes. *Proceedings of the National Academy of Sciences of the United States of America* 100:7195–7200.

85. Schwarz, D.S., H. Ding, L. Kennington, J.T. Moore, J. Schelter, J. Burchard et al. 2006. Designing siRNA that distinguish between genes that differ by a single nucleotide. *PLoS Genetics* 2, e140.

86. Girard, A., R. Sachidanandam, G.J. Hannon, and M.A. Carmell. 2006. A germline specific class of small RNAs bind mammalian piwi proteins. *Nature* 442:199–202.

87. Brennecke, J., A.A. Aravin, and A. Starkr. 2007. Discrete small RNA-generating loci as master regulator of transposon activity in *Drosophila*. *Cell* 128:1089–1103.

88. Yin, H., and H. Lin. 2007. An epigenetic activation role of Piwi and Piwi-associated piRNA in *Drosphila melanogaster*. *Nature* 450:304.

89. Berretta, J., and A. Morillon. 2009. Pervasive transcription constitutes a new level of eukaryotic genome regulation. *EMBO Reports* 10:973–982.

90. Mercer, T.R., M.E. Dinger, and J.S. Mattick. 2009. Long non-coding RNAs: Insight into function. *Nature Reviews Genetics* 10:155–159.

91. Bumgarner, S.L., R.D. Dowell, and P. Grisafi. 2009. Toggle involving cis-interacting noncoding RNAs controls variegated gene expression in yeast. *Proceedings of the National Academy of Sciences of the United States of America* 106:18321–18326.

92. Mercer, T.R., M.E. Dinger, S.M. Sunkin, M.F. Mehler, and J.S. Mattick. 2008. specific expression of long noncoding RNAs in the mouse brain. *Proceedings of the National Academy of Sciences of the United States of America* 105:716–721.

93. Sone, M., T. Hayashi, H. Tarui, K. Agata, M. Takeichi, and S. Nakagawa. 2007. The mRNA-like noncoding RNA Gomafu constitutes a novel nuclear domain in a subset of neurons. *Journal of Cell Science* 120:2498–2506.

94. Prasanth, K.V., and D.L. Spector. 2007. Eukaryotic regulatory RNAs: An answer to the "genome complexity" conundrum. *Genes & Development* 21:11–42.

95. Faghihi, M.A., F. Modarresi, and A.M. Khahil. 2008. Expression of a noncoding RNA is elevated in Alzheimer disease and drives rapid feed-forward regulation of β secretase. *Nature Medicine* 14:723–730.

96. Whitehead, J., G.K. Pandey, and C. Kanduri. 2009. Regulation of mammalian epigenome by long noncoding RNAs. *Biochimica et Biophysica Acta* 1790:936–947.

97. Wilusz, J.E., H. Sunwoo, and D.L. Spector. 2009. Long noncoding RNAs: Functional surprises from the RNA world. *Genes & Development* 23:1494–1504.

98. Abarrategui, I., and M.S. Krangel. 2007. Non coding transcription controls down-stream promoters to regulate T cell receptor α recombination. *The EMBO Journal* 26:4380–4390.

99. Schoeftner, S., and M.A. Blaso. 2008. Developmentally regulated transcription of mammalian telomeres by DNA-dependent RNA polymerase II. *Nature Cell Biology* 10:228–236.

100. Zhang, L.F., K.D. Huynh, and J.T. Lee. 2007. Perinucleolar targeting of the inactive X during S phase: Evidence for a role in the maintenance of silencing. *Cell* 129:693–706.

101. Wang, X.Q., J.L. Crutchley, and J. Dostie. 2011. Shaping the genome with non-coding RNAs. *Current Genomics* 125:307–321.

102. Rougeulle, C., and E. Heard. 2002. Antisense RNA in imprinting: Spreading silence through air. *Trends in Genetics* 18:434–437.

103. Koerner, M.V., F.M. Pauler, R. Huang, and D.P. Barlow. 2009. The function of non-coding RNAs in genomic imprinting. *Development* 136:1771–1783.

104. Chow, J., and E. Heard. 2009. X-inactivation and the complexities of silencing a sex chromosome. *Current Opinion in Cell Biology* 21:359–366.

105. Chaumeil, J., P. LeBaccon, A. Wutz, and E. Heard. 2006. *Xist* and the order of silencing. *Genes & Development* 20:2223–2237.

106. Stavropoulos, N., N. Lu, and J.T. Lee. 2001. A functional role for T six transcription in blocking Xist RNA accumulation but not in X-chromosome choice. *Proceedings of the National Academy of Sciences of the United States of America* 98:10232–10237.

107. Anderson, P., and N. Kedersha. 2009. RNA granules: Post-transcriptional and epigenetic modulators of gene expression. *Nature Reviews Molecular Cell Biology* 10:430–436.

108. Gallo, C.M., E. Munro, and D. Rasoloson. 2008. Processing bodies and germ granules are distinct RNA granules that interact in *C. elegans* embryos. *Developmental Biology* 323:76–87.

109. Bond, C.S., and A.H. Fox. 2009. Paraspeckles: Nuclear bodies built on long noncoding RNA *Journal of Cell Biology* 1865:637–644.

110. Chen, L.-L., and G.G. Carmichael. 2009. Altered nuclear retention of mRNAs containing inverted repeats in human embryonic stem cells: Functional role of a nuclear noncoding RNA. *Molecular Cell* 35(4):467–478.

111. Clemson, C.M., J.N. Hutchinson, and S.A. Sara. 2009. An architectural role for a nuclear noncoding RNA: NEAT1 RNA is essential for the structure of paraspeckles. *Molecular Cell* 33:717–726.

112. Calin, G.A., and C.M. Croce. 2006. MicroRNA signatures in human cancers. *Nature Reviews Cancer* 6:857–866.

113. Chin, L.J., and F.J. Slack. 2008. A truth serum for cancer—microRNAs have major potential as cancer biomarkers. *Cell Research* 18:983–984.

114. Mitchell, P.S., R.K. Parkin, and E.M. Kroh. 2008. Circulating microRNAs as stable blood-based markers for cancer detection. *Proceedings of the National Academy of Sciences of the United States of America* 105:10513–10518.

115. Marquez, R.T., and A.P. McCaffery. 2008. Advances in micro RNAs: Implications for gene therapists. *Human Gene Therapy* 19:27–38.

116. Weiler, J., J. Hanziker, and J. Hall. 2006. Anti-microRNA oligonucleotides (AMOs): Ammunition to target microRNAs implicated in human disease? *Gene Therapy* 13:496–502.

117. Lujambio, A. 2007. Genetic unmasking of an epigenetically silenced microRNA in human cancer cells. *Cancer Research* 67:1424–1429.

118. Lecellier, C.H., P. Dunoyer, K. Arar, J. Lehmann-Che, and S. Eyquem. 2005. A cellular micro RNA mediates antiviral defence in human cells. *Science* 308:557–560.

119. Tan, F.L., and J.Q. Yin. 2004. RNAi, a new therapeutic strategy against viral infection. *Cell Research* 146:460–466.

120. Nieth, C., A. Priebsch, A. Stege, and H. Lage. 2003. Modulation of classical multidrug resistance MDR. Phenotype by RNA interference RNAi. *FEBS Letters* 545:144–150.

8

Spliceomics: The OMICS of RNA Splicing

A. R. Rao, Tanmaya Kumar Sahu, and Nishtha Singh

CONTENTS

8.1 RNA Splicing Mechanism

Splicing is highly necessary for a typical eukaryotic mRNA before its translation to a correct protein. It involves a series of reactions that are catalyzed by the spliceosome. The spliceosome is a complex of small nuclear ribonucleoproteins (snRNPs) and protein subunits that removes introns from a transcribed pre-mRNA (hnRNA). There are also certain self-splicing introns that do not need the interference of spliceosomes. The RNA-splicing highly depends on the introns and they are classified on the basis of biochemical analysis of splicing reactions. At least four distinct classes of introns have been identified [1] and are discussed in the following subsections. A typical eukaryotic nuclear intron has consensus sequences defining important regions. Each intron has GU at its 5′ end, a branch site followed by a series of pyrimidines or a polypyrimidine tract (PPT), then by AG at the 3′ end [2]. Generally, the nucleotide *A* is always found at the branch point and the consensus around this sequence varies to some extent.

8.1.1 Classification of Introns

There are generally four types of introns based on the nature of splicing. A fifth type of intron also exists but the biochemical apparatus mediating its splicing has not been well investigated. The following are the five types of introns:

1. *Introns spliced by spliceosomes:* These types of introns are generally present on the nuclear protein-coding genes and are removed by spliceosomes.

2. *Introns spliced by proteins:* Certain enzymes like tRNA splicing enzymes also help in intron splicing. These enzymes splice the introns in nuclear and archaeal tRNA genes.

3. *Group I introns:* Large, self-splicing ribozymes that catalyze their own excision from mRNA, tRNA, and rRNA precursors in various organisms. The secondary structure of group I introns contain 10 paired elements (P1–P10) that fold to form two essential domains [3], that is, P4 to P6 are formed by the stacking of P5, P4, and P6 helices and P3 to P9 by P8, P3, P7, and P9 helices [4]. This group of introns frequently have long open reading frames inserted in loop regions and recognizes

the 5' exon sequence by a 4- to 6-nt base-paired internal guide sequence inter-action. They are spliced from precursor RNAs through a two-step ester-transfer reaction [5], which is discussed later.

i.   The exogenous guanosine (exoG) is docked with the active G-binding site of P7, and then its 3'-OH is aligned to break the phosphodiester bond at the 5' splice site present in P1, which results in a free 3'-OH group at the upstream exon and the exoG being attached to the 5' end of the intron.

ii. In the next step, the terminal G (omega G) of the intron swaps the exoG and occupies the G-binding site to catalyze the second ester-transfer reaction, in which the 3'-OH group at the upstream exon in P1 is aligned to break the 3' splice site present in P10, leading to the ligation of the adjacent upstream and downstream exons and release of the catalytic intron.

4. *Group II introns:* This is a huge class of self-catalytic ribozymes as well as mobile genetic elements found within the genes. Ribozyme activity can occur under high-salt conditions *in vitro*. These introns are found in the rRNA, tRNA, and mRNA of organelles in fungi, plants, and protists, and also in mRNA in bacteria. Group II introns have six structural domains, designated as dI to dVI. Additionally, a subset of group II introns encodes essential splicing proteins in intronic open reading frames. Splicing in this group occurs in a similar fashion as that found in nuclear pre-mRNA splicing and in the two-step transesterification of group I introns. In the first step of transesterification, the 2' hydroxyl group of a bulged adenosine of dVI attacks the 5' splice site and, in the second step, the nucleophilic attack occurs on the 3' splice site by the 3'-OH of the upstream exon. However, the protein machinery is required for *in vivo* splicing. Here, the intron excision occurs in the absence of Guanosine-5'-triphosphate (GTP), a substrate for RNA synthesis and involves the formation of a lariat, with an adenosine residue branch point strongly resembling that found in lariats formed during the splicing of nuclear pre-mRNA. Group II introns are further classified into IIA and IIB groups, which are distinguished from each other by the splice site consensus and the distance of the bulged adenosine in domain VI at the 3' splice site.

5. *Group III introns:* These are proposed to be a fifth family, but very little is known about the mechanisms and biochemical apparatuses that catalyze their splicing. These types of introns are generally found in the mRNA of genes from chloroplasts in euglenoid protists. They have a conventional group II type dVI with a bulged adenosine, a streamlined dI, the absence of dII-dV, and a relaxed splice site consensus. Splicing of these introns occurs through lariat and circular RNA formation. Here, the splicing also takes place through a two-step transesterification reaction: (i) formation of dVI bulged adenosine as the initiating nucleophile and (ii) the intron is excised as a lariat. They seem to be related to group II introns, and probably to spliceosomal introns [6]. The conserved 5' boundary sequences, lariat formation, lack of internal structure, and ability to use alternate splice boundaries are similar features between the group III and nuclear introns.

In addition, the introns are also classified based on the splicing machineries that catalyze the reaction [7].

1. *U2 snRNP-dependent introns:* U2 snRNP-dependent introns contribute to a majority of introns. These introns are excised by spliceosomes having U1, U2, U4, U5, and U6 snRNPs. U2-type introns are found in virtually all eukaryotes [8] and

comprise the vast majority of splice sites found in any organism, whereas U12-types have thus far been demonstrated only in vertebrates, insects, jellyfish, and plants [9]. In the beginning of splicing, the U2-type introns involve base-pairing of U1 snRNA to the 5′ splice site, and U2 snRNA to the branch point sequence (BPS) [10]. The base-pairing of U2 snRNA to the BPS is assisted by the two events: (1) binding of the large subunit of the U2 auxiliary factor (U2AF65) to the PPT, which is located immediately upstream of the intron 3′ terminus and (2) binding of the small subunit (U2AF35) to the 3′ terminal AG dinucleotide of the intron [11]. The initial recognition of the splice sites by both U1 and U2 snRNPs is followed by the U4/U6/U5 tri-snRNP, recruited to the splice site which leads to the two-step catalysis of splicing [12]. The U2-type splicing signals contain highly degenerate sequence motifs so that many different sequences can function as U2-type splice sites. Generally, in U2-type cases, the PPT is located immediately upstream of the AG but there are examples in alternatively spliced exons in *Drosophila melanogaster* with a longer PPT-AG distance, or even with the PPT positioned downstream of the AG [13]. Besides, the mammalian U2-type BPS can also be located far away (>100 nt) from the intron–exon junction sequence [14]. U12-type introns also lack an obvious PPT at the 3′ splice site. These introns consist of three subtypes based on their terminal dinucleotides:

  i.   GT–AG: GT at the 3′ end and AG at the 5′ end of the intron

  ii. GC–AG: GC at the 3′ end and AG at the 5′ end of the intron

  iii. AT–AC: AT at the 3′ end and AC at the 5′ end of the intron

 2. *U12 snRNP-dependent introns:* U12 snRNP-dependent introns belongs to a minor class of introns and are excised by spliceosomes having U11, U12, U4atac, U6atac and U5 snRNPs. In U12-type introns, the role of U1, U2, U4 and U6 snRNPs in U2-type introns are replaced by U11, U12, U4atac and U6atac snRNPs, respectively [15]. U12-type 5′ splice site and the BPS, which lies close to the 3′ end of the intron, are highly conserved [16]. Generally, U12 introns consist of two subtypes (GT–AG and AT–AC), but they do not have GC-AG subtypes such as in U2 introns.

8.1.2 Spliceosomal Biogenesis

The spliceosome is made up of five snRNA and a range of associated protein factors. In combination with the protein factors, the snRNA make an RNA–protein complex called the snRNP, the RNA component of which is rich in uridine. The splicing reactions for both the removal of introns and ligation of the flanking exons are catalyzed by spliceosomes. The snRNAs that build up nuclear spliceosome are named U1, U2, U4, U5, and U6, and participate in several RNA–RNA and RNA–protein interactions. U1 binds to 5′ GU and U2 binds to branch site (A) of the intron. This binding is facilitated by U2AF protein factors and the complex formed is known as the "spliceosome A complex." Formation of this complex is usually the key step in determining the intron ends for splicing and defining the exon ends to be retained [2]. Then, U4, U5, and U6 complexes bind with the spliceosome A complex and the U6 then replaces the U1. Finally, the U1 and U4 go away from the whole complex. The remaining complex then performs the two-step transesterification splicing reactions. A group of less abundant snRNAs, U11, U12, U4atac, and U6atac, together with U5, are subunits of the minor spliceosome that splices a rare class of pre-mRNA introns, denoted as U12 type. These snRNPs located in the cytosol form the U12 spliceosome [17].

Spliceosomes are classified into two categories (the major and minor) based on the snRNP content. The major spliceosome, splices the introns containing GU at the 5′ splice site and AG at the 3′ splice site. It is composed of the U1, U2, U4, U5, and U6 snRNPs and is active in the nucleus. In addition, a number of proteins including U2AF and SF1 are required for the assembly of the spliceosome [18]. This assembly is called the canonical assembly and the splicing is termed as canonical splicing or the lariat pathway, which accounts for more than 99% of splicing. Noncanonical splicing occurs when the intronic flanking sequences do not follow the GU-AG rule [19]. However, the minor spliceosome is very similar to the major spliceosome, but it splices out rare introns with different terminal nucleotides. The minor and major spliceosomes contain the same U5 snRNP and different, but functionally analogous, U1, U2, U4, and U6 snRNPs, which are called U11, U12, U4atac, and U6atac, respectively.

Normally, the exons of same RNA transcripts are joined to form a mature mRNA but, in some cases, exons from different RNA transcripts are also joined to form mature mRNAs and the type of splicing is termed as trans-splicing.

8.1.3 Alternate Splicing

The process of the recombination of exons takes place in multiple ways, during RNA splicing, to produce different mRNAs, which may be translated into different protein isoforms, is termed alternative splicing or differential splicing. Thus, alternative splicing is responsible for the encoding of multiple proteins from a single gene. This process occurs as a normal phenomenon in eukaryotes, which greatly increases the biodiversity of proteins that can be encoded by the small genic region of the genome. Exon skipping is one of the commonly occurring types of alternative splicing, in which a particular exon is included in mRNAs under some conditions or in specific tissues, and omitted from the mRNA in others.

Alternative splicing was first observed in adenoviruses [20]. In 1981, the first example of alternative splicing in a transcript was that of a normal endogenous gene encoding the thyroid hormone calcitonin, and was characterized in mammalian cells [21]. In humans, it is estimated that alternative splicing occurs in more than 60% of genes. However, recent analyses of vast amounts of transcript data in human and other organisms suggest that alternative splicing is widespread in mammalian genomes. Alternative splicing plays an important role in regulating gene expression. It is also helpful for the determination of binding properties, intracellular localization, enzymatic activity, protein stability, and posttranslational modifications of a large number of proteins. Alternative splicing of untranslated regions play an important role for the determination of mRNA localization and stability, as well as efficiency of translation. The events in alternative splicing are classified into four main subgroups:

1. *Exon skipping.* The exon is spliced out from the transcript together with its adjacent introns

2. *Alternative 5′ splice site selection.* Recognition of two or more 5′ splice sites at one end of the exon

3. *Alternative 3′ splice site selection.* Detection of two or more 3′ splice sites at one end of the exon

4. *Intron retention.* In which an intron can remain in the mature mRNA molecule [22]

There are many consequences to alternative splicing, which occur at both the protein and transcript levels. These consequences are categorized into two major types:

1. *Protein level alteration.* Alternative splicing generates splice variants that give rise to different protein products
2. *Transcript level alteration.* Alternative splicing produces splice variants that have different translation or stability profiles

For eukaryotic organisms, alternative splicing plays an important role as a small genic region of the genome codes for the entire proteome of the organism. Alternative splicing also provides evolutionary flexibility. A single point mutation may cause a given exon to be occasionally excluded or included from a transcript during splicing, allowing the production of a new protein isoform without loss of the original protein. Alternative splicing preceded multicellularity in evolution and this mechanism is believed to have been coopted to assist in the development of multicellular organisms [23].

8.2 Biomedical Aspects of mRNA Splicing

Abnormal variations in mRNA splicing have lead to many human genetic diseases. Mutational alterations in the RNA processing machinery often result in missplicing of multiple transcripts. Also, the single nucleotide polymorphism in splice sites or cis-acting splicing regulatory sites causes differences in the splicing of a single gene. These differences often cause the expression of unwanted proteins resulting in many lethal diseases. One of them is cancer [24], which often occurs due to abnormal splicing variants. Aside from cancer, many other genetic disorders are also caused by abnormal splicing variations and some of these diseases are discussed below.

8.2.1 Atypical Cystic Fibrosis

The malfunction of cystic fibrosis transmembrane conductance regulator (CFTR) gene causes cystic fibrosis, an autosomal recessive disorder. This is due to the occurrence of two polymorphisms located between the presumptive branch site for intron 8 and the AG-terminal dinucleotide in the CFTR gene, which directly affects the splicing of exon 9 [25]. This abnormal splicing alters the original domain confirmation, causing the loss of CFTR protein functionality.

8.2.2 Cancer

Cancer is sometimes caused by abnormal splicing. For example, a specific splicing variant of DNA methyltransferase (DNMT) genes is associated with cancer in humans. Several abnormally spliced DNMT3B mRNAs are found in tumors and cancer cell lines. The expression of two of these abnormally spliced mRNAs in mammalian cells was found to cause changes in DNA methylation patterns. The cells with one of the abnormal mRNAs also grew twice as fast as the control cells, indicating a direct and significant contribution to tumor development [26]. The production of an abnormally spliced transcript of another gene called *Ron* (*MST1R*) proto-oncogene has been found to be associated with increased levels of SF2/ASF in breast cancer cells. The abnormal isoform of the Ron protein encoded by this mRNA leads to cell motility [27].

8.2.3 Familial Isolated Growth Hormone Deficiency Type II

This disease is a dominantly inherited disorder caused by mutations in the single growth hormone gene (GH-1), in which one of the main symptoms is dwarfism [28]. GH-1 contains five exons and generates a small amount (5%–10%) of alternatively spliced mRNAs [29]. The isolated growth hormone deficiency type II mutations cause increased alternative splicing of the third exon by disrupting one of three splicing elements, that is, intronic splicing enhancers, exonic splicing enhancers, and 5′ splice site.

8.2.4 Frasier Syndrome

Frasier syndrome is caused by abnormalities in Wilms tumor suppressor gene (WT1). The disease is characterized by urogenital disorders involving kidney and gonad developmental defects. The alternative splice conserved among vertebrates in this gene is the use of two alternative 5′ splice site for exon 9, separated by nine nucleotides that encode lysine (K)–threonine (T)–serine (S), and a tripeptide [30]. The majority of patients with Frasier syndrome were found to have mutations that inactivate the downstream 5′ splice site, resulting in a shift to the KTS isoform.

8.2.5 Frontotemporal Dementia and Parkinsonism Linked to Chromosome 17

Frontotemporal dementia and Parkinsonism linked to chromosome 17 (FTDP-17) is an autosomal dominant disorder caused by mutations in the microtubule-associated protein tau (MAPT) gene that encodes tau protein. As a result of this mutation, the MAPT protein aggregates into the neuronal cytoplasmic inclusions. Silent mutations in exon 10 were identified to be linked to FTDP-17, ruling out the expression of a mutated protein as the pathogenic event [31]. The last four microtubule-binding domains are encoded by exon 10. Inclusion of exon 10 is determinative for the ratio of the 4R-tau and 3R-tau protein isoforms. The normal 4R/3R ratio is 1; this balance is required either for normal tau function or to prevent tau aggregation. Due to some FTDP-17 mutations, this ratio is altered by approximately twofold resulting in the disorder.

8.2.6 Myotonic Dystrophy

Myotonic dystrophy (DM) is also an autosomal dominant disorder and is the most common form of adult-onset muscular dystrophy, with a worldwide incidence of 1 in 8000. In this disease, the disease phenotype is directly linked to the disrupted regulation of alternative splicing. The two forms of the disease are DM1 and DM2. DM1 is the most common and is caused by a CTG expansion in the 3′ untranslated region of the DMPK gene, whereas DM2 is caused by a CCTG expansion in the first intron of the ZNF9 gene [32].

8.2.7 Spinal Muscular Atrophy

Spinal muscular atrophy is an autosomal recessive disorder that is responsible for a high childhood mortality rate. Survival motor neuron gene-2 (*SMN2*) mostly generates a shorter transcript due to skipping of the last exon (exon 7). This results in a truncated highly unstable SMN protein. The inability of *SMN2* to compensate for the loss of *SMN1* causes spinal muscular atrophy [33].

8.3 Prediction of Splice Site Efficiency

In general, splice site efficiency is predicted by position weight matrices (PWMs) calculated from the collection of splice sites. PWMs are constructed under the assumption of independence between the positions in the splice site motifs. However, there are few methods that consider the dependencies between positions in the splice site motifs that have proved to provide improvements over PWMs for estimating splice site efficiency. Also, the splice site efficiency and relative splice site strength are influenced by strong context effects, that is, base-pairing between two intronic regions close to the 5′ and 3′ splice sites and the presence of BPSs.

8.3.1 Position Weight Matrices

The sequences at the 5′ and 3′ splice site junctions for each intron, when separately aligned using the intron–exon junction as the anchor, lead to a matrix with four rows (one for each of A, C, G, and T) and the number of columns are equal to the length of the splice site motif. Such a matrix is known as a position frequency matrix (PFM). When each element of the frequency matrix is divided by its respective column total (total number of splice sites), the PFM becomes a PWM. It can be verified that each column total in PWM is equal to one. In case the number of splice site motifs is small, necessary transformations are applied on the elements of the PFM. Theoretically, the PWM can be formulated such that the likelihood of the observed set of splice site motifs is maximized. Here, splice sites are assumed as random independent observations drawn from a product multinomial distribution, from which it follows that each entry of the PWM will be proportional to the observed count of its corresponding nucleotide at the corresponding position. The main use of a PWM is to construct a consensus sequence by choosing the nucleotide with the highest frequency of occurrence at each position of the motif. The consensus motif for the PWM shown in Figure 8.1 is CAGGTAAGTATTT. Also, color-coded vertical bars stacked one over the other are used to represent the proportion of each nucleotide at a given position in the splice site motif (Figure 8.2).

Positions

	−3	−2	−1	+1	+2	+3	+4	+5	+6	+7	+8	+9	+10
A	0.334	0.637	0.099	0.000	0.000	0.597	0.699	0.089	0.181	0.296	0.226	0.224	0.227
C	0.362	0.107	0.027	0.000	0.000	0.027	0.071	0.054	0.149	0.194	0.250	0.262	0.238
G	0.185	0.115	0.805	1.000	0.000	0.350	0.119	0.782	0.193	0.295	0.237	0.242	0.255
T	0.119	0.141	0.069	0.000	1.000	0.026	0.111	0.075	0.477	0.215	0.287	0.272	0.280

FIGURE 8.1
PWM with four rows representing nucleotides A, C, G, and T, and 13 columns representing positions on the 5′ splice site motifs of *H. sapiens*.

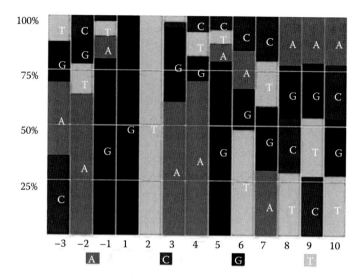

FIGURE 8.2
Graphical representation of PWM of 5′ splice site of *H. sapiens*.

Often, a log-odds (LOD) matrix is generated from PWM, where each *ij*th element of PWM (*i* representing the row or nucleotide and *j* representing the column or position in the motif) is replaced by logarithm to base 2 of the ratio of the weight at each position against a background proportion of 0.25 for each nucleotide. The background proportion, in general, is set to 0.25 under the assumption of equally, likely, or random occurrence of any observed nucleotide in a given position (as expectation of occurrence of one out of four nucleotides is 0.25). Using a different background proportion is not going to change the results but may change the scores of the motifs. Because the positions in the motif are assumed to be independent under PWM, the score for any motif can be calculated from the LOD matrix as the sum of the LOD scores of nucleotides at each position in the motif. In other words, a position on a motif is scored by aligning it with the LOD matrix and adding up the entries in the matrix that correspond to the nucleotides at each position on the motif [7]. The splice site motif score can be computed by $-\sum_{i=1}^{13} \log_2\left(\dfrac{p_i}{0.25}\right)$. For example, the splice motif score for CA**GGT**AAGTATTT is computed from PWM (Figure 8.1) as motif score

$$(\text{CA}\mathbf{GGT}\text{AAGTATTT}) = -\left[\log_2\frac{0.362}{0.250} + \log_2\frac{0.637}{0.250} + \log_2\frac{0.805}{0.250} + \log_2\frac{1.000}{0.250}\right.$$

$$\left. + \log_2\frac{1.000}{0.250} + \log_2\frac{0.597}{0.250} + \cdots + \log_2\frac{0.272}{0.250} + \log_2\frac{0.280}{0.250}\right]$$

or from the LOD matrix as the sum of the LOD scores of nucleotides at each position in the motif.

8.3.2 Information Content

Using information theory, the information content associated with each position on the splice site motif can be derived from the weights of four nucleotides in the PWM. This is of interest, as it says something about how different a given PWM is from a uniform distribution. If a random variable (each position on motif) takes four values (A, T, G, and C) with four probabilities (p_1, p_2, p_3, and p_4), then the self-information of observing a particular nucleotide at a particular position of the motif is $-\log p_i$ with the expected self-information as $-p_i \log p_i$. Hence, the total information content of a random variable taking four nucleotides with their corresponding probabilities become $-\sum_i p_i \log p_i$. As the information content and variance are inversely related, it is easy to study the variability, that is, the level of conservation at each position in the motif. The variability at each position can vary between 0 (one possibility is certain, $p = 1$) and 2 (all possibilities are likely equal, $p = 0.25$) when the logarithm in the said formula is taken with base 2. Often, it is more useful to calculate the information content with the background nucleotide frequencies of the sequences under consideration rather than assuming equal probabilities for each nucleotide. Under such situations, the information content at a given position becomes $-\sum_i p_i \log(p_i/p_b)$, where p_b is the probability of observing the nucleotide i in the background model.

8.3.3 Molecular Phylogeny

Molecular phylogenetics is the study of hereditary molecular differences to find out the evolutionary relationships of an organism by using its genomic sequences. It may be of great interest to understand the underlying mechanism involved in evolution through spliceomics. This is possible when the PWMs of different species for a given splice site type (5′ splice sites with GT-AG at intron terminals) are available. In probability and information theories, the Kullback–Leibler divergence, also known as information divergence, information gain, and relative entropy, is a nonsymmetric measure of the difference between two probability distributions (P and Q). In splice site analysis, given two PWMs, p and q, of equal motif length, a distance d can be calculated by Kullback–Leibler distance [34]. The distance calculation is given by the following formula:

$$d = -\sum_i \left[p_i \log_2(p_i/q_i) + q_i \log_2(q_i/p_i) \right]$$

where i is the index of position on PWM. The distance calculated between any two species is then used to construct the phylogenetic tree.

8.3.4 Dinucleotide Dependencies

The 5′ splice site motifs are normally considered with nine nucleotide lengths having three nucleotides in the exonic region (−1, −2, and −3) and six nucleotides in the intronic region (+1, +2, +3, +4, +5, and +6). The +1 and +2 positions are conserved at the 5′ intron terminal of the splice site. The nomenclature of positions in 5′ splice site is shown in Figure 8.3.

The expected occurrence of pairs of nucleotides at two positions, for example, A at +3 and G at +4, can be calculated from the PFM by assuming independence between the two

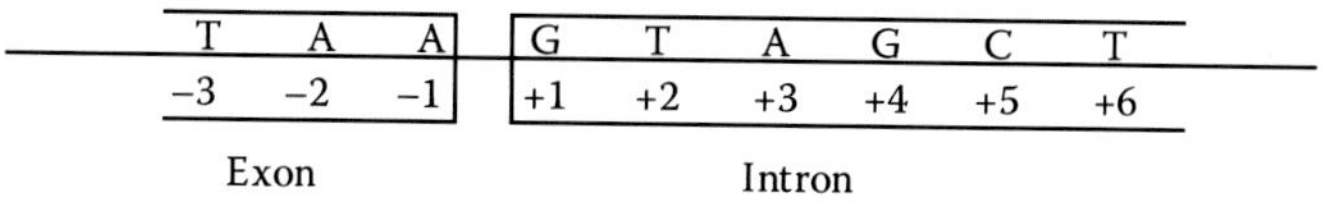

T	A	A	G	T	A	G	C	T
−3	−2	−1	+1	+2	+3	+4	+5	+6

Exon   Intron

FIGURE 8.3
Nomenclature of positions at 5′ splice site junction.

positions. The observed frequency of occurrence for such pairs can be measured from the genomic data, by counting the occurrence of pairs of nucleotides (16 possible combinations) at all possible combinations of positions (21 unique combinations, if the +1 and +2 invariant positions are excluded). Such observed frequencies can be put in the form of a matrix known as an association matrix with 21 rows and 21 columns. For example, the dinucleotide dependency between (−3T,−1A) is calculated by

dependency (−3T,−1A) = (observed frequency − expected frequency)/expected frequency.

Example:

Let

$$\text{Observed frequency} = O(-3T,-1A) = 2418$$

and

$$\text{Expected frequency} = E(-3T,-1A)$$
$$= p(-3T) * p(-1A) * \text{number of 5' splice site}$$
$$= 0.1114 * 0.0898 * 253268$$
$$= 2533$$

Then

$$\text{Dependency } (-3T,-1A) = (2418 - 2533)/2533 = -0.0454$$

8.4 Features of Splice Site Efficiency Derived from Mutation and SNPs

As discussed earlier, the dinucleotide dependencies or associations are strong determinants of splice site efficiency. Recently, Roca et al. [35] studied the features of 5′ splice site efficiency derived from disease-causing mutations at 5′ splice site, orthologous pairs of 5′ splice site between mouse and human, SNPs at 5′ splice site, and simulated SNPs.

8.4.1 Effect of SNPs and Mutations on Dinucleotide Dependencies

A general hypothesis is that the dinucleotide dependencies reflect the mechanism of 5′ splice site selection, and can be used to estimate subtle or second-order determinants of

5′ splice site strength, especially important for weak 5′ splice site, as measured by low motif scores based on PWM. However, the large number of pairwise combinations for each given 5′ splice site discourages the use of experimental testing to evaluate the global effect of dependencies on 5′ splice site strength.

Splicing mutations are very frequent among mutations that cause human disease, from genetic disorders to cancer. There are some disease-causing 5′ splice site mutations whose effects in splicing are not clear, such as those affecting the less-conserved positions, and very rare substitutions of a nonconsensus nucleotide by another nonconsensus nucleotide, in which nonconsensus nucleotides are the ones that are (other than the consensus nucleotides) obtained from the PWM. Because these mutations have a minor effect on the PWM-based scores, their deleterious consequences might be explained by the disruption of the dependencies at the wild-type 5′ splice site.

8.4.2 A Measure to Estimate the Effect of Nonconsensus SNPs and Mutations on Dinucleotide Dependencies

The following motif-to-weight (M–W) measure can be used to calculate the effect of change of nonconsensus nucleotides at different positions on splice sites. The M–W association ratio is defined as

$$\text{M} - \text{W association ratio} = \log_{10}\left[\frac{m(A1)/m(A2)}{w(A1)/w(A2)}\right]$$

where $m(A1)$, $m(A2)$, $w(A1)$, and $w(A2)$ are the frequencies of motifs with "nonconsensus nucleotide 1", frequencies of motifs with "nonconsensus nucleotide 2," frequencies of "nonconsensus nucleotide 1" and frequencies of "nonconsensus nucleotide 2," respectively, in the population of splice sites. The lower the value of the M–W association ratio, the lower is the extent of disruption of association/dependency between positions on splice sites.

8.4.2.1 Effect of "Real" versus "Simulated" Mutations/SNPs on Dinucleotide Associations

Roca et al. [35] collected 5′ splice site mutations from the Human Mutation Genome Database (http://www.hgmd.cf.ac.uk/ac/index.php) and sorted them into two groups, as substitution of a consensus nucleotide by another nonconsensus nucleotide and nonconsensus by another nonconsensus nucleotide. Those with substitutions of nonconsensus by another nonconsensus nucleotides (also, with small PWM score differences) are designated as "neutral" (N) 5′ splice site mutations. The hypothesis is that the N mutations substantially disrupt the associations between positions of the 5′ splice site. Before testing such hypotheses, mutations in splice sites are simulated (Figure 8.4).

Initially, for simulations, a random splice site sequence is drawn from the 5′ splice site of *Homo sapiens*, and the nonconsensus positions on the splice site are observed. Then, a random position is chosen. Because the other nonconsensus nucleotides at this position are known; a nonconsensus nucleotide is randomly chosen. In this way, the nonconsensus mutations are simulated. Indeed, it is observed that the N mutations severely disrupt the correlations at the wild-type 5′ splice site, much more than a simulated set of mutations (SIM) selected under the same criteria. This result not only highlights the importance of the correlations between 5′ splice site positions, but also shows that the correlations have predictive value to ascribe pathogenicity to mysterious 5′ splice site mutations.

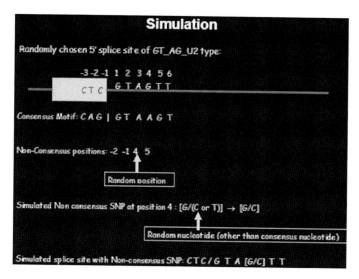

FIGURE 8.4
Procedure indicating simulation of nonconsensus SNPs in splice sites.

8.5 Visualization Tools for Splice Site Features

A heat map is a two-dimensional graphical representation of data in which the individual values contained in a matrix are represented by grayscale. To visualize the dinucleotide dependencies through heat maps, the dependencies are to be rescaled between −1 and +1 by equating minimum value, 0, median of positive values, median of negative values, and maximum value of dependencies to −1, 0, 0.5, −0.5, and +1, respectively, on the transformed scale. Based on the transformed scale, the associations plotted as heat maps are shown in Figure 8.5. In the case, where the observed frequency is greater than expected,

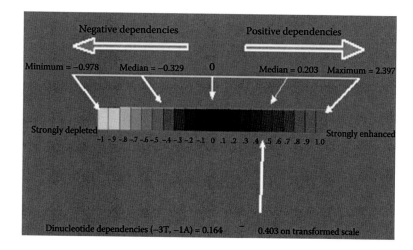

FIGURE 8.5
Transformation of correlations across species.

the dinucleotide dependency is strongly enhanced whereas it is depleted in the reverse case. For more details on dinucleotide dependencies, one may refer to Roca et al. [35].

The dinucleotide dependencies of human splice sites represented by a heat map is shown in Figure 8.6. Here, the largest positive value is represented by a gray color showing strong enhancement, whereas the smallest value with a white color indicates a strong depletion.

For a comparison of dinucleotide dependencies between two species, say between human and mouse, one heat map can be constructed with an upper diagonal map representing dinucleotide dependencies of the mouse and a lower diagonal map representing dinucleotide dependencies of the human species (Figure 8.7).

Associations between positions −2 or −1 and position +3, +4, +5, or +6 imply having non-consensus nucleotides on the exonic side causing a depletion of nonconsensus nucleotides

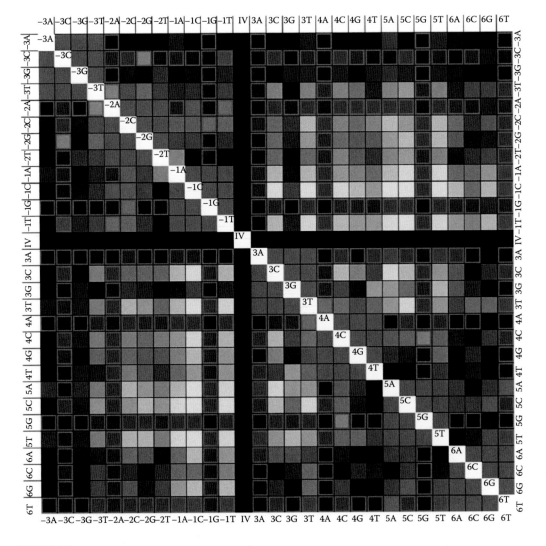

FIGURE 8.6
Heat map showing dinucleotide dependencies in human 5′ splice sites (lower and upper diagonals are symmetric).

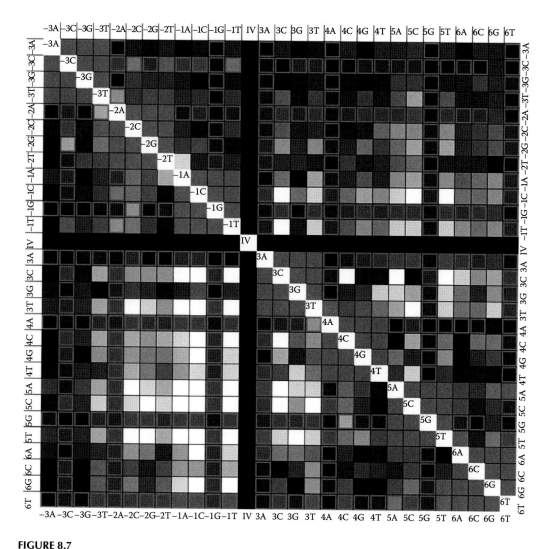

FIGURE 8.7

Dinucleotide correlation map with lower diagonal representing *H. sapiens* and upper diagonal representing *M. musculus*.

on the intronic side, and vice versa. Some features such as combinations of –3T, –2A and –2T, –1A are severely depleted, probably because these combinations are part of two of the three stop codons (TAG and TAA). The association at +5C, +6C is enriched, probably reflecting the gradual conversion of U12 type GT-AG introns into U2 type GT-AG introns.

8.6 Spliceomics Applications in Genome Annotation

Whole genome analysis of a single organism or comparison of genomes depends on accurate gene annotation. Gene annotation incorporates cDNA data, sequence similarity, and computational predictions based on the recognition of probable splice sites and coding regions [36]. Apart from gene prediction, gene structure prediction is an important part of genome annotation. Accurate prediction of gene structure depends on correct detection of the splicing signals and splice junctions. Splicing signals are elements in the DNA sequence of a gene that specify the accurate splicing of its primary RNA transcript. These are exonic splicing enhancers, intronic splicing enhancers, exonic splicing silencers, and intronic splicing silencers. Splicing enhancers are the small motifs to which the splicing activator proteins bind and enhance the probability of the nearby site to be a splice junction, whereas splicing silencers are the sites for binding of splicing repressor proteins which reduce this probability.

On the other hand, accurate splice site prediction is a critical component of eukaryotic gene prediction. The presence of the GT dinucleotide at the intron boundary of the donor site and the AG dinucleotide at the acceptor site is an important feature of the splice sites. However, in the donor sites, GC is observed in less than 1% of the cases. These donor sites are recognized by the U1 snRNA of the spliceosome through base-pairing with an ACUUACCU motif, and generally have the AG/GTAAGT pattern. The most common class of nonconsensus splice sites consists of 5′ splice sites with a GC dinucleotide [37]. The GC sites are assumed to be recognized by the U2-dependent spliceosome. In the U12 type of intron splice site sequences, a high variation from the standard consensus sequence is observed. U12 introns are recognized by the presence of highly conserved patterns, that is, RTATCCTY (where R = A or G; Y = C or T) at the 5′ splice site and TCCTRAY at the branch site.

Many reports of such variant splice sites can be traced to errors in annotation or interpretation, polymorphic differences between the sources of cDNA and genomic sequence, inclusion of pseudogene sequences, or failure to account for somatic mutation [38]. Some of the successful approaches for accurate splice site prediction include GRAIL [39], NetGene [40], GeneID [41], FGENEH [42], etc. The biostatisticians have successfully applied probabilistic modeling using hidden Markov models and neural networks for the splice site prediction. GeneParser [43] and Genie [44] are two of the successful applications of statistical modeling in splice junction prediction.

8.7 Statistical Modeling in Spliceomics

The question of what model should be used to distinguish the occurrence of non–splice site motifs/decoy splice site motifs from splice site motifs for a given set of aligned splice site motifs is common. It is important to identify signal nucleotides in splice site motifs and to recognize true motifs of RNA splicing from "decoy" splice sites in pre-mRNA sequences. A number of statistical models and data mining techniques, such as Markov models, hidden Markov models, maximum entropy model (MEM), decision trees, combinatorial models, neural networks, and support vector machines are quite extensively used in spliceomics. However, the discussion made here is confined to MEM, decision trees, and combinatorial models because they are most commonly used in the analysis of splice motifs.

8.7.1 Entropy Modeling

Although Markov and hidden Markov models are commonly used in gene structure prediction or splice site prediction, MEM is the most unbiased approximation for modeling splice site sequence motifs for a given set of constraints estimated from available splice site motifs. Yeo and Burge [45] showed that MEM has an attractive property, that is, it assumes nothing more about the distribution than that it is consistent with the features of empirical distribution that can be estimated from known signal sequences.

8.7.1.1 Maximum Entropy Method

Let X be a sequence of N random variables $X = \{X_1, X_2,..., X_N\}$ which can take values from $\{A,C,G, T\}$. Let $x = \{x_1, x_2,... x_N\}$ represents a specific splice site sequence. Let $p(X)$ be the joint probability distribution $p(X_1 = x_1, X_2 = x_2,..., X_N = x_N)$ and $P(X = x)$ denote the probability of a state in this distribution. The maximum entropy principle states that of all the possible distributions in the hypothesis space that satisfy a set of constraints, the distribution that is the best approximation of the true distribution, is the one with the largest Shannon entropy, H, given by

$$H(\hat{p}) = -\sum \hat{p}(x) \log_2(\hat{p}(x))$$

where the sum is taken over all possible sequences, x [46].

8.7.1.2 Constraints

There are mainly two categories of constraints, that is, complete constraints and specific constraints [45]. The complete constraints are those which specify sets of positional dependencies whereas specific constraints are those constraints on nucleotide frequencies at a subset of positions.

8.7.1.2.1 Complete Constraints

Let S_X be the set of all lower-order marginal distributions of the full distribution, $p(X_1 = x_1, X_2 = x_2,..., X_N = x_N)$. A lower-order marginal distribution is a joint distribution over a proper subset of X. For example, for $N = 3$,

$$S_X = \{p(X_1), p(X_2), p(X_3), p(X_1,X_2), p(X_2,X_3), p(X_1,X_3)\}$$

Let *mo* be the marginal-order of the marginal distributions and s be the skips of the marginal distribution. Then the first three elements of S_X are the first-order marginal distributions and the last three elements are second-order marginal distributions. Furthermore, out of three second-order marginal distributions, $p(X_1,X_2)$ and $p(X_2,X_3)$ are the second-order marginal distributions with skip 0, and $p(X_1,X_3)$ is the second-order marginal distribution with skip 1. In general, the first-order marginal distributions with skip 0 are included in the set of *m*th order marginal distributions with skip of order s [45], shown that for an aligned set of sequences of length N, the first-order constraints with skip 0 are the empirical frequencies of each nucleotide (A,C,G,T) at each position, and the maximum entropy distribution consistent with these constraints is the PWM. On the other hand, if second-order nearest-neighbor constraints are used, the solution is an inhomogeneous first-order Markov model. Consequently, different sets of constraints specify many different models.

8.7.1.2.2 Specific Constraints

These are the observed frequency values for a particular member of a set of "complete" constraints. For example, the specific constraints for $p(X_1, X_2)$ are {A A, AC, AG, A T, CA, CC, CG, CT, GA, GC, GG, GT, TA, TC, TG, TT}.

8.7.1.3 Maximum Entropy Model

The MEM is specified by a set of complete constraints and consists of two distributions, namely, the signal model and the decoy probability distribution. These distributions are generated by iterative-scaling over constraints from a set of aligned signals and a set of aligned decoys of the same sequence length N.

Let L be the likelihood ratio defined as

$$L(X = x) = \frac{P^{signal}(X = x)}{P^{decoy}(X = x)}$$

where $P^{signal}(X = x)$ and $P^{decoy}(X = x)$ are the probability of occurrence of sequence x from the distributions of signals and decoys, respectively.

Based on Neyman–Pearson lemma, the sequences for which $L(X = x) \geq C$, where C is a threshold that achieves the desired true-positive rate are predicted to be true signals [45]. Hence, the MEM can be used to distinguish true signals from decoys based on the likelihood ratio, L. For more details on the generation of MEDs by iterative-scaling over constraints, one may refer to Jaynes [46].

8.7.2 Decision Trees

The decision trees are computation/communication models in which an algorithm or communication process is considered as a decision tree. These are based on supervised machine-learning approaches that classify feature vectors hierarchically. Decision trees are generated by examining a collection of labeled vectors and statistically determining the feature containing most of the information relevant to classification [47]. If trained on high-quality data, decision trees can make very accurate predictions. Decision trees are expressive enough to model many partitions of the data that are not as easily achieved with classifiers that rely on logistic regression or support vector machines. Decision trees have been successfully applied for splice site prediction [48].

8.7.3 Combinatorial Model

The combinatorial model is based on the frequency of oligonucleotides for splice and splice like (decoy) signals. In this method, splice site sensors were built based on the Bayesian principle. It also outperforms the maximum entropy sensors for many representative test sets of genes when compared on receiver operating characteristic curve. In addition, it also presents combinatorial interaction of splice sites and related factors with the LOD matrix [49]. The combinatorial model of plant intron recognition is based on the assumption that splice site sequences as well as local intron and exon sequences contribute to splice site selection and splicing efficiency. Itoh et al. [50] suggested that alternative exon contain weak splice sites, regulated alternatively by potential regulatory sequences on the exons, in their combinatorial model of alternative splicing mechanisms.

8.8 Tools and Databases

8.8.1 Tools

8.8.1.1 Berkeley Drosophila Genome Project

The Berkeley Drosophila Genome Project is a neural network–based splice site prediction tool, in which a backpropagation feedforward neural network was trained with one layer of hidden units to recognize donor and acceptor sites, respectively, using a novel optimized representative data set. The genes having constraint consensus splice sites, that is, GT for the donor and AG for the acceptor sites, were considered in the tool. It gives scores for a potential splice site in the result. The training and test sets of human and *D. melanogaster* splice sites are also available for public use for testing splice site predictors (URL: http://www.fruitfly.org/seq_tools/splice.html).

8.8.1.2 FSPLICE

SPLICE provides the possibility to search for both donor and acceptor sites, and to define thresholds for them independently. The program allows the user to search minor variants of splicing donor site (GC sites) as well (URL: http://linux1.softberry.com/berry.phtml?topic=fsplice&group=programs&subgroup=gfind).

8.8.1.3 GeneSplicer

This is a fast, flexible system for detecting splice sites in the genomic DNA of various eukaryotes. The program uses decision tree method (maximal dependence decomposition) and enhances it with Markov models that capture additional dependencies among neighboring bases in a region around splice sites. The tool was executed successfully on *Plasmodium falciparum* (malaria), *Arabidopsis thaliana*, human, *Drosophila*, and rice. The training data sets for human and *A. thaliana* were also made available to the public. The overall splice site detection system is improved by a local score optimality feature (URL: http://www.cbcb.umd.edu/software/genesplicer/gene_spl.shtml).

8.8.1.4 HMMgene

The program is based on a hidden Markov model, which is a probabilistic model of the gene structure. HMMgene predicts several whole or partial genes in one sequence, splice sites, and start/stop codons. The program finds the best gene structure under constraints of known sequences having hits to ESTs, proteins, or repeat elements. Apart from reporting the best prediction, it also reports multiple best gene predictions for a given sequence to support the condition of the presence of several equally likely gene structures. The output is a prediction of partial or complete genes with information on the coding regions and scores for whole genes as well as exon scores (URL: http://www.cbs.dtu.dk/services/HMMgene/).

8.8.1.5 NetGene2 Server

The server works on an artificial neural network methodology and rule-based system to predict splice sites in human, *C. elegans*, and *A. thaliana*. The algorithm of this tool is based on the experimental evidence in the GenBank entries. It follows a two-step prediction scheme, in which global prediction of the coding potential regulates a cutoff level for a local prediction of splice sites and is refined by rules that are based on splice site confidence values, prediction scores, coding context, and distances between potential splice sites. The combined approach used in the tool significantly minimizes the false positives (URL: http://www.cbs.dtu.dk/services/NetGene2/).

8.8.1.6 SpliceMachine

The tool uses linear support vector machines to compute a linear classification boundary between actual and pseudo-splice sites. To achieve this, a candidate splice site is represented as a feature vector in which each feature contains some information about the candidate splice site and its context in the sequence. Splice sites are predicted by SpliceMachine based on the positional, compositional, and codon bias information that is extracted from a large local context around each candidate splice site (URL: http://bioinformatics.psb.ugent.be/webtools/splicemachine/).

8.8.1.7 SplicePort

SplicePort is a Web-based splice site analysis tool. It predicts the splice sites for submitted sequences and allows the browsing of predictive signals and motif exploration. In the tool, splice site prediction is done with the flexibility of exploring all the putative splice site locations, their scores, corresponding sensitivities, and false positive rate values. Moreover, it allows exploring the motifs for the requested location in the input sequence and browsing the complete collection of identified motifs for both acceptor and donor splice sites. SplicePort can also be readily implemented in other genomes other than human (URL: http://www.cs.umd.edu/projects/SplicePort).

8.8.2 Databases

8.8.2.1 Alternative Splicing and Transcript Diversity Database

The database has a collection of alternative transcripts that include transcription initiation, polyadenylation, and splicing variant data. Alternative transcripts are derived from the mapping of transcribed sequences to the complete human, mouse, and rat genomes using an extension of the computational pipeline developed for the Alternative Splicing Database and Alternative Transcript Diversity Database, which are now superseded by the Alternative Splicing and Transcript Diversity Database (URL: http://www.ebi.ac.uk/astd).

8.8.2.2 Drosophila Splice Site Database

The Drosophila Splice Site Database contains nucleotide sequences in the vicinity of 11,161 donor and acceptor splice sites collected from 3375 *Drosophila* cDNAs. This relational database is constructed by the Weir–Rice research group. The splice sites information was generated by a custom algorithm using *Drosophila* cDNA transcripts and genomic DNA. It also supports a set of procedures for analyzing splice site sequence space. The Web interface of the database permits the execution of procedures with a flexibility of parameter settings and custom-structured query languages (URL: http://igs.wesleyan.edu/).

8.8.2.3 HS³D (Homo sapiens Splice Sites Dataset)

HS³D is enriched with a collection of *H. sapiens* exon, intron, and splice regions extracted from GenBank Rel.123. The available data set gives standardized material to train and assess the prediction accuracy of computational approaches for gene discovery and characterization. Several statistics for exons and introns including overall nucleotides, average GC content, number of exons/introns including non-AGCT bases, number of exons/introns in which the annotated end is not found, exon/intron minimum length, exon/intron maximum length, exon/intron average length, exon/intron length standard deviation, number of introns in which the sequence does not start with GT, number of introns in which the sequence does not end with AG are reported in this database (URL: http://www.sci.unisannio.it/docenti/rampone/).

8.8.2.4 SpliceDB

SpliceDB is a public database of known mammalian splice site sequences. It holds information about verified splice site sequences for canonical and noncanonical sites with supporting evidence. It describes alternative introns supported by ESTs and noncanonical splice junctions (URL: http://linux1.softberry.com/spldb/SpliceDB.html).

8.8.2.5 SpliceRack

The SpliceRack database contains more than half a million splice sites from five species (*H. sapiens, M. musculus, D. melanogaster, C. elegans,* and *A. thaliana*). These splice sites were classified into four subtypes: U2-type GT-AG and GC-AG, and U12-type GT-AG and AT-AC. The database also includes some new examples of rare splice site categories, such as U12-type introns without canonical borders, and U2-dependent AT-AC introns (URL: http://katahdin.mssm.edu/splice/index.cgi?database=spliceNew).

8.8.2.6 TAndem Splice Site DataBase (TassDB)

This is a relational database which stores extensive data about alternative splice events at donors and acceptors, from both confirmed and unconfirmed cases. The database has two versions:

TassDB1. Holds information about GYNGYN donors and NAGNAG acceptors from eight species, based on the genome information available in 2006. It is enriched with 114,554 tandem splice sites, 5209 of which have EST/mRNA evidence for alternative splicing.

TassDB2. It contains data about GY(N)0-10GYN donors and NAG(N)0-10AG acceptors in human and mouse, in which the distance between the tandem splice sites ranges from 2 to 12. It includes 1,329,181 putative tandem splice sites, of which 14,312 are supported by EST/mRNA evidence for alternative splicing. This database is based on the genomic information available until 2008 (URL: http://www.tassdb.info/).

8.8.2.7 The Physcomitrella patens Sequences

This is a species-specific splice site database developed using hidden Markov models and support vector machines for the moss *Physcomitrella patens*. Both models minimize the false positives compared with NetPlantGene, an artificial neural network trained on *Arabidopsis* data. The software used for training and classification was SVMlight, libsvm, and svmsplice. The complete set of splice sites was divided into training/test sets of sizes 10% to 90% and three samples were drawn for each set. The set found to contain 90% of the sites for training proved to yield the best results (URL: http://www.cosmoss.org/ssp/).

8.9 Future Prospects

Spliceomics is very well studied for human and other model organisms although there is still a large scope to spliceomics applications in plant and animal sciences. Therefore, the location of splicing signals, splicing junctions, and their conservedness in plants and animals needs to be explored. Traditional data mining techniques have long been used for the prediction and analysis of splice sites. For more accurate prediction, the use of statistical models and data mining techniques such as random forest, boosting, least absolute shrinkage and selection, elastic net, least angle regression are needed, which will probably minimize the false-positive rates. Spliceomics can be applied for a better understanding of autosomal diseases in agriculturally important plants and microbes. Splice site efficiency derived from mutation and SNPs will also help us understand the underlying evolutionary mechanism. When PWM fails to explain the occurrence of subtle changes at splice junctions, the other measures, such as dinucleotide dependencies, are to be identified to explain the same. Thus, spliceomics play an important role in understanding the genetic inheritance and other complex biological phenomena.

Acknowledgments

The authors acknowledge World Bank–Funded National Agricultural Innovation Project (NAIP), ICAR Grants NAIP/Comp-4/C4/C-30033/2008-09 and 30(68)/2009/Bio Informatics/ NAIP/O&M for the resources utilized.

References

1. Bruce, A. 2008. *Molecular Biology of the Cell*. New York: Garland Science.
2. Matlin, A.J., F. Clark, and C.W.J. Smith. 2005. Understanding alternative splicing: Towards a cellular code. *Nature Reviews* 6(5):386–398.
3. Haugen, P., D.M. Simon, and D. Bhattacharya. 2005. The natural history of group I introns. *Trends Genetics* 21(2):111–119.
4. Woodson, S.A. 2005. Structure and assembly of group I introns. *Current Opinion in Structural Biology* 15(3):324–330.
5. Cech, T.R. 1990. Self-splicing of group I introns. *Annual Review of Biochemistry* 59:543–568.

6. Copertino, D.W., and R.B. Hallick. 1993. Group II and group III introns of twintrons: Potential relationships with nuclear pre-mRNA introns. *Trends in Biochemical Sciences* 18(12):467–471.

7. Sheth, N., X. Roca, M.L. Hastings, T. Roeder, A.R. Krainer, and R. Sachidanandam. 2006. Comprehensive splice-site analysis using comparative genomics. *Nucleic Acids Research* 34(14):3955–3967.

8. Collins, L., and D. Penny. 2005. Complex spliceosomal organization ancestral to extant eukaryotes. *Molecular Biology and Evolution* 22:1053–1066.

9. Burge, C.B., R.A. Padgett, and P.A. Sharp. 1998. Evolutionary fates and origins of U12-type introns. *Molecular Cell* 2:773–785.

10. Brow, D.A. 2002. Allosteric cascade of spliceosome activation. *Annual Review of Genetics* 36:333–360.

11. Reed, R. 2000. Mechanisms of fidelity in pre-mRNA splicing. *Current Opinion in Cell Biology* 12:340–345.

12. Staley, J.P., and C. Guthrie. 1998. Mechanical devices of the spliceosome: Motors, clocks, springs, and things. *Cell* 92:315–326.

13. Lallena, M.J., K.J. Chalmers, S. Llamazares, A.I. Lamond, and J. Valcarcel. 2002. Splicing regulation at the second catalytic step by sex-lethal involves 3′ splice site recognition by spf45. *Cell* 109:285–296.

14. Smith, C.W., and B. Nadal-Ginard. 1989. Mutually exclusive splicing of alpha-tropomyosin exons enforced by an unusual lariat branch point location: Implications for constitutive splicing. *Cell* 56:749–758.

15. Patel, A.A., and J.A. Steitz. 2003. Splicing double: Insights from the second spliceosome. *Nature Reviews Molecular Cell Biology* 4:960–970.

16. Abril, J.F., R. Castelo, and R. Guigo. 2005. Comparison of splice sites in mammals and chicken. *Genome Research* 15:111–119.

17. Konig, H., N. Matter, R. Bader, W. Thiele, and F. Müller. 2007. Splicing segregation: The minor spliceosome acts outside the nucleus and controls cell proliferation. *Cell* 13(4):718–729.

18. Black, D.L. 2003. Mechanisms of alternative pre-messenger RNA splicing. *Annual Reviews of Biochemistry* 72(1):291–336.

19. Ng, B., F. Yang, D.P. Huston, Y. Yan, Y. Yang, Z. Xiong, L.E. Peterson, H. Wang, and X.F. Yang. 2004. Increased noncanonical splicing of autoantigen transcripts provides the structural basis for expression of untolerized epitopes. *Journal of Allergy and Clinical Immunology* 114(6):1463–1470.

20. Chow, L.T., R.E. Gelinas, T.R. Broker, and R.J. Roberts. 1977. An amazing sequence arrangement at the 5′ ends of adenovirus 2 messenger RNA. *Cell* 12(1):1–8.

21. Leff, S.E., M.G. Rosenfeld, and R.M. Evans. 1986. Complex transcriptional units: Diversity in gene expression by alternative RNA processing. *Annual Review of Biochemistry* 55(1):1091–1117.

22. Sammeth, M., S. Foissac, and R. Guigo. 2008. A general definition and nomenclature for alternative splicing events. *PLoS Computational Biology* 4(8):e1000147.

23. Manuel, I., R. Jakob, P. David, and R. Scott. 2007. Functional and evolutionary analysis of alternatively spliced genes is consistent with an early eukaryotic origin of alternative splicing. *BMC Evolutionary Biology* 7:188.

24. He, C., F. Zhou, Z. Zuo, H. Cheng, and R. Zhou. 2009. A global view of cancer-specific transcript variants by subtractive transcriptome-wide analysis. *PLoS One* 4(3):e4732.

25. Strong, T.V., D.J. Wilkinson, M.K. Mansoura, D.C. Devor, K. Henze, Y.P. Yang, J.M. Wilson, J.A. Cohn, D.C. Dawson, R.A. Frizzell et al. 1993. Expression of an abundant alternatively spliced form of the cystic fibrosis transmembrane conductance regulator (CFTR) gene is not associated with a cAMP-activated chloride conductance. *Human Molecular Genetics* 2:225–230.

26. Fackenthal, J.D., and L.A. Godley. 2008. Aberrant RNA splicing and its functional consequences in cancer cells. *Disease Models & Mechanisms* 1(1):37–42.

27. Ghigna, C., S. Giordano, H., Shen, F. Benvenuto, F. Castiglioni, P.M. Comoglio, M.R. Green, S. Riva, and G. Biamonti. 2005. Cell motility is controlled by SF2/ASF through alternative splicing of the Ron proto-oncogene. *Molecular Cell* 20(6):881–890.

28. Cogan, J.D., J.A. Phillips, S.S. Schenkman, R.D. Milner, and N. Sakati. 1994. Familial growth hormone deficiency: A model of dominant and recessive mutations affecting a monomeric protein. *Journal of Clinical Endocrinology and Metabolism* 79:1261–1265.

29. Lecomte, C.M., A. Renard, and J.A. Martial. 1987. A new natural hGH variant—17.5 kD—produced by alternative splicing. An additional consensus sequence which might play a role in branchpoint selection. *Nucleic Acids Research* 15:6331–6348.

30. Miles, C., G. Elgar, E. Coles, D.J. Kleinjan, V. van Heyningen, and N. Hastie. 1998. Complete sequencing of the Fugu WAGR region from WT1 to PAX6: Dramatic compaction and conservation of synteny with human chromosome 11p13. *Proceedings of the National Academy of Sciences of the United States of America* 95:13068–13072.

31. D'Souza, I., P. Poorkaj, M. Hong, D. Nochlin, V.M. Lee, T.D. Bird, and G.D. Schellenberg. 1999. Missense and silent tau gene mutations cause frontotemporal dementia with Parkinsonism-chromosome 17 type, by affecting multiple alternative RNA splicing regulatory elements. *Proceedings of the National Academy of Sciences of the United States of America* 96:5598–5603.

32. Liquori, C.L., K. Ricker, M.L. Moseley, J.F. Jacobsen, W. Kress, S.L. Naylor, J.W. Day, and L.P. Ranum. 2001. Myotonic dystrophy type 2 caused by a CCTG expansion in intron 1 of ZNF9. *Science* 293:864–867.

33. Cifuentes-Diaz, C., T. Frugier, F.D. Tiziano, E. Lacene, N. Roblot, V. Joshi, M.H. Moreau, and J. Melki. 2001. Deletion of murine SMN exon 7 directed to skeletal muscle leads to severe muscular dystrophy. *Journal of Cell Biology* 152:1107–1114.

34. Kullback, S., and R.A. Leibler. 1951. On information and sufficiency. *Annals of Mathematical Statistics* 22(1):79–86.

35. Roca, X., A.J. Olson, A.R. Rao, E. Enerly, V.N. Kristensen, L. Borresen-Dale, B.S. Andresen, A.R. Krainer, and R. Sachidanandam. 2008. Features of 50-splice-site efficiency derived from disease-causing mutations and comparative genomics. *Genome Research* 18:77–87.

36. Stormo, G.D. 2000. Gene-finding approaches for eukaryotes. *Genome Research* 10(4):394–397.

37. Wu, Q., and A.R. Krainer. 1999. AT-AC pre-mRNA splicing mechanisms and conservation of minor introns in voltage-gated ion channel genes. *Molecular and Cellular Biology* 19(5):3225–3236.

38. Mount, S.M. 2000. Genomic sequence, splicing, and gene annotation. *American Journal of Human Genetics* 67(4):788–792.

39. Uberbacher, E., and R. Mural. 1991. Locating protein-coding regions in human DNA sequences by a multiple sensor-neural network approach. *Proceedings of the National Academy of Sciences of the United States of America* 88, 11261–11265.

40. Brunak, S., J. Engelbrecht, and S. Knudsen. 1991. Prediction of human mRNA donor and acceptor sites from the DNA sequence. *Journal of Molecular Biology* 220:49–65.

41. Guigo, R., S., Knudsen, N. Drake, and T.F. Smith. 1992. Prediction of gene structure. *Journal of Molecular Biology* 226:141–157.

42. Solovyev, V., A. Salamov, and C. Lawrence. 1995. Identification of human gene structure using linear discriminant functions and dynamic programming. *Proceedings International Conference on Intelligent Systems for Molecular Biology* 3:367–375.

43. Snyder, E.E., and G.D. Stormo. 1993. Identification of coding regions in genomic DNA sequences: An application of dynamic programming and neural networks. *Nucleic Acids Research* 21(3):607–613.

44. Kulp, D., D. Haussler, M.G. Reese, and F.H. Eeckman. 1997. Integrating database homology in a probabilistic gene structure model. *Pacific Symposium on Biocomputing* 1997:232–244.

45. Yeo, G., and C.B. Burge. 2004. Maximum entropy modeling of short sequence motifs with applications to RNA splicing signals. *Journal of Computational Biology* 11:377–394.

46. Jaynes, E. 1957. Information theory and statistical mechanics. *Physics Review* 106:620–630.

47. Patterson, D.J., K. Yasuhara, and W.L. Ruzzo. 2002. Pre-mRNA secondary structure prediction aids splice site prediction. *Pacific Symposium on Biocomputing* 7:223–234.

48. Kingsford, C., and S.L. Salzberg. 2008. What are decision trees? *Nature Biotechnology* 26:1011–1013.

49. Churbanov, A., I.B. Rogozin, J.S. Deogun, and H. Ali. 2006. Method of predicting splice sites based on signal interactions. *Biology Direct* 1:10.

50. Itoh, H., T. Washio, and M. Tomita. 2004. Computational comparative analyses of alternative splicing regulation using full-length cDNA of various eukaryotes. *RNA* 10(7):1005–1018.

9

Computational Regulomics: Information Theoretic Approaches toward Regulatory Network Inference

Vijender Chaitankar, Debmalya Barh, Vasco Azevedo, and Preetam Ghosh

CONTENTS

9.1 Introduction

A genetic regulatory network, at an abstract level, can be conceptualized as a network of interconnected genes/transcription factors that respond to internal and external conditions by altering the relevant connections within the network [1]; here, a connection represents the regulation of a gene by another gene/protein.

Babu et al. [2] organized the various approaches toward inferring the gene regulatory networks into three fundamental categories. These are:

1. Template-based methods
2. Inferring networks by predicting cis-regulatory elements
3. Reverse engineering using gene expression data

Although the first two approaches are more biologically oriented, the third one relies primarily on mathematical and computational-based approaches. Reverse engineering approaches using gene expression data scan for patterns in the time series microarray data sets generally known as the gene expression matrix [2]. The gene expression matrix is created after a series of steps are performed over the microarray chip, which includes image processing, transformation, and normalization. In a gene expression matrix, a row generally represents a gene and a column represents an external condition or a specific time point [3].

The advantage of the "reverse engineering approach using gene expression data" is that it does not require any prior knowledge to infer the regulatory networks, although incorporating such prior knowledge would yield even better inference accuracy. For example, because it is known that transcription factors regulate genes, giving a list of transcription factors as prior knowledge for these algorithms would greatly improve the results.

Understanding the global properties of regulatory networks has also resulted in improved inference results. One example of such a global property based on empirical studies on the yeast regulatory network is that a gene is generally regulated by a small number of transcription factors [4]. This property greatly reduces the computational complexity of popular methods such as dynamic Bayesian networks [5–7] and probabilistic Boolean networks [8,9], to name a few.

Figure 9.1 outlines the inference/reverse engineering process. The main input that a reverse engineering algorithm requires is the gene expression matrix, whereas other inputs such as prior knowledge (e.g., known interactions between genes) and other kinds of experimental data (e.g., gene knockout data) can also be utilized to obtain better results.

9.1.1 Microarray Experiments

Microarrays help biologists record the expression levels of genes in an organism at the genome level. A microarray is a glass plate that has thousands of spots; each of these spots

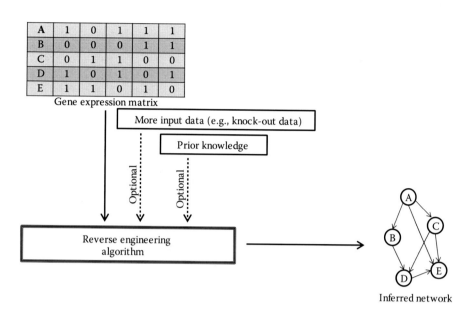

FIGURE 9.1
Inference/reverse engineering outline.

contains similar DNA molecules that uniquely identify a gene. In any microarray experiment, the RNA from the cells are first extracted; these extracted RNAs are then reverse transcribed into cDNA. The cDNAs are then labeled with colored dyes and allowed to hybridize on the microarray glass plate. At this stage, the labeled cDNAs will hybridize to the complementary sequences on the spots. If the concentration of a particular gene is high, then the corresponding spot on the microarray plate will show the dye color.

There are a number of ways in which a microarray experiment can record expression levels. The most popular one is to compare the expression level of a cell exposed to a particular external condition to that of a reference cell in a normal condition. In this approach, cells under the two different conditions are labeled with two different dye colors. If a particular gene is expressed in only one condition, then the corresponding spot on the microarray plate shows the color for that particular condition. If it is not expressed in any of the conditions, then the actual color of the spot (usually black) shows up. It is also possible that the genes may be expressed in both conditions; in such cases, a spot shows a variant color that is the combination of the two chosen colors. After hybridization, the colors on the microarray plate are scanned and recorded by a machine. This microarray data is then analyzed, which includes image processing, transformation, and normalization. For more details on these steps, refer to Babu [3]. Different microarray normalization and quantization methods might yield different networks, and hence, need to be carefully selected. The data after this analysis is used as the input by the reverse engineering algorithms that we discuss in this chapter. We next present some mathematical foundations on information theory.

9.1.2 Information Theoretic Metrics

This section introduces some of the basic metrics of information theory that are used in the algorithms for reverse engineering regulatory networks.

9.1.2.1 Entropy (H)

Entropy is the measure of the average uncertainty in a random variable. Entropy of a random variable X with probability mass function $p(x)$ is defined [10] by

$$H(X) = -\sum_{x \in X} p(x) \cdot \log p(x)$$
(9.1)

9.1.2.2 Mutual Information

Mutual information (MI) measures the amount of information that can be obtained about one random variable by observing another one [10]. MI is defined as

$$I(X;Y) = \sum_{x,y} \left[p(x,y) \log \frac{p(x,y)}{p(x) \cdot p(y)} \right]$$
(9.2)

MI can also be defined in terms of entropies [10] as

$$I(X;Y) = H(X) + H(Y) - H(X,Y)$$
(9.3)

9.1.2.3 Conditional Mutual Information

Conditional mutual information (CMI) is the reduction in the uncertainty of X due to knowledge of Y when Z is given [11]. The CMI of random variables X and Y given Z is defined [10,11] as

$$I(X;Y \mid Z) = \sum_{x,y,z} p(x,y,z) \cdot \log \frac{p(x,y \mid z)}{p(x \mid z) \cdot p(y \mid z)}.$$
(9.4)

CMI can also be expressed in terms of entropies as [11]:

$$I(X;Y \mid Z) = H(X,Z) + H(Y,Z) - H(Z) - H(X,Y,Z)$$
(9.5)

9.2 Methods/Algorithms

This chapter focuses on information theory–based inference of regulatory networks using expression data. Although approaches such as dynamic Bayesian networks [5–7] do infer quality networks, their capability is limited to inferring small regulatory networks (~300 nodes, i.e., 300 genes and transcription factors). As the microarray experiments provide data for thousands of genes, these approaches cannot utilize the full potential of the available data. Usually, a subset of the complete data set, such as differentially expressed genes, is filtered from the complete data set and is then given as the input to these approaches. However, information theory approaches are capable of inferring networks at the genome scale. In the following subsections, we will be discussing some of the popular information theory-based algorithms.

The algorithms discussed in this chapter are classified in the following four categories:

1. Relevance networks based algorithms
2. Minimum description length–based algorithms
3. Time-lag–based algorithms
4. REVEAL-based algorithms

9.2.1 Relevance Networks Based Algorithms

This section discusses the relevance network algorithm [12,13] and algorithms that were derived using relevance network algorithm. This family of algorithms is computationally very efficient and is capable of inferring very large networks.

9.2.1.1 Relevance Network Algorithm

A relevance network [12,13] is the simplest and computationally most effective information theory–based approach. This approach computes pairwise MI between every pair of genes from the given expression data. An edge between a pair of genes exists if the corresponding MI is greater than a user-selected threshold. The quality of inferred networks depends heavily on the threshold selected. A small threshold infers a network with a high number of true edges and false edges, whereas a high threshold infers networks with a low number of true and false edges. If a regulatory interaction inferred is correct, then the edge representing that interaction is a true edge; otherwise, if the regulatory interaction inferred is wrong, then the edge representing the interaction is a false edge. The selection of a proper threshold is an issue in this approach.

9.2.1.2 Algorithm for Reconstruction of Genetic Networks (ARACNE)

The relevance networks concept assumed that if the MI is low (lower than the user-defined threshold), then the genes are not connected, although they are connected if the MI is high. Based on the study of chemical kinetics, it has been found that the second assumption is not true [11]. Consider a network with three genes (X, Y, and Z). If gene X regulates gene Y, and gene Y regulated gene Z, then the MI between genes X and Z could be high, which infers a false edge. ARACNE [14] was the first inference algorithm to implement a method to identify and prune such false edges. ARACNE states that if the MI between two genes X and Y is less than or equal to the MI between genes X and Z or between Y and Z, that is, $I(X, Y) \leq [I(X, Z), I(Y, Z)]$, then there is no connectivity between X and Y.

9.2.1.3 Context Likelihood of Relatedness (CLR)

Various improvements in the relevance networks approach were achieved over the last decade, CLR [15] being the most popular among them today.

CLR applies an adaptive background correction step to eliminate false connections and indirect influences. In this approach, all pairwise MI values are computed similar to the relevance network approach and stored in a matrix. If the network inference is being performed over a data set that has expression data for N genes, then all pairwise MI values are stored in a matrix M, which has N rows and N columns. For every entry in this M matrix, z scores are computed over the row (z_1) and column (z_2). After z score computation is performed, the score for each pairwise edge is computed as $\sqrt{z_1^2 + z_2^2}$. Thus, the score for every edge is computed and stored in a matrix (e.g., Z), which also has N rows and N columns. The network is finally obtained by using a user-defined threshold over the Z matrix. For more information on computing z scores, refer to Faith et al. [15].

9.2.1.4 Direct Connectivity Algorithm

Like ARACNE and CLR, the direct connectivity algorithm [11] was also derived using relevance network algorithm. The false edge pruning process here is performed using the CMI metric. The algorithm initially builds a relevance network; then for every edge inferred, the algorithm computes the CMI between the two nodes of the inferred edge and every other node in the network. If the CMI metric is lower than a user-specified threshold, the edge is deleted. This algorithm, along with previously discussed algorithms, also suffers from the threshold selection problem.

9.2.2 Minimum Description Length–Based Algorithms

This section explains the minimum description length (MDL) principle [16–18] and the algorithms based on the MDL principle.

9.2.2.1 MDL Principle

The MDL principle states that if multiple theories exist, then the theory with the smallest model length is the best. This principle, along with entropy metric, can be used to exploit the robustness property of a regulatory network. It also solves the MI threshold selection problem that exists in relevance networks–based approaches. Description length for network inference was first formulated by Zhao et al. [19]. To understand the description length, let us first define the network formulation.

9.2.2.2 Genetic Network Formulation

The network formulation is similar to the one used by Zhao et al. [19]. A graph $G(V,E)$ represents a network in which V denotes a set of genes and E denotes a set of regulatory relationships between genes. If gene x shares a regulatory relationship with gene y, then there exists an edge between x and y. Genes can have more than one regulator. The notation $P(x)$ is used to represent a set of genes that share regulatory relationships with gene x. For example, if gene x shares a regulatory relationship with y and z, then $P(x) = \{y,z\}$. Also, every gene is associated with a function $f_x P(x)$, which denotes the expression value of gene x as determined by the values of the genes in $P(x)$.

Gene expression is affected by many environmental factors. Because it is not possible to incorporate all such factors, the regulatory functions are assumed to be probabilistic. Also, the gene expression values are assumed to be discrete and the probabilistic regulation functions are represented as lookup tables. If the expression levels are quantized to q levels and a gene x has n predecessors, then the lookup table has q^n rows and q columns and every entry in the table corresponds to a conditional probability.

TABLE 9.1

Conditional Probability

yz:x	0	1
00	0.6	0.4
01	0.3	0.7
10	0.5	0.5
11	0.8	0.2

Suppose that we have genes x, y, and z and the data is quantized to two levels, an example lookup table is given in Table 9.1. In this example, the entry 0.6 can be inferred as follows: if genes y and z are weakly expressed then the probability that x is also weakly expressed is 0.6.

9.2.2.3 Description Length Computation

The description length is the sum of the model length and the data length. The model length in Zhao et al. [19] was defined as the amount of memory consumed by the algorithm. The complex part of the description length computation is the data length. The following section gives a detailed explanation of the data length computation.

9.2.2.4 Data Length Computation

A gene can take any value when transformed from one time point to another due to the probabilistic nature of the network. The network is associated with a Markov chain, which is used to model state transitions. These states are represented as n gene expression vectors $X_t = (x_{1,t},...,x_{n,t})^T$ and the transition probability $p(X_{t+1}|X_t)$ can be derived as follows:

$$p(X_{t+1} \mid X_t) = \prod_{i=1}^{n} p\left(x_{i,t+1} \mid \mathbb{P}_t(x_i)\right). \tag{9.6}$$

The probability $p\left(x_{i,t+1} \mid \mathbb{P}_t(x_i)\right)$ can be obtained from the lookup table associated with the vertex x_i and is assumed to be time-invariant. It is estimated as follows:

$$p\left(x_{i,t+1} = j \mid \mathbb{P}_t(x_i)\right) = \frac{1}{m-1} \sum_{t=1}^{m-1} 1_{\{j\}}\left(x_{i,t+1} \mid \mathbb{P}_t(x_i)\right). \tag{9.7}$$

Each state transition brings some new information that is measured by the conditional entropy:

$$H(X_{t+1} \mid X_t) = -\log\left(p(X_{t+1} \mid X_t)\right). \tag{9.8}$$

The total entropy for the given m time series sample points $(X_1,...,X_m)$, is as follows

$$L_D = H(X_1) + \sum_{j=1}^{m-1} H(X_{j+1} \mid X_j). \tag{9.9}$$

As $H(X_1)$ is the same for all models, it is removed and thus the description length is given by

$$L_D = \sum_{j=1}^{m-1} H(X_{j+1} \mid X_j). \tag{9.10}$$

9.2.2.5 Network MDL Algorithm

In the network MDL approach [19], for every pairwise MI value that is greater than zero, the algorithm generates an initial network. To select the best network, the algorithm computes the MDL score for every network and chooses the network with the minimum score as the best network. The model length computation here is not standardized, and because a fine-tuning parameter is used in this approach to control the effects of the model length parameter, this method can be arbitrary. Another major disadvantage of this approach is that it is not scalable because it is computationally very complex. The complexity of this approach can be reduced by restricting the number of regulators inferred for each gene in the initial network to between three and six. Also, the network is biased to a very high MI threshold value. The initial network generated using a very high MI threshold value will result in a sparse network that is likely to have a very low data length. Such a network is not desired and this algorithm might not produce very good results. But considering certain special cases, such as with a restricted number of regulators and not using a very high or very low MI threshold, this algorithm can infer quality networks.

9.2.2.6 Predictive MDL Principle

The description length of the two-part MDL principle involves the calculation of the model length and the data length. As the length can vary for various models, the method is in danger of being biased toward the length of the model [16]. The normalized maximum likelihood model has been implemented in Dougherty et al. [20] to overcome this issue. Another such model based on the universal code length is the predictive MDL (PMDL) principle [18].

The description length for a model in PMDL [18] is given as:

$$L_D = -\sum_{t=0}^{m-1} \log\big(p(X_{t+1} \mid X_t)\big) \tag{9.11}$$

where $p(X_{t+1} \mid X_t)$ is the conditional probability or density function. In the PMDL algorithm, the description length is equivalent to the data length given in Zhao et al. [19]. The next section gives a detailed explanation of the PMDL algorithm.

9.2.2.7 PMDL Algorithm

Given the time series data, the data is first preprocessed, which involves filling missing values and quantizing the data. Then, the MI matrix $M_{n \times n}$ is evaluated to infer a connectivity matrix $C_{n \times n}$, which has two entries: 0 and 1. An entry of 0 indicates that no regulatory relationship exists between genes, but an entry of 1 at $C_{i \times j}$ indicates that gene i regulates j. The algorithm is given in Figure 9.2 [21–23].

```
1.  Input time series data
2.  Preprocess data
3.  Initialize M_{n×n}, C_{n×n} and P(x_j) ⇐ φ
4.  Calculate the cross-time mutual info
            between genes and fill M_{n×n}
5.  for i = 1 to n
6.    for j = 1 to n
7.      δ ⇐ M_{i×j}
8.      for k = 1 to n
9.        for l = 1 to n
10.         if M_{k×l} ≥ δ then
11.           C_{k×l} = 1, P(x_l) ⇐ P(x_l) ∪ x_k
12.         end if
13.       end for
14.     end for
15.     Compute probabilities using Eqn (7)
16.     Compute L_D using Eqn (10)
17.   end for
18. end for
19. Select the MI of model having least L_D as the MI threshold δ.
20. for i = 1 to n
21.   for j = 1 to n
22.     if C_{i×j} == 1 then
23.       for k = 1 to n and k ≠ i, j
24.         if CMI_{i,j,k} < Th then
25.           C_{i×j} ⇐ 0
26.           break;
27.         end if
28.       end for
29.     end if
30.   end for
31. end for
```

FIGURE 9.2
Pseudocode for PMDL algorithm.

From lines 5 to 18 (Figure 9.2), every value of the MI matrix is used as a threshold and a model is obtained. The conditional probabilities and the description lengths for each of these models are evaluated using Equations 9.7 and 9.10, respectively. Then, at line 19, the MI that was used to obtain the model with the shortest description length was used as the MI threshold to obtain the initial connectivity matrix. From lines 20 to 31, for every valid regulatory connection in the connectivity matrix, the CMI of the genes with every other gene was evaluated and if the value was lower than the user-specified threshold, the connection was deleted.

As the MDL principle selects the model with the smallest description length as the best one, in this case, the best model is the one that has the lowest entropy, that is, the lowest uncertainty. Thus, this approach exploits the robustness property of the biological networks.

9.2.2.7.1 *Time and Space Complexities of the PMDL Algorithm*

The performance of the PMDL algorithm depends on three factors. The number of genes, the number of time points in the expression matrix, and most importantly, the number of parents inferred for each gene by the algorithm. This section gives the time and space complexities of the algorithm.

Step 4 of the algorithm iterates n^2m times, where n is number of genes and m is the number of time points; from line 5 to line 18, the algorithm iterates n^4 times; lines 15 and 16 of the algorithm iterates n^3m times. Finally, from lines 20 to 31, the algorithm iterates n^3 times. Thus, the time complexity of the algorithm is $\Theta\ (n^4 + n^3m)$.

When it comes to space complexity, the conditional probability tables play a major role. If a gene has n parents, then the conditional probability tables take 2^n units of space. Thus, the amount of memory needed by the algorithm depends on the number of parents inferred by the network. As the space complexity grows exponentially based on the number of parents, it is possible that the algorithm may run out of memory for a data set with as few as 50 genes but run for as little as 5 min for a data set with several hundred genes. The memory limitation can be solved by restricting the number of regulators each gene can have.

9.2.3 Time-Lagged Information Theoretic Approaches

It was observed that the accuracy of information theory–based approaches saturate beyond specific data sizes [24,25] (in terms of the number of time points or columns in the expression matrix). Further analysis that showed why these approaches saturated was the motivation for conceiving time-lagged information theory approaches, which we discuss next.

9.2.3.1 Why Do Information Theory–Based Approaches Saturate?

Visual analysis about how the values of the basic metrics change according to the data size were performed by Chaitankar et al. [24,25]. For this analysis, biological synthetic data was created using the GeneNetWeaver tool [26,27]. A five-gene network and expression data with 100 time points was generated. The synthetic data was quantized to two levels and then the information theoretic quantities were calculated. Entropy of all the genes in the network across 100 time points was computed, whereas the conditional entropy and MI were computed for each pair of genes. It was observed that with more data (and correspondingly, with more time points) both the entropy and conditional entropies increased in the network (tending to unity), whereas the MI decreased (tending to zero).

This concludes that the saturation in the inference accuracy of the information theoretic approaches was due to the saturation in the MI quantity, which goes close to zero even though the entropy increases in the network. This would conceptually mean that there is room to improve on the inference accuracy (due to high entropy), yet the MI metric will not be able to point us to the right direction. Other information theoretic algorithms, like REVEAL [28], use the ratio of MI and entropy to infer the network for this purpose, which supposedly gives good performance.

However, as the entropy and MI metrics saturate, it is obvious that the ratio of MI and entropy will also saturate, hence, this ratio might also not be the right metric to achieve better accuracy by making use of more data. The directed mutual information metric [29] might be a better metric to use than the conventional MI-based algorithms.

The saturation in MI due to increasing number of time points would suggest that the MI should not be computed for the entire range of time points of microarray data available from the experiments. Gene Regulatory Networks (GRN) are inherently time-varying, and hence, the pairwise MI between any two genes needs to be computed over the time range in which the first gene will have substantial regulatory effect on the other one. This can be best approximated by estimating the regulatory time lags between each gene pair, and subsequently computing the MI between them for this particular time range. The time lag computation concept was initially proposed by Zou and Conzen [7] to predict potential regulators in their dynamic Bayesian network–based scheme.

9.2.3.2 Time Lags

The concept of time lags was first introduced by Zou and Conzen [7], where they proposed that the time difference between the initial expression change of a potential regulator (parent) and its target gene represents a biologically relevant period. Here, potential regulators are those sets of genes whose initial expression change happened before the target gene and the initial expression change is the upregulation or downregulation (ON or OFF) of genes.

Implementing Zou and Conzen's method of calculating time lags, for every pair of gene when the initial expression change is not at the same time point, one of the two time lags turns out to be negative. Figure 9.3 illustrates this problem. In the figure, *Ia* and *Ib* indicate the initial change in expression of gene *A* and gene *B* at time points 2 and 3, respectively. According to Zou and Conzen [7], gene *A* is the parent of gene *B*, and they do not consider the time lags between *B* and *A*. Time lag between *A* and *B* is *Ia* − *Ib* = 3 − 2 = 1. The time lag between *B* and *A* is *Ib* − *Ia* = 2 − 3 = −1, this is a negative time lag, which implies that a directed edge cannot exist between *B* to *A*. However, it is important to consider such time lags both in the forward and backward directions (i.e., from *A* to *B* and vice versa) as this can model the loops between two genes (i.e., *A* → *B* and *B* → *A* connections). Zou and Conzen's time lag computation cannot handle such cases.

One can argue that a gene can regulate another gene only when it is upregulated (ON). Based on the this argument, a new time lag computation approach was proposed in Chaitankar et al. [25], wherein time lags were defined as "the difference between initial upregulation of first gene and initial expression change of the second gene after the upregulation of first gene."

Figure 9.3 also illustrates the time lag computation approach proposed in Chaitankar et al. [25]. In the figure, *Ua* and *Ub* indicate the *initial upregulation* of genes *A* and *B* at time points two and three, respectively. *Ca* and *Cb* indicate that time points six and three are the

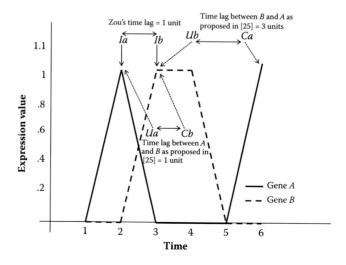

FIGURE 9.3
Time lag computation approaches.

time points in which the expression values of genes *A* and *B changed after the initial upregulation* of genes *B* and *A*. The time lag between *A* and *B* is calculated as $\tau 1 = Cb - Ua$ and the time lag between *B* and *A* is calculated as $\tau 2 = Ca - Ub$, respectively. In this example, the time lag between *X* and *Y* is one and time lag between *Y* and *X* is three.

9.2.3.3 Time-Lagged MI and Time-Lagged CMI

After implementing time lags, one no longer needs to compute the information theoretic metrics over the complete expression data. Suppose that the expression matrix has *E* columns of expression data. If a time lag τ is computed between two genes *A* and *B*, we remove the last τ columns from the row in the expression matrix representing gene *A* and the first τ columns of the row representing gene *B*. Computing the MI over the reduced expression data gives time-lagged MI (TLMI). Note that TLMI is not a symmetric quantity like MI, that is, $TLMI(A,B) \neq TLMI(B,A)$.

Considering a time lag, τ, between A and B, we compute time-lagged CMI (TLCMI) (A;B|C) by deleting the last τ columns of rows representing genes A and C, and the first τ columns of the row representing gene B. Computing the CMI over this reduced expression data gives TLCMI. Examples of time-lagged information theoretic quantities are given in Chaitankar et al. [25].

9.2.3.4 Time-Lagged Information Theory–Based Approaches

Virtually any information theory–based approach can be converted to its time-lagged equivalent [25,30] by replacing information theoretic metrics with corresponding time-lagged information theoretic quantities. A drawback of time-lagged approaches is that it can be implemented over time series data sets only.

9.2.4 REVEAL-Based Algorithms

The REVEAL [28] algorithm starts with pairwise combinations. If the *r*-score (ratio of the MI between the regulating gene and the entropy of the gene being regulated) equals one, REVEAL states that a regulatory interaction exists between these two genes and the algorithm proceeds to generate the corresponding wiring rules.

If not all genes are being regulated with a gene, REVEAL next proceeds to see if a combination of two genes regulate the remaining genes. If there are still other genes in which the MI to entropy ratio did not equal one, REVEAL will next consider a combination of three genes and so on. Thus, the algorithm iteratively considers more complex genetic interactions, thereby having exponential time and space complexities. Even for a small network of 50 genes, the algorithm runs out of memory on a modern 32-bit computer. The next subsections discuss some algorithms that are improvements/extensions over the REVEAL algorithm.

9.2.4.1 sREVEAL-1

By considering a set of known transcription factors, the number of iterations REVEAL has to execute for different input combinations is reduced by a huge margin, thus reducing both space and time complexities. The transcription factor prediction approach [31] can be used to identify transcription factors with high accuracy. As empirical studies in the past have shown that a gene is regulated by a small set of transcription factors, one can restrict the number of regulators to between three and six. This consideration further reduces the complexity of the algorithm. Unlike REVEAL, in which it keeps solving for a higher number of parents for a particular gene as it proceeds, this approach solves one gene at a time. If, for example, the number is restricted to three parents, the MI between all combinations of one, two, and three pairs of transcription factors are computed and the combination of transcription factors that produce the maximum MI to entropy ratio with the gene is chosen as the combination that regulates the gene. Figure 9.4 gives a pictorial presentation of REVEAL and sREVEAL [32] algorithms.

In REVEAL, all possible pairwise combinations are first checked to find the regulators of the genes; the genes wherein the ratio equals 1 have been properly characterized and taken out for future considerations. For the remaining genes, all combinations of two genes that can act as their parents are checked similarly, and if the ratio equals 1, they are taken out of consideration; the algorithm next looks into the combination of three genes to act as parents of the remaining genes and so on until all genes are characterized. The right-hand side in Figure 9.4 explains how sREVEAL-1 works. A maximum of two regulators for each gene is shown in this case. Hence, for every gene, all possible combinations of one and two genes are checked and the combination that gives the maximum MI is chosen as the potential parent set.

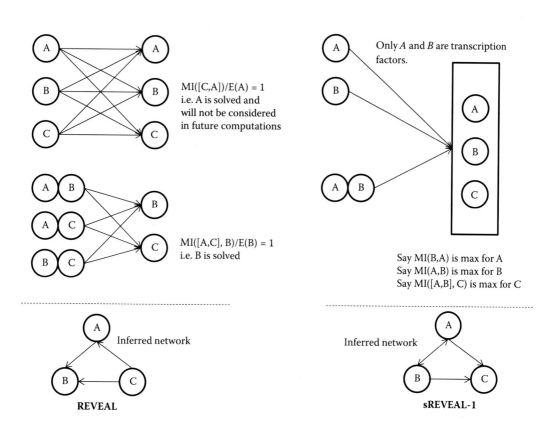

FIGURE 9.4
REVEAL and sREVEAL approaches.

9.2.4.2 sREVEAL-2

Although the abovementioned approach reduces complexity and can infer the network for several hundred genes, it can still run out of memory when the transcription factor list itself is large (>50) as in the case of the complete *Escherichia coli* transcriptional regulatory network. This issue can be solved by further reducing the potential set of transcription factors for each gene by using time lags. A transcription factor can regulate a gene if and only if the time lag between the transcription factor and the gene is greater than zero. In Chaitankar et al. [32], the authors suggest that very large time lags should be set to zero and hence not considered as potential regulators and higher priority must be given to smaller time lags. Thus, even after filtering the potential regulators using time lags, the list of potential regulators can still be large; further filtering can be performed by considering only small time lags between transcription factors and genes. This approach can be used to infer fairly large regulatory networks.

9.2.4.3 rREVEAL

rREVEAL removes the exit strategy (discarding cases for which the *r*-score ratio was evaluated as 1) implemented by REVEAL and uses a novel ranking scheme on the regulators to assess their likelihood of serving as the parents for any particular gene. rREVEAL starts by computing *r*-scores between every regulator and the gene. Then, it computes the *r*-score between every two combinations of regulators and the gene, and then it moves to three combinations of regulators, and so on. The scores of different combinations that produce maximum scores are stored in different bins, and regulators in each bin are given a normalized score. If the number of combinations that give the maximum score is C and the maximum number of regulators allowed in that bin is P, then the total number of regulators in the cell is $C*P$. If a regulator R occurs O times, then the normalized score for regulator R in bin P is given by:

$$N_R = \frac{O}{C \times P}.$$ (9.12)

Also, normalizing the scores for each regulator separately in every bin makes sure that the inference is not biased on the number of parents any gene can have. For example, as shown in Figure 9.5, for each gene, in bin one, we only have combination of one parent, whereas in bin two, we have combinations of two parents and so on. Also, assume that two cases in each bin achieve the maximum *r*-score: A and B in bin 1; (C,D) and (C,A) in bin 2. Here, the probability of the occurrence of A and B in bin 1 is 0.5 each, whereas that of C, D, and A in bin 2 is 0.5, 0.25, 0.25, respectively. If, instead of computing the normalized score in this fashion, we had simply counted the number of occurrences of the regulators in each bin, the results would be biased toward the higher bins because a regulator is more likely to occur a higher number of times in larger parent list combinations (regulator C in this example has a higher count). After this step, a final score is assigned to each regulator by summing the normalized scores across the individual bins. If P is the desired number of maximum regulators in the inferred network, a final score of N_S for each regulator is given by:

$$N_S = \sum_{i,j=1}^{R,P} N_{i,j}.$$ (9.13)

Regulator Combination	A	B	C	D
r-score	0.8	1	1	0.8

$$r\text{-score} = \frac{\text{MI between regulators}}{\text{entropy of gene}}$$

Regulator Combination	[A,B]	[A,C]	[A,D]	[B,C]	[B,D]	[C,D]
r-score	0.8	1	0	1	1	0.8

Regulator Combination	[A,B,C]	[A,B,D]	[A,C,D]	[B,C,D]
r-score	0.8	0.7	0.5	1

Regulator	A	B	C	D
Normalized score (N_R)	0	.5	.5	0

Bin 1

Regulator	A	B	C	D
Normalized score (N_R)	0.167	0.333	0.333	0.167

Bin 2

N_R = probability participating of in the regulator combination that achieves the maximum *r*-score.

Regulator	A	B	C	D
Normalized score (N_R)	0	0.333	0.333	0.333

Bin 3

Regulator	A	B	C	D
Final score (N_S)	0.055	0.388	0.388	0.167

N_S = Average N_R across all bins

FIGURE 9.5

rREVEAL example. To get the final regulator list for each gene, the *r*-scores, the normalized score (N_R), and the final scores (N_S) are computed as shown in the example above. Then the regulators are sorted based on the final scores and the desired number of regulators are inferred for the specific gene under consideration. In this fashion desired number of regulators are inferred for every gene. In the example above, we have four transcription factors as potential regulators, and the desired number of regulators for each gene is three, thus the final regulator list is [B, C, D].

Based on these scores, the regulator list for each gene is ranked, and depending on the desired number of regulators, the final regulator list for each gene is short-listed from the ranked list. Figure 9.5 gives a pictorial representation of rREVEAL. In the figure, the network has four transcription factors. The maximum number of desired parents is three; thus, scores are computed for all possible combinations of one, two, and three for all nodes in a network.

Although rREVEAL is more robust than REVEAL, there is a possibility that ordering of data might affect the resulting networks. If the maximum number of regulators allowed is *n*, and if *n*th and (*n* + 1)th ranked regulators have the same final scores, then ordering of data yields different networks. Such a scenario is less likely to occur and simple steps can be taken to tackle such scenarios. One strategy is to include all regulators beyond the (*n* + 1)th rank that have the same score as the *n*th ranked regulator. Also, normalization of scores is important to give each transcription factor a fair chance in the ranking process. Extensions of rREVEAL have been extensively studied in [33,34].

9.2.4.3.1 Complexity Analysis of rREVEAL

Entropy estimation and considering different combinations of regulators are the major players in time complexities of the rREVEAL algorithm. Entropy estimation for *n* variables and *q* level quantization has a time complexity of $O(q^n)$. As *q* is set to two and *n* is restricted to five, the entropy estimation complexity is a constant in the rREVEAL implementation. Now, considering the different regulator combinations in a network with *t* transcription factors and k being the maximum number of parents each gene can have, the total number of combinations to be considered is ${}^t C_5$, resulting in a worst-case run-time complexity of $O(t^k)$; as k is again restricted between three and five, the worst-case complexity of this step reduces to $O(t^5)$. For a network with s number of genes, the worst-case complexity of rREVEAL is hence $O(st^5)$.

9.2.5 Inference Using Knockout Data

Although the inference algorithms primarily work on the microarray data, other data sets can be incorporated to further improve the quality of inferred networks. One such data set that can be used is the knockout data. In this chapter, we will be looking at how the knowledge from the knockout data can be used to further improve the quality of inferred networks.

9.2.5.1 Knockout Data

Knockout data provides the expression levels of genes when a particular gene(s) is down-regulated. This data is effective in determining direct regulatory interactions [35]. These experiments are useful when certain genes of interests are targeted. External conditions that downregulate these genes are used to carry out the experiments. Because these experiments are expensive, they cannot be implemented on a large scale. However, the knowledge from these experiments can be used in network inference to improve the inference accuracy.

9.2.5.2 Knockout Network Generation

To generate the knockout data–based network, a simple model to detect fold change in gene expression with respect to the wild-type data (or expression data in normal condition) was used. If the fold change is greater than 1.2 or less than 0.7, then the gene has a direct regulatory relationship with the knocked out gene.

9.2.5.3 Number of Parents in the E. coli Genome

Studies on the yeast network [4] showed that a gene is usually regulated by a small number of transcription factors. This property can be used to greatly reduce the complexities of algorithms such as network MDL, PMDL, dynamic Bayesian network, etc. Chaitankar et al. [32] also performed similar studies on the genomic network of *E. coli*.

The gene regulatory network for *E. coli* was obtained from RegulonDB release 6.2 [36]. The network has 1502 genes and 3587 regulatory interactions. It was observed that genes usually have a small number of parents as observed in the yeast network. This observation indicates that a choice between three and six seems to be a good number of parents in GRN inference.

9.2.5.4 Network Generation Using Parent Restrictions

The parent restriction approach [37] for each gene takes the knockout data network, the network obtained from PMDL, the individual MI values, and time lags as inputs to compute the preference of parent node to be chosen in case a gene has more parents than the parent restriction constraint.

Figure 9.6 illustrates the parent selection process. In Figure 9.6, three networks (1, 2, and 3) are shown. Network 1 is the network obtained using MI and CMI thresholds. The

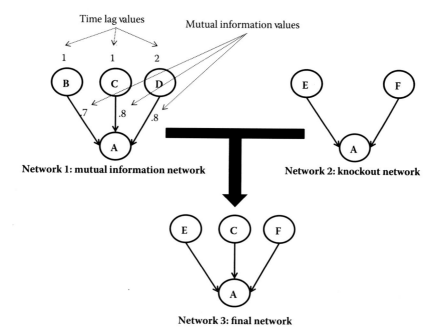

FIGURE 9.6
Knockout-based network inference approach.

edges in this network are labeled by the MI values and the time lags are given above the nodes. Network 2 is the knockout network, and network 3 is the final network obtained after the parent selection process is completed.

Next, the parent selection process is based on the following priorities:

- Knockout edges
- MI values
- Time lags

Irrespective of the MI or time lags, the knockout edges are added first, and then the parents from the MI network are chosen based on MI values; if there is a conflict in the MI values, the parent selection is based on the time lags. If the knockout edges are more than the desired number of parents, then the best of these are selected based on MI and time lags as discussed above. In Figure 9.6, the maximum number of parents that node A can have is three. Therefore, if genes E and F are chosen initially, and there is room for one more parent. There are two genes, C and D, both of which have 0.8 as an MI value, but as gene C has a shorter time lag, we choose gene C as the third parent.

9.2.5.5 Knockout Data Incorporation in rREVEAL

The knowledge from knockout data sets can be used to reduce the potential regulators for each gene, thus reducing the run-time of the algorithm. As knockout data has been shown to infer strong regulatory interactions [35], pruning potential parents based on knockout data not only improves the run-time but also the inference accuracy. We used a simple model to obtain the potential regulators based on knockout data. If the fold change of knocked out expression value between transcription factor A and gene B to steady state value of gene B is greater than or equal to a specific user-selected threshold, then we consider transcription factor A to be a potential regulator of gene B. In this fashion, for every gene, the complete transcription factor list is tested. In the worst-case, every transcription factor is a potential parent of every gene, but such a scenario is highly unlikely.

9.3 Applications

Disease diagnostics and drug development research teams can make use of these reverse engineering algorithms to generate a start-up regulatory network for their analysis. For example, during the drug development process, after the cell is exposed to the drug, the expression levels of the genes can be monitored using microarray data experiments. Using these observations across multiple time points, one can use a reverse engineering algorithm to build a network and start their analysis based on this network.

These algorithms can also be used for generating start-up networks for analysis by toxicologist, for example, the Engineer Research and Development Center of the U.S. Army is interested in knowing how chemicals being used in weapons affect the ecosystem. The Engineer Research and Development Center selects certain species and exposes their cells to a chemical, as in the previous example, again the expressions levels are recorded across multiple time points and a reverse engineering algorithm is used to build a start-up network for analysis.

9.4 World Wide Web Resources

Many of the abovementioned algorithms are available online. If one is interested in developing new algorithms, microarray data sets are also available online. Many *in silico* synthetic data generation tools are also available through which a person interested in developing newer algorithms can generate data sets to test the performance of the algorithm. We present the URLs to some of the important resources available online.

1. Yeast Cell Cycle Data set available from Dr. Spellman's laboratory: http://genome-www.stanford.edu/cellcycle/data/rawdata/
2. Yeast Cell Cycle Data set available from Dr. Chou's laboratory: http://genomics.stanford.edu/yeast_cell_cycle/full_data.html
3. GeneNetWeaver tool for *in silico* data generation: The GeneNetWeaver tool can be of immense help for one interested in developing newer reverse engineering algorithms. This tool generates data based on known biological functions. This tool also provides a way for generating subnetworks of very large networks. Specifically, this tool has the known yeast and *E. coli* regulatory networks. Using this tool, one can generate data sets for the desired network size and the desired number of microarray experiments. This tool is available for free at: http://gnw.sourceforge.net/

4. ARACNE and CLR: The popular ARACNE and CLR algorithms can be downloaded at the following link: http://gardnerlab.bu.edu/clr.html
5. rREVEAL implementation is available upon request. Please e-mail pghosh@vcu.edu to obtain a MATLAB® implementation of rREVEAL algorithm

References

1. Hartemink A.J. 2005. Reverse engineering gene regulatory networks. *Nature Biotechnology* 23: 554–555.
2. Babu, M.M., B. Lang, and L. Aravind. 2009. Methods to reconstruct and compare transcriptional regulatory networks. *Methods in Molecular Biology* 541:163–180.
3. Babu, M.M. 2004. Introduction to microarray data analysis. In *Computational Genomics: Theory and Application*, edited by Grant, R.P., 225–249. UK: Horizon Press.
4. Lee, T.I., N.J. Rinaldi, F. Robert, D.T. Odom, Z. Bar-Joseph, G.K. Gerber et al. 2002. Transcriptional regulatory networks in *Saccharomyces cerevisiae*. *Science* 298:799–804.
5. Imoto, S., T. Goto, and S. Miyano. 2002. Estimation of genetic networks and functional structures between genes by using Bayesian networks and nonparametric regression. *Pacific Symposium on Biocomputing* 7:175–186.
6. Murphy, K., and S. Mian. 1999. Modelling Gene Expression Data Using Dynamic Bayesian Networks. Technical report. Berkeley, CA: Computer Science Division University of California.
7. Zou, M., and S.D. Conzen. 2005. A new dynamic Bayesian network (DBN) approach for identifying gene regulatory networks from time course microarray data. *Bioinformatics* 21(1):71–79.
8. Schmulevich, I., E.R. Dougherty, S. Kim, and W. Zhang. 2002. Probabilistic Boolean networks: A rule-based uncertainty model for gene regulatory networks. *BMC Bioinformatics* 18(2):261–274.
9. Shmulevich, I., E.R. Dougherty, and W. Zhang. 2002. From Boolean to probabilistic Boolean networks as models of genetic regulatory networks. *Proceedings of the IEEE* 90(11):1778–1792.
10. Cover, T.M., and J.A. Thomas. 1991. *Elements of Information Theory*. New York: Wiley-Interscience.
11. Zhao, W., E. Serpedin, and E.R. Dougherty. 2008. Inferring connectivity of genetic regulatory networks using information-theoretic criteria. *IEEE Transactions on Computational Biology and Bioinformatics* 5(2):262–274.
12. Butte, A.J., and I.S. Kohane. 2000. Mutual information relevance networks: Functional genomic clustering using pairwise entropy measurements. *Pacific Symposium on Biocomputing* 418–429.
13. Eisen, M.B., P.T. Spellman, P.O. Brown, and D. Botstein. 1998. Cluster analysis and display of genome-wide expression patterns. *Proceedings of the National Academy of Sciences of the United States of America* 95:14863–14868.
14. Margolin, A.A., I. Nemenman, K. Basso, C. Wiggins, G. Stolovitzky, R. Dalla Favera, and A. Califano. 2006. ARACNE: An algorithm for reconstruction of genetic networks in a mammalian cellular context. *BMC Bioinformatics* 7:S7.
15. Faith, J.J., B. Hayete, J.T. Thaden, I. Mogno, J. Wierzbowski, G. Cottarel et al. 2007. Large-scale mapping and validation of *Escherichia coli* transcriptional regulation from a compendium of expression profiles. *PLoS Biology* 5.
16. Grünwald, P.D., I.J. Myung, and M.A. Pitt. 2005. *Advances in Minimum Description Length (Theory and Applications)*. Cambridge, USA: MIT Press.
17. Rissanen, J. 1978. Modeling by shortest data description. *Automatica* 18:465–471.
18. Rissanen, J. 1984. Universal coding, information, prediction and estimation. *IEEE Transactions on Information Theory* 30(4):629–636.
19. Zhao, W., E. Serpedin, and E.R. Dougherty. 2006. Inferring gene regulatory networks from time series data using the minimum description length principle. *Bioinformatics* 22(17):2129–2135.

20. Dougherty, J., I. Tabus, and J. Astola. 2008. Inference of gene regulatory networks based on a universal minimum description length. *EURASIP Journal on Bioinformatics and Systems Biology,* Article ID: 482090, 11 p.
21. Chaitankar, V., C. Zhang, P. Ghosh, E.J. Perkins, P. Gong, and P. Deng. 2009. Gene regulatory network inference using predictive minimum description length principle and conditional mutual information. *International Joint Conference on Bioinformatics, Systems Biology and Intelligent Computing* 487–490.
22. Chaitankar, V., C. Zhang, P. Ghosh, E.J. Perkins, P. Gong, and P. Deng. 2011. Predictive minimum description length principle approach to infer gene regulatory networks. *Advances in Experimental Biology and Medicine* 696:37–43.
23. Chaitankar, V., P. Ghosh, E.J. Perkins, P. Gong, P. Deng, and C. Zhang. 2010. A novel gene network inference algorithm using predictive minimum description length approach. *BMC Systems Biology* 4, Suppl 1:S7.
24. Chaitankar, V., C. Zhang, P. Ghosh, E.J. Perkins, and P. Gong. 2010. Effects of cDNA microarray time-series data size on gene regulatory network inference accuracy. *ACM International Conference on Bioinformatics and Computational Biology.*
25. Chaitankar, V., P. Ghosh, E.J. Perkins, P. Gong, and C. Zhang. 2010. Time lagged information theoretic approaches to the reverse engineering of gene regulatory networks. *BMC Bioinformatics* 11(Suppl 6):S19, doi:10.1186/1471-2105-11-S6-S19.
26. Marbach, D., T. Schaffter, C. Mattiussi, and D. Floreano. 2009. Generating realistic in silico gene networks for performance assessment of reverse engineering methods. *Journal of Computational Biology* 16(2):229–239.
27. Prill, R.J., D. Marbach, J. Saez-Rodriguez, P.K. Sorger, L.G. Alexopoulos, X. Xue et al. 2010. Towards a rigorous assessment of systems biology models: the DREAM3 challenges. *PLoS One* 5(2):e9202.
28. Shoudan, L. 1998. REVEAL: A general reverse engineering algorithm for inference of genetic network architectures. *Pacific Symposium on Biocomputing* 3:18–29.
29. Rao, A., A.O. Hero, 3rd, D.J. States, and J.D. Engel. 2006. Inference of biologically relevant gene influence networks using the directed information criterion. *ICASSP Proceedings.*
30. Chaitankar, V., P. Ghosh, M.O. Elasri, and M. Perkins. 2011. Poster: Gene regulatory network inference using time lagged context likelihood of relatedness. *International Conference on Computational Advances in Bio and Medical Sciences* 236.
31. Kummerfeld, S.K., and S.A. Teichmann. 2006. DBD: A transcription factor prediction database. *Nucleic Acids Research* 34:D74–78.
32. Chaitankar, V., P. Ghosh, M.O. Elasri, and E.J. Perkins. 2011. sREVEAL: Scalable extensions of REVEAL towards regulatory network inference. *Workshop on Computational Biology. 11th International Conference on Intelligent Systems Design and Applications.*
33. Chaitankar, V., P. Ghosh, M.O. Elasri, K.A. Gust, and E.J. Perkins. 2012. Genome scale inference of transcriptional regulatory networks using mutual information on complex interactions. To appear ACM BCB Workshop on Biological Network Analysis and Applications in Translational and Personalized Medicine (BNA-M).
34. Chaitankar, V., P. Ghosh, M.O. Elasri, K.A. Gust, and E.J. Perkins. 2012. sCoIn: A Scoring algorithm based on COmplex INteractions for reverse engineering regulatory networks. To appear in IEEE International Conference on BioInformaticsand BioEngineering (BIBE).
35. Yip, K., R. Alexander, K. Yan, and M. Gerstein. 2010. Improved reconstruction of in silico gene regulatory networks by integrating knockout and perturbation data. *PLoS One* 5(1):e8121.
36. Gama-Castro, S. 2008. RegulonDB (version 6.0): Gene regulation model of *Escherichia coli* K-12 beyond transcription, active (experimental) annotated promoters and Textpresso navigation. *Nucleic Acids Research* 36:D120–124.
37. Chaitankar, V., Ghosh P., Elasri, M., and E. Perkins. 2011. A scalable gene regulatory network reconstruction algorithm combining gene knockout data. *BiCob.*

10

Glycomics: Techniques, Applications, and Challenges

Veeraperumal Suresh, Chinnathambi Anbazhagan, and
Dakshanamurthy Balasubramanyam

CONTENTS

10.1 Introduction

Progress in genomic and proteomic research has elevated these fields to the forefront of scientific and biomedical research. These scientific endeavors have been facilitated by many of the modern laboratory techniques at the disposal of today's researcher. Automated synthesis of nucleic acids and peptides, rapid DNA and peptide sequencing, and gene expression profile analysis by using cDNA microarrays, protein expression systems, small interfering RNA (siRNA) gene silencing, and knockout organisms are widely used to elucidate the role of genes and proteins in biological systems. Until now, a complementary set of biophysical tools has remained out of reach to the growing discipline of glycomics and this void has greatly hindered the emergence of this field. Glycomics, a term analogous with genomics and proteomics, is the comprehensive study of glycomes—the entire complement of sugars, whether free or present in more complex molecules, of an organism, including genetic, physiologic, pathologic, and other aspects. Glycomics "is the systematic study of all glycan structures of a given cell type or organism" and is a subset of glycobiology. The term glycomics is derived from the chemical prefix for sweetness or a sugar, "glyco" and was formed to follow the naming convention established by genomics, which deals with genes, and proteomics, which deals with proteins. Glycomic analysis seeks to understand how a collection of glycans relate to a particular biological event. This includes carbohydrate–carbohydrate, carbohydrate–protein, and carbohydrate–nucleic acid interactions. Carbohydrates, in the form of glycopeptides, glycolipids, glycosaminoglycans, proteoglycans, or other glycoconjugates, have long been known to participate in a plethora of biological processes. These include viral entry, signal transduction, inflammation, cell–cell interactions, bacteria–host interactions, fertility, and development. Rapid advances in the field of glycomics, however, have been hindered by the complexity of the biomolecules involved. Due to their frequent branching and linkage diversity, oligosaccharides have greater structural complexity than nucleic acids and proteins. Furthermore, the difficulty in isolating, characterizing, and synthesizing complex oligosaccharides has been a significant challenge to progress in the field. Recent chemical advances, such as improved synthetic methods, including the development of an automated solid phase synthesizer, and methods for enzymatic synthesis, have opened new and exciting possibilities in obtaining pure, chemically defined carbohydrates. At the same time, the field has seen growing interest in the development of carbohydrate microarrays and neoglycoconjugates to facilitate otherwise laborious biological studies. By unifying synthetic advances and new biochemical tools, it is now possible to expand the tool-chest available to the glycomics researcher. This chapter illustrates the potential of some of these emerging technologies.

Over the past few years, there has been an increase in the application of systematic methods to the study of glycans and their interactions. In parallel with this trend, a revival in the number and scope of related databases, web sites, and both real and virtual glycan libraries has begun to address the information needs of glycobiologists and glycochemists. The term "glycoinformatics" refers to informatics tools available for assessing "primary data" (covalent and three-dimensional structures of glycans and glycoconjugates) and organizing these primary data into databases that can be used for speeding up the production of primary data, predicting new features, and characterizing structure/activity or structure/function relationships. There are several levels of glycan structural data for which this approach may be useful, from sequencing through three-dimensional structures to interactions (Figure 10.1). This chapter gives an overview of the status, methods, requirements and perspectives on the application of bioinformatics to the field of glycosciences. It covers

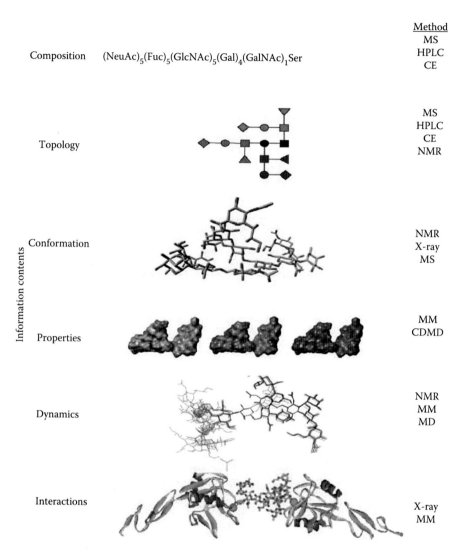

FIGURE 10.1
(See color insert.) Pictorial scheme of information content at several levels of glycan structural data.

the integration of computational methods in the elucidation of several of the levels of glycan structural organization, dynamics, and interactions, and provides a list of web tools and databases relevant to the field.

Carbohydrates are composed of monosaccharides that are covalently linked by glycosidic bonds, either in α or β form to formulate a standardized notation to glycans, the Consortium for Functional Glycomics proposed a standard symbolic representation for monosaccharides that are found in nature. Carbohydrates are most classically drawn as a tree in a two-dimensional plane, with the root monosaccharide placed at the rightmost position with the children branching toward the left. Each node represents a monosaccharide and each edge represents a glycosidic linkage, which includes the carbon numbers that are bond and the conformation. The International Union of Pure and Applied Chemistry-International Union of Biochemistry and Molecular Biology (IUPAC IUBMB) has specified the nomenclature of carbohydrates to uniquely describe complex oligosaccharides based on a three-letter code to represent monosaccharides. Each monosaccharide code is preceded by the anomeric descriptor and the configuration symbol. The ring size is indicated by an italic *f* for furanose and *p* for pyranose [1–4].

There are also other formats such as the Kyoto Encyclopedia of Genes and Genomes (KEGG) Chemical Function format, which represents glycans using a connected graph; LINCUS, which provides a unique and linear notation for glycans; and linear codes by Glycominds, which provides a commercial carbohydrate database. The encoding capabilities of all existing carbohydrate sequence formats and the content of publicly available structure databases as part of the EUROCarbDB project (http://www.eurocarbdb.org) [5]. Recently, a balancing technique to obtain more appropriate and robust models was created and its accuracy was compared with that of the conventional model [6]. To construct new models, they randomly sampled the same number of subjects with and without new dental caries and developed an efficient method for mining motifs or significant subtrees from glycans [7].

10.2 Sequencing Complex Carbohydrates

Despite recent progress in spectroscopy, the sequencing of structurally complex and diverse carbohydrates remains a challenging task. Mass spectrometry (MS) is the method of choice for sensitive identification and characterization, particularly for glycans released from glycoproteins, and automated high-throughput methodologies are currently being developed [8,9], some of which are also applicable to glycosaminoglycans [10]. Automatic MS spectrum interpretation is still a matter of active exploration [11,12], and the success of developed algorithms depends on the family of complex carbohydrates being investigated. Significant successes can be obtained with screening specific classes, such as N-linked glycoprotein glycans, for which databases of existing structural data are particularly complete and for which we have solid knowledge of biosynthesis pathways [13,14]. The combination of MS techniques with the use of accessible databases can also be used in the early stages of glycan analysis [15]. High-resolution MS is one of those biotechnologies that are highly promising for improved health outcomes. Proteomics biomarkers that can distinguish healthy patients from patients with cancer have been identified using MS data. MS studies that use glycomics to identify ovarian cancer have been previously demonstrated. MS is used for protein profiling in cancer on the basis of peptide or protein abundance from MS data. Recent literature on cancer classification using MS have identified some potential protein biomarkers in serum to distinguish cancer from normal samples. Because glycans play important roles in cell communication and signaling events, they may be implicated in cancer. Clinical glycomics is used to identify potential biomarkers for the early learning algorithms to classify high dimensional MS cancer data. Artificial neural networks are used to discriminate different tumor states. Decision tree–based assembled methods were proposed to identify biomarkers for inflammatory diseases. All these studies aim to discover the potential MS biomarkers that can distinguish one group from another and statistical methods are used to combine the high dimensional MS measurements into a single score to classify cancer status jointly with suitable preprocessing of the data. There are several studies on combining biomarkers.

Structure determination often also involves the use of nuclear magnetic resonance (NMR) spectroscopy with recent improvements in sensitivity; NMR provides complementary information to MS and the combination is powerful [16]. Assignment of ^{1}H and ^{13}C NMR spectra of complex carbohydrates can be achieved with an algorithm using an additive scheme, such as that developed in Computer-Aided Spectrum Evaluation of Regular Polysaccharides (CASPER) [17]. This allows both the simulation of spectra of known structures and the sequence determination of unknown structures using information from chemical analysis and unassigned NMR spectra.

10.3 Three-Dimensional Features of Glycan

Both NMR and x-ray diffraction can be used to assess three-dimensional features of complex carbohydrates. In solution, NMR is the method of choice to study conformation, through parameters such as chemical shifts, coupling constants, nuclear Overhauser effects, and relaxation time measurements [18]. However, a major difficulty arises from the flexibility of carbohydrates, especially the glycosidic links. When multiple conformations are present in solution, NMR data will represent a time-averaged conformation. Because the geometrical parameters are usually related in a nonlinear way to the experimental data, these data can be misleading. Almost all the available experimental data on three-dimensional structures come from x-ray diffraction. Molecular and crystal structures of small and medium-sized oligosaccharides can be found in the Cambridge Structural Database [19], which does not have a web interface. Out of the 4000 entries classified as carbohydrates, many are monosaccharides. Compounds of either structural or biological interest have been described in a recent review [20]. Among the crystal structures, no more than 10 disaccharide fragments of glycoconjugates and three trisaccharides are reported. The largest oligosaccharide crystal structure reported thus far is that of tricolorin, a glycolipid extracted from a Mexican plant used in traditional medicine [21]. For glycoprotein and protein–carbohydrate complexes, an increasing number of crystal structures have been reported, as the result of significant and rapid progress in the use of synchrotron radiation; sufficient data are available for useful statistical analyses [22]. In the case of glycan macromolecular structures, intrinsic conformational flexibility is mainly responsible for their inability to crystallize. Therefore, relatively few crystallographically solved structures are available from the Protein Data Bank [23]. A recent study identified 1562 entries in the Protein Data Bank, with a total of 5397 carbohydrate chains. Most of the chains correspond to *N*-glycosidically bound glycans, whereas *O*-glycans are a minority. Noncovalently bound carbohydrate ligands are also found; for example, more than 250 crystal structures of lectins have been solved, most of them as complexes with carbohydrate ligands. Many of these complexes involve biologically important oligosaccharides for which no structural information was available [24]. Polysaccharides form the most abundant family of biopolymers, offering a diversity of structures ranging from simple linear homopolymers to branched heteropolymers with repeat units that consist of octasaccharides. In contrast to other macromolecules, the three-dimensional structural data that can be obtained on polysaccharides come from x-ray fiber diffraction and are not sufficient to permit crystal structure determination. Modeling techniques that allow the calculation of diffraction intensities from various models must be used for comparison with the observed intensities. There are more than 100 apparently successful structure solutions for polysaccharides, counting all polymorphs, variants, derivatives, and complexes. A few structural models of polysaccharides have been deposited in the Protein Data Bank. A review by Chandrasekharan [25] gathers all three-dimensional models that have been established up to 1997.

10.4 Structure and Dynamics of Glycans

Elucidation of the three-dimensional structures and dynamic properties of complex carbohydrates is a prerequisite for understanding the biochemistry of recognition processes and for the design of carbohydrate-derived drugs. Procedures for the molecular modeling of carbohydrates have been devised to enable the characterization of their structural and dynamic properties [26]. The complexity of oligosaccharide and polysaccharide topology requires the design of dedicated molecular building procedures that can rapidly convert sequence information into a preliminary but reliable three-dimensional model. Particular procedures have been designed for such tasks [27,28] using libraries of constituent monomers. Once a set of atomic coordinates is available, a three-dimensional depiction of the structure is displayed and a molecular surface can be computed. This pictorial description can be extended to display features such as electrostatic potential, lipophilicity potential (from hydrophobicity to hydrophilicity), and hydrogen-bonding capacity. The conformational preferences of oligosaccharides are first determined using potential energy surfaces as a function of their glycosidic torsion angles (F and C). One can make the assumption that each glycosidic linkage displays conformational behavior that is independent of the structural features of the rest of the molecule. Several computational protocols have been proposed that systematically explore all F/C combinations, followed by energy calculations and subsequent minimization. In the best cases, the large number of possible conformations is "clustered" into conformational families that provide starting three-dimensional models. Several methods are available for search strategies, ranging from growing chain methods to exhaustive searching [28–30] via guided searches and stochastic searches such as Monte Carlo methods [31], genetic algorithms [32], and molecular dynamics [33]. The resulting three-dimensional models can be easily represented and used as starting points for further calculations, such as molecular dynamic simulations. This is important because, for complex carbohydrates, a rigid, well-defined conformation is not a good approximation, as molecular motions exist on the nanosecond timescale. Quite large complex carbohydrate structures have been computed, the largest being a plant mega-oligosaccharide [34]. Application to the structural characterization of a uromodulin tetradecasaccharide *O*-linked glycan suggested that a folding mechanism could occur in complex carbohydrates, and illustrated how the spatial arrangement of three sialyl Lewis x epitopes may favor its cooperative interaction with E- and P-selectins [35].

10.5 Interactions of Carbohydrates in Biological Systems

Progress with algorithms and increases in computational power have enabled the simulation of carbohydrates in their natural environment, that is, solvated in water [36,37], in concentrated solution (such as the insertion of a glycosylphosphatidylinositol anchor within a lipid environment) or in the binding site of a protein receptor. Tools are being developed to help design substrate analogues or inhibitors by submitting three-dimensional structures of proteins to virtual screening approaches [38]. Spatial structures of both ligands and protein are required to perform such explorations. The biological relevance of these interactions is best understood when structural features are interpreted together with kinetic and thermodynamic information, such as the results derived from isothermal microcalorimetric measurements, which yield information about the multivalency of the interaction and the entropic or enthalpic nature of the driving forces involved [39]. Other ambitious endeavors can be accomplished, such as the systematic exploration of how glycosaminoglycans such as heparan sulfate can affect the spatial orientation of chemokines [40]. In the absence of three-dimensional data, the biological roles of some complex carbohydrates and protein–carbohydrate interactions can be assessed through the use of quantitative structure–activity relationships, as exemplified by the elucidation of antibody and lectin recognition of histo-blood group antigens [41].

10.5.1 Carbohydrate Microarrays

There is great interest in developing microarray-based methods for probing the roles of nucleic acids, proteins, and carbohydrates in biology. The chip-based format offers many advantages over conventional methods. These include the ability to screen several thousand binding events in parallel and the fact that a minimal amount of analyte and ligand are required for study, making the most of precious synthetic or naturally procured materials. Many methods for preparing carbohydrate microarrays have been described to date: nitrocellulose-coated slides for noncovalent immobilization of microbial polysaccharides and neoglycolipid-modified oligosaccharides [42], polystyrene microtiter plates for presenting lipid-bearing carbohydrates [43], self-assembled monolayers modified by Diels–Alder mediated coupling of cyclopentadiene-derivatized saccharides [44], thiol-derivatized glass slides modified with maleimide-functionalized oligosaccharides [45], and thiol-functionalized carbohydrates immobilized on maleimide-derived gold and glass slides [46,47]. We adopted two surface chemistries for the preparation of our carbohydrate microarrays. Both involve maleimide functionalization of glass slides to form a stable bond between the slides and thiol-containing synthetic oligosaccharides. In one case, Bovine Serum Albumin (BSA)-derivatized aldehyde glass slides were functionalized with succinimidyl 4-(N-maleimidomethyl)cyclohexane-1-carboxylate to present a maleimide-reactive surface [46]. Alternatively, amine-derivatized Corning GAPSII slides were directly modified with succinimidyl 4-(N-maleimidomethyl)cyclohexane-1-carboxylate before incubation with thiol-presenting saccharides [48]. Microarrays were printed at high density by using standard DNA microarray robotic printers. These two methods of surface functionalization offer different advantages. BSA-derivatized slides present a relatively low density of immobilized oligosaccharide and excellent resistance to nonspecific binding of proteins to the surface. The GAPSII slides permit high-density immobilization of oligosaccharides, allowing the examination of carbohydrate clusters at the surface, and present the carbohydrate in a peptide-free context. These immobilization chemistries were developed, in part, to address the limitations inherent in existing methods for preparing carbohydrate microarrays. For instance, the microtiter method requires relatively large quantities of oligosaccharide and does not offer the same degree of high throughput available to robotically printed glass microarrays. In addition, the reliance on noncovalent, hydrophobic interactions to anchor carbohydrates to the microtiter wells places considerable limitations on the stringency of washes one may employ; the use of detergents to reduce nonspecific interactions invariably leads to the loss of carbohydrate from the microtiter wells. Nitrocellulose-based immobilization is limited to large polysaccharides or lipid-modified sugars. More sophisticated synthetic methods for immobilization have limited applications in the preparation of large oligosaccharides due to the sensitivity of the complex chemistries.

10.5.1.1 High-Mannose Microarrays

To establish the viability of the carbohydrate microarray, a panel of mannose-containing oligosaccharides was prepared. These structures were selected based on their relevance to the glycans found decorating viral surface envelope glycoproteins of HIV. Specifically, the arrays are composed of a series of closely related structural determinants of (Man)9-(GlcNAc). By presenting the various structural determinants of an important glycan on a single array, multiple proteins can be screened to determine their binding profiles. The carbohydrate-binding profiles of two potent HIV-inactivating proteins isolated from cyanobacterium, cyanovirin-N [49] and scytovirin [50].

10.5.1.2 Antibiotic Microarrays

Aminoglycosides are carbohydrate antibiotics that contain amino sugars and are composed of two to five monomers. Clinically, these compounds are used as broad-spectrum antibiotics against a variety of therapeutically important bacteria. Aminoglycosides exhibit their antibacterial effect by binding bacterial ribosomes and inhibiting protein synthesis. The most common binding site for this class of drugs is the A-site in the small ribosomal subunit, or 30S, portion of the bacterial ribosome. The therapeutic efficacy of aminoglycosides, however, has decreased recently due to antibiotic resistance. Resistance to aminoglycosides can be acquired either through the transfer of plasmid DNA or from the overexpression of endogenous enzymes. Several mechanisms cause resistance, including decreased uptake into cells, mutation of the target, binding to proteins, and covalent modification of the drug by enzymes [51]. Enzymatic modification is the most common aminoglycoside resistance mechanism. The result of aminoglycoside modification is a large decrease in binding affinity to the therapeutic target [52]. In recent years, the incidence of resistant bacteria has increased. To combat the growing threat that bacteria pose to human safety, new antibiotics must be identified. To facilitate the discovery of such compounds, high-throughput methods to identify compounds that weakly bind to resistance- and toxicity-causing proteins and strongly bind to therapeutic targets have been developed by using the microarray techniques described herein.

10.5.1.3 Antibiotic Microarrays to Interrogate Interactions to Therapeutic Targets and Resistance-Causing Enzymes

Aminoglycoside microarrays were constructed by random covalent immobilization of the antibiotics onto amine-reactive glass slides by using a DNA arraying robot. This approach provides a versatile platform for probing the interactions of these compounds with a variety of targets. Arrays were probed with an RNA mimic of the bacterial and human A-sites (Figure 10.2) [53]. These two different RNA sequences were used to establish this microarray method as a screen not only for tight binding to RNA but also for specific binding. Results from these studies showed that the antibiotic amikacin binds the tightest to both the bacterial and the human A-site mimics. These results do not exactly correlate with in-solution measurements of aminoglycoside binding affinity due to the nonspecific

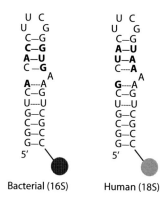

FIGURE 10.2
The oligonucleotide mimics of rRNA A-sites that were incubated with the aminoglycoside arrays. The bacterial oligonucleotide has been shown to be the binding site for some aminoglycosides in the ribosome. The human oligonucleotide has been tested for aminoglycoside binding by using MS experiments. Each oligonucleotide was fluorescently labeled; the bacterial RNA is labeled with TAMARA and the human with fluorescein.

immobilization of the compounds. Arrays were also incubated with two types of acetyl-transferase (AAC) resistance enzymes, an AAC(6′) from *Salmonella enterica* and an AAC(2′) from *Mycobacterium tuberculosis* [54,55]. Binding of these enzymes to the aminoglycosides correlated well with a previous calorimetric study of binding affinity.

10.5.1.4 Antibiotic Microarrays to Facilitate the Discovery of Inhibitors of Resistance-Causing Enzymes

A library of aminoglycoside mimetics was arrayed onto glass slides to find inhibitors of resistance. Guanidinoglycosides were chosen because:

1. They are easily synthesized from aminoglycosides [56].
2. The increased charge that guanidinoglycosides have relative to aminoglycosides might allow for tighter binding to the negatively charged aminoglycoside binding pocket in these enzymes [57].
3. The difference in pK_a between a guanidinino group and an amino group (12.5 versus 8.8, respectively) suggests that the guanidinoglycosides may not be substrates for these enzymes.

Screening this library revealed that each of the guanidinoglycosides exhibited higher affinity for the resistance enzymes than the corresponding aminoglycosides. Guanidinoglycosides were tested by using a spectrometric assay for their ability to serve as substrates for AAC(2′) and AAC(6′) [54,55]. The results demonstrate that guanidinoglycosides are not substrates and inhibit the acylation of several clinically important aminoglycosides [58]. Information from these studies will allow the development of new antibiotics that evade resistance.

10.5.1.5 Hybrid Carbohydrate/Glycoprotein Microarrays

A hybrid carbohydrate/glycoprotein microarray was developed to rapidly determine the contribution of protein–protein interactions in addition to carbohydrate–protein interactions in binding events. By arraying both the glycoprotein and the carbohydrate it displays, binding determinants can be rapidly identified. To develop these screens, a GAPSII glass slide was modified at the surface by using two chemistries: on one side, maleimide and, on the other, an NHS-activated ester. The carbohydrates and glycoproteins were printed on the maleimide and NHS-activated ester sides of a single chip, respectively. Hybridization with a carbohydrate-binding protein established whether the peptide context is required for binding. In the case of cyanovirin-N, both free carbohydrate and gp120 are bound. In contrast to a crude plant extract known to contain a high-mannose binding protein, free carbohydrate is not bound in the absence of the glycoprotein; this strongly suggests that either protein–protein contacts are required for glycan recognition or that protein conformation–dependent presentation of the high-mannose glycans influences recognition by this carbohydrate-binding protein.

10.5.1.6 Microsphere Arrays to Detect Protein–Carbohydrate Interactions

In contrast to the microarray systems described above, a system developed in collaboration with the Walt laboratory uses optically addressable, internally encoded microspheres to define the position and structure of a series of carbohydrates [59]. Although solid phase carbohydrate libraries have been employed previously [60], miniaturization of the assay, combined with fluorescently encoded microspheres, allows for rapid screening while requiring amounts of material comparable to or less than what is required by microarrays. To detect binding, the immobilized microsphere array is incubated with a fluorophore-labeled carbohydrate-binding protein. The binding profile is determined by measuring the fluorescence of beads that emit both at the wavelength of an internal code, which is used as a marker for the carbohydrate displayed on a microsphere (an entrapped fluorescent dye), and at the labeled protein. Fluorescence colocalization indicates a binding event.

10.5.2 Surface Plasmon Resonance to Study Protein–Carbohydrate Interactions

Immobilized carbohydrates are also used for surface plasmon resonance experiments to provide valuable insight into the binding of analytes to ligands in real time and to allow for both low- and high-affinity interactions to be measured [47]. In these experiments, a solution containing an analyte is washed over a surface. Binding is measured by the change in the refractive index of the surface upon accumulation of analytes. Because interactions are measured without the need for labeling, any influence of a label on the experimental results can be excluded. Surface plasmon resonance was completed by using self-assembled monolayers, which offer extensive control over the density of immobilized carbohydrate at the surface. By controlling the ratio of homogenously displayed maleimide in the self-assembled monolayer, it is possible to determine the concentration of thiol-modified oligosaccharide immobilized on the surface. Utilizing these precisely characterized self-assembled monolayers in concert with surface plasmon resonance, we established that complex synthetic carbohydrates can be used for detailed studies characterizing the activity of carbohydrate-binding proteins.

10.5.3 Fluorescent Carbohydrate Conjugates as Probes for Cell Biology

Although the microarray format yields a plethora of information regarding protein–carbohydrate interactions, such arrays may not be appropriate tools for studying cell surface receptors with presumed carbohydrate binding activity (i.e., lectins). One limitation of the arrays is the requirement for purified receptor. Furthermore, due to the high density of immobilized oligosaccharide, observation of binding to the surface is restricted to clustered or multivalent arrays of carbohydrate. Although it may be possible to immobilize carbohydrate at densities sufficiently low to approximate monovalent presentation of oligosaccharide, the microarray format is not ideal for examining monovalent protein carbohydrate interactions. To define the influence of oligosaccharide clustering on recognition by cell surface lectins, we have generated monovalent and multivalent fluorescent probes for applications in cell biology. These probes serve as reporters to enable an investigator to track receptor–carbohydrate interactions by fluorescence microscopy and flow cytometry.

10.5.3.1 Monovalent Oligosaccharide–Fluorophore Constructs for Receptor Studies

Most lectins have an increased affinity for a carbohydrate ligand that is proportional to the valency of the interactions [61]. Traditionally, this phenomenon has been investigated by solid phase assays that test the ability of carbohydrates to displace radioactively labeled lectin from binding to a high-affinity ligand (e.g., 10–8 m) [62]. Cell-based assays developed on the same principle of inhibitory concentrations have been employed as well. In these experiments, cells are incubated with fluorophore-labeled neoglycoproteins as the high-affinity ligand, then with potential unlabeled ligands, and the amount of cell-associated fluorescence is measured by flow cytometry. To establish a more direct method of detecting oligosaccharide–receptor interactions, we have generated monovalent oligosaccharide–fluorophore conjugates. These conjugates can be used to observe the approximate affinity of a cell surface lectin for the monomeric oligosaccharide in solution.

10.5.3.2 Multivalent Oligosaccharide Platforms for Cell Biology

To enable direct assessments of the effect of multivalency on oligosaccharide binding to cell surface lectins, we explored the use of semiconductor nanocrystal (quantum dot)–based systems as platforms to present multiple oligosaccharide monomers (>100) on a single particle.

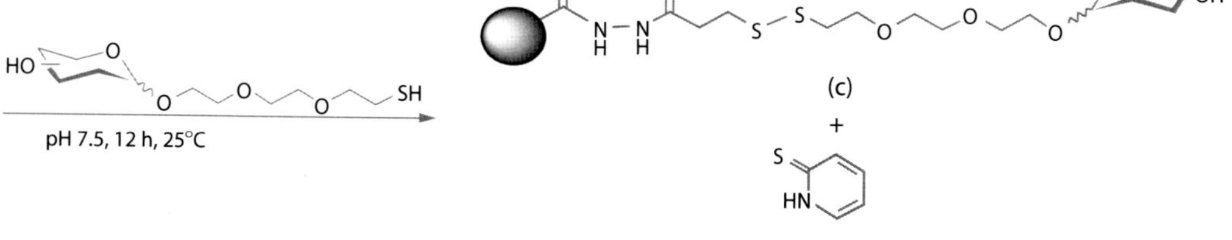

FIGURE 10.3
Modification of carboxylate-functionalized semiconductor quantum dots enables facile conjugation of thiol-bearing saccharides. The COOH-functionalized quantum dots (a) are treated in buffered solution with water-soluble 1-ethyl-3-(3 dimethylaminopropyl)carbodiimide (EDC) in the presence of 3-(2-pyridyldithio)propionyl hydrazide (PDPH) to generate a thiol-reactive nanoparticle (b). Coincubation of b with thiol-functionalized saccharides results in disulfide exchange between the quantum dot-anchored PDPH and the solution phase saccharide to yield carbohydrate–quantum dot conjugates (c) and liberated pyridine-2-thione. After purification of the carbohydrate–quantum dot conjugates away from liberated pyridine-2-thione, one can determine the amount of pyridine-2-thione in solution spectrophotometrically (UV/V is at 343 nm) and thereby determine the number of saccharides per particle.

Early nanoparticle-based multivalent platforms for evaluating multivalent oligosaccharide interactions were based on carbohydrate-modified gold nanoparticles [63]. The success of these nanoparticle studies led us to believe that quantum dots could enhance the utility of the nanoparticle platform. Given the high quantum yields in aqueous systems and unique photophysical properties (e.g., their lack of excitation-induced photobleaching, their extremely narrow and nonoverlapping emission spectra, and their ability to achieve multiple wavelength emissions after excitation from a single excitation source), quantum dot–carbohydrate conjugates will become a powerful tool in studying the cell biology of cell surface lectins [64]. We have developed a conjugation scheme using our linking chemistry to control the number of carbohydrates per quantum dot and methods to monitor the efficiency of conjugation (Figure 10.3). By using quantum dots bearing different amounts of saccharide, we are trying to further elucidate carbohydrate recognition by Dendritic Cell-Specific Intercellular adhesion molecule-3-Grabbing Non-integrin (DC-SIGN) and other mammalian lectins.

10.5.4 Carbohydrate Affinity Screening

Synthetic tools can facilitate the isolation and purification of carbohydrate-binding proteins from crude mixtures or biological extracts. Latex beads, magnetic particles, and agarose or sepharose resins modified to display a specific oligosaccharide can be used for affinity-based purification of carbohydrate-binding partners. Investigators have traditionally employed monosaccharide-derivatized matrices to identify and isolate carbohydrate-binding proteins. Although these matrices facilitate the isolation process, little information has been gleaned regarding the true structural specificity of the isolated protein. Matrices displaying more complex oligosaccharides will enable the simultaneous isolation of carbohydrate-binding proteins and the determination of their structural specificity.

10.6 Web Tools and Databases of Glycan Analysis

There are three major databases for complex carbohydrates, KEGG GLYCAN database developed by the Consortium for Functional Glycomics. All three databases are based on the CarbBank database developed in the 1990s. The major issue that was facing the glycolinformatics community was that each of these databases represented their glycan structures in different formats, a workshop was held at the U.S. National Institutes of Health. At this meeting, the GLYDE-2 XML format for glycans and glycoconjugates, developed by the Complex Carbohydrate Research Center (CCRC), was agreed upon as the standard format for exchanging carbohydrates.

Apart from the development of these various databases, several methods for analyzing the structures have been developed. They are classified into six types:

1. Structural analysis of carbohydrates databases
2. Three-dimensional structures of carbohydrate databases
3. Glycan graphical representations and nomenclature databases
4. Prediction of glycosylation positions in protein databases
5. Carbohydrate-active enzymes and other protein–glycan interaction databases
6. Carbohydrate databases

The Glycome Database is the integration of all available carbohydrate sequences into one central database. Approximately 100,000 database records with 73,341 sequences are accessible in the public domain. The biggest obstacle for data integration was the use of various sequence-encoding formats by different initiatives. Therefore, they found it necessary to develop a new sequence format called GlycoCT, which is a superset of the structural encoding capabilities inherent in all other formats developed thus far. They implemented a translation library which can read all of the carbohydrate sequence formats and translate them to GlycoCT. GlycoUpdateDB is a JAVA application program, which had to be designed to carry out the integration of the interpretable data obtained from the resources, depending on a PostgreSQL database that can be configured by an XML file. The configuration file contains settings for the local database and instructions for the download and data integration process. GlycomeDB consists of several database schemata with tables that store all downloaded and generated data sets. A software tool, GlycanBuilder for building and displaying glycan structures using a chosen symbolic notation, was reported by Ceroni et al. [65]. The tool uses an automatic rendering algorithm to draw the saccharide symbols and place them on the drawing board. Information about the symbolic notation is derived from a configurable graphical model as a set of rules governing the aspect and placement of residues and linkages [64]. The algorithm is able to represent a structure using only few traversals of the tree and is inherently fast. The tool uses an XML format for import and export of encoded structures.

Carbohydrates are involved in a variety of fundamental biological processes (cellular differentiation, embryonic development, and fertilization) and pathological situations (bacterial and viral infections, inflammatory diseases, and cancer). They therefore have a large pharmaceutical and diagnostic potential. Protein–carbohydrate interactions are intensively investigated using a variety of experimental methods. Among these, x-ray and NMR measurements provide a detailed three-dimensional picture of the spatial location of the ligand as well as the protein. It is estimated that more than half of all the proteins in the human body have carbohydrate molecules attached. Structural data taken from x-ray crystallography does not necessarily mean that a potential glycosylation site is unoccupied, but the presence of carbohydrate residues in the three-dimensional structures provides direct and unambiguous evidence for the occupancy of a glycosylation site. As illustrated above, the number of computer-based tools required to decipher the structures of complex carbohydrates has been notably increasing. Only a few carbohydrate-oriented resources are freely available on the Internet today, and these have been previously reviewed [65–67].

Web tools cover the areas of structural analysis (MS and NMR; Table 10.1), conformational analysis (Table 10.2), and graphical representations and nomenclature (Table 10.3). Others are oriented toward the prediction of glycosylation in proteins and glycosylphosphatidylinositol anchor sites (Table 10.4). It seems odd and unfortunate that databases of carbohydrate-interacting proteins (Table 10.5) are much more developed than carbohydrate databases (Table 10.6). Through an international effort, a first and ambitious enterprise to set up a carbohydrate database (CarbBank) was initiated as early as 1989 [68], but ceased in 1997. The last release contains approximately 50,000 records, related to bibliographical, biological, chemical, and physical data. Although other carbohydrate databases have since been implemented [69,70], efficient cross-linking and automated exchange of data are still under development [71]. This is because a universally accepted representation of carbohydrate structure is still lacking.

A canonical linear description of carbohydrate sequences has been developed in Linear Notation for Unique description of Carbohydrate Sequences (LINUCS) [72] and implemented in its Internet applications by the German Cancer Research Centre [73]. An

TABLE 10.1

Web Tools: Structural Analysis of Carbohydrates

GlycoMod	Predicts glycan structure from molecular mass	http://www.expasy.org/tools/glycomod/
GlycanMass	Calculates the mass of an oligosaccharide structure	http://www.expasy.org/tools/glycomod/glycanmass.html
Glyco-Search-MS	Suggests structures on the basis of peaks in MS spectra	http://www.dkfz.de/spec/glycosciences.de/sweetdb/start.php?action form_ms_search
GlycoMod	Predicts glycan structure from molecular mass	http://www.expasy.org/tools/glycomod/
GlycanMass	Calculates the mass of an oligosaccharide structure	http://www.expasy.org/tools/glycomod/glycanmass.html
GlyPeps	Glycopeptides detection by accurate mass measurement	http://www.dkfz-heidelberg.de/spec/glypeps/
CASPER	NMR assignment and simulation, structure prediction from NMR	http://www.casper.organ.su.se/casper/
SugaBase	Glycan ^{1}H, ^{13}C NMR data	http://www.boc.chem.uu.nl/sugabase/sugabase.html
NMR search	^{1}H, ^{13}C NMR search	http://www.glycosciences.de/sweetdb/nmr/

Source: Data from Mitchell, E.P. et al., *Proteins* 58:735–746, 2005; and Imberty, A. et al., *Bioeng. Med. Chem.* 4:1979–1988, 1996; and from http://www.cermav.cnrs.fr/glyco3d.

TABLE 10.2

Databases of Three-Dimensional Structures of Carbohydrates

CSD	Three-dimensional structures of small molecules	http://www.ccdc.cam.ac.uk/
XtalOligosaccharides	Crystal structures of oligosaccharides	http://www.cermav.cnrs.fr/cgi-bin/oligos/oligos.cgi
SWEET-II	Generation of three-dimensional structures from sequence	http://www.glycosciences.de/modeling/sweet2/doc/index.php
Monosaccharides	Three-dimensional structures of monosaccharide	http://www.cermav.cnrs.fr/cgi-bin/monos/monos.cgi
Disaccharides	Conformational maps of disaccharides	http://www.cermav.cnrs.fr/cgi-bin/di/di.cgi
Glydict Predicts	*N*-glycan structures from MD simulations	http://www.dkfz-heidelberg.de/spec/glydict/
GlycoMaps database	Conformational maps	http://www.dkfz-heidelberg.de/spec/glycomaps/
Dynamic Molecules	MD simulations	http://www.md-simulations.de/manager

Source: Data from Mitchell, E.P. et al., *Proteins* 58:735–746, 2005; Imberty, A. et al., *Bioeng. Med. Chem.* 4:1979–1988, 1996; and from http://www.cermav.cnrs.fr/glyco3d.
Note: CSD, Cambridge Structural Database; MD, molecular dynamics.

TABLE 10.3

Web Tools: Glycan Graphical Representations and Nomenclature

LINUCS	Linear encoding of carbohydrates	http://www.dkfz.de/spec/linucs/
LiGraph	Conversion between IUPAC and graphical representations	http://www.glycosciences.de/ligraph/
IUPAC	Nomenclature	http://www.chem.qmw.ac.uk/iupac/

Source: Data from Mitchell, E.P. et al., *Proteins* 58:735–746, 2005; and Imberty, A. et al., *Bioeng. Med. Chem.* 4:1979–1988, 1996; and from http://www.cermav.cnrs.fr/glyco3d.
Note: LINUCS, Linear Notation for Unique description of Carbohydrate Sequences.

TABLE 10.4

Web Tools: Prediction of Glycosylation Positions in Proteins

NetNGlyc	Prediction of *N*-glycosylation sites	http://www.cbs.dtu.dk/services/NetNGlyc/
NetOGlyc	Prediction of mucin-type *O*-glycosylation sites	http://www.cbs.dtu.dk/services/NetOGlyc/
YinOYang	Prediction of *O*-b-GlcNAc sites	http://www.cbs.dtu.dk/services/YinOYang/
Big PI Predictor	GPI anchor prediction	http://mendel.imp.univie.ac.at/sat/gpi/gpi_server.html
DGPI	GPI anchor prediction	http://129.194.185.165/dgpi/DGPI_demo_en.html

Source: Data from Mitchell, E.P. et al., *Proteins* 58:735–746, 2005; and Imberty, A. et al., *Bioeng. Med. Chem.* 4:1979–1988, 1996; and from http://www.cermav.cnrs.fr/glyco3d.

Note: GPI, glycosylphosphatidylinositol.

TABLE 10.5

Databases of Carbohydrate-Active Enzymes and Other Protein–Glycan Interactions

CAZy (carbohydrate-active enzymes)	Carbohydrate-active enzymes	http://afmb.cnrs-mrs.fr/CAZY/
KEGG Ligand	Glycans as ligands	http://www.genome.jp/kegg/ligand.html
KEGG Pathway	Carbohydrate metabolism	http://www.genome.jp/kegg/pathway.html
Lectins	Three-dimensional structures of lectins	http://www.cermav.cnrs.fr/lectines/
CTLD (C-type lectin-like domains)	Animal lectins	http://www.imperial.ac.uk/research/animallectins/
Glycosyltransferase database	Three-dimensional structures of glycosyltransferases	http://www.cermav.cnrs.fr/cgi-bin/rxgt/rxgt.cgi
GAGPRO	Three-dimensional structures of GAG binding proteins	http://www.cermav.cnrs.fr/cgi-bin/gag/gag.cgi
PDB2LINUCS	Glycan structures in the PDB	http://www.dkfz.de/spec/css/pdb2linucs/index.php
GlySeq	Analysis of glycoprotein sequences	http://www.dkfz.de/spec/css/glyseq/index.php
Consortium for Functional Glycomics	Glycan-binding proteins	http://www.functionalglycomics.org/glycomics/molecule/jsp/gbpMolecule-home.jsp
	Glycosylation pathways	http://www.functionalglycomics.org/static/gt/gtdb.shtml

Source: Data from Mitchell, E.P. et al., *Proteins* 58:735–746, 2005; and Imberty, A. et al., *Bioeng. Med. Chem.* 4:1979–1988, 1996; and from http://www.cermav.cnrs.fr/glyco3d.

Note: GAG, glycosaminoglycan; PDB, Protein Data Bank.

independent syntax, using a representation called Linear Code [74], was also developed. A simple one- to two-letter representation of monosaccharide units and linkages was used; the ordering of branches is assessed using a special lookup table. Despite the fact that both representations offer a unique description of complex structures and can be used efficiently in data processing, they are far from being intuitive to users. Dedicated graphic interfaces are being developed to bridge this gap. A systematic strategy has been reported for the identification of human glycans by MS alone [74]. An updated introduction to resources on the KEGG web site (see Tables 10.5 and 10.6) has also been published [75].

TABLE 10.6

Carbohydrate Databases

CarbBank	Glycan data	http://www.boc.chem.uu.nl/sugabase/carbbank.html
Glycosciences database	Databases and bioinformatics tools for glycobiology and glycomics	http://www.glycosciences.de
KCaM	Database searching by structure	http://www.genome.jp/ligand/kcam/
Consortium for Functional Glycomics	Glycan database	http://www.functionalglycomics.org/glycomics/molecule/jsp/carbohydrate/carbMoleculeHome.jsp
GlycoSuite	Glycoproteomics	http://glycosuite.com/
Glycominds	Glycomics	http://www.glycominds.com
SPECARB	Raman spectra of saccharides	http://www.models.kvl.dk/users/engelsen/specarb/specarb.html

Source: Data from Mitchell, E.P. et al., *Proteins* 58:735–746, 2005; and Imberty, A. et al., *Bioeng. Med. Chem.* 4:1979–1988, 1996; and from http://www.cermav.cnrs.fr/glyco3d.

10.7 Applications of Glycans

10.7.1 Drug Development with Glycans

The problem of how to exploit the vast potential of complex glycans in drug development involves overcoming fundamental challenges to clarify important structure–activity relationships. First, experimental procedures are needed that can accurately and reproducibly measure the structure or sequence of complex glycans. Second, there needs to be a framework for describing glycan sequences in terms of the properties that lead to their functions (Figure 10.2). The integration of these two approaches is especially important in the study of complex glycans as, in many cases, there seems to be a functional and structural overlap—a single structure might have many divergent functions depending on its spatial and temporal expression or, conversely, a given function might be elicited by several closely related glycan structures.

Although there are several systems with well-defined structure–function relationships that have facilitated drug discovery—such as the ATIII/pentasaccharide system [76], the mannose 6-phosphate pathway and lysosomal storage [77], and the sialic acid content of *N*- and *O*-linked glycoproteins controlling the clearance of glycoproteins from the blood [78], the integration of a bioinformatic framework with structural investigations in many other systems is lacking. Beyond this important limitation, from a drug development standpoint, several factors have hampered the development of glycan-based pharmaceuticals. First, glycans have remarkably high information content owing to the diversity in their primary chemical structures, which requires the application of various analytical techniques to discern the fine structure and sequence of the glycans. This information density derives from the fact that complex glycans not only contain information from diverse monosaccharides but also have linkage variability. There are several different families of complex glycans that are differentiated according to the types of linkage and exocyclic substitutions of the monosaccharides. Therefore, unlike chemical polymers, which often contain homogeneous repeating units in which the diversity of the mixture primarily arises from the degree of polymerization or cross-linking between these units, linear glycan mixtures are complex owing to the chemical diversity in their repeating units. Second, glycans are not synthesized *in vivo* by reading from a template, unlike proteins or DNA. Rather, complex glycans are created through the concerted action of several enzymes. This, together with a lack of proofreading machinery, results in complex glycans being heterogeneous and often polydisperse entities, particularly when compared with other biological polymers such as proteins and nucleic acids. Third, there is currently no mechanism to amplify glycan structures. The advent of amplification techniques for proteins and nucleic acids allowed numerous detailed structure–function studies to proceed for these biopolymers. Unfortunately, owing to the complexity of their biosynthesis, as well as the fact that they are created not in a template-based manner but rather by the concerted action of biosynthetic enzymes, there are no such strategies for complex glycans.

In addition, our ability to manipulate the structure of complex glycans through chemical modification is limited, although it is increasing. Fourth, glycans predominantly act at an extracellular or multicellular level and, as such, the screening of glycan activities using simple *in vitro* cell model systems does not always provide an avenue to determine the *in vivo* function accurately. Adding to the problem, many glycans interact only weakly as 1:1 complexes with targets; in fact, the strength of the interaction is typically provided through the multivalent interactions of the glycan with the targets [79]. As a result, screening for strong glycan binders assuming a 1:1 stoichiometry in a similar manner to screens that are used by the pharmaceutical industry to uncover small molecule lead compounds might overlook information that is crucial to understanding the role of complex glycans in a given biological process. An example of this is in the design of compounds to inhibit leukocyte homing and extravasation, in which multivalency and structural complexity are both important [80,81].

10.7.2 Glycans and Biotechnological Applications

Historically, a series of enzymes including imiglucerase (Cerezyme, Genzyme, Cambridge) and larodinase (Aldurazyme, Genzyme, Cambridge) that degrade or alter complex glycan structures have been successful in replacement therapy for individuals with glycan-related genetic disorders. Importantly, the development of the available drugs in this class required a detailed structural/functional understanding of their mechanism of action, their effect on glycan structure, and the role of specific glycan structures in various diseases such as Gaucher disease and mucopolysaccharidosis type I (Table 10.7).

TABLE 10.7

Representative Examples of Glycan-Based Therapeutics

Drug	Disease	Clinical Status	Manufacturer
Lovenox	Thrombosis	Market	Aventis
Fragmin	Thrombosis	Market	Pfizer
Aranesp	Anemia	Market	Amgen
Sepragel	Antiadhesive	Market	Genzyme
Healon	Cataracts	Market	Pfizer
Arixtra	Thrombosis	Market	Sanofi
Lovenox	Thrombosis	Market	Aventis
Aldurazyme	MPS I	Market	Genzyme
PI-88	Cancer	Phase II	Progen

Note: MPS I, mucopolysaccharidosis type I.

Glycans also have an important role in two distinct but related areas of drug development: either as part of glycoprotein therapeutics, modulating protein activity and stability, or alone as saccharide-based therapeutics. Given the importance of protein-based therapies, including antibodies and drugs such as growth hormone and erythropoietin, which together constitute a multibillion dollar per year industry, most of the initial focus has been on understanding and manipulating the complex glycan structures that are attached to glycoproteins. However, the development of structure–activity relationships for free glycans, specifically polysaccharides, has also allowed the development of new classes of therapeutics.

10.7.3 Complex Glycan Structures of Glycoproteins

Most therapeutic proteins, including antibodies, growth factors, and cytokines, are derived from endogenous glycoproteins. As a component of therapeutic glycoproteins, glycans modulate the activity, stability, serum half-life, and immunogenicity of biotechnologically derived proteins. The fact that complex glycans are important in protein function has been appreciated on a general level for many years—since the early days of the development of protein-based therapeutics. It has long been noted that different isoforms of a protein therapeutic, which carry distinct glycoforms, possess varied physiochemical and biological attributes. Therefore, it was concluded that the quality of a protein product could be influenced by the degree and differences in glycan or oligosaccharide structure [82–84].

Historically, the lack of analytical tools for understanding glycan structure and the dearth of structure–function relationships for specific therapeutic glycoproteins created a situation in which the focus was not on the optimization of glycosylation or on understanding how particular glycoforms impinge on biological functioning; rather, the focus was on making protein glycosylation as homogenous as possible. Given the apparent importance of glycosylation in the serum half-life and *in vivo* biological activity of therapeutic glycoproteins, this endeavor has sought to impinge on the glycosylation machinery in three ways: by the addition or subtraction of chemicals that influence the synthesis of complex polysaccharide structures, by the *in vitro* elaboration of glycan structures through the addition of exogenous glycosyltransferases and activated sugar substrates, and by the genetic manipulation of cells that produce therapeutic glycoproteins. In the first case, the effect of cell culture conditions on glycan structure has been investigated [85,86], as has the introduction of specific monosaccharides to cells, such as *N*-acetylmannosamine or glucosamine [83,84]. In the second case, it has been reported that specific saccharide structures can be introduced into recombinant glycoproteins using exogenous glycosyltransferases. A recent report illustrated that the addition of a specific saccharide structure (sialyl Lewis X) to a soluble human complement receptor, through the *in vitro* use of specific glycosyltransferases, resulted in the effective targeting of the complement receptor to activated endothelial cells that expressed E-selectin [87]. Finally, in the third case, the genetic manipulation of both mammalian and insect cells has been reported to result in the formation of fully elaborated, relatively homogenous glycoforms [88–90]. Recently, a new procedure has been developed in which a therapeutic glycoprotein (for example, epoetin; Epogen, Amgen, California) is manipulated at the genetic level to create further sites of potential asparagines or *N*-linked glycosylation [91]. As shown in Table 10.7, modifying the glycan coat of a protein therapeutic led to the creation of the second-generation antianemia drug, darbepoetin (Aranesp, Amgen, California), with new and improved therapeutic qualities [92].

10.7.4 Complex Glycans as Therapeutics

In addition to their role as conjugates of protein therapeutics, complex glycans, which are isolated from natural sources or are chemically synthesized, are themselves active biological and pharmaceutical agents. In this respect, several glycan drugs are now available clinically and many others are in various stages of clinical trials (Table 10.7). The development of glycan-based therapeutics has traditionally been accomplished through two separate approaches. The first approach involves the *de novo* synthesis of complex glycans by chemical means. This tactic has been successfully used to generate a synthetic version of a truncated heparin oligosaccharide for thrombotic indications: fondaparinux sodium (Arixtra, Sanofi, Paris). Solid phase synthetic procedures have recently been developed for glycans, which are analogous to the available peptide and nucleic acid synthetic routes, and hold promise for potential scale-up. Solid phase synthetic routes have already been developed for some drugs, such as carbohydrate-based vaccines [93].

The second manufacturing strategy uses the isolation of natural glycans, chemical modification, and degradation of the backbone structure followed by purification. This approach is in wider use owing to the ease of scale-up, as well as its ability to take advantage of the natural structural diversity of known complex glycans. The quintessential example of this approach is the family of low molecular-weight heparins (LMWHs) [94], which are described in more detail below. There are several practical reasons for glycans, either alone or as glycoconjugates, to have great potential for use as drugs. First, complex glycans are relatively small and intrinsically more stable than protein-based drugs. Second, they are more easily formulated for drug delivery. Third, and finally, sugars are highly specific and potentially less immunogenic than other natural products, such as proteins or RNA-based strategies [95]. At the same time, the large-scale production of specific biologically active glycans is still a technically demanding task. Owing to their nontemplate biosynthesis, glycans cannot be readily manufactured, unlike protein therapeutics in which cells can be used as "biotechnological factories." However, important advances have been made in several orthogonal approaches for their structural optimization and manufacture; these developments will enhance our ability to manufacture a wide range of complex polysaccharide biotherapeutics.

10.7.5 New Technologies to Study Complex Glycans

As our understanding of the structure–function relationships of complex glycans increases, so will our ability to harness their power in various disease processes. Glycan-based drugs have generated much excitement and provided important insight into the power of glycan-based therapeutics. However, the ultimate promise of glycans as drugs is only beginning to be exploited. In an effort to harness the promise of glycans as therapeutics, advances have been made in analyzing glycan structures in a rapid manner using a minimum of material, in synthesizing glycan structures *in vitro*, and in harnessing endogenous glycosylation pathways *in vivo* to create new reproducible glycan structures. Each of these is discussed.

10.7.6 Advances in Synthetic Strategies

In addition to advances in analytical methods, new techniques are also beginning to be designed for the synthesis of homogenous glycans and glycoconjugates; such approaches are crucial to develop and test structure–activity relationships. For example, solution and solid phase synthetic procedures are continuing to be evaluated for both linear and branched glycans [96,97]. The development of *in vitro* chemoselective and chemoenzymatic ligation procedures promises to create new glycoconjugates. This has most recently been illustrated by the creation of a long-lasting insulin through the enzymatic introduction of a trisaccharide to the amino-terminus of the polypeptide chain. Two additional techniques that can create homogenous protein/glycan conjugates in the cell in a precisely defined manner have been developed. The first involves the incorporation of nonnatural amino acids in the peptide backbone; these amino acids contain either a reactive chemical handle or an *N*-acetylglucosamine–modified amino acid that allows the elaboration of a polysaccharide chain through chemical or enzymatic routes [98]. The second approach involves the Herculean effort of recreating the mammalian glycosylation machinery in yeast [99]. Both have been shown to provide homogenous glycoforms that can be further elaborated and that hold promise for the routine creation of defined glycosylation patterns for structure–function relationships. Finally, an alternative metabolic engineering approach that has the potential to work both *in vitro* and *in vivo* has been reported, in which defined nonnatural monosaccharides are introduced into a cell, where they are incorporated into natural glycans through the action of endogenous polysaccharide synthetases [100–103]. After incorporation, these monosaccharides can be detected through unique chemical handles. Such an approach has shown some promise in terms of analyzing signaling pathways that are regulated by particular glycan posttranslational modifications [104,105] as well as in tailoring glycosylation *in vivo* for drug development efforts.

10.7.7 Improvements through Glycan Redesign

Advances in our understanding of the structure–function relationships of complex glycans have allowed three important developments: first, the creation of new therapeutic modalities for a wide range of diseases; second, the complete description of existing complex glycan therapeutics to potentially develop generics; and third, the redesign of known drugs to create new second-generation molecules with tailored biological characteristics. This last innovation, in particular, effectively illustrates the power of structure–function studies of complex glycans. Two examples of this approach are detailed below. The first involves the polysaccharide-based LMWHs; recent advances in our understanding of these molecules have created opportunities for the development of redesigned LMWHs as improved anticoagulants and, potentially, as effective agents in other diseases, such as cancer or inflammation. The second example is the anemia drug erythropoietin; understanding and systematically altering the glycan structure of this drug has allowed the development of second-generation molecules with optimized activity. This is of particular importance because understanding the glycan structure of a glycoprotein therapeutic, such as erythropoietin, and altering its glycan structure without affecting the protein backbone, makes it possible to create molecules with divergent properties and raises the possibility of the creation of tailored drugs.

10.7.7.1 LMWHs

Unfractionated heparin, which is the starting material for all LMWHs, has been used as an anticoagulant for more than 60 years [106]. The mechanism of action for heparin was identified through numerous structural and biological studies. This research revealed that the anticoagulant activity of heparin predominantly occurs through the inhibition of two components of the coagulation cascade: factor Xa and factor IIa. In both cases, a specific pentasaccharide sequence in heparin binds with high affinity to the serine protease inhibitor ATIII. This binding induces a conformational change in the protein that allows ATIII to inhibit factor Xa4. In addition, heparin chains that are longer than 18 saccharide units (which contain the ATIII-binding pentasaccharide motif) bridge factor IIa and ATIII—thus allowing ATIII to inhibit factor IIa. LMWHs, which are important antithrombotic drugs, are derived from unfractionated heparin by a range of chemical or enzymatic techniques (or both). Their structural complexity is therefore a consequence of the heterogeneity of the parent compound as well as the chemical differences in their preparation. Three main LMWHs are currently in clinical use: dalteparin (Fragmin, Pfizer, New York), tinzaparin (Innohep, DuPont Pharmaceuticals, New York), and enoxaparin (Lovenox, Aventis, New York). Given the understanding of the mechanisms by which heparin inhibits factor Xa and factor IIa, several strategies have been devised to design LMWHs (and mimetics) with tailored activity. *De novo* synthesis has generated molecules that have an ATIII-binding site separated from a thrombin-binding site by spacers with different properties. A series of related molecules that were created using this method were used to probe *in vivo* anticoagulant function, and were optimized for anti-Xa and anti-IIa activity [106].

To circumvent the challenges that are associated with chemical synthetic strategies for heparin-based polysaccharides, a chemoenzymatic synthesis has also been used for the synthesis of Arixtra. Finally, a practical methodology was recently developed to quantify the molar composition of a structural correlate (a tetrasaccharide, which is part of the pentasaccharide motif) in a given LMWH sample [107]. The abundance of this structural motif was found to strongly correlate with anticoagulant activity. Using this structural correlate, which is referred to as *glycorandomization* [108–111], a rational approach was applied to enhance the abundance of the correlate and to generate a new family of rationally designed LMWHs with increased anticoagulant activities [107]. These molecules were found to circumvent many of the shortcomings of known LMWHs, including the possession of an increased efficacy in arterial thrombotic settings. Finally, heparin, as a complex polysaccharide, has been shown to possess several activities that extend beyond anticoagulation, including anti-inflammatory and anticancer properties. Numerous scientific studies have identified a role for heparin oligosaccharide structures in mediating diseases, such as viral infection, cancer, Alzheimer disease, and inflammation [112–115]. There is great potential for the design of new treatments for these diseases through chemical, chemoenzymatic, and enzymatic routes using the heparin scaffold as a starting point.

10.7.8 Erythropoietin

One of the most important biotechnology products is the blockbuster antianemia drug erythropoietin. This 30-kDa protein contains three different sites of *N*-linked glycosylation and stimulates the production of red blood cells *in vivo* [116]. Investigations into the glycosylation of erythropoietin have highlighted the importance of specific capped structures in terms of the serum half-life of the protein [117]. Worldwide, there are two main versions of the drug: Procrit, which is marketed by Johnson & Johnson, New Jersey; and Epogen, which is marketed by Amgen. Recently, however, a new form of erythropoietin has been introduced that has generated much excitement in the nephrology and oncology fields. This new erythropoiesis-stimulating protein, Aranesp, differs from erythropoietin only in terms of the number of *N*-linked glycans that are attached to the peptide backbone [96].

In biological and clinical studies, Aranesp was found to possess higher activity and a substantially increased serum half-life compared with erythropoietin [97]. This has allowed the creation of a dosing schedule for Aranesp that supports higher patient compliance and so increases the overall effectiveness of the antianemia drug. Importantly, however, this is not the entire story. In marked contrast to sialyated erythropoietin, it was recently reported that erythropoietin that lacks the capping sialic acid (asialoEPO) possesses potent and broad neuroprotective activities in diseases ranging from cerebral ischemia to mechanical damage of the central nervous system [118]; in other words, it seems that the erythropoietic and tissue-protective effects of erythropoietin can be dissociated from one another. Taken together, these findings show that by understanding the glycosylation of a biomolecule therapeutic and rationally changing its glycosylation, it is possible to alter the efficacy and *in vivo* half-life of a protein or, alternatively, to change the *in vivo* effect of a drug. Therefore, it seems that, as with linear glycans, an array of sugar structures might be important for distinct functions and it is not simply a matter of homogenizing to a single glycoform. In fact, the development of complete integrated systems for understanding the structure–function activities of complex glycans is necessary to maximize the therapeutic potential of complex glycans.

10.8 Challenges and Opportunities Ahead

Given the structural and informational complexity of glycans, what are the main challenges ahead and how will they influence drug development in this field? We have, at our disposal, diverse techniques to characterize and synthesize glycans, yet it remains a challenge to accurately translate this into a description that reflects the function of a complex glycan pool and to allow rapid drug development. To add to the problem, although the arrangement of the monomer units or sequence is clearly an important characteristic of all polymers, including complex glycan mixtures, this is insufficient to completely describe a complex glycan pool. This is in contrast with other biopolymers, such as DNA, in which the sequence alone is sufficient to describe the molecule. Therefore, the descriptions of glycan structure–function relationships that are yielded by the burgeoning field of glycomics will undoubtedly differ from those produced by proteomics and genomics, which are used to describe protein and nucleic acid structure–function relationships. In the case of complex glycans, which exist as mixtures, there are several types of molecule present; therefore, beyond their identities, the relative abundance of each sequence in the mixture is required to accurately compare glycan pools to one another.

Traditionally, summary properties have been used to describe complex glycan mixtures. Although these are useful in describing certain attributes of the mixture, they often overlook other crucial properties. For instance, molecular weight averages and distributions, and monosaccharide composition, have been used to describe complex glycan mixtures such as LMWHs. Molecular weight distributions are sufficient to describe homopolymers and block polymers; however, they are not adequate to describe complex glycan mixtures. Clearly, molecular weight distributions and averages are important and necessary properties, but because of the structural complexity and heterogeneity of complex glycans, they do not capture many important aspects of the mixture. How should the task of characterizing such complex biological mixtures and applying this information to drug development be approached? The answer lies in the integration of experimentally derived data sets using a bioinformatics framework; that is, the development of strategies. It must be recognized that a single analytical tool, no matter how sophisticated, is insufficient to completely characterize the glycan mixtures that are currently used as drugs or that might be developed in the future. Instead of relying on one technique, several types of measurement that yield complementary information must be integrated to completely characterize complex glycan mixtures and yield important structural correlates to biological functions.

The methods generate different types of data sets, which often prove complementary to one another. Therefore, to complete a detailed characterization of complex glycans, it is necessary to integrate the diverse and orthogonal experimental measurements that are generated by many different methods. Data integration methodologies hold much promise for the analysis of complex glycans. The implementation of these techniques is expected to provide a more complete description of the structure–function relationships of complex glycans, as well as clarifying the biochemical pathways that are required to elicit specific responses. An effort currently being undertaken by the Consortium for Functional Glycomics seeks to understand the role of carbohydrate–protein interactions using such a strategy. In these ways, we might be able to develop a structure–function relationship model for complex glycans that will unlock the potential of these molecules as therapeutic and diagnostic agents. Because of their multifaceted responses, and the fact that many structures can "code" for a single function, such a model will capture the essence of the "glycome." This will allow the development of an "omics" approach to the study of glycan structure that is similar to, but in many ways distinct from, the way in which we have developed genomics for the analysis of the genetic programming of the cell, or proteomics for the analysis of the protein repertoire of the cell.

10.9 Summary

Historically, the study of carbohydrates in biology has been a significant challenge. The isolation of carbohydrates and glycoconjugates from natural sources is tedious, frequently yields heterogeneous products, and produces little material. Based on advances in synthetic chemistry, sensitive screening techniques for probing carbohydrate–protein interactions are being developed to facilitate discoveries in the emerging field of glycomics. Currently, access to this expanding set of tools remains limited, primarily due to the specialization required for preparing the synthetic oligosaccharides. Developments like the automated solid phase oligosaccharide synthesizer are likely to greatly expand access to these synthetically based advances. Until synthetic means are more widely available to the nonexpert, progress in the field is dependent on cross-discipline collaboration between glycobiologists and chemists with the synthetic capacity to generate structures of interest. With these new tools at the disposal of the glycobiologist, it is likely that previously unimagined roles for carbohydrates in cell biology will be discovered. These stand to be exciting years ahead, as revealing these new roles for complex glycans will illuminate fundamental cellular processes. In conjunction with genetic methods, biophysical tools of the kind described in this concepts paper will aid the growth of glycomics into a mature field, equal to genomics and proteomics.

In view of the various biochemical pathways and disease processes in which glycans are crucially engaged—angiogenesis, cancer, tissue repair, cardiovascular disease, immune system function, and microbial and viral pathogenesis, to name a few—the possibilities for glycans as therapeutics and diagnostics are numerous and exciting. As we further improve our ability to study glycans, we will gain a better understanding of the existing drugs, as well as developing new specific therapeutics. Important challenges lie ahead as we make our way in a postgenomic world in which an understanding of glycan structure and function will undoubtedly lead to a new generation of highly effective therapeutics.

Importantly, to overcome the challenges and to harness the excitement that this subject area has recently generated, the U.S. National Institutes of Health has funded the development of the Consortium for Functional Glycomics to advance the burgeoning field of glycomics. From our perspective, the future is exciting and the possibilities are numerous. Nowadays, complex glycan drugs are marketed in two main areas: the treatment of thrombosis (e.g., Lovenox and Fragmin) and as important structural components to promote healing [e.g., Sepragel (Genzyme) and Healon (Pfizer)]. The introduction of Aranesp, which is a hyperglycosylated form of an existing antianemia drug, has provided an intriguing hint to the potential of glycan-based drug improvement strategies. Furthermore, the potential of glycan-based strategies in multicellular diseases, such as cancer, is now beginning to be exploited. Finally, enzymes that act on glycoconjugates, such as Aldurazyme and Cerezyme, are now being used in enzyme-replacement strategies.

References

1. IUPAC-IUB Joint Commission on Biochemical Nomenclature (JCBN). 1982. Abbreviated terminology of oligosaccharide chains. Recommendations 1980. *European Journal of Biochemistry* 126:433–437.
2. IUB-IUPAC Joint Commission on Biochemical Nomenclature (JCBN). 1982. Abbreviated terminology of oligosaccharide chains. Recommendations 1980. *Journal of Biological Chemistry* 257:3347–3351.
3. IUPAC-IUB Joint Commission on Biochemical Nomenclature (JCBN). 1982. Polysaccharide nomenclature. Recommendations 1980. *European Journal of Biochemistry* 126:439–441.
4. Sharon, N., IUPAC-IUB Joint Commission on Biochemical Nomenclature (JCBN). 1986. Nomenclature of glycoproteins, glycopeptides and peptidoglycans. Recommendations 1985. *European Journal of Biochemistry* 159:1–6.
5. Herget, S., R. Ranzinger, K. Maass, and C.W. Lieth. 2008. GlycoCT—a unifying sequence format for carbohydrates. *Carbohydrate Research* 343:2162–2171.
6. Hashimoto, K., I. Takigawa, M. Shiga, M. Kanehisa, and H. Mamitsuka. 2008. Mining significant tree patterns in carbohydrate sugar chains. *Bioinformatics* 24:167–173.
7. Tamaki, Y., Y. Nomura, S. Katsumura, A. Okada, and H. Yamada. 2009. Construction of a dental caries prediction model by data mining. *Journal of Oral Science* 51:61–68.
8. Froesch, M., L.M, Bindila, G. Baykut, M. Allen, J. Peter-Katalinic, and A.D. Zamfir. 2004. Coupling of fully automated chip electrospray to Fourier transform ion cyclotron resonance mass spectrometry for high-performance glycoscreening and sequencing. *Rapid Communications in Mass Spectrometry* 18:3084–3092.
9. Cipollo J.F., A.M. Awad, C.E. Costello, and C.B. Hirschberg. 2005. N-glycans of *Caenorhabditis elegans* are specific to developmental stages. *Journal of Biological Chemistry* 280:26063–26072.
10. Zamfir, A, and J. Peter-Katalinic. 2004. Capillary electrophoresis-mass spectrometry for glycoscreening in biomedical research. *Electrophoresis* 25:1949–1963.
11. Goldberg, D., M. Sutton-Smith, J. Paulson, and A. Dell. 2005. Automatic annotation of matrix-assisted laser desorption/ionization N-glycan spectra. *Proteomics* 5:865–875.
12. Tang, H., Y. Mechref, and M.V. Novotny. 2005. Automated interpretation of MS/MS spectra of oligosaccharides. *Bioinformatics* 21(Suppl. 1):i431–i439.
13. Ethier, M., J.A. Saba, M. Spearman, O. Krokhin, M. Butler, W. Ens et al. 2003. Application of the StrOligo algorithm for the automated structure assignment of complex N-linked glycans from glycoproteins using tandem mass spectrometry. *Rapid Communications in Mass Spectrometry* 17:2713–2720.

14. Ethier, M., O. Krokhin, W. Ens, K.G. Standing, J.A. Wilkins, and H. Perreault. 2005. Global and site-specific detection of human integrin alpha 5 beta 1 glycosylation using tandem mass spectrometry and the StrOligo algorithm. *Rapid Communications in Mass Spectrometry* 19:721–727.

15. Lewandrowski, U., A. Resemann, and A. Sickmann. 2005. Laser-induced dissociation/high-energy collision-induced dissociation fragmentation using MALDI-TOF/TOF-MS instrumentation for the analysis of neutral and acidic oligosaccharides. *Analytical Chemistry* 77:3274–3283.

16. Voisin, S., R.S. Houliston, J. Kelly, J.R. Brisson, D. Watson, S.L. Bardy et al. 2005. Identification and characterization of the unique N-linked glycan common to the flagellins and S-layer glycoprotein of *Methanococcus voltae*. *Journal of Biological Chemistry* 280:16586–16593.

17. Stenutz, R., P.E. Jansson, and G. Widmalm. 1998. Computer-assisted structural analysis of oligo- and polysaccharides: An extension of CASPER to multibranched structures. *Carbohydrate Research* 306:11–17.

18. Hricovini, M. 2004. Structural aspects of carbohydrates and the relation with their biological properties. *Current Medicinal Chemistry* 11:2565–2583.

19. Allen, F.H., and R. Taylor. 2004. Research applications of the Cambridge Structural Database (CSD). *Chemical Society Reviews* 33:463–475.

20. Perez, S., C. Gautier, and A. Imberty. 2000. Oligosaccharide conformations by diffraction methods. In *Oligosaccharides in Chemistry and Biology*, edited by Ernst, B., G. Hart, and P. Sinay, 969–1001. Weinheim: Wiley-VCH.

21. Rencurosi, A., E.P. Mitchell, G. Cioci, S. Perez, R. Pereda-Miranda, and A. Imberty. 2004. Crystal structure of tricolorin A: Molecular rationale for the biological properties of resin glycosides found in some Mexican herbal remedies. *Angewandte Chemie (International ed. in English)* 43:5918–5922.

22. Petrescu, A.J., A.L. Milac, S.M. Petrescu, R.A. Dwek, and M.R. Wormald. 2004. Statistical analysis of the protein environment of N-glycosylation sites: Implications for occupancy, structure, and folding. *Glycobiology* 14:103–114.

23. Berman, H.M., T. Battistuz, T.N. Bhat, W.F. Bluhm, P.E. Bourne, K. Burkhardt et al. 2002. The Protein Data Bank. *Acta Crystallographica. Section D, Biological Crystallography* 58:899–907.

24. Lutteke, T., M. Frank, and C.W. der Lieth. 2004. Data mining the Protein Data Bank: Automatic detection and assignment of carbohydrate structures. *Carbohydrarate Research* 339:1015–1020.

25. Chandrasekharan, R. 1997. Molecular architecture of polysaccharide helices in oriented fibres. *Advanced Carbohydrate Chemistry and Biochemistry* 52:311–439.

26. Imberty, A., and S. Perez. 2000. Structure, conformation, and dynamics of bioactive oligosaccharides: Theoretical approaches and experimental validations. *Chemical Reviews* 100:4567–4588.

27. Bohne, A., E. Lang, and C.W. der Lieth. 1999. SWEET—WWW-based rapid 3D construction of oligo- and polysaccharides. *Bioinformatics* 15:767–768.

28. Engelsen, S.B., S. Cros, W. Mackie, and S. Perez. 1996. A molecular builder for carbohydrates: Application to polysaccharides and complex carbohydrates. *Biopolymers* 39:417–433.

29. Nyholm, P.G., L.A. Mulard, C.E. Miller, T. Lew, R. Olin, and C.P.J. Glaudemans. 2001. Conformation of the O-specific polysaccharide of *Shigella dysenteriae* type 1: Molecular modeling shows a helical structure with efficient exposure of the antigenic determinant α-L-Rhap-(1!2)-α-D-Galp. *Glycobiology* 11:945–955.

30. Koca, J. 1994. Computer program CICADA—travelling along conformational potential energy hyper surface. *Journal of Molecular Structure* 308:13–24.

31. Peters, T., B. Meyer, R. Stuike-Prill, R. Somorjai, J.R. Brisson, and A. Monte Carlo. 1993. Method for conformational analysis of saccharides. *Carbohydrate Research* 238:49–73.

32. Nahmany, A., F. Strino, J. Rosen, G.J.L.S. Kemp, and P.G. Nyholm. 2005. The use of a genetic algorithm search for molecular mechanics (MM3)-based conformational analysis of oligosaccharides. *Carbohydrate Research* 340:1059–1064.

33. Frank, M., P. Gutbrod, C. Hassayoun, and C.W. der Lieth. 2003. Dynamic molecules: Molecular dynamics for everyone. An internet-based access to molecular dynamic simulations: Basic concepts. *Journal of Molecular Modeling* 9:308–315.

34. Rodriguez-Carvajal, M.A., C. Herve du Penhoat, K. Mazeau, T. Doco, and S. Perez. 2003. The three-dimensional structure of the mega-oligosaccharide rhamnogalacturonan II monomer: A combined molecular modeling and NMR investigation. *Carbohydrate Research* 338:651–671.

35. Cioci, G., A. Rivet, J. Koca, and S. Perez. 2004. Conformational analysis of complex oligosaccharides: the CICADA approach to the uromodulin O-glycans. *Carbohydrate Research* 339:949–959.

36. Engelsen, S.B., C. Monteiro, C. Herve du Penhoat, and S. Perez. 2001. The diluted aqueous solvation of carbohydrates as inferred from molecular dynamics simulations and NMR spectroscopy. *Biophysical Chemistry* 93:103–127.

37. Almond, A., and J.K. Sheenan. 2003. Predicting the molecular shape of polysaccharides from dynamic interactions with water. *Glycobiology* 13:255–264.

38. Watson, P., M. Verdonk, and M.J. Hartshorn. 2003. A web-based platform for virtual screening. *Journal of Molecular Graphics & Modelling* 22:71–82.

39. Mitchell, E.P., C. Sabin, L. Snajdrova, M. Pokorna, S. Perret, C. Gautier et al. 2005. High affinity fucose binding of *Pseudomonas aeruginosa* lectin PA-IIL: 1.0 Å resolution crystal structure of the complex combined with thermodynamics and computational chemistry approaches. *Proteins* 58:735–746.

40. Lortat-Jacob, H., A. Grosdidier, and A. Imberty. 2002. Structural diversity of heparan sulfate binding domains in chemokines. *Proceedings of the National Academy of Sciences of the United States of America* 99:1229–1234.

41. Imberty, A., R. Mollicone, E. Mikros, P.A. Carrupt, S. Perez, and R. Oriol. 1996. How do antibodies and lectins recognize histo-blood group antigens? A 3D-QSAR study by comparative molecular field analysis (CoMFA). *Bioorganic & Medicinal Chemistry* 4:1979–1988.

42. Wang, D., S. Liu, B.J. Trummer, C. Deng, and A. Wang. 2002. Carbohydrate microarrays for the recognition of cross-reactive molecular markers of microbes and host cell. *Nature Biotechnology* 20:275–281.

43. Fazio, F., M.C. Bryan, O. Blixt, J.C. Paulson, and C.H. Wong. 2002. Synthesis of sugar arrays in microtiter plate. *Journal of the American Chemical Society* 124:14397–14402.

44. Houseman, B, and M. Mrksich. 2002. Carbohydrate arrays for the evaluation of protein binding and enzyme activity. *Biotechnology and Chemical Biology* 9:443–454.

45. Park, S., and I. Shin. 2002. Fabrication of carbohydrate chips for studying protein–carbohydrate interactions. *Angewandte Chemie (International ed. in English)* 114:3312.

46. Houseman, B.T., E.S. Gawalt, M. Mrksich, and M. Maleimide. 2003. Functionalized self-assembled monolayers for the preparation of peptide and carbohydrate biochips. *Langmuir* 19:1522.

47. Ratner, D.M., E.W. Adams, J. Su, B.R. O'Keefe, M. Mrksich, and P.H. Seeberger. 2004. Probing protein–carbohydrate interactions with microarrays of synthetic oligosaccharides. *ChemBioChem* 5:379–383.

48. Adams, E.W., D.M. Ratner, H.R. Bokesch, J.B. McMahon, B.R. O'Keefe, and P.H. Seeberger. 2004. Oligosaccharide and glycoprotein microarrays as tools in HIV-glycobiology: A detailed analysis of glycan dependent gp120/protein interactions. *Chemical Biology* 11:875–881.

49. Bolmstedt, A.J., B.R. O'Keefe, S.R. Shenoy, J.B. McMahon, and M.R. Boyd. 2001. Cyanovirin-N defines a new class of antiviral agent targeting N-linked, high-mannose glycans in an oligosaccharide-specific manner. *Molecular Pharmacology* 59:949–954.

50. Bokesch, H.R., B.R. O'Keefe, T.C. McKee, L.K. Pannell, G.M.L. Patterson, R.S. Gardella et al. 2003. A potent novel anti-HIV protein from the cultured cyanobacterium *Scytonema varium*. *Biochemistry* 42:2578–2584.

51. Walsh, C. 2000. Molecular mechanisms that confer antibacterial drug resistance *Nature* 406:775–781.

52. Llano-Sotelo, B., E.F. Azucena, L.P. Kotra Jr., S. Mobashery, and C.S. Chow. 2002. Aminoglycosides modified by resistance enzymes display diminished binding to the bacterial ribosomal aminoacyl-tRNA site. *Chemical Biology* 9:455.

53. Fourmy, D., M.I. Recht, S.C. Blanchard, and J.D. Puglisi. 1996. Structure of the A site of *Escherichia coli* 16S ribosomal RNA complexed with an aminoglycoside antibiotic. *Science* 274:1367–1371.

54. Magnet, S., T. Lambert, P. Courvalin, and J.S. Blanchard. 2001. Kinetic and mutagenic characterization of the chromosomally encoded Salmonella enterica AAC(6′)-Iy aminoglycoside N-acetyltransferase. *Biochemistry* 40:3700–3709.

55. Hegde, S.S., F. Javid-Majd, and J.S. Blanchard. 2001. Overexpression and mechanistic analysis of chromosomally encoded aminoglycoside 2′-N-acetyltransferase (AAC(2′)-Ic) from *Mycobacterium tuberculosis. Journal of Biological Chemistry* 276:45876–45881.

56. Baker, T.J., N.W. Luedtke, Y. Tor, and M. Goodman. 2000. Synthesis and anti-HIV activity of guanidinoglycosides. *Journal of Organic Chemistry* 65:9054–9058.

57. Vetting, M.W., S.S. Hegde, F. Javid-Majd, J.S. Blanchard, and S.L. Roderick. 2002. Aminoglycoside 2′-N-acetyltransferase from *Mycobacterium tuberculosis* in complex with coenzyme A and aminoglycoside substrates. *Nature Structural Biology* 9:653–658.

58. Disney, M.D., S.M. Magnet, J.S. Blanchard, and P.H. Seeberger. 2004. Aminoglycoside microarrays to study antibiotic resistance. *Angewandte Chemie (International ed. in English)* 43:1591–1594.

59. Adams, E.W., J. Ueberfeld, D.M. Ratner, B.R. O'Keefe, D.R. Walt, and P.D. Seeberger. 2003. Encoded fiber-optic microsphere arrays for probing protein–carbohydrate interactions. *Angewandte Chemie (International ed. in English)* 42:5317–5320.

60. Liang, R., L. Yan, J. Loebach, M. Ge, Y. Uozumi, K. Sekanina et al. 1996. Parallel synthesis and screening of a solid phase carbohydrate library. *Science* 274:1520.

61. Roseman, D.S, and J.U. Baenziger. 2001. The mannose/N-acetylgalactosamine-4-SO4 receptor displays greater specificity for multivalent than monovalent ligands. *Journal of Biological Chemistry* 276:17052–17057.

62. Iobst, S.T., and K. Drickamer. 1996. Selective sugar binding to the carbohydrate recognition domains of the rat hepatic and macrophage asialoglycoprotein receptors. *Journal of Biological Chemistry* 271(12):6686–6693.

63. Lin, C.C., Y.C. Yeh, C.Y. Yang, C.L. Chen, G.F. Chen, C.C. Chen, and Y.C. Wu. 2002. Selective binding of mannose-encapsulated gold nanoparticles to type 1 pili in *Escherichia coli. Journal of the American Chemical Society* 124:3508–3509.

64. Bruchez, M., M. Moronne Jr., P. Gin, S. Weiss, and A.P. Alivisatos. 1998. Semiconductor nanocrystals as fluorescent biological labels. *Science* 281:2013–2016.

65. Ceroni, A., A. Dell, and S.M. Haslam. 2007. The GlycanBuilder: A fast, intuitive and flexible software tool for building and displaying glycan structures. *Source Code for Biology and Medicine* 7:2–3.

66. Berteau, O., and R. Stenutz. 2004. Web resources for the carbohydrate chemist. *Carbohydrate Research* 339:929–936.

67. der Lieth, C.W. 2004. An endorsement to create open access databases for analytical data of complex carbohydrates. *Journal of Carbohydrate Chemistry* 23:277–297.

68. van der Lieth, C.W., A. Bohne-Lang, K.K. Lohmann, and M. Frank. 2004. Bioinformatics for glycomics: Status, methods, requirements and perspectives. *Brief Bioinformatics* 5:164–178.

69. Doubet, S., K. Bock, D. Smith, A. Darvill, and P. Albersheim. 1989. The complex carbohydrate structure database. *Trends in Biochemical Sciences* 14:475–477.

70. Cooper, C.A., H.J. Joshi, M.J. Harrison, M.R. Wilkins, and N.H. Packer. 2003. GlycoSuiteDB: A curated relational database of glycoprotein glycan structures and their biological sources. 2003 Update. *Nucleic Acids Research* 31:511–513.

71. Aoki, K.F., H. Mamitsuka, T. Akutsu, and M. Kanehisa. 2005. A score matrix to reveal the hidden links in glycans. *Bioinformatics* 21:1457–1463.

72. Bohne-Lang, A., E. Lang, T. Forster, and C.W. der Lieth. 2001. LINUCS: Linear Notation for Unique description of Carbohydrate Sequences. *Carbohydrate Research* 336:1–11.

73. Banin, E., Y. Neuberger, Y. Altshuler, A. Halevi, O. Inbar, D. Nir, and A. Dukler. 2002. A novel linear code nomenclature for complex carbohydrates. *Trends in Glycoscience and Glycotechnology* 14:127–137.

74. Kameyama, A., N. Kikuchi, S. Nakaya, H. Ito, T. Sato, T. Shikanai et al. 2005. A strategy for identification of oligosaccharide structures using observational multistage mass spectral library. *Analytical Chemistry* 77:4719–4725.

75. Hashimoto, K., S. Goto, S. Kawano, K.F. Aoki-Kinoshita, N. Ueda, M. Hamajima et al. 2005. KEGG as a glycome informatics resource. *Glycobiology* 16:63R–70R.
76. Desai, U.R., M. Petitou, I. Bjork, and S.T. Olson. 1998. Mechanism of heparin activation of antithrombin. Role of individual residues of the pentasaccharide activating sequence in the recognition of native and activated states of antithrombin. *Journal of Biological Chemistry* 273:7478–7487.
77. Kornfeld, S. 1992. Structure and function of the mannose 6-phosphate/insulin-like growth factor II receptors. *Annual Reviews in Biochemistry* 61:307–330.
78. Stockert, R.J. 1995. The asialoglycoprotein receptor: Relationships between structure, function, and expression. *Physiology Reviews* 75:591–609.
79. Yang, Z.Q., E.B. Puffer, J.K. Pontrello, and L.L. Kiessling. 2002. Synthesis of a multivalent display of a CD22-binding trisaccharide. *Carbohydrate Research* 337:1605–1613.
80. Gotte, M. 2003. Syndecans in inflammation. *FASEB Journal* 17:575–591.
81. van Zante, A., and S.D. Rosen. 2003. Sulphated endothelial ligands for L-selectin in lymphocyte homing and inflammation. *Biochemical Society Transactions* 31:313–317.
82. Gawlitzek, M., T. Ryll, J. Lofgren, and M.B. Sliwkowski. 2000. Ammonium alters N-glycan structures of recombinant TNFR-IgG: Degradative versus biosynthetic mechanisms. *Biotechnology and Bioengineering* 68:637–646.
83. Baker, K.N. 2001. Metabolic control of recombinant protein N-glycan processing in NS0 and CHO cells. *Biotechnology and Bioengineering* 73:188–202.
84. Yang, M., and M. Butler. 2002. Effects of ammonia and glucosamine on the heterogeneity of erythropoietin glycoforms. *Biotechnology Progress* 18:129–138.
85. Yang, M., and M. Butler. 2000. Effect of ammonia on the glycosylation of human recombinant erythropoietin in culture. *Biotechnology Progress* 16:751–759.
86. Senger, R.S., and M.N. Karim. 2003. Effect of shear stress on intrinsic CHO culture state and glycosylation of recombinant tissue-type plasminogen activator protein. *Biotechnology Progress* 19:1199–1209.
87. Thomas, L.J. 2004. Production of a complement inhibitor possessing sialyl Lewis X moieties by in vitro glycosylation technology. *Glycobiology* 14:883–893.
88. Weikert, S. 1999. Engineering Chinese hamster ovary cells to maximize sialic acid content of recombinant glycoproteins. *Nature Biotechnology* 17:1116–1121.
89. Zhang, X., S.H. Lok, and O.L. Kon. 1998. Stable expression of human α-2,6-sialyltransferase in Chinese hamster ovary cells: Functional consequences for human erythropoietin expression and bioactivity. *Biochimica Biophysica Acta* 1425:441–452.
90. Tomiya, N. 2003. Complex-type biantennary N-glycans of recombinant human transferrin from *Trichoplusia ni* insect cells expressing mammalian β-1,4-galactosyltransferase and β-1,2-*N*-acetylglucosaminyltransferase II. *Glycobiology* 13:23–34.
91. Egrie, J.C., and J.K. Browne. 2001. Development and characterization of novel erythropoiesis stimulating protein (NESP). *British Journal of Cancer* 84 (Suppl. 1):3–10.
92. Egrie, J.C., E. Dwyer, J.K. Browne, A. Hitz, and M.A. Lykos. 2003. Darbepoetin α has a longer circulating half-life and greater in vivo potency than recombinant human erythropoietin. *Experimental Hematology* 31:290–299.
93. Schofield, L., M.C. Hewitt, K. Evans, M.A. Siomos, and P.H. Seeberger. 2002. Synthetic GPI as a candidate anti-toxic vaccine in a model of malaria. *Nature* 418:785–789.
94. Linhardt, R.J, and N.S. Gunay. 1999. Production and chemical processing of low molecular weight heparins. *Seminars in Thrombosis and Hemostasis* 25 (Suppl. 3):5–16.
95. Sioud, M. 2004. Therapeutic siRNAs. *Trends in Pharmacological Sciences* 25:22–28.
96. Plante, O.J., E.R. Palmacci, and P.H. Seeberger. 2001. Automated solid-phase synthesis of oligosaccharides. *Science* 291:1523–1527.
97. Ritter, T.K., K.K. Mong, H. Liu, T. Nakatani, and C.H. Wong. 2003. A programmable one-pot oligosaccharide synthesis for diversifying the sugar domains of natural products: A case study of vancomycin. *Angewandte Chemie (International ed. in English)* 42:4657–4660.
98. Zhang, Z. 2004. A new strategy for the synthesis of glycoproteins. *Science* 303:371–373.

99. Hamilton, S.R. 2003. Production of complex human glycoproteins in yeast. *Science* 301: 1244–1246.

100. Wieser, J.R., A. Heisner, P. Stehling, F. Oesch, and W. Reutter. 1996. In vivo modulated N-acyl side chain of N-acetylneuraminic acid modulates the cell contact–dependent inhibition of growth. *FEBS Letters* 395:170–173.

101. Kayser, H. 1992. Biosynthesis of a nonphysiological sialic acid in different rat organs, using N-propanoyl-ᴅ hexosamines as precursors. *Journal of Biological Chemistry* 267:16934–16938.

102. Kiick, K.L., E. Saxon, D.A. Tirrell, and C.R. Bertozzi. 2002. Incorporation of azides into recombinant proteins for chemoselective modification by the Staudinger ligation. *Proceedings of the National Academy of Sciences of the United States of America* 99:19–24.

103. Luchansky, S.J., S. Goon, and C.R. Bertozzi. 2004. Expanding the diversity of unnatural cell-surface sialic acids. *Chembiochem* 5:371–374.

104. Vocadlo, D.J., H. Hang, E.J. Kim, J.A. Hanover, and C.R. Bertozzi. 2003. A chemical approach for identifying O-GlcNAc-modified proteins in cells. *Proceedings of the National Academy of Sciences of the United States of America* 100:9116–9121.

105. Hang, H.C., C. Yu, D.L. Kato, and C.R. Bertozzi. 2003. A metabolic labeling approach toward proteomic analysis of mucin-type O-linked glycosylation. *Proceedings of the National Academy of Sciences of the United States of America* 100:14846–14851.

106. Petitou, M., B. Casu, and U. Lindahl. 2003. A critical period in the history of heparin: The discovery of the antithrombin binding site. *Biochimie* 85:83–89.

107. Jin, L. 1997. The anticoagulant activation of antithrombin by heparin. *Proceedings of the National Academy of Sciences of the United States of America* 94:14683–14688.

108. Fu, X. 2003. Antibiotic optimization via in vitro glycorandomization. *Nature Biotechnology* 21:1467–1469.

109. Iozzo, R.V. 2001. Heparan sulfate proteoglycans: Intricate molecules with intriguing functions. *Journal of Clinical Investigation* 108:165–167.

110. Rudd, P.M., T. Elliott, P. Cresswell, I.A. Wilson, and R.A. Dwek. 2001. Glycosylation and the immune system. *Science* 291:2370–2376.

111. Petitou, M. 1999. Synthesis of thrombin-inhibiting heparin mimetics without side effects. *Nature* 398:417–422.

112. Sundaram, M. 2003. Rational design of low-molecular weight heparins with improved in vivo activity. *Proceedings of the National Academy of Sciences of the United States of America* 100:651–656.

113. Tiwari, V. 2004. A role for 3-O-sulfated heparan sulfate in cell fusion induced by herpes simplex virus type 1. *Journal of General Virology* 85:805–809.

114. Zacharski, L.R., and J.T. Loynes. 2002. The heparins and cancer. *Current Opinion in Pulmonary Medicine* 8:379–382.

115. Scholefield, Z. 2003. Heparan sulfate regulates amyloid precursor protein processing by BACE1, the Alzheimer's β-secretase. *Journal of Cell Biology* 163:97–107.

116. Malhotra, S., D. Bhasin, N. Shafiq, and P. Pandhi. 2004. Drug treatment of ulcerative colitis: Unfractionated heparin, low molecular weight heparins and beyond. *Expert Opinion in Pharmacotherapy* 5:329–334.

117. Goldwasser, E. 1984. Erythropoietin and its mode of action. *Blood Cells* 10:147–162.

118. Fukuda, M.N., H. Sasaki, and L. Lopez. 1989. Survival of recombinant erythropoietin in the circulation: The role of carbohydrates. *Blood* 73:84–89.

11

Breast Cancer Biomarkers

Rajneesh K. Gaur

CONTENTS

11.1 Introduction

Cancer is a class of diseases in which a group of cells undergo uncontrolled growth. The dividing cells may intrude on and destroy either adjacent or distant tissues via lymph or blood. Cancers are classified according to the resemblance of the tumor with a particular cell/tissue type. These types include carcinomas (epithelial origin such as breast, prostrate, lung, etc.), sarcomas (mesenchymal origin), lymphomas and leukemia (hematopoietic origin), reproductive organ cancers (pluripotent cells origin), and blastomas (embryonic tissue origin). There are more than 200 different types of cancer; fortunately, most of them are rare. Among all types of cancer, skin cancer is the most prevalent among men and women. In women, breast cancer is the second most common and lethal type of cancer.

The breast consists of milk-producing glands that are connected to the nipple opening through ducts. The gland and the respective ducts are surrounded by fat and fibrous tissues. Any form of lump either in the breast or in the underarm area is considered as a warning sign of breast cancer. Based on the anatomy of the breast, there are two kinds of breast cancer, namely, the *in situ* type and the invasive/infiltrating type. Cancer of the ducts and lobules of the breast is called *in situ* breast cancer, whereas when it starts invading the surrounding tissues, it is called invasive/infiltrating breast cancer. Most *in situ* breast cancers are of the ductal type and are curable. Multiple risk factors have been identified but the exact causes of breast cancer are unknown. There are two kinds of risk factors associated with breast cancer, namely, easily and non–easily modifiable risk factors. Easily modifiable risk factors are postmenopausal obesity, use of combined estrogen and progestin menopausal hormones, alcohol consumption, and physical inactivity, whereas non–easily modifiable breast cancer risk factors are age, family history, age at first full-term pregnancy, early menarche, late menopause, and breast density [1].

In general, cancer diagnosis is largely based on morphological examination of tumor biopsy, but this approach very poorly predicts the tumor's potential for progression and response to treatment. Therefore, the development of either chemical or biological markers (i.e., biomarkers) is highly desirable for the early detection, treatment planning, or targeting of the therapy. The current generation of breast cancer biomarker does not indicate the presence of breast cancer with certainty. The ideal biomarker must detect breast cancer at the earliest and most treatable stages, as well as improving patient outcome and the cost-effectiveness of therapy.

11.2 Biomarker Definition

In general, cancer biomarkers are defined as any measurable cellular, subcellular, or humoral factor that demonstrates the presence of malignancy or malignant potential, or predicts tumor behavior, prognosis, or response to treatment. According to the National Institutes of Health biomarkers definition, a biomarker (or biological marker or molecular marker or signature molecule) is a characteristic that can be objectively measured as an indicator of normal biological processes, pathogenic processes, or a pharmacological response to a therapeutic intervention [2]. These indicators could be any biochemical entities such as nucleic acids, proteins, sugars, lipids, and small metabolites as well as whole cells or biophysical characteristics of tissues.

11.3 Biomarker Development

Traditionally, biomarkers were discovered using conventional laboratory tools such as immunohistochemistry (IHC) and gel electrophoresis via identification of altered genes and changes in messenger RNA (mRNA) and protein expression [3]. Traditional methods are not suitable for the detection of early tumors because of their slow nature, highly limited approach, and requirement for a lot of raw material. Nowadays, biomarker discovery is heavily dependent on the available technology and established protocols (Table 11.1). The goal of the available and emerging technologies is to identify the link between the genetic variations or changes in gene or protein expression/activity with breast cancer or a response to therapeutic intervention. Recently, high-throughput technologies have made it possible to test hundreds to thousands of potential biomarkers at once without prior knowledge of the underlying biology and extensive experience in the relevant technology [4]. Biomarker discovery through genomic and proteomic profiling also suffers from a number of limitations including false findings, sample bias, and time-consuming and expensive evaluation with difficulties in validation.

Initially, the high-throughput methods to discover biomarker and expression patterns focused on a nucleic acid (i.e., DNA) because of the availability of more advanced and fully characterized methods for nucleic acids in comparison with other cellular components. RNA assays can detect dynamic changes in gene expression and identify upstream DNA level abnormalities, but the cost and difficulty of working with RNA in comparison with stable DNA are major hindrances in their adoption. Therefore, proteins can provide better alternatives as biomarkers than either type of nucleic acid because changes in DNA and RNA are not always directly linked to altered protein expression, modification, or function. But progress in the identification of protein biomarkers has lagged as a result of their sheer and unknown numbers, posttranslational modification, and partly due to current technological limitations [5]. Biomarkers from other routes such as metabolomics are even less characterized and developed. Metabolomics entails the study of metabolic responses to drugs, environmental changes, and diseases via the identification of small-molecule

TABLE 11.1

Examples of Biomarker Categories and High-Throughput Methods of Discovery

Biomarker Category	Examples of Methods
Genomics	
DNA based	
Copy number/loss of heterogeneity	Various DNA arrays
Sequence variation	Various sequencing methods
Epigenetic variation	
Genome rearrangements	
RNA based	
mRNA signatures	
miRNA signatures	
Proteomics	Mass spectrometry
Proteins	Liquid chromatography
Peptides	Protein arrays
Metabolomics	
Metabolites	Mass spectrometry
Lipids	Liquid chromatography
Carbohydrates	Protein arrays

Source: Derived from IOM (Institute of Medicine), Developing biomarker based tools for cancer screening, diagnosis and treatment. In *The State of the Science, Evaluation, Implementation and Economics. A Workshop.* Patlak, M., Nass, S., rappourteur, Washington, DC: The National Academic Press, 2006.

metabolite profiles, that is, it attempts to measure the metabolic consequences of altered gene and protein expressions.

To overcome the limitations of current biomarkers and to facilitate future research, a guideline named "Reporting Recommendations of Tumor Marker Prognostic Studies" was published, which recommends a description of the amount of information that should be provided when reporting the results of a biomarker study [6].

11.4 Biomarker Classification

Cancer biomarker–based diagnostics have applications for establishing disease predisposition, early detection, cancer staging, therapy selection, identifying whether or not a cancer is metastatic, therapy monitoring, assessing prognosis, and advances in the adjuvant setting. In the past decade, knowledge of cancer at the molecular level has been improved considerably in conjunction with the diagnostic tools. In general, biomarkers for breast cancer are classified as follows:

11.4.1 Risk Biomarkers

For individuals with no cancer, the risk biomarker predicts the onset of a clinical disease. For patients with cancer, it predicts the recurrence or progression of a disease. These biomarkers are associated with increased cancer risk and include mammographic abnormalities, proliferative breast disease with or without atypia, family clustering, and inherited germ line abnormalities. The examples of risk biomarkers are insulin-like growth factor 1 (IGF-1), the ratio of IGF-1 to its binding protein IGFBP-3, serum luteal phase progesterone and free testosterone in premenopausal women, prolactin, bioavailable estradiol, and testosterone in postmenopausal women.

11.4.2 Surrogate End Point Biomarkers

These biomarkers constitute tissue, cellular, or molecular alterations that occur between cancer initiation and progression. These biomarkers are used as end points in short-term chemoprevention trials. For a marker to be a surrogate end point biomarker, it should be reproducible, reliable, biologically plausible, and should have strong associations with ultimate outcome, for example, perillyl alcohol.

11.4.3 Prognostic Biomarkers

These biomarkers provide information about the patient's clinical outcome independent of treatment. Prognostic markers determine whether a patient requires treatment. Candidate prognostic biomarkers for breast cancer include increased proliferation indices such as Ki-67 and proliferating cell nuclear antigen; estrogen receptor (ER) and progesterone receptor (PR) overexpression; markers of oncogene overexpression such as c-erbB-2, transforming growth factor-α (TGF-α), and epidermal growth factor receptor (EGFR); indicators of apoptotic imbalance, including overexpression of bcl-2 and an increased bax/bcl-2 ratio; markers of disordered cell signaling such as p53 nuclear protein accumulation; alteration of differentiation signals, such as the overexpression of c-myc and related proteins; loss of differentiation markers such as Transforming Growth Factor (TGF) – β2 and retinoic acid receptor; and alteration of angiogenesis proteins such as vascular endothelial growth factor (VEGF) overexpression.

11.4.4 Predictive (or Response) Biomarkers

These biomarkers provide information about the response to therapy and are associated with tumor sensitivity or resistance to a specific treatment. Predictive markers are helpful in deciding which treatment will be best for the patient. Both predictive and prognostic biomarkers provide information about the future behavior of a tumor, whereas predictive markers help in selecting responsiveness or resistance to a specific treatment; prognostic factors provide information on outcome independent of systemic adjuvant therapy. These markers may be the target of a specific therapy itself, for example, the oncogene human EGFR-2 (HER-2) is the target of the monoclonal antibody "trastuzumab" and HER-2 amplification predicts good response toward anti–HER-2 therapy. A response biomarker is usually a reversible risk biomarker that predicts either disease prevention or eradication and temporary control in case of established cancer.

11.4.5 Pharmacodynamic Biomarkers

Pharmacodynamic biomarkers measure the near-term treatment effects of a drug on the tumor (or on the host) and can, in theory, be used to guide dose selection in the early stages of clinical development of a new anticancer drug. In cytotoxic chemotherapy, the dose that is used to determine antitumor activity in phase II clinical trials is usually the maximum tolerated dose discovered in a phase I dose escalation study. But this might be a less relevant end point for drugs that have been optimized to bind to a specific molecular target. An alternative way to determine an appropriate dose is to measure the effect of the drug on its target across a range of doses (known as a target-engagement study) and then to select a dose for phase II clinical trials on the basis of the magnitude of target modulation, for example, imatinib mesylate has been shown to block the protein-kinase activity of BCR-ABL in the tumor cells of patients with chronic myeloid leukemia at the same doses that induce clinical remission, which are well below those associated with toxicity. The utility of pharmacodynamic biomarkers might also extend beyond the clinical trial phase of drug development.

11.4.6 Novel Biomarkers

The lack of significant specificity and the number of shortcomings associated with each biomarker created a niche for the identification and commercialization of new biomarkers and related techniques.

1. Pharmaceutical companies classify biomarkers based on its type and how it will be used (Figure 11.1). Pharmaceutical categories of biomarkers include:
   a. Biomarkers that occur close to the actions of the target are termed target-engagement biomarkers. Target-engagement biomarkers help one understand how well a drug is acting on a target. These biomarkers undergo advanced validation but would not be subject to qualification assessments.

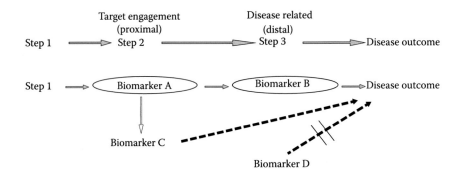

FIGURE 11.1
Target-engagement markers (biomarker A) versus disease-related markers (biomarker B). Pathophysiology is typically a multistep process. A putative biomarker may be (i) involved in one of the steps of the pathophysiology of a disease outcome (biomarker A), (ii) related to, but not directly involved in, the pathophysiology of a disease outcome (biomarker C), or (iii) not involved in the pathophysiology of a disease outcome (biomarker D). (Courtesy of Nass, S.J. and H.L. Moses, *Cancer Biomarkers: The Promises and Challenges of Improving Detection and Treatment*, Committee on Developing Biomaker-Based Tools for Cancer Screening, Diagnosis and Treatment, Institute of Medicine of the National Academies, The National Academies Press, Washington, D.C., 2007.)

b. Biomarkers that are closer to the disease outcome are called disease-related biomarkers. These biomarkers are used to assess the effect of a particular drug on a disease. These biomarkers usually undergo both validation and qualification assessments.

Hemoglobin is a useful biomarker, although it is not directly related to pathophysiology. Higher blood glucose levels, for example, in diabetes, leads to the glycation of more hemoglobin. Blood levels of hemoglobin A1c serve as an excellent surrogate end point in diabetes drug trials, yet this biomarker has nothing to do with the diabetes disease process, that is, the glycation of hemoglobin itself has no effect on the health of the patient. Pharmaceutical companies also classify biomarkers depending on the purpose of their use (Table 11.2).

2. Studies based on gene expression profiling have lead to the identification of four molecular classes of breast cancer [7], which are as follows:

a. *Luminal cancers.* Almost all are ER-positive and express cytokeratins 8 and 18. These are subdivided into "luminal A," which is mostly histologically low grade, and "luminal B," which tend to be of high grade with a worse prognosis.

b. *HER-2–positive cancers.* Characterized by the amplification and overexpression of the ERBB2 gene. They do not express hormone receptors and are of poor prognosis.

c. *Basal-like breast cancers.* These are triple-negative tumors [with ER(–), PR(–) and HER-2(–)] that show the expression of cytokeratins of the basal epithelial layer (CK 5/6, CK 17) with a poor prognosis.

d. *Normal-like breast cancers.* Recently published studies have used five Immuno-HistoChemical (IHC) surrogate markers (ER, PR, HER-2, CK5/6, and EGFR) for molecular class distinction, with luminal tumors being categorized by hormone receptor positivity, HER-2 expression (a feature of HER-2 tumors), and CK5/6 or EGFR indicative of basal-like tumors [8].

TABLE 11.2

Biomarker Types and Their Respective Definitions

Biomarker Types	Definition
Characterization	Known or established biomarker that often aids drug development decision making
Demonstration	A probable or emerging biomarker
Disease related	Used to assess the effect of a particular drug on a specific disease process
Dose stratifier	An indicator of the optimal dose of a specific drug for a specific patient
Early compound screening	Biomarker used early in drug development to detect likely effective drug candidates, that is, those that affect a specific drug target
Early response indicator	Biomarker that objectively indicates early in treatment whether the patient is responding to the treatment; for example, PET imaging of tumor size
Exploratory	Used to generate hypotheses; a research and development tool
Partial surrogate end point	Indicator of the effectiveness of treatment in early (phase I/II) clinical trials. Improvement of a partial surrogate end point is necessary for, but not sufficient to, ensure improvement of the primary clinical end point of interest. Partial surrogate end points serve as indicators of whether to continue the clinical testing of new drugs and progress to phase III trials
Patient classifier	Marker that classifies patients by disease subset
Pharmacodynamic	Marker that indicates drug activity and informs dose and schedule selection of a drug
Relapse risk stratifier	Indicator of the degree of risk for relapse after initial therapy
Response predictor	A measurement made before treatment to predict whether a particular treatment is likely to be beneficial
Risk management	Marker for patients or subgroups with high probability of experiencing adverse effects from their treatment, such as a marker for a drug metabolism subset
Risk stratifier	Indicator of the probability of an event (e.g. metastasis) or time to the event
Surrogate end point	An outcome measure that is thought to correlate with the primary clinical end point (outcome) of interest, and is used in place of the primary end point to determine whether the treatment is working
Target-engagement	Indicator of how well a drug is acting on a target
Tumor progression indicator	A measurement that provides early detection of tumor progression after treatment; for example, an increase in prostate-specific antigen levels can indicate progression of prostate cancer

Source: Courtesy of Nass, S.J. and H.L. Moses, *Cancer Biomarkers: The Promises and Challenges of Improving Detection and Treatment*, Committee on Developing Biomaker-Based Tools for Cancer Screening, Diagnosis and Treatment, Institute of Medicine of the National Academies, The National Academies Press, Washington, D.C., 2007.

11.5 Biomarker Types

Biomarkers of all types have been used by generations of epidemiologists, physicians, and scientists to study human diseases such as cancer. Recently, a large number of new biomarkers have been identified and are now in the pipeline for evaluation. The latest breast cancer biomarker report (up to May 2010) presents 200 utility-based breast cancer markers. The biomarkers include proteins, peptides, metabolites, genes and mutations, epigenetic species, and other molecules. These biomolecules cover different stages of cancer including diagnosis and staging (e.g. early detection), metastasis (e.g. to lymph nodes and bone), therapy directing (e.g. single and multiple drugs), response to therapy (e.g. favorable, resistance), prognosis (e.g. of relapse), potential drug targets and diagnostic opportunities (e.g. novel proteins upregulated in breast cancer tissues or increased in the circulation), biomarker source (e.g. blood, tissue, and urine), type (e.g. protein or gene), and function (e.g. of gene or protein).

11.6 Biomarkers of Clinical Importance

The number of molecules upregulated in breast cancer is used as a prognostic and predictive marker. Simultaneously, their overexpression makes them suitable targets for therapy. The distribution of these markers is highly varied in breast cancer cells. For clinical purposes, the American Society of Clinical Oncology (ASCO) recommends thirteen categories of breast tumor biomarkers, out of which six are new (Table 11.3). The following categories showed evidence of clinical utility and were recommended for use in practice: Soluble/Secreted form of MUC1 such as CA 15-3 & CA 27.29, carcinoembryonic antigen, ER, PR, HER-2, urokinase plasminogen activator, plasminogen activator inhibitor 1, and certain multiparameter gene expression assays.

The following categories demonstrated insufficient evidence to support routine use in clinical practice: DNA/ploidy by flow cytometry, p53, cathepsin D, cyclin E, proteomics, certain multiparameter assays, detection of bone marrow micrometastases, and circulating tumor cells.

11.6.1 Prognostic Biomarkers

Prognostic biomarkers help in making the decision of whom and how aggressively to treat. They provide the tools to distinguish "good outcome" tumors from "poor outcome" tumors. Traditional prognostic markers include the axillary lymph node status, tumor size, and nuclear and histologic grade. Unfortunately, the histologic information is clearly

TABLE 11.3

Molecular Biomarkers and Their Occurrence in Breast Cancer

Biomarker	Occurrence (%)
ER and PR	70–80
VEGFR	64 (invasive breast cancer)
EGFR	17
HER-2/neu	25
p53	25–30
p21	65
BRCA1 and BRCA2	2–4
Bcl2	80
Topoisomerase II α	>50
Caspase-3	75
TIP30/CC3	82
Ki67	56
uPA and PAI-1	46

not sufficient to accurately assess individual risk, therefore, a large number of molecular markers have been studied to determine their ability to predict prognosis or response to therapy (or both). Molecular markers routinely used to make treatment decisions in patients with early-stage breast cancer include markers of proliferation like Ki-67, mitosin, and hormone receptors such as growth factor and steroid receptors. Prognostic markers not used routinely include urokinase-like plasminogen activator/plasminogen activator inhibitor, cyclin D_1 and cyclin E, cathepsin D, p53, p21, proapoptotic and antiapoptotic factors, and BRCA1 and BRCA2. The estrogen, progesterone, and HER-2/neu receptor markers are recommended and incorporated into standard clinical practices. Some of the clinically detectable prognostic factors and techniques of their detection are discussed.

11.6.1.1 Clinicopathological Biomarkers

Most cancers are classified according to the type and extent of growth and spread in the body. Cancer classification has two essential parts: cancer staging and cancer grading. Cancer staging describes the tumor size and where it has spread (i.e., locally spread to the lymph nodes and distantly spread or metastasis), whereas cancer grading is based on the appearance and behavioral characteristics of the cancer cells. In general, cancer grading describes how the cancer might develop in the future. Both cancer staging and grading describe the type of cancer and the way it grows and spreads in the body. Pathological features, such as tumor size, histological subtype and grade, lymph node metastases, and lymphovascular invasion, derived from careful histological analysis of primary breast cancer samples, are good patient prognostic biomarkers. Tumor-node-metastasis (TNM) staging system integrates these features into cancer stages for early breast cancer detection [9]. The TNM staging system is described in Tables 11.4 and 11.5.

11.6.1.2 Receptor Biomarkers

Breast cancer is often characterized by higher expression of ER/PR and EGFR. It is also found that higher expression of vascular endothelial growth factor receptor-2 (VEGFR-2) is associated with metastasis and poor survival outcomes in patients with breast cancer. The overexpression of either growth factors or their receptors leads to abnormal signaling pathways that contribute to the progression, invasion, and malignant behavior of the tumor. The molecular markers currently used to predict treatment response in both early and metastatic breast cancer include ER, PR, and HER-2/neu receptor.

11.6.1.2.1 Estrogen Receptor

There are two types of ERs: ER-α and ER-β. Both are expressed separately but possess approximately 95% and 60% homology in their DNA and ligand binding domains, respectively, although less than 25% homology exists at the amino terminus. Both forms of receptors bind to the same DNA-responsive elements and possess similar ligand-binding characteristics.

ER-α is overexpressed in approximately 80% of breast cancers [10]. Estrogen binding to ER leads to a signaling cascade and finally initiates cell proliferation. ER-α is undoubtedly the most important biomarker in breast cancer because it provides the index for sensitivity to hormonal treatment. Generally, the ER in patients with ER(+) breast cancer is targeted with endocrine therapy. However, 40% of ER(+) patients do not respond to endocrine therapy. Multiple clinical studies have demonstrated that patients with ER(−) breast cancer are more likely to achieve a pathologically complete response with neoadjuvant chemotherapy

TABLE 11.4

TNM Classification for Breast Cancer (p = pathologically determined)

T (tumor size)	T_X	Primary tumor cannot be assessed
	T_0	No evidence of primary tumor
	T_{is}	Carcinoma *in situ*; intraductal carcinoma, lobular carcinoma *in situ*, or Paget's disease of the nipple with no associated tumor
	T_1	Tumor ≤2.0 cm at the greatest dimension T_{1mic}: Microinvasion ≤0.1 cm at the greatest dimension T_{1a}: Tumor >0.1 but ≥0.5 cm at the greatest dimension T_{1b}: Tumor >0.5 cm but ≥1.0 cm at the greatest dimension T_{1c}: Tumor >1.0 cm but ≥2.0 cm at the greatest dimension
	T_2	Tumor >2.0 cm but ≥5.0 cm at the greatest dimension
	T_3	Tumor >5.0 cm at the greatest dimension
	T_4	Tumor of any size with direct extension to T_{4a}: Extension to chest wall T_{4b}: Edema (including peau d'orange) or ulceration of the skin of the breast or satellite skin nodules confined to the same breast T_{4c}: Both of the above (T_{4a} and T_{4b}) T_{4d}: Inflammatory carcinoma
N (nodes)	N_X	Regional lymph nodes cannot be assessed (e.g. previously removed)
	N_0	No regional lymph node metastasis
	N_1	Metastasis to movable ipsilateral axillary lymph node(s) pN_{1a}: Only micrometastasis (≥0.2 cm) pN_{1b}: Metastasis to lymph node(s), any >0.2 cm pN_{1bi}: Metastasis in one to three lymph nodes, any >0.2 cm and all <2.0 cm at the greatest dimension pN_{1bii}: Metastasis to four or more lymph nodes, any >0.2 cm and all <2.0 cm at the greatest dimension pN_{1biii}: Extension of tumor beyond the capsule of a lymph node metastasis <2.0 cm at the greatest dimension pN_{1biv}: Metastasis to a lymph node ≥2.0 cm at the greatest dimension
	N_2	Metastasis to ipsilateral axillary lymph node(s) fixed to each other or to other structures
	N_3	Metastasis to ipsilateral internal mammary lymph node(s)
M (distant metastasis)	M_X	Presence of distant metastasis cannot be assessed
	M_0	No distant metastasis
	M_1	Distant metastasis present (includes metastasis to ipsilateral supraclavicular lymph nodes)

compared with ER(+) patients, with pathologically complete response rates of 7% to 8% versus 21% to 33%, respectively (Table 11.6) [11].

Clinically, the overexpression of ER is detected with ligand-binding assays such as dextran-coated charcoal assay. Ligand-binding assays for ER are technically challenging and expensive, require radioactive reagents and relatively large amounts of fresh-frozen tissue, and are insensitive and nonspecific in accounting for differences in the cellular composition of samples such as those with a low tumor cellularity or contaminating benign cells that might be ER positive. To avoid some of the shortcomings of ligand-binding assays, nowadays, IHC-based detection is used for ER status assessment. Although the detection of ER is easy, the most problematic aspect is the use of a wide variety of techniques for scoring and interpreting results arbitrarily rather than based on clinically calibrated definitions of ER positivity.

TABLE 11.5

Stages of Breast Cancer and TNM Staging System

Breast Cancer Stages	Tissue Involvement or Metastasis
Stage 0	T_{is}, N_0, M_0
Stage I	T_1, N_0, M_0
Stage IIA	T_0, N_1, M_0
	T_1, N_1, M_0
	T_2, N_0, M_0
Stage III	T_2, N_2, M_0
	T_3, N_0, M_0
	T_3, N_1, M_0
	T_3, N_2, M_0
	T_3, N_3, M_0
Stage IV	Any T, any N, M1; any breast cancer showing distant metastasis

TABLE 11.6

Hormone Receptor Frequency in Breast Cancer and Their Response to Hormone Therapy

Sr. No.	Tumor Phenotype	Phenotype Frequency (%)	Response to Hormonal Therapy (%)
1.	ER+/PR+	41	75–80
2.	ER+/PR−	30	20–30
3.	ER−/PR+	2	40–45
4.	ER−/PR−	27	<10

11.6.1.2.2 Progesterone Receptor

The PR is a member of the nuclear receptor superfamily and exists in two isoforms in most rodents and humans, PR-α and PR-β. PR-α is a truncated form of PR-β and lacks 164 amino acids at the amino terminus. Both isoforms are transcribed from the same gene under the control of different promoters. The PR receptor specifically regulates the expression of target genes in response to the hormonal stimulus. On binding with progesterone, the PR undergoes a series of events, including conformational changes, dissociation from heat shock protein complexes, dimerization, phosphorylation, and nuclear translocation, which enables its binding to progesterone-response elements within the regulatory regions of target genes. The binding of PR to the progesterone response elements leads to the upregulation of target gene. Both PR-A and PR-B are different functionally, as PR-A acts as a dominant repressor of both PR-B and ER in a promoter and cell type–specific manner. The expression of PR is strongly dependent on ER presence. Breast tumors expressing PR but not the ER are uncommon and represent only 1% cases [12]. As a result, patients with metastatic breast cancer respond better to endocrine therapy when expressing both PR and ER (Table 11.6).

Like ER, the upregulation of PR is detected routinely by IHC staining method. IHC measurements of PR, although not as important clinically as ER, can provide useful information and should also be performed on all samples of invasive breast cancer or ductal carcinoma *in situ*.

The recently released guidelines of the ASCO and the College of American Pathologists recommends ER and PR testing in all newly diagnosed cases as well as in any local or distant recurrence whenever appropriate [13].

11.6.1.2.3 HER-2/neu/c-erb-2 Receptor

HER-2 is a member of subclass 1 of the receptor tyrosine kinase superfamily; other members include EGFR (HER-1), HER-3, and HER-4. These receptors are made up of extracellular ligand-binding domains, membrane-spanning regions, and a cytoplasmic domain with tyrosine kinase activity. HER-2 is considered an orphan receptor because no binding ligand has been directly identified for this receptor. HER-2 heteromerizes with other HER family members and initiates intracellular signaling via mitogen-activated kinase, phosphatidylinositol-3-kinase, and phospholipase-C pathways. Overexpression of HER-2 leads to mitogenesis, malignant transformation, invasion, and metastasis. The HER-2 gene was found to be overexpressed in 10% to 34% cases of invasive breast cancer [14]. In human breast cancer, HER-2 gene amplification leads to an increased gene copy number of two copies of genes per cell up to 100 copies.

HER-2/neu status is assessed with both morphology (IHC, fluorescence *in situ* hybridization, and chromatin *in situ* hybridization) and molecular-based techniques (PCR and serum enzyme-linked immunosorbent assay). In contrast to IHC, fluorescence *in situ* hybridization provides a more objective scoring system. IHC staining is primarily used for HER-2 overexpression detection as a result of low cost, wide availability, and easy preservation of stained slides. For *in situ* hybridization techniques, two locus-specific probes, one for HER-2 and the other for the centromere of chromosome 17 (CEP17) are applied and the ratio of two is calculated. A ratio higher than 2.2 is considered unambiguously positive. Two commercially available HER-2/neu IHC kits, Dako Herceptest (DakoCytomation, Glostrup, Copenhagen, Denmark) and Ventana Pathway (Ventana Medical Systems, Inc., Tucson, Arizona) are in use to overcome the subjective errors of the IHC test and also to determine patients' eligibility to receive the anti–HER-2/neu therapeutic antibody "trastuzumab" (Herceptin; Genentech, South San Francisco, California). Real-time polymerase chain reaction (RT-PCR) is predominantly used to detect HER-2/neu mRNA in peripheral blood and bone marrow samples. An enzyme-linked immunosorbent assay is performed on the tumor cytosol of fresh tissues, but this is limited in use due to the small size of the breast cancer tissues available for screening procedures.

HER-2 amplification status is also a highly predictive biomarker. Patients with early and advanced breast cancer overexpressing HER-2 are more likely to develop resistance, suffer, or tend to have a shorter overall survival after treatment with hormonal therapy. The value of HER-2 in selecting for hormone resistance has not been validated in a level I evidence study. The ASCO states that "HER-2 may identify patients, who particularly benefit from anthracycline-based adjuvant therapy, but levels of HER-2 should not be used to exclude patients from this type of treatment." ASCO guidelines specify that "the use of HER-2 to decide whether to prescribe endocrine therapy either in the adjuvant or metastatic setting is not recommended."

11.6.1.2.4 Vascular Endothelial Growth Factor Receptor

A member of the receptor tyrosine kinase family, VEGFR consists of VEGFR-1, VEGFR-2, VEGFR-3, and a soluble form of VEGFR-1 (an intensive negative counterpart of the VEGFR family of receptors). There are common as well as unique ligands for VEGFRs. VEGF-A and VEGF-B bind to VEGFR-1. VEGF-A, VEGF-C, and VEGF-D bind to VEGFR-2, whereas VEGF-C and VEGF-D bind to VEGFR-3. VEGFR-1 and VEGFR-2 are responsible for angiogenesis, whereas VEGFR-3 promotes lymphogenesis. VEGFR-1 and VEGFR-2 are 43.2%

similar in sequence and are also structurally similar. Upon VEGF ligand binding, VEGFR receptors dimerize and undergo transphosphorylation, an increase in the intrinsic catalytic activity and creation of binding sites on the receptor tyrosine kinases to recruit cytoplasmic signaling proteins that activate the ERK1/2 pathway and the PI3K/Akt pathways. VEGFs/VEGFRs signaling are responsible for the survival and aggressive nature of tumor cells. Breast cancer cells mainly express VEGF-A, VEGF-B, VEGF-C, and their receptors VEGFR-A, VEGFR-B, and neuropilin (NP-1 and NP-2). VEGFR-1 is associated with poor prognosis in node-negative tumors with high risk of metastasis and relapse.

11.6.1.2.5 Epidermal Growth Factor Receptor

EGFR (c-erb-B-1 and Her-1 in humans) is a member of the ErbB family of receptors, a subfamily of four closely related receptor tyrosine kinases: EGFR (ErbB-1), HER-2/c-neu (ErbB-2), Her3 (ErbB-3), and Her4 (ErbB-4). Mutations affecting EGFR expression or activity could result in cancer. Because there is no standard test for EGFR, it is not considered a prognostic factor for routine use.

11.6.1.2.6 Transforming Growth Factor (TGF)-α

TGF-α is an activating ligand for EGFR and is found to be associated with breast cancer recurrence and adverse prognosis [15].

11.6.1.2.7 Breast Cancer Oncogene

Breast cancer oncogenes "BRCA1 and BRCA2" are responsible for 5% to 10% of inherited breast cancer cases. BRCA1 is a multifunctional protein that has been implicated in many normal cellular functions such as DNA repair, transcriptional regulation, cell cycle checkpoint control, and ubiquitination. Women who carry a germ line mutation in BRCA1 have a cumulative lifetime risk of 50% to 85% of developing breast cancer [16]. BRCA1 mutated breast tumors develop at an early age and are generally ER, PR, and HER-2/neu negative. These tumors are poorly differentiated and have a poor prognosis. DNA sequencing is used to detect the mutated BRCA1 gene but the method is not foolproof. Therefore, new methods such as quantitative PCR, multiplex ligation–dependent probe amplification, and quantitative multiplex PCR of short fluorescent fragments are in trials for increasing the accuracy of BRCA1 gene mutation detection.

11.6.1.3 Multigene Parameter-Based Markers (Based on DNA Microarrays)

The development of high-throughput technology allows the measurement of thousands of RNA transcripts using DNA microarrays. In general, DNA microarrays consists of multiple rows of oligonucleotides or cDNA lined up in an orderly manner on a small glass slide (usually 1–2 cm square). Each spot on the slide contains a unique DNA fragment with a known sequence. DNA microarrays are used to find differences in the expression of hundreds of genes, but the significance of most of those differences is unknown. DNA microarray–based genomic tests are developed on genomic profiling with the expectation that they might predict clinical outcome better than standard pathological and molecular markers. The various commercially available microarray based tests are as follows:

11.6.1.3.1 Oncotype DX (21-Gene Expression Assay; Genomic Health, Inc., Redwood City, California)

The Oncotype DX gene expression assay is intended for use by women with early-stage (stage I or II), node-negative, ER(+) invasive breast cancer. The test was designed to predict the risk of distant recurrences in patients with ER(+) early breast cancer receiving tamoxifen. The test also provides quantitative scores of ER, PR, and HER-2 for planning better treatment.

Oncotype DX gene expression assay is derived from 250 tumor-related genes selected from published literature, genomic databases, and experiments based on DNA microarray carried out on fresh-frozen tissues. Data were analyzed from three independent clinical studies involving 447 patients with breast cancer to test the relationship between the expression of 250 candidate genes and the recurrence of breast cancer. Three independent studies were used to select 21 genes for profiling. The test is based on the RT-PCR expression profile of 21 genes. The 21 genes include 16 breast cancer and five reference genes (Table 11.7). The genes are grouped by function: proliferation, invasion, hormone receptors (estrogen and progesterone), and growth factors.

A recurrence score (RS) is computed with a mathematical algorithm, which was developed and established using samples of the tamoxifen arm of the NSABP-B20 and B-1 trials [17]. The algorithm calculates the RS in the following manner:

RS = [(0.47) × HER-2 group score] + [(−0.34) × ER group score] + [(1.04) × Proliferation group score] + [(0.1) × Invasion group] + [(0.05) × CD68] + [(−0.08) × GSTM1] + [(−0.07) × BAG1]

The RS has been used to identify three groups with low (<18%), intermediate (18%–31%), and high (>31%) scores. These three groups have 6.8%, 18.6%, and 30.5% risk of distant recurrence in 10 years, respectively, among "tamoxifen"-treated patients. The RS predicts the overall survival and distant recurrence independent of age and tumor size. The U.S. Food and Drug Administration does not approve the Oncotype DX test whereas ASCO endorses it.

11.6.1.3.2 MammaPrint (70-Gene Expression Test; Agendia, Irvine, California)

MammaPrint is a DNA microarray–based prognostic, *in vitro* multivariate index assay for predicting the risk of distant metastasis in patients with breast cancer independent of ER status and any prior treatment. MammaPrint is based on the well-known "Amsterdam 70-gene breast cancer gene signature" from a group of 78 patients. These patients were selected retrospectively having node-negative breast cancer of less than 5 cm, no adjuvant chemotherapy, and younger than 55 years [18].

Out of the 70 genes used in MammaPrint (Table 11.8), 15 genes are well known to be involved in early embryonic development, two genes are known as development-related transcription factors, and three genes (TGFB3, FGF18, WISP1) represent well-characterized important epithelial–mesenchymal transitions mediating TGF-β, FGF, and Wnt family

TABLE 11.7

The 21 Genes Profiled in Oncotype DX Gene Expression Assay

Breast Cancer and Reference Genes Family	Genes Used for Profiling
Proliferation and hormone receptors	Ki-67, STK15, Survivin, ER, PR, Bl2
HER-2	GRB7, HER-2, Cyclin B1, MYBL2, Bcl2, SCUBE2
Invasion	Stromolysin 3, Cathepsin L2
GSTM1, CD68, BAG1	
Reference	β-Actin, GAPDH, RPLPO, GUS, TFRC

TABLE 11.8

The 70 Gene Signatures Used for Profiling in MammaPrint

Cellular Process or Molecule	Number of Genes
Metabolism	7
Cell cycle and DNA replication	12
Extracellular matrix adhesion and remodeling	5
Growth, proliferation, transformation, and cell death	17
Signal transduction and intracellular transport	3
Growth factor	7
Motility related	5
Intracellular hydrolase	1
Immune response	1
Neuropeptide	1
Predicted transmembrane protein	2
Predicted DNA-binding proteins	5
Unknown function	4
Metadherin or MTDH	1

proteins. The recently discovered metadherin or MTDH (associated with poor prognosis in ~40% of high-risk breast cancers) is also included in the 70-gene profiling. This test determines gene expression at the transcription level, which is in contrast with other studies in which the protein level is measured, for example, HER-2/neu and cyclin D1.

To assess the global gene expression pattern of a tumor sample, mRNA is extracted from the sample and transcribed into cRNA. Then, cRNA of tumor and reference labeled with a fluorescent dye before hybridization to MammaPrint (Agendia uses customized microarrays manufactured by Agilent). Each microarray contains three identical sets of the 70 genes to be analyzed. This enables three independent measurements of the 70-gene profile, increasing confidence in the test results. In addition, the customized arrays contain several hundred carefully selected normalization genes. Negative control genes are also present on each microarray; these are DNA sequences to which no human mRNA can bind and are used to monitor various technical aspects of the microarray process. The biological function of genes in the MammaPrint signature is annotated by using seven different algorithms and three associated databases and then interpreting these within the biological context. The values of the multiple variables are combined using the bioinformatics algorithm to produce patient-specific scores. The score determines low or high risk of recurrence according to the gene signature and clinical risk classifications (Figure 11.2).

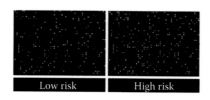

Low risk High risk

FIGURE 11.2
MammaPrint-based prediction of recurrence risk in patients with breast cancer without distant metastasis.

Patients are in the low-risk group if their 5-year probability of survival without distant metastasis is more than 90%. Patients were assigned to the clinicopathologic low-risk group if their 10-year probability of survival was more than 88% [for ER(+) patients] or more than 92% [for ER(–) patients] as estimated by Adjuvant software (Adjuvant Inc., San Antonio, Texas). MammaPrint has a 98.9% degree of accuracy in classifying patients as low risk or high risk and a technical reproducibility of 98.5%. In MammaPrint's low-risk patients, the risk can be reduced with endocrine therapy alone (e.g. tamoxifen), whereas for the management of high-risk patients with additional risk variables, more aggressive therapies including chemotherapy may be recommended. MammaPrint requires fresh tissue in contrast to the Oncotype DX assay, which works on both fresh and fixed tissues. The U.S. Food and Drug Administration approved the expensive "MammaPrint test" in February 2007 for patients with lymph node–negative breast cancer with tumor sizes of less than 5 cm and younger than 61 years.

11.6.1.3.3 Breast Cancer Index

This is a combination of two powerful indexes, HI and molecular grade index:

1. HI measures the ratio of two genes, HOXB13 to IL17BR (H/I; Aviara Dx) that stratifies ER(+) breast cancer according to low or high risk for recurrence and is predictive of benefit from endocrine therapy. A high mRNA expression ratio was associated with a high risk of recurrence in tamoxifen-treated patients.

2. Molecular grade index is an RT-PCR assay consisting of five proliferation-associated genes and identifies a subgroup of ER(+) patients with a worse outcome despite endocrine therapy. When combined with HI, it improves the accuracy of the test and helps in calculating the risk of recurrence and viable therapeutic options.

11.6.1.3.4 Genomic Grade Index

A genomic-grade signature consists of a 97-gene signature, which is able to discriminate grade 2 tumors into low and high genomic grade subgroups with outcomes comparable to histological low-grade and high-grade tumors. This signature enhances the importance of proliferation genes in ER(+) breast cancer. This signature was found to be better associated with outcome and prediction of relapse under endocrine treatment [19].

11.6.1.4 PCR-Based Markers

The main methods of breast cancer detection are mammography and tissue biopsies. Mammography may not detect small tumors and is often unsatisfactory for dense breast tissue. Tissue biopsies can be difficult due to small size of lump, lack of available medical facilities, and the patient's reluctance to undergo an invasive procedure. To address these empirical issues, diagnostic companies developed RT-PCR based tests for the detection of breast cancer in peripheral blood with minimum invasion. "BCtect" and copy number variation detection assays use the TaqMan RT-PCR assay (Applied Biosystems, Carlsbad, California).

Microarrays rely on the hybridization of the target to an array of probes anchored to a solid substrate. Several washing steps are required to observe the hybridization signals. Microarrays allow the detection of simultaneous expression of thousands of different genes but the range is limited to genes expressing at high levels. In contrast, TaqMan array relies on the hybridization of two primers and fluorescently labeled probe to the target followed by real-time detection of probe cleavage during standard PCR amplification. As a result,

TaqMan array is able to detect and quantitate the expression of all target genes. Additionally, microarrays are expensive and require more specialized equipment and efforts.

11.6.1.4.1 BCtect (DiaGenic, Oslo, Norway)

BCtect is a prognostic test for early stage detection of breast cancer. The test is based on the characteristic gene expression signature in the peripheral blood of women with breast cancer. RNA is extracted from the stored blood samples and converted to cDNA using a Food and Drug Administration approved protocol. The gene expression measurements are performed on 96 TaqMan Gene Expression Assays (Applied Biosystems) preloaded multiple times on a 384-well microfluidic card (array). Each array can be used to analyze the blood of four patients before data analysis by DiaGenic's proprietary statistical package. The test suffers from several limitations, including:

- Requires clinical evidence to confirm the presence and absence of breast cancer
- Intended for use in adult women
- Not for use with pregnant women
- Has not been documented to discriminate between breast cancer and other types of cancer
- Show reduced test specificity with acute inflammatory conditions

BCtect has been approved in India for the early detection of breast cancer.

11.6.1.4.2 Human Copy Number Variations

A copy number variation is defined as a deletion or duplication event involving less than 1 kb of DNA. In the human genome, approximately 20,000 copy number variations have been identified, which corresponds to more than 6000 unique regions/loci. Assays based on the analysis of copy number variations and smaller genomic regions in the human genome detect the target gene or genomic sequence of interest. The reference assay detects a sequence that is known to be present in two copies in a diploid genome. The results are specific, reproducible, and easy to interpret.

11.6.1.5 Cathepsin D

Cathepsin D is an estrogen-regulated lysosomal aspartyl protease, which facilitates cancer cell migration and promotes stromal invasion. Increased Cathepsin D levels act as an independent predictor of survival in breast cancer. Current immunoassay limitations on fresh tissue cytosolic preparations have resulted in a loss of interest in the use of Cathepsin D assessment as a prognostic factor and to guide therapy in breast cancer, although it is still performed in Europe.

11.6.1.6 Emerging Prognostic Biomarkers

The precise diagnosis of breast cancer will continue to challenge researchers as well as the diagnostic industry. As a result of continuous digging of cellular pathways, processes, and the availability of new and advanced technologies leading to the detection of even microquantities of a biomarker not only at the site of cancer but also peripherally. Therefore, the list of breast cancer biomarkers is constantly increasing and some of the emerging molecules will prove to be of clinical relevance.

11.6.1.6.1 Cell Cycle-Associated Markers

Traditional prognostic factors (lymph node status, tumor size) are not sufficiently accurate; therefore, better predictors of high risk and treatment response are needed. Breast cancer prognosticators are directly or indirectly related to proliferation, which strongly correlates with poor prognosis, irrespective of the methodology used. Some of the mitosis related prognosticators are promising alternatives, which require further development.

1. Ki67

   Ki stands for University of Kiel, Germany, and 67 refers to the number of clones on the 96-well plate [20]. Ki67 is a nuclear nonhistone protein, which is universally expressed among proliferating cells. It is expressed in G_1, S, and G_2 phases of the cell cycle with a peak during mitosis [21]. Ki67 have low proliferative activity in ER(+) tumors. Postneoadjuvant chemotherapy measurement of Ki67 is a strong predictor for recurrence free and overall survival. However, a high pretreatment score is associated with a good chance to achieve a pathologically complete response and this is a predictor of long-term outcome in these patients [22]. High levels of Ki67 in patients with cancer indicate an aggressive tumor. In invasive breast cancer, this marker is used to stratify good and poor prognostic categories. Although recent clinical studies indicate an association between Ki67 positivity and shorter overall survival, Ki67 staining is still not recommended as a prognostic marker for routine use [23].

2. Cyclin D1 (PRAD 1 or bcl-1)

   Cyclin D1 is localized on chromosome 11q13 and has been identified in 20% of clinical breast cancers [24]. Cyclin D1 is one of the most commonly overexpressed oncogenes in primary ductal breast carcinomas (45%–50%). After induction, cyclin D1 binds to cyclin-dependent kinases leading to the phosphorylation of various substrates including retinoblastoma proteins. This contributes the regulation of G_1-S phase transition in the cell cycle. Cyclin D1 has a strong correlation with ER and PR expression levels. Cyclin D1 has predictive value in hormone receptor–positive patients as its amplification and overexpression is a predictor of poor response to antiestrogen treatments [25].

3. Cyclin E

   Cyclin E is a positive regulator of cell cycle in the G_1 phase. The various isoforms of cyclin E have a correlation with increasing stage and grade of breast cancer. The role of cyclin E in the cell cycle suggests that increased levels may alter the response to chemotherapy and endocrine therapy, but these results need validation.

4. Oncogenes

   The c-myc gene is amplified in approximately 16% of breast cancers. C-myc is a 439–amino acid nuclear binding protein directly involved in cell differentiation, replication, and apoptosis. C-myc was found to be associated with decreased disease-free patient survival.

p53 is a cell cycle regulator having altered expression with adverse outcomes in breast cancer. p53 is mutated in approximately 50% of human cancers, although its mutation rate is lower in breast cancer. In general, breast carcinomas with p53 mutations are consistently associated high mitotic index, high cell proliferation rate and negative assays for ER and PR, and variable associations with the amplification of oncogenes such as HER-2/neu and c-myc. Different studies of metastatic disease have implicated p53 mutations with resistance to hormonal, adjuvant, neoadjuvant, and combination chemotherapy.

p21 (WAF1/Cip1) is an inhibitor of cyclin-dependent kinases and serves as a critical downstream effector in the p53-specific pathway of cell growth control. The relationship between p21-altered expression and breast cancer needs to be established clearly. p27 inactivates cyclin-dependent kinases and its low expression correlated with poor prognosis of breast cancer. S phase kinase–associated protein "Skp2" is involved in the degradation of p27 and is overexpressed in ER(–)/HER-2(–) breast carcinomas.

11.6.1.6.2 Circulating Tumor Cells

The circulating tumor cells were first reported in 1869 but their molecular detection was only recently facilitated in the blood of patients with cancer [26]. In the future, this biomarker can be used to predict therapy efficacy and resistance after initial exposure to therapy and may be beneficial in monitoring response to treatment. The identification of certain specific biomarkers in circulating tumor cells may help in further deciding the appropriateness of therapy. Although circulating tumor cells have been reported to be significant in the detection of cancer, the ASCO tumor marker group concluded that treatment decisions should not be influenced by circulating tumor cell counts.

11.6.1.6.3 Circulating Cell-Free DNA (cfDNA)

The genetic aberration of an individual's tumor, such as PIK3CA mutations in cell-free DNA of nuclear or mitochondrial origin, may be helpful in improving breast cancer management.

11.6.2 Predictive (or Response) Biomarkers

Predictive (or response) biomarkers determine the probability of a patient's benefit from a particular treatment. Currently, the recommended predictive biomarkers are ER and PR for selecting endocrine-sensitive breast cancers and HER-2 for identifying patients with breast cancer with metastatic disease. In addition to prediction factors, these three biomarkers also have strong prognostic value. Predictive biomarkers are also classified as the following.

11.6.2.1 Mutation Based

These biomarkers are characterized by a mutation which changes the biomolecular sensitivity toward a drug, for example, mutations in the genetic region encoding the kinase domain of EGFR predict the sensitivity of lung tumors to erlotinib or gefitinib-6. Similarly, in glioblastoma multiforme, distinct mutations in the genetic region encoding the extracellular domain of EGFR predict sensitivity to EGFR inhibitors but only in cases in which the tumor-suppressor protein phosphatase and tensin homologue is also intact.

11.6.2.2 Cytogenetic Analysis Based

These predictive biomarkers are of limited value and are normally disease specific. Although no such biomarker is used for breast cancer analysis, they are routinely used for leukemia assessment. Genotype-based cytogenetic analysis can be used to gain additional predictive information, for example, in patients with chronic myeloid leukemia, who develop resistance to imatinib mesylate, distinct mutations in the genetic region encoding the kinase domain of BCR-ABL predict differential sensitivity to the newer ABL inhibitors— dasatinib and nilotinib.

11.6.2.3 Gene Based

Three well-known predictive markers fall in this category: ER, PR, and HER-2/neu. These markers contribute as a result of gene amplification, translocation, or fusion. In patients with breast cancer in which the gene HER-2/neu (ERBB2) is amplified (i.e., extra copies are present), the patient responds to "trastuzumab" (Herceptin). However, in breast tumors that express both ER and HER-2, the patient responds to "tamoxifen" treatment. Similarly, patients with leukemia having promyelocytic leukemia–retinoic acid receptor-α translocation respond to all-trans retinoic acid, whereas patients with Philadelphia chromosome (containing BCR-ABL fusion genes) respond to "imatinib mesylate" (Gleevec or Glivec).

11.6.3 Promising Novel Biomarkers

The development of resistance against multiple drugs and their resulting ineffectiveness in breast cancer led to the search for new prognostic and predictive biomarkers. The number of potential markers is constantly increasing due to proteomic and genomic studies. Biomarkers poor sensitivity or specificity can be resolved by studying them simultaneously. The combined use of biomarkers will help in better prognosis and increasing sensitivity for chemotherapeutic agents.

Recently, a number of promising biomarkers have been identified and are under evaluation for their potential commercialization. These biomarkers include survivin, cadherin, epithelial cell adhesion molecule, integrins, cathepsin D, urokinase plasminogen activator and its receptor, plasminogen activators, matrix metalloproteins, Bcl-2 family members, transcription factors, epithelial mucins (Muc1 and Muc2), cell cycle regulators such as cyclin D1 and D2, telomerase, etc. Some of the newly developed and most promising novel breast cancer biomarkers include:

a. Changes in DNA methylation pattern (i.e. epigenetic changes) in human genome affect gene expression level and are often found to be associated with disease states. In vertebrate genome, the cytosine residue of DNA is methylated in CpG dinucleotides in DNA. Seventy to eighty percent to of CpG sites in the human genome are methylated [27]. Unlike genetic changes, epigenetic modifications are reversible. CpG islands are generally unmethylated in normal adult tissues with rare exceptions. Hypermethylation of various gene promoters is a common feature of malignant cells and these changes can occur early in the progression process. The DNA methylation changes can be detected with various techniques such as DNA array-based differential methylation hybridization and a PCR-based technology called methylation-sensitive restriction fingerprinting.

b. Micro-RNAs (miRNAs) are short 21- to 26-base-long nucleotides, noncoding RNA molecules that regulate gene expression by Watson–Crick base pairing to target mRNA. Therefore, they are responsible for regulating the mRNA level and protein expression. The miRNAs act as oncogenes and tumor suppressor genes are called "oncomirs." miR-21, miR-155, and miR-29b-2 are upregulated, whereas miR-143, miR-145, miR-155, and miR-200 are downregulated miRNAs in breast cancer. Most commercially available technologies for global miRNA expression profiling are based on DNA microarrays, in which the probes are spotted onto a glass surface. Such probes, however, are not optimal for miRNA detection because of cross-hybridization and lack of specificity. Exiqon (Vedbaek, Denmark) developed Locked Nucleic Acid (LNA) enhanced miRCURY microarrays for detecting miRNAs. LNAs are a class of conformationally restricted nucleotide analogs. LNAs incorporation in oligonucleotides increases their affinity for cRNA or cDNA target.

c. Osteopontin is a phosphorylated glycoprotein found in all body fluids but its overexpression in breast and other cancers has led to a potential biomarker and antimetastatic agent. The metastasis gene "osteopontin" exists in three isoforms—osteopontin-a, osteopontin-b, and osteopontin-c. Osteopontin-a is used by the immune system to help cells move and migrate; therefore, it is a crucial protein for cancer metastasis. Little is known about osteopontin-b. Osteopontin-c is selectively expressed in invasive, but not in noninvasive breast tumor cell lines [28]. In IHC staining, osteopontin-c was found to detect a higher fraction of breast cancers compared with ER, PR, or HER-2. Osteopontin reliably predicted grades 2 to 3 breast cancer in conjunction with ER and Her-2.

d. Chronic exposure to chemotherapeutic agents led to the development of resistant breast cancer. The best known molecular mechanism responsible for this type of multidrug resistance is the overproduction of membrane proteins known as ATP-dependent efflux pumps [29]. These proteins are characterized by an ATP-binding cassette or domain and are thus known as the ABC superfamily of transporters or ATP-dependent transporters. The prototype member, ABCB1 (also known as P-glycoprotein, P-170, PGP, or MDR1), is a broad-spectrum multidrug efflux pump that possesses 12 transmembrane domains and two ATP-binding sites. ABCB1 is thought to play a role in extruding neutral and cationic toxins out of cells. Another transporter, ABCC1 (MRP-1), is similar to P-glycoprotein in structure except for an amino-terminal extension that contains five transmembrane domains, giving a total of 17 transmembrane sequences. Overall, 42% of the breast tumors overexpressing P-glycoprotein are associated with lack of response to treatment.

e. Surface-enhanced laser desorption/ionization analysis of nipple aspirate fluid revealed several candidate biomarkers in 75% to 80% cases [30].

f. Immunohistochemical staining of breast cancer–associated glycoproteins including B72.3, α-lactalbumin, and milk fat globule has been proposed for identifying breast cancer metastasis in almost 75% of cases [31].

g. Mammaglobin and maspin have demonstrated promise as additional markers for early stage detection and occult metastasis [32].

h. Neubauer and colleagues [33] used a proteomic approach to identify proteins that were differentially regulated by ER in breast tumors. Their observations indicate that ER(–) tumors have increased levels of progesterone receptor membrane component-1, a hormone receptor component and binding partner for P450 proteins.

Author's Note

This chapter was compiled personally by the author and has no bearing/relation with his official position at the Department of Biotechnology, Ministry of Science and Technology, New Delhi, India.

References

1. American Cancer Society. 2010. Explore Research, Cancer Facts & Figures, Breast Cancer Facts and Figures 2009–2010. Retrieved on May 24, 2010 from http://www.cancer.org/acs/groups/content/@nho/documents/document/f861009final90809pdf.pdf.
2. Atkinson, A.J. 2001. Biomarkers Definitions Working Group. Biomarkers and surrogate end points: Preferred definitions and conceptual framework. *Clinical and Pharmacology and Therapeutics* 69:89–95.
3. Ross, J.S., K. Gray, and R. Mosher. 2004. Molecular techniques in cancer diagnosis and management. In *Cancer Diagnostics: Current and Future Trends*, edited by Nakamura, R.M., W.W. Grody, J.T. Wu, and R.B. Nagle, 325–360. Totowa, NJ: Humana Press Inc.
4. De Bortoli, M., and N. Biglia. 2006. Gene expression profiling with DNA microarrays: Revolutionary tools to help diagnosis, prognosis, treatment guidance, and drug discovery. In *Biomarkers in Breast Cancer: Molecular Diagnostics for Predicting and Monitoring Therapeutic Effect*, edited by Gasparini, G., and D.F. Hayes, 47–61. Totowa, NJ: Humana Press Inc.
5. Cottingham, K. 2006. Speeding up biomarker discovery. *Journal of Protein Research* 5:1047–1048.
6. McShane, L.M. 2005. Reporting recommendations for tumor marker prognostic studies. *Journal of Clinical Oncology* 23:9067–9072.
7. Perou, C.M. 2000. Molecular portraits of human breast tumours. *Nature* 406:747–752.
8. Carey, L.A. 2006. Race, breast cancer subtypes, and survival in the Carolina Breast Cancer Study. *Journal of the American Medical Association* 295:2492–2502.
9. Greene, F.L. 2002. AJCC Cancer Staging Manual, 6th edition, Prepared by American Joint Committee on Cancer, American Cancer Society and Published by Springer-Verlag, NY, USA.
10. Arpino, G., H. Weiss, A.V. Lee, R. Schiff, S. De Placido, C.K. Osborne, and R.M. Ellledge. 2005. Estrogen receptor–positive progesterone receptor–negative breast cancer: Association with growth factor receptor expression and tamoxifen resistance. *Journal of the National Cancer Institute* 97:1254–1261.
11. Colleoni, M. 2004. Chemotherapy is more effective in patients with breast cancer not expressing steroid hormone receptors: A study of preoperative treatment. *Clinical and Cancer Research* 10:6622–6628.
12. Viale, G. 2007. Prognostic and predictive value of centrally reviewed expression of estrogen and progesterone receptors in a randomized trial comparing letrozole and tamoxifen adjuvant therapy for postmenopausal early breast cancer: BIG 1-98. *Journal of Clinical Oncology* 25:3846–3852.
13. Hammond, M.E. 2010. American Society of Clinical Oncology/College of American Pathologists guideline recommendations for immunohistochemical testing of estrogen and progesterone receptors in breast cancer. *Journal of Clinical Oncology* 28:2784–2795.
14. Ross, J.S., and J.A. Flectcher. 1999. HER-2/neu (c-erb-B2) gene and protein in breast cancer. *American Journal of Clinical Pathology* 112:S53–S67.
15. Castellani, R., E.W. Visscher, S. Wkyes, F.H. Sarkar, and J.D. Crissman. 1994. Interaction of transforming growth factor-α and epidermal growth factor receptor in breast carcinoma. An immunohistologic study. *Cancer* 73:344–349.
16. King, M.C., J.H. Marks, and J.B. Mandell. 2004. Breast and ovarian cancer risks due to inherited mutations in BRCA1 and BRCA2. *Science* 302:643–646.
17. Paik, S. 2004. A multigene assay to predict recurrence of tamoxifen-treated, node negative breast cancer. *New England Journal of Medicine* 351:2817–2826.
18. van't Veer, L.J. 2002. Gene expression profiling predicts clinical outcome of breast cancer. *Nature* 415:530–536.
19. Desmedt, C. 2009. The gene expression grade index: A potential predictor of relapse for endocrine-treated breast cancer patients in the BIG 1-98 trial. *BMC Medical Genomics* 2:40.
20. Gerdes, J. 1983. Production of a mouse monoclonal antibody reactive with a human nuclear antigen associated with cell proliferation. *International Journal of Cancer* 31:13–20.

21. Lopez, F. 1991. Modalities of synthesis of Ki67 antigen during the stimulation of lymphocytes. *Cytometry* 12:42–49.
22. Jones, R.L. 2009. The prognostic significance of Ki67 before and after neoadjuvant chemotherapy in breast cancer. *Breast Cancer Research Treatment* 116:53–68.
23. Stuart-Harris, R. 2008. Proliferation markers and survival in early breast cancer: A systematic review and meta-analysis of 85 studies in 32,825 patients. *Breast* 17:323–334.
24. Wolman, S.R., R.J. Pauley, A.N. Mohmed, P.J. Dawson, D.W. Visscher, and F.H. Sarkar. 1992. Genetic markers as prognostic indicators in breast cancer. *Cancer* 70:1765–1774.
25. Jirstrom, K. 2005. Adverse effect of adjuvant tamoxifen in premenopausal breast cancer with cyclin D1 gene amplification. *Cancer Research* 65:8009–8016.
26. Bird, A.P. 1995. Gene number, noise reduction and biological complexity. *Trends in Genetics* 11:94–100.
27. Smith, B. 1991. Detection of melanoma cells in peripheral blood by means of reverse transcriptase and polymerase chain reaction. *Lancet* 338:1227–1229.
28. Weber, G.F., S. Zawaideh, V.A. Kumar, M.J. Glimcher, H. Cantor, and S. Ashkar. 2002. Phosphorylation-dependent interaction of osteopontin with its receptors regulates macrophage migration and activation. *Journal of Leukocyte Biology* 72:752–761.
29. Leonard, G.D., T. Fojo, and S.E. Bates. 2003. The role of ABC transporters in clinical practice. *Oncologist* 8:411–424.
30. Sauter, E.R., W. Zhu, X.J. Fan, R.P. Wassell, I. Chervoneva, and G.C. Du Bois. 2002. Proteomic analysis of nipple aspirates fluid to detect biologic markers of breast cancer. *British Journal of Cancer* 86:1440–1443.
31. Lee, A.K., R.A. DeLllis, P.P. Rosen, T. Herbert-Stanton, K. Tallberg, C. Garcia, and H.J. Wolfe. 1994. A-lactalbumin as an immunohistochemical marker for metastatic breast carcinomas. *American Journal of Surgery and Pathology* 8:93–100.
32. Maass, N., K. Nagasaki, M. Zeibart, C.M. Undhenke, and W. Jonat. 2002. Expression and regulation of tumor suppressor gene maspin in breast cancer. *Clinical Breast Cancer* 3:281–287.
33. Neubauer, H., S.E. Clare, W. Wozny, G.P. Schwall, S. Poznanovic, W. Stegmann et al. 2008. Breast cancer proteomics reveals correlation between estrogen receptor status and differential phosphorylation of PRMC1. *Breast Cancer Research* 10:R85.
34. Nass, S.J. and H.L. Moses. 2007. *Cancer Biomarkers: The Promises and Challenges of Improving Detection and Treatment.* Committee on developing Biomaker-based tools for cancer screening, diagnosis and treatment. Institute of medicine of the National Academies. Washington, DC: The National Academies Press.
35. IOM. 2006. Developing biomarker based tools for cancer screening, diagnosis and treatment. In *The State of the Science, Evaluation, Implementation and Economics. A Workshop,* edited by Patlak M., and S. Nass, rappourteur. Washington, DC: The National Academic Press.

12

Use of Protein Biomarkers for the Early Detection of Breast Cancer

Birendra Kumar, Purnmasi Ram Yadav, and Surender Singh

CONTENTS

12.1 Introduction

Breast cancer is a type of cancer originating from breast tissue, most commonly from the inner lining of the milk ducts or lobules that supply the ducts with milk [1]. Developments in breast cancer control will be greatly abetted by early discovery, thereby facilitating diagnosis and treatment of breast cancer in its preinvasive state before metastasis. Breast cancer is the most frequently occurring malignancy and is the second leading cause of cancer-related death for women in the United States [2]. The most efficacious screening modality utilized in the clinic is mammography, although lesions less than 0.5 cm in size remain undetectable by current machinery. Importantly, however, even though a breast lesion may be detected, given the low sensitivity/specificity of mammography, approximately fourfold more women have resultant biopsies. The 5-year survival rate for women with breast cancer is highly correlated with tumor stage, with tumor detection at very early stages (stages 0 and I) having an approximately 98% survival rate. The 5-year survival rate is approximately 85% for stage II tumors, approximately 60% for stage III tumors, and approximately 20% for stage IV tumors. Overall, women with breast cancer have an approximately 226,870 new cases and 39,510 deaths expected in the United States in 2010 [2]. Substantial progress has been made over the past three decades in our understanding of the epidemiology, clinical course, and basic biology of breast cancer and the integration of routine and molecular biomarkers into patient management [3]. Modern techniques designed to detect the disease at an earlier stage, combined with new methods of determining risk assessment and more optimized combined modality treatment, have enhanced our ability to manage, and in many cases, cure the disease. For more than 100 years, morphology has been the cornerstone for the assessment of breast cancer prognosis [4].

Early discovery of breast tumor does allow for increased treatment choices, including surgical resection, with a corresponding better patient response. Surgical resection may involve lumpectomy or mastectomy with removal of some of the axillary lymph nodes. After early discovery, radiation therapy, chemotherapy, and hormone therapy tamoxifen [5] and aromatase inhibitors [6–8] also have utility for therapeutic intervention. Targeted biological therapy with trastuzumab [9] or lapatinib [10,11] also has utility to treat HER2/ neu-positive breast tumors. Unfortunately, however, in the absence of good serum/plasma biomarkers, many patients with breast cancer are diagnosed too late in the disease process for surgical resection to be an effective option. Thus, these patients are typically offered several therapeutic treatment modalities dependent on tumor subtype (HER2+ or HER2– ER+ or ER–). The obtainable treatment modalities may include hormonal (antiestrogen), taxane (docetaxel or paclitaxel), or nontaxane chemotherapy. Usually, women with metastatic breast cancer are provided one therapeutic modality until treatment failure and are then switched to another therapeutic modality.

The basis of most breast cancer cases is not known. Although, several risk factors have been identified, including female gender, increasing patient age, family history of breast cancer at an early age, early menarche, late menopause, older maternal age at first live childbirth, prolonged hormone replacement therapy, exposure to therapeutic chest irradiation, benign proliferative breast disease, and genetic mutations in genes such as BRCA1/2 [12]. The overpowering majority of breast masses detected by palpation or by mammography are epithelial lesions, which include benign fibrocystic change, hyperplasia, carcinoma *in situ*, and infiltrating mammary carcinoma. Although several histologic types and subtypes of mammary carcinomas exist, more than 95% are either ductal or lobular carcinomas [13], with the majority (75%–80%) of mammary carcinomas being ductal carcinomas [14,15].

A number of hereditary alterations have been identified in breast cancers. The most frequent genomic aberrations identified are gains along chromosomes 1q, 8q, 17q, 20q, and 11q and losses along 8p, 13q, 16q, 18q, and 11q [16–21]. Interestingly, many of these chromosomal segments harbor known proto-oncogenes or tumor suppressor genes (or both) such as HER2-neu, BRCA1, BRCA2, C-MYC, and cyclin D1. Low-grade (grade 1) infiltrating ductal carcinomas have relatively few chromosomal alterations, with the highest frequency of aberrations occurring as losses on 16q and gains on 1q. It is generally accepted that estrogen receptor–positive (ER+) and ER-negative (ER–) breast cancers are two different disease entities. ER– tumors tend to be of high grade, have more frequent p53 mutations, and have worse prognosis compared with ER+ disease. Both ER+ and ER– tumors can be either HER2-positive or HER2-negative. Low-grade tumors are typically ER-positive, almost always HER2-nonamplified, and frequently overexpress cyclin D1 [13]. In contrast, high-grade (grade 3) tumors tend to be ER-negative, have frequent loss of p53 function, usually overexpress C-MYC, and commonly overexpress HER2 [16–18,22]. In high-grade tumors, loss of p53 function is usually due to 17p13 deletion, mutation, or inactivation, whereas overexpression of HER2 is usually because of 17q12 amplification [23–31]. Although early discovery of tumor has improved survival for a number of cancers, including breast cancer [32], colon cancer [33–35], prostate cancer [36,37], and cervical cancer [38], existing serum biomarkers for breast cancer are not adequate for early discovery. The possibility of early discovery of breast tumor may be realized through both noninvasive and invasive methods. To date, gains in the early discovery of breast tumor have been largely made due to routine mammography or by palpation (either self-examination or by physician or nurse practitioner). Imaging technologies (mammography, digital mammography, and magnetic

TABLE 12.1

Promising New Biomarkers for the Discovery of Breast Cancer

Biomarker's Name	Technology Used for Discovery	Kind	Study
RS/DJ-1	Humoral response	Autoantibody	[51]
p53	Humoral response	Autoantibody	[60]
HSP60	Humoral response	Autoantibody	[64]
HSP90	Humoral response	Autoantibody	[66]
Mucin-related	Humoral response	Autoantibody	[69]
CA 15-3	Serum profiling	Serum protein	[71]
RS/DJ-1	Serum profiling	Serum protein	[51]
HER-2/neu	Serum profiling	Serum protein	[87]
α-2-HS-glycoprotein	Nipple aspirate fluid profiling	Ductal protein	[107]
Lipophilin B	Nipple aspirate fluid profiling	Ductal protein	[107]
β-Globin	Nipple aspirate fluid profiling	Ductal protein	[107]
Hemopexin	Nipple aspirate fluid profiling	Ductal protein	[107]
Vitamin D–binding protein	Nipple aspirate fluid profiling	Ductal protein	[107]

resonance imaging) have been adopted clinically for mass screening purposes, but there is resistance for seeking such services on a yearly basis, given the relative complexity and high cost-to-benefit ratio of these imaging methodologies. As a result, there has been much interest in the growth and validation of serum-based biomarkers for the early discovery, risk stratification, prediction, and disease prognosis of breast cancer. This chapter will focus on current developments in the identification of new serum protein biomarkers with potential function for the early discovery of breast cancer (Table 12.1).

12.2 Autoantibodies in Breast Cancer

Sera from patients with cancer often react against their own tumors cells. These so-called tumor-specific antibodies have been shown to occur in a diversity of malignant conditions using cytotoxic techniques or immunofluorescence [39]. Apparently, the surface of the cancer cell contains antigens that are normal fetal or adult components but which become exposed to a greater degree or in a higher concentration than in normal cells and so induce the formation of these antibodies.

In 1971, Whitehouse and Holborow [40] showed that some other autoantibodies—those directed against smooth muscles—may occur in various types of cancer and, as they pointed out, their presence may further confound the detection of tumor-specific antigens. Smooth muscle autoantibodies in high and persistent titers occur mainly in the autoimmune type of chronic active hepatitis, but these antibodies may also be produced for 1 to 7 weeks by nonpredisposed subjects with viral hepatitis, infectious mononucleosis, or *Mycoplasma pneumonia* infections [41]. Some of these antibodies react with the actin microfilaments [42,43] present in many cells [44], and certain viruses may incorporate the antigen into their membrane, thus making it more immunogenic [45]. Farrow et al. [46] suggested that the conversion of a normal cell to a malignant form may lead to the exposure of this antigen, and Gabbiani et al. [42] showed that breast cancer cells contain more actin filaments than normal tissues, in keeping with the greater mobility of malignant growths. Subsequently, a higher frequency of antinuclear antibodies [47,48] was also shown to occur in malignant disease. A dissident report, however, came from Tannenberg et al. [49], who failed to observe any change in the incidence of these autoantibodies among 250 patients with various types of cancer. More recently, these "non–tumor-specific" antibodies have been studied in detail in patients with breast cancer by two groups of workers with somewhat conflicting results. Wasserman and his colleagues [50] observed a higher incidence of smooth muscle autoantibodies and antinuclear antibodies in patients with breast cancer at mastectomy than in matched controls. They noted that the patients who later had recurrence of their disease showed a significantly higher incidence of multiple antibodies compared with those who remained well. They proposed that in patients with breast cancer, the increased incidence of autoantibodies might reflect not only a change in the cell membrane but also tissue damage due to the presence of cancer or disordered immunological reactivity associated with deficient tumor surveillance. This theory might also explain the prognostic significance of these autoantibodies.

The humoral immune response to cancer in humans has been well demonstrated by the identification of autoantibodies to a number of different intracellular and surface antigens in patients with various tumor types [51–54]. A tumor-specific humoral immune response directed against oncoproteins [55,56], mutated proteins such as p53 [57,58], or other aberrantly expressed proteins have all been described. Although it is currently unknown whether the occurrence of such antibodies is beneficial, knowledge of potential tumor antigens that may evoke tumor-specific immune responses may have utility in early cancer diagnosis, in establishing prognosis, and in immunotherapy against the disease. Several approaches are currently available for the identification of tumor antigens. In contrast to the identification of tumor antigens based on analysis of recombinant proteins (which do not contain posttranslational modifications as found in tumors or tumor cell lines), it may be preferable to utilize a proteomics-based approach for the identification of tumor antigens. This may facilitate the identification of autoantibodies to naturally occurring proteins, such as in lysates prepared from tumors and tumor cell lines, and may uncover antigenicity associated with aberrant posttranslational modification of tumor cell proteins. Such a proteomics approach was implemented for the identification of breast tumor antigens that elicit a humoral response against proteins that are expressed in the SUM-44 breast cancer cell line. Two-dimensional polyacrylamide gel electrophoresis (PAGE) was used to simultaneously separate individual cellular proteins from the SUM-44 cell line. The separated proteins were transferred onto polyvinylidene difluoride (PVDF) membranes. Sera from patients with breast cancer were screened individually for antibodies that reacted against the separated proteins by Western blot analysis. Proteins specifically reacting with sera from patients with breast cancer were identified by mass spectrometry. Le Naour and colleagues [51] have shown that a humoral response directed against RS/DJ-1 occurred in 13.3% of patients with newly diagnosed breast cancer. None of the 25 healthy controls (0%) and none of the 46 patients (0%) with hepatocellular carcinoma exhibited autoantibodies to RS/DJ-1. Only 2 of 54 (3.7%) samples of sera from patients with lung adenocarcinoma demonstrated autoantibodies to RS/DJ-1. In breast cancer, besides RS/DJ-1 [51], autoimmunity has also been shown against a number of other cellular proteins. These proteins include p53 [59–62], heat shock protein 60 [63,64], heat shock protein 90 [65,66], and mucin-related antigens [64,67–69]. The presence of p53 autoantibodies has been observed in 15% of patients with breast cancer and was shown to be associated with a poor prognosis [59,60,62]. However, p53 autoantibodies have also been found in patients with other malignancies and inflammatory conditions [57,58], thus the humoral response to p53 is not specific to breast cancer. A humoral response to the 90 kDa heat shock protein has also been associated with poor survival in breast cancer [66]. In contrast, the presence of MUC1 autoantibodies has been associated with a reduced risk for disease progression in patients with breast cancer [68,69]. Although the antigenic epitope on MUC1 (or, for that matter, any of the other breast tumor antigens discussed above) is unknown, MUC1 has been shown to be aberrantly glycosylated frequently in breast cancer [69]. Currently, CA 15-3 (a soluble or secreted form of MUC1) has utility as a circulating marker for breast cancer [70,71]. Serial measurements of CA 15-3 have utility to detect recurrences and to monitor the treatment of metastatic breast cancer [70–72]. Additionally, the CA 15-3 concentration at initial presentation does have prognostic significance [73–77]. To circumvent many of the difficulties associated with two-dimensional PAGE (i.e., inadequate resolution, slow throughput, and limited dynamic range), protein microarrays were developed that have the capability to screen patient's sera for autoantibodies directed against tumor antigens [78–81]. In comparison to traditional enzyme linked immunosorbent assay (ELISA) that use single purified recombinant proteins, the protein microarrays are capable of presenting and analyzing more than 1000 tumor antigens simultaneously. In addition, as these tumor antigens are typically derived from diseased tissues or disease-related cells, they possess disease-related, potentially antigenic, posttranslational modifications not normally expressed by the particular cells or tissue. In this technology, proteins from diseased tissues or disease-related cell lines are separated by two-dimensional liquid chromatography (chromatofocusing or ion exchange high-performance liquid chromatography (HPLC) in the first dimension, followed by reverse phase high-performance liquid chromatography in the second dimension). After separation, all fractions (≥1700 fractions) from each separation are printed onto nitrocellulose-coated microscope slides and are subsequently probed with sera from patients or control subjects [78–81]. As each reactive fraction may contain a number of different proteins, each reactive fraction would need to be further assessed to determine the tumor antigen of interest. More recently, Ramachandran et al. [82,83] developed a novel protein microarray technology, termed nucleic acid protein programmable array. These arrays are generated by printing full-length cDNA encoding the target proteins at each feature of the array. The proteins are then transcribed and translated by a cell-free system and immobilized *in situ* using epitope tags fused to the proteins. Although this technology circumvents many of the difficulties of traditional protein microarrays (i.e., the need to resolve complex protein lysates), the printed proteins on the array lack all normal posttranslational modifications. Thus, any antigenicity resulting from aberrant modification of tumor proteins is not assessed. Anderson and colleagues [84] utilized the nucleic acid protein programmable array arrays to screen 4988 candidate tumor antigens with sera from patients with early stage breast cancer for autoantibodies. Twenty-eight of these antigens were confirmed using an independent serum cohort ($n = 51$ cases/38 controls, $p < 0.05$). Using all 28 antigens, a classifier was identified with a sensitivity of 80.8% and a specificity of 61.6% (AUC = 0.756). Although the sensitivity and specificity are not high, these 28 recombinant protein antigens may be considered as potential biomarkers for the early detection of breast cancer. It is not clear why only a subset of patients with a particular tumor type develop a humoral response to particular tumor antigens. Immunogenicity may depend on the level of expression, posttranslational modification, or other types of protein processing, the extent of which may be variable among tumors of a similar histological type. Other factors that may influence the immune response include variability among tumors and individuals in major histocompatibility complex molecules and in antigen presentation. Although a number of autoantibodies have been identified in breast cancer, in most cases, they occur in less than 50% of patients' sera. Therefore, they are not likely to be effective individually for the early detection of breast cancer but may show efficacy if utilized as a panel of biomarkers.

12.3 Discovery of Changed Plasma Protein Expression for Identification of Breast Cancer–Precise Biomarkers

Great interest has been shown in the hypothesis that tumor-specific proteins may be found in patients' circulation and may be useful in the early detection of cancer. For example, proteins such as CA125 in ovarian cancer and prostate-specific antigen (PSA) in prostate cancer have been used clinically as diagnostic markers of cancer. CA125 is a mucin commonly used as a diagnostic marker for epithelial ovarian cancer. PSA is secreted primarily by prostate epithelial cells into the seminal plasma and is one of the best characterized examples of a secreted glycoprotein used in cancer diagnostics. Several recent reports have described aberrantly expressed proteins in the serum of patients with breast cancer.

The most widely used serum marker in breast cancer diagnostics is CA 15-3, which detects soluble forms of the mucin MUC1. MUC1 is normally found in the apical membrane of normal secretory epithelium. After malignant transformation, however, MUC1 may be localized throughout the external surface of the entire plasma membrane. In addition, changes in MUC1 glycosylation have been reported during neoplastic transformation [85, 86]. Although MUC1 is expressed in normal and neoplastic breast epithelium, the clinical utility of MUC1 measurements is confined to measurements of shed or soluble forms (termed CA 15-3), released from the cell surface by proteolytic cleavage. Unfortunately, CA 15-3 is not suitable for early detection because serum levels are rarely increased in patients with early or localized breast cancer. The main utility for CA 15-3 is for monitoring therapy in patients with metastatic breast cancer. Catalona [36] evaluated RS/DJ-1 as a serum biomarker of breast cancer. In normal tissue, expression of RS/DJ-1 was observed in the epithelium, smooth muscle, blood vessels, and nerves. All 15 (100%) invasive ductal carcinomas and three (100%) invasive lobular carcinomas showed some level of cytoplasmic and nuclear reactivity in the neoplastic cells. Significantly increased levels of serum RS/DJ-1 were observed in the sera of 11 of 30 patients with newly diagnosed breast cancer, as compared with serum from 25 healthy subjects. However, serum RS/DJ-1 levels in patients with other types of breast lesions were not evaluated. Thus, it is unknown whether the increased serum RS/DJ-1 levels were cancer-specific.

In another study [87], significantly higher serum HER-2/neu levels were found in patients with tissue overexpression of HER-2/neu. Univariate analysis showed that HER-2/neu serum levels were prognostic factors in disease-free survival and overall survival only in patients with tissue overexpression. When only patients with HER-2/neu overexpression in tissue were studied, tumor size, nodal involvement, and tumor markers (at least one positive) were found to be independent prognostic factors for both disease-free survival and overall survival.

12.4 Use of Mass Spectrometric Methodologies for Identification of Breast Cancer–Precise Biomarkers

Mass spectroscopy approaches have recently been applied to breast cancer for the discovery of new and better biomarkers both in serum and nipple aspirate specimens [88,89]. Among these technologies, some, like surface-enhanced laser desorption and ionization (SELDI), are mass spectrometry–based. A number of investigators have used SELDI time-of-flight (TOF) mass spectrometry to investigate serum [90–98] and nipple aspirate/ductal lavage fluid [99–106] from patients with breast cancer. In one study, serum samples from women with or without breast cancer were analyzed using SELDI protein chip mass spectrometry [94]. Using a case-control study design, serum samples from 48 female patients with primary invasive breast cancer were compared with samples from 48 age- and sex-matched healthy controls. To increase the number of identifiable proteins, patients' serum were profiled on IMAC30 (activated with nickel) protein chip surfaces. Differences in protein intensity between breast cancer cases and controls were measured by the Mann–Whitney U test and adjusted for confounding variables in a multivariate logistic regression model. Three peaks, with mass-to-charge ratios (m/z) of 4276, 4292, and 8941 were found that showed significant decreased expression in cancer sera, as compared with control sera ($p < .001$). One drawback of the SELDI technology, however, is that given the limited dynamic range of SELDI, it is likely that distinctive features observed in serum with this approach represent relatively abundant proteins that are not necessarily specific to breast cancer. Furthermore, SELDI has difficulties in providing the identification of the distinctive proteins when used to directly profile complex protein mixtures.

12.5 Mass-Spectrometric Profiling of Nipple Aspirate Fluid or Ductal Lavage Fluid

Other mass spectrometric profiling methods have been utilized to profile proteins found in nipple aspirate fluid [107] and ductal lavage fluid to identify breast cancer–specific biomarkers. These investigators [107] analyzed paired nipple aspirate fluid samples from 18 women with stage I or stage II unilateral invasive breast cancer and four healthy volunteers using isotope-coded affinity tag labeling, followed by sodium dodecyl sulfate polyacrylamide gel electrophoresis (SDS-PAGE). Gel slices were cut from each sample, with subsequent analysis by liquid chromatography-tandem mass spectrometry. They identified 353 peptides from the tandem mass spectra. α2-HS-glycoprotein was found to be underexpressed in nipple aspirate fluid from tumor-bearing breasts, whereas lipophilin B, β-globin, hemopexin, and vitamin D–binding protein were all overexpressed. Unfortunately, these authors only identified abundant proteins whose overexpression or underexpression was somewhat modest. Moreover, these authors did not analyze nipple aspirate fluid from patients with inflammatory breast disease. Thus, conclusions cannot be drawn regarding breast cancer specificity of protein expression.

12.6 *N*-Linked Glycan Profiling for Biomarker Identification in Cancer Serum

Glycoproteins are the most heterogeneous group of posttranslational modifications known in proteins. Glycans show a high structural diversity reflecting inherent functional diversity. *N*- and *O*-oligosaccharide variants on glycoproteins (glycoforms) can lead to alterations in protein activity or function that may manifest itself as overt disease [108,109].

Many clinical biomarkers and therapeutic targets in cancer are glycoproteins [110–112], such as CA125 in ovarian cancer, HER2/neu in breast cancer, and PSA in prostate cancer. The human epidermal growth factor receptor 2 (HER2/neu) is a transmembrane glycoprotein, in which the presence of HER2 overexpression seems to be a key factor in malignant transformation and is predictive of a poor prognosis in breast cancer. CA125 is a mucin commonly employed as a diagnostic marker for epithelial ovarian cancer. Although CA125 has been used as an ovarian cancer marker for a long time, many of its O- and N-glycan structures have only recently been characterized [113]. PSA is secreted primarily by prostate epithelial cells into the seminal plasma. It is one of the best characterized examples of a secreted glycoprotein used in cancer diagnostics, and its glycoforms have been described [114]. The alteration in protein glycosylation that occurs through varying the heterogeneity of glycosylation sites or changing glycan structure of proteins on the cell surface and in body fluids has been shown to correlate with the development or progression of cancer and other disease states [115]. It has been reported that the glycosylation of PSA secreted by the tumor prostate cell line LNCaP differs significantly from that of PSA from seminal plasma (normal control). These carbohydrate differences allow a distinction to be made between PSA from normal and tumor origins and provide a valuable biochemical tool for the diagnosis of prostate cancer [116]. There is growing evidence that glycan structures on glycoproteins are modified in breast cancer [117–126]. Breast cancer–associated alterations have been demonstrated for fucosylation groups and for sialylations on the plasma protein α1-proteinase inhibitor [123]. Increased GlcNAc β1–6Man α1–6Man β-branching in asparagine-linked oligosaccharides has been observed in human tumor cells. The levels of the β1–6 branched oligosaccharides were evaluated in a series of benign and malignant human breast biopsies. Normal human breast tissue and benign lesions showed low expression, but 50% of the primary malignancies examined showed significantly increased β1–6 branching [124]. Subsequently, L-PHA (a lectin that binds specifically to the β1–6 branched oligosaccharides) lectin histochemistry was performed on paraffin sections of human breast tissues. All breast carcinomas and epithelial hyperplasia with atypia demonstrated significantly increased L-PHA staining as compared with fibroadenomas and hyperplasia without atypia [108]. More recently, L-PHA–reactive glycoproteins were identified from matched normal (nondiseased) and malignant tissue isolated from patients with invasive ductal breast carcinoma [126]. Comparison analysis of the data identified 34 proteins that were enriched by L-PHA fractionation in tumor relative to normal tissue for at least two cases of ductal invasive breast carcinoma. Of these 34 L-PHA tumor-enriched proteins, 12 were common to all four matched cases analyzed. Abd Hamid and coworkers [127] analyzed fluorescently tagged serum N-glycans of patients with advanced breast cancer using exoglycosidases and liquid chromatography-tandem mass spectrometry. They found that the expression of a trisialylated triantennary glycan containing an α1,3-linked fucose was increased in the presence of breast cancer. Kyselova and coworkers [128] profiled the permethylated N-glycans in sera of patients with breast cancer at different stages (stages I–IV) using matrix-assisted laser desorption ionization time-of-flight/time-of-flight mass spectrometry (MALDITOF/TOF MS) in one study. In a second study, they profiled reduced and methylated serum N-glycans of patients with late-stage breast cancer using nanoliquid chromatography chip/time-of-flight mass spectrometry (TOF MS) [129]. In both studies, they found an increase in fucosylation in both core and branched segments of N-glycans in the presence of breast cancer. In the latter study, they found a decrease in expression of a biantennary monosialylated N-linked glycan and an increase in expression of a fucosylated triantennary-trisialylated N-linked glycan in the presence of stage IV breast cancer. These glycosylation changes in a tumor-secreted protein may reflect fundamental activity changes in the enzymes involved in the glycosylation pathway, either through altered levels of enzymes or altered enzymatic activity. Importantly, the changes in glycan structure may serve as early detection biomarkers of breast cancer.

12.7 Summary

Early detection of breast cancer, so as to diagnose and treat cancer in its preinvasive state before metastasis, may greatly affect the treatment and prognosis of patients with this common, but deadly, malignancy. Unfortunately, suitable biomarkers have not yet been identified for the early detection of breast cancer. Biomarker discovery for this disease is still very much in its early phases. Multiple approaches have been developed, as described in previous sections, that hold promise for the identification of serum biomarkers. The protein biomarkers that have been identified to date do not possess the requisite sensitivity/specificity to have individual utility as biomarkers for the early detection of breast cancer but ultimately may have utility within a panel of protein biomarkers. Additionally, other emerging technologies, such as genetically engineered mouse models of breast cancer, may have utility to identify panels of serum biomarkers that can be further explored in human sera. To determine the utility of any promising protein biomarkers, the candidates will need to be tested and validated by multiple independent studies using an adequately sized test and training set of sera samples from very early-stage breast cancer. Development of such resources, including serum from patients with nonmalignant breast lesions and prospective serum collection from individuals at high risk of being diagnosed with breast cancer as well as serum from patients with other breast lesions and other types (non–breast cancer) of malignancies is of critical need for the identification of biomarkers with utility for the early detection of breast cancer. Up until now, serum/plasma collection has been primarily performed in individual laboratories, using heterogeneous sample collection methods. The Human Proteome Organization has conducted a study to assess efficacious serum collection methods. These findings have lead to efforts presently being made by the National Cancer Institute, through the Early Detection Research Network, to develop suitable serum resources for both the discovery phase and the subsequent validation phase of biomarkers for the early detection of cancer. With the ultimate development of these standardized resources, it is expected that suitable biomarkers would be validated and have utility for the early clinical detection of breast cancer within the next 5 to 10 years.

References

1. Sariego, J. 2010. Breast cancer in the young patient. *The American Surgeon* 76:1397–1401.
2. American Cancer Society. 2012. Cancer Facts & Figures. *Atlanta: American Cancer Society, Inc.*
3. Jemal A., T. Murray, A. Samuels et al. 2003. Cancer statistics. *CA: A Cancer Journal for Clinicians* 53:5–26.
4. Fitzgibbons, P.L., D.L. Page, D. Weaver et al. 2000. Prognostic factors in breast cancer. College of American Pathologists Consensus Statement 1999. *Archives of Pathology & Laboratory Medicine* 124:966–978.

5. Early Breast Cancer Trialist's Collaborative Group (EBCTCG). 2005. Effects of chemotherapy and hormonal therapy for early breast cancer on recurrence and 15-years survival: An overview of the randomised trials. *The Lancet* 365:1687–1717.

6. Nabholtz, J.M., A. Buzdar, M. Pollak et al. 2000. Anastrozole is superior to tamoxifen as first-line therapy for advanced breast cancer in postmenopausal women: Results of a North American multicenter randomized trial. *Journal of Clinical Oncology* 18:3758–3767.

7. Bonneterre, J., B. Thurlimann, J.F. Robertson et al. 2000. Anastrozole versus tamoxifen as first-line therapy for advanced breast cancer in 668 postmenopausal women: Results of the tamoxifen or arimidex randomized group efficacy and tolerability study. *Journal of Clinical Oncology* 18:3748–3757.

8. Bonneterre, J., A. Buzdar, J.M. Nabholtz et al. 2001. Anastrozole is superior to tamoxifen as first-line therapy in hormone receptor positive advanced breast carcinoma. *Cancer* 92:2247–2258.

9. Hudis, C.A. 2007. Trastuzumab—Mechanism of action and use in clinical practice. *New England Journal of Medicine* 357:39–51.

10. Bilancia, D., G. Rosati, A. Dinota, D. Germano, R. Romano, and L. Manzione. 2007. Lapatinib in breast cancer. *Annals of Oncology* 18:vi26–vi30.

11. Higa, G.M., and J. Abraham. 2007. Lapatinib in the treatment of breast cancer. *Expert Review of Anticancer Therapy* 7:1183–1192.

12. Carlson, R.W., B.O. Anderson, H.J. Burstein et al. 2007. Invasive breast cancer: Clinical practice guidelines in oncology. *Journal of the National Comprehensive Cancer Network* 5:246–312.

13. Yoder, B.J., E.J. Wilkinson, and N.A. Massoll. 2007. Molecular and morphologic distinctions between infiltrating ductal and lobular carcinoma of the breast. *Breast Journal* 13:172–179.

14. Tavassoli, F.A. 1999. Invasive lobular carcinoma. In *Pathology of the Breast,* edited by Tavassoli, F.A., 2nd ed. 426–436. Stamford, CN: Appleton & Lange.

15. Rosen, P.P. (ed.). 2001. Invasive lobular carcinoma. *Rosen's Breast Pathology,* 2nd ed., 627–652. Philadelphia, PA: Lippincott Williams & Wilkins.

16. Reis-Filho, J.S., P.T. Simpson, T. Gale, and S.R. Lakhani. 2005. The molecular genetics of breast cancer: The contribution of comparative genomic hybridization. *Pathology Research and Practice* 201:713–725.

17. Shackney, S.E., and J.F. Silverman. 2003. Molecular evolutionary patterns in breast cancer. *Advances in Anatomic Pathology* 10:278–290.

18. Simpson, P.T., J.S. Reis-Filho, T. Gale, and S.R. Lakhani. 2005. Molecular evolution of breast cancer. *Journal of Pathology* 205:248–254.

19. Stange, D.E., B. Radlwimmer, F. Schubert et al. 2006. High resolution genomic profiling reveals association of chromosomal aberrations on 1q and 16p with histologic and genetic subgroups of invasive breast cancer. *Clinical Cancer Research* 12:345–352.

20. Albertson, D.G., C. Collins, F. McCormick, and J.W. Gray. 2003. Chromosome aberrations in solid tumors. *Nature Genetics* 34:369–376.

21. Albertson D.G., and D. Pinkel. 2003. Genomic microarrays in human genetic disease and Cancer. *Human Molecular Genetics* 12:R145–R152.

22. Buerger, H., E.C. Mommers, R. Littmann et al. 2001. Ductal invasive G2 and G3 carcinomas of the breast are the end stages of at least two different lines of genetic evolution. *Journal of Pathology.* 194:165–170.

23. Tsuda, H., and S. Hirohashi. 1998. Multiple developmental pathways of highly aggressive breast cancers disclosed by comparison of histological grades and c-erbB-2 expression patterns in both the non-invasive and invasive portions. *Pathology International* 48:518–525.

24. Somerville, J.E., L.A. Clarke, and J.D. Biggart. 1992. c-erbB-2 overexpression and histological type of in situ and invasive breast carcinoma. *Journal of Clinical Pathology* 45:16–20.

25. Jain, A.N., K. Chin, A.L. Borresen-Dale et al. 2001. Quantitative analysis of chromosomal CGH in human breast tumors associates copy number abnormalities with p53 status and patient survival. *Proceedings of the National Academy of Sciences of the United States of America* 98:7952–7957.

26. Poller, D.N., C.E. Hutchings, M. Galea et al. 1992. p53 protein expression in human breast carcinoma: Relationship to expression of epidermal growth factor receptor, c-erbB-2 protein overexpression, and oestrogen receptor. *British Journal of Cancer* 66:583–588.

27. Barbareschi, M., E. Leonardi, F.A. Mauri, G. Serio, and P. Dalla Palma. 1992. p53 and c-erbB-2 protein expression in breast carcinomas: An immunohistochemical study including correlations with receptor status, proliferation markers, and clinical stage in human breast cancer. *American Journal of Clinical Pathology* 98:408–418.

28. Rudas, M., R. Neumayer, M.F.X. Gnant, M. Mittelböck, R. Jakesz, and A. Reiner. 1997. p53 protein expression, cell proliferation and steroid hormone receptors in ductal and lobular in situ carcinomas of the breast. *European Journal of Cancer* 33:39–44.

29. Eyfjord, J.E., S. Thorlacius, M. Steinarsdottir, R. Valgardsdottir, H.M. Ogmundsdottir, and K. Anamthawat-Jonsson. 1995. p53 abnormalities and genomic instability in primary human breast carcinomas. *Cancer Research* 55:646–651.

30. Wiltschke, C., I. Kindas-Muegge, A. Steininger, A. Reiner, G. Reinger, and P.N. Preis. 1994. Coexpression of HER-2/neu and p53 is associated with a shorter disease-free survival in node-positive breast cancer patients. *Journal of Cancer Research and Clinical Oncology* 120: 737–742.

31. Seshadri, R., A.S.Y. Leong, K. Mccaul, F.A. Firgaira, V. Setlur, and D.J. Horsfall. 1996. Relationship between p53 gene abnormalities and other tumour characteristics in Breast cancer Prognosis. *International Journal of Cancer* 69:135–141.

32. Goldhirsch, A., M. Colleoni, G. Domenighetti, and R.D. Gelber. 2003. Systemic treatments for women with breast cancer: Outcome with relation to screening for the disease. *Annals of Oncology* 14:1212–1214.

33. Lieberman, D. 1998. How to screen for colon cancer. *Annual Review of Medicine* 49:163–172.

34. Ferguson, J.A. 1993. Early detection of unsuspected colon cancers in asymptomatic people. *Diseases of the Colon and Rectum* 36:411.

35. Mandel, J.S., J.H. Bond, T.R. Church et al. 1993. Reducing mortality from colorectal cancer by screening for fecal occult blood, Minnesota Colon Cancer Control Study. *New England Journal of Medicine* 328:1365–1371.

36. Catalona, W.J. 1995. Management of cancer of the prostate. *New England Journal of Medicine* 331:996–1004.

37. Jacobsen, S.J., S.K. Katusic, E.J. Bergstralh et al. 1995. Incidence of prostate cancer diagnosis in the eras before and after serum prostate-specific antigen testing. *Journal of the American Medical Association* 274:1445–1449.

38. Lynge, E. 1989. Screening for cancer of the cervix uteri. *World Journal of Surgery* 13:71–78.

39. Southam, C.M. 1975. In *Immunological Disease*, edited by Samter, M. 1:743. Boston: Little, Brown and Co.

40. Whitehouse, J.M.A., and E.J. Holborow. 1971. Smooth Muscle Antibody in Malignant Disease. *British Medical Journal* 4:511–513..

41. Biberfeld, G., and G. Sterner. 1976. Smooth muscle antibodies in Mycoplasma pneumoniae infection. *Clinical and Experimental Immunology* 24:287–291.

42. Gabbiani, G., P. Trenchev, and E.J. Holborow. 1975. Increase of contractile protein in human cancer cell. *Lancet* 2:796–797.

43. Norberg, R., K. Lidman, and A. Fagraeus. 1975. Effects of cytochalasin B on fibroblasts, lymphoid cells, and platelets revealed by human anti-actin antibodies. *Cell* 6:507–512.

44. Pollard, T.D., R.R. Weihing. 1974. Actin and myosin and cell movement. *Critical Reviews in Biochemistry* 2:1–65.

45. Fagraeus, I.A. et al. 1974. Reaction of human smooth muscle antibodies with human blood lymphocytes and lymphoid cell line. *Nature* 252:246–247.

46. Farrow, L.J., E.J. Holborow, and W.D. Brighton. 1971. Reaction of human smooth muscle antibody with liver cells. *Nature New Biology* 232:186–187.

47. Burnham, T.K. 1972. Antinuclear antibodies in patients with malignancies. *Lancet* 2:436.

48. Zeromski, J.O., M.K. Gorny, and K. Jarczewska. 1972. Malignancy associated with antinuclear antibodies. *Lancet* 2:1035–1036.

49. Tannenberg, A.E.G. et al. 1973. Incidence of autoantibodies in cancer patients. *Clinical and Experimental Immunology* 15:153–156.

50. Wasserman, J., U. Glas, and H. Blomgren. 1975. Autoantibodies in patients with carcinoma of the breast. Correlation with prognosis. *Clinical and Experimental Immunology* 19:417–422.

51. Le Naour, F., D.E. Misek, M.C. Krause et al. 2001. Proteomics based identification of RS/DJ-1 as a novel circulating tumor antigen in breast cancer. *Clinical Cancer Research* 7:3328–3335.

52. Brichory, F.M., D.E. Misek, A.M. Yim et al. 2001. An immune response manifested by the common occurrence of annexins I and II autoantibodies and high circulating levels of IL-6 in lung cancer. *Proceedings of the National Academy of Sciences of the United States of America* 98:9824–9829.

53. Güre, A.O., N.K. Altorki, E. Stockert, M.J. Scanlan, L.J. Old, and Y.T. Chen. 1998. Human lung cancer antigens recognized by autologous antibodies: Definition of a novel cDNA derived from the tumor suppressor gene locus on chromosome 3p21.3. *Cancer Research* 58:1034–1041.

54. Stockert E., E. Jäger, Y.T. Chen et al. 1998. A survey of the humoral immune response of cancer patients to a panel of human tumor antigens. *Journal of Experimental Medicine* 187:1349–1354.

55. Ben-Mahrez, K., I. Sorokine, D. Thierry et al. 1990. Circulating antibodies against c-myc oncogene product in sera of colorectal cancer patients. *International Journal of Cancer* 46:35–38.

56. Pupa, S.M., S. Menard, S. Andreola, and M.I. Colnaghi. 1993. Antibody response against the c-erbB2 oncoprotein in breast carcinoma patients. *Cancer Research* 53:5864–5866.

57. Winter, S.F., J.D. Minna, B.E. Johnson, T. Takahashi, A.F. Gazdar, and D.P. Carbone. 1992. Development of antibodies against p53 in lung cancer patients appears to be dependent on the type of p53 mutation. *Cancer Research* 52:4168–4174.

58. Raedle, J., G. Oremek, M. Welker, W.K. Roth, W.F. Caspary, and S. Zeuzem. 1996. p53 autoantibodies in patients with pancreatitis and pancreatic carcinoma. *Pancreas* 13:241–246.

59. Lenner, P., F. Wiklund, S.O. Emdin et al. 1999. Serum antibodies against p53 in relation to cancer risk and prognosis in breast cancer: A population-based epidemiological study. *British Journal of Cancer* 79:927–932.

60. Soussi, T. 2000. p53 Antibodies in the sera of patients with various types of cancer. *Cancer Research* 60:1777–1788.

61. Angelopoulou, K., H. Yu, B. Bharaj, M. Giai, and E.P. Diamandis. 2000. p53 gene mutation, tumor p53 protein overexpression, and serum p53 autoantibody generation in patients with breast cancer. *Clinical Biochemistry* 33:53–62.

62. Kulić A., M. Sirotković-Skerlev, S. Jelisavac-Cosić, D. Herceg, Z. Kovac, and D. Vrbanec. 2010. Anti-p53 antibodies in serum: Relationship to tumor biology and prognosis of breast cancer patients. *Medical Oncology* 27:887–893.

63. Desmetz, C., F. Bibeau, F. Boissière et al. 2008. Proteomics-based identification of HSP60 as a tumor-associated antigen in early stage breast cancer and ductal carcinoma in situ. *Journal of Proteome Research* 7:3830–3837.

64. Desmetz, C., C. Bascoul-Mollevi, P. Rochaix et al. 2009. Identification of a new panel of serum autoantibodies associated with the presence of in situ carcinoma of the breast in younger women. *Clinical Cancer Research* 15:4733–4741.

65. Conroy, S.E., S.L. Gibson, G. Brunstrom, D. Isenberg, Y. Luqmani, and D.S. Latchman. 1995. Autoantibodies to 90 Kd heat-shock protein in sera of breast cancer patients. *The Lancet* 345:126.

66. Conroy, S.E., P.D. Sasieni, I. Fentiman, and D.S. Latchman. 1998. Autoantibodies to the 90 kDa heat shock protein and poor survival in breast cancer patients. *European Journal of Cancer* 34:942–943.

67. Von Mensdorff-Pouilly, S., M.M. Gourevitch, P. Kenemans et al. 1996. Humoral immune response to polymorphic epithelial mucin (MUC-1) in patients with benign and malignant breast tumours. *European Journal of Cancer A* 32:1325–1331.

68. Von Mensdorff-Pouilly, S., A.A. Verstraeten, P. Kenemans et al. 2000. Survival in early breast cancer patients is favorably influenced by a natural humoral immune response to polymorphic epithelial mucin. *Journal of Clinical Oncology* 18:574–583.

69. Blixt, O., D. Bueti, B. Burford et al. 2011. Autoantibodies to aberrantly glycosylated MUC1 in early stage breast cancer are associated with a better prognosis. *Breast Cancer Research* 13:R25.

70. Duffy, M.J. 1999. CA 15-3 and related mucins as circulating markers in breast cancer. *Annals of Clinical Biochemistry* 36:579–586.
71. Duffy, M.J., D. Evoy, and E.W. McDermott. 2010. CA 15-3: Uses and limitation as a biomarker for breast cancer. *Clinica Chimica Acta* 411:1869–1874.
72. De La Lande, B., K. Hacene, J.L. Floiras, N. Alatrakchi, and M.F. Pichon. 2002. Prognostic value of CA 15.3 kinetics for metastatic breast cancer. *International Journal of Biological Markers* 17:231–238.
73. Ebeling, F.G., P. Stieber, M. Untch et al. 2002. Serum CEA and CA 15-3 as prognostic factors in primary breast cancer. *British Journal of Cancer* 86:1217–1222.
74. Gion, M., P. Boracchi, R. Dittadi et al. 2002. Prognostic role of serum CA15.3 in 362 node-negative breast cancers. An old player for a new game. *European Journal of Cancer* 38:1181–1188.
75. Kumpulainen, E.J., R.J. Keskikuru, and R.T. Johansson. 2002. Serum tumor marker CA 15.3 and stage are the two most powerful predictors of survival in primary breast cancer. *Breast Cancer Research and Treatment* 76:95–102.
76. Martìn, A., M.D. Corte, A.M. Alvarez et al. 2006. Prognostic value of pre-operative serum CA 15.3 levels in breast cancer. *Anticancer Research* 26:3965–3971.
77. Molina, R., X. Filella, J. Alicarte et al. 2003. Prospective evaluation of CEA and CA 15.3 in patients with locoregional breast cancer. *Anticancer Research* 23:1035–1041.
78. Madoz-Gúrpide, J., H. Wang, D.E. Misek, F. Brichory, and S.M. Hanash. 2001. Protein based microarrays: A tool for probing the proteome of cancer cells and tissues. *Proteomics* 1:1279–1287.
79. Nam, M.J., J. Madoz-Gúrpide, H. Wang et al. 2003. Molecular profiling of the immune response in colon cancer using protein microarrays: Occurrence of autoantibodies to ubiquitin C-terminal hydrolase L3. *Proteomics* 3:2108–2115.
80. Bouwman, K., J. Qiu, H. Zhou et al. 2003. Microarrays of tumor cell derived proteins uncover a distinct pattern of prostate cancer serum immunoreactivity. *Proteomics* 3:2200–2207.
81. Qiu, J., J. Madoz-Gúrpide, D.E. Misek et al. 2004. Development of natural protein microarrays for diagnosing cancer based on an antibody response to tumor antigens. *Journal of Proteome Research* 3:261–267.
82. Ramachandran, N., E. Hainsworth, B. Bhullar et al. 2004. Self assembling protein microarrays. *Science* 305:86–90.
83. Ramachandran, N., J.V. Raphael, E. Hainsworth et al. 2008. Next-generation high-density self-assembling functional protein arrays. *Nature Methods* 5:535–538.
84. Anderson, K.S., S. Sibani, G. Wallstrom et al. 2011. Protein microarray signature of autoantibody biomarkers for the early detection of breast cancer. *Journal of Proteome Research* 10:85–96.
85. Hattrup, C.L., and S.J. Gendler. 2008. Structure and function of the cell surface (tethered) mucins. *Annual Review of Physiology* 70:431–457.
86. Kufe, D.W. 2009. Mucins in cancer: Function, prognosis and therapy. *Nature Reviews Cancer* 9:874–885.
87. Molina, R., J.M. Augé, J.M. Escudero et al. 2010. Evaluation of tumor markers (HER-2/neu oncoprotein, CEA, and CA 15.3) in patients with locoregional breast cancer prognostic value. *Tumour Biology* 31:171–180.
88. Li, J., Z. Zhang, J. Rosenzweig et al. 2002. Proteomics and bioinformatics approaches for identification of serum biomarkers to detect breast cancer. *Clinical Chemistry* 48:1296–1304.
89. Paweletz, C.P., B. Trock, M. Pennanen et al. 2001. Proteomic patterns of nipple aspirate fluids obtained by SELDI-TOF: Potential for new biomarkers to aid in the diagnosis of breast cancer. *Disease Markers* 17:301–307.
90. Fan, Y., J. Wang, Y. Yang et al. 2010. Detection and identification of potential biomarkers of breast cancer. *Journal of Cancer Research and Clinical Oncology* 136:1243–1254.
91. Gast, M.C., E.J. van Dulken, T.K. van Loenen et al. 2009. Detection of breast cancer by surface-enhanced laser desorption/ionization time-of-flight mass spectrometry tissue and serum protein profiling. *International Journal of Biological* Markers 24:130–141.
92. Gast, M.C., C.H. van Gils, L.F. Wessels et al. 2009. Serum protein profiling for diagnosis of breast cancer using SELDI-TOF MS. *Oncology Reports* 22:205–213.

93. Lebrecht, A., D. Boehm, M. Schmidt, H. Koelbl, and F.H. Grus. 2009. Surface-enhanced laser desorption/ionisation time-of-flight mass spectrometry to detect breast cancer markers in tears and serum. *Cancer Genomics and Proteomics* 6:75–84.

94. Van Winden, A.W., M.C. Gast, J.H. Beijnen et al. 2009. Validation of previously identified serum biomarkers for breast cancer with SELDI-TOF MS: A case control study. *BMC Medical Genomics* 2:4.

95. Belluco, C., E.F. Petricoin, E. Mammano et al. 2007. Serum proteomic analysis identifies a highly sensitive and specific discriminatory pattern in stage I breast cancer. *Annals of Surgical Oncology* 14:2470–2476.

96. Ricolleau, G., C. Charbonnel, L. Lodé et al. 2006. Surface enhanced laser desorption/ionization time of flight mass Spectrometry protein profiling identifies ubiquitin and ferritin light chain as prognostic biomarkers in node-negative breast cancer tumors. *Proteomics* 6:1963–1975.

97. Laronga, C., S. Becker, P. Watson et al. 2003. SELDI-TOF serum profiling for prognostic and diagnostic classification of breast cancers. *Disease Markers* 19:229–238.

98. Li, J., Z. Zhang, J. Rosenzweig, Y.Y. Wang, and D.W. Chan. 2002. Proteomics and bioinformatics approaches for identification of serum biomarkers to detect breast cancer. *Clinical Chemistry* 48:1296–1304.

99. Sauter, E.R., W. Davis, W. Qin et al. 2009. Identification of a β-casein-like peptide in breast nipple aspirate fluid that is associated with breast cancer. *Biomarkers in Medicine* 3:577–588.

100. Zhou, J., B. Trock, T.N. Tsangaris et al. 2010. A unique proteolytic fragment of alpha1-antitrypsin is elevated in ductal fluid of breast cancer patient. *Breast Cancer Research and Treatment* 123:73–86.

101. Noble, J.L., R.S. Dua, G.R. Coulton, C.M. Isacke, and G.P.H. Gui. 2007. A comparative proteinomic analysis of nipple aspiration fluid from healthy women and women with breast cancer. *European Journal of Cancer* 43:2315–2320.

102. He, J., J. Gornbein, D. Shen et al. 2007. Detection of breast cancer biomarkers in nipple aspirate fluid by SELDI-TOF and their identification by combined liquid chromatography-tandem mass spectrometry. *International Journal of Oncology* 30:145–154.

103. Li, J., J. Zhao, X. Yu et al. 2005. Identification of biomarkers for breast cancer in nipple aspiration and ductal lavage fluid. *Clinical Cancer Research* 11:8312–8320.

104. Pawlik, T.M., H. Fritsche, K.R. Coombes et al. 2005. Significant differences in nipple aspirate fluid protein expression between healthy women and those with breast cancer demonstrated by time-of-flight mass spectrometry. *Breast Cancer Research and Treatment* 89:149–157.

105. Sauter, E.R., S. Shan, J.E. Hewett, P. Speckman, and G.C. Du Bois. 2005. Proteomic analysis of nipple aspirate fluid using SELDI-TOF-MS. *International Journal of Cancer* 114:791–796.

106. Paweletz, C.P., B. Trock, M. Pennanen et al. 2001. Proteomic patterns of nipple aspirate fluids obtained by SELDI-TOF: Potential for new biomarkers to aid in the diagnosis of breast cancer. *Disease Markers* 17:301–307.

107. Pawlik, T.M., D.H. Hawke, Y. Liu et al. 2006. Proteomic analysis of nipple aspirate fluid from women with early-stage breast cancer using isotope-coded affinity tags and tandem mass spectrometry reveals differential expression of vitamin D binding protein. *BMC Cancer* 6:68.

108. Rudd, P.M., T. Elliott, P. Cresswell, I.A. Wilson, and R.A. Dwek. 2001. Glycosylation and the immune system. *Science* 291:2370–2376.

109. Kobata, A., and J. Amano. 2005. Altered glycosylation of proteins produced by malignant cells, and application for the diagnosis and immunotherapy of tumours. *Immunology and Cell Biology* 83:429–439.

110. Dube, D.H., and C.R. Bertozzi. 2005. Glycans in cancer and inflammation—Potential for therapeutics and diagnostics. *Nature Reviews Drug Discovery* 4:477–488.

111. Ørntoft, T.F., and E.M. Vestergaard. 1999. Clinical aspects of altered glycosylation of glycoproteins in cancer. *Electrophoresis* 20:362–371.

112. Semmes, O.J., G. Malik, and M. Ward. 2006. Application of mass spectrometry to the discovery of biomarkers for detection of prostate cancer. *Journal of Cellular Biochemistry* 98:496–503.

113. Wong, N.K., R.L. Easton, M. Panico et al. 2003. Characterization of the oligosaccharides associated with the human ovarian tumor marker CA125. *Journal of Biological Chemistry* 278:28619–28634.

114. Prakash, S., and P.W. Robbins. 2000. Glycotyping of prostate specific antigen. *Glycobiology* 10:173–176.
115. Block, T.M., M.A. Comunale, M. Lowman et al. 2005. Use of targeted glycoproteomics to identify serum glycoproteins that correlate with liver cancer in woodchucks and humans. *Proceedings of the National Academy of Sciences of the United States of America* 102:779–784.
116. Peracaula, R., G. Tabarés, L. Royle et al. 2003. Altered glycosylation pattern allows the distinction between prostate-specific antigen (PSA) from normal and tumor origins. *Glycobiology* 13:457–470.
117. Hayes, D.F., M. Abe, J. Siddiqui, C. Tondini, and D.W. Kufe. 1989. Clinical and molecular investigations of the DF3 breast cancer-associated antigen. *Immunological Approaches to the Diagnosis and Therapy of Breast Cancer II*, edited by R.L. Ceriani, 45–53. New York: Plenum.
118. Hull, S.R., A. Bright, K.L. Carraway, M. Abe, D.F. Hayes, and D.W. Kufe. 1989. Oligosaccharide differences in the DF3 sialomucin antigen from normal human milk and the BT-20 human breast carcinoma cell line. *Cancer Communications* 1:261–267.
119. Perey, L., D.F. Hayes, and D. Kufe. 1992. Effects of differentiating agents on cell surface expression of the breast carcinoma associated DF3-P epitope. *Cancer Research* 52:6365–6370.
120. Perey, L., D.F. Hayes, P. Maimonis, M. Abe, C. O'Hara, and D.W. Kufe. 1992. Tumor selective reactivity of a monoclonal antibody prepared against a recombinant peptide derived from the DF3 human breast carcinoma-associated antigen. *Cancer Research* 52:2563–2568.
121. Sewell, R., M. Bäckström, M. Dalziel et al. 2006. The ST6GalNAc-I sialyltransferase localizes throughout the golgi and is responsible for the synthesis of the tumor-associated sialyl-Tn O-glycan in human breast cancer. *Journal of Biological Chemistry* 281:3586–3594.
122. Burchell, J.M., A. Mungul, and J. Taylor-Papadimitriou. 2001. O-linked glycosylation in the mammary gland: Changes that occur during malignancy. *Journal of Mammary Gland Biology and Neoplasia* 6:355–364.
123. Goodarzi, M.T., and G.A. Turner. 1995. Decreased branching, increased fucosylation and changed sialylation of alpha-1-proteinase inhibitor in breast and ovarian cancer. *Clinica Chimica Acta* 236:161–171.
124. Dennis, J.W., and S. Laferte. 1989. Oncodevelopmental expression of -GlcNAcβ1–6Manα1–6Man β1-branched asparagines-linked oligosaccharides in murine tissues and human breast carcinomas. *Cancer Research* 49:945–950.
125. Fernandes, B., U. Sagman, M. Auger, M. Demetrio, and J.W. Dennis. 1991. B1–6 branched oligosaccharides as a marker of tumor progression in human breast and colon neoplasia. *Cancer Research* 51:718–723.
126. Abbott, K.L., K. Aoki, J.M. Lim et al. 2008. Targeted glycoproteomic identification of biomarkers for human breast carcinoma. *Journal of Proteome Research* 7:1470–1480.
127. Abd Hamid, U.M., L. Royle, R. Saldova et al. 2008. A strategy to reveal potential glycan markers from serum glycoproteins associated with breast cancer progression. *Glycobiology* 18:1105–1118.
128. Kyselova, Z., Y. Mechref, P. Kang et al. 2008. Breast cancer diagnosis and prognosis through quantitative measurements of serum glycan profiles. *Clinical Chemistry* 54:1166–1175.
129. Alley, W.R., M. Madera, Y. Mechref, and M.V. Novotny. 2010. Chip-based reversed-phase liquid chromatography-mass spectrometry of permethylated N-linked glycans: A potential methodology for cancer-biomarker discovery. *Analytical Chemistry* 82:5095–5106.

Part II

The World of Omics and Applications in Plant and Agricultural Sciences

13

Genomics, Proteomics and Metabolomics: Tools for Crop Improvement under Changing Climatic Scenarios

Hifzur Rahman, D. V. N. Sudheer Pamidimarri, Ramanathan Valarmathi, and Raveendran Muthurajan

CONTENTS

13.1 Introduction

The world's population is increasing at an alarming rate and will probably reach more than nine billion by the end of 2050; yet global food productivity shows a decreasing trend due to the accumulating negative effects of environmental stress originated by global climate change. Abiotic stresses such as drought, salinity, extreme temperatures, chemical toxicity, and oxidative stress are serious threats for future agriculture and result in the deterioration of the environment. Abiotic stress is the primary cause for most of the crop losses worldwide, reducing average yields by more than 50% [1]. Drought and salinity are becoming particularly widespread in many regions, and may cause serious salinization of more than 50% of all arable lands by 2050. Because of the increasing intensity of abiotic stresses caused by climate change, the gap between the potential yield of crops and the realized yield is increasing day by day (Figure 13.1). Minimizing these losses is a major concern for developing countries to improve the global food security.

Abiotic stresses lead to a series of morphological, physiological, biochemical, and molecular changes that adversely affect plant growth and productivity [2]. To activate and integrate plant stress responses, the expression of thousands of genes is altered by the onset of stress [3–5]. These genes encode proteins involved in numerous biological processes including stress responses as well as a large number of proteins of unknown function. Importantly, the expression of many genes with regulatory functions such as transcription factors, RNA-binding proteins, calcium-binding proteins, kinases, phosphatases, etc., is altered by stress. These genes are probably involved not only in regulating downstream stress responses but also in stress perception and signaling [6,7]. Indeed, evidence suggests that stress signals are transduced to a set of transcription factors that together activate numerous downstream genes involved in the various stress responses [8].

Developing genotypes with acceptable performance under drought, submergence, salinity, high temperatures, or low nutrient availability is essential for the sustainability of crop production in view of climate change [9,10]. Conventional breeding has been successful, but the pace of genetic progress must now increase to meet the projected demand for agricultural products [11].

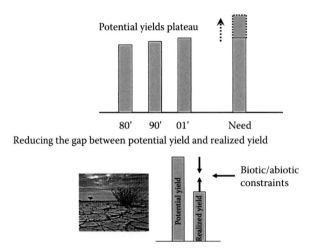

FIGURE 13.1
Constraints and strategies for increasing yield potential under changing climatic scenarios.

13.2 Functional Genomics

Large-scale genome sequencing projects have greatly changed the face of biology. Genomics has often been referred to as a new field, which has led to a paradigm shift in the way science has performed. Meanwhile, the postgenomic era has emerged by taking full advantage of the vast amount of genome sequence data. With the current improvements, it has become technically more convenient to handle/analyze biology and to answer the research/biological questions in nonhypothetical and nonbiased approaches [12,13]. Recent technological advances and the rapid development of novel genomic tools permits the investigation of a complete genome at once and in a single experiment. Currently, mass of the genome data is being converted into gene function data, providing functional value is added to the nucleotide sequence collections. Identifying the exact sequence and the location of all the genes of a given organism will be the first step towards understanding the functions of a biological system individually and together as a system [14]. In this respect, functional genomics is the key approach to transforming available data into functional information.

The term functional genomics can be referred to as the "development and application of global (genomewide or systemwide) experimental approaches to assess gene function by making use of the information and reagents provided by structural genomics" [15]. Gene "function" can be considered from several points of view: it can mean biochemical function (e.g., protein kinase), cellular function (e.g., a role in a signal transduction pathway), developmental function (e.g., a role in pattern formation), or adaptive function (the contribution of the gene product to the fitness of the organism). Having identified a new sequence, the simplest way to obtain (essentially biochemical) functional information is the comparison of sequence databases. Currently, approximately 50% of newly identified genes show sequence similarity to the functionally described known genes reported earlier. However, computerized analyses are generally not sufficient to define gene function with a high level of confidence, and thus experimental confirmation is needed in most cases. Indirect information on cellular or developmental function can be obtained from spatial and temporal expression patterns; for example, the presence of messenger RNA (mRNA) or protein (or both) in different cells/tissues, developmental stages, or under various types of biotic and abiotic stresses. The subcellular localization and posttranslational modifications of proteins can be informative as well. Knocking-out or overexpressing a gene could result in change(s) in the phenotype or development of the plant, which will allow us to deduce the function of the gene based on the changes observed. The function of mutants or natural variants of a gene for the adaptation/fitness of a plant can be identified by comparative analysis of wild-type with the plants under various environmental stresses [16].

Functional genomics is a general approach toward understanding how the genes of an organism work together by assigning new functions to unknown genes. The function of an unknown gene can be deduced from its sequence, structure, or location in the genome (on chromosome) using databases if genes with similar sequences are known in some other organisms. However, to pin down the exact function of unknown genes, it is necessary to understand each gene's role in the complex orchestration of all gene activities in the plant cell [14]. Gene function analysis therefore necessitates the analysis of temporal and spatial gene expression patterns. The most conclusive information about changes in gene expression levels can be gained from analysis of the varying qualitative and quantitative changes of mRNAs, proteins, and metabolites. These new developments have led to the creation of new fields of research within functional genomics, named comparative genomics, transcriptomics, proteomics, metabolomics, and phenomics. Together, these approaches will allow a comprehensive and systematic functional analysis of genomes with the potential to greatly accelerate the rate of gene function prediction [14].

13.2.1 Genomic DNA and cDNA Libraries

13.2.1.1 Creating a Genomic Library

Construction of the genomic library is the most classic yet most effective technique still in use for the execution of modern genomic experimentations. Technically, to create a genomic library, DNA is extracted from the desired cells/tissues and is cut into fragments by using a restriction enzyme for a limited amount of time (a partial digestion) so that only some of the restriction sites in each DNA molecule are cut. Because of random yet partial digestion, different DNA molecules will be cut at different loci, and a set of overlapping fragments will be produced. The fragments are then cloned to suitable vectors and transferred to bacteria. This technique produces a set of bacterial cells or phage particles containing the overlapping genomic fragments. A few of the clones contain the entire gene of interest; a few contain parts of the gene, but most contain fragments of unknown sequences. A genomic library contains all of the DNA sequences found in an organism's genome; therefore, a genomic library must contain a large number of clones to ensure that all DNA sequences in the genome are represented in the library.

13.2.1.2 Creating a cDNA Library

Although the genome libraries contain the most useful pool of consortia, researchers often might not need information from the nontranscribing regions, and fishing the transcribed target from the pool of consortia is labor-intensive. Therefore, a cDNA library consisting only of those DNA sequences that are transcribed into mRNA (called a cDNA library because the entire DNA in this library is *complementary* to mRNA) helps and provides much insight to the researcher for the target investigation. To create a cDNA library, mRNA is separated from a pool of RNA, that is, mRNA, rRNA, and tRNA, exploiting the poly(A) tail of mRNA by passing total cellular RNA through a column packed with oligo(dT) chains. As the RNA moves through the column, the poly(A) tails of mRNA molecules pair with the oligo(dT) chains and are retained in the column, whereas the rest of the RNA passes through. The mRNA can then be washed from the column by adding a buffer that breaks the hydrogen bonds between poly(A) tails and oligo(dT) chains. The mRNA molecules are then reverse-transcribed into cDNA using reverse transcriptase and short oligo(dT) primers. The resulting RNA–DNA hybrid molecule is then converted into a double-stranded cDNA molecule by one of several methods. One common method is to treat the RNA–DNA hybrid with RNase to partially digest the RNA strand. Partial digestion leaves gaps in the RNA–DNA hybrid, allowing DNA polymerase to synthesize a second DNA strand by using the short undigested RNA pieces as primers and the first DNA strand as a template. DNA polymerase eventually displaces all the RNA fragments, replacing them with DNA nucleotides, and nicks in the sugar–phosphate backbone are sealed by DNA ligase. These cDNA are then cloned into suitable vectors to produce a cDNA library.

Much of the eukaryotic DNA consists of repetitive (and other DNA) sequences that are not transcribed into mRNA and the sequences are not represented in a cDNA library. A cDNA library has two additional advantages. First, it is enriched with fragments from actively transcribed genes; second, introns do not interrupt the cloned sequences. Introns would pose a problem when the goal is to produce a eukaryotic protein in bacteria because most bacteria have no means of removing the introns.

The disadvantage of a cDNA library is that it contains only sequences that are present in mature mRNA. Introns and any other sequences that are altered after transcription are not present; sequences such as promoters and enhancers that are not transcribed into RNA are not present in a cDNA library. It is also important to note that the cDNA library represents only those gene sequences expressed in the tissue from which the RNA was isolated. Furthermore, the frequency of a particular DNA sequence in a cDNA library depends on the abundance of the corresponding mRNA in the given tissue. In contrast, almost all genes are present at the same frequency in a genomic DNA library.

13.2.1.3 Screening DNA Libraries

Technically, the construction of the libraries is very simple; however, the screening of the libraries is a massive task to identify and pick our target gene of interest. One common way to screen libraries is with probes. Probes can be used to find cloned fragments of DNA in bacteria or phages. To use a probe, replicas of the plated colonies or plaques in the library must first be made. The method for screening either cDNA or genomic DNA library is given in Figure 13.2.

How is a probe obtained when the gene has not yet been isolated? It is the major point of heddle; one option is to use a similar gene from another organism as the probe. For example, if we want to screen a pulse genomic library for the drought-responsive gene and the gene has already been isolated from rice, we could use a purified rice gene sequence as the probe to find the pulse gene for drought. Successful hybridization does not require perfect complementarity between the probe and the target sequence; therefore, a related sequence can often be used as a probe. The temperature and salt concentration of the hybridization reaction can be adjusted to regulate the degree of complementarity required for pairing

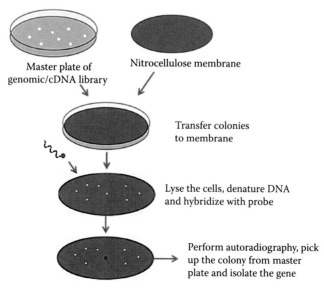

FIGURE 13.2
Screening of genomic and cDNA libraries using known/unknown probes.

to take place. Alternatively, synthetic probes can be created if the protein produced by the gene has been isolated and its amino acid sequence has been determined. With the use of the genetic code and the amino acid sequence of the protein, possible nucleotide sequences of a small region of the gene can be deduced and synthesized and used as a probe for hybridization. When part of the DNA sequence of the gene has been determined, a set of DNA probes can be synthesized chemically by using an automated machine known as an oligonucleotide synthesizer. The resulting probes can be used to screen a library for a gene of interest. Yet another method of screening a library is to look for the protein product of a gene. This method requires that the DNA library be cloned in an expression vector. The clones can be tested for the presence of the protein by using an antibody that recognizes the protein or by using a chemical test for the protein product. This method depends on the existence of a test for the protein produced by the gene. Almost any method used to screen a library will identify several clones, some of which will be false-positives that do not contain the gene of interest; several screening methods may be needed to determine which clones actually contain the gene.

13.2.2 Expressed Sequence Tags

Expressed sequence tags (ESTs) are the sequences obtained by single-pass sequencing of randomly chosen cDNA clones from libraries at all stages of plant growth and life cycle, and is currently the most efficient method for the large-scale discovery of many genes at a fast and affordable rate from crop plants with large genomes. ESTs represent nucleotide sequences of small portions of the expressed genes. These ESTs serve to tag and fish out the putative genes and also help in the quantification of their expression. Large EST programs for several crop and animal species are currently under way in many research laboratories worldwide, which lead to a steadily increasing number of entries in the EST database (http://www.ncbi.nlm.nih.gov/dbEST). Gene discovery via ESTs consists of several steps which include: (1) construction of cDNA libraries and single-pass sequencing of randomly selected clones, (2) EST quality check: the removal of vector and low-quality sequences, (3) alignment of ESTs to identify the number of represented genes, and (4) annotation of these genes or their partial sequences using available database searches.

13.2.2.1 EST Clustering/Gene Content

The assembly of gene sequences or parts thereof from a collection of ESTs to determine the number of represented genes is an important task. Special program packages such as the Phred/Phrap/Consed system, UniGene, Genexpress Index, TIGR_ASSEMBLER, STACK_PACK, CAP3, PCP/CAP4, HarvESTer, and others have been and continue to be developed for the assembly of large EST collections. The result of the assembly process can be divided into so-called singletons (sequences which do not assemble with any other sequence), groups of assembled sequences that might be called clusters, contigs, tentative consensus, tentative genes, unique gene, or unigenes, etc.

13.2.2.2 Applications of EST Clones and Sequences

The wealth of information generated by EST projects can be used in various ways. Currently, the most interesting uses are large-scale transcript profiling and the development of molecular markers. ESTs allow the efficient development of highly valuable molecular markers (functional markers) because genes often represent single or low copy sequences. These are of great value in the cereal genome, which consists of up to 80% of highly repetitive DNA. Classic hybridization-based restriction fragment length polymorphism (RFLP) markers have been developed from ESTs and used extensively for the construction of high-density genetic and physical maps in several crop species. Often, EST-based RFLP markers allow comparative mapping across different species because sequence conservation is high in the coding regions. Hence, marker development and map-based cloning in one species will profit directly from data that are available in any other species [17]. The available sequence information allows the design of primers, which can be used to screen cultivars of interest for length polymorphisms. In the future, expression cloning and the production of protein arrays may be added to this list. Protein arrays produced from expressed cDNA clones would facilitate functional studies of proteins with respect to enzymatic activity and ligand binding, including the search for interacting partners and the isolation of antibodies. Furthermore, sequence data generated through ESTs can be used to study gene families. In the future, one of the main challenges will be to couple expression data with metabolic and regulatory network models for better interpretation and access to the large amount of information that will soon be available in many crop species.

EST information is present in public databases for a variety of species, including a number of plants (Table 14.1 [18–21]). The database of the National Center for Biotechnology Information lists more than 39,000 *Arabidopsis* ESTs. The Institute for Genome Research very conveniently has ordered overlapping ESTs into tentative contigs. There are now more than 1 million cereal EST sequences in the public databases, with rice, wheat, and barley dominating. The large number of ESTs and the diversity of cDNA libraries that have been used to generate the sequences have made "Electronic Northerns" a useful method for assessing gene expression and this provides a good first measure of transcript abundance.

EST library has been constructed and utilized by many researchers to identify genes responsible for providing tolerance against various biotic and abiotic stresses. Nogueira et al. [22] found several cold-inducible genes, which have not been previously reported as being cold-inducible, including those for cellulose synthase, ABI3-interacting protein 2, *Os*NAC6 protein, and phosphate transporter after analyzing approximately 1500 ESTs. Pih et al. [23] identified 15 salt stress–inducible genes with early, late, or continuous expression patterns in *Arabidopsis* after sequencing 220 clones randomly from a cDNA library. Gorantla et al. [24], after sequencing and analyzing 7794 cDNA clones, identified 125 genes expressed under drought stress in leaf and panicle tissues.

A list of web sites and publicly available databases appears in Table 13.1. Aside from these, various large-scale EST projects are also under way for various crop species like maize (http://www.maizegdb.org/), wheat (http://wheat.pw.usda.gov/wEST/), and sugarcane (SUCEST; http://sucest-fun.org/en/projects/sucest/overview) for general and specific traits.

13.2.3 Differential Screening of cDNA Libraries/Subtractive Hybridization

Subtractive hybridization is a technique for identifying and characterizing differences between two populations of nucleic acids. It detects differences between the RNA in different cells, tissues, organisms, or sexes under normal conditions, or during different growth phases, after various treatments (i.e., hormone application, heat shock) or in diseased (or mutant) versus healthy (or wild-type) cells. Subtractive hybridization also detects DNA differences between different genomes or between cell types in which deletions or certain types of genomic rearrangements have occurred. Subtractive hybridization techniques have identified many differentially expressed sequences from a wide variety of

TABLE 13.1

Web Sites and Databases Relevant to Plant Functional Genomics

Database	Web Site
National Center for Biotechnology Information	http://www.ncbi.nlm.nih.gov/
REDB (Rice EST DataBase)	http://redb.ncpgr.cn/modules/redbtools/
Soybean EST database	http://soybase.org/EstDB/
SolEST Database (Solanaceae EST Database)	http://biosrv.cab.unina.it/solestdb/index.php
The Institute for Genome Research	http://www.jcvi.org
GRAMENE Database	http://www.gramene.org/
The Arabidopsis Information Resource (TAIR)	http://www.arabidopsis.org/
Rice Genome Project (RGP)	http://rgp.dna.affrc.go.jp/
EBI: The ArrayExpress Database	http://www.ebi.ac.uk/arrayexpress/
ApiEST-DB	http://www.cbil.upenn.edu/apidots/
NCBI EST Database	http://www.ncbi.nlm.nih.gov/dbEST/
Plant Genome Database	http://www.plantgdb.org/
UniGene Database at NCBI	www.ncbi.nlm.nih.gov/UniGene
The Rice Expression Profile Database (RiceXPro)	http://ricexpro.dna.affrc.go.jp/
The Database for Annotation, Visualization, and Integrated Discovery (DAVID)	http://david.abcc.ncifcrf.gov/
The Rice Annotation Project Database	http://rapdb.dna.affrc.go.jp/
ExPASy	http://expasy.org/
Mascot	http://www.matrixscience.com/home.html
Cold Spring Harbor Laboratory	http://genetrap.cshl.org/
Tomato Expression Database	http://ted.bti.cornell.edu/
GABI-Kat	http://www.gabi-kat.de/
SIGnAL	http://signal.salk.edu/cgi-bin/tdnaexpress
RIKEN	http://rarge.psc.riken.jp/dsmutant/index.pl
European Arabidopsis Stock Center	http://arabidopsis.info/

organisms. Subtractive hybridization requires two populations of nucleic acids; the tester (or tracer), which contains the target nucleic acid (the DNA or RNA differences that one wants to identify), and the driver, which lacks the target sequences. The two populations are hybridized with a driver-to-tester ratio of at least 10:1. Because of the large excess of driver molecules, tester sequences are more likely to form driver–tester hybrids than double-stranded testers. Only the sequences in common between the tester and the driver hybridize, however, leaving the remaining tester sequences either single-stranded or forming tester–tester pairs. The double stranded driver, driver-tester hybrid are removed by hydroxyapatite chromatography and single-stranded driver RNA can be removed chemically or enzymatically, leaving only single-stranded cDNA tester after the subtraction. (the "subtractive" step). Multiple rounds of subtractive hybridization are performed to identify truly tester-specific nucleic acid sequences. There are five basic steps to subtractive hybridization (Figure 13.3): (1) choosing the material for isolating tester and driver nucleic acids, (2) producing the tester and the driver, (3) hybridizing the tester and the driver, (4) removing driver–tester hybrids and excess driver (subtraction), and (5) isolating the complete sequence of the remaining target nucleic acid.

Variations are possible at each step, and the materials used and methods chosen depend on the desired results. It is necessary to further analyze isolated tester sequences by other expression profiling methods like Northern blotting, *in situ* hybridization, or polymerase

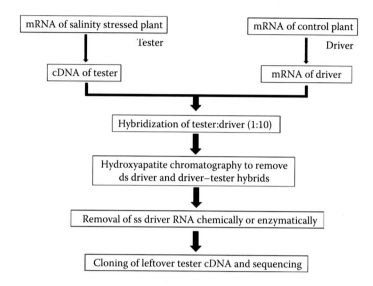

FIGURE 13.3
Schematic representation for the identification and isolation of genes using suppression subtractive hybridization.

chain reaction (PCR) methods to determine whether the sequences are truly tester-specific or not. Kumari et al. [25] constructed a subtractive cDNA library responsive to salinity tolerance in rice and reported 1194 salinity-regulated cDNAs that may serve as repositories for future individual gene-based functional genomics studies. Suppression subtractive hybridization was carried out to identify early salt stress responsive genes in tomato roots [26].

13.3 Expression Genomics/Transcriptomics—Gene Expression Analysis by mRNA Profiling

Expression profiling has become an important tool to investigate how an organism responds to environmental changes. Plants, being sessile, have the ability to dramatically alter their gene expression patterns in response to environmental changes such as temperature, water availability, or the presence of deleterious levels of ions [27]. Expression profiling of plant response to abiotic stresses are expected to lead to the identification of genes and regulators that will be useful to improve stress tolerance as well as to study the molecular basis of regulation of abiotic stresses in plants.

13.3.1 Techniques Used for Evaluating Gene Expression in Functional Genomics Studies

A fundamental step in any functional genomics study is the analysis of gene expression. One of the greatest strengths of genomics compared with other disciplines is the prospect of analyzing the expression of thousands of genes simultaneously; resulting in a more comprehensive picture of changes occurring in the transcriptome across different conditions [28].

The technology available for the analysis of gene expression can be divided into two categories: closed and open systems. By using a closed system, a definite number of genes can be assessed and taken for analysis. Therefore, the coverage of genes will be related to the completeness of the knowledge of the genome being studied, limiting this kind of analysis to the most well-characterized species or systems [28]. Typically, closed systems such as microarrays and real-time PCR have been extensively used in gene expression analysis in plants [29,30]. On the other hand, with open systems, the expression patterns of infinite number of genes can be assessed and there is also no need for previous knowledge of the genome or transcriptome of the organism. cDNA-Amplified Fragment Length Polymorphism (cDNA-AFLP), Massively Parallel Signature Sequencing (MPSS), and especially Serial Analysis of Gene Expression (SAGE) have been successfully used to quantify transcript abundance and generate expression data across different tissue types or developmental stages in higher plants [31–34]. Open and closed systems should not be considered as competitors, but rather as complementary technologies to be used depending on the subject to be analyzed and the objectives of the research.

13.3.1.1 cDNA-AFLP

The cDNA-AFLP is an RNA fingerprinting technique used to analyze differentially expressed genes grown under normal and stress conditions. It is evolved from AFLP and was first described by Vos et al. [35] for the fingerprinting of expressed genomic DNA. The classic cDNA-AFLP procedure [36] uses the standard AFLP protocol on a cDNA template. The technique involves five steps (Figure 13.4): (1) conversion of mRNA in double-stranded cDNA, (2) restriction digestion of cDNA with two restriction enzymes—a frequent cutter and a rare cutter, (3) ligation of synthetic oligonucleotide adapters to the cDNA ends,

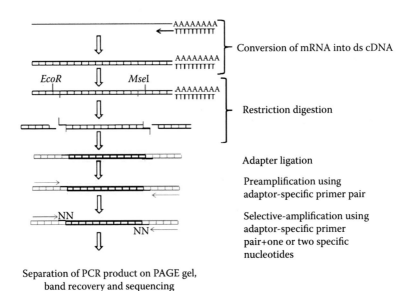

FIGURE 13.4
Schematic representation for identifying differentially expressed genes using cDNA-AFLP.

(4) preamplification of sets of restriction fragments using primers complementary to the adapter sequences, and (5) selective amplification of transcription-derived fragments using primer sequence complementary to adaptor sequence and small extensions of one, two, or three nucleotides to amplify fragments specific to those nucleotides. Gel analysis of the amplified fragments on sequencing polyacrylamide gels.

Restriction of plant cDNA with a combination of two restriction enzymes, a tetra-cutter and a hexa-cutter, allows a significant fraction of the cDNA population to be cleaved and represented as a discrete banding pattern on a sequencing polyacrylamide gel electro- phoresis. Specific fragments can be eluted from gels and sequenced to identify genes with differential expression [36].

The method needs only minute amounts of RNA due to the preamplification step per- formed with nonselective primers. Because stringent hybridization conditions are used in the amplification reactions, mismatched priming events are observed only in cases in which transcript levels are extremely high. This results in cDNA-AFLP banding patterns being highly reproducible and almost free of false-positives. Because band intensity is a direct function of template concentration, it is likely that transcript concentrations can be measured quite accurately by combining cDNA-AFLP with high resolution quantita- tive separation techniques. A significant disadvantage of the cDNA-AFLP method is the requirement for appropriate restriction sites on the cDNA molecules. cDNA-AFLP experi- ments using different enzymes are necessary to visualize every cDNA present in a plant cell with high probability.

cDNA-AFLP has been successfully employed in identifying transcription-derived frag- ments responsible against various types of abiotic stresses. Gui et al. [37] used cDNA-AFLP to analyze differentially expressed genes in contrasting wheat genotypes for salinity tol- erance, which lead to the identification of a large number of gene fragments related to salt stress; among which glycogen synthase kinase-shaggy kinase (TaGSK1) was strongly expressed in salt-tolerant genotypes, suggesting its role in salt stress response in wheat. Frank et al. [38] identified calcium-dependent protein kinase 2 (*CDPK2*), which was upreg- ulated in heat-stressed microspores of tomato.

13.3.1.2 Serial Analysis of Gene Expression

The SAGE method is a first-pass screening that relies on high-throughput sequencing of short fragments of cDNAs (tags) to estimate the relative concentration of mRNA in a tis- sue [39]. Among the various techniques used to assess transcript profiling, SAGE is one of the most powerful, and because of its reliance on DNA sequencing, SAGE can select transcripts not yet identified by high-throughput sequencing projects. However, it is dif- ficult to analyze large numbers of samples because SAGE is very expensive to perform. A prerequisite for the identification of the tags is the availability of large-sequence databases for the species under study. The technique is powerful but not very convenient for the comparison of many different samples and for the study of the rarer transcripts. The basic steps involved in SAGE are (Figure 13.5) the following:

- Double-stranded cDNA is synthesized from mRNA with a biotinylated oligo(dT) primer. The DNA is then cleaved with *Nla*III, which recognizes 4 bp and cuts DNA.

- The biotinylated 3' ends of the cleaved cDNA molecules are isolated by binding to streptavidin. The truncated cDNA is divided in half, and each half is ligated via the cohesive 3' end to one of two linkers containing a *Fok*I restriction site (*Fok*I is

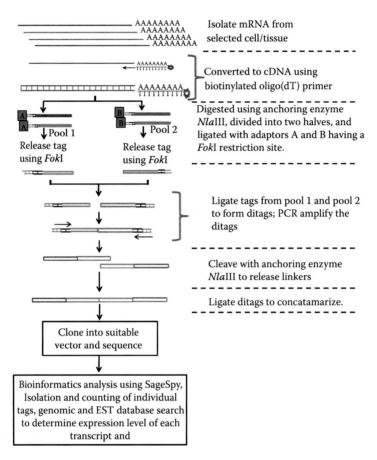

AAAAAAAA
AAAAAAAA
AAAAAAAA
AAAAAAAA   Isolate mRNA from
                selected cell/tissue

AAAAAAA
TTTTTTTT   Converted to cDNA using
AAAAAAAA   biotinylated oligo(dT) primer
TTTTTTTTTT

Pool 1   Pool 2   Digested using anchoring enzyme
Release tag   Release tag   NlaIII, divided into two halves, and
using FokI   using FokI   ligated with adaptors A and B having a
                                FokI restriction site.

                                Ligate tags from pool 1 and pool 2
                                to form ditags; PCR amplify the
                                ditags

                                Cleave with anchoring enzyme
                                NlaIII to release linkers

                                Ligate ditags to concatamarize.

Clone into suitable
vector and sequence

Bioinformatics analysis using SageSpy,
Isolation and counting of individual
tags, genomic and EST database search
to determine expression level of each
transcript and

FIGURE 13.5
(See color insert.) Schematic diagram showing steps involved in developing SAGE library for expression profiling.

a type II S-restriction endonuclease. These enzymes cleave at a defined sequence-independent distance away from their asymmetric recognition site).

- Cleavage of the ligation product with *FokI* results in the release of a linker with a short piece of cDNA called a tag.
- Cohesive ends are blunted, and the two pools of released tags are ligated to each other.
- Ligated tags then serve as templates for PCR amplification with primers specific for each linker. The resulting amplicons contain two tags (one "ditag") linked tail to tail, flanked by the *NlaIII* recognition site and a linker sequence at each end. Cleavage of the PCR product with *NlaIII* allows the isolation of ditags that are concatenated, inserted into a plasmid vector, and cloned.
- Clones containing 10 to 50 ditags are amplified by PCR and sequenced.
- In the sequence, ditags are counted easily because they are separated by the tetra-cutter recognition sequence, which serves as punctuation signal. The frequency of each tag in a SAGE library directly reflects the abundance of the corresponding mRNA in the tissue.

Two major advantages of SAGE are that uncloned cDNA can be analyzed and that no special device other than a sequencer is required. Because SAGE clones may contain 20 to 50 tags, a representative SAGE library is much smaller than the cDNA library. The SAGE technology has been employed to reveal changes in gene expression in *Arabidopsis* leaves and pollen undergoing cold stress [40,41]. Matsumura et al. [42] observed that among differentially expressed genes between anaerobically treated and untreated rice seedlings, metallothionein and globulin genes were highly expressed and prolamin gene was most highly inducible. Eight genes, including six showing no match to any rice EST, were anaerobically induced and six genes were repressed.

13.3.1.3 Differential Display Reverse Transcription PCR

The differential display reverse transcription PCR (DDRT-PCR) method was first described by Liang and Pardee [43]. The method provides an effective tool to detect individual mRNA that are differentially expressed in different eukaryotic cell types, and then allows the recovery and cloning of the corresponding cDNAs. The principle of the DDRT-PCR technique is to amplify specific subpopulations of mRNA, using reverse transcriptase and PCR to produce a population of PCR fragments of different lengths. To do this, mRNA is reverse-transcribed in subsets using anchored oligo(dT) primers that recognize different fractions of the total mRNA population. The resulting cDNA is then amplified with the same anchored oligo(dT) primer and a short arbitrary primer. Any one of the labeled dNTPs is introduced into the reaction and the labeled products are separated on a DNA sequencing gel and visualized by autoradiography (Figure 13.6).

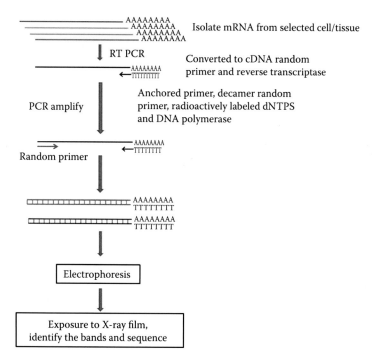

FIGURE 13.6

Flow chart of DDRT-PCR methodology for identifying differentially expressed genes.

Many refinements of the original differential display technology have been described. Instead of decameric upstream primers, elongated arbitrary primers have been used. This opens the possibility of including common sequence motifs to target particular classes or families of genes [44,45]. The DDRT-PCR technique has been applied to isolate several cDNAs induced in plants under various stress conditions, for example, salt stress [46], heat shock [47], ozone treatment [48], nutrient starvation [49], wounding [50], etc.

The DDRT-PCR technique suffers from several drawbacks. DDRT-PCR generates approximately 30% false-positives that do not represent differentially expressed genes [51]. This makes further confirmation of differentially expressed genes necessary, which is not only labor-intensive but also requires large amounts of RNA that often cannot be obtained without RNA amplification [52]. One other drawback of this technique is that because amplification starts from the 3' end of the mRNA, amplicons are from the untranslated 3' region of the gene, which often does not contain the information required to perform successful similarity searches in sequence databases. Finally, using DDRT-PCR quantitative measurement of transcript concentration cannot be done because there is no accurate relationship between signal strength and the initial concentration of the corresponding mRNA.

13.3.1.4 RFLP-Coupled Domain-Directed Differential Display

It is not always necessary to detect all differences in gene expression within a given system, as is the goal of differential display, but rather, the detection of differences within a certain gene family may be desired. Many genes and their protein products have a modular structure with the number of existing modules being quite limited [53]. In many transcription factors, in which the presence of certain domains (family-specific domains) defines membership in the different gene families, for example, the MADS-box family.

The presence of a family-specific domain in many gene families has been used in developing gene family–specific version of mRNA display, which is termed as "RFLP-coupled domain-directed differential display" (RC4D). By introducing RFLP, family members can easily be distinguished by size [54]. RC4D is a method specifically designed to analyze the expression of multigene families at different developmental stages, in diverse tissues or in different organisms [54]. RC4D combines cDNA-AFLP technology with a gene family–specific version of DDRT-PCR. In RC4D, instead of arbitrary decameric primers, longer primers directed against a family-specific domain are used, allowing cDNAs belonging to the same gene family to be selectively amplified. As the amplification products are relatively uniform in length, RFLP is introduced by digestion with a frequently cutting restriction enzyme. This reduces the amplicon size from approximately 1 kbp to several hundred base pairs, which is optimal for separation on acrylamide gels. Family members can thus be easily distinguished by size. The use of longer primers and stringent annealing temperatures makes the RC4D method more reproducible and far less parameter-dependent than DDRT-PCR, allowing semiquantitative determination of transcript concentration. The main limitation of the RC4D method is that it requires a conserved domain to be situated upstream of the genes' 3' end, which only allows genes belonging to the same gene family to be analyzed. Family members bearing a respective restriction site within the family-specific domain will be lost because cleavage of the amplicon resulting from the initial PCR amplification will destroy the domain-specific primer binding site.

RC4D was first used to analyze differential expression of MADS-box genes in male and female inflorescences of maize [54,55]. Tahtiharju et al. [56] identified several cDNAs coding for calcium dependent protein kinases involved in calcium signaling during cold induction of the kinase genes of *Arabidopsis thaliana* using RC4D.

13.3.1.5 Microarray

The microarray is also called DNA chips or Biochips made up of silicon or nylon or glass on which DNA fragments are fabricated. The sources of DNA fragments may be obtained from cDNA clones, EST clones, genomic clones, or DNA amplified from open reading frames. The size of a single DNA chip varies from 1 to 3.24 cm^2, which can display 25,000 to 44,000 genes and the chips are known as 25K or 44K based on the number of genes spotted on the slide. The high density and miniaturization make genomewide expression studies feasible by using either cDNA or oligonucleotide arrays.

DNA chip technologies utilize microscopic arrays (microarrays) of molecules immobilized on solid surfaces for biochemical analysis. Advanced arraying technologies such as photolithography, microspotting, and ink-jetting, coupled with sophisticated fluorescence detection systems and bioinformatics, permit molecular data gathering at an unprecedented rate [57]. Microarray technology evolved from Southern blotting but is the opposite to Southern blotting because microarray probe sequences are immobilized on a solid surface and allowed to hybridize with fluorescently labeled "target" mRNA. The intensity of fluorescence of a spot is proportional to the amount of target sequence that has hybridized to that spot and, therefore, to the abundance of that mRNA sequence in the sample. Microarrays allow for the identification of thousands of candidate genes involved in a given process based on variation between transcript levels for different conditions and shared expression patterns with genes of known function [57].

There are a number of methods for producing microarray slides, which are mainly done using robotic systems that have their own advantages and disadvantages. The three main methods for doing this are (1) spotting of DNA fragments directly onto the slide, (2) arraying of prefabricated oligonucleotides, and (3) *in situ* synthesis of oligonucleotides, done on the chip.

13.3.1.5.1 Preparation of the Probes

After the mRNA has been extracted from the cells or tissues under study, it is converted into cDNA. During this reaction, the DNA is labeled by the incorporation of fluorescently labeled dNTPs. The most commonly used fluorescent dyes in cDNA arrays are Cy3 and Cy5. This is due to their widely separated absorption and emission spectra, high molar absorption coefficients, and good fluorescence quantum yields. The steps involved in microarray analysis include (Figure 13.7): (1) probe preparation, (2) construction of array by using methods such as microspotting, piezoelectric printing/ink-jetting, photolithography, or by using a "pin and ring" device developed by Genetic Microsystems; (3) sample preparation and labeling with Cy3 and Cy5 dye (Cy3dUTP or Cy5dUTP); (4) hybridization (for 16–20 h at 45°C–60°C); (5) washing and drying of chip; and (6) scanning at 10 μm resolution, image acquisition, and analysis.

This high-throughput technology has been successfully used to analyze the regulation of genes at different stages of development [57,58] and in response to both abiotic and biotic stresses (Table 13.2). Gene expression profiling is the most widespread application of microarrays. Microarray assays may be directly integrated into functional genomic approaches aimed both at assigning function to identified genes, and to studying the organization and control of genetics. Using cDNA microarray, Seki et al. [59] identified genes related to drought, cold, or salinity, and examined the differences and cross-talk between three

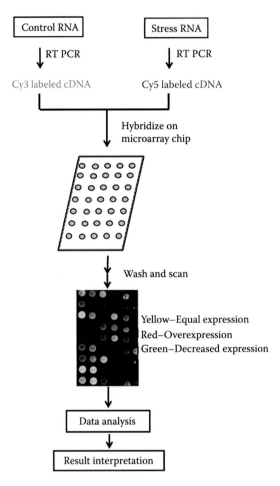

FIGURE 13.7
Flow diagram from microarray experimental principle.

TABLE 13.2

Use of Microarrays for Expression Profiling under Various Abiotic Stress Conditions in Different Crops

Trait under Study	Plant Species	Microarray Type	Reference
Drought and cold stresses	*Arabidopsis*	cDNA	[60]
Cold, drought, and high salinity	*Arabidopsis*	cDNA	[59]
Low-oxygen response	*Arabidopsis*	cDNA	[61]
Salt stress	Rice	cDNA	[62]
High-salinity stress	Ice plant	cDNA	[63]
Oxidative stress	*Arabidopsis*	cDNA	[64]
Water-deficit stress	*Arabidopsis*	cDNA	[65]
Salt, osmotic, and cold stress	*Arabidopsis*	Oligonucleotide	[4]
Drought and salinity	*Hordeum vulgare*	cDNA	[66]
Pollen under cold stress	*Arabidopsis*	cDNA	[40]
Rehydration after dehydration	*Arabidopsis*	cDNA	[67]

stresses. In total, 277 drought-inducible, 53 cold-inducible, and 194 salinity-inducible genes were identified. Only 22 genes were identified as drought-, cold-, and salinity-inducible genes, whereas 70% of salinity-inducible genes were also induced by drought stress, which indicates the strong correlation between the drought and salinity stress responses. Yu and Setter [68], using cDNA microarray, found that under water deficit conditions, heat shock proteins, chaperones, and major intrinsic proteins were upregulated in the placenta of maize.

13.4 Reverse Genetics as a Tool of Functional Genomics

"Reverse genetics" is an approach to discover the function of a gene by analyzing the phenotypic effects of specific gene sequences obtained by DNA sequencing. The reverse genetic process proceeds in the opposite direction of the so-called forward genetic screens of classic genetics. Although forward genetics seeks to find the genetic basis of a phenotype or trait, reverse genetics seeks to find what phenotypes arise as a result of particular genes. To learn, the influence a sequence has, "on phenotype," or to discover its biological function, researchers can engineer a change or disruption in the DNA. After this change has been made, a researcher can look for the effect of such alterations in the whole organism. There are several different methods of reverse genetics that have proved useful.

13.4.1 RNA Interference and Cosuppression

RNA interference (RNAi, also known as posttranscriptional gene silencing or cosuppression) is thought to be a key defense against viruses, as well as a way for regulating endogenous genes. RNAi has several advantages over other strategies of functional genomics; the primary advantage is that it specifically targets the chosen gene. As RNAi is a homology-dependent process, careful selection of a unique region of the target sequence is necessary to assure that whether the construct will specifically silence a member of a gene, gene family, or multiple members of the gene family (if conserved sequence domain is targeted). Besides, this RNAi can also silence the gene in specific tissues only by using tissue-specific promoter and also the expression of ihpRNAs using inducible promoters can control the extent and timing of gene silencing [69,70], such that essential genes are only silenced at chosen growth stages or in chosen plant organs. In these ways, RNAi provides the flexibility necessary for the characterization of genes of diverse functions. Genetic studies to "knock down" SCaBP and PKS levels using RNAi have revealed that *scabp5* and *pks3* mutants are impaired in their response to abscisic acid (ABA), and that the SCaBP5–PKS3 complex specifically senses and transduces ABA-specific Ca^{2+} signals [63]. ABA-induced expression levels of the cold- and drought-responsive genes *COR47, COR15A,* and *RD29A* were substantially higher in the *scabp5* and *pks3* mutants when compared with their expression in the wild-type, and these mutants expressed higher levels of *COR47* and *COR15A* even without exogenous ABA [71]. A collection of gene-specific sequence tags had been generated for at least 21,500 *Arabidopsis* genes [72], and as an application, hairpin RNA–expressing lines have been constructed for 8136 different gene-specific sequence tags. When combined with appropriate screening strategies, these resources will greatly improve our knowledge of plant stress tolerance.

13.4.2 Mutagenesis

A direct way of characterizing gene function is to mutate the gene and study the effect on the phenotype. Gene expression can be repressed by use of antisense, cosuppression, or RNAi strategies, but these approaches are largely based on a single-gene approach and are currently difficult to use on a large scale [73]. Chemical and insertional mutagenesis approaches have been extensively used for functional characterization of plant genes, and many successful examples of their uses have been published. However, the major disadvantage of this approach is the randomness of mutagenesis. For this reason, mutations need to be mapped to confirm their positions and must be sequenced to specifically identify the gene in which the mutation has happened. This requires large collections of mutant lines that are necessary to obtain good coverage of whole genomes. These limitations make the application of mutagenesis on a genomic scale very labor-intensive. Insertional mutagenesis is more commonly used in large-scale plant functional genomics because mobile DNA tags of known sequence can be utilized to retrieve flanking gene sequences, which helps in identifying the gene sequence in which the mutation has happened.

13.4.2.1 Insertional Mutagenesis

Insertion of a known segment of DNA into a gene of interest is a commonly used strategy for mutagenesis. Insertional mutagenesis offers a more rapid means to clone a gene as the insertion not only creates a mutation but also serves to tag the region, which helps in its identification [16]. The two most commonly used tools for insertional mutagenesis in plants are the use of transposons or T-DNA tags [74].

13.4.2.1.1 Transposon Tagging

Transposon tagging has been used to isolate numerous genes [75] since the first successful cloning of the bronze locus in maize by the *Ac/Ds* (activator–dissociation) transposon system [76]. For characterizing the MYB family of genes, three transposon tag collections [77–79] and one T-DNA tagged collection [80] were used. A total of 47 insertions were identified in 36 members of the R2R3 MYB family. Screening for an altered phenotype in 32 lines representing disruptions in 26 MYB genes showed no distinct phenotype in most of the lines analyzed using screens for stress among others with the exception of a UVB stress–tolerant line [78].

Mutant studies provide valuable data for functional validation of gene function, but because this is also a transgenic approach, environmental issues are naturally raised [81]. To overcome these concerns, endogenous transposons are desirable. *Tos17*, a rice retrotransposon, has been used for large-scale mutagenesis [82]. *Tos17* is activated only in tissue culture conditions and is not active under normal conditions. There are only two copies of *Tos17* under normal conditions in japonica rice and 5 to 30 transposed copies are found in plants regenerated from tissue culture.

Transposon tagging is often confined to either the self-fertilized diploid plants such as snapdragon or the cross-fertilized diploid crops like maize, both having well-characterized endogenous transposable elements. Although derivatives of maize Ac have been successfully used as transposable elements in crops to which they do not belong, their use is still limited and cannot be extended to all plant species. Therefore, T-DNA tagging involving random insertion of T-DNA into plant genomes using *Agrobacterium tumefaciens*–mediated transformation has been exploited.

13.4.2.1.2 T-DNA Insertion

T-DNA is a segment of the tumor-inducing (Ti) plasmid of *A. tumefaciens* and is delimited by short imperfect repeat border sequences. Any fragment cloned in between the left and right border can be transferred by *Agrobacterium* into the plants along with T-DNA, which gets inserted into the plant genome. The mutants are screened based on the unique sequence within the left and right borders, for instance, a selectable marker gene or a reporter gene [83]. T-DNA–tagged lines have been generated in *Arabidopsis* with the specific aim of isolating mutants altered in stress signaling. These tagged lines have been generated in the transgenic background of a stress-inducible *RD29A* promoter (responsive to cold, salt, dehydration, and ABA) fused to luciferase reporter gene. Several important genes in the abiotic stress pathway have been identified using these mutants. These lines have recently been used for identifying low temperature–responsive genes in rice. Lee et al. [84] analyzed 15,586 T-DNA–tagged rice lines, out of which 81 (0.52%) showed cold-responsive beta-glucuronidase (GUS) expression. Thirty-seven tagged genes were identified from these lines, two of which, *Os*RLK1 (putative LRR-type receptor-like protein kinase) and *Os*DMKT1 (putative demethyl menaquinone methyl transferase) were confirmed experimentally to be cold-responsive.

13.4.2.1.3 Activation Tagging

Insertional mutagenesis usually generates recessive loss-of-function mutations making them unsuitable for the functional analysis of redundant genes, are compensated by alternative metabolic or regulatory circuits, or have an additional role in early embryo or gametophyte development. Jeong et al. [85] have demonstrated that the CaMV 35S enhancer element efficiently increases the expression of nearby genes from either their 5′ or 3′ ends. This element enhances the expression of neighboring genes on either side of the randomly integrated T-DNA tag, resulting in gain-of-function in phenotypes. As the mutated gene contains a copy of T-DNA, it can be rapidly identified by either Thermal asymmetric interlaced PCR (TAIL-PCR) or plasmid rescue. Thus, if it is advantageous to do so, screens can be designed to simultaneously uncover both loss-of-function and gain-of function alleles. Weigel et al. [86], using T-DNA vectors that contain multimerized enhancers from the CaMV 35S gene and replacing kanamycin resistance gene with one conveying resistance against the herbicide glufosinate, facilitated the selection of transformed plants on soil rather than in culture. Lines identified from these initial populations have led to the discovery of a number of novel alleles and genes which undertake important functions in plant development, metabolism, and environmental interactions in *Arabidopsis* [87–89].

13.4.3 Gene and Enhancer Traps

Classic genetic approaches to gene identification rely on the disruption of a gene leading to a recognizable phenotype. However, the function of all genes cannot be uncovered by mutagenesis for two main reasons. First, many genes from multigene families are functionally redundant, sharing overlapping functions with other genes that may or may not be related at the sequence level. Mutation of a functionally redundant gene is not likely to lead to an easily recognizable phenotype because one or more other members of the family can provide the same function. Therefore, it is likely that the disruption of many plant genes will not result in an easily identifiable phenotype [90]. Second, many genes function at multiple stages of development. Mutations in these genes may lead to early lethality or may be highly pleiotropic, which can mask the role of a gene in a specific pathway.

To overcome such problems and to complement other efforts, modified tags containing a reporter gene have been developed.

Gene and enhancer trap insertions allow the identification of genes based on the expression pattern of a reporter gene; therefore, a mutant phenotype is not required. It requires the creation of a library which contains random genomic insertions of a reporter gene, the expression of which can be easily visualized. When the reporter gene is inserted within or nearby a chromosomal gene, the reporter gene used to be driven by promoter of chromosomal gene. Reporter gene expression in specific tissues or cell types can be identified, which shows gene expression in specific tissues or cells.

There are three basic kinds of traps: gene, enhancer, and promoter traps. Enhancer traps contain a reporter gene fused to a minimal promoter. The reporter gene will be expressed only when enhancer elements are close to the site of integration. Promoter traps contain a promoterless reporter gene, and expression will be seen only when the insertion is in an exon and only if that too is in the correct orientation in between the gene. On the other hand, a gene trap contains a reporter with one or more splice acceptor sites preceding it. In this case, expression will occur only if insertion occurs in an intron [90]. Later, the site of insertion of the trap is identified by performing TAIL-PCR using the sequences of the reporter gene. The bacterial *gus*A gene and the jellyfish green fluorescent protein gene are the two most commonly used reporter genes for traps. Sundaresan et al. [91] reported the use of gene and enhancer traps in the *Ac/Ds* family of maize transposons in *Arabidopsis* with GUS as a reporter. GUS expression could be detected in approximately 50% of enhancer trap lines and in 25% of gene trap lines. Several large collections of traps are now available for functional analysis. In rice, a total of 31,443 enhancer trap lines were generated using *GAL4/VP16*-UAS elements and GUS as a reporter [92]. A collection of 20,261 transgenic *Arabidopsis* lines were generated using a promoterless firefly *luciferase* reporter gene. Among these mutants, 3.7% showed altered luciferase expression upon exposure to stress. Detailed analysis of one of these lines showing salt-inducible luciferase expression helped identify an ABC transporter, *At1g*17840, the expression of which is regulated by sugar, salt, ABA, and GA [93]. It should, however, be mentioned that an expression in these mutants need not reveal involvement of gene in stress tolerance and further validation of gene function is required.

13.5 Proteomics as a Tool for Functional Genomics

The word "proteome" is derived from *prote*ins expressed by a gen*ome*, and it refers to all the proteins produced by an organism, much like the genome is the entire set of genes. Proteomics is the large-scale study of protein, particularly their structures and functions. The first protein studies that can be called proteomics began in 1975 with the introduction of the two-dimensional (2D) gel by O'Farrell in 1975 [94]. The major goal of proteomics is to obtain a more global and integrated view of biology by studying all the proteins of a cell rather than each one individually. After genomics, proteomics is considered the next step in the study of biological systems. The term proteomics was coined in the 1990s by Wilkins to make an analogy with genomics. Proteomics particularly focuses on the systematic and detailed analysis of the protein population in a cellular compartment, tissue, and whole organism for specific properties such as their identity, quantity, activity and molecular interaction.

After genomics, proteomics is considered the next step in the study of biological systems. It is much more complicated than genomics mostly because the genome is a rather constant entity, whereas the proteome differs from cell to cell and is constantly changing through its biochemical interactions with the genome and the environment. Protein expression differs in different parts of plants, in different stages of its life cycle, and under different environmental conditions. This is because distinct genes are expressed in distinct cell types. Proteomics tells us what fraction of the genome is functional and at what levels. Proteome analysis provides better annotation of genome sequences by taking into account very small open reading frames that are functional. It can also assign an unidentified open reading frame to a particular protein product. Furthermore, as mentioned, any one protein can undergo a wide range of posttranslational modifications to combat certain stress. Therefore, a "proteomics" study can become quite complex very quickly, even if the object of the study is very restricted.

13.5.1 Methods of Studying Proteins

The proteomics concept is based on high-quality separation of proteins. This science of proteins dates back to the late 1970s with the advent of 2D gel electrophoresis (2DGE) [94]. In proteomics, a measured quantity of proteins are separated by 2DGE, estimated using image analysis and then the mass spectrometer-based identification of gel separated proteins is carried out. 2D polyacrylamide gel electrophoresis is an important proteomics technique to characterize thousands of proteins from protein lysate sample, either from cells, tissues, or organisms. It is one of the driving forces for proteomics study that aims at separating, resolving, and detection of protein of interests from thousands of proteins in the cells.

2D-polyacrylamide gel electrophoresis separates protein according to charge (pI) by isoelectric focusing in the first dimension and according to size/mass (M_r) by SDS-polyacrylamide gel electrophoresis in the second dimension, has a unique capacity for the resolution of complex mixtures of proteins, permitting the simultaneous analysis of hundreds or even thousands of gene products (Figure 13.8). After separation of proteins, the gels are scanned using densitometric scanner and protein in the spots is quantified using various software like Melanie and Image Master 2D Platinum. Differentially regulated spots are cut from the gel and digested with trypsin and proceed to matrix-assisted laser desorption/ionization-time of flight analysis to obtain the charge-to-mass (m/z) ratio. This charge-to-mass ratio is searched against protein and EST databases using MASCOT to identify the protein. In *Arabidopsis*, using 2DGE and matrix-assisted laser desorption/ionization-time of flight, Bae et al. [95] characterized the nuclear proteome in response to cold stress. Forty proteins were induced and 14 had repressed responses to cold stress by more than two fold. Many of them were already known to be involved in stress, including heat-shock proteins, transcription factors (*AtMYB2* and *OBF4*), DNA binding proteins (DRT102 and Dr1), catalytic enzymes, syntaxin, calmodulin, and germin-like proteins. Similarly, following heat stress in wheat grains, a total of 37 proteins that were significantly changed by heat treatments were identified [96]. Similarly, in rice, out of 1000 quantified proteins from stressed and well-watered rice leaves, 42 proteins were found to be responsive to stress [97], these were identified using matrix-assisted laser desorption-ionization time-of-flight mass spectrometry (MALDI-TOF MS) and electro spray ionisation-time-of-flight mass spectrometry (ESIQ-TOF MS). Jagadish et al. [98] identified the proteins responsive to heat tolerance from the anthers of rice, Muthurajan et al. [99] identified the proteins responsible for panicle exsertion during drought stress in rice peduncles.

Nowadays, protein arrays are rapidly becoming established as a powerful means to detect proteins, monitor their expression levels, and investigate protein interactions and

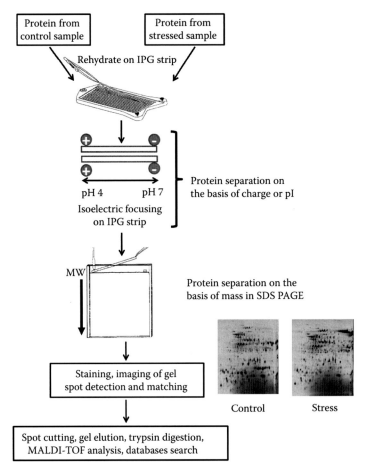

FIGURE 13.8
A schematic diagram for protein separation and identification using 2DGE and mass spectrometry.

functions. Protein chips are particularly useful for assessing and analyzing protein diversity on a large scale, including hydrophobic and large proteins that are difficult to visualize using a 2D method. Protein arrays are solid-phase ligand-binding assay systems using immobilized proteins on surfaces which include glass, membranes, etc. Protein samples are labeled with different florescent dyes and probed against the protein chip. The bound proteins can be identified by a color change as in the case of microarrays, which are later eluted and identified by mass spectrometry. Proteomics-based research is proving helpful in the characterization of stress-related proteins and the identification of corresponding novel stress-responsive genes for the production of stress-tolerant plants.

Many types of information cannot be obtained from the study of genes alone because proteins, not genes, are functional in cells and are responsible for the phenotypes of the cells. After translation, the protein would undergo several modifications that could decide their function and localization, which may not be predicted from the gene sequence. Proteomics helps in annotating the genome, differential expression of protein under various conditions, predicting protein function, protein modification, localization and compartmentalization and in protein–protein interactions.

13.6 Metabolomics

The most recent addition to the array of methodologies used in functional genomics is metabolomics, in which a set of analytical technologies (such as gas chromatography–mass spectrometry (GC-MS), liquid chromatography-photodiode array detection-mass spectrometry (LC-DAD-MS), fourier transform infrared spectroscopy (FTIR), and nuclear magnetic resonance (NMR)) are used for comprehensive, nonbiased, high-throughput analyses of complex metabolite mixtures typical of plant extracts. Metabolomics has emerged as a functional genomics methodology that contributes to our understanding about complex molecular interactions in biological systems [100]. The enormous biochemical diversity displayed in the plant kingdom is estimated to exceed 200,000 different metabolites [101].

Upon exposure to osmotic stress as a result of low temperature, drought, or high salinity, plants accumulate a range of osmolytes with the primary function of turgor maintenance. The solutes accumulated vary among species and include sugars (i.e., sucrose, glucose, fructose, and trehalose), polyols, betaines, and amino acids such as proline [102,103]. For some compounds, such as proline and glycine betaine, the exogenous application of the molecule or the enhancement of their biosynthesis through ectopic expression has resulted in a stress tolerance improvement [104–107]. Overexpression of transcription factors involved in stress-specific gene regulation, such as DREB or MYB factors, in particular, are involved in regulating the synthesis of osmoprotectants [108]. Thus, the importance of metabolite changes during abiotic stress suggests that detailed metabolite profiling and its correlation with transcript profiling may provide valuable insights into stress response mechanisms and will enlighten us about the plant machinery and how it is operating to combat stress tolerance. The possibility of monitoring a complete set of metabolites could largely improve the understanding of the adaptation mechanisms. This systematic study significantly contributes to the study of stress biology in plants. The metabolomic changes that have been observed in plants subjected to stress conditions depend on different causes; therefore, they have different significances and are expected to correlate differently with tolerance/sensitive phenotypes. The main goal of studying metabolic changes during stress responses is to identify the metabolites involved in the restoration of homeostasis and normal metabolic fluxes and how the synthesis or accumulation of compounds are involved in mediating tolerance mechanisms.

Examining metabolomics, or changes in metabolic profiles, can be an important part of an integrative approach for assessing gene function and relationships to phenotypes [109]. Metabolomics, a science still in its infancy, provides a better understanding of the correlation between genes and the functional phenotype of an organism, which is the true goal of all functional genomics strategies. Unlike transcriptomic and proteomic strategies measuring changes in mRNA and protein levels, metabolic profiling allows us to get closer to the specific role that a certain gene performs in a biochemical process. The co-occurrence principle of transcripts and metabolites, particularly transcriptome coexpression network analysis, is powerful for decoding functions of genes not only in a model plant such as *Arabidopsis* but also in crops and medicinal plants.

13.7 Phenomics

The term phenomics is an emerging trans-discipline that stands for the large-scale analysis of plant diversity with respect to plant morphology. Plant phenomics is the study of plant growth, performance, and composition. Phenomics is performed by combining novel technologies such as noninvasive imaging, spectroscopy, image analysis, robotics, and high-performance computing [110].

To connect the wealth of genomic information for agricultural application, it has to be carefully and comprehensively linked to phenotypes in a "real-world" environment. A clear goal of phenomics is to bridge the gap between genomics, plant function, and agricultural traits [110]. Such phenotypic profiling approaches aim to measure the physical and chemical properties of an organism at specific time points during its life cycle. There are abiotic stresses which produce similar phenotypic effects (manual phenotyping used to be quite difficult). Under such conditions, phenomics tools like carbon isotope discrimination, infrared thermography, chlorophyll fluorescence analysis, digital growth analysis, etc., can be used. Phenomics provides the opportunity to study previously unexplored areas of plant science, and it provides the opportunity to bring together genetics and physiology to reveal the molecular genetic basis of a wide range of previously intractable plant processes [110].

13.8 Conclusions

From the day the first plant genome sequence became available, a new path was presented to researchers who wished to further develop plant varieties with an emphasis on abiotic stress tolerance. Abiotic tolerance has come to the forefront in the topic of plant biology because of the drastic climatic changes that have led to decreases in crop yields. In this scenario, the future of human food security is becoming uncertain as the currently available germplasm is slowly losing its yield viability due to climatic changes leading to abiotic stress. With the recent developments in "omics" technologies, global genome scanning has allowed researchers to quickly identify novel alleles of target genes and, with ease of manipulation, to also create mutants or transgenics that will help increase the production of agricultural crops along with an enhanced ability to resist the coming global climatic changes and abiotic stresses.

Because of the increased knowledge of omics in particular crop species, the development of novel methods, and modification of the existing resistant germplasm, we will be able to raise varieties/species with improved crop qualities and desired characteristics such as enhanced yields that are resistant to abiotic stresses. Novel germplasm collections with the desired characteristics, and screening of the genetic background using high-throughput omics technology, will provide novel strategies toward improvement of other species by further genetic manipulations and will advance the goals to include future prospective climatic changes. Although a considerable amount of knowledge has been achieved, there is, however, a need to make the technology more available, for expertise in a single platform combining all omics technologies, and for infrastructure to be developed to enhance the productivity of crops with improved characteristics to fight against the forthcoming global climatic changes and to give food security to our future generations.

The omics revolution is creating intrinsic pressure on budding researchers to ignore classic approaches, such as plant breeding, which should not be neglected. The promotion of all the innovative approaches in combination with the fundamental technologies is important in addressing the need for enhancing agricultural productivity and sustainability, and to put them to use for the public good.

References

1. Bray, E.A., J. Bailey-Serres, and E. Weretilnyk. 2000. Responses to abiotic stresses. In *Biochemistry and Molecular Biology of Plants*, edited by Buchanan, B.B., W. Gruissem, and R.L. Jones, 1158–1203. Rockville, MD: American Society of Plant Physiologists.
2. Wang, W.X., B. Vinocur, O. Shoseyov, and A. Altman. 2001. Biotechnology of plant osmotic stress tolerance: Physiological and molecular considerations. *Acta Horticulturae* 560:285–292.
3. Chen, W., N.J. Provart, J. Glazebrook et al. 2002. Expression profile matrix of *Arabidopsis* transcription factor genes suggests their putative functions in response to environmental stresses. *Plant Cell* 14:559–574.
4. Kreps, J.A., Y. Wu, H.S. Chang, T. Zhu, X. Wang, and J.F. Harper. 2002. Transcriptome changes for *Arabidopsis* in response to salt, osmotic and cold stress. *Plant Physiology* 130:2129–2141.
5. Hannah, M.A., A.G. Heyer, and D.K. Hincha. 2005. A global survey of gene regulation during cold acclimation in *Arabidopsis thaliana*. *PLoS Genetics* 1(2):179–196.
6. Shinozaki, K., K. Yamaguchi-Shinozaki, and M. Seki. 2003. Regulatory network of gene expression in the drought and cold stress responses. *Current Opinion in Plant Biology* 6:410–417.
7. Bartels, D., and R. Sunkar. 2005. Drought and salt tolerance in plants. *Critical Reviews in Plant Science* 24:23–58.
8. Vinocur, B., and A. Altman. 2005. Recent advances in engineering plant tolerance to abiotic stress: Achievements and limitations. *Current Opinion in Biotechnology* 16:123–132.
9. Battisti, D.S., and R.L. Naylor. 2009. Historical warnings of future food insecurity with unprecedented seasonal heat. *Science* 323:240–244.
10. Lobell, D.B., M.B. Burke, C. Tebaldi, M.D. Mastrandrea, W.P. Falcon, and Naylor R.L. 2008. Prioritizing climate change adaptation needs for food security in 2030. *Science* 319:607–610.
11. FAO. 2050. Increased investment in agricultural research essential. Published online September 25, 2009, http://www.fao.org/news/story/en/item/35686/icode/2009 (accessed March 22, 2012).
12. Rounsley, S., and S. Briggs. 1999. The paradigm shift of genomics—a complement to traditional plant science. *Current Opinion in Plant Biology* 2:81–82.
13. Brent, R. 2000. Genomic biology. *Cell* 100:169–183.
14. Holtfort, H., M.C. Guitton, and R. Reski. 2002. Plant functional genomics. *Naturwissenschaften* 89:235–249.
15. Hieter, P., and M. Boguski. 1997. Functional genomics: It's all how you read it. *Science* 278:601–602.
16. Bouchez, D., and H. Hofte. 1998. Functional genomics in plants. *Plant Physiology* 118:725–732.
17. Semagn, K., A. Bjornstad, and M.N. Ndjiondjop. 2006. An overview of molecular marker methods for plants. *African Journal of Biotechnology* 5(25):2540–2568.
18. Hofte, H., T. Desprez, J. Amselem, H. Chiapello, M. Caboche, A. Moisan et al. 1993. An inventory of 1152 expressed sequence tags obtained by partial sequencing of cDNAs from *Arabidopsis thaliana*. *The Plant Journal* 4:1051–1061.
19. Newman, T., F.J. de Bruijn, P. Green, K. Keegstra, H. Kende, L. McIntosh et al. 1994. Genes galore: A summary of methods for accessing results from large-scale partial sequencing of anonymous *Arabidopsis* cDNA clones. *Plant Physiology* 106:1241–1255.
20. Cooke, R., M. Raynal, M. Laudié, F. Grellet, M. Delseny, P.C. Morris et al. 1996. Further progress towards a catalogue of all *Arabidopsis* genes: Analysis of a set of 5000 non-redundant ESTs. *The Plant Journal* 9:101–124.
21. Yamamoto, K., and T. Sasaki. 1997. Large-scale EST sequencing in rice. *Plant Molecular Biology* 35:135–144.
22. Nogueira, F.T., V.E. De Rosa Jr., M. Menossi, E.C. Ulian, and P. Arruda. 2003. RNA expression profiles and data mining of sugarcane response to low temperature. *Plant Physiology* 132:1811–1824.

23. Pih, K.T., H.J. Jang, S.G. Kang, H.L. Piao, and I. Hwang. 1997. Isolation of molecular markers for salt stress responses in *Arabidopsis thaliana*. *Molecules and Cells* 7:567–571.

24. Gorantla, M., P.R. Babu, V.B. Reddy Lachagari, A.M.M. Reddy, W. Ramakrishna, L.J. Bennetzen et al. 2007. Identification of stress-responsive genes in an indica rice (*Oryza sativa* L.) using ESTs generated from drought-stressed seedlings. *Journal of Experimental Botany* 58(2):253–265.

25. Kumari, S., V.P. Sabharwal, H.R. Kushwaha, S.K. Sopory, S.L. Singla-Pareek, and A. Pareek. 2009. Transcriptome map for seedling stage specific salinity stress response indicates a specific set of genes as candidate for saline tolerance in *Oryza sativa* L. *Functional & Integrative Genomics* 1:109–123.

26. Ouyang, B., T. Yang, H. Li, L. Zhang, Y. Zhang, J. Zhang et al. 2007. Identification of early salt stress response genes in tomato root by suppression subtractive hybridization and microarray analysis. *Journal of Experimental Botany* 58(3):507–520.

27. Hazen, S.P., Y. Wu, and J.A. Kreps. 2003. Gene expression profiling of plant responses to abiotic stress. *Functional & Integrative Genomics* 3(3):105–111.

28. Green, C.D., J.F. Simons, B.E. Taillon, and D.A. Lewin. 2001. Open systems: Panoramic views of gene expression. *Journal of Immunological Methods* 250:67–79.

29. Monroy, A., A. Dryanova, B. Malette, D. Oren, M. R. Farajalla, W. Liu et al. 2007. Regulatory gene candidates and gene expression analysis of cold acclimation in winter and spring wheat. *Plant Molecular Biology* 64:409–423.

30. Fernandez, P., J.D. Rienzo, L. Fernandez, H.E. Hopp, N. Paniego, and R.A. Heinz. 2008. Transcriptomic identification of candidate genes involved in sunflower responses to chilling and salt stresses based on cDNA microarray analysis. *BMC Plant Biology* 8:11.

31. Fizames, C., S. Munos, C. Cazettes, P. Nacry, J. Boucherez, F. Gaymard et al. 2004. The *Arabidopsis* root transcriptome by serial analysis of gene expression. Gene identification using the genome sequence. *Plant Physiology* 134:67–80.

32. Meyers, B.C., S.S. Tej, T.H. Vu, C.D. Haudenschild, V. Agrawal, S.B. Edberg et al. 2004. The use of MPSS for whole-genome transcriptional analysis in *Arabidopsis*. *Genome Research* 14:1641–1653.

33. Calsa, T., and A. Figueira. 2007. Serial analysis of gene expression in sugarcane (*Saccharum* spp.) leaves revealed alternative C4 metabolism and putative antisense transcripts. *Plant Molecular Biology* 63:745–762.

34. Song, S., H. Qu, C. Chen, S. Hu, and J. Yu. 2007. Differential gene expression in an elite hybrid rice cultivar (*Oryza sativa*, L.) and its parental lines based on SAGE data. *BMC Plant Biology* 7:49

35. Vos, P., R. Hogers, M. Bleeker, M. Reijans, T. van de Lee, M. Hornes et al. 1995. AFLP: A new technique for DNA fingerprinting. *Nucleic Acids Research* 23(21):4407–4414.

36. Bachem, C.B.W., R.S. van der Hoeven, S.M. de Brujin, D. Vreugdenhil, M, Zabeau, and R.G.F. Visser. 1996. Visualization of differential gene expression using novel method of RNA fingerprinting based on AFLP: Analysis of gene expression during potato tuber development. *The Plant Journal* 9:745–753.

37. Gui Ping, Ch., W.S. Maa, Z.J. Huanga, T. Xu, Y.B. Xueb, and Y.Z. Shen. 2003. Isolation and characterization of *TaGSK1* involved in wheat salt tolerance. *Plant Science* 165:1369–1375.

38. Frank, G., E. Pressman, R. Ophir, L. Althan, R. Shaked, M. Freedman et al. 2009. Transcriptional profiling of maturing tomato (*Solanum lycopersicum* L.) microspores reveals the involvement of heat shock proteins, ROS scavengers, hormones, and sugars in the heat stress response. *Journal of Experimental Botany* 60:3891–3908.

39. Velculescu, V.E., L. Zhang, B. Vogelstein, and K.W. Kinzler. 1995. Serial analysis of gene expression. *Science* 270:484–487.

40. Lee, J.Y., and D.H. Lee. 2003. Use of serial analysis of gene expression technology to reveal changes in gene expression in *Arabidopsis* pollen undergoing cold stress. *Plant Physiology* 132:517–529.

41. Jung, S.H., J.Y. Lee, and D.H. Lee. 2003. Use of SAGE technology to reveal changes in gene expression in *Arabidopsis* leaves undergoing cold stress. *Plant Molecular Biology* 52(3):553–567.

42. Matsumura, H., S. Nirasawa, and R. Terauchi. 1999. Transcript profiling in rice (*Oryza sativa* L.) seedlings using serial analysis of gene expression (SAGE). *The Plant Journal* 20:719–726.

43. Liang, P., and A.B. Pardee. 1992. Differential display of eukaryotic messenger RNA by means of the polymerase chain reaction. *Science* 257:967–971.
44. Donohue, P.J., G.F. Alberts, Y. Guo, and J.A. Winkles. 1995. Identification by targeted differential display of an immediate early gene encoding a putative serine/threonine kinase. *Journal of Biological Chemistry* 270(10):351–357.
45. Johnson, S.W., N.A. Lissy, P.D. Miller, J.R. Testa, R.F. Ozols, and T.C. Hamilton. 1996. Identification of zinc finger mRNAs using domain-specific differential display. *Analytical Biochemistry* 236:348–352.
46. Muramoto, Y., A. Watanabe, T. Nakamura, and T. Takabe. 1999. Enhanced expression of a nuclease gene in leaves of barley plants under salt stress. *Gene* 234:315–321.
47. Visioli, G., E. Maestri, and N. Marmiroli. 1997. Differential display-mediated isolation of a genomic sequence for a putative mitochondrial LMW *HSP* specifically expressed in conditions of induced thermotolerance in *Arabidopsis thaliana* (L.) Heynh. *Plant Molecular Biology* 34:517–527.
48. Kiiskinen, M., M. Korhonen, and J. Kangasjarvi. 1997. Isolation and characterization of cDNA for a plant mitochondrial phosphate translocator (*Mpt*1): Ozone stress induces *Mpt*1 mRNA accumulation in birch (*Betula pendula* Roth). *Plant Molecular Biology* 35:271–279.
49. Petrucco, S., A. Bolchi, C. Foroni, R. Percudani, G.L. Rossim, and S. Ottonello. 1996. A maize gene encoding an NADPH binding enzyme highly homologous to isoflavone reductases is activated in response to sulfur starvation. *Plant Cell* 8:69–80.
50. Titarenko, E., E. Rojo, J. Leon, and J.J. Sanchez-Serrano. 1997. Jasmonic acid dependent and independent signaling pathways control wound-induced gene activation in *Arabidopsis thaliana*. *Plant Physiology* 115:817–826.
51. Malhotra, K., L. Foltz, W.C. Mahoney, and P.A. Schueler. 1998. Interaction and effect of annealing temperature on primers used in differential display RT-PCR. *Nucleic Acids Research* 26:854–856.
52. Poirier, G.M., J. Pyati, J.S. Wan, and M.G. Erlander. 1997. Screening differentially expressed cDNA clones obtained by differential display using amplified RNA. *Nucleic Acids Research* 25:913–914.
53. Campbell, I.D., and A.K. Downing. 1994. Building protein structure and function from modular units. *Trends in Biotechnology* 12:168–172.
54. Fischer, A., H. Saedler, and T. Gunter. 1995. Restriction fragment length polymorphism-coupled domain directed differential display: A highly efficient technique for expression analysis of multigene families. *Proceedings of the National Academy of Sciences of the United States of America* 92:5331–5335.
55. Huang, H., M. Tudor, T. Su, Y. Zhang, Y. Hu, and H. Ma. 1996. DNA binding properties of two *Arabidopsis* MADS domain proteins: Binding consensus and dimer formation. *Plant Cell* 8:81–94.
56. Tahtiharju, S., V. Sangwan, A.F. Monroy, R.S. Dhindsa, and M. Borg. 1997. The induction of kin genes in cold-acclimating *Arabidopsis thaliana*. Evidence of a role for calcium. *Planta* 203:442–447.
57. Lemieux, B., A. Aharoni, and M. Schena. 1998. Overview of DNA chip technology. *Molecular Breeding* 4:277–289.
58. Kehoe, D.M., P. Villand, and S. Somerville. 1999. DNA microarrays for studies of higher plants and other photosynthetic organisms. *Trends in Plant Science* 4:38–41.
59. Seki, M., M. Narusaka, J. Ishida, T. Nanjo, M. Fujita, Y. Oono et al. 2002. Monitoring the expression profiles of 7000 *Arabidopsis* genes under drought, cold and high-salinity stresses using a full-length cDNA microarray. *The Plant Journal* 31:279–292.
60. Seki, M., M. Narusaka, H. Abe, M. Kasuga, K. Yamaguchi-Shinozaki, P. Carninci, Y. Hayashizaki, and K. Shinozakia. 2001. Monitoring the expression pattern of 1300 *Arabidopsis* genes under drought and cold stresses by using a full-length cDNA microarray. *The Plant Cell* 13(1):61–72.
61. Klok, E.J., I.W. Wilson, D. Wilson, S.C. Chapman, R.M. Ewing, S.C. Somerville, W.J. Peacock, R. Dolferus, and E.S. Dennis. 2002. Expression profile analysis of the low-oxygen response in *Arabidopsis* root cultures. *The Plant Cell* 14(10):2481–2494.
62. Kawasaki, S., C. Borchert, M. Deyholos, H. Wang, S. Brazille, K. Kawai, D. Galbraith, and H.J. Bohnerta. 2001. Gene expression profiles during the initial phase of salt stress in rice. *The Plant Cell* 13(4):889–905.

63. Bohnert, H.J., P. Ayoubi, C. Borchert, R.A. Bressan, R.L. Burnap, J.C. Cushman, M.A. Cushman, M. Deyholos, R. Fischer, D.W. Galbraith et al. 2001. A genomics approach towards salt stress tolerance. *Plant Physiology and Biochemistry* 39:295–311.

64. Desikan, R., S.A.H. Mackerness, J.T. Hancock, and S.J. Neill. 2001. Regulation of the *Arabidopsis* transcriptome by oxidative stress. *Plant Physiology* 127(1):159–172.

65. Bray, E.A. 2002. Classification of genes differentially expressed during water-deficit stress in *Arabidopsis thaliana*: An analysis using microarray and differential expression data. *Annals of Botany* 89(7):803–811.

66. Ozturk, Z.N., V. Talame, M. Deyholos, C.B. Michalowski, D.W. Galbraith, N. Gozukirmizi, R. Tuberosa, and H.J. Bohnert. 2002. Monitoring large scale changes in transcript abundance in drought- and salt-stressed barley. *Plant Molecular Biology* 48:551–573.

67. Oono, Y., M. Seki, T. Nanjo, M. Narusaka, M. Fujita, R. Satoh, M. Satou, T. Sakurai, J. Ishida, K. Akiyama, K. Iida, K. Maruyama, S. Satoh, K. Yamaguchi-Shinozaki, and K. Shinozaki. 2003. Monitoring expression profiles of *Arabidopsis* gene expression during rehydration process after dehydration using ca. 7000 full-length cDNA microarray. *The Plant Journal* 34:868–887.

68. Yu, L.X., and T.L. Setter. 2003. Comparative transcriptional profiling of placenta and endosperm in developing maize kernels in response to water deficit. *Plant Physiology* 131:568–582.

69. Chen, S., D. Hofius, U. Sonnewald, and F. Bornke. 2003. Temporal and spatial control of gene silencing in transgenic plants by inducible expression of double-stranded RNA. *The Plant Journal* 36:731–740.

70. Guo, H.S., J.F. Fei, Q. Xie, and N.H. Chua. 2003. A chemical regulated inducible RNAi system in plants. *The Plant Journal* 34:383–392.

71. Guo, Y., L. Xiong, C.P. Song, D. Gong, U. Halfter, and J.K. Zhu. 2002. A calcium sensor and its interacting protein kinase are global regulators of abscisic acid signalling in *Arabidopsis*. *Development Cell* 3:233–244.

72. Hilson, P., J. Allemeersch, T. Altmann, S. Aubourg, A. Avon, J. Beynon et al. 2004. Versatile gene-specific sequence tags for *Arabidopsis* functional genomics: Transcript profiling and reverse genetics applications. *Genome Research* 14:2176–2189.

73. Parinov, S., and V. Sundaresan. 2000. Functional genomics in *Arabidopsis*: Large-scale insertional mutagenesis complements the genome sequencing project. *Current Opinion in Biotechnology* 11:157–161.

74. Krysan, P.J., J.C. Young, and M.R. Sussman. 1999. T-DNA as an insertional mutagen in *Arabidopsis*. *Plant Cell* 11:2283–2290.

75. Sundaresan, V. 1996. Horizontal spread of transposon mutagenesis: New uses of old elements. *Trends in Plant Science* 1:184–191.

76. Fedoroff, N.V., S. Wessler, and M. Shure. 1983. Isolation of the transposable maize controlling elements *Ac* and *Ds*. *Cell* 35:235–242.

77. Baumann, E., J. Lewald, H. Saedler, B. Schulz, and E. Wisman. 1998. Successful PCR-based reverse genetic screens using an *En-1*-mutagenised *Arabidopsis thaliana* population generated via single-seed descent. *Theoretical and Applied Genetics* 97:729–734.

78. Meissner, R.C., H. Jin, E. Cominelli, M. Denekamp, A. Fuertes, R. Greco et al. 1999. Function search in a large transcription factor gene family in *Arabidopsis*: Assessing the potential of reverse genetics to identify insertional mutations in R2R3 MYB genes. *Plant Cell* 11:1827–1840.

79. Tissier, A.F., S. Marillonnet, V. Klimyuk, K. Patel, M.A. Torres, G. Murphy et al. 1999. Multiple independent defective suppressor-mutator transposon insertions in *Arabidopsis*: A tool for functional genomics. *Plant Cell* 11:1841–1852.

80. Bouchez, D., C. Camilleri, and M. Caboche. 1993. A binary vector based on Basta resistance for in planta transformation of *Arabidopsis thaliana*. *Comptes Rendus de l'Académie des Sciences. Série III, Sciences de la Vie* 316:1188–1193.

81. Hirochika, H. 2001. Contribution of the *Tos17* retro-transposon to rice functional genomics. *Current Opinion in Plant Biology* 4:118–122.

82. Miyao, A., K. Tanaka, K. Murata, H. Sawaki, S. Takeda, K. Abe et al. 2003. Target site specificity of the *Tos17* retrotransposon shows a preference for insertion within genes and against insertion in retrotransposon-rich regions of the genome. *Plant Cell* 15:1771–1780.

83. Azpiroz-Leehan, R., and K.A. Feldmann. 1997. T-DNA insertion mutagenesis in *Arabidopsis*: Going back and forth. *Trends in Genetics* 13:152–156.

84. Lee, S., S.H. Kim, S.J. Kim, K. Lee, and S.K. Han. 2004. Trapping and characterization of cold responsive genes from T-DNA tagging lines in rice. *Plant Science* 166:69–79.

85. Jeong, D.H., S. An, H.G. Kang, S. Moon, J.J. Han, S. Park et al. 2002. T-DNA insertional mutagenesis for activation tagging in rice. *Plant Physiology* 30:1636–1644.

86. Weigel, D., J.H. Ahn, M.A. Blazquez, J.O. Borevitz, S.K. Christensen, C. Fankhauser et al. 2000. Activation tagging in *Arabidopsis*. *Plant Physiology* 122:1003–1013.

87. Kardailsky, I., V.K. Shukla, J.H. Ahn, N. Dagenais, S.K. Christensen, J.T. Nguyen et al. 1999. Activation tagging of the floral inducer FT. *Science* 286:1962–1965.

88. Zhao, Y., S.K. Christensen, C. Frankhauser, J.R. Cashman, J.D. Cohen, D. Weigel et al. 2001. A role for flavin mono-oxygenase like enzymes in auxin biosynthesis. *Science* 291:306–309.

89. Borevitz, J.O., Y. Xia, J. Blount, R.A. Dixon, and C. Lamb. 2000. Activation tagging identifies a conserved MYB regulator of phenylpropanoid biosynthesis. *Plant Cell* 12:2383–2393.

90. Springer, P.S. 2000. Gene traps: Tools for plant development and genomics. *The Plant Cell* 12:1007–1020.

91. Sundaresan, V., P. Springer, T. Volpe, S. Haward, J.D. Jones, C. Dean et al. 1995. Patterns of gene action in plant development revealed by enhancer trap and gene trap transposable elements. *Genes & Development* 9:1797–1810.

92. Wu, C., X. Li, W. Yuan, G. Chen, A. Kilian, J. Li et al. 2003. Development of enhancer trap lines for functional analysis of the rice genome. *The Plant Journal* 35:418–427.

93. Alvarado, M.C., L.M. Zsigmond, I. Kovacs, A. Cseplo, C. Koncz, and L.M. Szabados. 2004. Gene trapping with firefly luciferase in *Arabidopsis*. Tagging of stress-responsive genes. *Plant Physiology* 134:18–27.

94. O'Farrell, P.H. 1975. High resolution two-dimensional electrophoresis of proteins. *The Journal of Biological Chemistry* 250:4007–4021.

95. Bae, M.S., E.J. Cho, E.Y. Choi, and O.K. Park. 2003. Analysis of the *Arabidopsis* nuclear proteome and its response to cold stress. *The Plant Journal* 36:652–663.

96. Majoul, T., E. Bancel, E. Triboi, J. Ben Hamida, and G. Branlard. 2003. Proteomic analysis of the functional genomics of stress tolerance effect of heat stress on hexaploid wheat grain: Characterization of heat-responsive proteins from total endosperm. *Proteomics* 3:175–183.

97. Salekdeh, G.H., J. Siopongco, L.J. Wade, B. Ghareyazie, and J. Bennett. 2002. Proteomic analysis of rice leaves during drought stress and recovery. *Proteomics* 2:1131–1145.

98. Jagadish, S.V.K., R. Muthurajan, R. Oane, T.R. Wheeler, S. Heuer, J. Bennett et al. 2010. Physiological and proteomic approaches to address heat tolerance during anthesis in rice (*Oryza sativa* L.). *Journal of Experimental Botany* 61:143–156.

99. Muthurajan, R., Z.S. Shobbar, S.V. Jagadish, R. Bruskiewich, A. Ismail, H. Leung et al. 2011. Physiological and proteomic responses of rice peduncles to drought stress. *Molecular Biotechnology* 48(2):173–182.

100. Hall, R. et al. 2002. Plant metabolomics: The missing link in functional genomics strategies. *Plant Cell* 14:1437–1440.

101. Pichersky, E., and D. Gang. 2000. Genetics and biochemistry of secondary metabolites: An evolutionary perspective. *Trends in Plant Science* 5:439–445.

102. Shulaev, V., D. Cortes, G. Miller, and R. Mittler. 2008. Metabolomics for plant stress response. *Physiologia Plantarum* 132(2):199–208.

103. Smirnoff, N. 1998. Plant resistance to environmental stress. *Current Opinion in Biotechnology* 9:214–219.

104. Chen, T.H., and N. Murata. 2008. Glycinebetaine: An effective protectant against abiotic stress in plants. *Trends in Plant Science* 13(9):499–505.

105. Kishor, P.B.K., Z. Hong, G.H. Hu, C.A.A. Miao, and D.P.S. Verma. 1995. Overexpression of Δ1-pyrroline-5-carboxylate synthetase increases proline production and confers osmotolerance in transgenic plants. *Plant Physiology* 108(4):1387–1394.
106. Quan, R., M. Shang, H. Zhang, Y. Zhao, and J. Zhang. 2004. Improved chilling tolerance by transformation with *bet*A gene for the enhancement of glycine betaine synthesis in maize. *Plant Science* 166(1):141–149.
107. Szabados, L., and A. Savouré. 2010. Proline: A multifunctional amino acid. *Trends in Plant Science* 15(2):89–97.
108. Gosal, S.S., S.H. Wani, and M.S. Kang. 2009. Biotechnology and drought tolerance. *Journal of Crop Improvement* 23(1):19–54.
109. Thomas, G.H. 2001. Metabolomics breaks the silence. *Trends in Microbiology* 9:158.
110. Furbank, R.T., and M. Tester. 2011. Phenomics—technologies to relieve the phenotyping bottleneck. *Trends in Plant Science* 16(12):635–644.

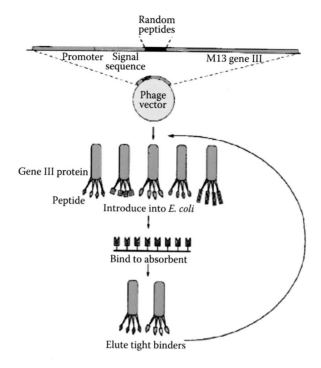

FIGURE 1.6
A peptide library in a filamentous phage vector.

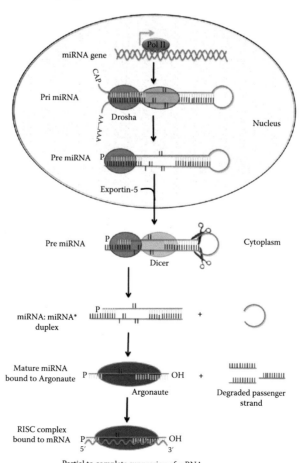

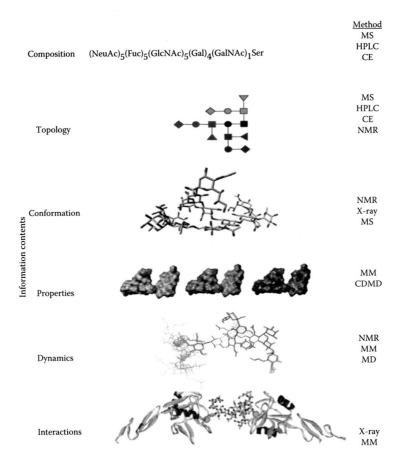

	Composition	$(NeuAc)_5(Fuc)_5(GlcNAc)_5(Gal)_4(GalNAc)_1Ser$	Method MS HPLC CE
Information contents	Topology		MS HPLC CE NMR
	Conformation		NMR X-ray MS
	Properties		MM CDMD
	Dynamics		NMR MM MD
	Interactions		X-ray MM

FIGURE 10.1

Pictorial scheme of information content at several levels of glycan structural data.

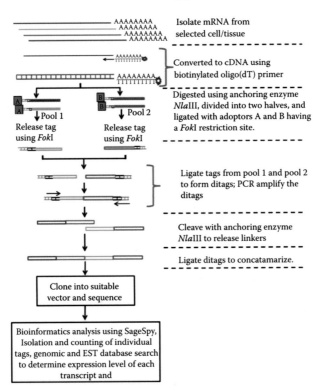

Isolate mRNA from selected cell/tissue

Converted to cDNA using biotinylated oligo(dT) primer

Digested using anchoring enzyme *Nla*III, divided into two halves, and ligated with adaptors A and B having a *Fok*I restriction site.

Pool 1   Pool 2
Release tag   Release tag
using *Fok*I   using *Fok*I

Ligate tags from pool 1 and pool 2 to form ditags; PCR amplify the ditags

Cleave with anchoring enzyme *Nla*III to release linkers

Ligate ditags to concatamarize.

Clone into suitable vector and sequence

Bioinformatics analysis using SageSpy, Isolation and counting of individual tags, genomic and EST database search to determine expression level of each transcript and

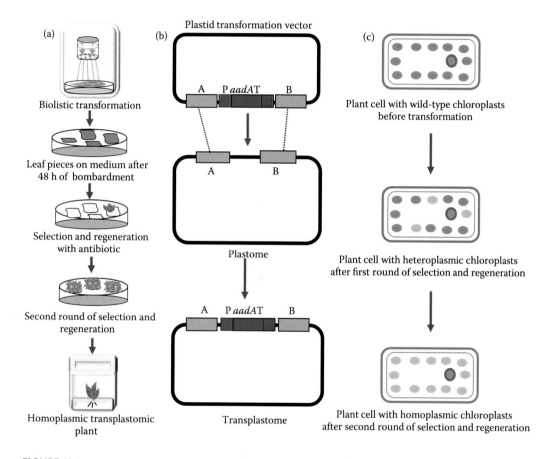

FIGURE 16.1

Stepwise chloroplast transformation in tobacco: biolistic bombardment, transgene integration into the chloroplast genome via two events of homologous recombination and sorting at the genome and organelle level. (a) A tobacco leaf is bombarded with vector DNA-coated metal particles. Sectioning of leaves into small pieces after 48 h of bombardment and placing the leaf sections onto the surface of regeneration medium (RMOP), containing spectinomycin because the transformation vector carries *aadA* genes. Leaves from recovered spectinomycin-resistant shoots were used for the second round of selection and regenerated homoplasmic shoots are transferred to boxes. (b) The chloroplast transformation vector, harboring the expression cassette that carries tails of sequences homologous to plastid DNA on both sides, targets the expression cassette in the plastid genome. Recombination between homologous sequences, in the vector and the plastome, allows the integration of expression cassette into the plastome. (c) A plant cell with a homogenous population of wild-type chloroplasts before bombardment. Cells from a shoot regenerated during the first round of selection and regeneration are generally heteroplasmic, whereas a shoot from the second round are homoplasmic.

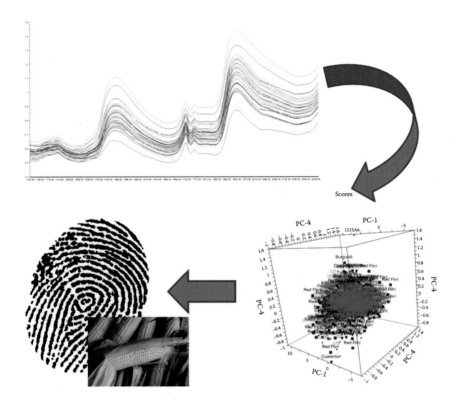

FIGURE 22.3
Example of combining NIR analysis and PCA to fingerprint corn samples for high oil content.

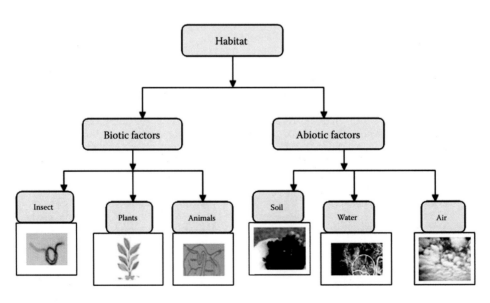

FIGURE 23.1
Habitats for metagenomic analysis.

14

Application of Next-Generation Sequencing for Abiotic Stress Tolerance

Tapan Kumar Mondal and Keita Sutoh

CONTENTS

14.1 Introduction

Crop production is highly vulnerable under the unprecedented climate changes that have led to heavy crop yield loss, and is a serious concern worldwide. Environmental extremes such as high and low temperatures, insufficient water supply, salinity, heavy metals, radiation, high and low nutrient content in the soil, etc., are the consequences of a rapidly changing climate, wreaking havoc on crop yield and productivity. Despite rapid advancements in techniques and research strategies, few crops are suitable for the changing climate, which will account for 15% to 32% of the yield reduction in the next 50 years, a situation that is just the reverse of what we actually need. Thus, there is an urgent need to accelerate our research efforts to develop new cultivars both by widening the genetic base of the available germplasm as well as by improving existing cultivars through integrated breeding approaches. However, "genomics" is considered to be an important component of integrated breeding approaches as the development of advanced molecular markers requires prior knowledge of genome sequences. Therefore, the application of "omics" techniques for the varietal improvement of plants demands sequenced genomes, so much so that it has become an era of "individual" genomics, which is only achievable with the assistance of high-throughput sequencing techniques that are discussed in the following sections.

14.2 A Brief Account of the Different High-Throughput Sequencing Platforms

All biological experiments that are based on DNA sequencing have changed drastically after the development of high-throughput sequencing. In particular, plant science, in which genetic diversity is the central theme of the research, needs robust high-throughput sequencing techniques to harness the benefits of genetic diversity. Although first-generation sequencing has been the predominant sequencing technology since 1974, there are several shortfalls with regard to the use of this technique, that is, low-throughput, involvement of bacterial cloning of the "DNA sequence," lengthy procedure, and labor intensiveness. Despite these facts, "Sanger" sequencing was and is suitable for the sequencing of individual genes due to its lower error rate. Simultaneously, another method known as "Maxam–Gilbert sequencing" was also developed. However, due to its technological simplicity, use of nonradioactive materials, and amenability to higher scale-up, Sanger sequencing moved forward so much so that it has become a highly automated technique [1].

The cloning of the DNA sequence into the bacterial plasmid before sequencing was eliminated by various second-generation sequencing technologies, even though they differ in chemistry of sequencing among themselves. For example, instead of cloning, a highly efficient *in vitro* DNA amplification method known as "emulsion PCR" was used in Roche 454 sequencing, the first next-generation sequencing (NGS) technology, which was developed in 2005. The 454 chemistry is known as "pyrosequencing," which is based on a sequencing-by-synthesis technique that measures the release of inorganic pyrophosphate by chemiluminescence during the sequencing reaction. The sequence of the DNA template is determined from a "pyrogram," which corresponds to the order of correct nucleotides that have been incorporated. Because chemiluminescent signal intensity is proportional to the amount of pyrophosphate released and hence the number of bases incorporated, the pyrosequencing approach is prone to errors that result from incorrectly estimating the length of homopolymeric sequence stretches (i.e., indels). The advantage of this technique is that it can perform the sequencing for longer read lengths but the disadvantage being the higher cost per unit of sequence. Nevertheless, due to larger sequencing length, this is suitable for *de novo* sequencing [2].

In Illumina/Solexa, the DNA templates are sequenced in a massively parallel fashion using a sequencing-by-synthesis approach, which employs reversible terminators with removable fluorescent moieties and special DNA polymerases that can incorporate these terminators into growing oligonucleotide chains. Its major advantage is that it can generate millions of reads but with the disadvantage of having very high initial costs for the instruments. Nevertheless, this technique is the most popular and is most suitable for resequencing, which has greater applications for studying epigenetic changes, ChIP-seq (identification of genomewide protein binding sequence of DNA), and the discovery of small noncoding RNA, among others [3].

Later, another second-generation sequencing platform, the supported oligonucleotide ligation and detection system was developed, the chemistry of which is based on massively parallel sequencing by ligation which is quite similar that of 454 sequencing but differs in the sense that after amplification, the sequencing is done in sequential rounds of hybridization and ligation with 16 dinucleotide combinations labeled with four different fluorescent dyes (each separate dye is used to label four dinucleotides). The advantage of this technology is that it has higher sequencing accuracy with the generation of larger amounts of sequencing data at a minimum amount of time. Due to the generation of a large amount of sequencing data, it gives better coverage, a term which is often used to describe the number of times a region of the genome has been sequenced. This method is often used in the resequencing of a gene [4], transcriptomic sequencing, or in genomic sequencing alongside other technologies [5].

"Ion torrent" is another sequencing technology that is based on the fact that the incorporation of one nucleotide in the template DNA chain always releases one H^+ in the medium, which causes a change in the pH that is then detected by an ion sensor—somewhat like a pH meter. This technique is also suitable for *de novo* sequencing of the genome with higher accuracy [6].

In 2011, PacBio, a revolutionary third-generation sequencer that works on single-molecule real-time technology to perform DNA sequencing was developed. It has the potential of sequencing up to 20,000-bp-long DNA within 1 h. Therefore, this technique is highly recommended for *de novo* sequencing. Another third-generation technology, known as "Helicos," is gaining popularity and has the advantage of using direct RNA sequencing. Therefore, the cDNA library construction step (the collection of all converted RNA molecules to its complementary DNA) is eliminated, which ultimately reduces the cost of sample preparation [7].

However, based on the principles of nanotechnology, there are several "future-generation sequencing" or "fourth-generation sequencing" technologies, such as Visigene [8] and Oxford Nanopore Technology [9], which will soon be available for commercial use. These techniques are based on single-molecule real-time sequencing but the sequencing length will be potentially unlimited, which differs from third-generation sequencing.

14.3 Applications

Because of rapid climate change, agriculture becoming more and more intensified, and location specifics, crop improvement objectives are also becoming more and more target/trait oriented. Therefore, it is necessary to discover new alleles in cultivated genotypes as well as in their wild relatives. Importantly, the conservation, utilization, and exploration of present genetic resources in the backdrop of their genetic erosion as well as eventual wipeout can be enhanced by genomic resources. Because of advancements in high-throughput sequencing techniques, the generation of genomic resources has become much easier and is taking place at a much faster pace. Thus, the different approaches of NGS are described in the next section and depicted in Figure 14.1.

14.3.1 *De novo* Sequencing or Whole-Genome Sequencing

Plants are sessile and hence evolved to be more diverse. Thus, in many instances, agronomically important traits are found in noncrop plants or otherwise not-so-useful plants. Perhaps the best example is the "mangrove," a halophyte that grows in salty, coastal regions. Thus,

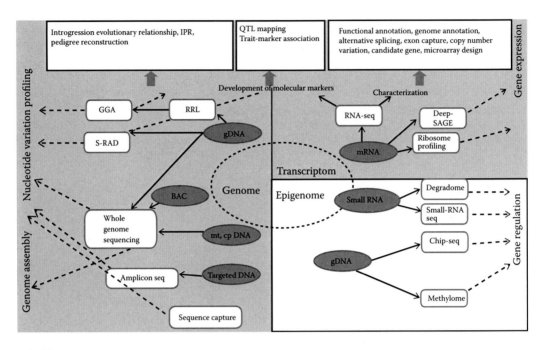

FIGURE 14.1

Application of NGS to the abiotic stresses of plant sciences. There are three different domains in a plant cell, that is, the genome, transcriptome, and epigenome, which are sources of variation (described by dotted central circle of the diagram). These sources, along with the help of NGS, are applied in the area of gene expression, gene regulation, or to study nucleotide diversity. Genomic DNA (gDNA), mitochondrial or chloroplast DNA (mtDNA, cpDNA), vectors such as BAC, mRNA, noncoding small RNAs, or target regions of the genome (targeted DNA) are the samples and are considered as the starting material (marked with filled circles). Different steps that can be performed with NGS are shown within the dotted arrows. gDNA samples, for example, can be used to produce reduced representation libraries and these can be either sequenced directly or used to generate sequenced restriction site–associated DNA (S-RAD) markers. Targeted DNA can be generated by PCR for amplicon sequencing or through sequence capture (NimbleGen, Roche, Basal, Switzerland; SureSelect, Agilent Technologies, Santa Clara, USA). Similarly, mRNA can be used in a deep-serial analysis of gene expression approach, RNAseq, or riboprofiling can be performed. gDNA can be also used in epigenetic studies of DNA–protein interactions through ChIP-seq or for studying methylation patterns through BS-seq (high-throughput bisulfite sequencing). The main application of genomes and transcriptomes are described in the top square boxes with some overlap applications that are described in the top middle square boxes, which includes quantitative trait loci (QTL) mapping, and trait marker association.

there is a need to sequence the genomes of such nonmodel, noncrop (even wild) plant species. Until recently, the sequencing of entire genomes, even those with small sizes, was time-consuming, tedious, and required substantial funds from multi-institutional collaborations. The development of NGS technology has increased the throughput and reduced the cost as well as time of sequencing by several orders of magnitude. Therefore, several crops have recently been sequenced *de novo* and are shown in Table 14.1. It is noteworthy to mention here that using this technology, neem [10], pigeon pea [11], and an indigenous medicinal plant (*Picrorhiza*) [12] have recently been sequenced *de novo* by Indian scientists for the first time. Several factors determine the *de novo* sequencing. For example, due to the shorter read lengths in Solexa/Illumina compared with 454 sequencing, *de novo* assembly can be easier using the latter (although coverage is deeper). Furthermore, for nonmodel plants, the absence of a reference genome, trait-specific transcriptome expression, tissue sampling, bioinformatics algorithms, and data analysis are some of the issues that are important for *de novo* sequencing.

TABLE 14.1

Summary of *De novo* Sequencing of Some Economically Important Plant Species

Plant	Approximate Genome Size	Purpose of Sequencing	Platform	Remarks	Algorithms Used	Reference
Eucalyptus grandis	600 Mbp	Transcriptome characterization, SNP identification	454	10% coverage of total transcript	Newbler and Paracel Transcript Assembler	[13]
Pachycladon enysii	550 Mb	Reference-guided assembly using diverged reference	Illumina	22,438 unigenes identified	ELAND and SOAP, Velvet	[14]
California poppy (*Eschscholzia californica*) and magnoliid avocado (*Persea americana*)	1115 Mbp	Comparison of two platforms for transcriptome characterization	454 and Illumina	Total identified unigenes 115,000	Newbler assembler	[15]
Sugarcane (*Saccharum officinarum*)	3961 Mbp	Transcriptome characterization for SNP detection	454	1632 SNPs were detected	CAP3	[16]
Red mangrove (*Rhizophora mangle*)	—	Transcriptome characterization, comparative gene expression	454	13,000 distinct gene models for each species were identified	Newbler/gsAssembler, Phrap	[17]
Grain amaranth (*Amaranthus hypochondriacus*)	675 Mbp	Transcriptome characterization	454	21,207 unigenes were identified	Newbler	[18]
Artemesia annua	823 Mbp	Glandular trichome transcriptome characterization	454	28,573 protein coding unigenes	SeqClean and CAP3	[19]
American chestnut (*Castanea dentata*) and Chinese chestnut (*Castanea mollissima*)	800 Mbp	Transcriptome comparison between canker tissues from *C. mollissima* and *C. dentata*	454	40,039 and 28,890 unigenes from *C. mollissima* and *C. dentata*, respectively	Newbler	[20]
Looking-glass mangrove (*Heritiera littoralis*)	—	Transcriptome characterization, comparative gene expression	Roche GSFLX	7.1% coverage for *H. littoralis* and 6.9% coverage for *R. mangle*	Phrap and Newbler assembler	[17]

(continued)

TABLE 14.1 (Continued)

Summary of *De novo* Sequencing of Some Economically Important Plant Species

Plant	Approximate Genome Size	Purpose of Sequencing	Platform	Remarks	Algorithms Used	Reference
Olive (*Olea europaea* cv. Coratina and Tendellone)	1907 Mbp	Transcriptome characterization between two genotypes with contrasting phenolic accumulation in fruits	454	26,563 Unigenes were detected	ParPEST	[21]
Amaranthus tuberculatus	757 ± 257 Mbp	Transcriptome characterization, comparative gene expression	454	44,469 unigenes identified	CAP3 and EGassembler	[22]
Cucumber (*Cucumis sativus*)	880 Mbp	Transcriptome characterization, comparative gene expression of two near-isogenic lines with male and female flowers	454	81,401 unigenes identified with 2X coverage	iAssembler and Splan	[23]
Eucalyptus grandis × *Eucalyptus urophylla*	650 Mbp	*De novo* assembly	Illumina	37X coverage of the transcript	Velvet, Mosaik and BWA	[24]
Wild oat (*Avena barbata*)	8729 Mbp	Differentially expressed transcriptome under water stress and different nitrogen treatment in root as well as leaf	454	8319 SSR were detected	MALIGN, CAP3, and BLAT	[25]
Garden pea (*Pisum sativum*)	4768 Mbp	Transcriptome characterization, comparative gene expression	454	81,449 unigenes identified	MIRA, TGICL, and BWASW	[26]

Species	Genome size	Application	Platform	Results	Software	Reference
Bracken fern (*Pteridium aquilinum*)	6259 Mbp	Transcriptome characterization	454	56,256 unigenes, 548 potentially amplifiable SSRs	MIRA and CAP3	[27]
Scabiosa columbaria	1174 Mbp	Transcriptome characterization, SNP identification	454 and illumina	29,676 unigenes identified, 4320 putative simple sequence repeats and 75,054 putative SNPs detected	CLC bio	[28]
Medicinal Chinese herb (*Salvia miltiorrhiza*)	534 ± 139 Mbp	Transcriptome sequencing to identify genes involved in the biosynthesis of active ingredients	Illumina	34,340 unigenes identified	CAP3	[29]
Neem (*Azadirachta indica*)	384 Mbp	*De novo* assembly of transcriptome from mature fruit	Illumina	26,908 unigenes identified	Trinity	[10]
Pigeon pea (*Cajanus cajan*)	858 Mbp	Whole genome sequence	454	47,004 unigenes identified with 10× coverage	CLC Bio	[11]
Picrorhiza kurroa	1720 Mbp	*De novo* transcript sequencing and identification of picroside containing gene under different temperature treatments	Illumina	74,336 unigenes identified	SOAPdenovo	[12]

14.3.2 Whole-Genome Resequencing

Once the genome of a plant is sequenced, it can act as a reference genome for the alignment of other sequences of the same species or related species, which speeds up the detection of genetic variations for a large number of accessions within a limited time. Thus, rapid developments in high-throughput sequencing have made whole-genome resequencing in several individuals, or targeted resequencing of "core germplasm" collections in a reality. This has also eliminated the important difficulty of ascertainment bias (i.e., the presence of rare alleles) obtained through biparental mapping populations in the estimation of linkage disequilibrium and genetic relationships between accessions [30,31]. The 1001 Genomes Project, perhaps the largest plant resequencing project, was launched at the beginning of 2008 and had the goal of discovering whole-genome sequence variations in 1001 accessions of the model plant *Arabidopsis thaliana*. Several *Arabidopsis* lines have been sequenced since then [32,33], describing the majority of common small-scale polymorphisms as well as many larger insertions and deletions in the *A. thaliana* pan-genome, their effects on gene function, and the patterns of local as well as global linkage among these variants. The actions of processes other than spontaneous mutation are identified by comparing the spectrum of mutations that have accumulated since *A. thaliana* diverged from its closest relative 10 million years ago with the spectrum observed in the laboratory. Recent species-wide selective sweeps are rare, and potentially deleterious mutations are more common in marginal populations. Subsequently, whole-genome resequencing in several crop species initiated primarily to detect the single nucleotide polymorphisms (SNPs) in maize [34,35], rice [36], *Medicago truncatula* [37], and soybean [38] will provide very valuable information for developing crop-specific markers for breeding programs.

14.3.3 Global Expression of Differentially Expressed Abiotic Stress-Related Genes

Abiotic stresses are polygenic in nature and therefore it is important to detect the entire gene pool including the rare or minor genes. Although differentially expressed genes can be identified in a number of ways, such as differential display of reverse transcriptase, micro-arrays, serial analysis of gene expression, selective subtractive hybridization, massive parallel sequence signatures, and cDNA-amplified fragment length polymorphisms [39], each of them, however, present advantages and disadvantages. Nevertheless, newer techniques always become more superior than older techniques. Due to the larger dynamic range and sensitivity of RNAseq, several additional factors have contributed to the rapid uptake of sequencing for differential expression analysis. For example, microarrays are simply not available for many nonmodel organisms (e.g., Affymetrix offers microarrays for ~30 organisms). Additionally, sequencing gives unprecedented details about transcriptional features that arrays cannot, such as novel transcribed regions, allele-specific expression, RNA editing, and a comprehensive capability to capture alternative splicing. Thus, for a number of reasons, RNAseq has become the popular choice to identify differentially expressed genes. Prolonged exposure to cold temperatures (vernalization) in the range of 6°C to 12°C induces reproductive growth, leading to bolting (rapid elongation of the main stem) and flowering in the sugar beet. Thus, to find the differentially expressed genes during vernalization in sugar beet, RNAseq was conducted. Several differentially expressed genes were found, which leads to the conclusion that RAV1-like AP2/B3 domain–containing genes are important for vernalization [40]. Using the same technology, that is, RNAseq, the genes responsible for cold acclimatization in blueberry (*Vaccinium corymbosum*) have been identified and their expressions were validated through Q-PCR analysis. Furthermore, they developed genic/simple sequence repeat markers for mapping some candidate genes [41]. The combination of SuperSAGE (a high-throughput transcriptome profiling technology), with whole-genome sequencing revealed the deep molecular understanding of salinity tolerance in chickpea [42]. Cross-validation with recent reports enriched the information about the salt stress dynamics of more than 9000 chickpea expressed sequence tags, and enlarged their pool of alternative transcript isoforms. Similarly, drought-responsive genes of *Gossypium herbaceum* were identified using RNAseq [43]. Several important pathways and transcription factors were detected in the drought tolerance cultivar, which are either not expressed or downregulated in susceptible cultivars of cotton. Based on the RNAseq data, an initial assemblage of sorghum genes and gene networks regulated by osmotic stress was made. This offers a preliminary look into the cascade of global gene expression patterns that arise in a drought-tolerant crop subjected to abiotic stress [44].

14.3.4 Discovery of Novel Small RNA

Small RNAs are a recently emerging group of noncoding RNAs, which plays an important role in gene expression under diverse stress conditions including various abiotic stresses. Although several abiotic stress–specific miRNAs have been discovered in various plant species through conventional cloning [45], the number of novel small RNAs discovered has increased significantly after the introduction of NGS due to the greater depth of sequencing. The use of NGS technology has also uncovered an interesting fact; until recently, it was believed that miRNAs were produced from a noncoding RNA that undergoes various steps to process 21- to 26-bp-long miRNA. As an exception to this, based on high-throughput sequencing analysis, it has been discovered that miRNAs are also produced from the intron region of some genes in *A. thaliana* and rice [46]. Furthermore, to investigate their biological roles, the targets of these mirtrons were predicted and validated based on degradome sequencing data. The results indicated that the mirtrons could guide target cleavages to exert their regulatory roles posttranscriptionally.

Several trait-specific miRNAs have been discovered. In *Populus euphratica*, 197 conserved miRNAs have been identified from the young leaves by small RNAseq analysis, out of which 104 miRNA sequences were found to be upregulated, whereas 27 were downregulated under osmotic stress [47]. In a separate experiment, 32 conserved and 8 new miRNAs that were responsive to drought stress were discovered after high-throughput sequencing of small RNAs from *M. truncatula* [48]. Subsequently, drought-specific miRNA in soybean [49], rice [45], barley [50], and cowpea [51] were identified, and their expressions later validated through Q-PCR under osmotic stress. Later, using illumine sequencing, it was found that 71, 50, and 45 miRNAs are either uniquely or differently expressed under drought, salinity, and alkalinity, respectively, in *Glycine max* [52]. In *Populus tomentosa*, 17 conserved and 9 novel miRNAs were found to be responsive to drought stress, whereas 7 conserved and 5 novel miRNAs were found to be responsive to flooding stress [53]. They found that both miRNA and miRNA*s (the complementary stand of miRNA) were involved in the regulation of these stresses.

The sterility of thermosensitive genic male lines of wheat (*Triticum aestivum*) is strictly controlled by temperature. Thus, through small RNA followed by degradome sequencing, a total of seven miRNAs and eight *trans*-acting small interfering RNAs in wheat have been discovered [54]. Deep sequencing of *Brachypodium* small RNAs at the global genome level identifies 27 conserved miRNAs, as well as 129 predicted miRNAs. They also reported that genome organization of the miR395 family in *Brachypodium* was quite different from that in rice, which is an exception to the conventional belief that miRNAs and their loci are conserved across the plant species. The expression of three conserved miRNAs and 25 predicted miRNAs showed significant changes in response under cold stress [55]. Low temperatures are also necessary to initiate stalk development in the orchid, *Phalaenopsis aphrodite* subsp. *formosana*. Thus, small RNA sequencing was performed to identify miRNAs in *Phalaenopsis* responding to low temperatures. Comparing sequencing data and small RNA Northern hybridization results, several miRNAs such as miR156, miR162, miR528, and miR535 were identified as low temperature–induced miRNAs [56]. Twenty-nine cold-responsive miRNA of *P. tomentosa* were identified by deep sequencing [57]. Among them, 21 miRNAs were downregulated and 8 miRNAs were upregulated in response to cold stress.

Several studies recently demonstrated that plant miRNAs play an important role in nutrient uptake under deficiency conditions. To elucidate the molecular mechanism underlying nitrogen signaling in maize, four small RNA libraries and one degradome library from maize seedlings exposed to nitrogen deficiency were constructed, which revealed a total of 99 absolutely new loci belonging to 47 miRNA families [58]. A small RNA library of *Arabidopsis* was sequenced to reveal miRNAs which were differentially expressed in response to phosphate deficiency. It was found that the expression of miR156, miR399, miR778, miR827, and miR2111 were upregulated, whereas the expression of miR169, miR395, and miR398 were repressed upon phosphate deprivation. Aluminum toxicity is a major factor limiting plant growth in acidic soils. To understand the miRNAs that are associated with aluminum uptake, a high-throughput sequencing analysis of small RNA library from the root of *M. truncatula* was made, which identified 326 known and 21 new miRNAs [60]. Among these, only the expression of 23 miRNAs was upregulated under aluminum stress [60].

Low-oxygen (hypoxia) stress associated with natural phenomena, such as water-logging, results in widespread transcriptome changes and a metabolic switch from aerobic respiration to anaerobic fermentation. Deep sequencing of small RNA libraries obtained from hypoxia-treated and control root tissues identified a total of 65 unique miRNAs and 14 *trans*-acting RNAs in *Arabidopsis* that were responsive to hypoxia [61]. In another study, it was found that miR169, miR397, miR528, miR1425, miR827, miR319, and miR408 were found to be responsive to hydrogen peroxide stress in rice. The predicted targets were experimentally validated and found to be involved in different cellular responses and metabolic processes, indicating the importance of diverse miRNAs in response to oxidative stress [62].

Topping or removal of the floral bud is an important cultural operation that increases the yield of tobacco. There are three major conserved miRNA families (nta-miR156, nta-miR172, and nta-miR171) and two new major miRNA families (nta-miRn2 and nta-miRn26), which were found to be associated with mechanical injury (topping) in tobacco [63]. Furthermore, a series of conserved and novel miRNAs that responded to heat stress were detected in *Brassica rapa* by deep sequencing [64]. In a separate study, it was found that Chinese cabbage shares at least 35 conserved miRNA families with *A. thaliana*, among them; five miRNA families were responsive to heat stress.

Salinity is another important abiotic stress that occurs worldwide. Using NGS technology, miRNA expressed under saline conditions has also been detected in barley [50] and rice [45]. The plant growth regulator, ethylene, and its precursor, 1-aminocyclopropane-1-carboxylic acid, inhibit root growth. However, there has been no information about the microRNAs that mediate ethylene-dependent physiological processes in plants. It was reported that eight miRNAs that are associated with 10 μM of 1-aminocyclopropane-1-carboxylic acid in *M. truncatula* have been identified [65].

14.3.5 Sequence Information–Based Conservation and Utilization of Plant Genetic Resources

NGS technologies are useful for the conservation and characterization of plant genetic resources. Often, a high degree of redundancy is found among the different *ex situ* germplasm collections. Most attempts to identify duplicated samples suffer from the difficulty of agreeing on a common panel of markers for a given species and the reproducibility of DNA marker data between different laboratories. NGS data do not suffer from such shortcomings and therefore represent an ideal information platform to tackle the issue of redundancy. Sequencing *ex situ* collections just for the sake of eliminating redundancy would practically be too expensive and it would be impossible to sequence every individual in a large collection of crops. Therefore, developing the "core collection set" (a set of genotypes that are true representatives of the genetic resources of a crop for its genetic diversity) is a better alternative [66]. Every core collection set will serve as a public, standardized, and well-characterized resource for the scientific community. Well-characterized, multiplied, isolated core reference set (CRS) have to be maintained for reference purposes, comparative studies, future reanalysis, and integrative genomic analysis [67].

NGS methods are suitable for low-cost sequencing of core collection sets with low coverage to develop genomewide markers that will facilitate the rejection of duplication across genome populations [68,69]. There are several methods in this regard such as reduced-representation libraries [70], complexity reduction of polymorphic sequences [71,72], restriction site–associated DNA sequencing [73], and low-coverage sequencing for genotyping [74–76]. These are applicable for the genetic analysis of crop plants, particularly of nonmodel species or species with high levels of repetitive DNA or for breeding germplasms with low levels of polymorphism without the need for prior sequence information. These methods can be applied to compare SNP diversity within and between closely related plant species or within wild natural populations [77,78].

14.3.6 Genomewide SNP or Haplotype Discovery

One of the central themes in the genomics of plant genetic resources is to study allelic differences or variations that can be detected by SNPs or haplotypes (i.e., groups of SNPs that are linked to a particular trait). SNPs are the most abundant type of DNA sequence polymorphism. Due to higher availability and stability compared with other robust markers, such as simple sequence repeats, it provides enhanced possibilities for studying genetic diversity, genetic resource management, cultivar identification, construction of genetic maps, assessment of genetic diversity, detection of genotype/phenotype associations, or marker-assisted breeding [79]. Furthermore, the efficiency of these activities can be improved by the automation of SNP genotyping. Using NGS technology, large-scale SNPs have been discovered in several crops. Traditionally, sequence variations are compared among the large number of plant genetic resources comprising diverse genetic backgrounds. Genomewide SNPs have been discovered using both transcriptomic [80] as well as whole-genome [81] sequencing techniques. Several abiotic trait-specific SNPs have been discovered in crops such as wheat (for drought and salt) [82], rye (for frost) [83], maize (for drought tolerance) [84,85], and *Arabidopsis* (for drought) [86], although they are mostly allele mining in nature. However, NGS has the highly likely potential of discovering SNPs that are specific to various abiotic stresses. These trait-specific SNPs could be converted into functional markers and then used for crop improvement by marker-assisted selection.

14.3.7 Allele Mining

Although the allelic diversity of genes has been shown to contribute to many phenotypic variations associated with different physiological processes in plants, information on the allelic diversity of abiotic stress–responsive genes is limited [87]. Plant accessions from wild or locally adapted landrace gene pools, conserved either *in situ* or *ex situ*, contain a rich repertoire of alleles that have been left behind by the selective processes of domestication, selection, and cross-breeding, which paved the way for today's elite cultivars. These resources stored in various gene banks remain underexplored owing to a lack of efficient strategies to screen, isolate, and transfer important alleles. The most effective strategy for determining allelic richness at a given locus is currently to determine its DNA sequence in a representative or core collection of a species. Although it is currently being done mostly with Sanger sequencing, nevertheless, NGS will soon be a cost-effective measure for allele mining [88]. Recently, the NGS were applied to study the allelic diversity of 10 candidate genes related to osmotic stress from 336 sessile oak (*Quercus robur*) and pedunculate oak (*Quercus petraea*). Out of these, eight genes from individual oaks growing in Austria have been detected with a total number of 158 polymorphic sites [89].

14.3.8 Genomewide Association Studies

Association mapping is an alternative technique to the traditional linkage mapping, which uses recombination events from many lineages [90]. Association mapping uses unstructured populations that have been subjected to many recombination events [91]. Thus far, linkage mapping based on biparental progenies has proven useful in detecting major genes and QTLs [92,93]. Although this approach has been successful in many analyses, it suffers from several drawbacks [30,31,94]. To overcome the shortfall of biparental-based linkage mapping, association genetics has started to supplement these efforts in several crops [95]. Furthermore, an improved approach, nested association mapping, which combines the advantages of linkage analysis and association mapping in a single unified mapping population, is also being used for the genomewide dissection of complex traits [96]. Several QTLs related to abiotic traits have been mapped, cloned, and transferred to elite lines. Alternatively, genomewide association studies in diverse germplasm collections offer new perspectives toward gene and allele discovery for traits of agricultural importance and for dissecting the genetic basis of complex quantitative traits in plants [95,97]. However, genomewide association studies require a genomewide assessment of genetic diversity (preferably based on a reference genome sequence and resequenced parts thereof), patterns of population structure, and the decay of linkage disequilibrium. For this, effective genotyping techniques for plants, high-density marker maps, phenotyping resources, and, if possible, a high-quality reference genome sequence are required [98]. However, in many cases, the results of genomewide association studies need confirmation by linkage analysis.

In maize, water stress at the flowering stage causes loss of kernel-set, which leads to low production. Therefore, an association mapping approach was used to identify loci involved in the accumulation of carbohydrates and abscisic acid (ABA) metabolites during stress. A panel of SNPs in genes from these metabolic pathways was used to genotype 350 tropical and subtropical inbred maize that were well-watered or water-stressed at the flowering stage. Association mapping with 1229 SNPs in 540 candidate genes identified an SNP in the maize homologue of the *Arabidopsis* MADS-box gene, PISTILLATA, which was significantly associated with phaseic acid in ears of well-watered plants, and an SNP in pyruvate dehydrogenase kinase, a key regulator of carbon flux into respiration, that was associated with silk sugar concentration. A SNP in an aldehyde oxidase gene was significantly associated with abscisic acid levels in silks of water-stressed plants [99].

14.3.9 Generation of Mutation for Novel Genetic Variation

Although natural variation is the basis for selecting the parent for mapping populations, it always remains a challenge for the plant breeder to produce new mutants. Besides, it is easy to search for mutants among the controlled or structured population rather than identifying the natural variant among the vast genetic resources. Although physical or chemical mutagenesis has remained the primary choice for decades, several techniques consisting of chemical mutagenesis coupled with molecular breeding have been developed recently.

Targeting-induced local lesions in genomes (TILLING) is a technique that can identify polymorphisms (more specifically, point mutations) resulting from induced mutations in a target gene by heteroduplex analysis [100]. Mutants have long been a valuable resource in plant breeding as well as in plant genomics research [100–102]. However, the method employed (irradiation or chemical) to induce a mutated population will affect its usefulness and application for genomics research [103]. It allows for genotypic screening for allelic variations before commencing with the more costly and labor-intensive phenotyping [102]. TILLING is fast becoming a mainstream technology for mutation characterization [103] and for analyzing SNPs [104]. To further speed up the process, a very sensitive high-throughput screening method based on capillary electrophoresis has been developed [105] using endonucleolytic mutation analysis by internal labeling to greatly improve the effectiveness of this new reverse genetics approach to crop improvement. A variant of this technique, EcoTILLING, allows us to assign haplotypes, thus reducing the number of accessions to be sequenced, becoming a cost-effective, time-saving, and high-throughput method, ideal for use in laboratories with limited financial resources. Although EcoTILLING techniques have been used to identify the salt tolerance–associated SNPs [106] as well as drought tolerance–associated SNPs in rice [107], and drought-associated SNP haplotype in barley [108], a high-throughput screening of mutation using NGS has not yet been done for identifying abiotic stress–related mutations.

Although TILLING has not yet been widely used due to certain limitations such as the use of endonuclease CEL I, which has poor cleavage efficiency and 5′-3′ exonuclease activity, this method reduced signal or noise levels and limits analyzing more than eight samples per pool [109]. Moreover, being a prescreening method, it does not reveal information about the nature of sequence changes and their possible effect on gene function. Thus, there is a need for robust and low-cost amplicon sequencing methods applicable to large populations. To overcome this, KeyPoint technology, a high-throughput mutation/polymorphism discovery technique based on 454 sequencing technology of target genes amplified from mutant or natural populations, was developed [110]. Using this technique, two mutants in the tomato eIF4E gene were identified after screening more than 3000 M2 families in a single sequencing run. Nevertheless, NGS will soon be useful for the rapid identification of abiotic stress tolerance genes.

14.4 Conclusions

Abiotic stresses are a major bottleneck in agricultural production and productivity today. There is a need to develop various abiotic stress-tolerant plants, which can be achieved by an understanding of the molecular mechanisms of stress tolerance, and by generating of the genomic resources associated with abiotic stress. Simultaneously, marker-assisted breeding is extremely powerful technique for developing new cultivars or improving an existing cultivar, or for the introduction of a gene from an alien species, but needs the information from a genome sequence primarily for two purposes: (1) to develop a marker that spans the entire genome and (2) to develop a reference genome that can be used in several ways to detect polymorphisms. Therefore, sequencing the genome for crop and noncrop plants, as well as alien species, becomes very important. Although a handful of nonmodel plants have been sequenced, efforts to sequence the genomes of several other crop species are ongoing. These will become reality only with the lowering of sequencing costs per nucleotide base, which will soon facilitate the development of need-based cultivars to combat rapid climate change.

References

1. Schadt, E.E., S., Turner, and A. Kasarskis. 2010. A window into third-generation sequencing. *Human Molecular Genetics* 19:227–240.
2. Zalapa, J.E., H. Cuevas, H. Zhu, S. Steffan, D. Senalik, E. Zeldin et al. 2012. Using next-generation sequencing approaches for the isolation of simple sequence repeat (SSR) loci in the plant sciences. *American Journal of Botany* 99:193–208.
3. Bentley, D.R. 2006. Whole-genome re-sequencing. *Current Opinion in Genetics and Development* 16:545–552.
4. Ashelford, K., M.E. Eriksson, C.M. Allen, R.D. Amore, M. Johansson, P. Gould et al. 2011. Full genome re-sequencing reveals a novel circadian clock mutation in *Arabidopsis*. *Genome Biology* 12:28–40.
5. Shulaev, V., D.J. Sargent, R.N. Crowhurst, T.C. Mockler, O. Folkerts, A.L.D. Elcher et al. 2011. The genome of woodland strawberry (*Fragaria vesca*). *Nature Genetics* 43:109–116.
6. Rothberg, J.M., W. Hinz, T.W. Rearick, J. Schultz, W. Mileski, M. Davey et al. 2011. An integrated semiconductor device enabling non-optical genome sequencing. *Nature* 475:348–352.
7. Levene, M.J., J. Korlach, S.W. Turner, M. Foquet, H.G. Craighead, and W.W. Webb. 2003. Zero-mode waveguides for single-molecule analysis at high concentration. *Science* 299:682–686.
8. Hardin, S.H. 2008. Real-time DNA sequencing. In *Next Generation Genome Sequencing: Towards Personalized Medicine*, edited by Janitz, M., 97–102, Weinheim, Germany: Wiley-VCH Verlag GmbH & Co. KGaA.
9. Lieberman, K.R., G.M. Cherf, M.J. Doody, F. Olasagasti, Y. Kolodji, and M. Akeson. 2010. Possessive replication of single DNA molecules in a nanopore catalyzed by phi29 DNA polymerase. *Journal of the American Chemical Society* 132:17961–17972.
10. Krishnan, N.M., S. Pattnaik, S.A. Deepak, A.K. Hariharan, P. Gaur, R. Chaudhary et al. 2011. *De novo* sequencing and assembly of *Azadirachta indica* fruit transcriptome. *Current Science* 10:1533–1561.
11. Singh, N.K., D.K. Gupta, P.K. Jayaswal, M.K. Mahato, S. Dutta, S. Singh et al. 2012. The first draft of the pigeon pea genome sequence. *Journal of Plant Biochemistry Biotechnology* 1:98–112.
12. Gahlan, P., H.R. Singh, R. Shankar, N. Sharma, A. Kumari, V. Chawla et al. 2012. *De novo* sequencing and characterization of *Picrorhiza kurrooa* transcriptome at two temperatures showed major transcriptome adjustments. *BMC Genomics* 13:126–138.
13. Novaes, E., D.R. Drost, W.G. Farmerie, J.P. Pappas, D. Grattapaglia, R.R. Sederoff et al. 2008. High-throughput gene and SNP discovery in *Eucalyptus grandis*, an uncharacterized genome. *BMC Genomics* 9:312–319.
14. Collins, L.J., P.J. Biggs, C. Voelckel, and S. Joly. 2008. An approach to transcriptome analysis of non-model organisms using short-read sequences. *Genome Informatics* 21:3–14.

15. Wall, P.K., J. Leebens-Mack, A.S. Chanderbali, A. Barakat, E. Wolcott, H. Liang et al. 2009. Comparison of next generation sequencing technologies for transcriptome characterization. *BMC Genomics* 10:347–366.

16. Bundock, P.C., F.G. Eliott, G. Ablett, A.D. Benson, R.E. Casu, K.S. Aitken et al. 2009. Targeted single nucleotide polymorphism (SNP) discovery in a highly polyploid plant species using 454 sequencing. *Plant Biotechnology* 7:347–354.

17. Dassanayake, M., J.S. Haas, H.J. Bohnert, and J.M. Cheeseman. 2009. Shedding light on an extremophile lifestyle through transcriptomics. *New Phytologist* 183:764–775.

18. Lee, R.M., J. Thimmapuram, A. Kate, T.G. Gong, A.G. Hernandez, C.L. Wright et al. 2009. Sampling the waterhemp (*Amaranthus tuberculatus*) genome using pyrosequencing technology. *Weed Science* 57:463–469.

19. Wang, W.Y., Q. Wang, Q. Zhang, and D. Guo. 2009. Global characterization of *Artemisia annua* glandular trichome transcriptome using 454 pyrosequencing. *BMC Genomics* 10:465–475.

20. Barakat, A., D.S. Diloreto, Y. Zhang, C. Smith, K. Baier, W.A. Powell et al. 2009. Comparison of the transcriptomes of American chestnut (*Castanea dentata*) and Chinese chestnut (*Castanea mollissima*) in response to the chestnut blight infection. *BMC Plant Biology* 9:51–62.

21. Alagna, F., N.D. Agostino, L. Torchia, M. Servili, R. Rao, M. Pietrella et al. 2009. Comparative 454 pyrosequencing of transcripts from two olive genotypes during fruit development. *BMC Genomics* 10:399–414.

22. Riggins, C.W., Y. Peng, J.R.C.N. Stewart, and P.J. Tranel. 2010. Characterization of *de novo* transcriptome for waterhemp (*Amaranthus tuberculatus*) using GS-FLX 454 pyrosequencing and its application for studies of herbicide target-site genes. *Pest Management Science* 66:1042–1052.

23. Guo, S., Y. Zheng, J. Joung, S. Liu, Z. Zhang, O.R. Crasta et al. 2010. Transcriptome sequencing and comparative analysis of cucumber flowers with different sex types. *BMC Genomics* 11:384–395.

24. Mizrachi, E., C.A. Hefer, M. Ranik, F. Joubert, and A.A. Myburg. 2010. *De novo* assembled expressed gene catalog of a fast-growing Eucalyptus tree produced by Illumina mRNA-Seq. *BMC Genomics* 11:681–687.

25. Swarbreck, S.M., E.A. Lindquist, D.D. Ackerly, and G.L. Andersen. 2011. Analysis of leaf and root transcriptomes of soil-grown *Avena barbata* plants. *Plant and Cell Physiology* 52:317–332.

26. Franssen, S.U., R.P. Shrestha, A. Bräutigam, E. Bornberg-Bauer, and A.P. Weber. 2011. Comprehensive transcriptome analysis of the highly complex *Pisum sativum* genome using next generation sequencing. *BMC Genomics* 12:227–245.

27. Der, J.P., M.S. Barker, N.J. Wickett, C.W. Depamphilis, and P.G. Wolf. 2011. *De novo* characterization of the gametophyte transcriptome in bracken fern, *Pteridium aquilinum*. *BMC Genomics* 12:99–113.

28. Angeloni, F., C.A.M. Wagemaker, M.S.M. Jetten, H.J.M. Opden, I. Camp, E.M. Janssen-Megens et al. 2011. *De novo* transcriptome characterization and development of genomic tools for *Scabiosa columbaria* L. using next generation sequencing techniques. *Molecular Ecology Research* 11:662–674.

29. Wenping, H., Z. Yuan, S. Jie, Z. Lijun, and W. Zhezhi. 2011. *De novo* transcriptome sequencing in *Salvia miltiorrhiza* to identify genes involved in the biosynthesis of active ingredients. *Genomics* 98:272–279.

30. Moragues, M., J. Comadran, R. Waugh, I. Milne, A.J. Flavell, and J.R. Russell. 2010. Effects of ascertainment bias and marker number on estimations of barley diversity from high-throughput SNP genotype data. *Theoretical and Applied Genetics* 120:1525–1534.

31. Cosart, T., A. Beja-Pereira, S. Chen, S.B. Ng, J. Shendure, and G. Luikart. 2011. Exome-wide DNA capture and next generation sequencing in domestic and wild species. *BMC Genomics* 12:347–355.

32. Cao, J., K. Schneeberger, S. Ossowski, T. Günther, S. Bender, J. Fitz et al. 2011. Whole-genome sequencing of multiple *Arabidopsis thaliana* populations. *Nature Genetics* 43:956–963.

33. Lister, R., and J.R. Ecker. 2009. Finding the fifth base: Genome-wide sequencing of cytosine methylation. *Genome Research* 19:959–966.

34. Lai, J., R. Li, X. Xu, W. Jin, M. Xu, H. Zhao et al. 2010. Genome-wide patterns of genetic variation among elite maize inbred lines. *Nature Genetics* 42:1027–1030.

35. Xu, Z., S. Zhong, X. Li, W. Li, S.J. Rothstein, S. Zhang et al. 2011. Genome-wide identification of microRNAs in response to low nitrate availability in maize leaves and roots. *PLoS One* 6:28009–28018.

36. Xu, X., X. Liu, S. Ge, J.D. Jensen, F. Hu, X. Li et al. 2011. Resequencing 50 accessions of cultivated and wild rice yields markers for identifying agronomically important genes. *Nature Biotechnology* 30:105–111.

37. Branca, A., T.D. Paape, P. Zhou, R. Briskine, A.D. Farmer, J. Mudge et al. 2011. Whole-genome nucleotide diversity, recombination, and linkage disequilibrium in the model legume *Medicago truncatula*. *Proceedings of the National Academy of Sciences of the United States of America* 108:864–870.

38. Lam, H.M., X. Xu, X. Liu, W. Chen, G. Yang, F.L. Wong et al. 2010. Resequencing of 31 wild and cultivated soybean genomes identifies patterns of genetic diversity and selection. *Nature Genetics* 42:1053–1059.

39. Nobuta, K., K. Vemaraju, and B.C. Meyers. 2007. Methods for analysis of gene expression in plants using MPSS. *Methods in Molecular Biology* 406:387–408.

40. Mutasa-Göttgens, E.S., A. Joshi, H.F. Holmes, P. Hedden, and B. Göttgens. 2012. A new RNASeq-based reference transcriptome for sugar beet and its application in transcriptome scale analysis of vernalization and gibberellins responses. *BMC Genomics* 13:99–117.

41. Rowland, L.J., N. Alkharouf, O. Darwish, E.L. Ogden, J.J. Polashock, N.V. Bassil et al. 2012. Generation and analysis of blueberry transcriptome sequences from leaves, developing fruit, and flower buds from cold acclimation through deacclimation. *BMC Plant Biology* 12:46–61.

42. Molina, C., M. Zaman-Allah, F. Khan, N. Fatnassi, R. Horres, B. Rotter et al. 2011. The salt-responsive transcriptome of chickpea roots and nodules via deep SuperSAGE. *BMC Plant Biology* 11:31–57.

43. Ranjan, A., D. Nigam, M.H. Asif, R. Singh, S. Ranjan, S. Mantri et al. 2012. Genome wide expression profiling of two accession of *G. herbaceum* L. in response to drought. *BMC Genomics* 13:94–112.

44. Dugas, D.V., M.K. Monaco, A. Olsen, R.R. Klein, S. Kumari, D. Ware et al. 2011. Functional annotation of the transcriptome of *Sorghum bicolor* in response to osmotic stress and abscisic acid. *BMC Genomics* 12:514–535.

45. Sunkar, R., X. Zhou, Y. Zheng, W. Zhang, and J.K. Zhu. 2008. Identification of novel and candidate miRNAs in rice by high throughput sequencing. *BMC Plant Biology* 8:25–42.

46. Meng, Y., and C. Shao. 2012. Large-scale identification of mirtrons in *Arabidopsis* and rice. *PLoS One* 7:31163–31170.

47. Li, B., Y. Qin, H. Duan, W. Yin, and X. Xia. 2011. Genome-wide characterization of new and drought stress responsive microRNAs in *Populus euphratica*. *Journal of Experimental Botany* 62:3765–3779.

48. Wang, T., L. Chen, M. Zhao, Q. Tian, and W.-H. Zhang. 2011. Identification of drought-responsive microRNAs in *Medicago truncatula* by genome-wide high-throughput sequencing. *BMC Genomics* 12:367–378.

49. Kulcheski, F.R., L.F.V. Oliveira, L.G. Molina, M.P. Almerão, F.A. Rodrigues, J. Marcolino et al. 2011. Identification of novel soybean microRNAs involved in abiotic and biotic stresses. *BMC Genomics* 12:307–324.

50. Lv, S., X. Nie, L. Wang, X. Du, S.S. Biradar, X. Jia et al. 2012. Identification and characterization of microRNAs from barley (*Hordeum vulgare* L.) by high-throughput sequencing. *International Journal of Molecular Science* 13:2973–2984.

51. Blanca, E.B.-F., L. Gao, N.N. Diop, Z. Wu, J.D. Ehlers, P.A. Roberts et al. 2011. Identification and comparative analysis of drought-associated microRNAs in two cowpea genotypes. *BMC Plant Biology* 11:127–138.

52. Li, H., Y. Dong, H. Yin, N. Wang, J. Yang, X. Liu et al. 2011. Characterization of the stress associated microRNAs in *Glycine max* by deep sequencing. *BMC Plant Biology* 11:170–182.

53. Ren, Y., L. Chen, Y. Zhang, X. Kang, Z. Zhang, and Y. Wang. 2012. Identification of novel and conserved *Populus tomentosa* microRNA as components of a response to water stress. *Functional and Integrative Genomics* 10.1007/s10142-012-0271-6.

54. Tang, Z., L. Zhang, C. Xu, S. Yuan, F. Zhang, Y. Zheng et al. 2012. Uncovering small RNA-mediated responses to cold stress in a wheat thermo-sensitive genic male sterile line by deep sequencing. *Plant Physiology* 159(2):721–738.

55. Zhang, J., Y. Xu, Q. Huan, and K. Chong. 2009. Deep sequencing of *Brachypodium* small RNAs at the global genome level identifies microRNAs involved in cold stress response. *BMC Genomics* 10:449–465.

56. An, F.-M., S.-R. Hsiao, and M.-T. Chan. 2011. Sequencing-based approaches reveal low ambient temperature-responsive and tissue-specific microRNAs in *Phalaenopsis* orchid. *PLoS One* 6:18937–18948.

57. Chen, L., Y. Zhang, Y. Ren, J. Xu, Z. Zhang, and Y. Wang. 2012. Genome-wide identification of cold-responsive and new microRNAs in *Populus tomentosa* by high-throughput sequencing. *Biochemistry Biophysics Research Communication* 417:892–896.

58. Zhao, M., H. Tai, S. Sun, F. Zhang, Y. Xu, and W.-X. Li. 2012. Cloning and characterization of maize miRNAs involved in responses to nitrogen deficiency. *PLoS One* 7:29669–29680.

59. Hsieh, L.-C., S.-I. Lin, A.C. Shih, J.-W. Chen, W.-Y. Lin, C.-Y. Tseng et al. 2009. Uncovering small RNA-mediated responses to phosphate deficiency in *Arabidopsis* by deep sequencing. *Plant Physiology* 151:2120–2132.

60. Chen, L., T. Wang, M. Zhao, Q. Tian, and W.-H. Zhang. 2012. Identification of aluminum-responsive microRNAs in *Medicago truncatula* by genome-wide high-throughput sequencing. *Planta* 235:375–386.

61. Moldovan, D., A. Spriggs, J. Yang, J.B. Pogson, E.S. Dennis, W. Iain et al. 2010. Hypoxia-responsive microRNAs and trans-acting small interfering RNAs in *Arabidopsis*. *Journal of Experimental Botany* 61:165–177.

62. Li, T., H. Li, Y.-X. Zhang, and J.-Y. Liu. 2011. Identification and analysis of seven H_2O_2-responsive miRNAs and 32 new miRNAs in the seedlings of rice (*Oryza sativa* L. ssp. *indica*). *Nucleic Acid Research* 39:2821–2833.

63. Guo, H., Y. Kan, and W. Liu. 2011. Differential expression of miRNAs in response to topping in flue-cured tobacco (*Nicotiana tabacum*) roots. *PLoS One* 6:28565–28580.

64. Yu, X., H. Wang, Y. Lu, M.D. Ruiter, M. Cariaso, M. Prins et al. 2012. Identification of conserved and novel microRNAs that are responsive to heat stress in *Brassica rapa*. *Journal of Experimental Botany* 63:1025–1038.

65. Chen, L., T. Wang, M. Zhao, and W. Zhang. 2012. Ethylene-responsive miRNAs in roots of *Medicago truncatula* identified by high-throughput sequencing at whole genome level. *Plant Science* 184:14–19.

66. Glaszmann, J.C., B. Kilian, H.D. Upadhyaya, and R.K. Varshney. 2010. Accessing genetic diversity for crop improvement. *Current Opinion of Plant Biology* 13:167–173.

67. Hawkins, R.D., G.C. Hon, and B. Ren. 2010. Next-generation genomics: An integrative approach. *Nature Review Genetics* 11:476–486.

68. Bansal, V., O. Harismendy, R. Tewhey, S.S. Murray, N.J. Schork, E.J. Topol et al. 2010. Accurate detection and genotyping of SNPs utilizing population sequencing data. *Genome Research* 20:537–545.

69. Davey, J.W., P.A. Hohenlohe, P.D. Etter, J.Q. Boone, J.M. Catchen, and M.L. Blaxter. 2011. Genome-wide genetic marker discovery and genotyping using next generation sequencing. *Nature Review Genetics* 12:499–510.

70. You, F.M., N. Huo, K.R. Deal, Y.Q. Gu, M.C. Luo, P.E. McGuire et al. 2011. Annotation-based genome-wide SNP discovery in the large and complex *Aegilops tauschii* genome using next-generation sequencing without a reference genome sequence. *BMC Genome* 12:59–78.

71. van Orsouw, N.J., R.C.J. Hogers, A. Janssen, F. Yalcin, S. Snoeijers, E. Verstege et al. 2007. Complexity reduction of polymorphic sequences (CRoPS™): A novel approach for large-scale polymorphism discovery in complex genomes. *PLoS One* 2:1172–1182.

72. Mammadov, J.A., W. Chen, R. Ren, R. Pai, W. Marchione, F. Yalçin et al. 2010. Development of highly polymorphic SNP markers from the complexity reduced portion of maize (*Zea mays* L.) genome for use in marker-assisted breeding. *Theoretical and Applied Genetics* 121:577–588.

73. Baxter, S.W., J.W. Davey, J.S. Johnston, A.M. Shelton, D.G. Heckel, C.D. Jiggins et al. 2011. Linkage mapping and comparative genomics using next-generation RAD sequencing of a non-model organism. *PLoS One* 6:19315–19326.

74. Huang, X., Q. Feng, Q. Qian, Q. Zhao, L. Wang, A. Wang et al. 2009. High-throughput genotyping by whole-genome resequencing. *Genome Research* 19:1068–1076.

75. Andolfatto, P., D. Davison, D. Erezyilmaz, T.T. Hu, J. Mast, T. Sunayama-Morita et al. 2011. Multiplexed shotgun genotyping for rapid and efficient genetic mapping. *Genome Research* 21:610–617.

76. Elshire, R.J., J.C. Glaubitz, Q. Sun, J.A. Poland, K. Kawamoto, E.S. Buckler et al. 2011. A robust, simple genotyping-by-sequencing (GBS) approach for high diversity species. *PLoS One* 6:19379–19389.

77. Ossowski, S., K. Schneeberger, J.I. Lucas-Lledó, N. Warthmann, R.M. Clark, R.G. Shaw et al. 2010. The rate and molecular spectrum of spontaneous mutations in *Arabidopsis thaliana*. *Science* 327:92–94.

78. Pool, J.E., I. Hellmann, J.D. Jensen, and R. Nielsen. 2010. Population genetic inference from genomic sequence variation. *Genome Research* 20:291–300.

79. Ganal, M.W., T. Altmann, and M.S. Röder. 2009. SNP identification in crop plants. *Current Opinion of Plant Biology* 12:211–217.

80. Barbazuk, W.B., and P.S. Schnable. 2011. SNP discovery by transcriptome pyrosequencing. *Methods in Molecular Biology* 729:225–246.

81. Ahmad, R., D.E. Parfitt, J. Fass, E. Ogundiwin, A. Dhingra, T.M. Gradziel et al. 2011. Whole genome sequencing of peach (*Prunus persica* L.) for SNP identification and selection. *BMC Genomics* 12:569–576.

82. Mondini, L., M. Nachit, E. Porceddu, and M.A. Pagnotta. 2012. Identification of SNP mutations in DREB1, HKT1, and WRKY1 genes involved in drought and salt stress tolerance in durum wheat (*Triticum turgidum* L. var *durum*). *Omics: A Journal of Integrative Biology* 16:178–187.

83. Li, Y., G. Haseneyer, C.-C. Schön, D. Ankerst, V. Korzun, P. Wilde et al. 2011. High levels of nucleotide diversity and fast decline of linkage disequilibrium in rye (*Secale cereale* L.) genes involved in frost response. *BMC Plant Biology* 11:6–20.

84. Hao, Z., X. Li, C. Xie, J. Weng, M. Li, D. Zhang et al. 2011. Identification of functional genetic variations underlying drought tolerance in maize using SNP markers. *Journal of Integrated Plant Biology* 53:641–652.

85. Lu, Y., S. Zhang, T. Shah, C. Xie, Z. Hao, X. Li et al. 2010. Joint linkage-linkage disequilibrium mapping is a powerful approach to detecting quantitative trait loci underlying drought tolerance in maize. *Proceedings of the National Academy of Sciences of the United States of America* 107:19585–19590.

86. Hao, G.P., Z.Y. Wu, M.Q. Cao, G. Pelletier, D. Brunel, C.L. Huang et al. 2004. Nucleotide polymorphism in the drought induced transcription factor CBF4 region of *Arabidopsis thaliana* and its molecular evolution analyses. *Yi Chuan Xue Bao* 31:1415–1425.

87. Tao, Z., Y. Kou, H. Liu, X. Li, J. Xiao, and S. Wang. 2011. OsWRKY45 alleles play different roles in abscisic acid signalling and salt stress tolerance but similar roles in drought and cold tolerance in rice. *Journal of Experimental Botany* 62:4863–4874.

88. Kumar, G.R., K. Sakthive, R.M. Sundaram, C.N. Neeraja, S.M. Balachandran, R.N. Shobha et al. 2010. Allele mining in crops: Prospects and potentials. *Biotechnology Advances* 28:451–461.

89. Homolka, A., T. Eder, D. Kopecky, M. Berenyi, K. Burg, and S. Fluch. 2012. Allele discovery of ten candidate drought-response genes in Austrian oak using a systematically informatics approach based on 454 amplicon sequencing. *BMC Research Notes* 5:175–185.

90. Abdurakhmonov, I., and A. Abdukarimov. 2008. Application of association mapping to understanding the genetic diversity of plant germplasm resources. *International Journal of Plant Genomics* 5:1–18.

91. Zhao, K., C.W. Tung, G.C. Eizenga, M.H. Wright, M.L. Ali, A.H. Price et al. 2011. Genome-wide association mapping reveals a rich genetic architecture of complex traits in *Oryza sativa*. *Nature Communication* 2:467–475.

92. Frary, A., T.C. Nesbitt, A. Frary, S. Grandillo, E. van der Knaap, B. Cong et al. 2000. fw2.2: A quantitative trait locus key to the evolution of tomato fruit size. *Science* 289:85–88.

93. Komatsuda, T., M. Pourkeirandish, C. He, P. Azhaguvel, H. Kanamori, D. Perovic et al. 2007. Six-rowed barley originated from a mutation in a homeodomainleucine zipper I–class homeobox gene. *Proceedings of the National Academy of Sciences of the United States of America* 104:1424–1429.

94. Schuenemann, V.J., K. Bos, S. DeWitte, S. Schmedes, J. Jamieson, A. Mittnik et al. 2011. Targeted enrichment of ancient pathogens yielding the pPCP1 plasmid of *Yersinia pestis* from victims of the Black Death. *Proceedings of the National Academy of Sciences of the United States of America* 108:746–752.

95. Hall, D. 2010. Using association mapping to dissect the genetic basis of complex traits in plants. *Brief Functional Genomics* 9:157–165.

96. Yu, J., J.B. Holland, M.D. McMullen, and E.S. Buckler. 2008. Genetic design and statistical power of nested association mapping in maize. *Genetics* 178:539–551.

97. Mackay, T.F.C., E.A. Stone, and J.F. Ayroles, 2009. The genetics of quantitative traits: Challenges and prospects. *Nature Review Genetics* 10:565–577.

98. Rafalski, J.A. 2010. Association genetics in crop improvement. *Current Opinion of Plant Biology* 13:174–180.

99. Setter, T.L., J. Yan, M. Warburton, J.M. Ribaut, Y. Xu, M. Sawkins et al. 2010. Genetic association mapping identifies single nucleotide polymorphisms in genes that affect abscisic acid levels in maize floral tissues during drought. *Journal of Experimental Botany* 62:701–716.

100. Till, B.J., S.H. Reynolds, E.A. Greene, C.A. Codomo, L.C. Enns, J.E. Johnson et al. 2003. Large-scale discovery of induced point mutations with high-throughput TILLING. *Genome Research* 13:524–530 (http://genome.cshlp.org/content/13/3/524.abstract-target-2).

101. Henikoff, S., and L. Comai. 2003. Single-nucleotide mutations for plant functional genomics. *Annual Review of Plant Biology* 54:375–401.

102. Henikoff, S., B.J. Till, and L. Comai. 2004. TILLING: Traditional mutagenesis meets functional genomics. *Plant Physiology* 135:630–636.

103. Comai, L., and S. Henikoff. 2006. TILLING: Practical single-nucleotide mutation discovery. *The Plant Journal* 45:684–694.

104. Cordeiro, G.M., F. Eliott, C.L. McIntyre, R.E. Casu, and R.J. Henry. 2006. Characterization of single nucleotide polymorphisms in sugarcane ESTs. *Theoretical and Applied Genetics* 113:331–343.

105. Cross, M., S.L. Lee, and R.J. Henry. 2007. A novel detection strategy for scanning multiple mutations using CEL I. Paper presented at the Plant and Animal Genomes Conference XV, Mutation Screening Workshop, San Diego, CA. January 13–17, 2007.

106. Negrão, S., C. Almadanim, I. Pires, K.L. McNally, and M.M. Oliveira. 2011. Use of EcoTILLING to identify natural allelic variants of rice candidate genes involved in salinity tolerance. *Plant Genetic Resources* 9:300–304.

107. Naredo, Ma. E.B., J. Cairns, H. Wang, G. Atienza, M.D. Sanciangco, R.J.A. Melgar et al. 2009. EcoTILLING as a SNP discovery tool for drought candidate genes in *Oryza sativa* germplasm. *Philippine Journal of Crop Science* 34:10–16.

108. Cseri, A., M. Cserháti, M. von Korff, B. Nagy, G.V. Horváth, A. Palágyi et al. 2011. Allele mining and haplotype discovery in barley candidate genes for drought tolerance. *Euphytica* 181:341–356.

109. Till, B.J., T. Zerr, L. Comai, and S. Henikoff. 2006. A protocol for TILLING and ecotilling in plants and animals. *Nature Protocols* 1:2465–2477.

110. Rigola, D., J.V. Oeveren, A. Janssen, A. Bonne, H. Schneiders, H.J.A. van der Poel et al. 2009. High-throughput detection of induced mutations and natural variation using KeyPoint™ technology. *PLoS One* 4:4761–4770.

15

Next-Generation Sequencing and Assembly of Bacterial Genomes

Artur Silva, Rommel Ramos, Adriana Carneiro, Sintia Almeida,
Vinicius De Abreu, Anderson Santos, Siomar Soares, Anne Pinto,
Luis Guimarães, Eudes Barbosa, Paula Schneider, Vasudeo Zambare,
Debmalya Barh, Anderson Miyoshi, and Vasco Azevedo

CONTENTS

15.1 Introduction

One of the most important advances in biology has been our capacity to sequence the DNA of organisms. However, long after the conclusion of human genome sequencing, there are still regions of the genome that are unworkable; that is, they are difficult to mount and remain incomplete. Answers may come from second-generation sequencing, which has produced large volumes of data, generating millions of short reads per run, a reality that was unimaginable with Sanger sequencing.

Although we can now generate a high degree of sequencing coverage (Figure 15.1), the *de novo* assembly of short reads is more complex compared with reference assembly. Various algorithms and bioinformatics tools have been developed to take care of these new

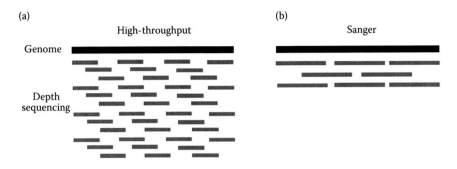

FIGURE 15.1
Qualitative comparison between the sequences generated by Sanger and those from the NGS platforms. There is a higher abundance and depth of coverage with the short reads, but they are also significantly shorter, with little overlap available for assurance. (a) Depth of coverage in the high-throughput platforms. (b) Depth of coverage in the Sanger platforms.

problems and computational challenges, such as identification of repeat regions, sequencing errors, and simultaneous manipulation of short reads [1,2].

After reads are generated by the sequencer, it is necessary to join them in a logical fashion to mount the final sequence. Over the years, various tools have been developed to resolve this issue, for example, the assemblers PHRAP (http://www.phrap.org), ARACHNE [3], and Celera [4]. They have a paradigm in common, often referred to as overlap-layout-consensus (OLC) [5]. This approach is quite similar to that used to resolve a jigsaw puzzle, as described below.

The first step consists of aligning the reads, two by two, exhaustively; the pairs of reads should present a consistent overlap from one read to another, similar to the search for pieces in a jigsaw puzzle that fit each other and have colors that match. Especially in eukaryotic genomes, the main difficulty is in distinguishing inexact overlaps due to sequencing errors and similarities within the genome, such as highly conserved repeat regions [6]. Sequence alignment is a widely studied area of bioinformatics, which consists of supplying the ideal alignment between two sequences as a function of an evaluation "score." Most of these methods are based on the Needleman–Wunsch algorithm [7], which uses the spatial dynamics of the possible alignments between the sequences. Many extensions have been conceived, for example, for multiple alignments [8], local alignments [9], or rapid research in large data banks [10].

Current techniques can be executed rapidly in parallel processors. To process short reads generated by second-generation sequencing platforms, one of the solutions found for simultaneously manipulating thousands of sequences has been the use of computing clouds [11]. The assemblers detect a group of reads with consistent alignments with each other, forming contiguous sequences (contigs). This would be equivalent to partially forming an image by putting together pieces of a jigsaw puzzle. In both montages, genome and jigsaw puzzle, the process can be interrupted in ambiguous regions, where various continuations or holes are possible, and where no connecting piece was found [12].

Finally, the assembler tries to order and orient the contigs with each other in an *de novo* manner; that is, without the help of a reference sequence. Returning to the metaphor of a jigsaw puzzle, this would correspond to identifying corners and different parts of the image that relate to each other. In the final mounting phase, a scaffolded group of contigs will become available. Nevertheless, it is desirable to remove most possible gaps, eventually converging to a group of integral chromosomes, that is, those that do not include breaks. This phase, called finishing, can be expensive and could take considerable time, depending on the strategy used to close the occasional holes in the genome scaffold [13].

15.2 Treatment of the Data

Preprocessing of the data, involving a quality filter and correction of the sequencing errors, is essential to increase the accuracy of the assemblies, as it prevents incorrect or low-quality reads from becoming part of the genome assembly process.

15.2.1 Base Quality

In 1998, Ewing and collaborators developed the PHRED algorithm, with the objective of determining the probability of occurrence of the one of the four nucleotides (A, C, G, or T) for each base of a DNA sequence during the base-calling process; the intensity of the wavelength that is obtained through the fluorescence produced by ligations between nucleotides is used to calculate the PHRED quality value (Q), which is logarithmically related to the observed probability of error for each base (P), according to the formula presented in Equation 15.1.

$$Q = -10\log_{10}P \tag{15.1}$$

In Table 15.1, we can observe examples of PHRED quality associated with the probability of being incorrect and the precision of the identification of the base.

The sequences obtained from automatic sequencers are not considered to be reliable due to their low quality at the extremities (Figure 15.2) and because of contaminants. Consequently, working on the quality of the data was fundamental so that the following phases of processing biological information would not be compromised [15]. In the case of the sequences obtained from next-generation sequencing (NGS), despite the high degree of coverage, base quality should be evaluated. In this way, the reads can be trimmed and

TABLE 15.1

Error Probability and Precision Based on PHRED Quality Values

PHRED Quality Score	Error Probability	Accuracy (%)
10	1/10	90
20	1/100	99
30	1/1000	99.9
40	1/10,000	99.99
50	1/100,000	99.999

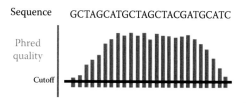

Sequence   GCTAGCATGCTAGCTACGATGCATC

Phred quality

Cutoff

FIGURE 15.2

Low quality of the ends of the reads obtained from automatic sequencers, which use the dideoxynucleotide method.

quality filters applied. As examples of tools that do this quality treatment, we can cite Quality Assessment [16], Galaxy [17], ShortRead [18], and PIQA [19]; the latter being used exclusively for Illumina data.

Analysis of sequence quality, followed by data treatment, makes it possible to reduce alignment errors because it provides precise alignment parameters according to the data that are the objects of study [20]. In the assembly of genomes, software such as Quality-value guided Short Read Assembler (QSRA) [21] propose assembling genomes with the extension of sequences through analysis of base quality, giving better results than the assembler on which it was based (VCAKE, which does not take base quality into account to extend sequences) [22]. In transcriptome studies with NGS platforms using RNAseq, evaluation of the quality of the data is extremely important because the coverage represents the level of expression; consequently, quality filters using stringent parameters can provoke variations in the expression levels that are found [23].

15.2.2 Error Correction (Tools)

Despite the high degree of accuracy provided by NGS platforms, due to the extensive coverage generated by this equipment, sequencing errors can cause problems in the assembly of genomes when using an *de novo* approach because the generation of contigs is very sensitive to these errors [24]. Consequently, to obtain better results, it is necessary to correct the errors before mounting the genomes, which will make the data more reliable [25].

In re-sequencing projects, in which the reads obtained from the sequencers are aligned against a reference genome, error correction can avoid elimination (trimming) of the 3' extremities of the read due to the low quality observed when one tries to improve the alignments [26].

Some genome assemblers have already included error correction procedures: SHARCGS only considers reads that have been produced by the sequencer n times, a parameterized value, and those that present overlap with other reads [25]. In 2001, Pevzner and collaborators used the spectral alignment method to correct errors, which consisted of a given string S in spectrum T, formed by all of the continuous strings of fixed size (T string), a search is made for the smallest number of modifications that need to be made in S to transform it into a T string. This method of correction is implemented by the assembler EULER-SR [27] before the process of mounting the genome.

As examples of independent tools that can correct errors, we can cite Short-Read Error Correction (SHREC) [25] for SOLEXA/Illumina data, which uses a generalized tree of suffixes to process the data, the SOLiD Accuracy Enhancement Tool for SOLiD data, which is available in LifeScope Genomic Analysis Software (http://www.lifetechnologies.com/lifescope), using an approach similar to that of EULER-SR, and Hybrid SHREC [24], which is based on the SHREC algorithm, but can process files from various sequencing platforms.

15.3 Strategies for Assembling Genomes

Reads from the sequencer should be submitted to preprocessing, where base quality and sequencing errors are evaluated with software, commonly specific to corresponding sequencing platforms; they are then submitted to *de novo* assembly and then oriented and ordered to produce the scaffold (Figure 15.3). If the assembly is done with a reference

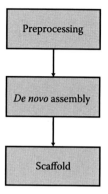

FIGURE 15.3
Steps used for *ab initio* assembly of genomes. After data treatment in the preprocessing stage, *ab initio* assembly is run, generating the contigs, which then are oriented and ordered to generate the scaffold.

genome, after preprocessing, the reads are mapped against this reference, and after alignment is finished, a consensus sequence is produced.

15.3.1 Reference Mounting

Basically, reference assembly consists of mapping the reads obtained from sequencing against a reference genome (Figure 15.4), preferentially, of a phylogenetically closely related organism, making it possible to align a large part of the reads. However, the alignment configurations will also influence the quantity of reads that is utilized; consequently, the parameters such as depth of coverage and the number of mismatches that are permitted should be defined based on the sequencing information: estimated coverage and PHRED quality of the bases [20]. Mapping using a reference sequence provides the identification of the nucleotide substitutions as well as the indels, principally with the use of NGS platforms, due to the high degree of sequence coverage [28].

After mounting, regions of the reference genome that are not covered are observed, representing gaps, which can occur as a function of the presence of a nucleotide sequence in the reference that does not occur in the sequenced organism, or because this region was not sequenced. Among the problems with sequence mounting with the use of a reference, we can cite the representation of repeated regions, for example, the case of a reference genome that has two such regions and the sequenced organism that has only one; during mapping against a reference, the two reference regions will be covered, which can result in mounting errors (Figure 15.5). For mapping reads using a reference genome, one can use software such as SHRiMP [29] and SOLiD BioScope (Applied Biosystems, Foster City, CA); both align in color-space SOLiD, SOAP2 [30], MAQ [31], RMAP [26], and ZOOM [32]. The program SOLiD BioScope is a Java-based application that has various integrated tools in a

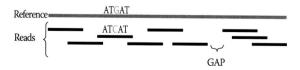

FIGURE 15.4
Alignment of reads against a reference genome, showing mismatches and a gap.

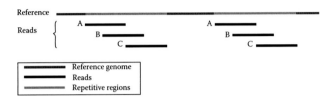

FIGURE 15.5
Double mapping of reads A, B, and C in the reference genome, because it involves a repetitive region. However, in the sequenced genome, the number of repeats can be different from that observed in the reference genome.

web interface for resequencing and transcriptome analyses. The reference assembly pipeline permits mapping of reads based on reference genomes, identifying single nucleotide polymorphisms, indels (insertions and deletions), and inversions, as well as generating a consensus genome (Applied Biosystems).

15.3.2 *De Novo* Assemblers

This consists of reconstructing genome sequences without the aid of any other information aside from the reads produced by the sequencing process. With this strategy, similarity alignments can be made among the reads themselves, or through the overlap of *k*-mers. This allows, at the end of the alignment process, the formation of contiguous sequences (contigs) as seen in Figure 15.6. Most assemblers is based on graph theory, in which vertices and edges can represent overlap, a *k*-mer or a read varies according to the strategy that is used, in which the contigs are the paths formed in the graph. Thus, the assemblers can be divided into greedy, OLC, or de Bruijn graph (DBG) algorithms; the latter uses a Eulerian path [28].

DBG is the approach that is mostly widely used by assemblers of short reads because it works better with large numbers of reads, typical of NGS sequencers. The main programs that adopt this approach are AllPaths [33], Euler-SR [34], SOAPdenovo [35], and Velvet [36]. Among these programs, Velvet is the only one that mounts short sequences in the color-space format.

15.3.3 Challenges and Difficulties for *De Novo* Assembly

The limitations of *de novo* assembly approaches are directly associated with the technological limitations and the features of the data generated by second-generation sequences as well as the sizes of the reads and the volume of data that is generated, which exponentially increases the processing time and sometimes makes mounting unviable. Within this

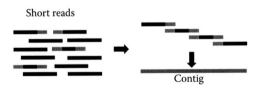

FIGURE 15.6
Alignment between the reads generated by the sequencing, finally obtaining scaffolded contigs.

context, various problems can occur, such as the grouping of repeat regions; there are also regions in which sequences are of low quality, base compression in the sequencing, and even regions with a low degree of coverage due to the random character of the sequencing [37,38].

One of the classic examples of problems with *de novo* assembly is finding a path in the overlap graph that passes through each of the vertices only once (Hamiltonian path) or each edge only once (Eulerian path); this often results in the loss of connectivity between very distant sequences, showing that strategies based on graphs, especially the de Bruijn strategy, are extremely sensitive to sequencing errors [39]. These problems are more complex and common in assemblies made with short reads, as the number of reads is larger than with Sanger sequences because the lengths are much shorter, which exponentially increases the size of the problem. The large sizes of the conserved repeated regions also make the process of the genome assembly difficult and as it involves Eukaryotic genomes, which have very large repeat regions, this task is sometimes a problem that is hard to resolve [5]. Despite the problems cited above, studies show that up to 96.29% of a gene can be reconstructed using short sequences, with sizes starting at 25 nucleotides [40].

15.4 Tools for Assembling Genomes

Second-generation sequencers are capable of generating thousands of reads, providing a high degree of coverage and accuracy. As examples of these platforms, we can cite SOLiD, Illumina, and 454 FLX Titanium. Despite the reduction in sequencing costs, among other advantages, the reduction in the sizes of the reads, along with the increase in the number of reads, results in computational challenges for the processing of this data, principally for genome assemblies [28]. Assembly a genome consists of overlapping based on the similarity of reads generated by a sequencer, to produce contiguous sequences (contigs), which in turn are aligned and oriented with each other to construct the scaffold. This is called a reference assembly when it involves mapping reads in comparison with a reference sequence, whereas mounting reads without such a reference is called *de novo* assembly [31].

15.4.1 Tools for Reference Assembly

For alignment against a reference genome, there are two approaches: using hash tables and using prefix/suffix trees [20]. Many software that use hash tables define the subsequences obtained from the search sequence as the key. The program tries to map identical sequences, known as seeds from the reference, so that the sequences can be subsequently extended. However, the use of templates with spaced seeds gives better results because it considers internal mismatches (Figure 15.7). Even so, independent of whether seeds or

Seed	11111111111
Spaced Seed	11101010101101011

FIGURE 15.7
Templates using seed in which an exact match of 11 bases is necessary to initiate the extension, and spaced seeds in which a match of 11 bases in required, but permitting the existence of internal mismatches.

spaced seeds are used, the alignments do not accept gaps; identification of such gaps is made in a step after the extension of the alignment.

The mapping algorithms that use prefix/suffix trees search for exact alignments, represented by suffix trees, enhanced suffix array, and Ferragina-Manzini index (FM index) [41]; then they extend the alignments considering mismatches. Among the tools that use suffix trees, we have MUMmer [42] and OASIS [43]. Among the alignment software based on enhanced suffix arrays, we can cite Vmatch [44] and Segemehl [45]. The FM index method uses a small amount of memory (from 0.5 to 2 bytes per nucleotide), which can vary as a function of implementation and parameters that are used [20]; examples of such programs include Bowtie [46], BWA [47], BWA-SW [48], SOAP2 [30], and BWT-SW [49].

15.4.2 Tools for *De Novo* Assembly

According to Miller et al. [28], the *de novo* assembly of genomes consists of aligning the reads with each other to produce contiguous sequences (contigs). The principal methodologies for NGS data are based on graphs, these being:

- OLC
- DBG
- Greedy

15.4.2.1 Tools that Use OLC

This is the most widely used approach for large sequences, such as those produce by Sanger; nevertheless, there are also applications based on this method for short reads, such as Edena [50]. The OLC method can be divided into three phases: overlap, layout, and consensus. In the overlap phase, each read is compared with all of the others to identify overlaps, considering the minimum size of overlap and *k*-mer, which will affect the accuracy of the contigs. Among the types of overlaps that are recorded, four categories are possible: containment, normal fitting, prefix, and suffix fitting, as shown in Figure 15.8 [51].

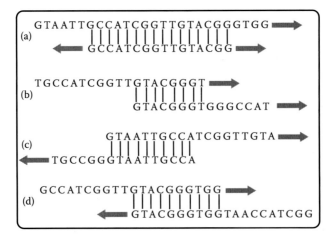

FIGURE 15.8
Containment (a), partial overlap (b), prefix overlap (c), and suffix overlap (d).

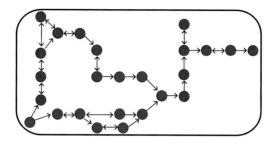

FIGURE 15.9
Layout graph: G graph valid based on graph construction theory. First draft of what could be the genome.

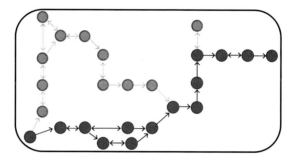

FIGURE 15.10
Consensus graph: using progressive alignment guided by pairs.

In the layout phase, the information obtained from the previous phase is used to construct a graph (Figure 15.9), which is reconstructed at each actualization. At the end of this phase, the first draft representing the genome will be generated, taking into account that in this phase, various methods are used to simplify the pathways and remove errors detected in the graph, such as bubbles and linear extensions, known as dead paths [28]. In the consensus phase, multiple alignments are made of the fragments, progressively, to develop a consensus sequence [28], as shown in Figure 15.10. As examples of the assemblers that use the OLC strategy, we have Celera Assembler [4], ARACHNE [3], CAP, and PCAP [52]. Edena is the only program for the platforms Solexa and SOLiD that uses OLC [50].

15.4.2.2 Tools that Use the DBG

In 1995, Idury and Waterman introduced the use of a graph to represent a sequence assembly. Their method consisted of creating a vertex for each word. Then, the vertices that correspond to the overlap of k-mers are connected; k can be represented by a sequence with a specific number of bases. The original vertex corresponds to prefix k-1 of the corresponding overlap region (k-mer), and the vertex destination of suffix k-1 of the same region, providing a reconstruction of the sequence through a path that traverses each edge exactly once. Pevzner and Tang [39] proposed a representation that was slightly different from the graph of the sequence, the so-called DBG, which uses a Eulerian path; that is, a pathway that visits each edge exactly once, through which the k-mers are represented as arcs or edges, and overlapping of the k-mers join their ends.

The classic method of assembling fragments is based on the notion of a graph of over-laps. Each read corresponds with a vertex in the overlap graph and two vertices are con-nected by an arc, if the corresponding reads overlap [53]. Different from the problem of the Hamiltonian path, the Eulerian path is less complex and is resolved even with graphs with millions of vertices, as there are linear-time algorithms that can provide a solution for them [54]. It is important to emphasize that a de Bruijn graph is centered on the k-mer, which means that its topology is not affected by the fragmentation of the reads [12]. And compared with the overlap phase in OLC, the computational cost is much smaller because the overlaps are not performed against all reference genomes [28]. Operationally, the DBG containing the vertices with length k is constructed with the result of the division of the k-mer, being linked in exact, identical overlapping for the previously defined k-mer values. After construction of the graph, it is generally possible to simplify it without any loss of information. The reads of the vertices are interrupted and initiated again at each simpli-fication. Simplification of two vertices is similar to the concatenation of two strands of characters [51]. As in other approaches, the assemblers add to their main algorithm, acces-sory algorithms to help remove assembly errors, such as reduction of redundant pathways, removal of bubbles or pathway loops, and linear extensions; that is, those that do not pos-sess defined pathways. The main programs that adopt the use of the DBG are AllPaths [33], Euler-SR for short sequences [34], SOAPdenovo [54], and Velvet [36], which is specialized in the localization of the use of paired reads. Velvet is the only one that assembles short sequences in a color-space. Other assembly programs, such as ABySS [55,56], were success-ful in constructing DBGs, eliminating the limitations in the use of memory that are com-mon during assemblies; Table 15.2 shows the principal characteristics of each software [28].

In 2009, inspired by the ideas of Pevzner et al. [5], Zerbino implemented the program Velvet, the structure of which differs in various aspects. Among these, maps of k-mers are generated for the vertices and not for the arcs, and there can be reverse complementary-associated sequences to obtain a bidirectional graph [57]. In this way, the vertices can be connected by a directed edge or an arc. Due to the symmetry of the blocks, an arc goes from vertex A to B, a symmetrical arc goes from B to A. With any alteration of an arc, it is implicit that the same change will be made symmetrically in the paired arc.

Each vertex can be represented by a single rectangle, which represents a series of k-mer overlaps (in this case, $k = 5$) listed directly above or below. The only nucleotide of each k-mer is colored red. The arcs are represented as arrows between knots. The last k-mer has overlaps of an arc of origin with the first of its destination arcs. Each arc has a symmetrical arc [12].

15.4.2.3 Tools that Use Greedy Graph

The greedy algorithms were widely implemented in assembly programs for Sanger data, such as PHRAP, TIGR Assembler, and CAP3 [1]. When new sequencing technologies became available, other software were developed for assembling the NGS data (short reads) using different greedy strategies, such as SSAKE [58], SHARCGS [59], and VCAKE [22]. These algorithms can use an OLC approach or a DBG, applying a basic function (Figure 15.11); starting with any read of a group of data, add another, and in this way numerous interactions are run until all possible operations are tested and the overlaps are identified, in which a suffix of a read overlaps a prefix of another (Figure 15.11a). Each operation uses the overlap of a major score, measured by the size of the overlap between reads to make the next junction [1,2,28].

TABLE 15.2

Feature Comparison between *De Novo* Assemblers for Whole-Genome Shotgun Data from NGS Platforms

	Algorithms Feature	**Greedy Assemblers**	**OLC Assemblers**	**DBG Assemblers**
Modeled features of reads	Base substitutions	—	—	Euler, AllPaths, SOAP
	Homopolymer miscount	—	CABOG	—
	Concentrated error in 3′ end	—	—	Euler
	Flow space	—	Newbler	—
	Color space	—	Shorty	Velvet
Removal of erroneous reads	Based on *k*-mer frequencies	—	—	Euler, Velvet, AllPaths
	Based on *k*-mer frequency and quality value	—	—	AllPaths
	For multiple values of *k*	—	—	AllPaths
	By alignment to other reads	—	CABOG	—
	By alignment and quality value	SHARCGS	—	—
Correction of erroneous base calls	Based on *k*-mer frequencies	—	—	Euler, SOAP
	Based on *k*-mer frequencies quality value	—	—	AllPaths
	Based on alignments	—	COBOG	—
Approaches to graph construction	Implicit	SSAKE, SHARCGS, VCAKE	—	—
	Reads as graph nodes	—	Edena, CABOG, Newbler	—
	k-mer as graph nodes	—	—	Euler, Velvet, AbySS, SOAP
	Simple path as graph nodes	—	—	AllPaths
	Multiple values of *k*	—	—	Euler
	Multiple overlap stringencies	HARCGS	—	—
Approaches to graph reduction	Filter overlaps	—	CABOG	—
	Greedy contig extension	SSAKE, SHARCGS, VCAKE	—	—
	Collapse simple paths	—	CABOG, Newbler	Euler, SOAP, Velvet
	Erosion of spurs	—	CABOG, Edena	Euler, AllPaths, SOAP, Velvet
	Transitive overlap reduction	—	Edena	—
	Bubble smoothing	—	Edena	Euler, SOAP, Velvet
	Bubble detection	—	—	AllPaths
	Reads separate tangled paths	—	—	Euler, SOAP
	Break at low coverage	—	—	SOAP, Velvet
	Break at high coverage	—	CABOG	Euler
	High coverage indicates repeat	—	CABOG	Velvet
	Special use of long reads	—	Shorty	Velvet

Source: Miller, J.R. et al., *Genomics* 95: 315–327, 2010. With permission.

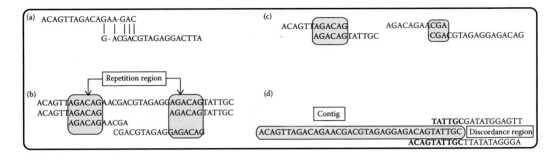

FIGURE 15.11
(a) Overlap between two reads, in which the overlapping region does not need to be a perfect match; (b) example of correct assembly of a region of the genome that has two repetitive regions (box) using four reads; (c) assembly generated by the greedy approach. Two overlaps were identified, but the second has a greater alignment which must be considered. (d) discordance between two reads (discordance region) that could extend a contig (bold sequence). Extension of the contig could be finalized to avoid misassembles.

The quality of the overlaps is measured by the size and the identity (percentage bases shared between two reads in the overlap region). Also, simplification of the graph is based only on the size of the overlaps between reads, it being necessary to implement mechanisms to prevent misassemblies [1,28]. The term "greedy" refers to the fact that the decisions taken by the algorithm occur as a function of a local quality (in the case of assembly, the quality of the overlaps between the reads), which may not be an optimal global solution; in this way, assemblers based on greedy can generate numerous misassemblies (Figure 15.11b and 15.11c) [1].

During the assembly process, where the reads are added by iteration, the fragments are considered in descending order according to their quality, as explained previously. Consequently, to avoid missassemblies, the extension process is finalized when conflicting information is identified, for example, when two or more reads extend a contig, but with no overlap between them (Figure 15.11d) [1].

The approaches based on greedy algorithms need a mechanism to avoid incorporating false-positive overlaps into contigs. The overlaps induced by repeat sequences can have high scores more than the overlaps of regions without repetitions; also, an assembly that generates a false-positive overlap will unite unrelated sequences to the ends of a repetition and produce a chimera [28]. Some assemblers use other algorithms to avoid including errors; SHARCGS, for example, includes a preprocessing step that filters out erroneous reads. The parameters of this filter can be modified by the user of the program [59].

15.5 Tools for Visualizing NGS Data and Producing a Scaffold

The development of NGS platforms, also named high-throughput sequencing machines has opened new opportunities for biological applications, including resequencing of genomes, sequencing of the transcriptome, ChIP-seq, and discovery of miRNA [60]. These NGS technologies created a necessity to develop new tools to visualize the results of the assemblies and alignments of short reads [61]. Consequently, new challenges arose: a need to rapidly and efficiently process an enormous quantity of reads, a need for high-quality interpretation of data, a user-friendly interface, and the capacity to accept various formats of files produced by different sequencers and assemblers [62]. Visualization of the sequences generated from the process of mounting genomes can be done, for example, by the software Consed [63], which allows the data to be edited, and Hawkeye [64]. Various programs have been developed for NGS reads, including EagleView [65], Tablet [62], MapView [66], MaqView [31], SAMtools [48,55], and Integrative Genomics Viewer (http://www.broadinstitute.org).

The main differences between the visualizers are in the interfaces for presentation to the user, data processing velocity, as well as the different formats of the data entry files and development of the scaffold. Loading NGS data in programs such as Consed and Hawkeye, for example, requires a large amount of memory, which is normally not available to users of desktop computers. EagleView is a visualization tool developed only for NGS, but it does not permit visualization of paired reads and it has memory limitations. MapView permits analysis of genetic variation, supports paired-end data and single-end reads, and various different entry and output file formats [66].

The scaffold is made of DNA sequences that are reconstructed after sequencing; it can be composed of contigs, which should be ordered and oriented with each other with the help of a reference genome, and by gaps: regions where the DNA sequence is not recognized because it does not exist in the genome or because it is not covered in the sequencing or assembly [67]. Options for software for generating genome scaffolds using a reference genome include Bambus [68] from the software package AMOS, which can be used as entry and exit by the software Mummer [42]. Genscaff [69] uses contiguous sequences generated by an assembly program without the help of a reference genome, through implementation of graph theory. CLCBio Workbench (http://www.clcbio.com) and the software package Lasergene (http://www.dnastar.com), besides other functionalities, produce and edit the scaffold; however, they are commercial software packages.

15.6 Closing Gaps

An artifact related to the assembly of a genome is the formation of gaps (holes or spaces). Usually, the strategy used to resolve these spaces would be to design specific primers for this region, and posterior alignment of the amplified sequences by the primers, thereby closing the gap. However, for large gaps (2 kb or longer), new primers are needed. Therefore, this process requires considerable time and becomes expensive [70]. Given this situation, we describe here an *in silico* strategy to resolve this type of artifact, a solution consisting of the use of short reads generated by SOLiD (NGS) that were not mapped during the assembly process.

15.7 Description of the Gap Closing Strategy

Align the short sequences in the flanking regions of the mapped genes. Then, the nucleotides that have a PHRED quality of 20 or more and a minimum of 10× coverage should be added manually (Figure 15.12; steps 1 and 2). This extension will close the small gaps (1–100 bp).

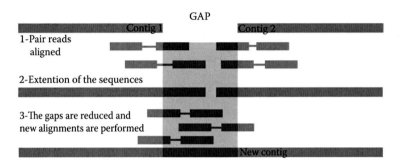

FIGURE 15.12
Description of the strategy for closing the gaps: step 1, short reads are aligned in the initial assembly; step 2, short reads that align in terminal contigs are mounted in new contigs; step 3, the short reads are aligned against the updated sequence and the process is repeated until the gap is closed.

If there are still gaps, the short reads should be realigned in relation to the reference genome (Figure 15.12; step 3), because with the production of the new contigs, new short reads align in the flanking regions of the gaps, forming what we call a merged contig. In this way, the genome could possibly be closed completely *in silico*, without using PCR. We emphasize that this strategy was used during the mounting of the genome of a strain of *Corynebacterium pseudotuberculosis*, which was sequenced with the SOLiD platform, which generated 19,091,361 reads (140× coverage). This system mapped 590 gap regions, closing 100% of the gaps [71].

References

1. Pop, M. 2009. Genome assembly reborn: Recent computational challenges. *Brief Bioinformatics* 10: 354–366.
2. Pop, M., and S.L. Salzberg. 2008. Bioinformatics challenges of new sequencing technology. *Trends in Genetics* 24: 142–149.
3. Batzoglou, S., D.B. Jaffe, K. Stanley et al. 2002. ARACHNE: A whole genome shotgun assembler. *Genome Research* 12: 177–189.
4. Myers, E.W., G.G. Sutton, A.L. Delcher et al. 2000. A whole-genome assembly of drosophila. *Science* 287: 2196–2204.
5. Pevzner, P.A., H. Tang, and M.S. Waterman. 2001. An eulerian path approach to DNA fragment assembly. *Proceedings of the National Academy of Sciences of the United States of America* 98.
6. Phillippy, A.M., M.C. Schatz, and M. Pop. 2008. Genome assembly forensics: finding the elusive misassembly. *Genome Biology* 9: R55.
7. Needleman, S., and C. Wunsch. 1970. A general method applicable to the search for similarities in the amino acid sequence of two proteins. *Journal of Molecular Biology* 48: 443–453.
8. Higgins, D., and P. Sharp. 1988. CLUSTAL: a package for performing multiple sequence alignment on a microcomputer. *Gene* 73: 237–244.
9. Smith, T., and M. Waterman. 1981. Identification of common molecular subsequences. *Journal of Molecular Biology* 147: 195–197.
10. Altschul, S., W. Gish, W. Miller, E. Myers, and D. Lipman. 1990. Basic local alignment search tool. *Journal of Molecular Biology* 215: 403–410.
11. Bateman, A., and M. Wood. 2009. Cloud computing. *Bioinformatics* 2512: 1474.

12. Zerbino, D. 2009. Genome Assembly and Comparison Using de Bruijn Graphs. PhD thesis, University of Cambridge.

13. Cole, C.G., O.T. McCann, J.E. Oliver et al. 2008. Finishing the finished human chromosome 22 sequence. *Genome Biology* 9: R78.

14. Sasson, S.A. 2010. From Millions to One: Theoretical and Concrete Approaches to De novo Assembly Using Short Read DNA Sequences. PhD thesis, Graduate School-New Brunswick Rutgers, The State University of New Jersey.

15. Chou, H.H., and M.H. Holmes. 2001. DNA sequence quality trimming and vector removal. *Bioinformatics* 1712: 1093–1104.

16. Ramos, R.T., A.R. Carneiro, J. Baumbach, V. Azevedo, M.P. Schneider, and A. Silva. 2011. Analysis of quality raw data of second generation sequencers with quality assessment software. *BMC Research Notes* 4: 130.

17. Blankenberg, D., A. Gordon, G. Von Kuster, N. Coraor, J. Taylor, and A. Nekrutenko A. 2010. Manipulation of FASTQ data with Galaxy. *Bioinformatics* 26: 1783–1785.

18. Morgan, M., S. Anders, M. Lawrence, P. Aboyoun, H. Pagès, and R. Gentleman. 2009. ShortRead: a bioconductor package for input, quality assessment and exploration of high-throughput sequence data. *Bioinformatics* 25: 2607–2608.

19. Martínez-Alcántara, A., E. Ballesteros, C. Feng et al. 2009. PIQA: Pipeline for Illumina G1 genome analyzer data quality assessment. *Bioinformatics* 25: 2438–2439.

20. Li, H., and N. Homer. 2010. A survey of sequence alignment algorithms for next-generation sequencing. *Briefings Bioinformatics* 11: 181–197.

21. Bryant, D.W., W.K. Wong, and T.C. Mockler. 2009. QSRA—a quality-value guided de novo short read assembler. *BMC Bioinformatics* 10: 69.

22. Jeck, W., J. Reinhardt, D. Baltrus et al. 2007. Extending assembly of short DNA sequences to handle error. *BMC Bioinformatics* 23: 2942–2944.

23. Marioni, J.C., C.E. Mason, S.M. Mane, M. Stephens, and Y. Gilad. 2008. RNA-seq: An assessment of technical reproducibility and comparison with gene expression arrays. *Genome Research* 18: 1509–1517.

24. Salmela, L. 2010. Correction of sequencing errors in a mixed set of reads. *Bioinformatics* 26: 1284–1290.

25. Schroder, J., H. Schroder, S.J. Puglisi, R. Sinha, and B. Schmidt. 2009. SHREC: a short-read error correction method. *Bioinformatics* 25: 2157–2163.

26. Smith, A.D., Z. Xuan, and M.Q. Zhang. 2008. Using quality scores and longer reads improves accuracy of Solexa read mapping. *BMC Bioinformatics* 9: 128.

27. Chaisson, M.J., and P.A. Pevzner. 2008. Short read fragment assembly of bacterial genomes. *Genome Research* 18: 324–330.

28. Miller, J.R., S. Koren, and G. Sutton. 2010. Assembly algorithms for next-generation sequencing data. *Genomics* 95: 315–327.

29. Rumble, S.M., P. Lacroute, A.V. Dalca, M. Fiume, A. Sidow, and M. Brudno. 2009. SHRiMP: Accurate mapping of short color-space reads. *PLoS Computational Biology* 5: 5.

30. Li, R., C. Yu, Y. Li et al. 2009. SOAP2: an improved ultrafast tool for short read alignment. *Bioinformatics* 25: 1966–1967.

31. Li, H., J. Ruan, and R. Durbin. 2008. Mapping short DNA sequencing reads and calling variants using mapping quality scores. *Genome Research* 18: 1851–1858.

32. Lin, H., Z. Zhang, M.Q. Zhang, B. Ma, and M. Li. 2008. ZOOM! Zillions of oligos mapped. *Bioinformatics* 24: 2431–2437.

33. Butler, J., I. MacCallum, M. Kleber et al. 2008. ALLPATHS: De novo assembly of whole-genome shotgun microreads. *Genome Research* 18: 810–820.

34. Chaisson, M., P. Pevzner, and H. Tang. 2004. Fragment assembly with short reads. *Bioinformatics* 20: 2067–2074.

35. Li, Y., Y. Hu, L. Bolund, and J. Wang. 2010. State of the art de novo assembly of human genomes from massively parallel sequencing data. *Human Genomics* 44: 271–277.

36. Zerbino, D.R., and E. Birney. 2008. Velvet: Algorithms for de novo short read assembly using de Bruijn graphs. *Genome Research* 18: 821–829.
37. Ewing, B., and P. Green. 1998. Base-calling of automated sequencer traces using PHRED. II. Error probabilities. *Genome Research* 83: 186–194.
38. Flicek, P., and E. Birney. 2009. Sense from sequence reads: Methods for alignment and assembly. *Nature Methods* 6: S6–S12.
39. Pevzner, P.A., and H. Tang. 2001. Fragment assembly with double-barreled data. *Bioinformatics* 17: 225–233.
40. Kingsford, C., M.C. Schatz, and P. Pop. 2010. Assembly complexity of prokaryotic genomes using short reads. *BMC Bioinformatics* 111: 21.
41. Ferragina, P., and G. Manzini. 2000. Opportunistic data structures with applications. In *Proceedings of the 41st Symposium on Foundations of Computer Science FOCS 2000*. California: Redondo Beach 390–398.
42. Kurtz, S., A. Phillippy, A.L. Delcher et al. 2004. Versatile and open software for comparing large genomes. *Genome Biology* 5: R12.
43. Meek, C., J.M. Patel, and S. Kasetty. 2003. OASIS: An online and accurate technique for local-alignment searches on biological sequences. In *Proceedings of 29th International Conference on Very Large Data Bases VLDB 2003*, Berlin: 910–921.
44. Abouelhoda, M.I., S. Kurtz, and E. Ohlebusch. 2004. Replacing suffix trees with enhanced suffix arrays. *Journal of Discrete Algorithms* 2: 53–86.
45. Hoffmann, S., C. Otto, S. Kurtz et al. 2009. Fast mapping of short sequences with mismatches, insertions and deletions using index structures. *PLoS Computational Biology* 5: 1–10.
46. Langmead, B., C. Trapnell, M. Pop, and S.L. Salzberg. 2009. Ultrafast and memory-efficient alignment of short DNA sequences to the human genome. *Genome Biology* 10: R25.
47. Li, H., and R. Durbin. 2010. Fast and accurate long-read alignment with Burrows–Wheeler transform. *Bioinformatics* 265: 589–595.
48. Li, H. and R. Durbin. 2009. Fast and accurate short read alignment with Burrows–Wheeler transform. *Bioinformatics* 25: 1754–1760.
49. Lam, T.W., W.K. Sung, S.L. Tam, C.K. Wong, and S.M. Yiu. 2008. Compressed indexing and local alignment of DNA. *Bioinformatics* 24: 791–797.
50. Hernandez, D., P. François, L. Farinelli, M. Osterås, and J. Schrenzel. 2008. De novo bacterial genome sequencing: Millions of very short reads assembled on a desktop computer. *Genome Research* 18: 802–809.
51. Myers, E.W. 1995. Towards simplifying and accurately formulating fragment assembly. *Journal of Computational Biology* 2: 1–21.
52. Huang, X., and S. Yang. 2005. Generating a genome assembly with PCAP. In *Current Protocols in Bioinformatics* Unit 11.3.
53. Lemos, M., A. Basílio, and A. Casanova. 2003. Um Estudo dos Algoritmos de Montagem de Fragmentos de DNA. PUC Rio, Rio de Janeiro.
54. Fleishner, H. 1990. *Eulerian Graphs and Related Topics*. London: Elsevier Science.
55. Li, H., B. Handsaker, A. Wysoker et al. 2009. 1000 Genome Project Data Processing Subgroup. The Sequence Alignment/Map format and SAMtools. *Bioinformatics* 16: 2078–2079.
56. Simpson, J.T., K. Wong, S.D. Jackman, J.E. Schein, S.J.M. Jones, and I. Birol. 2009. ABySS: A parallel assembler for short read sequence data. *Genome Research* 19: 1117–1123.
57. Medvedev, P., K. Georgiou, G. Myers, and M. Brudno. 2007. Computability of models for sequence assembly. In *Proceedings of Workshop on Algorithms in Bioinformatics WABI* 289–301.
58. Warren, R.L., C.G. Sutton, S.J. Jones, and R.A. Holt. 2007. Assembling millions of short DNA sequences using SSAKE. *Bioinformatics* 15: 234.
59. Dohm, J.C., C. Lottaz, T. Borodina, and H. Himmelbauer. 2007. SHARCGS, a fast and highly accurate short-read assembly algorithm for de novo genomic sequencing. *Genome Research* 17: 1697–1706.
60. Shendure, J., and H. Ji. 2008. Next-generation DNA sequencing. *Nature Biotechnology* 26: 1135–1145.

61. Magi, A., M. Benelli, A. Gozzini, F. Girolami, and M.L. Brandi. 2010. Bioinformatics for next generation sequencing data. *Genes* 1: 294–307.

62. Milne, I., M. Bayer, L. Cardle et al. 2010. Tablet—Next generation sequence assembly visualization. *Bioinformatics* 3: 401–402.

63. Gordon, D., C. Abajian, and P. Green. 1998. Consed: a graphical tool for sequence finishing. *Genome Research* 8: 195–202.

64. Schatz, M.C., A.M. Phillippy, B. Shneiderman, and S.L. Salzberg. 2007. Hawkeye: an interactive visual analytics tool for genome assemblies. *Genome Biology* 8: R34.

65. Huang, W. and G. Marth. 2008. EagleView: A genome assembly viewer for next-generation sequencing technologies. *Genome Research* 9: 1538–1543.

66. Bao, H., H. Guo, J. Wang, R. Zhou, X. Lu, and S. Shi. 2009. MapView: Visualization of short reads alignment on a desktop computer. *Bioinformatics* 12: 1554–1555.

67. Schuster, S.C. 2008. Next-generation sequencing transform today's biology. *Nature Methods* 5: 16–18.

68. Pop, M., D.S. Kosack, and S.L. Salzberg. 2004. Hierarchical scaffolding with Bambus. *Genome Research* 14: 149–159.

69. Setúbal, J.C., and R. Werneck. 2001. A program for building contig scaffolds in double-barrelled shotgun genome sequencing. Campinas Instituto de Computação, Unicamp.

70. Tsai, I.J., D.T. Otto, and M. Berriman. 2010. Improving draft assemblies by iterative mapping and assembly of short reads to eliminate gaps. *Genome Biology* 11: R41.

71. Silva, A., M.P. Schneider, L. Cerdeira et al. 2011. Complete genome sequence of *Corynebacterium pseudotuberculosis* I19, a strain isolated from a cow in Israel with bovine mastitis. *Journal of Bacteriology* 1931: 323–324.

16

Towards Engineering Dark-Operative Chlorophyll Synthesis Pathways in Transgenic Plastids

Muhammad Sarwar Khan

CONTENTS

16.1 Chloroplasts

Chloroplasts are green organelles in a plant cell, which develop either by differentiation of proplastids in the meristematic cells or by dedifferentiation of already developed plastids such as chromoplasts, amyloplasts, and leucoplasts. The transformation of proplastids into chloroplasts is an important change that occurs during development, leading to the formation of an extensive membrane system (outer and inner membranes) [1]. The inner membrane encloses a fluid-filled region called the stroma, which contains enzymes for the light-independent reactions of photosynthesis. The folding of the inner membrane forms interconnected stacks of disc-like sacs called thylakoids, and is often arranged in stacks known as grana. The thylakoid membrane, which encloses a fluid-filled thylakoid interior space, contains the pigments, chlorophyll and carotenoids, as well as enzymes and the electron transport chain used in photosynthesis.

The number of plastids, from a proplastid in a meristematic cell to a mature chloroplast in a developed leaf mesophyll cell, varies from 10 to 100, respectively. Genome copy number increases with the differentiation of proplastids to chloroplasts. Although the plastid genome is very small with respect to the nuclear genome, it constitutes approximately 10% to 20% of the total cellular DNA content. This is because only two copies of the nuclear genome are present in a diploid plant cell, whereas it contains thousands of copies of the plastid genome, thus extraordinarily increasing the ploidy level of the plastid genome [2,3]. In land plants, plastome copy number is usually highest in photosynthetically active cells, in which plastids are present as green chloroplasts. In contrast, non–green plastid types often possess fewer plastomes [4]. The plastome in flowering plants is a double-stranded and circular DNA molecule, and is divided into small and large single-copy regions separated by two inverted repeats [5–7]. The two inverted repeats are identical in their nucleotide sequence but differ in their relative orientation. It has been observed that legumes lack these inverted repeats.

Most plastome-encoded genes are grouped into two major classes [8], the genetic system genes and photosynthesis genes. The first group encodes genes for ribosomal RNAs, transfer RNAs, ribosomal proteins, and RNA polymerase subunits, whereas the second group encodes genes for subunits of photosystem I, photosystem II, cytochrome b_6f complex, and ATP synthase. In addition, the plastid genome contains a number of open reading frames called hypothetical chloroplast open reading frames (ycfs) that have unknown functions [9,10]. In addition, several polypeptides encoded by the nuclear genome are synthesized on cytoplasmic ribosomes and are imported into chloroplasts.

16.2 Chloroplast Genome Organization and Omics Approaches to Characterize Transcriptomes and Proteomes

Plastids have a self-replicating genome, which is a predominantly circular DNA molecule that varies in size from 120 to 220 kb, encoding approximately 120 protein and RNA genes in various species [5,11], suggesting that most of their ancestral genes have been lost or transferred to the nuclear genome during evolution. When it comes to the copy number of the plastid genome, it reaches up to 10,000 copies per cell because a single leaf mesophyll cell carries 100 chloroplasts, and each chloroplast contains 100 plastome (the plastid genome) copies. Duplication of a large region of plastome in an inverted orientation in most plant species results in doubling the number of genes encoded by the inverted repeat region. Furthermore, plastids have their own transcription and translation machinery because they are descendents of an ancient cyanobacterial endosymbiont [12], and many of its functions have been conserved. The availability of complete genome sequences, a large number of expressed sequence tags, and high-throughput methodologies such as transcriptomics and proteomics has enabled the analyses of gene expression in photosynthetic organisms, including *Chlamydomonas*, cyanobacteria, maize, and *Arabidopsis*, during different developmental stages or in response to environmental cues [13–16].

From the transcriptomics point of view, DNA microarray technology is one of the key tools used to functionally analyze the transcriptomes. Typically, DNA microarrays are constructed by arranging thousands of target DNA fragments, which includes genomic or cDNA clones (or oligonucleotides) in a grid format on slides that are subsequently probed with fluorescently labeled fragments derived from poly(A) messenger RNA (mRNA) or total RNA [17]. Recently, customized DNA microarrays have also been developed for studying photosynthesis-related gene expression profiles [18] because genes with similar functions often display similar transcriptional profiles. Using the concept of "guilt-by-association" coupled with a reverse genetic approach, a number of genes of unknown function and exhibiting photosynthesis gene–like transcription profiles were characterized. For example, PGRL1, a central component of cyclic electron flow around photosystem II was identified [19]. More recently, a deep sequencing approach, "RNAseq" has successfully been used to investigate the transcriptional control of ectopic chloroplast development in green curd cauliflower and, as a result, gene expression between green and white curds was compared [20]. The approach is considered a powerful tool to compare gene expression on a genomewide scale in a species, without a reference genome, and helps in dissecting the genetic basis of naturally occurring variations in crops. Kahlau and Bock [21] have recently characterized transplastomic tomatoes using systematic transcriptomic analysis of the tomato plastid genome during fruit development and chloroplast-to-chromoplast conversion, and found that plastid genes related to photosynthesis compared with genetic systems are strongly downregulated in fruits due to developmental changes rather than the chloroplast-to-chromoplast conversion. Nevertheless, the *accD* gene encoding fatty acid biosynthesis is upregulated during chromoplast development.

With the introduction of new high-throughput technologies, a significant fraction of chloroplast proteomes is characterized. The most commonly used techniques are two-dimensional gel electrophoresis, chromatography, mass spectrometry, and bioinformatics. Using these techniques, protein identification, expression profiling, posttranslational modifications, compartmentalization, and protein–protein interactions are recorded. However, a number of problems, including the detection of low-abundance proteins and capturing the dynamics of plastid proteomes, are underlined in the literature [22]. Nevertheless, proteome dynamics during plastid differentiation in rice has been reported by Kleffmann et al. [23]. The analysis of proteome dynamics was carried out using quantitative two-dimensional gel electrophoresis and tandem mass spectrometry protein identification during light-induced development of chloroplasts from etioplasts of rice. Additionally, a number of proteomic studies on chloroplasts have been performed that include analyses of the stroma [24,25], thylakoid membranes [24–27], envelope membrane [25,28], and the thylakoid lumen [29,30].

16.3 Chloroplast Genome Engineering Technology

When it comes to transformation, the nuclear genome gets the attention. One of the major drawbacks of nuclear genome engineering, in addition to position effects due to random insertion of transgenes into the genome, is the possibility of causing genetic pollution in other cultivated or wild species [9]. Recently, plastid genome is engineered to express transgenes of agronomic and medical importance. Successful plastid transformation relies on the totipotency of the explant, choice of selectable marker gene and antibiotics, potential regeneration medium, choice of expression cassettes, and mode of transformation. The tobacco plant has been used for the expression of genes of medical importance, examining the suitability of the transgene/protein, and investigating the functions of the endogenous genes or ycfs. However, plastid transformation technology has been extended to other plant species including potato [31], rice [32], cotton [33], carrot [34], rapeseed [35], cabbage [36], sugarcane [37], *Arabidopsis* [38], and other species. Although plastid transformation has been achieved using polyethylene glycol treatment of protoplasts [39], it is restricted to tobacco only. Biolistic particle bombardment is the most widely used and effective method of transforming plastids. Tiny gold or tungsten particles are coated with DNA molecules

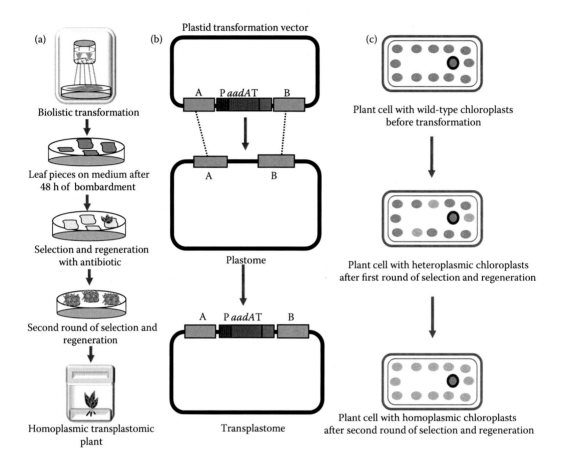

FIGURE 16.1

(See color insert.) Stepwise chloroplast transformation in tobacco: biolistic bombardment, transgene integration into the chloroplast genome via two events of homologous recombination and sorting at the genome and organelle level. (a) A tobacco leaf is bombarded with vector DNA-coated metal particles. Sectioning of leaves into small pieces after 48 h of bombardment and placing the leaf sections onto the surface of regeneration medium (RMOP), containing spectinomycin because the transformation vector carries *aadA* genes. Leaves from recovered spectinomycin-resistant shoots were used for the second round of selection and regenerated homoplasmic shoots are transferred to boxes. (b) The chloroplast transformation vector, harboring the expression cassette that carries tails of sequences homologous to plastid DNA' on both sides, targets the expression cassette in the plastid genome. Recombination between homologous sequences, in the vector and the plastome, allows the integration of expression cassette into the plastome. (c) A plant cell with a homogenous population of wild-type chloroplasts before bombardment. Cells from a shoot regenerated during the first round of selection and regeneration are generally heteroplasmic, whereas a shoot from the second round are homoplasmic.

to transform plastids of fully expanded tobacco leaves or embryogenic cells. In the case of tobacco, leaves are cut into small pieces after 48 h of bombardment and placed on antibiotic-containing regeneration medium (RMOP). The most commonly used antibiotic is spectinomycin because the *aadA* gene is predominantly used to transform plastids and confers resistance to both spectinomycin and streptomycin. Transformed shoots (which are green) are selected on bleached leaf sections because chlorophyll synthesis is inhibited by the antibiotic [9,32]. During the process, transgenic and wild-type plastids are sorted out. Approximately 20 to 25 cell divisions are required to develop a homoplasmic transgenic shoot, a shoot in which each and every cell and each plastid carries a transgenome [32]. The biolistic approach, including the selection of bombarded leaves on antibiotic-containing RMOP medium and sorting of chloroplasts to develop homoplasmic clones is explained (Figure 16.1).

The first successful plastid transformation using this method was achieved in *Chlamydomonas reinhardtii* [40], and the first successful chloroplast transformation in a higher plant tobacco was carried out by introducing *aadA* gene as a selectable marker [41]. Plastid transformation has some advantages over conventional nuclear transgenic technologies, including increased transgene containment, lack of position effects, and high levels of expression that yield proteins with a bona fide structure [42], which are explained in the subsequent sections.

16.3.1 Biological Containment and Coexistence

From the biological containment perspective, the technology significantly increases transgene containment because plastids are inherited uniparentally in a strictly maternal fashion. Although pollen from such plants contains metabolically active plastids, the plastid DNA itself is lost during the process of pollen maturation and hence is not transmitted to the next generation [3,43]. This makes plastid transformation a valuable tool for developing genetically modified plants that are biologically contained, thus posing lower environmental risks [3,44]. This gene containment strategy is therefore suitable for establishing the coexistence of most conventional and genetically modified cultivated crops. The strategy is further refined to improve transgene containment by engineering a β-ketothiolase–based system that promises the development of sterile male plants with fertility restoration and a greatly reduced risk of genetic outcrossing [45,46]. Male sterility is achieved by metabolic engineering of plastids of tobacco in such a way that functional β-ketothiolase accumulates in the leaves and anthers. In chloroplasts, β-ketothiolase utilizes the acetyl-CoA pool, destined for *de novo* fatty acid biosynthesis, to synthesize acetoacetyl-CoA. Thus, depletion of the acetyl-CoA pool by β-ketothiolase interferes in the fatty acid biosynthesis pathway and may affect pollen development, resulting in a sterile male phenotype.

16.3.2 High-Level Gene Expression and Transprotein Accumulation

Increased ploidy at plastid and genome level results in very high copy number of plastid-encoded genes in each cell, which offers an enormous potential for expressing foreign genes to maximum levels [3]. Plastid transformation has the ability to produce and accumulate large amounts of foreign protein [up to 46% of total soluble protein (TSP)] when the transgene is stably integrated and expressed in plastids [47]. In addition to ploidy, the expression of transgene(s) can be enhanced by developing expression cassettes that include a promoter along with translation control sequences (leader sequences) and a terminator. Generally, transgenes are expressed under the strong plastid ribosomal RNA operon (*rrn*) promoter (*Prrn*). The stability of the transgenic mRNA is ensured by the 5′ untranslated region (5′-UTR) and 3′-UTR sequences flanking the transgenes. The most commonly used 5′-UTR and 3′-UTR is psbA/TpsbA [10,48,49]. The translation control sequences may be the mRNA 5′-UTR or the 5′ translation control region, which includes the 5′-UTR and the coding region's N terminus. Details are described elsewhere [32,50–52]. The terminator or the 3′ regulatory region encodes the mRNA 3′-UTR, which typically harbors a stem–loop-type RNA secondary structure. The expression levels in chloroplasts are thus several-fold higher than those obtained upon expression of the same transgenes in the nuclear compartment.

16.3.3 Homologous Recombination and Absence of Position Effects

Transgene integration into the plastid genome occurs through homologous recombination, which facilitates the targeting of a foreign gene to a specific location in the plastid DNA [3,10,53]. Due to the targeted insertion of the transgene, a uniform population of transgenic lines is recovered in antibiotic-containing regeneration medium. In contrast, the transforming DNA integrates into the nuclear genome predominantly by nonhomologous recombination, which leads to transgenic lines with largely different copy numbers and integration sites of the transgene. In practice, the inserted transgene has short DNA sequence tails added at each end, the tails are homologous to sequences on the chloroplast target gene, which thus initiates homologous recombination.

16.3.4 Gene Stacking in Operons and Pathway Engineering

Generally, nuclear mRNAs from plants are monocistronic, which translates into a single protein, causing a serious drawback when engineering multiple genes or pathways. Plastid possesses its own transcription and translation machinery, which allows the integration and expression of multiple genes in operons [54]. Multiple genes cotranscribe as polycistronic RNAs, which are subsequently processed to form translatable transcripts [55]. Therefore, introduction of multiple transgenes arranged in an operon should allow the expression of an entire pathway in a single transformation event. A number of operons, including the *Bt cry*2Aa2 and the *mer* operons, were expressed from chloroplasts of tobacco [56]. Recently, a photorespiration pathway based on the stepwise incorporation of genes into the nuclear genome and posttranslational targeting of proteins to the chloroplasts was engineered in *Arabidopsis thaliana* [57]. The pathway works independently in chloroplasts, in which it converts glycolate directly to glycerate and thus increases productivity by improving photosynthesis in transgenic plants [57,58].

16.4 Chlorophyll Biosynthesis Pathway and Recent Developments

Photosynthetic organisms contain one of the most important molecules, chlorophyll, which converts captured solar energy into biochemical energy. Chlorophyll biosynthesis is a multireaction pathway in which the reduction of protochlorophyllide (PChlide) to chlorophyllide (Chlide), which is subsequently converted into chlorophyll by phytylation, is a major regulatory step. Integration of the newly synthesized chlorophyll into the developing thylakoid membranes is tightly coupled with the biogenesis of the photosynthetic apparatus and the development of chloroplasts [59]. The reduction of PChlide to Chlide is catalyzed by two different enzymes: a light-dependent, nuclear-encoded, plastid-localized single subunit enzyme (light-dependent protochlorophyllide oxidoreductase; LPOR) that requires light for its activation in angiosperms, and a light-independent, plastid-encoded enzyme (dark-operated protochlorophyllide oxidoreductase; DPOR) that is composed of three subunits.

From an evolutionary perspective, light-dependent chlorophyll biosynthesis is assumed to be a universal feature of oxygenic photosynthetic organisms [60]; however, genes (*chlL*, *chlN*, and *chlB*) encoding three protein subunits of DPOR are highly conserved in their nucleotide sequence in the plastid genome of most gymnosperms [61], Bryophyta [62],

Pteridophyta [63] and algae, Chlorophyta [64,65] and Rhodophyta [66], and in the chromosomal DNA of cyanobacteria [67,68] and anoxygenic photosynthetic bacteria [69]. These photosynthetic organisms can green and assemble functional chloroplasts in the dark and are ready for photosynthesis upon exposure to light [70,71]. Nevertheless, darkness promotes etiolation in the seedlings of angiosperms because they lack DPOR genes. Consequently, the distribution of DPOR and LPOR entails that the two systems coexisted throughout evolution from the cyanobacteria to the gymnosperms but that the genes encoding DPOR subunits were lost during the evolution from gymnosperms to angiosperms and, as a result, lost their ability to green in the dark [72–74]. Hence, the reduction of PChlide to Chlide in the absence of light in most dark-grown organisms, other than etiolated angiosperms, correlates with the presence of the *chlL, chlN,* and *chlB* genes [73]. To date, the plastid genomes sequenced have revealed that the plastid genomes either encode all three subunits or completely lack these genes. Most plastid-encoded chlorophyll-binding proteins like *D2, CP43, CP47, PsaA,* and *PsaB* are unstable in the absence of chlorophyll [75,76]. On the other hand, chlorophyll and its intermediates like Chlide are present in a free state, can be lethal to cells by the formation of highly reactive oxygen radicals and singlet oxygen upon illumination. Hence, the stability of chlorophyll-binding proteins demands a coordinated synthesis of chlorophyll and these proteins, produced in a light-dependent fashion in angiosperms and in a light-independent manner in gymnosperms [77].

16.4.1 Regulation of Expression of DPOR Coding Genes

As described previously, DPOR is a three-subunit enzyme encoded by *chlL, chlN,* and *chlB* genes. The regulation of these three genes in oxygenic photosynthetic organisms suggests that the pattern of accumulation of the mRNAs of three subunits is organism specific [78]. Interestingly, light seems to increase the accumulation of mRNAs for some of these genes in certain organisms. The mRNA of *chlB* accumulates to several-fold higher levels in continuously light-grown versus dark-grown *Chlamydomonas* cell cultures [79]. However, *chlN* is constitutively expressed under both light-grown and dark-grown algal cultures [80]. Furthermore, *chlB, chlL,* and *chlN* mRNAs accumulate independent of light in cell cultures of liverwort, which are equally green in the dark or in the light. Among gymnosperms, the *chlL* and *chlN* genes are expressed constitutively in light-grown and dark-grown pine cotyledons, but the accumulation of these mRNAs is light-induced from a low basal level in dark-grown stems [81]. In the larch species, *chlL* and *chlN* are expressed in both dark-grown and light-grown seedlings [82]. It has been suggested that inefficient processing of *chlLN* polycistronic transcripts could be responsible for the limited ability of this conifer to synthesize chlorophyll in the dark. In addition, the *chlB* chloroplast mRNAs of the larch species and two species of pine have been found to be edited posttranscriptionally [83]; whereas in pine, this editing alters the amino acid specified at two locations within the *ChlB* polypeptide. Thus, the impaired greening of dark-grown larch seedlings could conceivably reflect not only inefficient *ChlLN* transcript splicing but also the synthesis of a partially inactive *ChlB* polypeptide.

16.4.2 Engineering DPOR Pathways in Heterologous Plant Species

To ascertain how DPOR operates in oxygenic photosynthetic organisms, disruption of endogenous genes to develop null allele mutants and complementation experiments were carried out. Disruption of the chloroplast-encoded *chlB* from *C. reinhardtii* resulted in the development of a yellow phenotype of mutants in the dark, which was indistinguishable from mutants for *chlL* or *chlN*, suggesting that *chlB* is essential for the activity of DPOR in chloroplasts [84]. Later on, complementation experiments were performed on cyanobacterium mutants that lack *chlL* and *chlB* genes, which were incapable of reducing PChlide to Chlide. Two mutants—one carrying a disruption in the *chlB* gene with a kanamycin resistance cassette and is named YFB14 [85], and the other carrying a *chlL*-defective gene in which the gene is disrupted by *neo* gene cassette and is named YFC2 [86]—were transformed with two shuttle vectors for overexpression of strep-tagged L-protein (ChlL) and strep-tagged NB-protein (ChlN–ChlB), respectively, by electroporation. Two transformants, YFC2/L2 and YFB14/NB2, were recovered and the amounts of ChlL and ChlB subunits were confirmed by Western blot analysis. Both transformants restored the ability to produce chlorophyll in the dark [15]; however, further biochemical experiments suggested that the DPOR from an oxygenic photosynthetic organism did not acquire oxygen tolerance mechanisms during evolution but that cyanobacteria had developed a mechanism to protect DPOR from oxygen. The ChlB subunit was immunochemically detected in membrane fractions of cyanobacteria, showing either cytoplasmic or thylakoid membranes as the site of light-independent reduction of PChlide. Similarly, cyanobacterium was used for the evaluation of the physiological roles of maize ferredoxin isoforms by overexpression [87]. A mutant lacking the *petF* gene encoding cyanobacterial ferredoxin was successfully isolated from parental strains expressing maize ferredoxin isoforms. The functional differentiation of maize ferredoxin isoforms is demonstrated by this heterologous expression system.

In our laboratories, we developed a number of expression cassettes carrying DPOR coding genes (*chlL, chlN* and *chlB*). The genes were isolated from the plastid genome of *Pinus thubergii*, Japanese black pine, and was introduced into the plastome of *Nicotiana tabacum*. The genes were cloned either singly or in operon form in tobacco chloroplast transformation vector with FLARE-S as a selectable marker [88]. The transformed tobacco plants, recovered on spectinomycin-containing regeneration medium, were tested using polymerase chain reaction and showed successful integration of expression cassettes into the chloroplast genome and that homoplasmy levels were determined through Southern blot analysis. DPOR comprises two protein subunits, that is, L-protein and NB-protein encoded by *chlL* and *chlNB* genes, respectively. In the reduction process of PChlide to Chlide, six cysteine residues are involved in 4Fe-4S cluster, three of which are contributed by *ChlN*, one by *ChlB*, and two by *ChlL*. Likewise, nitrogenase, an L-protein, is suggested to act as an ATP-dependent electron donor to NB-proteins. This NB-protein has been proposed to be an important catalytic component of the DPOR. To identify the individual role of these subunits, the transformed tobacco plants were subjected to different morphological and physiological experiments. Because these genes are involved in the photosynthetic performance of the plants, especially in chlorophyll biosynthesis, thus the homoplasmic transgenic plants carrying *chlB* gene were first investigated. As expected, shoots developed from the cuttings of wild-type plants grown in the dark showed etiolated growth with no roots, whereas shoots from the cuttings of transgenic plants developed early and had more roots. Upon shifting from dark to light growing conditions, the leaves of the transgenic shoots showed early development of chlorophyll pigments compared with the wild-type shoots. Furthermore, photosynthetically indistinguishable transgenic shoots also showed significant differences in root development from untransformed wild-type shoots when cuttings were grown in the light. Therefore, it may be concluded that the *chlB* gene is involved, directly or indirectly, in the root development of tobacco. Furthermore, the gene promotes the early development of chlorophyll pigments upon illumination from dark growing conditions [88]. Two more sets of chloroplast transgenic plants were developed by engineering plastomes with expression cassettes; one carrying *chlL* and *chlN* genes and the other carrying *chlLN* and *chlB* genes in an operon to be expressed under a constitutive promoter from a tobacco chloroplast ribosomal RNA operon. Plants were screened and investigated to generate physiological and biochemical data. Preliminary data from these plants supports the early findings on gymnosperms, in which a three-subunit enzyme is required for greening in the dark. It also signifies that the enzyme is capable of greening angiosperms in the dark and thereby developing photosynthetically competent chloroplasts.

16.5 Perspective

Protochlorophyllide oxidoreductase (LPOR) catalyzes the reduction of PChlide to Chlide in the light, whereas DPOR carries out the reduction of PChlide to Chlide in the dark; hence, photosynthetic organisms harboring DPOR are capable of greening in the dark. There are three genes, namely, *chlL*, *chlN*, and *chlB*, encoding three subunits that constitute DPOR, which could be introduced into the genomes of plastids of angiosperms, particularly, cereal crop plants, to improve photosynthesis by increasing their chlorophyll contents and thereby developing photosynthetically competent chloroplasts. Our attempts are the first known efforts to engineer the three-subunit pathway in the plastid genome of tobacco, a representative model plant. This will help in understanding the molecular biology of transgenic angiosperms.

Acknowledgments

The author thanks the Ministry of Science and Technology, Islamabad and the Higher Education Commission, Islamabad for providing funds and supporting PhD students to carry out the research in my group.

References

1. Ruhlman, T., and H. Daniell. 2007. Plastid pathways. Molecular engineering via the chloroplast genome. In *Applications of Plant Metabolic Engineering*, edited by R. Verpoorte et al. Springer, Netherlands 79–108.
2. Bendich, A. 1987. Why do chloroplast and mitochondria contain so many copies of their genome. *BioEssays* 6: 279–282.
3. Daniell, H., M.S. Khan, and L. Allison. 2002. Milestones in chloroplast genetic engineering: an environmentally friendly era in biotechnology. *Trends in Plant Science* 7:84–91.
4. Isono, K., Y. Niwa, K. Satoh, and H. Kobayashi. 1997. Evidence for transcriptional regulation of plastid photosynthesis genes in *Arabidopsis thaliana* roots. *Plant Physiology* 114:623–630.
5. Sugiura, M. 1992. The chloroplast genome. *Plant Molecular Biology* 19:149–168.

6. Martin, W., and R.G. Hermann. 1998. Gene transfer from organelle to the nucleus: How much, what happens and why? *Plant Physiology* 118:9–17.
7. Turmel, M., C. Otis, and C. Lemieux. 1999. The complete chloroplast DNA sequence of the green alga *Nephrselmis olivacea*: Insights into the architecture of ancestral chloroplast genomes. *Proceedings of the National Academy of Sciences of the United States of America* 96:10248–10253.
8. Shimada, H., and M. Sugiura. 1991. Fine structural features of the chloroplast genome: Comparison of the sequenced chloroplast genomes. *Nucleic Acids Research* 19:983–995.
9. Daniell, H., and M.S. Khan. 2003. Engineering the chloroplast genome for biotechnology applications. In *Transgenic Plants: Current Innovations and Future Trends*, edited by Stewart, N. Horizon Press, UK. 83–110.
10. Khan, M.S., W. Hameed, M. Nozoe, and T. Shiina. 2007. Disruption of the *psbA* gene by the copy correction mechanism reveals that the expression of plastid-encoded genes is regulated by photosynthesis activity. *Journal of Plant Research* 120:421–430.
11. Yang, M., X. Zhang, G. Liu, Y. Yin, K. Chen, Q. Yun et al. 2010. The complete chloroplast genome sequence of date palm (*Phoenix dactylifera* L.). *PLoS One.* 5:e12762.
12. Gruissem, W., and J.C. Tonkyn. 1993. Control mechanisms of plastid gene expression. *Critical Review of Plant Science* 12:19–55.
13. Duggan, D.J., M. Bittner, Y. Chen, P. Meltzer, and J.M. Trent. 1999. Expression profiling using cDNA microarrays. *Nature Genetics* 21:10–14.
14. Cahoon, A.B. and M.P. Timko. 2000. Yellow-in-the-dark mutants of *Chlamydomonas* lack the ChlL subunit of light-independent protochlorophyllide reductase. *Plant Cell* 12:559–568.
15. Yamamoto, H., S. Kurumiya, R. Ohashi, and Y. Fujita. 2009. Oxygen sensitivity of a nitrogenase-like protochlorophyllide reductase from the cyanobacterium *Leptolyngbya boryana*. *Plant Cell Physiology* 50:1663–1673.
16. Armbruster, U., P. Pesaresi, M. Pribil, A. Hertle, and D. Leister. 2011. Update on chloroplast research: New tools, new topics, and new trends. *Molecular Plant* 4:1–16.
17. Schaffer, R., J. Landgraf, M. Perez-Amador, and E. Wisman. 2000. Monitoring genome-wide expression in plants. *Current Opinion in Biotechnology* 11:162–167.
18. Kurth, J., C. Varotto, P. Pesaresi, A. Biehl, E. Richly, F. Salamini et al. 2002. Gene-sequence-tag expression analyses of 1800 genes related to chloroplast functions. *Planta* 215:101–109.
19. DalCorso, G., P. Pesaresi, S. Masiero, E. Aseeva, D. Schunemann, G. Finazzi et al. 2008. A complex containing PGRL1 and PGR5 is involved in the switch between linear and cyclic electron flow in Arabidopsis. *Cell* 132:273–285.
20. Zhou, X., Z. Fei, T.W. Thannhauser, and L. Li. 2011. Transcriptome analysis of ectopic chloroplast development in green curd cauliflower (*Brassica oleracea* L. var. *botrytis*). *BMC Plant Biology* 11:169.
21. Kahlau, S., and R. Bock. 2008. Plastid transcriptomics and translatomic of tomato fruit development and chloroplast-to-chromoplast differentiation: Chromoplast gene expression largely serves the production of a single protein. *Plant Cell.* 20:856–874.
22. Beginsky, S., and W. Gruissem. 2004. Chloroplast proteomics: Potentials and challenges. *Journal of Experimental Botany* 55:1213–1220.
23. Kleffmann, T., A. von Zychlinski, D. Russenberger, M. Hirsch-Hoffmann, P. Gehrig, W. Gruissem et al. 2007. Proteome dynamics during plastid differentiation in rice. *Plant Physiology* 143:912–923.
24. Rutschow, H., A.J. Ytterberg, G. Frisco, R. Nilsson, and K.J. Van Wijk. 2008. Quantitative protoeomics of a chloroplast SRP54 sorting mutant and its genetic interactions with CLCP1 in *Arabidopsis. Plant Physiology* 148:156–175.
25. Ferro, M., S. Brugière, D. Salvi, D. Seigneurin-Berny, M. Court, L. Moyet et al. 2010. AT_ CHLORO, a comprehensive chloroplast proteome database with subplastidial localization and curated information on envelope proteins. *Molecular Cell Proteomics* 9:1063–1084.
26. Peltier, J.B., G. Friso, D.E. Kalume, P. Roepstorff, F. Nilsson, I. Adamska et al. 2000. Proteomics of the chloroplast: Systematic identification and targeting analysis of lumenal and peripheral thylakoid proteins. *Plant Cell* 12:319–341.

27. Giacomelli, L., A. Andrea Rudella, and K.J. van Wijk. 2006. High light response of the thylakoid proteome in *Arabidopsis* wild type and the ascorbate-deficient mutant *vtc2-2*. A comparative proteomics study. *Plant Physiology* 141:685–701.

28. Froehlich, J.E., C.G. Wilkerson, W.K. Ray, R.S. McAndrew, K.W. Osteryoung, D.A. Gage et al. 2003. Proteomic study of the *Arabidopsis thaliana* chloroplastic envelope membrane utilizing alternatives to traditional two-dimensional electrophoresis. *Journal of Proteome Research* 2:413–425.

29. Peltier, J.B., O. Emanuelsson, D.E. Kalume, J. Ytterberg, G. Friso, A. Rudella et al. 2002. Central functions of the lumenal and peripheral thylakoid proteome of *Arabidopsis* determined by experimentation and genome-wide prediction. *Plant Cell* 14:211–236.

30. Schubert, M., U.-A. Petersson, B.-J. Haas, C. Funk, W.-P. Schroder, and T. Kieselbach. 2002. Proteome map of the chloroplast lumen of *Arabidopsis thaliana*. *Journal of Biological Chemistry* 277:8354–8365.

31. Sidorov, V.A., D. Kasten, S.Z. Pang, P.T.J. Hajdukiewicz, J.M. Staub, and N.S. Nehra. 1999. Stable chloroplast transformation in potato: Use of green fluorescent protein as a plastid marker. *Plant Journal* 19:209–216.

32. Khan, M.S., and P. Maliga. 1999. Fluorescent antibiotic resistance marker for tracking plastid transformation in higher plants. *Nature Biotechnology* 17:910–915.

33. Kumar, S., A. Dhingra, and H. Daniell. 2004. Stable transformation of the cotton plastid genome and maternal inheritance of transgenes. *Plant Molecular Biology* 56:203–216.

34. Kumar, S., A. Dhingra, and H. Daniell. 2004. Plastid-expressed betaine aldehyde dehydrogenase gene in carrot cultured cells, roots, and leaves confer enhanced salt tolerance. *Plant Physiology* 136:2843–2854.

35. Cheng, L., H.P. Li, B. Qu, T. Huang, J.X. Tu, T.D. Fu et al. 2010. Chloroplast transformation of rapeseed (*Brassica napus*) by particle bombardment of cotyledons. *Plant Cell Reports* 29:371–381.

36. Liu, C.W., C.C. Lin, J.J. Chen, and M.J. Tseng. 2007. Stable chloroplast transformation in cabbage (*Brassica oleracea* L. var. *capitata* L.) by particle bombardment. *Plant Cell Reports* 26:1733–1744.

37. Mustafa, G. 2011. Development of Plastid Transformation of Sugarcane. PhD diss. Quaid-I-Azam University, Pakistan.

38. Sikdar, S.R., G. Serino, S. Chaudhuri, and P. Maliga. 1998. Plastid transformation in *Arabidopsis thaliana*. *Plant Cell Reports* 18:20–24.

39. Golds, T., P. Maliga, and H.U. Koop. 1993. Stable plastid transformation in PEG-treated protoplasts of *Nicotiana tabacum*. *Biotechnology* 11:95–97.

40. Boynton, J.E., N.W. Gillham, E.H. Harris, J.P. Hosler, A.M. Johnson, A.R. Jones et al. 1988. Chloroplast transformation in *Chlamydomonas* with high velocity microprojectiles. *Science* 240:1534–1538.

41. Svab, Z., and P. Maliga. 1993. High-frequency plastid transformation in tobacco by selection for a chimeric *aadA* gene. *Proceedings of the National Academy of Sciences of the United States of America* 90:913–917.

42. Khan, M.S., A.M. Khalid, and K.A. Malik. 2005. Intein-mediated protein trans-splicing and transgene containment in plastids. *Trends in Biotechnology* 23:217–220.

43. Hagemann, R. 2004. In *Molecular Biology and Biotechnology of Plant Organelles*, edited by Daniell, H., and C. Chase. 87–108.

44. Verma, D., and H. Daniell. 2007. Chloroplast vector systems for biotechnology applications. *Plant Physiology* 145:1129–1143.

45. Khan, M.S. 2005. Engineered male sterility. *Nature* 436:783–784.

46. Ruiz, O.N., and H. Daniell. 2005. Engineering cytoplasmic male sterility via the chloroplast genome by expression of β-ketothiolase. *Plant Physiology* 138:1232–1246.

47. DeCosa, B., W. Moar, S.B. Lee, M. Miller, and H. Daniell. 2001. Overexpression of the Bt *cry2Aa2* operon in chloroplasts leads to formation of insecticidal crystals. *Nature Biotechnology* 19:71–74.

48. Zoubenko, O.V., L.A. Allison, Z. Svab, and P. Maliga. 1994. Efficient targeting of foreign genes into the tobacco plastid genome. *Nucleic Acids Research* 22:3819–3824.

49. Fernandez-San Millan, A., A. Mingo-Castel, M. Miller, and H. Daniell. 2003. A chloroplast transgenic approach to hyperexpress and purify human serum albumin, a protein highly susceptible to proteolytic degradation. *Plant Biotechnology* 1:71–79.

50. Eibl, C., Z. Zou, A. Beck, M. Kim, J. Mullet, and H.U. Koop. 1999. In vivo analysis of plastid *psbA*, *rbcL* and *rpl32* UTR elements by chloroplast transformation: Tobacco plastid gene expression is controlled by modulation of transcript levels and translation efficiency. *Plant Journal* 19:333–345.

51. Kuroda, H., and P. Maliga. 2001. Complementarity of the 16S rRNA penultimate stem with sequences downstream of the AUG destabilizes the plastid mRNAs. *Nucleic Acids Research* 29:970–975.

52. Kuroda, H., and P. Maliga. 2001. Sequences downstream of the translation initiation codon are important determinants of translation efficiency in chloroplasts. *Plant Physiology* 125:430–436.

53. Cerutti, H., M. Osman, P. Grandoni, and A.T. Jagendorf. 1992. A homolog of *Escherichia coli* RecA protein in plastids of higher plants. *Proceedings of the National Academy of Sciences of the United States of America* 89:8068–8072.

54. Sugita, M., and M. Sugiura. 1996. Relaxation of gene expression in chloroplasts of higher plants. *Plant Molecular Biology* 32:315–326.

55. Bogorad, L. 2000. Engineering chloroplasts: An alternative site for foreign genes, proteins, reactions and products. *Trends in Biotechnology* 18: 257–263.

56. Ruiz, O.N., H.S. Hussein, N. Terry, and H. Daniell. 2003. Phytoremediation of organomercurial compounds via chloroplast genetic engineering. *Plant Physiology* 132:1344–1352.

57. Kebeish, R., M. Niessen, K. Thiruveedhi, R. Bari, H.-J. Hirsch, R. Rosenkranz et al. 2007. Chloroplastic photorespiratory bypass increases photosynthesis and biomass production in *Arabidopsis thaliana*. *Nature Biotechnology* 25:593–599.

58. Khan, M.S. 2007. Engineering photorespiration in chloroplasts: A novel strategy for increasing biomass production. *Trends in Biotechnology* 25:437–440.

59. Philippar, K., T. Geis, I. Ilkavets, U. Oster, S. Schwenkert, J. Meurer et al. 2007. Chloroplast biogenesis: The use of mutants to study the etioplast–chloroplast transition. *Proceedings of the National Academy of Sciences of the United States of America* 104:678–683.

60. Eckhardt, U., B. Grimm, and S. Hortensteiner. 2004. Recent advances in chlorophyll biosynthesis and breakdown in higher plants. *Plant Molecular Biology* 56:1–14.

61. Wakasugi, T., J. Tsudzuki, S. Ito, K. Nakashima, T. Tsudzuki, and M. Sugiura. 1994. Loss of all *ndh* genes as determined by sequencing the entire chloroplast genome of the black pine *Pinus thunbergii*. *Proceedings of the National Academy of Sciences of the United States of America* 91:9794–9798.

62. Ohyama, K., H. Fukuzawa, T. Kohchi, H. Shirai, T. Sano, S. Sano et al. 1986. Chloroplast gene organization deduced from complete sequence of liverwort *Marchantia polymorpha* chloroplast DNA. *Nature* 322:572–574.

63. Yamada, K., M. Matsuda, Y. Fujita, H. Matsubara, and M. Sugai. 1992. A frxC homolog exists in the chloroplast DNAs from various pteridophytes and in gymnosperms. *Plant Cell Physiology* 33:325–327.

64. Suzuki, J.Y., and C.E. Bauer. 1992. Light-independent chlorophyll biosynthesis: Involvement of the chloroplast gene chlL (frxC). *Plant Cell* 4:929–940.

65. Li, J., M. Goldschmidt-Clermont, and P. Timko. 1993. Chloroplast-encoded ChlB is required for light-independent protochlorophyllide reductase activity in *Chlamydomonas reinhardtii*. *Plant Cell* 5:1817–1829.

66. Reith, M., and J. Munholland. 1993. A high-resolution gene map of the chloroplast genome of the red alga *Porphyra purpurea*. *Plant Cell* 5:465–475.

67. Fujita, Y. 1996. Protochlorophyllide reduction: a key step in the greening of plants. *Plant Cell Physiology* 37:411–421.

68. Ogura, Y., M. Takemura, K. Oda, K. Yamato, E. Ohta, H. Fukuzawa et al. 1992. Cloning and nucleotide sequence of afrxCORF469 gene cluster of Synechocystis PCC6803: Conservation with liverwort chloroplast frxC-ORF465 and nif operon. *Bioscience Biotechnology and Biochemestry* 56:788–793.

69. Burke, D.H., M. Alberti, and J.E. Hearst. 1993. bchFNBH bacteriochlorophyll synthesis genes of *Rhodobacter capsulatus* and identification of the third subunit of light-independent protochlorophyllide reductase in bacteria and plants. *Journal of Bacteriology* 175:2414–2422.

70. Yamazaki, S., J. Nomata, and Y. Fujita. 2006. Differential operation of dual protochlorophyllide reductases for chlorophyll biosynthesis in response to environmental oxygen levels in the cyanobacterium *Leptolyngbya boryana*. *Plant Physiology* 142:911–922.

71. Kusumi, J., A. Sato, and H. Tachida. 2006. Relaxation of function constraints on light-independent protochlorophyllide oxidoreductase in *Thuja*. *Molecular Biology and Evolution* 23:941–948.

72. Shi, C., and X. Shi. 2006. Characterization of three genes encoding the subunits of light-independent protochlorophyllide reductase in *Chlorella protothecoides* CS-41. *Biotechnology Progress* 22:1050–1055.

73. Shi, C., and X. Shi. 2006. Expression switching of three genes encoding light-independent protochlorophyllide oxidoreductase in *Chlorella protothecoides*. *Biotechnology Letters* 28:261–265.

74. Demko, V., A. Pavlovib, D. Valkova, L. Slovakova, B. Grimm, and J. Hudak. 2009. A novel insight into the regulation of light-independent chlorophyll biosynthesis in *Larix decidua* and *Picea abies* seedlings. *Planta* 230:165–176.

75. Mullet, J.E., P.G. Klein, and R.R. Klein. 1990. Chlorophyll regulates accumulation of the plastid-encoded chlorophyll apoproteins CP43 and D1 by increasing apoprotein stability. *Proceedings of the National Academy of Sciences of the United States of America* 87:4038–4042.

76. Eichacker, L., H. Paulsen, and W. Riidiger. 1992. Synthesis of chlorophyll a regulates translation of chlorophyll a apoproteins P700, CP43, CP47 and D2 in barley etioplasts. *European Journal of Biochemistry* 205:17–24.

77. Forreiter, C., and K. Apel. 1993. Light-independent and light-dependent protochlorophyllide-reducing activities and two distinct NADPH-protochlorophyllide oxidoreductase polypeptides in mountain pine (*Pinus mugo*). *Planta* 190:536–545.

78. Nomata, J., T. Ogawa, M. Kitashima, K. Inou, and Y. Fujita. 2008. NB-protein (BCHN-BchB) of dark-operative protochlorophyllide reductase is the catalytic component containing oxygen-tolerant Fe-S clusters. *FEBS Letters* 582:1346–1350.

79. Li, J., M. Goldschmidt-Clermont, and P. Timko. 1993. Chloroplast-encoded ChlB is required for light-independent protochlorophyllide reductase activity in *Chlamydomonas reinhardtii*. *Plant Cell* 5:1817–1829.

80. Choquet, Y., M. Rahire, J. Girard-Bascou, J. Erikson, and J.D. Rochaix. 1992. A chloroplast gene is required for the light-independent accumulation of chlorophyll in *Chlamydomonas reinhardtii*. *The EMBO Journal* 11:1697–1704.

81. Spano, A.J., Z. He, and M.P. Timko. 1992. NADPH:protochlorophyllide oxidoreductases in white pine (*Pinus strobus*) and loblolly pine (*P. teada*). *Molecular General Genetics* 236:86–95.

82. Yamada, A. 1996. The study of ecological community on ectomycorrhizal fungi in *Pinus densiflora* forests. PhD diss., University of Tsukuba, Japan.

83. Karpinska, B., S. Karpinski, and J.E. Hällgren. 1997. The *chlB* gene encoding a subunit of light-independent protochlorophyllide reductase is edited in chloroplasts of conifers. *Current Genetics* 31:343–347.

84. Liu, X.Q., H. Xu, and C. Huang. 1993. Chloroplast chlB is required for light independent chlorophyll accumulation in *Chlamydomonas reinhordhi*. *Plant Molecular Biology* 23:297–308.

85. Fujita, Y., H. Takagi, and T. Hase. 1996. Identification of the *chlB* gene product essential for light-independent chlorophyll biosynthesis in the cyanobacterium *Plecronema boyanum*. *Plant Cell Physiology* 37:313–323.

86. Kada, S., H. Koike, K. Satoh, T. Hase, and Y. Fujita. 2003. Arrest of chlorophyll synthesis and differential decrease of photosystems I and II in a cyanobacterial mutant lacking light-independent protochlorophyllide reductase. *Plant Molecular Biology* 51:225–235.

87. Kimata-Ariga, Y., T. Matsumura, S. Kada, H. Fujimoto, Y. Fujita, T. Endo et al. 2000. Differential electron flow around photosystem I by two C(4)-photosynthetic-cell-specific ferredoxins. *The EMBO Journal* 19:5041–5050.
88. Nazir, S. 2012. Engineering novel pathways in chloroplast for improving the crop yield. PhD diss., Quaid-I-Azam University, Pakistan.

17

Utilization of Omics Technology to Analyze Transgenic Plants

Marisela Rivera-Domínguez and Martín-Ernesto Tiznado-Hernández

CONTENTS

17.1 Introduction

Gene manipulation allows the isolation of discrete pieces of a genome from its host organism, and its propagation in the same or different host, a technique known as cloning. This in turn enables the DNA segment to be sequenced [1]. A typical cloning experiment requires the isolation of the DNA of interest (sometimes called foreign, passenger, or target DNA), a cloning vector, restriction endonucleases, DNA ligase, and a prokaryotic or eukaryotic cell to serve as the biological host. Once the vector and foreign DNA have been isolated, they are treated with the same restriction endonuclease to produce site-specific scission in the DNA. Because both vector and foreign DNA have sticky ends when they are generated by the same restriction enzyme, they can be ligated, thus producing a recombinant molecule. This recombinant molecule can be physically inserted into the appropriate host by transformation [2]. Once a gene has been isolated and cloned (multiplied in a bacterial vector), it must undergo several modifications to create a construct that can then be inserted into a plant cell, as described in the following:

1. A promoter sequence must be added for the gene to be expressed (i.e., translated into a protein product). The promoter is the on/off switch that controls how much, when, and in what plant tissues the gene will be expressed. To date, most promoters in transgenic crop varieties have been "constitutive," that is, triggering gene expression throughout the life cycle of the plant and in most of the tissues. The most commonly used constitutive promoter is CaMV35S, which was isolated from the cauliflower mosaic virus. This promoter generally results in a high degree of expression in plants. Other promoters are more specific and respond to cues in the plant's internal or external environment. An example of this is the light-inducible promoter from the cab gene, encoding the major chlorophyll *a/b* binding protein. On the other hand, the cloned gene is sometimes modified to achieve a large expression in a plant. For example, the genes encoding toxins from the bacteria *Bacillus thuringiensis*, which confers insect resistance in plants, have a higher percentage of A–T nucleotide pairs compared with plants in which coding regions show a higher degree of G–C nucleotide pairs. In a clever modification, researchers substituted A–T nucleotides with G–C nucleotides in the Bt gene without inducing a significant change in the protein amino acid sequence. The result was an enhanced production of the gene product in plant cells.

2. A termination sequence, which signals to the cellular machinery that the end of the gene sequence has been reached.

3. A selectable marker gene is added to the gene "construct" to identify plant cells or tissues in which the integration of the transgene was successful. This is necessary because achieving incorporation and expression of transgenes in plant cells is a rare event, occurring in just a small percentage of the targeted tissues or cells. Selectable marker genes encode proteins that provide resistance to agents that are normally toxic to plants, such as antibiotics or herbicides. As explained below, only plant cells that have integrated the selectable marker gene will survive on a medium containing the appropriate antibiotic or herbicide. As for the other inserted genes, marker genes also require promoter and termination sequences for proper function.

17.2 Techniques for Transforming Plants: Gene Gun and *Agrobacterium*

Transformation is the heritable change in a cell or organism brought about by the uptake and establishment of introduced DNA. There are two main methods for transforming plant cells and tissues:

1. The *Agrobacterium* method, successfully used in dicots (broadleaf plants like soybeans and tomatoes) for many years, but only recently in monocots (grasses and their relatives). In general, the *Agrobacterium* method is preferentially used instead of the gene gun because of the greater frequency of single-site insertions of the foreign DNA, making it easier to monitor the transgene segregation.

2. The "gene gun" method (also known as microprojectile bombardment or biolistics), has been especially useful in transforming monocot species like corn and rice.

17.2.1 Creation of Transgenic Plants

17.2.1.1 Agrobacterium Method of Plant Transformation

Agrobacterium tumefaciens is a soil phytopathogen, Gram-negative bacterium that can genetically transform cells of numerous dicot plant species as well as some monocots and gymnosperms [3]. It is a remarkable species of soil-dwelling bacteria that has the ability to infect plant cells and insert a piece of its DNA into the plant cell genome. When the bacterial DNA is integrated into a plant chromosome, it effectively hijacks the plant's cellular machinery and uses it to ensure the proliferation of the bacterial population. *Agrobacterium* can additionally transform numerous fungal species, including yeasts, ascomycetes, and basidiomycetes, as well as human cells [4,5]. *A. tumefaciens* reproduces on susceptible plants by transferring its T-DNA, a portion of its tumor-inducing plasmid (pTi), into plant cells [6–8]. This T-DNA is translocated to the nucleus, integrated into plant chromosomal DNA, and expressed.

The DNA in an *A. tumefaciens* cell is contained in the bacterial chromosome as well as in the Ti (tumor-inducing) plasmid. The Ti plasmid contains:

- A stretch of DNA termed T-DNA (~20 kb long) that is transferred to the plant cell in the infection process
- A series of *vir* (virulence) genes that direct the infection process

A. tumefaciens can only infect a plant that had been wounded. When a plant root or stem is wounded, it gives off certain chemical signals, which can diffuse into the rhizosphere, inducing the expression of *vir* genes [6]. The activated *vir* genes direct a series of events necessary for the transfer of the T-DNA from the Ti plasmid to the plant's chromosome, as follows:

- Copying of the T-DNA
- Attachment of a protein product to the copied T-DNA strand to act as a leader
- Addition of proteins along the length of the T-DNA, possibly as a protective mechanism
- Opening a channel in the bacterial cell membrane, through which the T-DNA passes

Phenolic molecules (e.g., acetosyringone) and acidic pH are essential for *vir* gene expression [9–11], whereas aldose monosaccharides (e.g., arabinose) enhance sensitivity to phenolic inducers but are not essential for *vir* gene induction [12,13]. The gene expression of the integrated T-DNA in the plant genome leads to the accumulation of plant growth hormones responsible for the formation of crown gall tumors. The wild-type T-DNA carries two types of genes: oncogenes and genes for the biosynthesis of amino acid derivatives (opines). Expression of the integrated T-DNA, therefore, results in uncontrolled cell division and the formation of tumors [14,15]. These transformed cells produce and secrete opines to be utilized by *A. tumefaciens* and several other microorganisms as sources of carbon and nitrogen [16]. The T-DNA itself does not encode the genes required for its transfer and integration and is defined only by two 25-bp direct repeats, termed T-DNA left and right borders [17]. Thus, the entire wild-type T-DNA sequence between the borders can be replaced with a gene of interest that will be transferred to plants and integrated into the plant genome, representing the molecular basis for plant genetic engineering [8,18,19]. The path that the T-DNA follows during translocation to the nucleus and integration into the plant chromosomal DNA to be expressed has already been shown [20].

Several *A. tumefaciens chv* (chromosomal virulence) genes and a set of *vir* (virulence) genes are present in the pTi plasmid code for the protein machinery of the T-DNA transport [8,21,22]. The VirA/VirG two-component signal transduction system senses the secretion of phenolics and specific sugar compounds from the wounded susceptible plant tissues and induces the expression of other *vir* genes [23]. The VirD2/VirD1 complex nicks the bottom strand of the Ti plasmid at the T-DNA borders and, with the assistance of the bacterial DNA synthesis and repair machinery, a single-stranded copy of the T-DNA, the T strand, is liberated from the Ti plasmid [24–26]. VirD1 is then released, whereas VirD2 remains covalently attached to the 5' end of the T strand [27–29]. The resulting protein–DNA complex is exported to the host cell through the type IV secretion system encoded by the *virD4* gene and *virB* operon including 11 open reading frames [30,31]. The VirD4/VirB channel also mediates the export of VirE2 [32], a single-stranded DNA-binding protein [33–35] that is thought to associate cooperatively with the T-strand in the plant cell cytoplasm [36,37], producing the T-complex [7,38], a semirigid, hollow, cylindrical filament composed of a T-strand molecule attached to one molecule of VirD2 and coated by multiple VirE2 monomers [39]. Both VirD2 and VirE2 are thought to facilitate the import of the T-complex into the host cell nucleus [40–43] to integrate the T-DNA into the plant genome by illegitimate recombination [44,45].

Two different models have been suggested for T-DNA integration, double-stranded break (DSB) repair and single-stranded gap repair [45]. The first model predicts that unwound ends of a double-stranded T-DNA molecule anneal with single-stranded overhangs of a DSB in the plant DNA, the residual 5′ and 3′ overhangs are removed, and the inserted T-DNA is ligated. The second model suggests that integration is initiated with a nick, which leads to a gap in the plant DNA, both ends of a single-stranded T-DNA molecule then anneal to this gap, the residual 5′ and 3′ overhangs are removed, and the integration is completed by repair synthesis of the second T-DNA strand. Both models also suggest a role for VirD2 in the recognition of nicks or gaps by interaction with plant factors [45].

Additionally, direct evidence for the role of DSBs and double-stranded T-DNA intermediates in T-DNA integration was presented [46]. They used transient expression of an intron-encoded endonuclease I-*Sce*I to create DSBs in the genomes of plants transgenic for the I-*Sce*I recognition site [47]. Afterward, transgenic plants were retransformed with two *A. tumefaciens* strains to induce transient expression of I-*Sce*I and DSB with one strain, whereas the other strain was carrying a T-DNA with the I-*Sce*I site. Results indicate that T-DNA is preferentially targeted and integrated into the DSB at their I-*Sce*I sites. Furthermore, the invading T-DNA with the I-*Sce*I recognition site was often integrated after being digested by I-*Sce*I, indicating its conversion to a double-stranded form before integration.

17.2.1.2 Biolistic Method of Plant Transformation

Particle bombardment technology is valuable for both gene expression [48] and stable transformation research [49]. The biolistic process represents a completely new approach to the problem of how to deliver DNA into cells and tissue. High-velocity microprojectiles are used to carry DNA, or other substances, past the plant cell wall and membranes. Because DNA is being shot into the cells, it represents a type of biological ballistics, hence the term biolistics [50]. The basis of biolistics is the acceleration of small DNA-coated particles (gold or tungsten) toward cells, resulting in the penetration of the protoplasm by the particle and subsequent expression of the introduced DNA [51]. The physical nature of DNA introduction permits reliable and effective gene transfer to major agronomic monocotyledonous and dicotyledonous species, and into intact plant and animal tissues [51]. In addition, it had been also applied to transforming species recalcitrant to conventional *Agrobacterium* or protoplast methods [52]. The first experiments in microprojectile bombardment for delivering DNA to living cells was done in 1987 [53]. DNA was delivered by discharging a 0.22 caliber cartridge to accelerate tungsten microprojectiles carrying DNA through an evacuated chamber and into the cell. A particle inflow gun was designed followed by the flowing helium gun to make it more efficient and compatible with biological targets. This equipment used a vacuum chamber to reduce the drag on the particle as well as to diminish the tissue damage, and also a timer relay–driven solenoid to provide more consistent acceleration by permitting better control of the amount of helium released [51].

The efficiency of this technique requires the optimization of several parameters for each species or tissue to be transformed [54].

- The particle size
- Helium pressure of the shot
- Helium source and rupture disc distance
- Distance between rupture disc and macrocarrier membrane
- Distance from macrocarrier membrane to target tissue

17.2.2 Selection of Transformed Tissue: Selectable and Nonselectable Marker Genes

17.2.2.1 Selectable Marker

Selectable marker genes confers resistance to selected chemical agents (e.g., antibiotics and herbicides), and have been used in the development of plant transformation technologies, mainly because the marker genes allow for the selection of rare transgenic cells among the nontransgenic cells in the original explant tissue. The use of marker genes for selecting transgenes has three major pitfalls: (1) the negative effects of selective chemicals can limit the ability of transgenic cells to proliferate and differentiate into transgenic plants, (2) the recent increase in public concern caused by the release of transgenic plants carrying antibiotic-resistant genes, and (3) for repeated transformation (transgene stacking) is difficult to increase the number of desired genes using the same selectable marker [55].

Selectable marker genes are used to discriminate between transgenic and nontransgenic cells in systems that generally display low transformation efficiencies. With high cotransformation frequencies, they facilitate the identification of plants containing cotransformed transgenes. The utility of individual selectable marker genes is a function of both the properties of the respective resistance protein they encode and the relative sensitivity of the target tissue to their corresponding selective agent. Selectable marker genes act by expressing: an enzyme that inactivates the selective agent (detoxification) and a resistant variant of a selective agent's target enzyme (tolerance). The features of a particular transformation system should be considered when choosing the resistance mechanism and the individual marker gene to be employed in any selection scheme [56,57]. The most common selectable markers are listed in Tables 17.1 and 17.2.

TABLE 17.1

Toxic Antibiotics and Selectable Marker Genes Used for the Conditional-Positive Selection of Transgenic Plants

Antibiotics	Genes	Enzymes	Genetic Sources
Neomycin	Neo, nptII	Neomycin	Escherichia coli Tn5
Kanamycin	aphA2	Phosphotransferases	
Paramomycin, G418	nptII (aphA1)		E. coli Tn601
Aminoglycosides	aaC3	Aminoglycoside-N-acetyl transferases	Serratia marcesens
	aaC4		Klebsiella pneumonia
	6'gat		Shigella sp.
Spectinomycin	aadA	Aminoglycoside-3″-adenyl transferase	Shigella sp.
Spectinomycin	SPT	Streptomycin phosphotransferase	Tn5
Streptomycin			
Hygromycin B	hph (aphIV)	Hygromycin phosphotransferase	E. coli
Bleomycin	Ble	Bleomycin resistance	E. coli Tn5
Phleomycin			Streptoalloteichus hindustanus
Sulfonamides	sulI	Dihydropteroate synthase	E. coli pR46
Streptothricin	sat3	Acetyl transferase	Streptomyces sp.
Chloramphenicol	Cat	Chloramphenicol acetyl transferase	E. coli Tn5
			Phage p1cm

Source: Miki, B. and McHugh, S., J. Biotechnol. 107:193–232, 2004. With permission.

TABLE 17.2

Toxic Herbicides and Selectable Marker Genes Used for the Conditional-Positive Selection of Transgenic Plants

Herbicides	Genes	Enzyme	Genetic Source
Phosphinothricin	*pat, bar*	Phosphinothricin acetyltransferase	*Streptomyces hygroscopicus* *Streptomyces viridochromogenes* Tu494
Glyphosate	EPSP synthase *aro*A *cp4 epsps* *gox*	5-Enolpyruvylshikimate-3-phosphate synthase Glyphosate oxidoreductase	*Petunia hybrida, Zea mays* *Salmonella typhimurium* *E. coli* *A. tumefaciens* *Ochrobactrum anthropi*
Sulfonylureas	*csr1-1*	Acetolactate synthase	*A. thaliana*
Imidazolinones	*Csr1-2*	Acetolactate synthase	*A. thaliana*
Oxynils	*Bnx*	Bromoxynil nitrilase	*K. pneumoniae* subsp. *ozanaenae*
Gabaculine	*hemL*	Glutamate-1-semialdehyde aminotransferase	*Synechococcus* PCC6301
Cyanamide	*Cah*	Cyanamide hydratase	*Myrothecium verrucaria*

Source: Miki, B. and McHugh, S., *J. Biotechnol.* 107:193–232, 2004. With permission.

17.2.2.2 Nonselectable Markers: Reporter Genes

Reporter genes are used in plant transformation for analyzing gene function, monitoring selection efficiency in both transformed tissue and transgenic plants, and to follow the inheritance of foreign genes in subsequent plant generations. Transient expression analysis using constructs with promoter and reporter genes may be used to analyze gene regulation and function. The utility of different reporter genes in plant transformation is a function of the properties of the respective protein they encode.

Good reporter genes should have the following characteristics:

- Expression in plant cells
- Low background activity in transgenic plants
- No detrimental effects on plant metabolism
- Only moderate stability *in vivo* so as to detect downregulation of gene expression as well as gene activation
- Easy to quantify in an assay system—nondestructive, quantitative, sensitive, versatile, simple to carry out, and inexpensive

The most commonly used reporter genes are listed in Table 17.3.

17.2.3 Selection of Transformed Plant Tissues

After the gene insertion process, plant tissues are transferred into a selective medium containing an antibiotic or herbicide, depending on which selectable marker was used. Only plants expressing the selectable marker gene will survive, and it is assumed that these plants are also harboring the transgene of interest. Thus, subsequent steps in the tissue culture process will be focused on these surviving plants.

TABLE 17.3

Nonselectable Markers or Reporter Genes Used to Create Transgenic Plants

External Substrates	Genes	Enzymes	Genetic Source
O-Nitrophenol-β-D-galactoside 5-Bromo-4-chloro-3-indolyl-β-D-galactoside	*lacZ*	β-Galactosidase	*E. coli*
5-Bromo-4-chloro-3-indoxyl-β-D-glucuronide cyclohexylammonium salt; 4-Methylumbelliferyl β-D-glucuronide	*quidA* (gusA)	β-Glucuronidase	*E. coli*, *Bacillus* sp.
Luciferin	*Luc*	Luciferase	*Photinus pyralis*
Decanal	*luxA, B lux F*	alkanal,reduced FMN:oxygen oxidoreductase	*Vibrio harveyi*
None	*Gfp*	Green fluroescence protein	*Aequorea victoria*
None		Phytoene synthase	*Erwinia herbicola*
None	R, C1, B	Anthocyanin pathway regulatory factors	Maize
None		Thaumatin II	*Thaumatococcus danielli* Benth
Oxalic acid		Oxalate oxidase	Wheat

Source: Miki, B. and McHugh, S., *J. Biotechnol.* 107:193–232, 2004. With permission.

17.2.4 Regeneration of Whole Plant Tissue: Tissue Culture Techniques

To obtain whole plants from transgenic tissues such as immature embryos, they are grown under controlled environmental conditions in a series of media containing nutrients and hormones, a process known as tissue culture. Once whole plants are generated and produce seed, evaluation of the progeny begins. This regeneration step has been a stumbling block in producing transgenic plants in many species, but specific varieties of most crops can now be transformed and regenerated. Tissue culture is not a theoretical prerequisite for plant transformation, but it is employed in almost all current practical transformation system to archive a workable efficiency of gene transfer, selection, and regeneration of transformed individuals [58].

A high multiplication ratio with a micropropagation system does not necessarily indicate a large number of regenerable cells accessible for gene transfer [59]. Gene transfer into potentially regenerable cells may not allow the recovery of transgenic plants if the capacity for efficient regeneration is short-lived [60]. There does not seem to be any specific reason to prefer embryogenic or organogenic plant regeneration. The choice can be made based on features affecting convenience or efficiency, including ready availability of explants and minimal time in tissue culture. Somaclonal variation, once considered a potentially useful source of genetic variation for plant improvement in gene transfer programs, is not desirable. On the contrary, in this system, it is desirable to minimize somaclonal variation by minimizing the tissue culture phase [61].

17.3 Plant Breeding and Testing

Intrinsic to the production of transgenic plants, there is an extensive evaluation process to verify whether the inserted gene has been stably incorporated without detrimental effects on other plant functions, product quality, or the intended agroecosystem in which the transgenic is going to be grown. Initial evaluation includes analysis of:

- Activity of the introduced gene
- Stable inheritance of the gene
- Unintended effects on plant growth, yield, and quality

If a plant passes these tests, it will most likely not be used directly for crop production, but rather it will be crossed with improved varieties of the crop. This is because only a few varieties of a given crop can be efficiently transformed, and these generally do not possess all the producer and consumer qualities required from modern cultivars. The initial genetic cross to the improved variety must be followed by several cycles of repeated crosses to the improved parent, a process known as backcrossing. The goal is to recover as much of the improved parent's genome as possible, with the addition of the transgene from the transformed parent. The next step in the process is multilocation and multiyear evaluation trials in greenhouse and field environments to test the effects of the transgene and overall performance.

17.4 Regulatory Measures Adopted for Transgenic Plants: National and International Agreements

Risk management related to transgenic plants was developed on the basis of the "precautionary principle," formulated in 1998 at the Wingspread Conference sponsored by the Science and Environmental Health Network, which basically states that it is better to take precautionary measures now rather than face serious harm to human health as a consequence later on. However, it has been mentioned that this principle, in all its forms, is loaded with vagueness and ambiguity [62]. There are other risk management approaches; one is called the "trial and error" method, whereas the other is called the "precaution through experience" method, which has been suggested to be a hybrid of the precautionary approach along with adjustments of the trial and error approach after commercialization. This approach is thought to be better because the development of a new technology usually takes into account previous experiences with similar technologies or technologies with the same applications [63].

Aside from the abovementioned principles, the concept known as substantial equivalence [64] was developed to evaluate the safety of food derived from transgenic plants. Both the substantial equivalence and the precautionary principle are subjects of hard scientific and public controversies as mentioned in other sections of the present review. However, the same controversies that led to the development of substantial equivalence and precautionary principles as approaches to evaluate risk in different countries led to the development of laws to reduce the potential risks associated with transgenic plants. Furthermore, in general, the risks that the laws are trying to control are related with human health and the environment [65]. By the year 2000, 52 countries were known to have biosafety legislation worldwide [66]. In this review, due to lack of space, only some of the laws developed in some of the countries will be described. In Canada, the Canadian Food Inspection Agency regulates the importation, environmental release, and livestock feed use of plants with novel traits, which include, but are not limited to, transgenic plants. Importation, confined release, and unconfined environmental release and the use of plants with novel traits as livestock feed are regulated under the Plant Protection Act, the Seeds Act, and the Feeds Act, respectively. Health Canada regulates novel foods for human consumption, including food products derived from transgenic plants under the Food and Drugs Act [67].

17.4.1 National Agreements

17.4.1.1 United States

In the United States, the Animal and Plant Health Inspection Service of the U.S. Department of Agriculture regulates the importation, interstate movement, and environmental release of transgenic plants that contain plant pest components. These activities are regulated under the Federal Plant Protection Act. The regulatory authority for food and livestock feed use in the United States lies with the Food and Drug Administration. The U.S. Environmental Protection Agency regulates the sale, distribution, production, and use of pesticides, including those produced by transgenic plants, and establishes tolerances for the pesticides expressed in transgenic plants intended for food or feed [68,69].

17.4.1.2 Mexico

In Mexico, the Secretariat of Agriculture, Livestock, Rural Development, Fisheries and Food regulates the importation, interstate movement, and environmental release of transgenic plants under the Federal Plant Health Law and the Law for the Production Certification and Commercialization of Seeds. The Intersecretariat Committee for Biosafety and Genetically Modified Organisms conducts risk assessments and provides biosafety recommendations in support of the Secretariat of Agriculture, Livestock, Rural Development, Fisheries and Food's regulatory decisions. The Secretariat of Agriculture, Livestock, Rural Development, Fisheries and Food also regulates the use of transgenic plants as livestock feed under the Federal Animal Health Law. Food products derived from transgenic plants are regulated by the Secretary of Health under the General Health Law [70].

17.4.1.3 Brazil

Brazil approved its biosafety law in 1995 (Law 8974/95), providing a regulatory framework in the areas of agriculture, health, and the environment. Although the biosafety regulatory framework was still being developed, the scientific work in research and development related to biotechnology had to follow three sets of regulations: biosafety laws, pesticide laws, and National Environmental Council resolutions [71]. In this same country, a group of scientists were reportedly developing the scientific guidelines for "transgenic organisms in integrated pest management and biological control," forming the basic principles for biosafety testing of transgenic crops to avoid negative environmental effects [72].

17.4.1.4 Latin American Countries and Caribbean Region

The Latin American and Caribbean Region is made up of 30 countries: Antigua and Barbuda, Argentina, Belize, Bolivia, Brazil, Chile, Colombia, Costa Rica, Dominica, Dominican Republic, Ecuador, El Salvador, Grenada, Guatemala, Guyana, Haiti, Honduras, Jamaica, Mexico, Nicaragua, Organization of the Eastern Caribbean States, Panama, Paraguay, Peru, Saint Kitts–Nevis, Saint Lucia, Saint Vincent and the Grenadines, Suriname, Trinidad and Tobago, Uruguay, and Venezuela. Out of these countries, there are several with legislation like Argentina (Resolution 124/91), Brazil (Law 8974/95), Cuba, Peru (Law 27104), Bolivia, Costa Rica, Colombia (Resolution 3492/98), Chile (Resolution 1027/93), Paraguay, México, and Uruguay. On the other hand, the countries without legislation are Dominican Republic, Ecuador, El Salvador, Guatemala, Honduras, Nicaragua, Panama, Venezuela, and the majority of the Caribbean countries. In all countries analyzed, the legislation covers research and development for greenhouse and field trials, excluding Colombia and Uruguay, in which the law does not consider laboratory research [66].

17.4.1.5 Russian Federation

In the Russian Federation, by 1996, a new law was adopted by the Duma on the state regulation of genetic engineering activity. The law is rather similar to the European legislation, described below. It basically establishes standards related with the safe conduct of genetic engineering. In general, the law induced the creation of a framework favorable to genetic engineering activity and biotechnology. New Russian legislation for approval and labeling of all food prepared from genetically modified (GM) organisms is expected to come into force soon. In April 1997, the Russian government established an Interdepartmental Commission on Problems in Genetic Engineering Activity, with the main task of ensuring the full implementation of Russia's genetic engineering law [73].

17.4.1.6 New Zealand

In New Zealand, the Hazardous Substances and New Organisms Act of 1996 regulates the research and release of any living organisms not present by evolution in New Zealand, which includes the GM ones. In order to import, field test, or release into the environment, the approval of the Environmental Risk Management Authority is needed. Because each organism is different, and therefore the risks related, the analysis is done on a case-by-case basis [74,75].

17.4.1.7 African Countries

In Africa, several countries are exerting efforts to develop a special legislation for controlling transgenic plants. In Ghana, efforts have been carried out to develop biosafety guidelines in genetic engineering and biotechnology by a national biosafety committee with financial support from the United Nations Environment Program–Global Environment Facility to basically regulate all works related to gene manipulation with recombinant DNA technology, including the development of transgenic plants, animals, and microorganisms, as well as the production of vaccines and the industrial production of GM organisms. Furthermore, the guidelines also aim to regulate the release into the environment for field trials and commercial purposes and the export and import of GM organisms [76]. Lesotho also developed a National Biosafety Policy by the year 2001. Other countries in Africa with regulations for the control of transgenic organisms are Cameroon (Law

2003/006), Kenya, Malawi, Mauritius (Law XLIV), Republic of Mozambique, South Africa, Burkina Faso, Central African Republic, Côte d'Ivoire, Namibia, Nigeria, Rwanda, Senegal, Seychelles, and Zambia [77].

17.4.1.8 China

China issued its first regulation in 1993, under the name of the Safety Administration Regulation on Genetic Engineering; however, labeling of GM products and possible trade barriers resulting from biotechnology concerns in countries that follow precautionary and preventive policies have begun to take an effect on the commercialization of GM organisms in China. Because of this, along with the United States of America's decision in 2001 to ban Chinese soy sauce imports produced with GM soybean, led the State Council to decree a new and general rule on Safety Administration of Agricultural GM organisms [78].

17.4.2 International Agreements

17.4.2.1 North American Plant Protection Organization

For example, in the United States, Canada, and México (member countries of the North American Plant Protection Organization), the importation into contained facilities and release into the environment of transgenic plants is regulated through the North American Plant Protection Organization Regional Standards for Phytosanitary Measures number 14, which was created in 2003 and is subject to review in 2008 [63]. The Regional Standard for Phytosanitary Measures is based on the laws existing in the countries mentioned above.

17.4.2.2 European Union Agricultural Policy

The European Union is an intergovernmental and supranational group of 25 democratic member states: Belgium, France, West Germany, Italy, Luxembourg, the Netherlands, Denmark, Ireland, United Kingdom, Greece, Portugal, Spain, unified Germany, Austria, Finland, Sweden, Cyprus, Czech Republic, Estonia, Hungary, Latvia, Lithuania, Malta, Poland, Slovakia, Slovenia, Bulgaria, and Romania (members by January 1, 2007). The union has a common agricultural policy, which also regulates transgenic plants and food derived from them. According to this legislation, fresh GM plants and the protein encoded by the genes are not considered substantially equivalent to conventional food. Therefore, there needs to be a permit for the cultivation or import (Directive 2001/18) as well as for the use of GM material as a novel food (Regulation 258/97). Also, manufacturers have to include data on the genetic modification, and the safety of the GM plant in the environment and for consumers. Furthermore, data on toxicity, nutritive value, and estimated consumption (Recommendation 618/97) need to be provided. A special authorization exists for food elaborated from GM plants, which does not contain transgenic DNA or protein (Food Regulation 258/97). In the European Union, food containing 1% and higher levels of ingredients prepared with GM material must be labeled (Regulation 49/2000). Since April 2000, all food additives prepared from GM material must be labeled under Regulation 50/2000 [79].

17.4.2.3 Southern Africa Regional Biosafety Programme

The Southern Africa Regional Biosafety Programme is a group of seven African Countries: Malawi, Mauritius, Mozambique, Namibia, South Africa, Zambia, and Zimbabwe, with the goal of technical training in biosafety regulatory implementation [77].

17.4.2.4 *Cartagena Protocol on Biosafety*

The Cartagena Protocol on Biosafety was adopted on January 29, 2000 and was in force by September 11, 2003. As of October 19, 2006, 135 parties (listed below) had already ratified the protocol. Latin America and the Caribbean: Antigua and Barbuda, Bahamas, Barbados, Belize, Bolivia, Brazil, Colombia, Cuba, Dominica, Dominican Republic, Ecuador, El Salvador, Grenada, Guatemala, Mexico, Nicaragua, Panama, Paraguay, Peru, Saint Kitts and Nevis, Saint Lucia, Saint Vincent and the Grenadines, Trinidad and Tobago and Venezuela. Africa: Algeria, Benin, Botswana, Burkina Faso, Cameroon, Cape Verde, Congo, Democratic Republic of the Congo, Djibouti, Egypt, Eritrea, Ethiopia, Gambia, Ghana, Kenya, Lesotho, Liberia, Libyan Arab Jamahiriya, Madagascar, Mali, Mauritania, Mauritius, Mozambique, Namibia, Niger, Nigeria, Rwanda, Senegal, Seychelles, South Africa, Sudan, Swaziland, Togo, Tunisia, Uganda, United Republic of Tanzania, Zambia, and Zimbabwe.

Asia and Pacific: Bangladesh, Bhutan, Cambodia, China, Cyprus, Democratic People's Republic of Korea, Fiji, India, Indonesia, Iran (Islamic Republic of), Japan, Jordan, Kiribati, Kyrgyzstan, Lao People's Democratic Republic, Malaysia, Maldives, Marshall Islands, Mongolia, Nauru, Niue, Oman, Palau, Papua New Guinea, Philippines, Samoa, Solomon Islands, Sri Lanka, Syrian Arab Republic, Tajikistan, Thailand, Tonga, Vietnam, and Yemen. From Central and Eastern Europe: Albania, Armenia, Azerbaijan, Belarus, Bulgaria, Croatia, Czech Republic, Estonia, Hungary, Latvia, Lithuania, Poland, Republic of Moldova, Romania, Serbia, Slovakia, Slovenia, The Former Yugoslav Republic of Macedonia, and Ukraine. Western Europe and other groups: Austria, Belgium, Denmark, European Community, Finland, France, Germany, Greece, Ireland, Italy, Luxembourg, Netherlands, New Zealand, Norway, Portugal, Spain, Sweden, Switzerland, Turkey, and the United Kingdom of Great Britain and Northern Ireland.

The main objective of the protocol is to ensure a good level of protection during the transference, manipulation, and safe utilization of living organisms created with tools from modern biotechnology that can have adverse effects on conservation, sustainable utilization of biodiversity, and human health. It also seeks to regulate the transboundary movement of GM organisms for intentional introduction into the environment. It obliges exporters to notify importers in advance, in such a way that importing countries can have the opportunity to assess the environmental and human health risks associated with the particular genetic modified organism to take a decision based on the precautionary principles. An important article of the protocol states that it cannot be used to restrict a party from taking harder measures to protect the conservation and sustainable utilization of biodiversity as long as the decision is compatible with the protocol itself. In the case of an illegal transboundary movement, the affected party will be able to request to the party of origin the elimination of the living organism either by taking into the country of origin or by destroying it.

The protocol launched an information center on biotechnology safety to ease the exchange of information, scientific experience, technical expertise, environmental knowledge and juridical experience related with living modified organisms. Also, the biotechnology information center provides technical support during the execution of protocols to developing countries or transitioning economies as well as to centers of biodiversity or centers of origin. The biotechnology information center will also make information for the execution of the protocol accessible and it will facilitate, whenever possible, other international information exchange on biotechnology safety. The protocol includes annexes specifying the information required of living modified organisms to be used for food, feed, or industrial processing and the general principles for risk evaluation [80].

17.5 Risks and Concerns Related with Transgenic Plants

17.5.1 Environmental Risks Associated with the Release of Transgenic Plants

The analysis of risk associated with the release of transgenic plants has to take into account that even traditional agricultural practices have been altering the natural environment to some degree, including the presence of artificial chemicals like fungicides, herbicides, fertilizers, etc., increasing of salt concentrations, negative effects on biological beings of both plant species and pests, and changes in the microbial community, among others. Therefore, to assess the risk of growing in the field, GM organisms must take into account the factors mentioned because transgenic plants are being grown under the agricultural practices developed for normal plants. In this context, the utilization of genetic engineering in agriculture can either accelerate the damaging effects of agriculture or contribute to more sustainable agricultural practices, including the conservation of natural resources for future generations. Furthermore, it had been suggested that the effects of different transgenic plants must be evaluated case-by-case due to intrinsic differences in the transgenic organism, the specific characteristics of the environment in which it is growing, and the agricultural practices that will be used in the process. There are several risks thought to be associated with the release of GM organisms into the environment and these are described in the following section.

17.5.1.1 Gene Flow

One of the most important risks is the possibility of gene flow from the transgenic organism to the wild type. Although it had been mentioned that the possibility of gene flow depends on the interfertility of the two populations, overlapping in flowering phonologies, and density of interbreeding populations [81], there is enough evidence to conclude that gene flow can occur and it has occurred from at least 47 domesticated plants to their wild types [82].

In experiments carried out in greenhouses and open fields, it was found that gene flow can occur between wheat (*Triticum aestivum* L.) and jointed goatgrass (*Aegilops cylindrica* Host) with any of the two species acting as a pollen donor. This is another example of interspecific gene flow and is an important concern because jointed goatgrass has been a weed for wheat in the United States since the early 1900s [83]. A study to evaluate the crop to crop gene flow in *Sorghum bicolor* subsp. *bicolor* was carried out in South Africa. The study used a male-fertile sorghum field surrounded by male-sterile recipients. The results showed an average of outcrossing rate of 2.54% at 13 m, less than 1% at 26 m, and 0.06% at 158 m. In the same study, a mathematical model predicted a maximum gene flow distance of 200 to 700 m [84]. A transgenic barley (*Hordeum vulgare* L.) line carrying the neomycin phosphotransferase II (*npt*II) gene was used as a pollen donor and a cytoplasmically male-sterile barley line as a recipient to analyze the *npt*II flow at distances of 1, 2, 3, 6, 12, 25, 50, and 100 m from the donor plots. Analysis revealed that all seeds from plants located at 1 m were transgenic, whereas only 3% of plants located at 50 m showed that characteristic. This study suggested that adequate isolation distance should be carried out to cultivate transgenic barley [85].

However, it was found that in some studies, the protocols used to estimate gene flow could be overestimating this rate, and the use of normal plants instead of male-sterile recipients had been suggested [86]. Aside from the abovementioned studies, it had been found that wild relatives can fertilize transgenic plants, as found in an open-field experiment in which a male-sterile plant of *Brassica napus* was found to contain phenotypic characteristics of *Raphanus raphanistrum*, a wild radish [87]. The possibility of gene flow had been found even in the case of autogamous crop plants in which the percentage of outcrossing was low. This was shown in an experiment with Foxtail millet (*Setaria italica*). This is a species showing less than 2% outcrossing. Pollen dispersal from a fertile Foxtail millet into normal and male-sterile plants found that gene flow was possible up to 24 and 40 m, respectively [88]. In sunflower, it was found that pollinators could transfer crop pollen to wild plants as far as 1000 m away [89].

Grass species used for turf, forage, rangeland, and bioremediation often have wild relatives growing sympatric and therefore the risk of spreading transgenic DNA in these species is particularly high. An experiment was done with creeping bent grass (*Agrostis stolonifera*) transformed with the bar gene and growing in a nursery. The experiment included 250 nontransgenic bent grass planted around the nursery. Transgenic weeds were reportedly at a distance of 426 m from the plantation. Southern blot analysis confirmed the presence of the bar gene in the genome. Furthermore, it was found that pollen from creeping bent grass can fertilize *A. canina*, *A. capillaries*, *A. castellana*, *A. gigantea*, and *A. pallens*, demonstrating the possibility of interspecific gene flow [90]. It was reported that traditional maize landraces at Oaxaca, México were contaminated with transgenic DNA [91], although the results were severely questioned [92,93] and the article was finally retracted from the periodical in which it was published.

17.5.1.2 Consequences of Gene Flow

The consequences of gene flow to close relatives growing nearby are variable depending on the type of construct that the transgenic organism is expressing. In this context, a gene conferring resistance to a either biotic (insect or fungal attack) or abiotic (drought, heavy metals, or salt) stress will bring some advantages of the organisms over either other individuals of the same specie, other species lacking the gene, or both [94]. This situation will increase the percentage of individuals in the population with the transgene and at the same time, a decrease in biodiversity will take place. In this context, an experiment done to evaluate the performance in the field of *Helianthus annuus* expressing the Bt toxin Cry1Ac found that the transgenic showed lower damage from four native insects that attack the plant, namely, *Cochylis hospes* (Chchylidae), *Isophrictis similiella* (Gelechidae), *Plagiomimicus spumosum* (Noctuidae), and *Suleima helianthana* (Tortricidae). The results from the study suggested that other native lepidopterans can be affected as well, namely, *Homeosoma electellum* and *Ecusoma womonana*. In the study, it was concluded that the advantage conferred by the Bt toxin could lead to an increase of the percentage in the population of transgenic plants [95].

Another concern is the development of a resistant phenotype in the target organism, which can be for the control of insects, fungi, or weeds. In this case, the constant selective pressure could induce mutations in the organisms, which would eliminate the advantage conferred by the introduced gene. Indeed, it was reported that the diamondback moth (*Plutella xylostella*) evolved resistance to four *B. thuringiensis* toxins (Cry1Aa, Cry1Ab, Cry1Ac, and Cry1F) with one gene, suggesting that insects can develop resistance much faster than previously thought [96]. In the case of genes conferring resistance to herbicides, the situation is less complicated, considering that the advantage provided by the phenotypic trait of resistance to a particular herbicide will be important under conditions in which the herbicide is present and this will most likely happen within a commercial field.

However, since the first case reported in 1970, 258 weed species have evolved resistance to one or more of 18 herbicides available. Furthermore, one of the most recent cases is the resistance to glyphosate (which is the primary ingredient in Monsanto's popular herbicide, Roundup) in the United States and in four other countries [97].

Also related with this issue is the potential effect of the transgene product on nontarget organisms like soil microorganisms playing a fundamental role in crop residues' degradation and in biogeochemical cycles [98]. In this case, the transgene can alter the normal population of other species not directly interacting with the transgenic plant and even others that are distantly related. An example of this was reported when monarch larvae were fed *in vitro* with pollen from the transgenic corn event Bt11 expressing the *B. thuringiensis* toxic protein cry1Ab. This study found lower weights in larvae fed the transgenic pollen as compared with controls [99]. This is important, considering that the dispersal of recombinant DNA by pollen and its persistence in soil for at least a year had already been investigated [100]. Aside from this, the high-dose refuge strategy recommended to delay pest adaptation in transgenic crops expressing *B. thuringiensis* toxin was shown to be suboptimal within the context of sustainable agricultural development [101]. Therefore, it is imperative to develop better management strategies to avoid the effects of recombinant DNA in the biosphere.

17.5.1.3 Horizontal Gene Transfer

Horizontal gene transfer [102] is a phenomenon in which genetic exchange between evolutionarily distant organisms, such as plants and bacteria, occurs [103]. Analysis of the horizontal gene transfer between plants and bacteria present in soil was done by detection of spreading intact DNA from plants (pollen) as well as during growth or decay, by studies of DNA persistence of free DNA in soil, and potential of bacteria to take up the DNA present in soil. It was concluded that the probability of a successful event of this kind was very low [104]. However, it had been found that the techniques to monitor horizontal gene transfer are too insensitive to detect these phenomena in field trials, even if it occurred at high rates [105]. In agreement with this statement, an experiment was done to evaluate the effect on the microbial population of soil amended with transgenic and nontransgenic papaya. We recovered higher colony forming units of bacteria, actinomycetes, and fungi resistant to kanamycin in soils with transgenic papaya [106] as compared with nontransgenic papaya, suggesting the transference of the kanamycin-resistant genes to the bacteria, although it was not experimentally probed.

In agreement with these results, it was shown that it is possible to obtain kanamycin-resistant clones by treating *Acinetobacter* sp. BD413 carrying a plasmid containing an inactivated *npt*II gene, with DNA from either GM papaya or potato. However, no transformation event was observed when *Acinetobacter* was not carrying the plasmid [107]. Also, the *in vitro* transfer of *npt*II gene from the disrupted leaves of six species of donor plants into the soil bacterium *Acinetobacter* spp. BD413 was shown. However, this transference was not detected in the absence of a homologous *npt*II gene in the bacteria [108]. Furthermore, the transfer of DNA segments from the leaves and roots of transplastomic tobacco into naturally transformable cells of *Acinetobacter* sp. was investigated. These experiments showed that plant DNA integration into bacterial genome could probably occur through hotspots of illegitimate recombination [109]. However, in disagreement with this experiment, the natural transformation of *Pseudomonas stutzeri* soil bacteria by transgenic plant DNA was shown to be strongly dependent on the presence of a homologous sequence in the bacterial genome. Also, the results of the experiments on *Acinetobacter* sp. transformation described previously supports the need for the presence of a homologous sequence in the bacteria for the event to take place. It suggests a low probability of nonhomologous DNA fragment from being integrated into the bacterial genome by illegitimate recombination events [110]. Therefore, this suggests that it is highly improbable that gene transfer occurs with uncommon genes. In agreement with this, an experiment done with extracts of transgenic tobacco plants did not alter haploidization, mitotic crossing over, mutation rate, or chromosomal alteration in soil bacteria [111].

The possibility of transference of DNA from plant tissue into bacteria depends on the persistence of the molecules in the soil. In this regard, some experiments found that the recombinant DNA can be spread intro nearby areas through pollen and actively persist in the soil for years [100]. The spread of DNA, including the *npt*II gene, by roots and pollen into the rhizosphere of the transgenic and nontransgenic plants growing nearby was investigated. Furthermore, it was shown that the recombinant DNA persisted for up to 4 years in soil at 4°C. Besides, it was shown that this DNA can transform *Acinetobacter* sp. strain BD413 by homologous recombination [112]. In agreement with these results, an experiment was done to analyze the presence of recombinant DNA in soil from fields with sugar beet transgenic plants. The recombinant DNA was found in soil at a distance of 50 m from pollen-producing transgenic plants. Recombinant DNA was found in the form of extracellular molecules in the soil close to the transgenic plants' roots. It was concluded that recombinant DNA is deposited in the soil during the growth of transgenic sugar beets and that a major mechanism of recombinant DNA spread in the environment is pollen. This mechanism of dispersal allows the DNA to persist in the field for at least a year [100]. As can be seen, the survival of the recombinant DNA allows for a transformation event to take place; however, the presence of a homologous sequence in the bacterial genome is still needed for this phenomenon to take place.

17.5.1.4 Unpredictable Effects

Using modern genetic engineering tools, it is possible to exchange genes between different species. Although this seems to be a common phenomenon in bacteria [113,114], this is not for the case for most eukaryotic organisms, and shows another risk to take into account. By breaking the natural barriers, which avoid indiscriminate gene exchange among species, genetic engineering is altering the natural gene pool available for a particular species [115]. Because the reasons for the presence of natural barriers controlling genetic exchange among species are largely unknown, it will be hard to understand or predict the consequences of breaking such barriers. That is why it had been suggested that for three moral reasons (respect for the opinions of people, staying within species borders, and respect for nature), we should maintain the modification of plants intragenically [116].

Another risk comes from the fact that we are releasing into the environment species with a genetic background not developed through the natural selection phenomena. The introduction of constructs into a genome cannot be seen as an isolated incident, but rather an event that can bring large changes in the normal genome regulation [102]. For instance, the disruption of a coding region of a gene encoding for a transcription factor can completely eliminate the possibility of the transgenic organism from using a set of changes controlled by that transcription factor. Also, if the disruption happens to occur in the promoter of that gene, this can induce alterations in the normal regulation of the promoter region controlling the expression of the gene, either eliminating the expression under certain conditions, turning an otherwise controlled promoter into a constitutive one, or reducing the level of transcripts for that gene. On the other hand, the disruption could occur in a gene important for the response to a variety of biotic or abiotic stresses, which in turn will result in an individual being unable to adapt, resist, and survive particular stresses. These situations have been previously observed during the induction of random mutagenesis in plants.

17.5.2 Reviews of Environmental Risks Related with Transgenic Plants

A review sponsored by the Food and Agriculture Organization of the United Nations evaluating the effect produced by the use of transgenic plants in agriculture production concluded that there are no verifiable reports of transgenic organisms causing any significant environmental harm. Furthermore, some important environmental benefits are emerging from the tendency of farmers to use fewer pesticides with low level of toxicity. The decreased use of these chemicals produces less contamination in water and farm workers. It is expected that a more generalized use of techniques to eliminate the marker genes introduced during the transformation of the organism (usually proteins conferring resistance to antibiotics), as well as the use of methodologies to prevent gene flow, will render the transgenic organisms potentially less dangerous for biodiversity. Finally, it was stated that "science cannot declare any technology completely risk-free." Genetically engineered crops can reduce some environmental risks associated with conventional agriculture, but will also introduce new challenges that must be addressed, and society will have to decide when and where genetic engineering is safe enough [117]. Indeed, several authors have suggested the importance of public education, familiarization, and participation to enable them to understand and accept the technology [63,102,118,119].

In agreement with the conclusions mentioned previously, a bibliographical review of the effects of transgenic plants on nontarget species, invasiveness, potentiality of gene escape by pollen or horizontal gene transfer, and adverse effects on soil biota concluded that the utilization of transgenic insecticide-tolerant and herbicide-tolerant plants do not show negative effects on the environment, rather they can improve both the environment and human health by reducing the utilization of chemical insecticides and herbicides [120]. Moreover, scientists do not have full knowledge of the risks and benefits of the utilization of Bt plants; however, the expectation is that the risks would be lower than that of current or alternative technologies and that the benefits would be greater, which seems to be the case [121]. Aside from the abovementioned reasons, transgenic Bt cotton plants were first commercialized in 1996, and eight years later, Bt cotton plants have been grown in an area reaching more than 80 million ha worldwide. However, no resistance to Bt proteins have yet been found, suggesting that resistance management strategies have been effective [122].

17.6 Improving the Environment by Using Transgenic Plants

From the above, it can be clearly seen that there is a lot of controversy about the risks of releasing transgenic plants into the environment. However, transgenic plants can be used to have positive effects on the environment through bioremediation. Bioremediation has recently attracted attention as an alternative technology for the elimination of pollutants in soil and water. This is an effective, environmentally nondestructive, and cheap method based on the use of green plants to remove, contain, or transform toxic chemicals into harmless ones [123,124]. This technique has been tested, with some success, to remove polynitrated aromatic compounds, which are components of explosives [125–127], toxic heavy metals [128,129], and arsenic [130]. However, the same risks to the environment as discussed for other transgenic plants also apply for this approach, and it has been suggested that the importance of avoiding gene flow from transgenic plants to be used as an environmental cleaner [131].

17.7 Risks to Human Health Associated with Transgenic Plants

Evaluating the safety of food derived from transgenic organisms has proved to be very difficult, and as a criteria, besides the "classic" toxicological safety evaluation, a concept known as substantial equivalence has been suggested [64], which basically states that "if a new food or food component is found to be substantially equivalent to an existing food or food component, it can be treated in the same manner with respect to safety. No additional safety concerns would be expected. Where substantial equivalence is more difficult to establish because the food or food component is either less well-known or totally new, then the identified differences, or the new characteristics, should be the focus of further safety considerations" (Organisation for Economic Cooperation and Development, 1993:13). Many objections concerning the scientific validity for risk assessment of the substantial equivalence concept had been mentioned aside from the fact that its application has several practical difficulties, such as the availability of isogenic or at least nearly isogenic lines to compare the modified food, few methodologies to detect unintended effects resulting from the genetic modification, and limited information about the natural variations in the crop components due to changes in the environment [132], differences by variety [133], and effects due to field management [134]. Because of this, the need to take into account an environmental assessment, aside from food analysis, to have what is called an integrated safety assessment, has been suggested [135]. Furthermore, substantial equivalence suggests that it is possible to qualify the food from a substantial basis only, without taking into account the legitimate factors involved, namely, substance, quality, and ethics [136,137].

The risks to human health associated with the consumption of food derived from transgenic plants includes the presence of allergenic proteins, toxic compounds, horizontal gene transfer associated with antibiotic resistance genes (which are used as selectable markers during transformation) [138,139], the pollination of plants intended for human consumption by plants transformed to create industrial biochemicals by gene flow [82], and the consumption of chemicals to which the transgenic plant is tolerant and used for weed control in the field, such as the presence of phosphinothricin in food derived from phosphinothricin-tolerant plants [140].

17.7.1 Horizontal Gene Transfer Involving Transgenic Food

The horizontal gene transfer of an antibiotic-resistant gene, from food prepared from a transgenic plant to a bacteria present in the human gastrointestinal tract, will confer resistance to the bacteria from that particular antibiotic [141]. This will give to the bacteria an antibiotic-resistant phenotype [142,143]. Because bacteria can exchange genetic information, this gene can be passed on to other bacteria present, and during an infection, this particular situation will make it difficult to eliminate from the human body.

Several experiments were carried out to evaluate, and describe, this possibility. It was found that the survival of free DNA encoding the antibiotic resistance gene lasts less than 1 min in biologically active ovine saliva and ovine rumen fluid [144]. Also, an experiment was conducted to evaluate the transfer of the *npt*II gene from transgenic plants to *Acinetobacter* BD413 in the gut of *Manduca sexta*. However, neither the entire gene nor pieces of the gene were detected in the bacteria [145]. Other experiments done did not find evidence of DNA transference from GM food to bacteria [146–148]. Furthermore, a report from the Working Party of the British Society for Antimicrobial Chemotherapy concluded that the transfer of antibiotic resistance genes from GM plants to bacteria is remote, and the hazard from that event is, at worst, slim [149]. The transference of DNA from bacterial cells to mammalian cells has also been hypothesized, although such an event is considered very remote [150]. In strong disagreement with previous statements and findings, the transference of transgenic DNA from bacteria in the human gut has been demonstrated. Although this was an experiment involving people with ileostomies, it still suggests that the event can take place [151]. In agreement, an experiment was carried out with 36 laying hens and three groups of 94 broiler chickens fed diets containing either 60% conventional or 60% Bt176 transgenic corn expressing a Bt toxin conferring resistance against the European corn borer. Polychain reaction analysis of chicken tissues from muscle, liver, and spleen showed the presence of the *ivr* gene from chloroplast corn. In this study, it was concluded that the transference of DNA fragments into the body is a constant phenomenon. However, it was not possible to find the presence of the Bt toxin responsible for the resistance of the transgenic corn to the attack of the European corn borer [152]. Also, an experiment with piglets fed with either GM maize (event MON810) or conventional maize was carried out. After 35 days, an analysis of blood, spleen, liver, kidney, and muscle tissues was done to evaluate the presence of plant DNA. The presence of maize genes (Zein, Sh-2) as well as fragments of the *B. thuringiensis* toxin, Cry1Ab, was recorded in the blood, liver, spleen, and kidney, except in the muscle [153].

Analysis of the CaMV35S promoter characteristic, which is a promoter present in almost all transgenic plants released, had suggested a high possibility of horizontal gene transfer. This is because of the presence of several recombination hotspots in this promoter, similar to the recombination hotspots present in other organisms—elements in common with promoters of both plants and animals—its ability to recombine with infecting viruses and the particular recombination mechanism of this piece of double-stranded DNA break repair, which does not need the presence of DNA sequence homologies [154]. Concerns about horizontal gene transfer led to the development of techniques to eliminate the marker gene from the transgenic organism by manipulating the recombination level surrounding the locus of the construct [155,156].

Although the potential risks associated with the consumption of food derived from transgenic organisms is generally agreed upon, no scientific evidence based on seriously designed and conducted experiments published in peer-reviewed scientific journals showing negative effects on human health have been found in the literature research. In agreement with this, it has been mentioned that there is no reason to question the safety of food derived from transgenic crops [157]. Also, a review sponsored by the Food and Agriculture Organization of the United Nations, evaluating the effect produced by the use of transgenic plants in agricultural production, concluded that no verifiable toxic or nutritionally deleterious effect resulting from the consumption of food prepared from GM organisms had been discovered anywhere in the world [117]. However, because of multiple concerns about the potential risks to human health with regard to the consumption of food derived from GM organisms, it was recommended that the information about how the safety of the food supply is ensured must be available to the general public. Although experiments with humans are not available, at least one study done using experimental animals showed some nutritional effects due to the consumption of food derived from transgenic plants, suggesting that the possibility of negative effects on human health due to the consumption of food derived from transgenic plants is real.

Rats fed with potato GM expressing the bar gene did not show any effect related to genetic modifications and the results suggest that GM crops have no adverse effects on the multigeneration reproductive–developmental ability [158]. Also, rats were fed with either a potato GM containing viral genome sequences of the necrotic strain of potato virus Y or the parental line cv Irga as well as the cultivars Maryna and Ania. No effects were observed in the biological response of rats, concluding that the transgenic line was substantial and had nutritional equivalence to the normal cultivar [159].

Pigs were fed with the parental line and the corresponding transgenic expressing a gene from the *B. thuringiensis* to confer resistance against the European corn borer (*Ostrinia nubilalis*). The digestibility of crude protein, the amount of nitrogen-free extract, and the amount of metabolizable energy for pigs fed with the normal and transgenic lines were analyzed in the feces. No differences were found in the nutritional assessment and the conclusion of the study was that the two plants can be considered substantially equivalent [160]. Furthermore, no differences were found in the protein efficiency ratio determined in rats fed with conventional and transgenic maize Roundup Ready resistance to the herbicide glyphosate. From here, it was concluded that parental and transgenic maize lines could be regarded as equivalent in terms of nutritional qualities [161].

In contrast with the results showed by the experiments described previously, broiler chickens fed with diets including 250 or 300 gr/kg of transegenic peas (*Pisum sativum L.*) expressing the bean (*Phaseolus vulgaris*) α-amylase inhibitor showed a reduction in body weight and an increase in feed intake suggesting a lower conversion efficiency. It was demonstrated that this was due to a reduction in starch digestibility rather than effects on protein digestibility [162]. In one experiment, an effect on food was observed due to a genetic modification on the vegetable source of that food.

Based on the results mentioned, it will be advisable to take into account the final recommendation of a report prepared under the auspices of the Royal Society of London, the U.S. National Academy of Sciences, the Brazilian Academy of Sciences, the Chinese Academy of Sciences, the Indian National Science Academy, the Mexican Academy of Sciences, and the Third World Academy of Sciences: the regulatory systems in every country must monitor any potential adverse human health effects of transgenic plants taking into account the possibility of long-term adverse effects on human health [163].

17.8 Presence of Novel Proteins or New Metabolites in Transgenic Plants

The presence of new proteins with novel properties such as allergenicity, toxicity, or ability to induce responses similar to some hormones or other properties is real if we consider the protocols used to create the transgenic plants currently available in the market. Basically, all of the transgenic organisms released were created using techniques that forcibly introduced a piece of DNA into the plant genome. Because of the random nature of the phenomenon, the new piece of DNA can be introduced in the middle of a gene-coding region. If that is the case, it will have created a new chimeric protein containing a combination of the natural DNA sequence of the gene and the DNA sequence of the introduced construct. Furthermore, it is also known that more than one copy of the gene can be introduced, which means that there is a higher probability for the event mentioned to take place. Although multiple copies inserted into the genome can be identified by Mendelian segregation analysis, sometimes the inserted piece of DNA cannot be active and, in this case, it will be hard to locate it unless expensive and time-consuming protocols of recombinant DNA technology are used. An example of this situation is the case of the Roundup Ready Soybean Event 40-3-2 created by Monsanto Corporation. In this case, analysis of the transgenic soybean genome, including the creation of a genomic library, found a 72-bp piece from the gene CP4 EPSPS (5-enol-pyruvyl-shikimate-3-phosphate synthase isolated from *Agrobacterium* sp. strain CP4) besides the introduced construct which in turn was found to have an additional 250-bp segment of the CP4 EPSPS gene at the 3' end of the nopaline synthase transcriptional termination sequence. The study of the company on the transgenic line concluded that "no unexpected gene expression product (either mRNA or protein) or unexpected changes in other components of nutrition or safety importance are produced as a result of the presence of the 72 bp secondary insert nor the presence of the 250 bp segment of CP4 EPSPS DNA at the 3' end of the NOS termination sequence of the primary insert" [164]. In agreement with these results, no evidence of allergenicity was found when Soya Roundup Ready was tested. On the other hand, the same results were recorded for the transgenic maize MON810, Bt11, T25, and Bt176 [165], suggesting that the risks of new allergenic reaction due to the consumption of GM food are small, although the experiments done do not rule out the possibility for the case of other new transgenic organisms to soon develop or for long-term effects.

In some cases, the induction of the expression of a protein can produce allergenic reactions during the consumption of the food created from a transgenic plant because of the intrinsic characteristics of the protein, as was the case when a Brazilian nut protein was introduced into soybean [166]. Aside from the abovementioned case, if the gene which is inactivated by the construct insertion is a transcription factor, large changes in the genome regulation can be expected, which in turn could lead to many alterations in metabolite concentration and composition. Because of this fact, along with the concept of substantial equivalence, the use of metabolomic, proteomic, and transcriptomic techniques had been suggested to analyze alterations in transgenic plants [167] as a better approach compared with the analysis of the major nutrients such as ash, total solid, protein, lipid, crude fiber, dietary fiber, total carbohydrate, and minerals [160,168–172], or the analysis of specific components of the plant such as glycoalkaloids, vitamin C, total nitrogen, fatty acids, and trypsin inhibitor activity [173], or the proximate composition as well as essential amino acids such as lysine and tryptophan [161].

17.8.1 Identification of Novel Proteins or Metabolites in Transgenic Plants

The transcriptome of conventional wheat lines and those expressing additional genes encoding a high molecular weight of glutelin were analyzed and rather small differences were found in the transcriptome profile of untransformed and isogenic transgenic lines. In one study, it was concluded that transgenic plants can be considered substantially equivalent with the untransformed parental line [174]. The protein expression was analyzed using two-dimensional gel electrophoresis of two isogenic tomato plants, with the exception of the virus resistance characteristic introduced by genetic engineering. No qualitative or quantitative differences were found [175]. Comparison of the proteome among different potato varieties and between normal and GM potatoes found that natural variations between potato varieties were higher compared with variations observed between GM and normal potato varieties [176]. The protein profile of kiwifruit (*Actinidia deliciosa*) expressing the transcription factor OSH1 was compared with the parental line, and was found to have 11 proteins decreased in the transgenic line. From these results, the importance of analyzing substantial equivalence at the level of the proteins between transgenic and normal plants has been suggested when the introduction of a transcription factor is involved [177].

Using two-dimensional gel electrophoresis, the seed proteome of 12 transgenic plants of *Arabidopsis thaliana*, expressing three different genes and three different promoters, were compared with the isogenic lines as well as with 12 *Arabidopsis* ecotype lines. It was found that the differences induced by any of the genetic modifications tested fell within the range of natural variability among the 12 different ecotypes of *Arabidopsis*, suggesting that the introduced construct did not cause pleiotropic effects in the plant [178,179]. A metabolome analysis of three transgenic wheat lines expressing high molecular weight subunit genes and the corresponding parental lines growing in two different areas of the United Kingdom was analyzed. The results of the experiment showed that the differences observed between parental and transgenic lines are within the range observed due to the effects from the environment [132]. Also, GM potatoes were compared with conventional potato crops by using metabolomics and were found to have a similar composition [180]. The results of the experiments mentioned previously clearly showed the importance of knowing the effects that different factors could have in the composition of different crops, and suggest the need to know these effects before reaching any conclusion about the possible pleiotropic effects associated with the creation of transgenic plants.

17.9 Economic Effect of Transgenic Crops

17.9.1 Economic Benefits

Modern biotechnology seems to offer the potential for substantial yield gains, cost reductions, and better characteristics of agricultural crops. However, there is a large price associated with the development of these new technologies, which has led to the concentration of intellectual property rights for GM varieties in large multinational biotechnology companies. This situation makes it difficult for poor countries to develop transgenic crops from isogenic lines adapted to local environments and weather conditions. Specifically, depending on transgenic crops developed by large companies raises concerns about the control of prices for these varieties, which can lead to a few benefits for producers. However, although there is a large price to pay for transgenic technology, it is expected that the reduction in costs associated with the application of chemicals in the field such as pesticides and insecticides for the transgenic crops [181], lower losses due to varieties that are virus resistant, and in general, lower costs associated with field management will compensate for the initial investment needed. In this work, we will analyze mainly the economic effects of insect-resistant transgenic cotton, which is the most widely adopted transgenic crop in both developed and underdeveloped countries. We will also analyze the economic effect of other transgenic crops that have been grown in small areas of the world.

Transgenic cotton expressing a gene from the bacterium *B. thuringiensis* (Bt), which renders it resistant to some insect attacks, was first grown in Australia, Mexico, and the United

States in 1996 and, after that, in six other countries: Argentina, China, Colombia, India, Indonesia, and South Africa. The global area planted with Bt and Bt herbicide–tolerant (Bt/Ht) cotton varieties increased from less than 1 million ha in 1996 to 4.6 million ha in 2002 [117]. Farmers from the United States adopted Bt cotton very quickly and it had been estimated that the average annual use of pesticides for cotton in the United States has been reduced by approximately 1000 MT [182]. This situation brought a higher net income of $105 million per year. On the other hand, the industry earned approximately $80 million from sales of Bt technology. As a consequence, consumer prices were reduced, with a gain of approximately $45 million per year for consumers in the United States and elsewhere. The average shares of the benefits were 46% to United States farmers, 35% to industry, and 19% to cotton consumers [117,183].

In the case of developing countries, studies had been carried out to evaluate the effect of the Bt transgenic cotton in Argentina, Mexico, China, India, and South Africa. In general, although benefits in specific amounts of money are not available as in the case of the United States, it was found that the net income was higher for growers producing Bt cotton as compared with non-Bt cotton. This income varied widely, from very little in the case of Argentina up to large benefits in the case of China. All cases also reported a large reduction in the use of pesticides to control insect attacks. The study concluded that there is a potential for economic benefits in developing countries adopting the transgenic technology. However, these benefits depend on the presence of a high level of national institutional capacity [185]. Although the economic benefits from growing Bt cotton in Argentina were low, in this country, the growth of herbicide-tolerant soybean had been a success. The aggregate welfare benefits from herbicide-tolerant soybean production was estimated to be $1.2 billion. Of this amount, 53% went to consumers, 34% to biotechnology companies, and 13% to farmers [184].

In India, work was done with the objective of exploring the effect of insect-resistant *B. thuringiensis* (Bt) cotton on costs and profit over two seasons after its commercial release in three regions of the Maharashtra State. Data was collected from 7793 cotton plots in 2002 and 1577 plots in 2003. It was found that although the cost of cotton seed was much higher for Bt cotton as compared with non-Bt cotton, the costs of insecticides were lower. Also, the yields and revenue were much higher from Bt cotton plots, accounting for 39% and 63% for 2002 and 2003, respectively. From here, the gross margins of Bt cotton plots were 43% and 73% higher than non-Bt cotton plots for 2002 and 2003, respectively. Altogether, these results suggest that growing Bt cotton provided substantial benefits for farmers in India. However, there still remains the question of whether or not these benefits will be sustainable [186]. An evaluation was carried out to know the economic benefits of growing herbicide-resistant transgenic rice in Uruguay. The stochastic simulation technique included effects on potential benefits after changes in technology, yield, operational costs, and adoption parameters. The results indicated a profit of $1.82 million for the producers and 0.55 million for the company that owns the transgenic crop, suggesting that there was a real economic benefit for the producers instead of the owner of the patent being the only one who would benefit [187].

The economic effect of introducing transgenic potatoes into the Netherlands is still uncertain basically because both the efficacy and the price of transgenic potatoes are still unknown. Also, it is evident that potato breeders will have to increase their investments in research and development, and whether this investment will yield profits is pretty much uncertain, mainly because the return on investment in genetic engineering research is also dependent on the system of intellectual property protection. Furthermore, it is evident that potato varieties with improved resistance, storage, or processing characteristics may lead to lower costs for pesticides, storage control, and industrial inputs, which in turn can lead to higher profits for potato farmers. However, the costs of the starting material will be much higher and the economic effect can only be stated in hypothetical terms [188].

A study was carried out to analyze the global economic effects on farm income of the use of transgenic crops in the period from 1996 to 2004. The study was done by compiling existing data from studies done in farms growing GM crops. It was found that recombinant DNA technology has had, in general, a positive effect on farm income as shown by the enhanced productivity and efficiency gains. In 2004, the direct global farm income benefit from GM crops was $4.8 billion. The largest gains were reported for the soybean sector, in which an additional $4.14 billion worth of income was generated. In the case of cotton, in 2004, cotton farm income levels increased by $1.62 billion, and since 1996, the sector has benefited from an additional $6.5 billion. Significant increases in farm incomes have also resulted from maize and canola. The combination of insect resistance and herbicide-resistant technology in maize has increased farm incomes by more than $2.5 billion since 1996. In the North American canola sector, it has produced an additional income of $713 million [189].

From a review of the different works done analyzing the economic benefits of transgenic technology, it is fairly clear that an economic benefit most likely will be present in countries adopting the transgenic technology. However, it is important to mention that all these studies were done in a relatively short period and it still remains to be seen whether the economic benefits observed during these studies will be sustainable for future generations.

17.10 Public Perception

The development of recombinant DNA technology has raised suspicions among consumers who are afraid that this technology can bring unexpected and unwanted side effects. It has been observed that among the major concerns of the public sector are the cross-pollination between transgenic crops and wild species, the use of antibiotic resistance marker genes, and the possible expression of proteins that are able to induce allergies [190]. This technology has also been related with the idea of "playing God." In general, it has been found that every new technology produces the same effect in consumers in the sense of being afraid of new things that are difficult to understand, and because of that, it is difficult to realize the consequences of its use.

Public attitudes to biotechnology are playing an important role in determining how widely genetic engineering techniques will be adopted in food and agriculture. Public opinion has been studied extensively in Europe and North America but not in other countries, and therefore, internationally comparable data are very limited. The different studies that have been conducted in many countries on the public acceptance of GM food products are difficult to compare because of differences in the adoption rates of GM crops and differences in national legislation related with the regulation of GM crop production and consumption. In most countries, we can say that public awareness and acceptance have been shaped primarily by mixed messages from action groups like Greenpeace and the industry. Because of this situation, in this part of the review, we will discuss the studies done in several parts of the world and the largest study in comparable public opinion conducted on agricultural biotechnology.

A doctoral work in Sweden with the objective of studying the consumer attitudes to and perceptions of organic and GM foods concluded that the attitude toward GM foods was generally negative and that these foods were considered not healthy as compared with organic foods. Also, consumers felt that tampering with nature can bring unexpected consequences in the ecological system [191]. A perception survey conducted in South African stakeholders involved in national public debates on the risks and benefits of GM crops found that academia, government, producers, consumer organizations, and industry strongly believe in the benefits of GM crops, although nongovernment organizations and churches do not. In fact, the latter emphasizes the potential risks. Because of that, South Africa has become a leader in promoting as well as in opposing to the use of recombinant DNA technology in agriculture [192].

A study done in three cities in the United States (Orono, Maine; Columbus, Ohio; and Phoenix, Arizona) concluded that there was a relatively low level of awareness and understanding of the issues surrounding GM foods. Also, although the research suggests that there is a consumer demand to have GM food labeled, in reality, this demand does not seem to exist, which suggests that there is no need for such a labeling program to be instituted [193]. A study done in 607 Tennessee residents showed that concerns about GM ornamental plants are similar to the ones related with GM food products. Moreover, one of the main concerns was the possibility of the invasiveness of the plant growing in the fields [194].

A survey was done in both the United States and Italy to identify the consumers' risk perceptions, knowledge, and awareness of GM foods. The survey included 3450 randomly selected households in the metropolitan areas of Denver, Chicago, Atlanta, Los Angeles, New Orleans, New York, and Houston. From this survey, 509 usable questionnaires were returned. In Italy, personal interviews of 500 Italian consumers were done outside large supermarkets near the metropolitan area of Piacenza, which is in northern Italy. In this case, 459 usable data sheets were collected. It was found that higher levels of perceived risk decreased the likelihood of purchase in both countries. These surveys also concluded that confidence in government agencies is important to Italian and US consumers' willingness to purchase GM foods. However, Italians were more sensitive to the potential risks that GM foods may pose to human health and the environment, relative to the US consumer. Furthermore, in general, Italians were less prone to eat GM foods as compared with US consumers [195]. This negative attitude toward GM organisms found in the case of Italian people is prevalent in Europe. Because of this, the organization of consensus conferences has been proposed as a good approach to stimulate public information and public debate in Europe, with adaptations to accommodate the differences in the cultural context of each country [196].

A survey of consumer awareness and acceptance of GM food products was conducted in Beijing in four supermarkets and included almost 1000 customers. The overall majority of respondents (71%) showed a little knowledge of GM food products. Also, the survey showed a relatively high percentage of people willing to consume GM food (40%). On the other hand, a substantial proportion of the consumers were undecided (51%) or unwilling (9%) to consume GM food. However, after the authors provided both positive and negative background information about GM food risks, the percentage of willingness decreased by 25%. The results clearly indicate that the attitudes of the Chinese consumers can change depending on the information about the GM crops and suggests the importance of giving accurate and updated information [197].

A recent survey of 257 American students and 319 Korean students from undergraduate courses was done with the goal of evaluating the consumer's attitudes toward GM foods. It was found that concern about health risks from GM foods differs by gender, race, urbanization, and frequency of exercising. In general, females are more concerned about health risks from GM food than males. Furthermore, Koreans are more concerned about health risks from GM foods than Americans. Suburban residents are less concerned about health risk from GM foods than those who live in urban and small town/rural areas, and students who exercised frequently for good health were more concerned about health risks from GM foods [198]. The results of this survey agree with the fact that the public acceptance of products developed from transgenic plants is biased by the perception of direct or indirect risks, and the potential benefits or credibility of government regulatory agencies that evaluate food and environmental safety. In North America, the acceptance of foods derived from transgenic organisms is holding steady, and knowledge about them remains generally low [199].

All of the surveys mentioned above were done in industrialized countries. However, two studies were done in underdeveloped countries, one in Mexico and the Philippines and the other in Costa Rica. The study carried out in Mexico and the Philippines had the objective of comparing attitudes toward the risks and benefits of using recombinant DNA technology in agriculture in underdeveloped countries. With this purpose, a nearly identical questionnaire was sent to stakeholders in the two countries. In general, the results indicate that the participants in Mexico and Philippines believed genetic engineering to be an important tool to address agricultural, nutritional, and environmental problems. Furthermore, they do not regard transgenic foods as risky for consumers. However, they are concerned about the potential effect of such transgenic crops on their countries' rich biological diversity and do not believe that national biosafety guidelines will be created and implemented properly [200]. In Costa Rica, a survey was done of 750 university students from the three public universities available, with the goal of determining the perceptions and knowledge about biotechnology and GM organisms. It was found that 88% showed a satisfactory level of knowledge and 79% of them have a favorable opinion about this modern biotechnology. It is interesting that students attending social disciplines showed a higher rejection of this technology as compared with students from life sciences and technology [201].

The most extensive international study of public perceptions of recombinant DNA technology was done with a survey of approximately 35,000 people in 34 countries in Africa, Asia, the Americas, Europe, and Oceania. Approximately 1000 people in each country were asked whether they agreed or disagreed with the following statement: "The benefits of using biotechnology to create genetically modified food crops that do not require chemical pesticides and herbicides are greater than the risk." People in the Americas, Asia, and Oceania agreed more than Africans or Europeans. Approximately 20% and 33% of people in Africa and Europe, respectively, were undecided in their responses, compared with approximately 13% of people in the Americas, Asia, and Oceania.

In general, people in higher-income countries tend to be more skeptical of the benefits of biotechnology and more concerned about the potential risks. However, there were some exceptions to this rule. In this context, in Asia, higher-income countries such as Japan and the Republic of Korea were more skeptical of the benefits and more concerned about the potential risks associated with biotechnology compared with people from lower-income countries such as the Philippines and Indonesia. On the other hand, in Latin America, people in higher-income countries such as Argentina and Chile were more skeptical than people from lower-income countries such as the Dominican Republic and Cuba. Therefore, it is clear that factors other than income levels are important in determining attitudes toward biotechnology. In general, people in developing countries were more likely to support the application of genetic engineering to reduce the use of chemical pesticides and herbicides. The countries with the highest percentage of agreement were those in which genetically engineered crops were being grown: Canada, Mexico, and the United States. In a second question, the study asked whether people would support or oppose the use of biotechnology to develop different applications. It was found that the public supported the use of this new technology depending on the specific biotechnology application. In this way, use of biotechnology to be applied in human health or environmental concerns was far more supported than applications with the goal of increase agricultural productivity. More than 70% supported the use of biotechnology to protect or repair the environment. Biotechnology applications related to animals received considerably less support than crop or bacterial applications, suggesting that people are less comfortable with animal biotechnology, perhaps because it involves more complex ethical issues [117].

Before describing the studies done about the risks related with the use of the new technology of DNA manipulation, it is important to define what risk analysis is. Risk analysis consists of three steps: risk assessment, risk management, and risk communication. Risk assessment evaluates and compares the scientific evidence regarding the risks associated with alternative activities, risk management develops strategies to prevent and control risks until acceptable limits, and risk communication involves an ongoing dialogue between regulators and the public about the risks and options to manage risk so that appropriate decisions can be made. Risk is often defined as "the probability of harm." A hazard, by contrast, is anything that might conceivably go wrong. A hazard does not in itself constitute a risk [117].

A study was done to analyze the risks related with recombinant DNA technology, and was evaluated based on the definition of risk as a property of an activity and risk as a property of an unwanted consequence. The study concluded that risk perception, with respect to gene technology, was related to the possibility of an unwanted side effect even without knowing the type of effect [202]. In this case, considering the short explanation about risk mentioned previously, it seems that risk communication is part of the risk analysis that has to be done so that people can get a better perception of risk. Furthermore, a study investigating the American and European public reaction toward GM organisms found that consumer and public response to risks can be volatile. This was found to be due, at least partially, to strong judgment from memorable events which are common when individuals do not have a good understanding of the risks associated with the technology. This study suggested that risk communication, by way of public dialogue and action, must be used to stimulate a deeper public evaluation about the benefits and costs of the use of GM organisms [203].

On the other hand, it had been mentioned that the only way society would benefit from the use of transgenic crops is by moving away from polarized positions, which had basically shaped the discussion in the past, to open a rational discussion of the risks and benefits of recombinant DNA technology [199]. In agreement with this statement, a study that sought to analyze the issues about the use of GM crops concluded that there is little risk, if any, and there are several potential benefits to the consumer in both developed and developing countries with the improvement in food quality and a reduction in the environmental effects of agriculture with the decreased use of less toxic agrochemicals [138]. Therefore, it seems that the perception of risk related with the use of recombinant DNA technology is not based on a careful analysis of the benefits against the risks involved in the use of this new technology.

It had been suggested that for the discussion about the risks of this technology, it is important to include both technical and social issues. Furthermore, scientists need to be heard in this dialogue because environmental activist groups like Greenpeace and the Environmental Defense Fund are being used increasingly as the sources of news instead of the opinion of university scientists. In addition to the responsibility of the scientific community, there is also a responsibility for society to educate itself in biotechnology. From the surveys mentioned previously and others, it is clear that the scientific illiteracy had led to a growing public distrust of science and technology [121].

17.11 Conclusion

The far-reaching consequences of the artificial introduction of portions of DNA into the plant genome are largely unknown. The few works that have been carried out with the help of omics technology showed that there are no large changes in the plant's metabolism. However, it is clear that more experimentation is needed with the techniques of genomics, proteomics, and metabolomics to more clearly assess the real effects and, as a consequence, the risks related with the creation of transgenic plants. Aside from that, other risks related with the creation of transgenic plants, such as effects on the environment and human health, need to be studied with the modern tools of science. Undoubtedly, the manipulation of plant genomes offers a great potential to solve problems that the human society is already facing. However, without more experimentation, the questions about the risks related with the creation of transgenic plants will stay unanswered and therefore the consequences of their use will be rather difficult to predict. Many laws have been created to control the growth and utilization of products derived from transgenic plants, and hopefully, those laws will stay in force while science elucidates the risks related with transgenic plants or develops technology for their safe containment. It is important to remember that planet earth is not disposable, it is the only home we humans have.

References

1. Old, R.W., and S.B. Primrose. 1994. *Principles of Gene Manipulation: An Introduction to Genetic Engineering*. 5th ed., 474. Cambridge, MA: Blackwell Scientific Publications.
2. Rodriguez, R.L., and R.C. Trait. 1983. *Recombinant DNA Techniques: An Introduction*. p. 236. Boston: Addison-Wesley Publishing Company.
3. DeCleene, M., and J. DeLey. 1976. The host range of crown gall. *The Botanical Review* 42:839–466.
4. Bundock, P., A. den Dulk-Ras, A. Beijersbergen, and P.J. Hooykass. 1995. Trans-kingdom T-DNA transfer from *Agrobacterium tumefaciens* to *Saccharomyces cerevisiae*. *The EMBO Journal* 14:3206–3214.
5. de Groot, M.J., P.J. Bundock, and A.G. Beijersbergen. 1998. *Agrobacterium tumefaciens*–mediated transformation of filamentous fungi. *Nature Biotechnology* 16:839–842.
6. Zhu, J., P.M. Oger, B. Schrammeijer, P.J.J. Hooykaas, S.K. Farrand, and S.C. Winans. 2000. The bases of crown gall tumorigenesis. *Journal of Bacteriology* 182:3885–3895.
7. Zupan, J., and P.C. Zambryski. 1997. The *Agrobacterium* DNA transfer complex. *Critical Reviews in Plant Science* 16:279–295.
8. Gelvin, S.B. 2003. *Agrobacterium*-mediated plant transformation: The biology behind the "gene-jockeying" tool. *Microbiology and Molecular Biology Reviews* 67:16–37.
9. Stachel, S.E., E. Messens, M. Van Montagu, and P. Zambryski. 1985. Identification of the signal molecules produced by wounded plant cells that activate T-DNA transfer in *Agrobacterium tumefaciens*. *Nature (London)* 318:624–629.

10. Stachel, S.E., E.W. Nester, and P.C. Zambryski. 1986. A plant cell factor induces *Agrobacterium tumefaciens* vir gene expression. *Proceedings of the National Academy of Sciences of the United States of America* 83:379–383.

11. Winans, S.C., R.A. Kerstetter, and E.W. Nester. 1988. Transcriptional regulation of the virA and virG genes of *Agrobacterium tumefaciens*. *Journal of Bacteriology* 170:4047–4054.

12. Ankenbauer, R.G., and E.W. Nester. 1990. Sugar-mediated induction of *Agrobacterium tumefaciens* virulence genes: Structural specificity and activities of monosaccharides. *Journal of Bacteriology* 172:6442–6446.

13. Shimoda, N., A. Toyoda-Yamamoto, J. Nagamine, S. Usami, M. Katayama, Y. Sakagami, and Y. Machida. 1990. Control of expression of *Agrobacterium* vir genes by synergistic actions of phenolic signal molecules and monosaccharides. *Proceedings of the National Academy of Sciences of the United States of America* 87:6684–6688.

14. Gaudin, V., T. Vrain, and L. Jouanin. 1994. Bacterial genes modifying hormonal balances in plants. *Plant Physiology and Biochemistry* 32:11–29.

15. Das, A. 1998. DNA transfer from *Agrobacterium* to plant cells in crown gall tumor disease. *Subcellular Biochemistry* 29:343–363.

16. Savka, M.A., Y. Dessaux, P. Oger, and S. Rossbach. 2002. Engineering bacterial competitiveness and persistence in the phytosphere. *Molecular Plant–Microbe Interactions* 15:866–874.

17. Zambryski, P.C., A. Depicker, K. Kruger, and H.M. Goodman. 1982. Tumor induction by *Agrobacterium tumefaciens*: Analysis of the boundaries of T-DNA. *Journal of Molecular and Applied Genetics* 1:361–370.

18. Gelvin, S.B. 1998. The introduction and expression of transgenes in plants. *Current Opinion in Biotechnology* 9:227–232.

19. Potrykus, I., R. Bilang, J. Futterer, C. Sautter, M. Schrott, and G. Spangenberg. 1998. *Genetic Engineering of Crop Plants*, 773. New York: Marcel Dekker.

20. Nair, G.R., Z. Liu, and A.N. Binns. 2003. Reexamining the role of the accessory plasmid pAtC58 in the virulence of *Agrobacterium tumefaciens* strain C581. *Plant Physiology* 133:989–999.

21. Sheng, J., and V. Citovsky. 1996. *Agrobacterium*–plant cell interaction: Have virulence proteins, will travel. *The Plant Cell* 8:1699–1710.

22. Tzfira, T., and V. Citovsky. 2000. From host recognition to T-DNA integration: The function of bacterial and plant genes in the *Agrobacterium*–plant cell interaction. *Molecular Plant Pathology* 1:201–212.

23. Winans, S.C., N.J. Mantis, C.Y. Chen, C.H. Chang, and D.C. Han. 1994. Host recognition by the VirA, VirG two-component regulatory proteins of *Agrobacterium tumefaciens*. *Research in Microbiology* 145:461–473.

24. Wang, K., S.E. Stachel, B. Timmerman, M. Van Montagu, and P.C. Zambryski. 1987. Site-specific nick occurs within the 25 bp transfer promoting border sequence following induction of vir gene expression in *Agrobacterium tumefaciens*. *Science* 235:587–591.

25. Filichkin, S.A., and S.B. Gelvin. 1993. Formation of a putative relaxation intermediate during T-DNA processing directed by the *Agrobacterium tumefaciens* VirD1, D2 endonuclease. *Molecular Microbiology* 8:915–926.

26. Scheiffele, P., W. Pansegrau, and E. Lanka. 1995. Initiation of *Agrobacterium tumefaciens* T-DNA processing: Purified proteins VirD1 and VirD2 catalyze site- and strand-specific cleavage of superhelical T-border DNA *in vitro*. *Journal of Biological Chemistry* 270:1269–1276.

27. Ward, E., and W. Barnes. 1988. VirD2 protein of *Agrobacterium tumefaciens* very tightly linked to the 5′ end of T-strand DNA. *Science* 242:927–930.

28. Young, C., and E.W. Nester. 1988. Association of the VirD2 protein with the 5′ end of T-strands in *Agrobacterium tumefaciens*. *Journal of Bacteriology* 170:3367–3374.

29. Howard, E.A., B.A. Winsor, G. De Vos, and P.C. Zambryski. 1989. Activation of the T-DNA transfer process in *Agrobacterium* results in the generation of a T-strand protein complex: Tight association of VirD2 with the 5′ ends of T-strands. *Proceedings of the National Academy of Sciences of the United States of America* 86:4017–4021.

30. Christie, P.J. 1997. *Agrobacterium tumefaciens* T-complex transport apparatus: A paradigm for a new family of multifunctional transporters in eubacteria. *Journal of Bacteriology* 179:3085–3094.

31. Christie, P.J., and J.P. Vogel. 2000. Bacterial type IV secretion: Conjugation systems adapted to deliver effector molecules to host cells. *Trends in Microbiology* 8:354–360.
32. Vergunst, A.C., B. Schrammeijer, A. den Dulk-Ras, C.M.T. de Vlaam, T.J. Regensburg-Tuink, and P.J.J. Hooykaas. 2000. VirB/D4-dependent protein translocation from *Agrobacterium* into plant cells. *Science* 290:979–982.
33. Citovsky, V., G. De Vos, and P.C. Zambryski. 1988. Single-stranded DNA binding protein encoded by the *virE* locus of *Agrobacterium tumefaciens*. *Science* 240:501–504.
34. Citovsky, V., M.L. Wong, and P.C. Zambryski. 1989. Cooperative interaction of *Agrobacterium* VirE2 protein with single stranded DNA: Implications for the T-DNA transfer process. *Proceedings of the National Academy of Sciences of the United States of America* 86:1193–1197.
35. Sen, P., G.J. Pazour, D. Anderson, and A. Das. 1989. Cooperative binding of *Agrobacterium tumefaciens* VirE2 protein to single-stranded DNA. *Journal of Bacteriology* 171:2573–2580.
36. Citovsky, V., J. Zupan, D. Warnick, and P.C. Zambryski. 1992. Nuclear localization of *Agrobacterium* VirE2 protein in plant cells. *Science* 256:1802–1805.
37. Gelvin, S.B. 1998. *Agrobacterium* VirE2 proteins can form a complex with T strands in the plant cytoplasm. *Journal of Bacteriology* 180:4300–4302.
38. Howard, E.A., V. Citovsky, and P.C. Zambryski. 1990. The T-complex of *Agrobacterium tumefaciens*. *UCLA Symposia on Molecular & Cellular Biology* 129:1–11.
39. Citovsky, V., B. Guralnick, M.N. Simon, and J.S. Wall. 1997. The molecular structure of *Agrobacterium* VirE2-single stranded DNA complexes involved in nuclear import. *Journal of Molecular Biology* 271:718–727.
40. Howard, E., J. Zupan, V. Citovsky, and P.C. Zambryski. 1992. The VirD2 protein of *A. tumefaciens* contains a C-terminal bipartite nuclear localization signal: Implications for nuclear uptake of DNA in plant cells. *Cell* 68:109–118.
41. Rossi, L., B. Hohn, and B. Tinland. 1993. The VirD2 protein of *Agrobacterium tumefaciens* carries nuclear localization signals important for transfer of T-DNA to plant. *Molecular and General Genetics* 239:345–353.
42. Citovsky, V., D. Warnick, and P.C. Zambryski. 1994. Nuclear import of *Agrobacterium* VirD2 and VirE2 proteins in maize and tobacco. *Proceedings of the National Academy of Sciences of the United States of America* 91:3210–3214.
43. Ziemienowicz, A., T. Merkle, F. Schoumacher, B. Hohn, and L. Rossi. 2001. Import of *Agrobacterium* T-DNA into plant nuclei: Two distinct functions of VirD2 and VirE2 proteins. *The Plant Cell* 13:369–384.
44. Gheysen, G., R. Villarroel, and M. Van Montagu. 1991. Illegitimate recombination in plants: A model for T-DNA integration. *Genes and Development* 5:287–297.
45. Mayerhofer, R., Z. Koncz-Kalman, C. Nawrath, G. Bakkeren, A. Crameri, K. Angelis et al. 1991. T-DNA integration: A mode of illegitimate recombination in plants. *The EMBO Journal* 10:697–704.
46. Tzfira, T., L.R. Frankman, M. Vahadilla, and V. Citovsky. 2003. Site-specific integration of *Agrobacterium tumefaciens* T-DNA via double-stranded intermediates. *Plant Physiology* 133:1011–1023.
47. Monteilhet, C., A. Perrin, A. Thierry, L. Colleaux, and B. Dujon. 1990. Purification and characterization of the in vitro activity of I-*Sce* I, a novel and highly specific endonuclease encoded by a group I intron. *Nucleic Acids Research* 18:1407–1413.
48. Ludwig, S.R., B. Bowen, L. Beach, and S.R. Wessler. 1990. A regulatory gene as novel visible marker for maize transformation. *Science* 247:449–450.
49. Christou, P., D.E. McCabe, and W.F. Swain. 1988. Stable transformation of soybean callus by DNA coated gold particles. *Plant Physiology* 87:671–674.
50. Sanford, J.C. 1989. Biolistic plant transformation. *Physiologia Plantarum* 79:206–209.
51. Vain, P., M.D. McMullen, and J.J. Finer. 1993. Osmotic treatment enhances particle bombardement–mediated transient and stable transformation of maize. *Plant Cell Reports* 12:84–88.
52. Songstad, D.D., D.A. Somers, R.J. Griesbach. 1995. Advances in alternative DNA delivery techniques. *Plant Cell Tissue and Organ Culture* 40:1–15.

53. Klein, T.M., E.D. Wolf, R. Wu, and J.C. Sanford. 1987. High-velocity microprojectile for deliver-ing nucleic acids into living cells. *Nature* 327:70–73.
54. Agius, F., I. Amaya, M.A. Botella, and V. Valpuesta. 2005. Functional analysis of homolo-gous and heterologous promoters in strawberry fruits using transient expression. *Journal of Experimental Botany* 56:37–46.
55. Yoder, J.I., and A. Goldsbrough. 1994. Transformation systems for generating marker-free transgenic plants. *Biotechnology* 12:263–267.
56. McElroy, D., and R. Wu. 1997. Rice actin gene and promoter. US Patent 5641876.
57. Miki, B., and S. McHugh. 2004. Selectable marker genes in transgenic plants: Applications, alternatives and biosafety. *Journal of Biotechnology* 107:193–232.
58. Birch, R.G. 1997. Plant transformation. Problems and strategies for practical application. *Annual Review of Plant Physiology and Plant Molecular Biology* 48:297–326.
59. Livingstone, D.M., and R.G. Birch. 1995. Plant regeneration and microprojectile-mediate gene transfer in embryonic leaflets of peaneaut (*Arachis hypogaca* L.). *Australian Journal of Plant Physiology* 22:585–591.
60. Ross, A.H., J.M. Manners, and R.G. Birch. 1995. Embryogenic callus production, plant regen-eration, and transient gene expression following particle bombardement, in the pasture grass *Cenchrus ciliarus* (*Gramineacea*). *Australian Journal of Botany* 43:193–199.
61. Karp, A. 1995. Somaclonal variation as a tool for crop improvement. *Euphytica* 85:295–305.
62. Turner, D., and L. Hartzell. 2004. The lack of clarity in the precautionary principle. *Environmental Values* 13:449–460.
63. Welsh, R., and D.E. Ervin. 2006. Precaution as an approach to technology development: The case of transgenic crops. *Science Technology and Human Values* 31:153–172.
64. Siekel, P. 2005. Safe production of genetically modified food. *Biologia* 60:157–160.
65. World Health Organization (2005). *Modern Food Biotechnology, Human Health and Development: An Evidence-Based Study.* WHO Library Cataloguing-in-Publication Data. Geneva, Switzerland: WHO.
66. Artunduaga-Salas, R. 2000. Biosafety Regulations Related with Transgenic Plants in Latin American and the Caribbean Region: The Andean Countries as a Model. 6th International Symposium on the Biosafety of Genetically Modified Organisms. Saskatoon, Saskatchewan, Canada.
67. North American Plant Protection Organization–Regional Standards for Phytosanitary Measures No. 14. 2000. Importation and Release into the Environment of Transgenic Plants in NAPPO Member Countries, 30. Ottawa, Ontario, Canada.
68. Belson, N.A. 2000. US regulation of agricultural biotechnology: An overview. *Journal of Agrobiotechnology Management & Economics* 3:268–280.
69. Sayre, P., and R.J. Seidler. 2005. Application of GMOs in the US: EPA research and regulatory considerations related to soil systems. *Plant and Soil* 275:77–91.
70. CIBIOGEM Comisión Intersecretarial de Bioseguridad de los Organismos Genéticamente Modificados. 2002. Regulatory framework of genetically modified organisms (Marco regulato-rio de organismos genéticamente modificados). pp. 39.
71. Fontes, E.M.G. 2003. Legal and regulatory concerns about transgenic plants in Brazil. *Journal of Invertebrate Pathology* 83:100–103.
72. Capalbo, D.M.F., A. Hilbeck, D. Andow, A. Snow, B.B. Bong, F.H. Wan et al. 2003. Brazil and the development of international scientific biosafety testing guidelines for transgenic crops. *Journal of Invertebrate Pathology* 83:104–106.
73. European Federation of Biotechnology. 1999. Biotechnology legislation in central and eastern Europe. Briefing Paper 9. *Task Group on Public Perceptions of Biotechnology*, 4.
74. Brent, P., D. Bittisnich, S. Brooke-Taylor, N. Galway, L. Graf, M. Healy, and L. Kelly. 2003. Regulation of genetically modified foods in Australia and New Zealand. *Food Control* 14:409–416.
75. Ministry for the Environment. 2004. *Genetic Modification. The New Zealand Approach.* http://www.gm.govt.nz.

76. United Nations Environment Programme–Global Environment Facility. 2003–2004. Guidelines for Risk Assessment of Genetically Modified Organisms in Ghana. Project on the Development of National Biosafety Frameworks. National Progress Report Submitted to the Third Series of Subregional Workshops.

77. Mayet, M. 2005. Biosafety in Africa: A complex web of interests. *African Centre for Biosafety,* 9.

78. Huang, J., and Q. Wang. 2002. Agricultural biotechnology development and policy in China. *Journal of Agrobiotechnology Management & Economics* 5:122–135.

79. Kleter, G.A., W.M. van der Krieken, E.J. Kok, D. Bosch, W. Jordi, and L.J.W.J. Gilissen. 2001. Regulation and exploitation of genetically modified crops. *Nature Biotechnology* 19:1105–1110.

80. Secretariat of the Convention on Biological Diversity. 2000. Cartagena Protocol on Biosafety to the Convention on Biological Diversity: Text and annexes, 19. Montreal: Secretariat of the Convention on Biological Diversity.

81. Ellstrand, N.C., H.C. Prentice, and J.F. Hancock. 1999. Gene flow and introgression from domesticated plants into their wild relatives. *Annual Review of Ecology and Systematics* 30:539–563.

82. Ellstrand, N.C. 2002. Gene flow from transgenic crop to wild relatives: What have we learned, what do we know, what to we need to know. In *Proceedings of the Ecological and Agronomic Consequences of Gene Flow from Transgenic Crops to Wild Relatives,* 39–46. Ohio State University.

83. Zemetra, R.S., C.A. Mallory-Smith, J. Hansen, Z. Wang, J. Snyder, A. Hang et al. 2002. The evolution of a biological risk program: Gene flow between wheat (*Triticum aestivum* L.) and jointed goatgrass (*Aegilops cylindrica* Host). In *Proceedings of the Ecological and Agronomic Consequences of Gene Flow from Transgenic Crops to Wild Relatives,* 169–178. Ohio State University.

84. Schmidt, M., and G. Bothma. 2006. Risk assessment for transgenic sorghum in Africa: Crop-to-crop gene flow in *Sorghum bicolor* (L.) Moench. *Crop Science* 46:790–798.

85. Ritala, A., A.M. Nuutila, R. Aikasalo, V. Kauppinen, and J. Tammisola. 2002. Measuring gene flow in the cultivation of transgenic barley. *Crop Science* 42:278–285.

86. Wang, J.M., X.S. Yang, Y. Li, and P.F. Elliott. 2006. Pollination competition effects on gene-flow estimation: Using regular vs. male-sterile bait plants. *Agronomy Journal* 98:1060–1064.

87. Darmency, H., E. Lefol, and A. Fleury. 1998. Spontaneous hybridizations between oilseed rape and wild radish. *Molecular Ecology* 7:1467–1473.

88. Wang, T.Y., H.B. Chen, X. Reboud, and H. Darmency. 1997. Pollen-mediated gene flow in an autogamous crop: Foxtail millet (*Setaria italica*). *Plant Breeding* 116:579–583.

89. Arias, D.M., and L.H. Rieseberg. 1994. Gene flow between cultivated and wild sunflower. *Theoretical and Applied Genetics* 89:655–660.

90. Wipff, J.K. 2002. Gene flow in turf and forage grasses (*Poaceae*). In *Proceedings of the Ecological and Agronomic Consequences of Gene Flow from Transgenic Crops to Wild Relatives,* 134–152. Ohio State University.

91. Quist, D., and I.H. Chapela. 2001. Transgenic DNA introgressed into traditional maize landraces in Oaxaca, México. *Nature* 414:541–543.

92. Metz, M., and J. Futterer. 2001. Suspect evidence of transgenic contamination. *Nature* 416:600–601.

93. Kaplinsky, N., D. Braun, D. Lisch, A. Hay, S. Hake, and M. Freeling. 2002. Maize transgene results in Mexico are artifacts. *Nature* 416:601–602.

94. Pilson, D., and H.R. Prendeville. 2004. Ecological effects of transgenic crops and the escape of transgenes into wild populations. *Annual Review of Ecology and Evolution and Systematics* 35:149–174.

95. Pilson, D., A. Snow, L. Rieserberg, and A. Helen. 2002. Fitness and population effects of gene flow from transgenic sunflower to wild *Helianthus annuus.* In *Proceedings of the Ecological and Agronomic Consequences of Gene Flow from Transgenic Crops to Wild Relatives,* 58–70. Ohio State University.

96. Tabashnik, B.T., Y.B. Liu, N. Finson, L. Masson, and D.G. Heckel. 1997. One gene in diamondback moth confers resistance to four *Bacillus thuringiensis* toxins. *Proceedings of the National Academy of Sciences of the United States of America* 94:1640–1644.

97. Holt, J. 2002. Prevalence and management of herbicide-resistant weeds. In *Proceedings of the Ecological and Agronomic Consequences of Gene Flow from Transgenic Crops to Wild Relatives*, 47–57. Ohio State University.

98. Giovannetti, M., C. Sbrana, and A. Turrini. 2005. The impact of genetically modified crops on soil microbial communities. *Rivista di Biologia-Biology Forum* 98:393–417.

99. Hellmich, R.L., B.D. Slegfried, M.K. Sears, D.E. Stanley-Horn, M.J. Daniels, H.R. Mattila et al. 2001. Monarch larvae sensitivity to *Bacillus thuringensis* purified proteins and pollen. *Proceedings of the National Academy of Sciences of the United States of America* 98:11925–11930.

100. Meier, P., and W. Wackernagel. 2003. Monitoring the spread of recombinant DNA from field plots with transgenic sugar beet plants by PCR and natural transformation of *Pseudomonas stutzeri*. *Transgenic Research* 12:293–304.

101. Vacher, C., D. Bourguet, F. Rousset, C. Chevillon, and M.E. Hochberg. 2003. Modelling the spatial configuration of refuges for a sustainable control of pests: A case study of Bt cotton. *Journal of Evolutionary Biology* 16:378–387.

102. Natarajan, S., V. Renczesova, M. Kukuckova, S. Stuchlik, and J. Turna. 2005. Genetically modified organisms from the point of view of horizontal gene transfer. *Biologia* 60:633–639.

103. Nielsen, K.M., A.M. Bones, K. Smalla, and J.D. van Elsas. 1998. Horizontal gene transfer from transgenic plants to terrestrial bacteria—a rare event? *FEMS Microbiology Reviews* 22:79–103.

104. de Vries, J., and W. Wackernagel. 2004. Microbial horizontal gene transfer and the DNA release from transgenic crop plants. *Plant and Soil* 266:91–104.

105. Heinemann, J.A., and T. Traavik. 2004. Problems in monitoring horizontal gene transfer in field trials of transgenic plants. *Nature Biotechnology* 22:1105–1109.

106. Wei, X.D., H.L. Zou, L.M. Chu, B. Liao, C.M. Ye, and C.Y. Lan. 2006. Field released transgenic papaya effect on soil microbial communities and enzyme activities. *Journal of Environmental Sciences—China* 18:734–740.

107. Iwaki, M., and Y. Arakawa. 2006. Transformation of *Acinetobacter* sp. BD413 with DNA from commercially available genetically modified potato and papaya. *Letters in Applied Microbiology* 43:215–221.

108. Tepfer, D., R. Garcia-Gonzales, H. Mansouri, M. Seruga, B. Message, F. Leach, and M.C. Perica. 2003. Homology-dependent DNA transfer from plants to a soil bacterium under laboratory conditions: Implications in evolution and horizontal gene transfer. *Transgenic Research* 12:425–437.

109. de Vries, J., T. Herzfeld, and W. Wackernagel. 2004. Transfer of plastid DNA from tobacco to the soil bacterium *Acinetobacter* sp. by natural transformation. *Molecular Microbiology* 53:323–334.

110. de Vries, J., P. Meier, and W. Wackernagel. 2001. The natural transformation of the soil bacteria *Pseudomonas stutzeri* and *Acinetobacter* sp. by transgenic plant DNA strictly depends on homologous sequences in the recipient cells. *FEMS Microbiology Letters* 195:211–215.

111. Azevedo, J.L., and W.L. Arauno. 2003. Genetically modified crops: Environmental and human health concerns. *Mutation Research* 544:223–233.

112. de Vries, J., M. Heine, K. Harms, and W. Wackernagel. 2003. Spread of recombinant DNA by roots and pollen of transgenic potato plants, identified by highly specific biomonitoring using natural transformation of an *Acinetobacter* sp. *Applied and Environmental Microbiology* 69:4455–4462.

113. Krzywinska, E., J. Krzywinski, and J.S. Schorey. 2004. Naturally occurring horizontal gene transfer and homologous recombination in *Mycobacterium*. *Microbiology* 150:1707–1712.

114. Call, D.R., M.S. Kang, J. Daniels, and T.E. Besser. 2006. Assessing genetic diversity in plasmids from *Escherichia coli* and *Salmonella enterica* using a mixed-plasmid microarray. *Journal of Applied Microbiology* 100:15–28.

115. Michaud, D. 2005. Environmental impact of transgenic crops. I. Transgene migration. *Phytoprotection* 86:93–105.

116. Myskja, B.K. 2006. The moral difference between intragenic and transgenic modification of plants. *Journal of Agricultural & Environmental Ethics* 19:225–238.

117. Raney, T. 2004. The State of Food and Agriculture 2003–2004. Agricultural Biotechnology: Meeting the needs of the poor? *FAO Agriculture Series* 145.

118. Myhr, A.I., and T. Traavik. 2003. Genetically modified (GM) crops: Precautionary science and conflicts of interests. *Journal of Agricultural & Environmental Ethics* 16:227–247.

119. Karlsson, M. 2003. Biosafety principles for GMOs in the context of sustainable development. *International Journal of Sustainable Development and World Ecology* 10:15–26.

120. Velkov, V.V., A.B. Medvinsky, M.S. Sokolov, and A.I. Marchenko. 2005. Will transgenic plants adversely affect the environment? *Journal of Biosciences* 30:515–548.

121. Shelton, A.M., J.-Z. Zhao, and R.T. Roush. 2002. Economic, ecological, food safety and social consequences of the deployment of Bt transgenic plants. *Annual Review of Entomology* 47:845–881.

122. Bates, S.L., J.Z. Zhao, R.T. Roush, and A.M. Shelton. 2005. Insect resistance management in GM crops: Past, present and future. *Nature Biotechnology* 23:57–62.

123. Tengerdy, R.P., and G. Szakacs. 1998. Perspectives in agrobiotechnology. *Journal of Biotechnology* 66:91–99.

124. Suresh, B., and G.A. Ravishankar. 2004. Phytoremediation—A novel and promising approach for environmental clean-up. *Critical Reviews in Biotechnology* 24:97–124.

125. French, C.E., S.J. Rosser, G.J. Davies, S. Nicklin, and N.C. Bruce. 1999. Biodegradation of explosives by transgenic plants expressing pentaerythritol tetranitrate reductase. *Nature Biotechnology* 17:491–494.

126. Rosser, S.J., C.E. French, and N.C. Bruce. 2001. Engineering plants for the phytodetoxification of explosives. *In Vitro Cellular & Developmental Biology-Plant Journal of the Tissue Culture Association* 37:330–333.

127. Ramos, J.L., M.M. Gonzalez-Perez, A. Caballero, and P. van Dillewijn. 2005. Bioremediation of polynitrated aromatic compounds: Plants and microbes put up a fight. *Current Opinion in Biotechnology* 16:275–281.

128. Kramer, U., and A.N. Chardonnens. 2001. The use of transgenic plants in the bioremediation of soils contaminated with trace elements. *Applied Microbiology and Biotechnology* 55:661–672.

129. Singh, O.V., S. Labana, G. Pandey, R. Budhiraja, and R.K. Jain. 2003. Phytoremediation: An overview of metallic ion decontamination from soil. *Applied Microbiology and Biotechnology* 61:405–412.

130. Alkorta, I., J. Hernandez-Allica, and C. Garbisu. 2004. Plants against the global epidemic of arsenic poisoning. *Environment International* 30:949–951.

131. Davison, J. 2005. Risk mitigation of genetically modified bacteria and plants designed for bioremediation. *Journal of Industrial Microbiology & Biotechnology* 32:639–650.

132. Baker, J.M., N.D. Hawkins, J.L. Ward, A. Lovegrove, J.A. Napier, P.R. Shewry, and M.H. Beale. 2006. A metabolomic study of substantial equivalence of field-grown genetically modified wheat. *Plant Biotechnology Journal* 4:381–392.

133. Reynolds, T.L., M.A. Nemeth, K.C. Glenn, W.P. Ridley, and T.D. Astwood. 2005. Natural variability of metabolites in maize grain: Differences due to genetic background. *Journal of Agricultural and Food Chemistry* 53:10061–10067.

134. Kuiper, H.A., G.A. Kleter, H.P.J.M. Noteborn, and E.J. Kok. 2002. Substantial equivalence—An appropriate paradigm for the safety assessment of genetically modified foods? *Proceedings of the IXth. International Congress of Toxicology*. Brisbane, Australia, 427–431.

135. Haslberger, A.G. 2006. Need for an "integrated safety assessment" of GMOs, linking food safety and environmental considerations. *Journal of Agricultural and Food Chemistry* 54:3173–3180.

136. Pouteau, S. 2000. Beyond substantial equivalence: Ethical equivalence. *Journal of Agricultural & Environmental Ethics* 13:273–291.

137. Pouteau, S. 2002. The food debate: Ethical versus substantial equivalence. *Journal of Agricultural & Environment Ethics* 15:291–303.

138. Phipps, R.H., and D.E. Beever. 2000. New technology: Issues relating to the use of genetically modified crops. *Journal of Animal and Feed Sciences* 9:543–561.

139. Goldstein, D.A., B. Tinland, L.A. Gilberston, J.M. Staub, G.A. Bannon, R.E. Goodman et al. 2005. Human safety and genetically modified plants: A review of antibiotic resistance markers and future transformation selection technologies. *Journal of Applied Microbiology* 99:7–23.

140. Metz, P.L.J., W.J. Stiekema, and J.P. Nap. 1998. A transgene-centered approach to the biosafety of transgenic phosphinothricin-tolerant plants. *Molecular Breeding* 4:335–341.

141. Jelenic, S. 2003. Controversy associated with the common component of most transgenic plants—Kanamycin resistance marker gene. *Food Technology and Biotechnology* 41:183–190.

142. Malik, V.S., and M.K. Saroha. 1999. Marker gene controversy in transgenic plants. *Journal of Plant Biochemistry and Biotechnology* 8:1–13.

143. Droge, M., A. Puhler, and W. Selbitschka. 1998. Horizontal gene transfer as a biosafety issue: A natural phenomenon of public concern. *Journal of Biotechnology* 64:75–90.

144. Duggan, P.S., P.A. Chambers, J. Heritage, and J.M. Forbes. 2000. Survival of free DNA encoding antibiotic resistance from transgenic maize and the transformation activity of DNA in ovine saliva, ovine rumen fluid and silage effluent. *FEMS Microbiology Letters* 191:71–77.

145. Deni, J., B. Message, M. Chioccioli, and D. Tepfer. 2005. Unsuccessful search for DNA transfer from transgenic plants to bacteria in the intestine of the tobacco horn worm *Manduca sexta*. *Transgenic Research* 14:207–215.

146. Mercer, D.K., K.P. Scott, W.A. Bruce-Johnson, L.A. Glover, and H.J. Flint. 1999. Fate of free DNA and transformation of oral bacterium *Streptococcus gordonii* DL1 plasmid DNA in human saliva. *Applied and Environmental Microbiology* 65:6–10.

147. Einspanier, R., A. Klotz, J. Kraft, K. Aulrich, R. Poser, F. Schwägele et al. 2001. The fate of forage plant DNA in farm animals: A collaborative case-study investigating cattle and chicken fed recombinant plant material. *European Food Research and Technology* 212:129–134.

148. Martin-Orue, S., A. O'Donnell, J. Arino, T. Netherwood, H. Gilbert, and J. Mathers. 2002. Degradation of transgenic DNA from genetically modified soya and maize in human intestinal simulations. *British Journal of Nutrition* 87:533–542.

149. Bennett, P.M., C.T. Livesey, D. Nathwani, D.S. Reeves, J.R. Saunders, and R. Wise. 2004. An assessment of the risks associated with the use of antibiotic resistance genes in genetically modified plants. *Journal of Antimicrobial Chemotherapy* 53:418–431.

150. Thomson, J.A. 2001. Horizontal transfer of DNA from GM crops to bacteria and to mammalian cells. *Journal of Food Science* 66:188–193.

151. Netherwood, T., R. Bowden, P. Harrison, A.G. O'Donnell, D.S. Parker, and H.J. Gilbert. 1999. Gene transfer in the gastrointestinal tract. *Applied and Environmental Microbiology* 65:5139–5141.

152. Aeschbacher, K., R. Messikommer, L. Meile, and C. Wenk. 2005. Bt176 corn in poultry nutrition: Physiological characteristics and fate of recombinant plant DNA in chickens. *Poultry Science* 84:385–394.

153. Mazza, R., M. Soave, M. Morlacchini, G. Piva, and A. Marocco. 2005. Assessing the transfer of genetically modified DNA from feed to animal tissues. *Transgenic Research* 14:775–784.

154. Ho, M.W., A. Ryan, and J. Cummins. 2000. Hazards of transgenic plants containing the cauliflower mosaic viral promoter. *Microbial Ecology in Health and Disease* 12:6–11.

155. Zubko, E., C. Scutt, and P. Meyer. 2000. Intrachromosomal recombination between attP regions as a tool to remove selectable marker genes from tobacco transgenes. *Nature Biotechnology* 18:442–445.

156. Zhang, Y., H. Li, O. Bo, Y. Lu, and Z. Ye. 2006. Chemical-induced autoexcision of selectable markers in elite tomato plants transformed with a gene conferring resistance to lepidopteran insects. *Biotechnology Letters* 28:1247–1253.

157. Chassy, B.N. 2002. Food safety evaluation of crops produced through biotechnology. *Journal of the American College of Nutrition* 21:166s–173s.

158. Rhee, G.S., D.H. Cho, Y.H. Won, J.H. Seok, S.S. Kim, S.J. Kwack, and R.D. Lee. 2005. Multigeneration reproductive and developmental toxicity study of bar gene inserted into genetically modified potato on rats. *Journal of Toxicology and Environmental Health* 68:2263–2276.

159. Zdunczyk, Z., S. Frejnagel, J. Fornal, M. Flis, M.C. Palacios, B. Flis, and W. Zagorski-Ostoja. 2005. Biological response of rat fed diets with high tuber content of conventionally bred and transgenic potato resistant to necrotic strain of potato virus (PVYN) Part I. Chemical composition of tubers and nutritional value of diets. *Food Control* 16:761–766.

160. Reuter, T., K. Aulrich, A. Berk, and G. Flachowsky. 2002. Investigations on genetically modified maize (Bt-maize) in pig nutrition: Chemical composition and nutritional evaluation. *Archives of Animal Nutrition* 56:23–31.

161. Chrenkova, M., A. Sommer, Z. Ceresnakova, S. Nitrayova, and M. Prostredna. 2002. Nutritional evaluation of genetically modified maize corn performed on rats. *Archives of Animal Nutrition* 56:229–235.

162. Li, X.H., T.J.V. Higgins, and W.L. Bryden. 2006. Biological response of broiler chickens fed peas (*Pisum sativum* L.) expressing the bean (*Phaseolus vulgaris* L.) alpha-amylase inhibitor transgene. *Journal of the Science of Food and Agriculture* 86:1900–1907.

163. Transgenic Plants and World Agriculture. 2000. Report prepared under the auspices of the Royal Society of London, the U.S. National Academy of Sciences, the Brazilian Academy of Sciences, the Chinese Academy of Sciences, the Indian National Science Academy, the Mexican Academy of Sciences and the Third World Academy of Sciences, 39. Washington, D.C.: National Academic Press.

164. Monsanto Company. 2000. Updated molecular characterization and safety assessment of Roundup Ready® soybean event 40-3-2. Confidential Report MSL-16712. 19.

165. Batista, R., B. Nunes, M. Carmo, C. Cardoso, H.S. José, A.B. de Almeida et al. 2005. Lack of detectable allergenicity of transgenic maize and soya samples. *Journal of Allergy and Clinical Immunology* 116:403–410.

166. Nordlee, J.A., S.L. Taylor, J.A. Townsend, and L.A. Thomas. 1996. Identification of a Brazil nut allergen in transgenic soybean. *New England Journal of Medicine* 334:688–692.

167. Cellini, F., A. Chesson, I. Colquhoun, A. Constable, H.V. Davies, K.H. Engel et al. 2004. Unintended effects and their detection in genetically modified crops. *Food and Chemical Toxicology* 42:1089–1125.

168. Aulrich, K., H. Bohme, R. Daenicke, I. Halle, and G. Flachowsky. 2001. Genetically modified feeds in animal nutrition 1st communication: *Bacillus thuringensis* (Bt) corn in poultry, pig and ruminant nutrition. *Archives of Animal Nutrition* 54:183–195.

169. El Sanhoty, R., A.A. Abd El-Rahman, and K.W. Bogl. 2004. Quality and safety evaluation of genetically modified potatoes Spunta with Cry V gene: Compositional analysis, determination of some toxins, antinutrients compounds and feeding study in rats. *Nahrung-Food* 48:13–18.

170. Lee, S.H., H.J. Park, S.M. Cho, H.K. Chun, D.H. Kim, T.H. Ryu, and M.C. Cho. 2004. Comparison of major nutrients and mineral contents in genetically modified herbicide-tolerant red pepper and its parental cultivars. *Food Science and Biotechnology* 13:830–833.

171. Han, J.H., Y.X. Yang, S.R. Chen, Z. Wang, X.L. Yang, G.D. Wang, and J.H. Men. 2005. Comparison of nutrient composition of parental rice and rice genetically modified with cowpea trypsin inhibitor in China. *Journal of Food Composition and Analysis* 18:297–302.

172. Jonnala, R.S., N.T. Dunford, and K. Chenault. 2005. Nutritional composition of genetically modified peanut varieties. *Journal of Food Science* 70:254–256.

173. Shepherd, L.V., J.W. McNicol, R. Razzo, M.A. Taylor, and H.V. Davies. 2006. Assessing the potential for unintended effects in genetically modified potatoes perturbed in metabolic and developmental processes.Targeted analysis of key nutrients and anti-nutrients. *Transgenic Research* 15:409–425.

174. Baudo, M.M., R. Lyons, S. Powers, G.M. Pastori, K.J. Edwards, M.J. Holdsworth, and P.R. Shewry. 2006. Transgenesis has less impact on the transcriptome of wheat grain than conventional breeding. *Plant Biotechnology Journal* 4:369–380.

175. Corpillo, D., G. Gardini, A.M. Vaira, M. Basso, S. Aime, G.R. Accotto, and M. Fasano. 2004. Proteomics as a tool to improve investigation of substantial equivalence in genetically modified organisms: The case of a virus-resistant tomato. *Proteomics* 4:193–200.

176. Lehesranta, S.J., H.V. Davies, L.V.T. Shepherd, N. Nunan, J.W. McNicol, S. Auriola et al. 2005. Comparison of tuber proteomes of potato varieties, landraces, and genetically modified lines. *Plant Physiology* 138:1690–1699.

177. Kita, M., C. Honda, S. Komatsu, S. Kusaba, Y. Fujii, and T. Moriguchi. 2006. Protein profile in the transgenic kiwifruit overexpressing a transcription factor gene, OSH1. *Biologia Plantarum* 50:759–762.

178. Ruebelt, M.C., M. Lipp, T.L. Reynolds, J.D. Astwood, K.H. Engel, and K.D. Jany. 2006. Application of two-dimensional gel electrophoresis to interrogate alterations in the proteome of genetically modified crops. 2. Assessing natural variability. *Journal of Agricultural and Food Chemistry* 54:2162–2168.

179. Ruebelt, M.C., M. Lipp, T.L. Reynolds, J.J. Schmuke, J.D. Astwood, D. Della Penna et al. 2006. Application of two-dimensional gel electrophoresis to interrogate alterations in the proteome of gentically modified crops. 3. Assessing unintended effects. *Journal of Agricultural and Food Chemistry* 54:2169–2177.

180. Catchpole, G.S., M. Beckmann, D.P. Enot, M. Mondhe, B. Zywicki, J. Taylor et al. 2005. Hierarchical metabolomics demonstrates substantial compositional similarity between genetically modified and conventional potato crops. *Proceedings of the National Academy of Sciences of the United States of America* 102:14458–14462.

181. Fanxi, M. 1999. Impact of transgenic crops in developing world agriculture. *Proceeding of 99 International Conference on Agricultural Engineering*. Beijing, China.

182. Gianessi, L.P., C.S. Silvers, S. Sankula, and J.E. Carpenter. 2002. *Plant Biotechnology: Current and Potential Impact for Improving Pest Management in U.S. Agriculture: An Analysis of 40 Case Studies*. National Center for Food and Agricultural Policy, Washington, D.C.

183. Falck-Zepeda, J.B., G. Traxler, and R.G. Nelson. 2000. Surplus distribution from the introduction of a biotechnology innovation. *American Journal of Agricultural Economics* 82:360–369.

184. Raney, T. 2006. Economic impact of transgenic crops in developing countries. *Current Opinion in Biotechnology* 17:1–5.

185. Trigo, E., and E. Cap. 2003. The impact of the introduction of transgenic crops in Argentinean agriculture. *Journal of Agrobiotechnology Management & Economics* 6:87–94.

186. Morse, S., R.M. Bennett, and Y. Ismael. 2005. Genetically modified insect resistance in cotton: Some farm level economic impacts in India. *Crop Protection* 24:433–440.

187. Hareau, G.G., B.F. Mills, and G.W. Norton. 2006. The potential benefits of herbicide-resistant transgenic rice in Uruguay: Lessons for small developing countries. *Food Policy* 31:162–179.

188. Bijman, W.J. 2002. The development and introduction of genetically modified potatoes in the Netherlands. Agricultural Economics Research Institute (LEI-DLO). The Netherlands.

189. Brookes, G., and P. Barfoot. 2005. GM crops: The global economic and environmental impact—the first nine years 1996–2004. *Journal of Agrobiotechnology Management & Economics* 8:187–196.

190. Halford, N.G., and P.R. Shewry. 2000. Genetically modified crops: Methodology, benefits, regulation and public concerns. *British Medical Bulletin* 56:62–73.

191. Magnusson, M. 2004. Consumer Perception of Organic and Genetically Modified Foods. Health and Environmental Considerations. Acta Universitatis Upsaliensis. Comprehensive Summaries of Uppsala Dissertations from the Faculty of Social Sciences, 137. Uppsala, Sweden.

192. Aerni, P. 2005. Stakeholder attitudes towards the risks and benefits of genetically modified crops in South Africa. *Environmental Science & Policy* 8:464–476.

193. Teisl, M.F., L. Halverson, K. O'Brien, B. Roe, N. Ross, and M. Vayda. 2002. Focus group reactions to genetically modified food labels. *Journal of Agrobiotechnology Management & Economics* 5:6–9.

194. Klingeman, W., B. Babbit, and C. Hall. 2006. Master gardener perception of genetically modified ornamental plants provides strategies for promotings research products through outreach and marketing. *HortScience* 41:1263–1268.

195. Harrison, R.W., S. Boccaletti, and L. House. 2004. Risk perceptions of urban Italian and United States consumers for genetically modified foods. *Journal of Agrobiotechnology Management & Economics* 7:195–201.

196. Ruibal-Mendieta, N.L., and F.A. Lints. 1998. Novel and transgenic food crops: Overview of scientific versus public perception. *Transgenic Research* 7:379–386.

197. Ho, P., and E.B. Vermeer. 2004. Food safety concerns and biotechnology: Consumers' attitudes to genetically modified products in urban China. *Journal of Agrobiotechnology Management & Economics* 7:158–175.

198. Finke, M.S., and H. Kim. 2003. Attitudes about genetically modified foods among Korean and American college students. *Journal of Agrobiotechnology Management & Economics* 6:191–197.

199. Byrne, P.F. 2006. Safety and public acceptance of transgenic products. *Crop Science* 46:113–117.
200. Aerni, P. 2002. Stakeholder attitudes toward the risks and benefits of agricultural biotechnology in developing countries: A comparison between Mexico and the Philippines. *Risk Analysis* 22:1123–1137.
201. Valdez, M., I. Rodriguez, and A. Sittenfeld. 2004. Perception about biotechnology in university students in Costa Rica. *Revista de Biologia Tropical* 52:745–756.
202. Sjoberg, L. 2002. Attitudes toward technology and risk: Going beyond what is immediately given. *Policy Sciences* 35:379–400.
203. Nelson, C.H. 2001. Risk perception, behavior, and consumer response to genetically modified organisms—Toward understanding American and European public reaction. *American Behavioral Scientist* 44:1371–1388.

18

Microalgal Omics and Their Applications

Shanmugam Hemaiswarya, Rathinam Raja, Ramanujam Ravikumar,
Annamalai Yogesh Kumar, and Isabel S. Carvalho

CONTENTS

18.1 Introduction

Microalgae constitute a large and diverse group of unicellular phototrophic and hetero-trophic organisms, which comprise the base of the food chain and are evolutionarily dis-tinct from other species. They have emerged as a promising group in the production of bioproducts and biofuel, as well as for the remediation of effluents. Indigenous popula-tions have used microalgae for centuries and the commercial application of microalgae has been extensively reviewed [1–4]. The efficiency of the microalgal production process depends on higher biomass, yield, productivity, and process robustness. These param-eters highly depend on the host microorganism. Natural screening, mutagenesis, selec-tion, bioprocess development, genetic engineering, and metabolic engineering strategies have been adopted to increase the metabolic capabilities of the host microorganisms [5]. Nevertheless, problems such as the accumulation of toxic intermediates or metabolic stress resulting in a decreased cellular fitness need to be solved. The lack of knowledge about the regulatory mechanisms of key enzymes and the complex relationships between genotype and phenotype are still barriers to the development of efficient cell factories. The over-expression, deletion, or introduction of heterologous genes in specific metabolic pathways does not always result in the desired phenotype. Recent remarkable innovations in plat-forms for omics-based research and application development have provided crucial solu-tions to these problems. A combinatorial approach using multiple omics platforms and the integration of their outcomes is now an effective strategy for clarifying the molecular systems that are integral to improving algal productivity.

18.2 Algal Genomics

The algal genome sequences revealed that individual organisms possess unique features in terms of primary sequence structures and gene compositions, making each useful for understanding basic physiological aspects and the evolution of photosynthetic eukaryotes. Algae have contributed to research into a variety of other phenomena in cell biology. Studies of eukaryote sexual reproduction and life cycles have often been investigated using algae. In addition, research into flagellar movement [6] or organelle multiplication, applicable to eukaryotic organisms in general, has been developed using unicellular algae with simple morphologies. One of the most important topics in algal research is photosynthesis. The best studied algal species is *Chlamydomonas reinhardtii*, an excellent organism for basic genetic studies. Being a unicellular microorganism, *Chlamydomonas* grow quickly with a doubling time of less than 10 h, and the cells behave homogeneously in terms of physiological and biochemical characteristics. Because *Chlamydomonas* is haploid and has a controlled sexual cycle with the possibility of tetrad analysis, it is an excellent genetic model. *Chlamydomonas* is the only known eukaryote in which the nuclear, chloroplast, and mitochondrial genomes can all be transformed. Furthermore, it has a specialized organelle called the eyespot, which enables it to sense light and respond accordingly. The unicellular green alga, *C. reinhardtii*, offers significant advantages for the genetic dissection of photosynthesis. Recent experiments with *Chlamydomonas* have substantially advanced our understanding of several aspects of photosynthesis including chloroplast biogenesis, structure–function relationships in photosynthetic complexes, human diseases, metabolism, nutrient acquisition, circadian rhythm, and environmental regulation [7–9].

In the case of algae, several organellar genomes have been sequenced, and for a limited number of species, the whole genome sequence has been generated or is currently being generated in one of the algal genome projects. *Cyanidioshyzon merolae* was the first alga to provide a complete genome sequence [10]. *C. merolae* has been very useful as a model organism for studies of the division and multiplication of cell organelles [11,12]. Table 18.1 presents a list of microalgae whose genomes have been sequenced. The complete genome of eukaryotic marine phytoplankton species (centric diatom) *Thalassiosira pseudonana* has been published [13], followed 2 years later by the green alga, *Ostreococcus tauri* [14] and *Ostreococcus lucimarinus* [15], the pennate diatom, *Phaeodactylum tricornutum* [16], and subsequently two other Prasinophyceae of the genus *Micromonas* [17]. Both *Ostreococcus* genomes encode a relatively large number of putative selenoproteins containing selenocysteine. The completed genome sequences have provided important clues about the

TABLE 18.1

Genomes of a Few Microalgal Species

Microalgae	Class	Genome Size (Mbp)	Reference
C. merolae	Rhodophyta	16.5	[10]
O. tauri	Chlorophyta	12.6	[14]
O. lucimarinus	Chlorophyta	12.6	[15]
C. reinhardtii	Chlorophyta	120	[85]
T. pseudonana	Heterokontophyta	31.3	[13]
P. tricornutum	Heterokontophyta	27.4	[16]
Aureococcus anophagefferens	Pelogophyte	56.7	[86]
Micromonas pusilla	Prasinophyceae	22	[17]

course of events during the evolution of algae [18]. Phytoplankton genome analysis also revealed a higher genetic divergence even between related species. Similar morphological species such as *O. tauri* and *O. lucimarinus* were very divergent, with an average amino acid identity of only 70% [15]. Analyzing algal genomes provides important information for research into not only the simple acquisition of nucleotide sequences, but also the physiology, environmental adaptation mechanisms, morphogenesis, and evolution of organisms.

18.3 Algal Transcriptomics

The study of the functional importance of all genes that are expressed is called transcriptomics. This study is required to achieve at least a few of the following goals, such as the identification of individual transcriptional units, the determination of the level of expression for each gene expressed, the expression of unique transcripts restricted to cell conditions or cell types, the precise assessment of transcript diversity at each transcriptional locus (i.e., splice isoforms, alternative transcription start sites, and polyadenylation sites), and the examination of the protein factors that control the transcriptional cassettes of a cell and their mechanisms of action. Currently, technologies exist that can comprehensively study all aspects of this problem. These technologies are separated into approaches that primarily assess levels of expression, to technologies that provide precise transcript boundaries and other transcript-focused characteristics, and to technologies that look at transcription factor–DNA interactions.

Transcriptomics techniques range from Northern blot hybridization and quantitative reverse transcriptase polymerase chain reaction approaches to microarrays, serial analysis of gene expression, and massively parallel signature sequencing [19]. Expressed sequence tags (ESTs) have become valuable tools for these technologies. ESTs are small pieces of the DNA sequence (usually 200 to 500 nucleotides long) that are generated by sequencing either one or both ends of an expressed gene. EST collections derived from cells grown under different conditions have proven to be a good tool for transcriptomics studies and genome annotation. EST studies [20,21], together with the first whole-genome sequences from diatoms, *T. pseudonana* [13] and *P. tricornutum* [16], have shown that less than 50% of diatom genes can be assigned a putative function using homology-based methods due to the lack of genomic information from well-studied taxonomically related organisms. Similar observations were also made in a pilot study of ESTs derived from the polar diatom, *Fragilariopsis cylindrus*, grown at low temperature [22]. The diatom EST database [21] enabled comparative studies of eukaryotic algal genomes and revealed some interesting differences in the genes involved in basic cell metabolism [23,24]. It also aided the study of key signaling and regulatory pathways [25], silica metabolism [23,26], nitrogen metabolism [27], and carbohydrate metabolism [28,21].

The study of messenger RNAs (mRNAs) expressed under different conditions can provide a systematic exploration of the molecular adaptations of a cell by differential gene expression and has proven to be a good tool for transcriptomics studies and genome annotation. A microarray has been developed for the green unicellular freshwater alga, *C. reinhardtii*, containing 10,000 oligonucleotide sequences—each sequence representing a unique gene and covering nearly the full genome of the alga (http://www.chlamy.org). Using this microarray, the transcription of light-regulated genes and the involvement of phototropin therein have been studied in *C. reinhardtii* [29], along with the specificity of a chloroplast RNA stability mutant [30].

C. reinhardtii was exposed to copper and a microarray analysis was carried out to investigate the differential gene transcription upon exposure. The mRNA levels of both the glutathione peroxidase gene and a probable glutathione-S transferase gene were found to be upregulated. Moreover, several genes in the thioredoxin system were differentially transcribed [31]. These are all genes involved in oxidative stress defense mechanisms, indicating that the microarray analysis was able to identify those genes playing a role in the metabolic processes that are involved in defense against copper toxicity. Microarrays were likewise used for *C. reinhardtii* to test the alga for its ability to acclimate to specific forms of oxidative stress [32], to understand the response of phosphorus deprivation [33] and survival during sulfur starvation [34], to profile the mRNA expression patterns of genes after exposure to the explosive 2,4,6-trinitrotoluene [35], to study anoxic mRNA expression of genes after anaerobic acclimation [36], and to survey the plastid and mitochondrial transcriptomes for changes in RNA profiles as a response to certain biotic and abiotic stimuli [30,37]. Microarray analyses were used to obtain global expression profiles of mRNA abundance in the green algae *C. reinhardtii* at different time points before the onset and during the course of sulfur-depleted hydrogen production.

Photobiological hydrogen production using microalgae is being developed into a promising clean fuel stream for the future. It has provided new insights into photosynthesis, sulfur acquisition strategies, and carbon metabolism–related gene expression during sulfur-induced hydrogen production. A general trend toward the repression of transcripts encoding photosynthetic genes was observed. During the three phases (anaerobic, hydrogen production, and termination phases), in addition to S starvation, cells also experience a lack of oxygen [38]. As a result, carbon metabolism shifts from respiration to fermentation and reduced hydrogen is released as hydrogen gas to circumvent nicotinamide adenine dinucleotide phosphate (NAD(P)$^+$) reduction [39]. Therefore, a large number of sulfur starvation and anoxia-responsive genes mask the process of hydrogen production.

The mRNA expression profiles of genes in *Euglena gracilis* were studied under different stress conditions [40], and they developed a microarray by constructing a nonnormalized cDNA library from *E. gracilis* and sequencing 1000 cDNAs. Microarrays were generated by spotting those ESTs whose sequences showed similarity to either Plantae or Protista genes. Similarly, a cDNA microarray was developed to screen differentially transcribed genes in the unicellular green algae, *Haematococcus pluvialis*, under astaxanthin-inducive culture conditions [41]. A cDNA library was constructed using *Nannochloropsis* algal cells grown in the exponential growth phase and EST analysis was carried out. A total of 1960 nonredundant sequences were found, a large proportion of which could be unique to this microalga. Several sequences with similarity to proteins involved in lipid transport and metabolism were obtained including long-chain acyl-CoA synthetases, long chain acyl-CoA transporter, fatty acid desaturase, and so forth [42]. They presumably participate in the biosynthesis of poly unsaturated fatty acids (PUFAs) in *Nannochloropsis oculata*. Considering that this microalgae is rich in eicosapentaenoic acid (EPA), the identification of these genes will be very important. For the unicellular red algae, *Cyanidioschyzon merolae,* a plastid DNA microarray was created to investigate the transcriptional regulation of the red algal plastid genome [43]. Microarray technology is more frequently used than the other techniques such as differential display and serial analysis of gene expression [44]. Long serial analysis of gene expression for gene discovery and transcriptome profiling was used in the marine coccolithophore, *Emiliana huxleyi,* whereas differential display was used to identify and study the transcriptional induction of genes by high light and cadmium stress in *C. reinhardtii,* respectively [29,45].

Microalgae in marine and brackish waters cause "harmful effects," considered from the human perspective, in that they threaten public health and cause economic damage to fisheries and tourism. Cyanobacteria also cause similar problems in freshwater. These episodes encompass a broad range of phenomena collectively referred to as harmful algal blooms. They include the discoloration of waters by mass occurrences of microalgae and the presence of toxin-producing species that may be harmful even in low cell concentrations. The sandwich hybridization assay has been used for monitoring the toxic microalgal species, *Heterosigma akashiwo* [46,47], *Cochlodinium polykrikoides* [48], and *Alexandrium* sp. [49,50] as well as for the development of multiprobe chips for the simultaneous detection of several toxic species [51]. Ahn et al. [52] developed a fiber-optic microarray for the simultaneous detection of multiple harmful algal bloom species. They describe a specific sandwich hybridization assay that combines fiber-optic microarrays with oligonucleotide probes to detect and enumerate different harmful algal bloom species. Along the same line, a low-density oligonucleotide array for the simultaneous detection of several harmful algal bloom species has been developed [53].

18.4 Algal Proteomics

During the last few years, proteomics has been established as a powerful tool for understanding various biological problems in several organisms. Proteins can be identified down to the femtomole and even the attomole range with modern mass spectrometry. The integration of multidimensional chromatograph and mass spectrometry, and the proteomic informatics, multidimensional shotgun proteomics has proven to be a powerful tool to separate and identify proteins from complex protein mixtures and be complementary to two-dimensional electrophoresis–based analysis [54]. Compared with traditional proteomic two-dimensional electrophoresis, shotgun proteomics serves as a powerful approach for high-throughput analysis of complex protein mixtures and is highly efficient, time saving, and labor saving [55]. Proteomic studies may provide a platform for the discovery of some unidentified ciliary/flagella genes and proteins [56]. With the development of proteomics, two-dimensional electrophoresis has been the primary method of separation and comparison for complex protein mixtures in *Dunaliella salina* [57]. A shotgun proteomics strategy was used to identify flagellar proteins after flagella were released and collected from *D. salina*. A total of 520 groups of proteins were identified under a stringent filter condition [58]. In addition to six kinds of known flagella proteins, the putative flagella proteins of *D. salina*, identified by one or more peptides, are abundant in signaling, cell division, metabolism, etc. A large number of different mutant loci have also been shown to affect the assembly and function of the flagellar apparatus in *C. reinhardtii*. Biochemical analyses of wild-type and mutant type flagella have demonstrated that the flagellar axoneme alone is composed of more than 200 proteins [6].

Proteomic analyses have been performed on the polypeptide components of the flagella and basal body [59–62]. Vener recently published a review focusing on environmentally modulated phosphorylation, flagellar proteins, and dynamics of proteins in photosynthetic membranes in plants and *C. reinhardtii* [63,64]. In *C. reinhardtii*, proteome analyses is generally used to understand the construction of the photosynthetic apparatus [65–69], light and circadian programs [70], the composition of mitochondrial protein components [71], and so on. Symbiosomes are specific intracellular membrane-bound vacuoles containing microalgae in a mutual Cnidaria (host)–dinoflagellate (symbiont) association. The symbiosome membrane is originally derived from host plasma membranes during phagocytosis of the symbiont; however, its molecular components and functions are not clear. A total of 17 proteins were identified. Based on their different subcellular origins and functional categories, it indicates that symbiosome membranes serve as the interface for interaction between host and symbiont by fulfilling several crucial cellular functions such as those of membrane receptors/cell recognition, cytoskeletal remodeling, adenosine-5′-triphosphate (ATP) synthesis/proton homeostasis, transporters, stress responses/chaperones, and antiapoptosis [72].

Haematococcus is the most widely used resource of astaxanthin for human consumption. Because it has considerable potential and promising applications in human health and nutrition, a molecular genetics approach to assess the genes involved in astaxanthin biosynthesis was performed [41,73]. The identification of differentially expressed proteins in response to stress is critical for understanding the mechanism(s) associated with astaxanthin biosynthesis in the microalgae, *H. lacustris*, to investigate changes in protein expression associated with astaxanthin accumulation triggered by sodium orthovanadate as a stress inducer, or to mimic stress. The proteomic and bioinformatic analyses showed global expression patterns of soluble cellular proteins revealing multiple, concerted enzymatic reactions along with a complex battery of proteins involved in fatty acid biosynthesis, central metabolism, signal transduction, and stress response that underwent dynamic changes in response to oxidative stress. For a better understanding of the full mechanism of astaxanthin biosynthesis under various stress conditions, additional studies including cDNA-chip array and genomic analysis are currently underway.

18.5 Algal Metabolomics

Metabolomics (in a pharmacological context) describes the quantitative measurement of the dynamic metabolic response of a living system to a toxic or physiological challenge [74,75]. Metabolomics shares two distinct advantages with proteomics in terms of the elucidating gene function. That is, the total complement of proteins or metabolites changes according to the physiological, developmental, or pathological state of a cell, tissue, organ, or organism, and unlike transcript (mRNA) analysis, proteins and metabolites are functional entities within the cell [76]. A third advantage that is distinct to metabolomics is that there are far fewer metabolites than genes or gene products to be studied. The majority of metabolomic studies use nuclear magnetic resonance (NMR)-based technologies to complement the information provided by measuring the transcriptomic and proteomics responses to contaminant exposure [74,77–81]. These H-NMR–based technologies are usually coupled to pattern recognition, expert systems, and related bioinformatic tools to interpret and classify complex NMR-generated data sets [74].

In algae, most metabolic analyses have thus far been focused on the quantification and identification of secondary metabolites with economic value in food science, pharmaceutical industry, and public health among others. Fatty acids, steroids, carotenoids, polysaccharides, lectins, polyketides, and algal toxins are among the algal products being studied (for an extensive coverage of species with high potential and their valuable secondary products; see Richmond [81]). Environmental metabolomic studies—in contrast with the application of metabolomics to characterize the metabolic response of an organism to environmental stimuli or stressors—have thus far only rarely been carried out in algae.

One study reports on the metabolite levels of *C. reinhardtii* under nutrient deprivation [82]. They describe a procedure for cell preparation and metabolite extraction. In chromatogram extracts of algae grown in standard conditions (sufficient nutrients), more than 800 metabolites could be detected with Ala, pyruvate, Glu, glycerolphosphate, and adenosine 5-monophosphate among the most prominent peaks. When cells were grown under nutrient-deficient conditions (depletion of nitrogen, phosphorus, sulfur, or iron), highly distinct metabolic phenotypes were observed. Metabolomic studies of the response of *C. reinhardtii* to sulfur depletion and anoxia, as well as to how such responses underpin hydrogen production, were studied. Metabolomics employs a nontargeted profiling approach and can potentially detect and quantify hundreds of metabolites, particularly the more abundant intermediates of primary metabolism. Nuclear magnetic resonance (NMR), gas chromatography–mass spectrometry (GC-MS), and thin layer chromatography (TLC) were used in parallel to obtain as much information as possible about key metabolites. Hydrogen production does not slow down due to depletion of energy reserves but rather due to loss of essential functions resulting from sulfur depletion or due to a build-up of the toxic fermentative products formate and ethanol [83].

For an integrated analysis of the molecular repertoire of *C. reinhardtii* under reference conditions, bioinformatics annotation methods combined with Two-dimensional gas chromatography coupled with mass spectrometric detection (GC × GC/MS)–based metabolomics and liquid chromatography–mass spectrometry (LC/MS)-based shotgun proteomics profiling technologies were applied to characterize abundant proteins and metabolites, and resulted in the detection of 1069 proteins and 159 metabolites. Among the proteins measured, 204 currently do not have EST sequence support; thus, a significant portion of the proteomics-detected proteins provide evidence for the validity of *in silico* gene models. Furthermore, the peptide data generated lends support to the validity of a number of proteins currently in the proposed model stage. By integrating genomic annotation information with experimentally identified metabolites and proteins, we constructed a draft metabolic network for *Chlamydomonas*. Computational metabolic modeling allowed the identification of missing enzymatic links [84].

18.6 Conclusion

Despite the growing number of completed microalgae genome sequences, only a few examples of genetic engineering of the metabolism for the production of by-products have been reported. Nevertheless, metabolic engineering in model strains of microalgae will likely provide important leads for proof-of-principle studies in the future. Furthermore, the complete genome sequences provide a valuable framework for the huge amounts of marine metagenome projects. Comparative genomics is a powerful tool to unravel previously unknown gene functions. Each algal genome project provides enough genome sequence data to allow comparative analysis of each genome. It is based on small genome repertoires, particularly for unicellular algae; algal genome information should accelerate studies on, for example, the establishment of cellular components, and it will allow us to elucidate cellular and molecular properties that they have in common with land plants or other eukaryotes. Principally omic projects should aim at different microalgae features such as evolution, adaptation, and divergence compared with other species, gene, protein and metabolite information, and their interaction. This would facilitate an understanding of the biology of microalgae in detail and the application of these concepts in the production of valuable products.

Acknowledgments

The authors wish to express their sincere gratitude to Dr. Velu Subramani, Senior Scientist, Refining Technology, The British Petroleum Company Ltd., Chicago, and to Prof. R. Manivasakan, Indian Institute of Technology, Madras, India, for their critical review of this chapter.

References

1. Cardozo, K.H.M., G. Thais, P.B. Marcelo, R.F. Vanessa, P.T. Angela, P.L. Norberto et al. 2007. Metabolites from algae with economical impact. *Comparative Biochemistry and Physiology Part C* 146:60–78.
2. Raja, R., S. Hemaiswarya, N. Ashok Kumar, S. Sridhar, and R.A. Rengasamy. 2008. Perspective on the biotechnological potential of microalgae. *Critical Reviews in Microbiology* 34:77–88.
3. Plaza, M., H. Miguel, C. Alejandro, and I. Elena. 2009. Innovative natural functional ingredients from microalgae. *Journal of Agriculture and Food Chemistry* 57:7159–7170.
4. Greenwell, H.C., L.M.L. Laurens, R.J. Shields, R.W. Lovitt, and K.J. Flynn. 2010. Placing microalgae on the biofuels priority list: A review of the technological challenges. *Journal of Royal Society Interface* 7:703–726.
5. Beer, L.L., E.S. Boyd, J.W. Peters, and M.C. Posewitz. 2009. Engineering algae for biohydrogen and biofuel production. *Current Opinion in Biotechnology* 20:264–271.
6. Dutcher, S.K. 1995. Flagellar assembly in two hundred and fifty easy-to-follow steps. *Trends Genetics* 11:398–404.
7. Grossman, A.R. 2005. Paths toward algal genomics. *Plant Physiology* 137:410–427.
8. Harris, E.H. 2001. *Chlamydomonas* as model organism. *Annual Review in Plant Physiology and Plant Molecular Biology* 52:363–406.
9. Mittag, M., and V. Wagner. 2003. The circadian clock of the unicellular eukaryotic model organism, *Chlamydomonas reinhardtii. Biological Chemistry* 384:689–695.
10. Matsuzaki, M., O. Misumi, I.T. Shin, S. Maruyama, M. Takahara, S.Y. Miyagishima et al. 2004. Genome sequence of the ultrasmall unicellular red alga, *Cyanidioschyzon merolae* 10D. *Nature* 428:653–657.
11. Kuroiwa, T. 1998. The primitive red algae, *Cyanidium caldarium* and *Cyanidioschyzon merolae* as model system. *Bioessays* 20:344–354.
12. Kuroiwa, T., K. Nishida, Y. Yoshida, T. Fujiwara, T. Mori, H. Kuroiwa et al. 2006. Structure, function and evolution of the mitochondrial division apparatus. *Biochimica et Biophysica Acta* 1763:510–521.
13. Armbrust, E.V., C.B. Berges, R.G. Beverley, M. Diego, H.P. Nicholas, S. Zhou et al. 2004. The genome of the diatom, *Thalassiosira pseudonana*: Ecology, evolution and metabolism. *Science* 306:79–86.
14. Derelle, E., C. Ferraz, S. Rombauts, P. Rouzé, A.Z. Worden, S. Robbens et al. 2006. Genome analysis of the smallest free-living eukaryote, *Ostreococcus tauri* unveils many unique features. *Proceedings of the National Academy of Sciences of the United States of America* 103:11647–11652.

15. Palenik, B., J. Grimwood, A. Aerts, P. Rouzé, A. Salamov, N. Putnam et al. 2007. The tiny eukaryote *Ostreococcus* provides genomic insights into the paradox of plankton speciation. *Proceedings of the National Academy of Sciences of the United States of America* 104:7705–7710.

16. Bowler, C., B. Chris, E.A. Andrew, H.B. Jonathan, G. Jane, J. Kamel et al. 2008. The *Phaeodactylum* genome reveals the evolutionary history of diatom genomes. *Nature* 456:239–244.

17. Worden, A.Z., J.H. Lee, T. Mock, P. Rouzé, M.P. Simmons, A.L. Aerts et al. 2009. Green evolution and dynamic adaptations revealed by genomes of the marine picoeukaryotes *Micromonas*. *Science* 324:268–272.

18. Parker, M.S., M.E. Thomas, and A. Virginia. 2008. Genomic insights into marine microalgae. *Annual Review of Genetics* 42:619–645.

19. Pariset, L., C. Giovanni, B. Silvia, R.S. Vincenzo, and V. Alessio. 2009. Microarrays and high-throughput transcriptomic analysis in species with incomplete availability of genomic sequences. *New Biotechnology, Biotechnology Annual Review* 25:272–279.

20. Scala, S., N. Carels, A. Falciatore, M.L. Chiusano, and C. Bowler. 2002. Genome properties of the diatom, *Phaeodactylum tricornutum*. *Plant Physiology* 129:993–1002.

21. Maheswari, U., A. Montsant, J. Goll, S. Krishnaswamy, K.R. Rajyashri, V.M. Patell et al. 2005. The diatom EST database. *Nucleic Acids Research* 33:D344–D347.

22. Mock, T., A. Krell, G. Glockner, U. Kolukisaoglu, and K. Valentin. 2006. Analysis of expressed sequence tags (ESTs) from the polar diatom, *Fragilariopsis cylindrus*. *Journal of Phycology* 42:78–85.

23. Montsant, A., K. Jabbari, U. Maheswari, and C. Bowler. 2005. Comparative genomics of the pennate diatom, *Phaeodactylum tricornutum*. *Plant Physiology* 137:500–513.

24. Herve, C., T. Tonon, J. Collen, E. Corre, and C. Boyen. 2006. NADPH oxidases in eukaryotes: Red algae provide new hints! *Current Genetics* 49:190–204.

25. Montsant, A., A.E. Allen, S. Coesel, A. De Martino, A. Falciatore, M. Mangogna et al. 2007. Identification and comparative genomic analysis of signaling and regulatory components in the diatom, *Thalassiosira pseudonana*. *Journal of Phycology* 43:585–604.

26. Lopez, P.J., J. Descles, A.E. Allen, and C. Bowler. 2005. Prospects in diatom research. *Current Opinion Biotechnology* 16:180–186.

27. Allen, A.E., A. Vardi, and C. Bowler. 2006. An ecological and evolutionary context for integrated nitrogen metabolism and related signaling pathways in marine diatoms. *Current Opinion in Plant Biology* 9:264–273.

28. Kroth, P.G., A. Chiovitti, A. Gruber, V. Martin-Jezequel, T. Mock, M.S. Parker et al. 2008. A model for carbohydrate metabolism in the diatom, *Phaeodactylum tricornutum* deduced from comparative whole genome analysis. *PLoS One* 3:e1426.

29. Im, C.S., and A.R. Grossman. 2001. Identification and regulation of high light-induced genes in *Chlamydomonas reinhardtii*. *Plant Journal* 30:301–313.

30. Erickson, B., D.B. Stern and D.C. Higgs. 2005. Microarray analysis confirms the specificity of a Chlamydomonas reinhardtii chloroplast RNA stability mutant. *Plant Physiology* 137:534–544.

31. Jamers, A., K. van der Van, L. Moens, J. Robbens, G. Potters, Y. Guisez et al. 2006. Effect of copper exposure on gene expression profiles in *Chlamydomonas reinhardtii* based on microarray analysis. *Aquatic Toxicology* 80:249–260.

32. Ledford, K., B.L. Chin, and K.K. Niyogi. 2007. Acclimation to singlet oxygen in *Chlamydomonas reinhardtii*. *Eukaryotic Cell* 6:919–930.

33. Moseley, J.L., C.W. Chang, and A.R. Grossman. 2006. Genome-based approaches to understanding phosphorus deprivation responses and PSR1 control in *Chlamydomonas reinhardtii*. *Eukaryotic Cell* 5:26–44.

34. Zhang, Z., J. Shrager, M. Jain, C.W. Chang, O. Vallon, and A.R. Grossman. 2004. Insights into the survival of *Chlamydomonas reinhardtii* during sulfur starvation based on microarray analysis of gene expression. *Eukaryotic Cell* 3:1331–1348.

35. Patel, N., V. Cardoza, E. Christensen, B. Rekapalli, M. Ayalew, and C.N. Stewart Jr. 2004. Differential gene expression of *Chlamydomonas reinhardtii* in response to 2,4,6,-trinitrotoluene (TNT) using microarray analysis. *Plant Science* 167:1109–1122.

36. Mus, F., A. Dubini, M. Seibert, M.C. Posewitz, and A.R. Grossman. 2007. Anaerobic acclimation in *Chlamydomonas reinhardtii*: Anoxic gene expression, hydrogenase induction, and metabolic pathways. *Journal of Biology Chemistry* 282:25475–25486.

37. Lilly, J.W., J.E. Maul, and D.B. Stern. 2002. The *Chlamydomonas reinhardtii* organellar genomes respond transcriptionally and post-transcriptionally to abiotic stimuli. *Plant Cell* 14:2681–2706.

38. Nguyen, A.V., H. Thomas, R. Skye, A. Malnoe, M. Timmins, J.H. Mussgnug et al. 2008. Transcriptome for photobiological hydrogen production induced by sulfur deprivation in the green alga, *Chlamydomonas reinhardtii*. *Eukaryotic Cell* 7:1965–1979.

39. Hemschemeier, A., and T. Happe. 2005. The exceptional photofermentative hydrogen metabolism of the green alga, *Chlamydomonas reinhardtii*. *Biochemical Society Transition* 33:39–41.

40. Ferreira, V.d.-S., I. Rocchetta, V. Conforti, S. Bench, R. Feldman, and M.J. Levin. 2007. Gene expression patterns in Euglena gracilis: insights into the cellular response to environmental stress. *Gene* 389:136–145.

41. Eom, H., C.G. Lee, and E. Jin. 2006. Gene expression profile analysis in astaxanthin-induced, *Haematococcus pluvialis* using a cDNA microarray. *Planta* 223:1231–1242.

42. Juan, S., K. Pan, J. Yu, B. Zhu, G. Yang, W. Yu, and X. Zhang. 2008. Analysis of expressed sequence tags from the marine microalga *Nannochloropsis oculata* (Eustigmatophyceae). *Journal of Phycology* 44:1529–8817.

43. Minoda, A., K. Nagasawa, M. Hanaoka, M. Horiuchi, H. Takahashi, and K. Tanaka. 2005. Microarray profiling of plastid gene expression in a unicellular red alga, *Cyanidioschyzon merolae*. *Plant Molecular Biology* 59:375–385.

44. Dyhrman, S.T., S.T. Haley, S.R. Birkeland, L.L. Wurch, M.J. Cipriano, and A.G. McArthur. 2006. Long serial analysis of gene expression for gene discovery and transcriptome profiling in the widespread marine coccolithophore *Emiliania huxleyi*. *Applied Environmental Microbiology* 72:252–260.

45. Rubinelli, P., S. Siripornadulsil, F. Gao-Rubinelli, and R.T. Sayre. 2002. Cadmium- and iron-stress inducible gene expression in the green alga, *Chlamydomonas reinhardtii*: Evidence for H43 protein function in iron assimilation. *Planta* 215:1–13.

46. Tyrrell, J.V., L.B. Connell, and C.A. Scholin. 2002. Monitoring for *Heterosigma akashiwo* using a sandwich hybridization assay. *Harmful Algae* 1:205–214.

47. Ayers, K., L.L. Rhodes, J. Tyrrell, and M. Gladstone. 2005. International accreditation of sandwich hybridisation assay format DNA probes for micro-algae. *New Zealand Journal of Marine and Freshwater Research* 39:1225–1231.

48. Mikulski, C.M., Y.T. Park, K.L. Jones, C.K. Lee, W.A. Lim, Y. Lee et al. 2008. Development and field application of rRNA targeted probes for the detection of *Cochlodinium polykrikoides* Magalef in Korean coastal waters using whole cell and sandwich hybridization formats. *Harmful Algae* 7:347–359.

49. Metfies, K., S. Huljic, M. Lange, and L.K. Medlin. 2005. Electrochemical detection of the toxic dinoflagellate, *Alexandrium ostenfeldii* with a DNA biosensor. *Biosensor and Bioelectron* 20:1349–1357.

50. Diercks, S., L.K. Medlin, and K. Metfies. 2008. Colorimetric detection of the toxic dinoflagellate, *Alexandrium minutum* using sandwich hybridization in a microtiter plate assay. *Harmful Algae* 7:137–145.

51. Diercks, S., K. Metfies, and L.K. Medlin. 2008. Development and adaptation of a multiprobe biosensor for the use in a semi-automated device for the detection of toxic algae. *Biosensor and Bioelectron* 23:1527–1533.

52. Ahn, S., D.M. Kulis, D.L. ErdnerL, D.M. Anderson, and D.R. Walt. 2006. Fiber-optic microarray for simultaneous detection of multiple harmful algal bloom species. *Applied Environmental Microbiology* 72:5742–5749.

53. Ki, S., and M.S. Han. 2005. A low-density oligonucleotide array study for parallel detection of harmful algal species using hybridization of consensus PCR products of LSU rDNA D2 domain. *Biosensor and Bioelectron* 21:1812–1821.

54. McDonald, W.H., and J.R. Yates. 2003. Shotgun proteomics: Integrating technologies to answer biological questions. *Current Opinion Molecular Therapy* 5:302–309.

55. Swanson, S.K., and M.P. Washburn. 2005. The continuing evolution of shotgun proteomics. *Drug Discovery Today* 10:719–725.

56. Snell, W.J., J. Pan, and Q. Wang. 2004. Cilia and flagella revealed: From flagellar assembly in *Chlamydomonas* to human obesity disorders. *Cell* 117:693–697.

57. Katz, A., P. Waridel, A. Shevchenko, and U. Pick. 2007. Salt-induced changes in the plasma membrane proteome of the halotolerant alga, *Dunaliella salina* as revealed by blue native gel electrophoresis and nano-LC-MS/MS analysis. *Molecular Cell Proteomics* 6:1459–1472.

58. Jia, Y., L. Xue, J. Li, and H. Liu. 2010. Isolation and proteomic analysis of the halotolerant alga *Dunaliella salina* flagella using shotgun strategy. *Molecular Biology Reproduction* 37:711–716.

59. Li, J.B., J.M. Gerdes, C.J. Haycraft, Y. Fan, T.M. Teslovich, H. May-Simera et al. 2004. Comparative genomics identifies a flagellar and basal body proteome that includes the BBS5 human disease gene. *Cell* 117:541–552.

60. Keller, L.C., E.P. Romijn, I. Zamora, I.I.I. Yates, Jr., and W.F. Marshall. 2005. Ciliary trafficking: CEP290 guards a gated community. *Current Biology* 15:1090–1098.

61. Pazour, G.J., N. Agrin, J. Leszyk, and G.B. Witman. 2005. Proteomic analysis of a eukaryotic cilium. *Journal of Cell Biology* 170:103–113.

62. Yang, P., D.R. Diener, C. Yang, T. Kohno, G.J. Pazour, J.M. Dienes et al. 2006. Radial spoke proteins of *Chlamydomonas* flagella. *Journal of Cell Science* 15:1165–1174.

63. Vener, A.V. 2007. Environmentally modulated phosphorylation and dynamics of proteins in photosynthetic membranes. *Biochimica Biophysica Acta* 1767:449–457.

64. Yamaguchi, K., M.V. Beligni, S. Prieto, P.A. Haynes, W.H. McDonald, I.I.I. Yates Jr., and S.P. Mayfield. 2003. Proteomic characterization of the *Chlamydomonas reinhardtii* chloroplast ribosome: Identification of proteins unique to the 70S ribosome. *Journal of Biology Chemistry* 278:33774–33785.

65. Rolland, N., A. Ariane, D. Paulette, G. Jerome, H. Michael, K. Georg et al. 2009. *Chlamydomonas* proteomics. *Current Opinion in Microbiology Ecology and Industrial Microbiology Technology* 12:285–291.

66. Stauber, E.J., A. Fink, C. Markert, O. Kruse, U. Johanningmeier, and M. Hippler. 2003. Proteomics of *Chlamydomonas reinhardtii* light-harvesting proteins. *Eukaryotic Cell* 2:978–994.

67. Turkina, M.V., J. Kargul, A. Blanco-Rivero, A. Villarejo, J. Barber, and A.V. Vener. 2006. Environmentally modulated phosphoproteome of photosynthetic membranes in the green alga, *Chlamydomonas reinhardtii*. *Molecular Cell Proteomics* 5:1412–1425.

68. Gillet, S., P. Decottignies, S. Chardonnet, and P. Le Marechal. 2006. Cadmium response and redoxin targets in *Chlamydomonas reinhardtii*: A proteomic approach. *Photosynthesis Research* 89:201–211.

69. Schmidt, M., G. Gessner, M. Luff, I. Heiland, V. Wagner, M. Kaminski et al. 2006. Proteomic analysis of the eyespot of *Chlamydomonas reinhardtii* provides novel insights into its components and tactic movements. *Plant Cell* 18:1908–1930.

70. Wagner, V., M. Fiedler, C. Markert, M. Hippler, and M. Mittag. 2004. Functional proteomics of circadian expressed proteins from *Chlamydomonas reinhardtii*. *FEBS Letters* 559:129–135.

71. Lis, R., A. Atteia, G. Mendoza-Hernandez, and D. Gonzalez-Halphen. 2003. A proteomic approach. *Plant Physiology* 132:318–330.

72. Peng, S.E., Y.B. Wang, L.H. Wang, W.N. Chen, C.Y. Lu, L.S. Fang, and C.S. Chen. 2010. Proteomic analysis of symbiosome membranes in Cnidaria–dinoflagellate endosymbiosis. *Proteomics* 10:1002–1016.

73. Steinbrenner, J., and H. Linden. 2001. Regulation of two carotenoid biosynthesis genes coding for phytoene synthase and carotenoid hydroxylase during stress-induced astaxanthin formation in the green alga *Haematococcus pluvialis*. *Plant Physiology* 125: 810–817.

74. Nicholson, J.K., J.C. Lindon, and E. Holmes. 1999. Metabonomics: Understanding the metabolic responses of living systems to pathophysiological stimuli via multivariate statistical analysis of biological NMR spectroscopic data. *Xenobiotica* 29:1181–1189.

75. Bundy, J.G., E.M. Lenz, N.J. Bailey, C.L. Gavaghan, C. Svendsen, D. Spurgeon et al. 2002. Metabonomic assessment of toxicity of 4-fluoroaniline, 3-5-difluoraniline and 2-fluoro-4-methylaniline to the earthworm *Eisenia veneta* (Rosa): Identification of new exogenous biomarkers. *Environmental Toxicology Chemistry* 21:1966–1972.

76. Raamsdonk, L.M., B. Teusink, D. Broadhurst, N. Zhang, A. Hayes, M. Walsh et al. 2001. A functional genomics strategy that uses metabolome data to reveal the phenotype of silent mutations. *Nature Biotechnology* 19:45–50.

77. Nicholson, J.K., M.J. Buckingham, and P.J. Sadler. 1983. High resolution proton NMR studies of vertebrate blood and plasma. *Biochemical Journal* 221:605–615.

78. Nicholson, J.K., D. Higham, J.A. Timbrell, and P.J. Sadler. 1985. Quantitative $_1$H-NMR urinalysis studies on the biochemical effects of cadmium exposure in the rat. *Molecular Pharmacology* 36:125–132.

79. Bales, J.R., D.P. Higham, I. Howe, J.K. Nicholson, and P.J. Sadler. 1984. Use of high resolution proton nuclear magnetic resonance spectroscopy for rapid multi-component analysis of urine. *Clinical Chemistry* 20:426–432.

80. Nicholson, J.K., and I.D.Wilson. 1989. High resolution proton NMR spectroscopy of biological fluids. *Progress in Nuclear Magnetic Resonance Spectroscopy* 21:449–501.

81. Richmond, A. 2004. Part III. Economic applications of micro algae. In *Handbook of Microalgal Culture: Biotechnology and Applied Phycology*, 255–353. Oxford, UK: Blackwell Publishing.

82. Bölling, C., and O. Fiehn. 2005. Metabolite profiling of *Chlamydomonas reinhardtii* under nutrient deprivation. *Plant Physiology* 139:1995–2005.

83. Matthew, T., Z. Wenxu, R. Jens, L. Lysha, R. Skye, H. Thomas et al. 2009. The metabolome of *Chlamydomonas reinhardtii* following induction of anaerobic H_2 production by sulfur depletion. *Journal of Biological Chemistry* 284:23415–23425.

84. May, P., S. Wienkoop, S. Kempa, B. Usadel, N. Christian, J. Rupprecht et al. 2008. Metabolomics- and proteomics-assisted genome annotation and analysis of the draft metabolic network of *Chlamydomonas reinhardtii*. *Genetics* 179:157–166.

85. Grossman, A.R., E.E. Harris, C. Hauser, P.A. Lefebvre, D. Martinez, D. Rokhsar, and J. Shrager. 2003. *Chlamydomonas reinhardtii* at the crossroads of genomics. *Eukaryotic Cell* 2:1137–1150.

86. Gobler C.J., D.L. Berry, S.T. Dyhrman, S.W. Wilhelm, A. Salamov, A.V. Lobanov et al. 2011. Niche of harmful alga, *Aureococcus anophagefferens* revealed through ecogenomics. *Proceedings of the National Academy of Sciences of the United States of America* 108:4352–4357.

19

Plant Metabolomics: Techniques, Applications, Trends, and Challenges

Shoaib Ahmad

CONTENTS

19.1 Introduction

Metabolomics basically deals with cellular metabolites. It is widely recognized as a branch of postgenomic science. Plant metabolomics helps in phenotyping even in the absence of any visible marked changes, either in the plant or in its cells. Plant metabolomic studies are apparently complex [1]. The plant metabolome represents all metabolites and other molecules that are involved in and produced during a plant cell's metabolism. In other words, plant metabolome is a fingerprint impression of all the processes taking place in a cell [2–4]. The usual focus is on small molecules [5,6]. Hence, plant metabolomics can be considered as a relatively newer branch of science dealing exclusively with plant metabolomes. Metabolomics is sometimes known as metabonomics [7]. Metabolomics is different from the traditional studies on metabolites in the sense that it focuses on cellular processes instead of just some metabolites or enzymes [8].

Many people consider metabolomics as an offshoot of systems biology [9]. There are two routes for metabolome studies [10]:

1. Quantitative analysis of metabolites
2. Chemometrics (chemical pattern) methods

The first route involves the identification and quantification of metabolites present in a plant specimen. It can be applied for a wide range of specimen extracts and could be useful for the identification and subsequent quantification of a large number of metabolites belonging to diverse chemical classes. The second route has a restricted applicability in comparison to the first route. This route cannot be used to identify or quantify the metabolites. It can be used for suggesting typical metabolite patterns in particular conditions. Despite contrasting features, both routes can be used for identification of diagnostic chemical patterns [10].

19.1.1 Advantages of Metabolomics Studies Over Other Omic Techniques

Metabolomics offers several advantages over the other omic applications [11]:

1. It is a downstream analysis and depends on only the analysis of metabolites and not the influxes
2. It does not require whole or large genome sequences
3. It is much cheaper than other "omics" technologies
4. The same analytical method can be used in a number of plant species if they produce the same category of chemicals
5. Metabolic networks are simpler and easier to understand compared with the complicated signaling networks
6. Metabolomic methods have proven their efficacy and worth in numerous studies

19.1.2 Relation between Genomics, Proteomics, and Metabolomics

Genomics, proteomics, and metabolomics are interrelated sciences. Genomics deals with genes, proteomics deals with proteins (mainly enzymes), and metabolomics deals with metabolites [11–13]. In the context of plant sciences, genomics deals with genes controlling the genetic setup of the plant. These genes are responsible for particular phenotypes or individual taxa characteristics. These genes control the expression of various proteins of which enzymes form one major type. Many of the enzymes control key steps in the primary and secondary metabolite synthesis pathways. The presence or absence of a particular gene may be associated with a particular enzyme, and this enzyme may further be associated with complicated metabolite synthesis pathways or networks in plants. As we go up in the hierarchy series from genes to enzymes to metabolites, the number of study subjects (genes/enzymes/metabolites) decreases. Any change in the genetic setup (induced or natural) will result in a corresponding change in protein expression and may cause alteration in the enzyme structure and properties. As expected, such a change will lead to a qualitative or quantitative change in metabolite production patterns. As compared with genomics and proteomics, the metabolite data generated by metabolomics is less expensive, and requires less sophisticated and relatively cheaper instrumentation with wide adaptability. The metabolomics data has a lower degree of complexity and can be rapidly converted into useful information. It is easier to use for dereplication purposes.

19.1.3 Plant Metabolomics and System Biology

Biology has undergone paradigm changes in the last few years. Studies in diverse fields such as plant anatomy, physiology, biochemistry, and molecular biology (at the cellular and subcellular levels) have resulted in a huge pile of data that needs to be converted into meaningful information—particularly for application to the production of metabolites and their utilization. Often, computational technologies are required for acquiring and analyzing volumes of data. Complex mathematical modeling is required to understand and mimic the processes within plant structures and cells in normal and altered growth conditions. Some times, a combination of omic technologies and *in silico* methods is required for the rapid transformation of data into utilizable information. As a result, a new branch of science—systems biology—has emerged. It is very difficult to think of metabolomics without system biology tools [5,14–16].

Systems biology plays an important role in plant metabolomics, as depicted by the following examples:

1. Storage of metabolomic studies data
2. Construction of databases for acquisition and storage of data
3. Construction of data retrieval systems for analysis
4. Establishment of plant metabolome size
5. Study of organization and control of metabolic networks
6. Mathematical models for the study of plant anabolic and catabolic reactions
7. Development of computational models for physiological processes
8. Prediction of plant cell and foreign protein interaction

Systems biology has contributed much to the development of plant metabolomics. It has helped in the identification of biomarkers, acquisition of metabolite-related data from phytochemistry-oriented studies (when hyphenated techniques are used), or biochemical data from genomic studies. Systems biology can also provide links to connect several metabolic pathways with some common point (chemical reactant, catalyst, or intermediary in biosynthesis) to get a larger picture of metabolism in plant systems. In the future, systems biology may help in *in silico* prediction of the effect of chemical agents such as herbicides on metabolite production.

19.2 Development of Plant Metabolomics

Metabolomics started developing from metabolic profiling. It owes much to the technique of gas chromatography-mass spectroscopy (GC-MS), which was being used in the early 1970s for the analysis of steroids and related compounds in plant specimens [17]. Until the 1980s, metabolic profiling remained a stable field, and in the 1990s, it received recognition as one of the diagnostic techniques useful for metabolomic studies. Many scientists have favored the integration of complementary DNA-amplified fragment length polymorphism (cDNA-AFLP) based transcript profiling and metabolomics [18]. Systems biology has also been a contributor to the development of plant metabolomics.

19.2.1 Methodologies

Metabolomics is largely considered a postgenomic science. Plant metabolomics assumes significance from the huge number of known metabolites (more than 200,000) produced by plants. This number is larger than the metabolites produced by animals and microorganisms taken together [1]. The strength of metabolic machinery in plants can be estimated from the fact that a single plant (*Arabidopsis thaliana*) produces more than 5000 metabolites. Hence, *A. thaliana* is a model system for all types of metabolic studies. Metabolomics is entirely different from plant metabolite profiling, which deals with the qualitative and quantitative studies on metabolites (particularly, secondary plant metabolites). On the other hand, plant metabolomics deals with the entire range of metabolites produced in the course of cellular events taking place in plants.

Plants such as rose, lavender, chamomile, calendula, and lemongrass produce a range of terpenoids, which are mainly responsible for their characteristic fragrance. The exact chemical nature can be determined to a large extent using hyphenated GC-MS technology. Similarly, DNA microarrays can detect genes and gene functions. If the data from GC-MS and DNA microarray experiments are pooled together, it is very easy to locate the gene responsible for the production of a typical fragrance constituent in a particular plant part (for example, in rose petals).

Metabolomic analysis usually generates huge data, which requires proper correlation with the existing body of knowledge, proper interpretation, and above all, utmost utilization. A new branch of chemometrics has recently come into play. Metabolome data requires integration with omics data (e.g., genomics and transcriptomics data). For this purpose, *in silico* or computerized methods are adopted. Software tools are available to depict gene–protein–metabolite relationships [1].

19.2.2 Instrumentation

Due to the diverse nature of metabolites produced by plants, different types of equipment are required for investigational studies (Table 19.1). It can be easily understood why

TABLE 19.1

Indicative List of Instruments Used in Plant Metabolomics

Instrument	Instrument Abbreviation	Application(s)
Mass spectrometer	MS	Determination of types of compounds produced Identification of chemical nature of compounds Molecular mass determination
Gas chromatography	GC	Identification of volatile compounds
High-performance liquid chromatography	HPLC	Chromatographic identification of metabolites Metabolite profiling Quantitative determination of metabolites
High-performance thin layer chromatography	HPTLC	Chromatographic identification of metabolites Metabolite profiling Quantitative determination of metabolites
Gas chromatography coupled with MS	GC-MS	Combined applications of GC and MS with better sensitivity
High-performance liquid chromatography coupled with MS	HPLC-MS	Combined applications of HPLC and MS with better sensitivity
High-performance thin layer chromatography coupled with MS	HPTLC-MS	Combined applications of HPTLC and MS with better sensitivity
Nuclear magnetic resonance spectrometer	NMR	Analysis of small and large molecules Spatial orientation of metabolite subparts
Capillary electrophoresis	CE	Analysis of charged metabolites
Capillary electrophoresis coupled with MS	CE-MS	Analysis of charged metabolites with better sensitivity

a single instrument is not sufficient for plant metabolite analysis. These analytical technologies are based on chromatographic [e.g., high-performance liquid chromatography (HPLC) and high-performance thin layer chromatography (HPTLC)], spectroscopic [e.g., MS, nuclear magnetic resonance (NMR)], or hyphenated (a combination of the chromatographic and spectroscopic methods, e.g., HPLC-MS, HPTLC-MS) techniques. For plant metabolomics, all analytical methods used must have accuracy, sensitivity, and reproducibility. The sensitivity has to be quite higher so that even the chemical compounds that are produced in trace quantities are not missed in metabolomic analysis. One will find variations in the existing analytical instrumentation or newer instrumental methods available with the progress of time. Such change will definitely keep benefiting plant metabolomics and contribute to the understanding of complex metabolic pathways in plants.

19.2.3 Spectroscopy as an Aid to Metabolomics

Near-infrared spectroscopy has been suggested and applied to work related to microbial metabolomics by some researchers. This spectroscopic technique utilizes true fingerprints in the 110 to 250 nm region of the electromagnetic spectrum. Data can be captured from microbial cultures in a very short period, that is, 30 s only. This spectral data can then be manipulated using standard chemometric tools. Near-infrared spectroscopy does not cause damage to the microbes in culture and, hence, does not affect their metabolic profiles. These spectra correspond to the vibrational energy of chemical bonds and functional groups, making it easier to look for the particular types of chemical transformations taking place in the cultures. The ease of data capture and manipulation makes near-infrared spectroscopy a favorable tool for high throughput screening (HTS) type metabolomics [19]. Proton NMR spectroscopic analysis has been used for metabolic profiling of ginkgolic acids in ginkgo products. These acids are the main chemical constituents of ginkgo. Multiplicity of substituent chains makes analysis of ginkgolic acids by conventional chromatographic methods more complex and difficult. This method can be used for profiling of various species of ginkgo and their products. The quantitative results from this study are comparable to those obtained with routine GC analysis. Clearly, this method can be adopted for studying metabolomics in ginkgo [20].

^{1}H-NMR spectroscopy has been successfully utilized for metabolomic analysis of 12 *Cannabis sativa* cultivars. These cultivars have been differentiated on the basis of δ-9-tetrahydrocannabinolic acid and cannabidiolic acid. A simpler (although less reliable) differentiation can be made using water extracts containing carbohydrates and amino acids. Sucrose, glucose, asparagine, and glutamic acid levels are cultivar-specific. They are, however, not a replacement for δ-9-tetrahydrocannabinolic acid and cannabidiolic acid profiling for accurate metabolomic studies [21]. Ten species of *Echinacea* grow in North America. The proton spectroscopy has been used for studying metabolomics in *Echinacea*. This analysis has also found a clear-cut distinction among some of the species, for example, *Echinacea purpurea*, *E. pallid*, and *E. angustifolia* on the basis of metabolic profiling [22].

19.3 Applications of Plant Metabolomics

The following section gives details of applications of plant metabolomics.

19.3.1 Plant Physiology

Plant metabolomics has several applications in the study of plant physiology and related processes.

19.3.2 Hormonal Interactions

Plants have a positive or negative interaction with many different organisms. The positive interaction may be exemplified by the interaction with pollinators, whereas a negative interaction may be exemplified by interaction with pathogens. Plants have a built-in capacity to regulate responses to single or multiple biotic interactions. These interactions are largely governed by signaling hormone interactions. These interactions may have a positive or negative effect on physiological processes in plants. Ecogenomics and metabolomics may help in understanding the hormonal interactions of plants with their biotic environment [23,24].

The NCED3 gene in *Arabidopsis* mutant (nc3-2) is a key factor in dehydration-inducible biosynthesis of abscissic acid (ABA) in higher plants. Raffinose levels are independently regulated by the ABA under dehydration stress. The biosynthesis of amino acids, saccharopine, proline, and polyamine depends on ABA-dependent transcriptional regulation. Amino acid accumulation depends on ABA production. This type of metabolomic analysis, based on the utilization of mass spectrometry (MS) systems [gas chromatography–time of flight–MS (GC-TOF-MS) and capillary electrophoresis-MS (CE-MS)] can be used for the characterization of metabolome of plants with reference to hormonal interaction with abiotic environment condition, that is, drought [25]. Benzothiadiazole is a functional analogue of salicylic acid (SA). SA is a hormone-like compound present naturally in plants. SA is needed for the induction of plant defense genes. Benzothiadiazole-treatment of *Arabidopsis* reduces the levels of phenolic metabolites, induced generally by other signaling molecules [26]. Carbon/nitrogen assimilation in plants is dependent, to some extent, on hormonal signals. Carbon/nitrogen assimilation affects the composition of minor amino acids [27].

19.3.3 Plant Breeding

Metabolomics allows the estimation of a wide range of metabolites and is helpful in phenotyping as well as diagnostic analysis in plant crops. Natural variance in the metabolite composition of crops has been the subject of metabolomic analyses. Metabolomics is one of the current genomics-assisted selection tools for crop improvement strategies. Metabolomics has been used in the assessment of broad genetic variance among different varieties of crops. This approach should be continued and extended beyond the limitations of costs and extent of heritability. There has been a shift from the classic concept of single metabolite measurement to the use of metabolite study platforms providing information on a huge array of metabolites. Better models have been developed to provide/create links between metabolism and yield-associated traits. The high levels of metabolites are ensured in the crops by using hybrids. This process does not compromise on the yield levels of the crop and hence the market value of the crop remains steady even if prices do not increase. A high content of these metabolites makes these crops more popular and acceptable in commerce and trade. The metabolic response to biotic and abiotic stress usually results in higher metabolite contents and hence indicate the need for metabolomics-assisted breeding. It has been asserted that the application of postgenomic tools will accelerate the selection, and the combined use of metabolomics, genome sequencing, and high-throughput reverse genetics may shorten the time needed for producing elite lines. Metabolomics-assisted breeding is definitely applicable to crop species. This strategy can increase crop resistance to diseases, and herbicide or salinity tolerance. It is a viable option for crop improvement programs [28].

19.3.4 Plant Ontology

Plant ontology refers to the development of individual plants following the same set of strategies followed by the systematic or classification category to which they belong. "Ontogeny repeats phylogeny" is a common and popular statement to the effect that a plant undergoes all the developmental stages through which its "parents or ancestors" passed. It is very difficult to use the ontology terminologies with uniformity across the disciplines (or discipline subsets) of plant sciences, biotechnology, genomics, etc. The Plant Ontology Consortium is a collaborative effort among several plant databases and experts in these disciplines toward developing simple, robust, and extensible controlled vocabularies reflecting plant structure biology and developmental stages. These efforts have resulted in a network of vocabularies linked by relationships (ontologies) and allow queries across data sets within a database or between multiple databases. Such ontological vocabularies provide descriptions of *Arabidopsis*, maize, and rice (*Oryza* sp.) anatomy, morphology, and developmental stages.

The databases for these three species are:

1. The *Arabidopsis* Information Resource for *A. thaliana* and other species of *Arabidopsis*
2. Gramene for rice (*Oryza sativa* and other species of *Oryza*)
3. MaizeGDB for maize (*Zea mays*)

More than 3500 gene annotations exist in three species-specific databases, and these annotations can be queried using the ontology browser and relevant information can be retrieved [29].

19.3.5 Plant Stress

Environmental stress in plants may be defined as any change in growth condition(s) within the plant's natural habitat, which changes or tends to change the state of metabolic homeostasis. Overcoming this change requires an adjustment of metabolic pathways and the whole process of adjustment to the new growth conditions is termed as acclimation. Acclimation involves several phases. Initially, the plant senses the change in environmental conditions and this sensing activates a complicated network of signaling pathways. Later, different proteins and phytochemical compounds are produced. These compounds are termed as phytoalexins and are produced generally in response to damaging stimuli from physical, chemical, and microbiological factors affecting the growth of normal plants. The stress is termed abiotic stress if it involves nonliving physical factors (such as mechanical injuries, extreme thermal conditions, and ultraviolet radiation exposure) or chemical factors (such as extreme salinity, heavy metals, loss of minerals due to soil erosion, and nonavailability of trace elements). On the other hand, stress may be defined as biotic stress if caused by living organisms or as a result of their presence in the vicinity (such as the presence of *Salmonella* and *Escherichia coli*, growth of *Phytophthora* or *Aspergillus*

TABLE 19.2

Compounds Produced during Biotic or Abiotic Stress Conditions

Compound Type	Example
Polyols	Mannitol and sorbitol
Dimethylsulfonium compounds	Dimethylsulfoniopropionate, glycine, and betaine
Sugars	Sucrose, trehalose, and fructan
Amino acids	Proline and ectoine
Antioxidants or free radical scavengers	Ascorbic acid, glutathione, tocopherols, anthocyanins, and carotenoids

fungus, and excess growth of weeds competing for nutrition). A large number of molecules are formed during the conditions of abiotic or biotic stress (Table 19.2). From the metabolomics viewpoint, three types of compounds are involved in these processes resulting from stress [30]:

1. Compounds involved in the acclimation (e.g., antioxidants or osmoprotectants)
2. Compounds/by-products produced as a result of stress due to alteration(s) in growth conditions
3. Signal transduction molecules acting as mediators of acclimation response

The first category of compounds includes antioxidants such as flavonoids and osmoprotectants such as amino acid–proline. The second type of molecules may include reactive oxygen species or oxidized compounds such as phenolic compounds, for example, SA or antioxidants. The last category of molecules may include:

1. Compounds synthesized *de novo*
2. Compounds freed from their conjugated form(s), for example plant hormones and SA
3. By-products of stress metabolism, for example phytoalexins

The plant defense response in stress leads to phytoalexin production, phenylpropanoid pathway activation, and lignin biosynthesis induction. Small molecules such as SA, methyl salicylate, jasmonic acid, methyl jasmonate, etc., may signal the activation of responses related to systemic defense and acclimation.

19.3.5.1 Effects of Stress on Metabolite Production in Selected Plants

From the metabolomic studies in *Arabidopsis*, it has been concluded that regulatory processes play a vital role in the metabolic adjustments during cold acclimation [30]. These processes operate independent of transcript abundance. The *Arabidopsis* metabolome undergoes extensive reconfiguration in low temperatures and C-repeat/dehydration–responsive element-binding factor plays an important role in the typical responses. *Arabidopsis* produces higher quantities of sucrose and proline molecules when exposed to salt stress. Chilling-tolerant and chilling-sensitive genotypes have been identified in rice. The former genotype accumulates galactose and raffinose under stress, whereas the latter genotype loses these sugars under cold stress. The former genotype has been found to have a more active reactive oxygen species scavenging system.

Glucose, malate, and proline are found in higher concentrations in plants exposed to water deficit stress than in the plants exposed to salt stress. Amino acids, polyols, and sugars are increased in the roots of bean plants when grown in phosphorous-deficient conditions. Cadmium stress in some plants may increase the levels of acetates and malate while decreasing the levels of glutamate and branched-chain amino acids. In case of drought and heat stress or a combination thereof, plants accumulate sucrose and maltose. Zwitterionic quaternary ammonium compounds, glycine and betaine have been found to act as osmoprotectants in members of Plumbaginaceae. They decrease the effects of salinity and drought [31].

19.4 Primary and Secondary Plant Metabolism

Plant metabolomics can be utilized in the detailed studies on primary and secondary metabolism.

19.4.1 Metabolic Control

Several genetic and physicochemical factors affect metabolic activities in plant tissues and hence have a variable and definitive degree of metabolic control over plant processes. At the molecular level, posttranslational modifications in the proteins (e.g., chemical modifications via phosphorylation, acetylation, phenylation, glycosylation, or methylation) contribute a lot to control the metabolism in plants. A protein is located at a particular cellular compartment and it is active only within a specified location. The functionality of a protein may also require interaction with other protein components under the influence of several factors. The synthesized messenger RNA and protein can be stored within the cells for use at a later stage. The gene expression levels are dependent on the levels of synthesis, degradation, and storage of DNA, messenger RNA, and proteins, chemical modifications in protein structures, protein–protein interactions, and proper spatial localization. All these conditions are a must for a gene to perform its classified function [32,33].

Light and temperature remain the leading physical factors that affect metabolite production in medicinal plant species. The production of the anticancer drug camptothecin in *Camptotheca acuminata* increases with a decrease in light reaching the plant [34–40]. Periodic flooding also increases camptothecin production in this plant [41]. Temperature is another factor that controls metabolite production. Plant enzymes are inactive at 0°C. They resume their activities as the temperature increases up to 40°C, beyond which they lose their catalytic properties.

19.4.2 Metabolic Network Interpretation

Nontargeted metabolite profiling techniques lead to the generation of huge quantities of data needed for constructing dynamic metabolomic networks. Recently, comparative correlation analysis has been suggested as a complementary approach for the characterization of metabolites associated with different physiological states of plant parts or tissue types. Several correlations are induced by uncontrollable fluctuations in environmental conditions (e.g., light intensity, temperature, or nutrient uptake). This leads to an induced variability of metabolites affecting the enzymatic reaction network and to an emergent pattern of correlations. Metabolite correlations can be associated with the physiologies of different plant parts. These covariance and correlation matrices form the basis of current methods of data analysis (clustering algorithms and principal component analysis). Comparative analysis of correlations is a newer approach and allows us to identify key points of alteration in metabolic regulation. This approach can be readily extended to include data originating from transcriptomic and protein expression studies. Recently, a process for the integrated extraction and quantification of metabolites, proteins, and RNA from complex plant mixtures has been proposed to simplify the otherwise difficult-to-understand coregulation in complex biochemical networks [13].

19.4.3 Metabolite Identifier Visualization

The nonexistence of a definite system for metabolite notation in general, and in plant metabolites in particular, leads to ambiguity in plant metabolomic studies. Such ambiguity usually hampers studies and research in plant metabolism and related sciences. The same metabolite may be referred to differently by different systems including the Chemical Abstract Service, the Kyoto Encyclopedia of Genes and Genomes, the Chemical Entities of Biological Interest, and the Comprehensive species metabolite relationship database (KNApSAcK; Table 19.3).

Often, the same classification system may be using two different identifiers for the isomeric or optical forms of the same metabolite. There may be another case when a system may be using only one identifier in this condition. It necessitates a technical improvement process regarding the metabolite identifier information without any ambiguity and incompatibility with other identifier systems. Proper management of metabolite identifiers is understandably a key technology for metabolomics data analysis, which may include data set integration and contextual interpretation. Recently, the software called MetMask has been developed for automatic integration, mapping, and conversions of different metabolite identifiers. This software is capable of constructing consensus metabolite mappings from variable sources such as reference libraries and external databases. These metabolite identifier mappings are required to be so accurate that each identifier mapping corresponds only to exactly one metabolite. MetMask utilizes a graph-based approach for meeting its objectives. In this approach, database entries for a particular compound are used to create a small graph mentioning the name of the metabolite and identifier entries in other databases. Several graphs for the same compound from different systems are then correlated to form a larger merged network usually using the platform of RIKEN metabolomics or

TABLE 19.3

Systems and Metabolite Identifiers

Systems Using Metabolite Identifiers	System Abbreviation	Metabolite Identifiers
Chemical Abstract Service	CAS	Accession codes or registration numbers
Kyoto Encyclopedia of Genes and Genomes	KEGG	Accession codes or registration numbers
Chemical Entities of Biological Interest	ChEBI	Accession codes or registration numbers
Comprehensive species metabolite relationship database	KNApSAcK	Species–metabolite relationship

PRIMe lists. The information from ontology–metabolite or species–metabolite relation-ships can be easily incorporated into a larger network to make it meaningful and more practical-oriented. A graph-based approach is more suitable for quick and accurate data mining in the domain of plant metabolomics. Application of MetMask has suggested that piperidine and quinazoline moieties in several plant species in Fabaceae originate from the same precursor, cadaberine [42].

19.4.4 Phenotype and Genotype Gap-Filling

Metabolomics has been claimed as a science that bridges the gap between phenotypes and genotypes. Metabolome analysis, instead of focusing on a single metabolite, represents the summary of all upstream activities in a cell. It also takes into account the external influence on metabolic regulatory processes. Any difference in the chemometric patterns of metabolites automatically means a difference in the genetic constitution. A study of the metabolomic information and its correct interpretation provides a narrowed down approach to pinpointing a single gene or a group of genes that are possibly responsible for particular variance in the chemometric pattern, metabolic profiling, or biological profiling. In this sense, metabolomics can be considered as being successful in filling the hitherto unfilled gap between phenotype and genotype [43].

19.4.5 Product Dereplication

Dereplication is basically the process of detection of previously known compounds pres-ent in a plant-based extract. Dereplication is aimed at rapid drug discovery from drugs of plant origin. This process utilizes basic and sophisticated techniques of chromatogra-phy and spectroscopy. Of late, dereplication has become largely dependent on the com-bination of chromatography and spectral techniques, which are frequently referred to as hyphenated techniques. This process ensures that efforts are not wasted in the discovery of known compounds. Obvious advantages of this method include minimal wastage of efforts and resources (manpower, machine time, and money) in the discovery of known compounds. It accelerates the drug discovery process—a factor which is quite important particularly in bioassay-guided fractionation procedures, and in commercial research where the registration of intellectual property rights gives an edge to a particular com-pany over its competitors. Practically speaking, NMR spectroscopy is the most versatile tool in the dereplication procedures. This instrumental technique is quite helpful in the analysis of small molecules. Here, a scientist checks the status of an isolated compound (pre-existing compound or novel compound) after recording its fingerprint molecular spectra. The search relies on matching spectral chemical shifts of the compound isolated. Usually, one has to be dependent on large chemical databases such as NMR-ShiftDB. If the database searches are not successful, one has to go for rather sophisticated computer-aided structure elucidation [19,44–47].

NMR is frequently favored as a dereplication tool because unlike MS, it is a nonde-structive technique and allows even a very small quantity of plant extract to be analyzed rapidly and with considerable accuracy. Dereplication is usually done in conjunction with bioactivity-guided fractionation. The usual bioactivities range from simple enzyme-based assays to *in vitro* and *in vivo* biological activities. Discovery of new bioactive molecules (with the potential of becoming new leads or new drugs) from plant sources essentially requires extracts to be subjected simultaneously to chemical screening processes and to various biological or pharmacological targets. Metabolite profiling using hyphenated analytical techniques (e.g., LC/UV, LC/MS, and LC/NMR) provides important structural information within a considerably short period. This information often has vital clues and leads to a partial or a complete on-line *de novo* chemical compound structure determination. Bioassays after extract LC/microfractionation allow easy and efficient identification of bioactive LC peaks in the experimental chromatograms. This combination of metabolite profiling and LC/bioassays provides help in differentiating between known bioactive compounds and new molecules directly in crude plant extracts. This approach helps in the isolation of new bioactive molecules and sometimes in constituents with novel or unusual spectroscopic features.

Utilization of LC/UV/NMR/MS and LC/bioassay methods is a new development in plant metabolomics [48]. Examples of dereplication in the context of plant metabolomics are as follows:

1. Phorbol esters in *Lyngbya majuscula* and *Croton cuneatus* have been dereplicated by utilization of HPLC and UV–visible spectrophotometer and online phorbol ester receptor binding activity [49].

2. Pseudoguaianolides from *Parthenium hispitum* have activity against hepatitis C virus. These compounds have been dereplicated using capillary-scale NMR probes [50].

3. Three cytotoxic withanolides have been isolated from the leaves of Acnistus arborescens. Dereplication analysis of the ethyl ether extract performed using a combination of NMR and MS has led to the identification of the constituents responsible for significant cytotoxic activity [51].

4. The chloroform extract of Kielmeyera albopunctata stem bark has been subjected to a bioassay-related LC-MS dereplication using the KB cell lines [52]. This dereplication procedure has led to the identification of new coumarins, namely,

    i. 4-(1-methylpropyl)-5,7-dihydroxy-8-(4-hydroxy-3-methylbutyryl)-6-(3-methylbut-2-e nyl)chromen-2-one

    ii. 5,7-dihydroxy-8-(4-hydroxy-3-methylbutyryl)-6-(3-methylbut-2-enyl)-4-phenylchromen-2-one

    iii. 9-(1-methylpropyl)-4-hydroxy-5-(4-hydroxy-3-methylbutyryl)-2-(1-hydroxy-1-methyle thyl)-2,3-dihydrofuro[2,3-f]chromen-7-one

        Coumarins (A and B) have exhibited moderate cytotoxicity against KB cell lines, whereas the last compound has not shown any cytotoxicity.

5. Phytoestrogens are present in seed oil, juice, fermented juice, and peel extract of pomegranate fruit (*Punica granatum*) and have been suggested to exert anticancer effects. This claim has been experimentally proven on *in vitro* grown human breast cancer cells. On-line biochemical detection coupled to mass spectrometry (LC-BCD-MS) has been used for profiling estrogenic activity in the pomegranate peel extract. This typical dereplication procedure has involved the use of HPLC, on-line β-estrogen receptor (ER) bioassay, and the use of MS/MS fingerprint. A total of three estrogenic compounds (i.e., luteolin, quercetin, and kaempferol) belonging to a flavonoid subclass of phenolics have been identified. The dereplication procedure in this typical study involved repeated cycles of HPLC fractionation and biological screening. It offers convincing evidence that LC-BCD-MS markedly accelerates the time required for compound description and identification processes [53].

19.4.6 Quantitative Trait Locus Analysis

Quantitative trait locus (QTL) analysis is one of the mainstream techniques of plant genetics and metabolomics [54,55]. QTL analysis refers to a specific DNA sequence, which is related to a particular known trait in plants, for example total yield of fodder leaves. QTL analyses specifically deal with quantitative genetics. These analyses can provide information on the following:

1. Number of loci controlling each trait in the plant
2. Basis of the trait control
3. Presence or absence of directionality in parents with reference to QTLs

Combined application of QTL mapping and metabolomics makes it easier to understand key aspects of metabolic relationships as well as finding answers to the certain questions considered fundamental to the field of quantitative genetics.

19.4.7 Prerequisites for Undertaking QTL Mapping and Analysis

There are four essential requirements [56], which are to be fulfilled for undertaking QTL analysis and applying it to plant metabolomics. These requirements are as follows:

1. Population produced from parents with phenotypically contrasting features
2. Molecular marker–based linkage map
3. Reliable phenotypical screening tool
4. A special software that can integrate phenotypic information with genotypic information for the purpose of QTL detection.

19.4.8 Methods for QTL Analysis

QTL analysis is largely dependent on statistical sciences [57], and QTL analysis methods include:

1. Regression methods
2. Maximum likelihood estimation
3. Precision of mapping and hypothesis testing

These methods have several subcategories.

19.4.8.1 Regression Methods

These methods are relatively simple, require less computing, and are faster. These methods do not require much sophisticated software applications.

Subcategories
   a. ANOVA analysis using single marker genotypes
   b. ANOVA analysis using multiple marker genotypes
   c. Regression on QTL probability, conditional on marker haplotypes

   d. Haley–Knott regression

   e. Regression of phenotype on marker type

19.4.8.2 Maximum Likelihood Estimation

These methods are relatively complex, require more computing work, and are time-consuming. These methods need specific tailor-made software requiring considerable practice and expertise.

   Subcategories

   a. Comparison of likelihood and regression procedures

   b. Multiple regression on marker genotypes

   c. Interval mapping with marker cofactors (composite interval mapping)

19.4.8.3 Precision of Mapping and Hypothesis Testing

These methods are used for authenticating the first two methods (regression and likelihood estimation methods).

   Subcategories

   a. Permutation testing

   b. Bootstrapping

   c. Accounting for multiple testing

19.4.9 Molecular Markers in QTL Analysis of Crop Plants

Most of the crop plant quantitative traits are under the control of polygenes. These gene loci (QTL) can be detected using certain molecular markers, segregating in a Mendelian fashion. Molecular markers include DNA-based markers [58]. Classic examples of QTL analysis in metabolomics include the cases of canola and tomato. Oilseed rape/canola (*Brassica napus*) is a rich source of vegetable proteins because of its essential amino acid content. Rapeseed proteins are considered not suitable for human nutrition because of the presence of antinutritive components and compounds (dietary fibers, dark-colored tannins, and bitter-tasting sinapate esters). Yellow colored seeds have a thin seed coat with low contents of dietary fiber and condensed tannins. Condensed tannins accumulate mainly in the cells found between the outer integument and the aleuronic layer in rapeseed. These tannins impart a dark color to the seed coat. Seed cotyledons contain 0.1% to 0.5% (dry weight) condensed tannins as compared with up to 6% tannins in the seed coat. A study was conducted for the identification of QTL for seed color, individual and total condensed tannins in a winter rapeseed doubled haploid population in Germany. QTL mapping was largely based on seed analyses of doubled haploid lines grown in field trials. A specialized QTL software (PLABQTL) was used in the study. This population was found to have three QTLs each for seed color and total seed flavonoids. The doubled haploid population also contained 10 and 4 QTLs, respectively, for oligomeric proanthocyanidins and polymeric proanthocyanidins [59].

   Tomatoes are a rich source of antioxidants, which are generally claimed to have a preventive role in chronic diseases (e.g., cancer, arthritis, and heart disease). An introgression

TABLE 19.4

QTLs in Introgression Lines of Tomato

Trait	QTLs Identified
Total antioxidant capacity of the water-soluble fraction	5
Ascorbic acid	6
Total phenolics	9
Lycopene	6
β-Carotene	2

line of tomato producing the wild, green-fruited species (*Lycopersicon pennellii*) and the domesticated tomato (*Lycopersicon esculentum*) was subjected to QTL analysis for nutritional and antioxidant component molecules. The results of these studies are presented in Table 19.4.

Some of these QTLs exhibited increased levels of ascorbic acid and phenolics as compared with the parental line (*L. esculentum*). Although four QTLs were detected for lycopene, no increase in lycopene levels relative to *L. esculentum* was observed. In contrast, two QTLs were detected for β-carotene and they were found to have increased β-carotene levels in ripe fruits. This study has also highlighted the fact that the traits being investigated are largely affected by environmental conditions. It was evident from experiments that only one-third of the water-soluble antioxidant QTL was consistent over at least two crop seasons [60].

19.4.10 Metabolite Production

All metabolic reactions in plants start with the process of carbon fixation in photosynthesis, in which carbon dioxide and solar energy are utilized. Metabolic pathways are located in one or more cellular compartments (e.g., cell walls, cell or vacuole membrane, and cytosol) within tissues specialized for biosynthetic processes. Most metabolites remain inside the body of the producer plant and serve several functions including plant defense using bitter alkaloids. Occasionally, the plant releases some of its secondary metabolites such as terpenes (produced from isoprene units and modified under existing circumstances, including the presence of molecular oxygen). These released chemical compounds attract many insects due to their characteristic aroma or smell and help in pollination or seed dispersal processes, which are vital for the continuation of the plant species. Specific enzymes required for catalytic reactions are encoded in nuclear, chloroplast, and mitochondrial genomes by specific genes.

The *Arabidopsis* PAD3 gene is required for the biosynthesis of a specific phytoalexin (camalexin) in response to bacterial infection caused by *Pseudomonas syringae*. Although the mutation of this gene leads to defective camalexin production, mutation of the PAD3 gene does not affect resistance to *P. syringae*. This mutation, however, decreases plant resistance to the fungus, *Alternaria brassicicola* [61]. Tdc and Str genes in the periwinkle plant (*Catharanthus roseus*) are responsible for the production of indole alkaloids, which have cytotoxic and anticancer effects [62–65].

19.4.11 Metabolite Manipulation

Metabolite production in plants can be manipulated by using a number of approaches. These approaches include the following:

1. Insertion of a gene responsible for production of a particular metabolite
2. Controlling the expression of specific genes
3. Increasing or decreasing the key chemical substances required for particular biosynthetic reaction
4. Experimental induction of mutation
5. Experimental induction of foreign gene(s)
6. Introduction of stress to a plant using physical, chemical, or biological methods

A classic example in metabolite production and manipulation is the use of Bax for indole alkaloid production in *C. roseus*. Bax is a mammalian proapoptotic member of the Bcl-2 family. It is responsible for triggering hypersensitivity reactions when expressed in plants. In a study, transgenic *C. roseus* cells were produced that overexpressed a mouse Bax protein under the β-estradiol–inducible promoter. Expression of mouse Bax was found to induce the transcriptional activation of two key genes (Tdc and Str) involved in terpenoid indole alkaloid biosynthesis. Bax was also found to stimulate the accumulation of defense-related protein PR1 in the cells and activate the indole alkaloid biosynthesis pathway. Bax can be used as a potential regulator for secondary metabolite production in plants. This unique strategy can be used for increasing the metabolic flux to natural products with the activation of an entire biosynthetic pathway. This strategy is far less complicated than the strategy of manipulating single structural genes within the pathways [65].

19.4.12 Integrated Metabolite Engineering

Plants are excellent producers of chemical compounds—many of which find use in food and medicines. The compounds produced by plants are diverse in phytochemistry and chemical structures. As indicated by the shortage of artemisinin from plant sources in recent years, an alternative approach for metabolite production through biotechnological tools is currently needed. Metabolic engineering of plants is an important technique that can be used for this purpose [66].

Metabolite engineering can be utilized toward one or more of the following objectives:

1. Production of a desirable/beneficial chemical constituent in higher quantities
2. Decreased production of toxic compound(s)
3. Production of novel chemical compounds

The usual strategy of plant secondary metabolism engineering has one of the following three approaches [67]:

1. Transformation of single/multiple enzyme gene(s) or a whole metabolic pathway
2. Reduced target gene expression or blockade of the competitive metabolic pathway
3. Manipulation of transcription factors required for multiple point metabolic regulation

Plant secondary metabolism engineering is a feasible and challenging area. It requires a sound knowledge of the biosynthetic pathways involved in the production of particular chemical constituent(s). Plant secondary metabolism engineering can utilize one of the following strategies for increasing secondary metabolite production:

1. Overcoming rate-limiting steps
2. Reducing flux through competitive pathways
3. Reducing catabolism
4. Overexpression of regulatory genes

Apart from overexpression of plant genes in plants, overexpression of plant genes in microbes can be used toward the bioconversion of readily available precursors into valuable chemical constituents. Limited knowledge of secondary metabolic pathways and the genes involved represent some of the major hurdles in the realization of the full potential of plant secondary metabolism engineering [68,69].

Artemisinin is an important sesquiterpene endoperoxide lactone. Commercially, it is produced from a herb (*Artemisia annua*) of the Compositae family. Artemisinin therapy is a preferred treatment against the falciparum malaria caused by *Plasmodium falciparum*. Optimum clinical usage of artemisinin is hampered by two factors [70]:

1. Low *in vivo* production
2. High cost of chemical synthesis

Shikonin is a well-known industrial product from *Lithospermum erythrorhizon*. The effect of expression of the bacterial ubiA gene on genetic engineering of shikonin biosynthesis has been studied, and it has been found that no significant correlation between ubiA enzyme activity and shikonin accumulation exists. Overexpression of ubiA gene alone cannot enhance shikonin formation [71]. Efforts are being made by scientists to increase flux through engineering of metabolic pathways in both *E. coli* and *Saccharomyces cerevisiae* using combinations of precursor pathway engineering and downstream gene expression optimization for flavonoid and taxoid production, respectively [71–73]. Baccatin III is an intermediate in biosynthesis of anticancer molecule taxol produced in yew (*Taxus*) trees. Its production involves a sequence of 15 enzymatic steps from primary metabolism. Efforts are being made to reconstruct the early steps of taxane diterpenoid metabolism in yeast (*S. cerevisiae*). Genes encoding five sequential pathway steps leading from primary isoprenoid metabolism to the taxadien-5α-acetoxy-10β-ol were successfully installed in a single yeast host. Yeast isoprenoid precursors can be utilized in the reconstituted pathway [74].

19.4.13 Candidate Metabolite Prediction through GC-MS

GC-MS is an important technique for the identification of plant metabolites having a volatile character. The GC-MS analysis involves six basic steps (Table 19.5).

The readouts or results of GC-MS pose one problem—proper and rapid identification of peaks corresponding to different compounds. Assigning spectral peaks to metabolites is a very tedious job and requires a thorough survey of the old as well as new research presented in journals and databases. This peak matching is a time-consuming process. Indirectly, it poses a hindrance and threat to the high-throughput screening required in present-day industry and research organizations [3,75]. Recently, a team of Japanese scientists have developed a database called KNApSAcK for this purpose [8]. This database is presently composed of 49,165 species–metabolite relationships, 24,847 metabolites, and approximately 13,094 plant species. This database has been used for elucidating structures of nearly 47% of the 50,000 known secondary metabolites. This database is particularly

TABLE 19.5

Basic Steps in GC-MS Analysis

Steps	Brief Information on Step
Extraction	Extraction of metabolites from plant specimen. Due care to prevent degradation of metabolites
Derivatization	Making the metabolites volatile, usually by trimethylsylation
Separation	Separation of derivatized metabolites on GC column under standard reproducible conditions. Temperature programming most crucial
Ionization	Ionization of metabolites eluted from GC column for better fragmentation studies
Detection	Detection of ionized metabolites. GC-TOF-MS most popular
Evaluation	Process of matching retention times of peaks and MS fragment pattern with reference to existing literature. *In silico* matching quite popular

Source: Desbrosses, G. et al. in *Lotus japonicus* Handbook, edited by Marquez, A.J., Springer-Verlag, New York, 165–174, 2005. With permission.

aimed at bringing systemization to the plethora of chemical compound structures and for the retrieval of information on molecules of therapeutic or industrial significance. There are approximately 26 families having nearly 100 metabolites each. There are several plant families for which 500 or more metabolites have been reported. These families include Fabaceae (2260 metabolites), Asteraceae (1757 metabolites), and Brassicaceae (764 metabolites).

This database is searchable according to organism name, metabolite, molecular formula, and molecular weight. The whole procedure from peak detection to metabolite prediction involves three simple steps:

1. Detection of peak in GC-MS is an important task and requires proper baseline filtering, peak deconvolution, and alignment of multiple charts. Any error in the procedural operations can lead to missing an important metabolite for the second step.

2. Search for possible compounds in the NIST (Nara Institute of Science & Technology, Ikoma, Nara, Japan) database having more than 200,000 mass spectra. Assessment of natural products involves a fair prediction of whether the metabolites are natural products or not. For this purpose, species of origin of compounds in the NIST database are surveyed before due confirmation by KNApSAcK database.

3. Finally, the KNApSAcK database can predict whether the metabolite identified can be biochemically synthesized or not.

19.4.14 Metabolite Profiling through CE-MS

Primary metabolic pathways in plants are represented by glycolysis, Krebs tricarboxylic acid cycle, pentose phosphate pathway, Calvin cycle, and biosynthetic pathways involved in the production of amino acids. These pathways produce primary metabolites which are usually ionic and nonvolatile in nature, and belong to different chemical classes. Analyzing them in plant cells or tissues by routine methods such GC-MS or LC-MS poses a number of difficulties. CE-MS is a relatively new analytical technique. It is specifically applicable to the analysis of ionic compounds. The working principle of this technique involves injecting sample solution to a capillary, followed by the application of a high voltage (in the range of ~30 kV) after which the ionic compounds move to a cathode and anode under the influence of electrostatic force. These compounds are then separated and identified on the basis of ratios of charge to mass [76].

CE-MS can be utilized for two types of analyses:

1. Cationic metabolite analysis
2. Anionic metabolite analysis

Cationic metabolite analysis is usually used for the analysis of amino, alcoholic, phenolic, carboxylic, thiol, and phosphate/sulfate groups. It is also used for the analysis of amino acids, amines, peptides, and nucleosides. This method generally uses fused silica capillaries. CE-Electron Spray Ionization (ESI)-MS is quite popular in such analyses. Anionic metabolite analysis is successful in the analysis of sugar phosphates, organic acids, CoA compounds, and nucleotides such as ATP, NAD, or NADPH. This method relies on the use of sulfonated glass capillaries.

19.5 Applications

19.5.1 Cationic Metabolite Analysis

Nontargeted metabolite profiling of T-DNA–inserted knockout *Arabidopsis* mutants using CE-ESI-TOF-MS has been reported. The physiological roles of the β-substituted alanine synthase (Bsas) gene family in plants were not clear. γ-Glutamyl-β-cyanoalanine levels were found to have decreased in *Bsas3;1* mutant. *Bsas3;1* plays a key role in the synthesis of β-cyanoalanine and subsequent production of γ-glutamyl-β-cyanoalanine *in vivo*.

19.5.2 Anionic Metabolite Analysis

CE-ESI-MS has been used to study metabolic responses in glycolysis and the pentose phosphate cycle. The response of *E. coli* to genetic and environmental perturbations has been studied using this technique. Metabolic enzyme gene disruptant metabolite variation levels were smaller in comparison to messenger RNA and protein level variations. *E. coli* possibly uses some complementary strategies against perturbations for protecting the metabolic machinery. CE-MS metabolite profiling is used for investigating the photosynthetic metabolism in plants. The addition of glucose to the cultures of *Synechocystis* sp. PCC 6803 led to changed metabolite pattern in glycolysis, pentose phosphate pathway, and Calvin cycle. The culture changed from photoautotrophic to photomixotrophic conditions. Anionic metabolite profiling of *Arabidopsis*, rice, and *C. roseus* has been done.

19.5.2.1 Advantages of CE-MS

CE-MS offers several advantages. Some of them are as follows:

1. Separation of structural isomers is possible
2. Use of MS offers high sensitivity and selectivity in analytical procedures
3. Chemical derivatization is not involved

4. Rapid recovery of analytes is possible

5. Accuracy of quantification is not affected

19.5.2.2 CE-MS and Other Methods of Analysis

CE-MS is a high-sensitivity and high-resolution analysis method and operates in a high-throughput manner. Still, it is not as popular as GC-MS or LC-MS because of reduced stability and comparatively low sensitivity. However, CE-MS has an advantage over these popular methods as it can successfully analyze a variety of ionic metabolites.

19.5.3 Biomarkers

The term biomarker, in the context of plant metabolomics, refers to proteins, chemicals including primary and secondary metabolites, genes, and biological events that are indicative of specific biological conditions in the plants.

19.5.3.1 Biomarker Identification

Metabolomics provides a tool for the comprehensive and quantitative analysis of a variety of metabolites in biological samples. Mass spectrometry methods are one important technique used for discovering biomarkers [77]. Biomarkers are required to be specific to the biological conditions. In a general sense, a normal small molecule can be considered a biomarker in the plant kingdom when its level in the plant metabolic processes either decreases or increases due to a particular reason (for example, water deficit stress [78] or soil salinity stress). Small molecule biomarkers can be identified using routine analytical methodologies such as HPLC-MS, GC-MS, or HPTLC-MS. The sophisticated infrared and NMR spectrometers can also be utilized in the identification of small as well as medium-sized biomarkers. The large-sized biomarkers, such as particular genes or proteins, can be identified using pregenomic technologies (such as gel electrophoresis or SDS-PAGE) or postgenomic technologies (such as microarrays). Similarly, camalexin can be used as a biomarker for pathogenic attack conditions in *A. thaliana* [79]. Chlorogenic acid can be used as a biomarker to identify thrips attack on resistant *Chrysanthemum* species [80].

19.5.3.2 Biomarker Discovery and Validation

The process of discovery of biomarkers is a complicated one and is largely dependent on the condition of the plant. The biomarkers discovered from normal and stressed plants may differ. Usually, the small molecules are preferred as biomarkers rather than the large-sized biomarkers because their estimation and use in metabolomic studies requires less time and less sophisticated, cheaper instrumentation. Accordingly, their validation is easier and faster. These molecules are several times more difficult to integrate with a complex network of biosynthetic pathways. To overcome this, metabolic profiling for secondary plant metabolites is resorted to in metabolomics. Restriction fragment length polymorphisms can be used as biomarkers for rice genotypes from different geographic areas. Thirty-four such restriction fragment length polymorphisms are being used as markers [81].

19.5.4 Pharmaceutical and Medicine Aspects

Plant metabolomics can be easily utilized toward drug development from natural sources, particularly plants. It is also applied in research on herbal remedies.

19.5.4.1 Metabolomics in Drug Development

Metabolomics can be utilized very well in drug development, particularly the lead molecule discoveries from plants and traditional remedies existing in various systems of medicine [8,82–84]. The overcomplicated postgenomic technologies can be used for the following purposes:

1. Determination of chemical composition of a traditional remedy (having plants as ingredients) or its formulation
2. Development of phytochemical and biological spectrum profiles for plant-based drugs
3. Prediction of safety profiles for plant-based drugs
4. Prediction of toxicity profiles for plant-based drugs
5. Detection of adulteration in plant drugs and their formulations
6. Establishing the true botanical identity of plant drugs used for therapeutic and quasitherapeutic purposes.

Recent years have witnessed much work in this direction on *Echinacea, Oryzae,* and *Nerium* species. Galanthamine is an example of a bioprospecting drug discovery from natural plant sources. This benzazepine alkaloid has been frequently used for symptomatic relief of Alzheimer disease. Traditionally, it has been obtained from the plant *Leucojum aestivum* (Amaryllidaceae) or produced by synthetic chemical means. Limited supplies of *L. aestivum* and high costs of synthesis forced scientists to bioprospect Amaryllidaceae for galanthamine. The bulbs of *Narcissus pseudonarcissus* (Amaryllidaceae) were identified as potential alternatives to *L. aestivum* for galanthamine. ^{1}H-NMR and principal component analysis were performed for determining the effect of phytogeographical constraints on galanthamine [85].

19.5.4.2 Metabolomics in Phytomedicine Research

The systems biology approach (based on genomics, proteomics, and metabolomics) has been applied to phytomedical research [86,87]. This approach may help in evidence-based phytotherapeutics and contribute to a paradigm change in the use of complex plant/phytocompound mixtures in orthodox medicine. Metabolomics is capable of providing vital clues to the molecular pharmacology of complex mixtures used in oriental medicine (particularly in traditional Chinese medicine). Phytochemical-specific signatures in gene or protein expression profiles can be used in "biological fingerprinting" of medicinal plant extracts and hence contribute to the holistic standardization of plant origin or plant-based drugs used in many of the herbal medicine traditions of the Eastern and Western worlds. Apart from this, metabolomics can also help in the prediction of safety and toxicity profiling of these medicines. Hyphenated NMR techniques may be utilized for determining the chemical composition of a complex, bioactive fraction from plants or plant remedies.

The applications of metabolomics in phytomedicine can very well be exemplified by the case of *Echinacea*. Recently, a comparative metabolomics approach has been successfully developed and applied for the characterization of three *Echinacea* species, that is, *E. purpurea*, *E. pallid*, and *E. angustifolia*, which are known and used for immunomodulatory action with many clinical implications. The metabolomics approach has resulted in the development of unique secondary plant metabolite profiles for each of the three species. These profiles can be used for better quality control over crude *Echinacea* and its extracts. These profiles have also been important in forming a basis for the typical pharmacological activities possessed by *Echinacea* [84,88].

19.5.5 Special Cases

The special cases in plant metabolomics include the study of *Arabidopsis*, food and potato metabolomes.

19.5.6 Arabidopsis Metabolome

A. thaliana has served as a model system for almost all types of metabolomic studies. Its genome sequencing was completed in the year 2000. The biochemical diversity in *A. thaliana* can be judged from the fact that it alone contains 30 terpene synthases, 272 cytochrome P450s, 130 oxogluterate-dependent dioxygenases, 64 acyltransferases, and 130 transporter proteins. This plant alone produces approximately 5000 metabolites and 4000 proteins, which constitute nearly one-third of all known proteins. As mentioned in the foregoing text, this plant forms the focus of The *Arabidopsis* Information Resource database. Ontological vocabularies providing a description of *Arabidopsis* for all databases have been constructed [29,89].

Nontargeted metabolite profiling of T-DNA–inserted knockout *Arabidopsis* mutants using CE-ESI-TOF-MS has shown that β-substituted alanine (Ala) synthase (Bsas) *Bsas3;1* plays a key role in the synthesis of β-cyanoalanine and γ-glutamyl-β-cyanalanine *in vivo* [90]. This plant has been the subject of numerous studies related in metabolite production, particularly under the influence of abiotic stress factors. Regulatory processes play a vital role in the metabolic adjustments in *Arabidopsis* during cold acclimation. These processes remain independent of transcript abundance. *Arabidopsis* metabolome undergoes extensive reconfigurations in extreme low temperature conditions and C-repeat–binding factor plays an important role in eliciting the typical responses. Upon exposure to salt stress, *Arabidopsis* has been shown to produce higher quantities of sucrose and proline molecules as an indication of abiotic type of stress [30]. An exclusive *Arabidopsis* Metabolomics Consortium exists and operates under the National Science Foundation in the United States [91].

19.5.7 Food Metabolomics

Food is one of the basic requirements for living organisms. Human beings value food for its nutritional value, taste, and fragrance [92]. Many of the foods provide primary and secondary metabolites, which have therapeutic value and find an important role in the maintenance of physical and mental health, and prevent the occurrence of many diseases. It is this biochemical profile which determines the value, stability, and shelf-life of food items. Some of the foods may contain substances that can exert harmful effects. Food policies across the globe (particularly in the European Union) demand a stricter control over food

TABLE 19.6

Important Metabolites of Commercial Crops Under META-PHOR

Crop	Chemical Constituents	Significance
Broccoli	Phytosterols	Prevention of cancer, heart disease, and postmenopausal syndrome
	Glucosinolates	Prevention of thyroid gland disorders (e.g., hyperthyrotoxicosis)
	Flavonoids	Free-radical scavengers
Melon	Isoprenoids	Characteristic fragrance and taste
	Flavonoids	Free radical scavengers
	Sugars	Taste
Rice	Micronutrients	Nutritional value
	Fragrant components	Characteristic aroma desirable for high market value
	Vitamins	Nutritional value

items whether used in the crude form, in semifinished products, or in finished products such as nutraceuticals. The metabolomics approach has been tried, in the case of food, by the EU Research Division and has led to the initiation of many EU funded projects. One such project started in October 2006 is called the metabolomics for plants, health, and outreach (META-PHOR). Presently, this project focuses on three commercial crops: broccoli, melon, and rice (Table 19.6).

META-PHOR is determined to ensure the quality and safety of food in the EU. It is also aiming at meeting particular local nutritional demands, particularly in developing countries, and makes efforts to generate biofortified crops—with benefits beyond nutrition alone [92].

19.5.8 Potato Metabolomics

Potato is the fourth largest global food crop. It is a rich source of dietary carbohydrates, commercial starch, and glycoalkaloids of pharmaceutical interest. Recent years have seen an increase in usage of genetic modification of various food crops. Similar increases were expected for safety concerns in the usage of products of such GM crops. This safety concern has led to numerous studies on the metabolomic perspective of potato. The main work in this regard has been conducted at the Scottish Crop Research Institute and the Institute of Food Research located at Dundee and Norwich, respectively, in United Kingdom. The potato varieties, normal landraces and GM transgenic crops, tissue cultures and vector-only control were subjected to metabolomic studies. The effects of genetic modification and tissue culture were studied on the glycoalkaloid levels and ratios of α-chaconine and α-solanine. These processes were found to increase the relative contents of the reportedly more toxic α-chaconine [93,94].

19.5.9 Plant Research

Metabolomics technologies find a large number of applications in the field of in-plant sciences and research. These applications include the following (indicative list only):

1. Study of plant developmental biology stages
2. Phenotyping of genetically modified plants

3. Quantitative trait analysis

4. Systems biology of plants

5. Discovery and identification of disease and stress markers in plants

6. Determination of the effects of genetic alterations on plant performance

7. Investigation of altered gene effect on the cellular composition of the plant

8. Monitoring the effects of transgenesis under normal and stress conditions

9. Assessment of potential risks linked with transgenesis

10. Production of new crop varieties with better adaptability to stresses

11. Determination of altered levels of osmolytes and osmoprotectants

Acknowledgments

The author is thankful to the Rayat and Bahra Group of Institutions, Mohali for providing computing and literature search facilities for the present work. No words can ever be sufficient to acknowledge the authors/researchers whose works have formed the basis of this work. Support and constant encouragement from my family members (especially Zeba) and teachers (particularly Dr. Shibli Jameel of Hamdard University, New Delhi) are also acknowledged.

References

1. Oksman-Caldentey, K., and K. Saito. 2005. Integrating genomics and metabolomics for engineering plant metabolic pathways. *Current Opinion in Biotechnology* 16:174–179.

2. Fiehn, O., J. Kopka, P. Dormann, T. Altmann, R.N. Trethewey, and L. Willmitzer. 2000. Metabolite profiling for plant functional genomics. *Nature Biotechnology* 18:1157–1161.

3. Desbrosses, G., D. Steinhauser, J. Kopka, and M. Udvardi. 2005. Metabolome analysis using GC-MS. In *Lotus japonicus* Handbook, edited by Marquez, A.J., 165–174. Springer-Verlag, New York.

4. Deuschle, K., M. Fehr, M. Hilpert, I. Lager, S. Lalonde, L.L. Looger, and S. Okumoto. 2005. Genetically encoded sensors for metabolites. *Cytometry A* 64:3–9.

5. Weckwerth, W. 2003. Metabolomics in system biology. *Annual Reviews in Plant Biology* 54: 669–689.

6. Last, R.L., A.D. Jones, and Y. Shachar-Hill. 2007. Towards the plant metabolome and beyond. *Nature Reviews Molecular Cell Biology* 8:167–174.

7. Huang, Q., P. Yin, X. Lu, H. Kong, and G. Xu. 2009. Applications of chromatography-mass spectrometry in metabonomics. *Se Pu* 27:566–572.

8. Oishi, T., K. Tanaka, T. Hashimoto, Y. Shinbo, K. Jumtee, T. Bamba et al. 2009. An approach to peak detection in GC-MS chromatograms and application of KNApSAcK database in prediction of candidate metabolites. *Plant Biotechnology* 26:167–174.

9. Dan, M., X.F. Gao, G.X. Xie, Z. Liu, A.H. Zhao, and W. Jia. 2007. Application of metabolomics in research of plant metabolites. *Zhongguo Zhong Yao Za Zhi* 32:2337–2341.

10. Wishart, W. 2010. Applications of Metabolomics in Nutritional Science, University of Alberta. Available from: http://www.metabolomics.ca/News/lectures/UNC-metabolomics-final.pdf.

11. Dunn, W.B., and D.I. Ellis 2005. Metabolomics: Current analytical platforms and methodologies. *Trends in Analytical Chemistry* 24:285–294.
12. Bino, R.J., R.D. Hall, O. Fiehn, J. Kopka, K. Saito, J. Draper et al. 2004. Potential of metabolomics as functional genomics tool. *Trends in Plant Science* 9:418–425.
13. Morgenthal, K., W. Weckwerth, and R. Steuer. 2006. Metabolomic networks in plants: Transition from pattern recognition to biological interpretation. *BioSystems* 83:108–117.
14. Ekins, S., A. Bugrim, and Y. Nikolsky. 2005. System biology: Applications in drug discovery. In *Drug Discovery Handbook*, edited by Gad, S.C., 123–184. New Jersey: Wiley Interscience.
15. Kell, D.B. 2004. Metabolomics and systems biology: Making sense of the soup. *Current Opinion in Microbiology* 7:296–307.
16. Huang, L., W. Gao, J. Zhou, and R. Wang. 2010. Systems biology applications to explore secondary metabolites in medicinal plants. *Zhongguo Zhong Yao Za Zhi* 35:8–12.
17. Maloney, V. 2004. Plant metabolomics. *BioTeach Journal* 2:92–99.
18. Wang, Y., Z. Liu, A. Zhao, M. Su, G. Xie, and W. Jia. 2009. Functional genomic approaches to explore secondary metabolites in medicinal plants. *Zhongguo Zhong Yao Za Zhi* 34:6–10.
19. Peric-Concha, N., and P.F. Long. 2003. Mining the microbial metabolome: A new frontier for natural product lead discovery. *Drug Discovery Today* 23:1078–1084.
20. Choi, Y.H., H.K. Choi, A.M. Peltenburg-Looman, A.W. Lefeber, and R. Verpoorte. 2004. Quantitative analysis of ginkgolic acids from ginkgo leaves and products using 1H-NMR. *Phytochemical Analysis* 15:325–330.
21. Choi, Y.H., H.K. Kim, A. Hazekamp, C. Erkelens, A.W. Lefeber, and R. Verpoorte. 2004. Metabolomic differentiation of *Cannabis sativa* cultivars using 1H NMR spectroscopy and principal component analysis. *Journal of Natural Products* 67:953–957.
22. Frederich, M., C. Jansen, P. de Tullio, M. Tits, V. Demoulin, and L. Angenot. 2010. Metabolomic analysis of *Echinacea* spp. by 1H nuclear magnetic resonance spectrometry and multivariate data analysis technique. *Phytochemical Analysis* 21:61–65.
23. Kant, M.R., and I.T. Baldwin. 2007. The ecogenetics and ecogenomics of plant–herbivore interactions: Rapid progress on a slippery road. *Current Opinion in Genetic Development* 17:519–524.
24. Van-Dam, N.M. 2009. How plants cope with biotic interactions. *Plant Biology (Stuttg)* 11:1–5.
25. Urano, K., K. Maruyama, Y. Ogata, Y. Morishita, M. Takeda, N. Sakurai et al. 2009. Characterization of the ABA-regulated global responses to dehydration in *Arabidopsis* by metabolomics. *Plant Journal* 57:1065–1078.
26. Hien Dao, T.T., R.C. Puig, H.K. Kim, C. Erkelens, A.W. Lefeber, H.J. Linthorst et al. 2009. Effect of benzothiadiazole on the metabolome of *Arabidopsis thaliana*. *Plant Physiology and Biochemistry* 47:146–152.
27. Foyer, C.H., M. Parry, and G. Noctor. 2003. Markers and signals associated with nitrogen assimilation in higher plants. *Journal of Experimental Biology* 54:585–593.
28. Fernie, A.R., and N. Schauer. 2008. Metabolomics-assisted breeding: A viable option for crop improvement? *Trends in Genetics* 25:39–48.
29. Jaiswal, P., S. Avraham, K. Ilic, E.A. Kellogg, S. McCouch, A. Pujar et al. 2005. Plant ontology (PO): A controlled vocabulary of plant structures and growth stages. *Computational and Functional Genomics* 6:388–397.
30. Shulaev, V., D. Cortes, G. Miller, and R. Mittler. 2008. Metabolomics for plant stress response. *Physiologia Plantarum* 132:199–208.
31. Hanson, A.D., B. Rathinasabapathi, J. Rivoal, M. Burnet, M.O. Dillon, and D.A. Gage. 1994. Osmoprotective compounds in the Plumbaginaceae: A natural experiment in metabolic engineering of stress tolerance. *Proceedings of the National Academy of Sciences of the United States of America* 91:306–310.
32. Fiehn, O. 2010. Study of Metabolic Control in Plants by Metabolomics. Available from: http://fiehnlab.ucdavis.edu/publications/Fiehn%202005%20Plaxton-ed%20Blackwell_book%20chapter_Study%20of%20metabolic%20control%20in%20plants%20by%20metabolomics.pdf.
33. Kannicht, C., and B. Fuchs. 2008. Post-translational modifications of proteins. In *Molecular Biomethods Handbook*, edited by Walker, J.M., and R. Raplay, 427–449. New York: Springer.

34. Liu, W.Z., and Z.F. Wang. 2004. Accumulation and localization of camptothecin in young shoot of *Camptotheca acuminata*. *Journal of Plant Physiology and Molecular Biology* 30:405–412.
35. van Hengel, A.J., M.P. Harkes, H.J. Wichers, P.G.M. Hessleink, and R.M. Buitelaar. 1992. Characterization of callus formation and camptothecin production by cell lines of *Camptotheca acuminata*. *Plant Cell Tissue and Organ Culture* 28:11–18.
36. López-Meyer, M., C.L. Nesler, and T.D. McKnight. 1994. Sites of accumulation of the antitumor alkaloid camptothecin in *Camptotheca acuminata*. *Planta Medica* 60:558–560.
37. Lu, H., and T.D. McKnight. 1999. Tissue specific expression of the beta subunit of tryptophan synthase in *Camptotheca acuminata*, an indole alkaloid producing plant. *Plant Physiology* 120:43–52.
38. Sakato, K., H. Tanaka, N. Mukai, and M. Misawa. 1974. Isolation and identification of camptothecin from cells of *Camptotheca acuminata* suspension cultures. *Agricultural and Biological Chemistry* 38:217–218.
39. Wiedenfeld, H., M. Furmanowa, E. Roeder, J. Guzewska, and H. Gustowski. 1997. Camptothecin and 10-hydroxycamptothecin in callus and plantlets of *Camptotheca acuminata*. *Plant Cell, Tissue and Organ Culture* 49:213–218.
40. Lorence, A., F. Medina-Bolivar, and C.L. Nessler. 2004. Camptothecin and 10-hydroxycamptothecin from *Camptotheca acuminata* hairy roots. *Plant Cell Reports* 22:437–441.
41. Li, Z., and Z. Liu. 2005. Plant regeneration from leaf petioles in *Camptotheca acuminata*. *In vitro Cellular & Developmental Biology. Plant* 41:262–265.
42. Matsuda, F., H. Redestig, Y. Sawada, Y. Shinbo, M.Y. Hirai, S. Kanaya, and S. Kazuki. 2009. Visualization of metabolite identifier information. *Plant Biotechnology* 26:479–483.
43. McConville, M. 2009. Metabolomics: Bridging the phenotype–genotype gap. *Australian Biochemist* 40:15.
44. Mendonça-Filho, R.R. 2006. Bioactive phytocompounds: New approaches in the phytosciences. In *Modern Phytomedicine. Turning Medicinal Plants into Drugs*, edited by Ahmad, I., F. Aqil, and M. Owais, 1–24. Weinheim: Wiley-VCH.
45. Wolf, D., and K. Siems. 2007. Burning the hay to find the needle—Data mining strategies in natural product dereplication. *Chimia* 61:339–345.
46. Bruno, D. 2005. Drug discovery, natural substances & pharmaceutical industry. *11th NAPRECA Symposium Book of Proceedings, Antananarivo, Madagascar, 27–34. Feb 10, 2010.* http://www .napreca.net/publications/11symposium/pdf/D-27-34-Bruno.pdf.
47. Steinbeck, C. 2004. NMRShiftDB—A free information system for organic molecules and their spectral data. *CDK News* 1:2–3.
48. Wolfender, J.L., K. Ndjoko, and K. Hostettmann. 2003. Liquid chromatography with ultraviolet absorbance-mass spectrometric detection and with nuclear magnetic resonance spectroscopy: A powerful combination for the on-line structural investigation of plant metabolites. *Journal of Chromatography A* 1000:437–455.
49. Beutler, J.A., A.B. Alvarado, D.E. Schaufelberger, P. Andrews, and T.G. McCloud. 1990. Dereplication of phorbol bioactives: *Lyngbya majuscula* and *Croton cuneatus*. *Journal of Natural Products* 53:867–874.
50. Hu, J.F., R. Patel, B. Li, E. Garo, G.W. Hough, M.G. Goering et al. 2007. Anti-HCV bioactivity of pseudoguaianolides from *Parthenium hispitum*. *Journal of Natural Products* 70:604–607.
51. Minguzzi, S., L.E. Barata, Y.G. Shin, P.F. Jonas, H.B. Chai, E.J. Park et al. 2002. Cytotoxic withanolides from *Acnistus arborescens*. *Phytochemistry* 59:635–641.
52. Scio, E., A. Ribeiro, T.M. Alves, A.J. Romanha, Y.G. Shin, G.A. Cordell, and C.L. Zani. 2003. New bioactive coumarins from *Kielmeyera albopunctata*. *Journal of Natural Products* 66: 634–637.
53. van Elswijk, D.A., U.P. Schobel, E.P. Lansky, H. Irth, and J. van der Greef. 2004. Rapid dereplication of estrogenic compounds in pomegranate (*Punica granatum*) using on-line biochemical detection coupled to mass spectrometry. *Phytochemistry* 65:233–241.
54. Borevitz, J.O., and J. Chory. 2004. Genomics tools for QTL analysis and gene discovery. *Current Opinion in Plant Biology* 7:132–136.

55. Kliebenstein, D.J. 2007. Metabolomics and plant quantitative trait locus analysis—the optimum genetical genomics platform? In *Concepts in Plant Metabolomics*, edited by Nikolau, B.J., and E.S. Wurtele, 29–45. Dordrecht: Springer.

56. Prasanna, B.M. 2010. QTL analysis and its applications in crop plants. Available from: http://www.iasri.res.in/ebook/EB_SMAR/e-book_pdf%20files/Manual%20IV/10-QTL.pdf.

57. Werf, J. 2010. Methods for QTL analysis. Available from: http://www-personal.une.edu.au/~jvanderw/Models_for_QTL_analysis.pdf.

58. Gupta, P.K. 2002. Molecular markers and QTL analysis in crop plants. *Current Science* 83:113–114.

59. Lipsa, F.D., R.J. Snowdon, and W. Friedt. 2009. QTL analysis of condensed tannins content in *Brassica napus* L. *Research Journal of Agricultural Science* 41:274–278.

60. Rousseaux, M.C., C.M. Jones, and D. Adams. 2005. QTL analysis of fruit antioxidants in tomato using *Lycopersicon penellii* L. *Theorotical and Applied Genetics* 111:1396–1408.

61. Böttcher, C., L. Westphal, C. Schmotz, E. Prade, D. Scheel, and E. Glawischnig. 2009. The multifunctional enzyme CYP71B15 (PHYTOALEXIN DEFICIENT3) converts cysteine-indole-3-acetonitrile to camalexin in the indole-3-acetonitrile metabolic network of *Arabidopsis thaliana*. *Plant Cell* 21(6):1830–1845.

62. Hong, S.B., C.A. Peebles, J.V. Shanks, K.Y. San, and S.I. Gibson. 2006. Expression of the *Arabidopsis* feedback-insensitive anthranilate synthase holoenzyme and tryptophan decarboxylase genes in *Catharanthus roseus* hairy roots. *Journal of Biotechnology* 122:28–38.

63. Hughes, E.H., S.B. Hong, S.I. Gibson, J.V. Shanks, and K.Y. San. 2004. Metabolic engineering of the indole pathway in *Catharanthus roseus* hairy roots and increased accumulation of tryptamine and serpentine. *Metabolic Engineering* 6:268–276.

64. Goklany, S., R.H. Loring, J. Glick, and C.W. Lee-Parsons. 2009. Assessing the limitations to terpenoid indole alkaloid biosynthesis in *Catharanthus roseus* hairy root cultures through gene expression profiling and precursor feeding. *Biotechnology Progress* 25:1289–1296.

65. Xu, M., and J. Dong. 2007. Enhancing terpenoid indole alkaloid production by inducible expression of mammalian Bax in *Catharanthus roseus* cells. *Science China C - Life Sciences* 50:234–241.

66. Liu, C., Y. Zhao, and Y. Wang. 2006. Artemisinin: Current state and perspectives for biotechnological production of an antimalarial drug. *Applied Microbiology and Biotechnology* 72:11–20.

67. Yang, Z.R., X. Mao, R.Z. Li. 2005. Research progress in genetic engineering of plant secondary metabolism. *Journal of Plant Physiology and Molecular Biology* 31:11–18.

68. Verpoorte, R., R. van der Heijden, and J. Memelink. 2000. Engineering the plant cell factory for secondary metabolite production. *Transgenic Research* 9:323–343.

69. Verpoorte, R., and J. Memelink. 2002. Engineering secondary metabolite production in plants. *Current Opinion Biotechnology* 13:181–187.

70. Arsenault, P.R., K.K. Wobbe, and P.J. Weathers. 2008. Recent advances in artemisinin production through heterologous expression. *Current Medicinal Chemistry* 15:2886–2896.

71. Boehm, R., S. Sommer, S.M. Li, and L. Heide. 2000. Genetic engineering on shikonin biosynthesis: Expression of the bacterial ubiA gene in *Lithospermum erythrorhizon*. *Plant Cell and Physiology* 41:911–919.

72. Fowler, Z.L., W.W. Gikandi, and M.A. Koffas. 2009. Increased malonyl coenzyme A biosynthesis by tuning the *Escherichia coli* metabolic network and its application to flavanone production. *Applied and Environmental Microbiology* 75:5831–5839.

73. Dejong, J.M., Y. Liu, A.P. Bollon, R.M. Long, S. Jennewein, D. Williams, and R.B. Croteau. 2006. Genetic engineering of taxol biosynthetic genes in *Saccharomyces cerevisiae*. *Biotechnology and Bioengineering* 93:212–224.

74. Brunakova, K., and J. Kosuth. 2009. Gene expression profiling in *Taxus baccata* L. seedlings and cell cultures. *Methods in Molecular Biology* 547:249–262.

75. Sinbo, Y., Y. Nakamura, H. Asahi, Md. Altaf-Ul-Amin, K. Kurokawa, and S. Kanaya. 2004. Species-specific diversity of metabolites: A natural product database system. In *Proceedings of 10th International Congress for Culture Collections*, edited by Watanabe, M.M., K. Suzuki, and T. Seki, 61–64.

76. Harada, K., and E. Fukusaki. 2009. Profiling of primary metabolite by means of capillary electrophoresis-mass spectrometry and its application for plant science. *Plant Biotechnology* 26:47–52.

77. Wolfender, J.L., G. Glauser, J. Boccard, and S. Rudaz. 2009. MS-based plant metabolomic approaches for biomarker discovery. *Natural Products Communications* 4:1417–1430.

78. Shao, H.B., L.Y. Chu, C.A. Jaleel, P. Manivannan, R. Panneerselvam, and M.A. Shao. 2009. Understanding water deficit stress-induced changes in the basic metabolism of higher plants—biotechnologically and sustainably improving agriculture and the ecoenvironment in arid regions of the globe. *Critical Reviews in Biotechnology* 29:131–151.

79. Zhou, N., T.L. Tootle, and J. Glazebrook. 1999. Arabidopsis PAD3, a gene required for camalexin biosynthesis, encodes a putative cytochrome P450 monooxygenase. *Plant Cell* 11:2419–2428.

80. Leiss, K.A., F. Maltese, Y.H. Choi, R. Verpoorte, and P.G. Klinkhamer. 2009. Identification of chlorogenic acid as a resistance factor for thrips in *Chrysanthemum*. *Plant Physiology* 150:1567–1575.

81. Kell, D.B. 2007. Metabolomic biomarkers: Search, discovery and validation. *Expert Reviews in Molecular Diagnosis* 7:329–333.

82. van der Kooy, F., F. Maltese, Y.H. Choi, H.K. Kim, and R. Verpoorte. 2009. Quality control of herbal material and phytopharmaceuticals with MS and NMR based metabolic fingerprinting. *Planta Medica* 75:763–775.

83. Lao, Y.M., J.G. Jiang, and L. Yan. 2009. Application of metabonomic analytical techniques in the modernization and toxicology research of traditional Chinese medicine. *British Journal of Pharmacology* 157:1128–1141.

84. Shyur, L., and N. Yang. 2008. Metabolomics for phytomedicine research and drug development. *Current Opinion in Chemical Biology* 12:66–71.

85. Lubbe, A., B. Pomahacova, Y.H. Choi, and R. Verpoorte. 2009. Analysis of metabolic variation and galanthamine content in *Narcissus* bulbs by 1H NMR. *Phytochemical Analysis* 21:66–72.

86. Wang, M., R.J. Lamers, H.A. Korthout, J.H. van Nesselrooij, R.F. Witkamp, R. van der Heijden et al. 2005. Metabolomics in the context of systems biology: Bridging traditional Chinese medicine and molecular pharmacology. *Phytotherapy Research* 19:173–182.

87. Wu, B., S.K. Yan, Z.Y. Shen, and W.D. Zhang. 2007. Metabonomic technique and prospect of its application in integrated traditional Chinese and Western medicine research. *Journal of Plant Physiology and Molecular Biology* 5:475–480.

88. Gilroy, C.M., J.F. Steiner, T. Byers, H. Shapiro, and W. Georgian. 2003. *Echinacea* and truth in labeling. *Achieves of Internal Medicine* 163:699–704.

89. Saito, K. 2010. Phytochemical Genomics for Manipulation of Plant Secondary Metabolites. http://www.psc.riken.go.jp/english/group/function/index.html and http://www.p.chiba-u.ac.jp/lab/idenshi/index-e.html.

90. Watanabe, M., M. Kusano, A. Oikawa, A. Fukushima, M. Noji, and K. Saito. 2008. Physiological roles of the β-substituted alanine synthase gene family in *Arabidopsis*. *Plant Physiology* 146:310–320.

91. Anonymous. 2010. Available from: http://www.plantmetabolomics.org.

92. Hall, R.D. 2007. Food metabolomics: META-PHOR: A new European research initiative. *Agro FOOD Industry Hi-tech* 18:14–16.

93. Stewart, D., I. Colquhoun, J. McNicol, and H. Davies. 2010. Plant Metabolomics: Addressing the Scientific Concerns and Uncertainties of Genetic Modification. Available from: http://www.scri.ac.uk/staff/derekstewart.

94. Shepherd, T., G. Dobson, R. Marshall, S.R. Verrall, S. Conner, D.W. Griffiths et al. 2004. Profiling of metabolites and volatile flavour compounds from *Solanum* species using gas chromatography-mass spectrometry. *Proceedings of the Third International Congress on Plant Metabolomics, Ames, Iowa, USA, June*. Available from: http://www.scri.ac.uk/staff/derekstewart.

20

Nutraceuticals: Applications in Biomedicine

Vellingiri Vadivel

CONTENTS

20.1 Introduction

Foods are primarily evaluated according to their nutritional value. In other words, the most important factor for the evaluation of foods is their "primary" function, that is, the role of providing standard nutrient components. This primary function is particularly important for people who are suffering from a shortage of food. The "secondary" function of foods, which is defined in terms of sensory properties such as taste, flavor, appearance, and texture, is also critical for the food industry in many countries. In addition to these functions, much attention has recently been paid to the "tertiary" function of foods, that is, the role of food components in preventing diseases by modulating physiological systems (e.g., immune, endocrine, nervous, circulatory, and digestive systems).

Examples of the tertiary functional properties of foods that have been studied are anticarcinogenic, antimutagenic, antioxidant, and antiaging activities [1]. Due to increasing concerns about health in recent years, efforts have been made by the food industries to develop new foods with tertiary functions called "functional foods." Numerous food components showing tertiary functions have been isolated and characterized. For example, many vegetables have been shown to contain a variety of biologically active phytochemicals. There has been an accumulation of scientific findings in recent years regarding the roles of such components in the prevention of diseases. Rapid progress has been made in the development of functional foods based on the results of studies on food components providing positive health benefits over and above normal nutritional benefits.

Diet is a major focus of public health strategies aimed at maintaining optimal health throughout life, preventing early onset of chronic diseases such as gastrointestinal disorders, cardiovascular disease, diabetes, obesity, cancer, osteoporosis, as well as promoting healthier aging. Although the highly complex relationship between food and health is still poorly understood, recent advances in research in a variety of different disciplines provide promising new approaches to improve our understanding. The increasing consumer health consciousness and the growing demand for healthy foods are stimulating innovative new product developments in the food industry internationally, and it is also responsible for the expanding worldwide interest in functional foods. In general, it is proposed that a food can be regarded as functional if, beyond its inherent nutritional effects, it satisfactorily demonstrates a beneficial effect on one or more target functions in the body in a way that is relevant to either the state of well-being and health or to the reduction of risk of a disease. Other definitions shortly state that a functional food is any food that may provide a health benefit beyond the traditional nutrients it contains.

That food might provide therapeutic benefit is clearly not a new concept. The tenet, "Let food be thy medicine and medicine be thy food" was embraced 2500 years ago by Hippocrates, the father of medicine. However, this "food as medicine" philosophy fell into relative obscurity in the nineteenth century with the advent of modern drug therapy. In the 1900s, the important role of diet in disease prevention and health promotion came to the forefront once again. During the first 50 years of the twentieth century, micronutrients such as minerals and vitamins received more attention. Nonetheless, today scientific focus is on the identification of various bioactive compounds and their role(s) in the prevention of various chronic diseases.

The development of novel functional foods is a great opportunity to improve the quality of foods available to consumers to benefit their health and well-being, and these added-value food products are also of growing industrial and social interest in most modern societies. This is a result of consumers being aware of serious nutritional imbalances in their diets. Through several nutrition surveys across Europe, it has been stated that several population subgroups do not reach nationally recommended daily allowances for several important nutrients [2]. Moreover, in many cases, the gap between actual and recommended intakes is significant and the prevalence of suboptimal intakes is high, so the development of such novel foods is highly desirable. In a workshop organized by the International Life Sciences Institute in Lisbon in 1997, experts from academia, regulatory agencies, industry, and consumer groups from key countries in Europe came to the conclusion that the addition of nutrients to foods can provide an effective and safe strategy to improve actual micronutrient intake and status by restoring amounts lost, by providing key nutrients in foods that replace traditional products, and by extending the range of foods rich in relevant micronutrients.

Functional foods offer significant sales and marginal growth opportunities. The growth value of functional foods is expected to be higher, that is, up to five times at its highest, over the next few years compared with that of total packaged foods. In most commercial functional foods, a number of bioactive components are added that are considered to be beneficial to the health of consumers. It is reckoned that an important aspect of these functional foods is to provide an appropriate dose of these bioactive components to have a beneficial rather than a toxic effect on human health. In connection with this, this chapter mainly emphasizes the effect of certain dietary bioactive compounds such as polyphenolics from berries, catechins from black tea, flavonoids from fruits and vegetables, tannins from sorghum grains, lycopene from tomato, protease inhibitors from legume seeds, and organosulfur from *Allium* vegetables on the prevention/treatment of different types of cancer. This chapter will give a better understanding to the readers about the importance of these phytochemicals in maintaining a healthy life and is also helpful in formulating functional foods to prevent the incidence of cancer.

20.2 Cancer

Cancer is the second leading cause of death in many developed countries next to cardio-vascular diseases, accounting for approximately one-quarter of all deaths. Among women aged 40 to 79 and among men aged 60 to 79, cancer is the leading cause of death in the United States. The lifetime probability of developing cancer is approximately 46% for men and approximately 38% for women. Among men, cancers of the prostate, lung, and colon–rectum account for approximately 56% of all newly diagnosed cancers and for approximately 51% of all causes of cancer death in the United States. Among women, cancers of the breast, lung, colon, and uterine corpus account for approximately 61% of all new cancer cases and for approximately 55% of all causes of cancer death in the United States [3].

Although genetic inheritance influences the risk of cancer, most of the variation in cancer risk across populations and among individuals is due to environmental and lifestyle factors. Evidence that lifestyle factors (e.g., unhealthy diets, excessive adiposity, and smoking) play a key role in promoting cancer comes from several sources. First, studies show that the chances of identical twins developing cancer at the same site are generally less than 10% [3]. Second, studies of migrants moving from a low- to a high-risk area have shown that they acquire the cancer pattern of the host country within a single generation. Finally, data from epidemiological studies strongly suggest that excessive calorie intake and adiposity, and low intake of vegetables, fruits, beans, and whole grains are key players in the pathogenesis of the most common types of cancer. Data from several large epidemiological studies indicate that excessive adiposity, especially abdominal adiposity, is a major contributor to the increased incidence or death from adenocarcinoma of the esophagus, colon cancer, postmenopausal breast cancer, endometrial, kidney, liver, gallbladder, and pancreatic cancers.

Excessive adiposity due to excessive energy intake and minimal physical activity is associated with insulin resistance, low-grade inflammation, and changes in hormone and growth factor levels that likely play a central role in the pathogenesis of many cancers. Chronic positive energy balance promotes adipose tissue hypertrophy, adipokine-mediated insulin resistance, compensatory hyperinsulinemia, and increased sex hormone availability. Insulin, estrogens, and androgens are strong mitogens for cells and stimulate the growth and development of several tumor types. Interestingly, the development of two of the most common cancers affecting men and women in the Western world (i.e., prostate and premenopausal breast cancer) are not directly associated with adiposity or chronic hyperinsulinemia, suggesting that other metabolic factors are also major causes in their pathogenesis.

Several epidemiological studies have found a strong association between plasma levels of insulin-like growth factor-I (IGF-I) and the risk of developing prostate cancer, premenopausal breast cancer and colon cancer. Nutrient intake is a major regulator of circulating IGF-I, which promotes tumor development by stimulating cell proliferation and inhibiting cell death. Recent data from observational studies indicated that long-term protein intake, but not calorie intake, regulates serum IGF-I concentration in humans, suggesting that long-term protein intake is an important cancer risk factor. Several other factors have also been hypothesized to increase the risk of cancer, including the lack of adequate consumption of vegetables, fruits, beans, and whole grains that are rich in antioxidant vitamins and protective phytochemicals, consumption of animal foods rich in fat and genotoxic heterocyclic amines and polycyclic aromatic hydrocarbons, hypovitaminosis D, and exposure to tobacco smoke, pollutants, and pesticides. So, it is urgently necessary to recognize the link between dietary habits and incidence of cancer and identify the protective effect of phytochemicals from common foods toward cancer. In connection with this, the protective/curative effects of certain well-proven dietary phytochemical compounds against various types of cancers are discussed in Sections 21.3 through 21.9.

20.3 Polyphenolics from Berries

The phenolic compounds constitute one of the most numerous and ubiquitously distributed groups of secondary plant metabolites, which range from simple molecules (e.g., simple phenols, phenolic acids, phenyl-propanoids, and flavonoids) to highly polymerized compounds (e.g., lignins, lignans, tannins, suberins, and cutins). The phenolic compounds have been demonstrated to prevent the development of many chronic diseases, which might be associated with their powerful antioxidant and free radical scavenging properties. Hence, food technologists are keen to harness the nutritional benefits of phenolics, namely, their antioxidant or free radical scavenging, food preservative, antimicrobial, antimutagenic, therapeutic, and pharmaceutical properties.

Ellagic acid (Figure 20.1) is a natural phenolic antioxidant found in numerous fruits and vegetables including blackberry, raspberry, strawberry, cranberry, wolfberry, walnut, pecan, pomegranate, and other plant foods. *In vivo* studies have shown that dietary intake of ellagic acid inhibits chemically induced tumorigenesis in the lung [4], skin [5], and esophagus [6]. In addition, dietary administration of ellagic acid has been found to inhibit the development of azoxymethane-induced small intestinal adenocarcinomas in rats [7]. The most active components in raspberry were the ellagitannins, but these components break down readily and the resultant products, including ellagic acid, may be the actual active components. Joint research conducted between the Scottish Crop Research Institute and the University of Ulster has shown that berry extracts can inhibit the initiation, progression, and invasiveness of colon cancer cells (Figure 20.2) [8]. Ellagic acid has been shown to cause G_1 phase cell cycle arrest, inhibit cell growth, increase the expression of cyclin-dependent kinase inhibitor p21, and induce apoptosis [9]. Ellagic acid inhibited

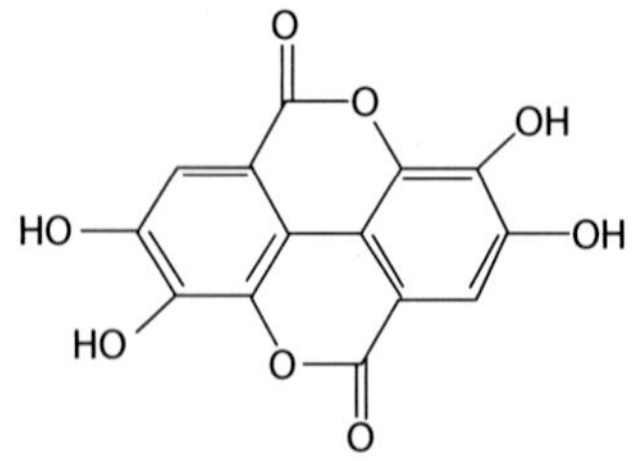

FIGURE 20.1
Structure of ellagic acid.

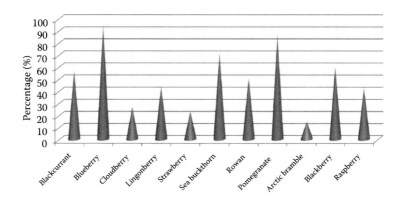

FIGURE 20.2
Inhibition of growth of GeLa cancer cells by berry extract. (From Battino, M. et al., *Nutr. Rev.* 67:S145–S150, 2009. With permission.)

N-nitrosomthylbenzylamine metabolism and DNA binding in cultured rat esophagus. Ferulic acid and β-sitosterol have been shown to inhibit the growth of premalignant and malignant human oral cavity cell lines but not normal oral cavity cells [10].

Dietary administration of phenolics from berries caused the inhibition of *N*-nitrosomthylbenzylamine-induced esophageal tumor incidence and multiplicity in rats [11,12]. Suppression of azoxymethane-induced colon tumorigenesis in rats by administration of berry phenolics has also been reported [13]. This suppression was associated with a reduction in the levels of 8-hydroxy-20-deoxyguanosine in the urine of azoxymethane-treated rats, suggesting that the phenolic compounds in berries reduce oxidative stress in carcinogen-treated animals. The phenolic compounds from berries also influence multiple signaling pathways through modulation of key regulatory transcription factors (e.g., nuclear factor-κB), kinases (e.g., Akt), and mitogen-activated protein kinases leading to effects on downstream genes such as cyclooxygenase-2 and inducible nitric oxide synthase.

Anthocyanins are glycosidically bound anthocyanidins present in many flowers and fruits. The cyanidin glycosides (Figure 20.3) found in black raspberries was demonstrated to be effective inhibitors of nuclear factor-κB expression in mouse epidermal JB-6 cells [14]. It is interesting to speculate that the inability of phenolic compounds from blueberries to inhibit *N*-nitrosomthylbenzylamine-induced tumorigenesis in the rat esophagus might be because they contain different anthocyanins than black raspberries and have very low levels of ellagitannins.

Topical application of a 10% black raspberry gel to the surface of oral dysplastic lesions in 17 patients for a period of 6 weeks led to histologic regression in a subset of patients, as well as a significant reduction in loss of heterozygosity at three tumor suppressor gene

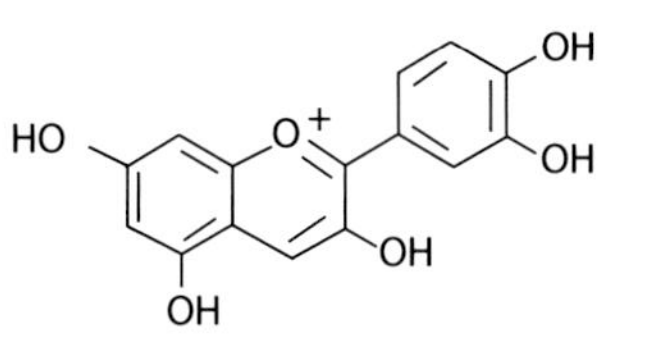

FIGURE 20.3
Structure of cyanidine.

loci. In phase II clinical trials, freeze-dried black raspberries were proven to be effective in decreasing the risk of oral, esophageal, and colon cancers [12]. Oral administration of 32 or 45 g/day (female and male, respectively) of freeze-dried black raspberries containing appreciable phenolic levels for 6 months to 10 patients with Barrett's esophagus did not result in a reduction in the length of the Barrett's lesion; however, it led to reductions in the urinary excretion of two markers of oxidative stress, 8-epi-prostaglandin F2a and 8-OHdG [15]. A clinical trial on the progression of esophageal dysplasia to squamous cell carcinoma was conducted in China, and the results indicate that freeze-dried black raspberries will modulate the expression of colon cancer genes or regress rectal polyps in patients with familial adenomatous polyposis [16].

20.4 Catechins from Black Tea

The primary sources of polyphenols in black tea are catechins (35%–52%) (Figure 20.4), which include epicatechin, epicatechin-3-gallate, epigallocatechin, and epigallocatechin-3-gallate. In addition, tea contains smaller concentrations of quercetin and theaflavins.

Black tea extract was found to inhibit 7,12-dimethylbenz[*a*]anthracene–induced carcinogenesis in the hamster buccal pouch carcinogenesis model. The efficiency of black tea polyphenols was found to be more than that of green tea polyphenols, which may be due to the presence of theaflavins and thearubigins. Administration of black tea extract had a protective effect against oxidative stress and reduced the incidence of bone marrow micronuclei. Neoplastic lesions and dimethylbenz[*a*]anthracene-induced hamster buccal pouch tumors also significantly decreased under the effect of black tea. Black tea was also found to interfere with the activity of carcinogen-metabolizing enzymes. Similarly, black tea was also found to inhibit esophageal tumors in male Sprague-Dawley rats, induced by nitrosomethyl benzamine. These rats (65%) developed esophageal tumors after 39 weeks. In rats given black tea extract as the only source of drinking fluid, a 70% reduction in the incidence of tumors as well as a decline in tumor growth was observed. When tea extract was administered after the nitrosomethyl benzamine treatment period was over, a 50%

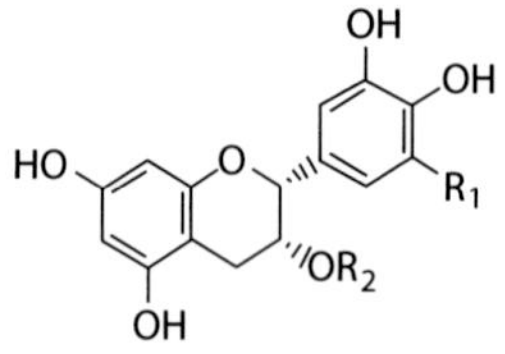

Catechin	R$_1$	R$_2$
(-)-Epicatechin (EC)	H	H
(-)-Epicatechin gallate (ECG)	H	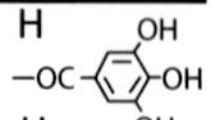 —OC—⟨⟩—OH
(-)-Epigallocatechin (EGC)	OH	H
(-)-Epigallocatechin gallate (EGCG)	OH	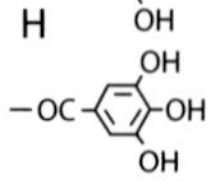 —OC—⟨⟩—OH

FIGURE 20.4
Structure of catechins (epicatechin, epicatechin-3-gallate, epigallocatechin and epigallocatechin-3-gallate).

reduction in esophageal papilloma incidence as well as a decrease in tumor multiplicity was seen. The volume of tumors was also significantly lower in black tea–treated rats. Thus, black tea could prevent esophageal tumors both during carcinogen exposure period and subsequent molecular events leading to tumor formation.

The risk/cases of gastric/esophageal cancer was found to be inversely related to urinary excretion of tea polyphenols epigallocatechin, epicatechin, M4, and M6 (M4 and M6 are metabolites of epigallocatechin and epicatechin, respectively). The inverse relationship stood even after adjustment for smoking, alcohol consumption, level of serum carotenes, and *Helicobacter pylori* seropositivity. The protective effect was found to be increased in individuals having low serum carotene levels. Therefore, tea polyphenols may protect against gastric and esophageal cancer formation [17]. There is plenty of literature available (and often conflicting) on the properties of various polyphenol-rich foods, especially tea and red wine/grapes in relation to various types of cancer in humans (Table 20.1).

In another study, cancer induced in the intestines of male F3-44 by using azoxymethane was suppressed by the administration of powdered black tea extract. Here, the animals

TABLE 20.1

Anticancer Effect of Selected Functional Nutrients and Their Main Food Sources

Functional Nutrients	Food Sources		
	Fruits	**Vegetables**	**Legumes**
Ascorbic acid	Citrus fruit, kiwi	Tomatoes, potatoes, broccoli	—
Carotenes	Tangerine, orange, yellow fruits, grapefruit, watermelon	Carrot, tomatoes, pumpkin, maize, potatoes, spinach, kale, chard, turnip, beet, broccoli, romaine lettuce	—
Catechins	Fruits in general, berries, grapes	Fresh tea leaf and vegetables in general	—
Fructo-oligosaccharides	Fruits in general, especially bananas	Onion, garlic, asparagus, tomatoes, leeks, artichoke	—
Isoflavones	—	Celery	Various legumes bean, soybeans, and chickpea
Lignans	Strawberries, apricots	Cabbage, brussels sprouts	Soybean, beans
Phytic acid	—	—	Fiber of legumes, soybeans
Polyphenols	Berries, grapes, pomegranates, and fruit skin	Vegetables	Most legumes and soybeans
Resveratrol	Grapes	—	—
Saponins	—	Most vegetables	Legumes in general, especially soybeans and peas
Tannins	Pomegranates, berries, apple juice, grape juices	—	Most legumes, red-colored beans, chickpeas
Tocopherol	Kiwi	Green leafy vegetables, spinach, carrot, avocado	Soybean

Source: Crujeiras, A.B. et al., in *Bioactive Foods in Promoting Health*, 359–380, 2010. With permission.

were killed after 26 weeks and it was found that the number of colorectal tumors decreased from 2.54 to 1.54. The incidence of adenoma was also reduced from 86% in the control group to 59% in the black tea–treated group. Black tea also increased the apoptotic index of tumors from 2.92 in controls to 4.13 in the black tea–treated rats. The inhibitory effect may be due to the polyphenols present in black tea. Black tea extract was found to be more effective against adenomas than cancer initiation [18]. These studies indicate that black tea extract was found to be effective in controlling the initiation of cancer for a limited period only.

Two large prospective cohort studies of men and women found no association between the consumption of tea containing caffeine and colon or rectal cancer [19]. The inhibition of the arachidonic acid metabolism was found to decrease the risk of colorectal cancer. Cyclooxygenases, lipoxyenases, and cytochrome P-450 enzymes are the major enzymes involved in the metabolism of arachidonic acid. Cyclooxygenase-1 and cyclooxygenase-2 convert arachidonic acid to prostaglandin G_2, which ultimately forms thromboxane and 12-hydroxyl heptadecatrienoate in case of cyclooxygenase-1 and prostaglandin E_2 and prostaglandin I_2 in case of cyclooxygenase-2. Prostaglandin E_2 is involved in rapid cell proliferation in tumors, mitogenesis, invasiveness and angiogenesis, and inhibition of apoptosis. Cyclooxygenase-2 is overexpressed in tumor cells. Theaflavins were found to increase the formation of prostaglandin E_2 in tumors by stimulating cyclooxygenase-2. Catechins, however, reduced the activity of both cylooxygenase-1 and cylooxygenase-2 and thus had anticancer activity [20].

In an epidemiological study on the effect of black tea consumption and the prevalence of rectal cancer in Moscow, black tea consumption was found to be associated with lower incidence of rectal cancer. The effect of black tea was found to be more pronounced for women than for men, possibly because of high alcohol intake by men. Tea consumption and the incidence of rectal cancer were inversely correlated in a dose-dependent manner and a minimum consumption level of approximately 80 g/month was required to have any significant effect [21].

Decaffeinated black tea was found to reduce the number of tumors in the lungs of male C-3H mice treated with diethylnitrosamine, a constituent of tobacco. Administration of black tea reduced the number of lung tumors [22]. Theaflavins were found to inhibit cell proliferation and tumor formation in the lungs of female A/J mice induced by 4-(methylnitrosamino)-1-(3-pyridyl)-1-butanone. A single dose of 4-(methylnitrosamino)-1-(3-pyridyl)-1-butanone (103 mg/kg body weight) caused cell proliferation in bronchiolar cells, which was inhibited when 0.3% theaflavin solution was administered. 4-(Methylnitrosamino)-1-(3-pyridyl)-1-butanone–induced tumor multiplication was decreased by 23% and tumor volume by 34% when 0.1% theaflavin solution was given as the sole source of drinking fluid for a period of 16 weeks starting 2 days after 4-(methylnitrosamino)-1-(3-pyridyl)-1-butanone administration [23]. Administration of black tea to F3-44 rats with adenocarcinomas and adenosquamous carcinomas induced by 4-(methylnitrosamino)-1-(3-pyridyl)-1-butanone was found to reduce tumorigenesis in lungs. Treatment with caffeine alone was also found to suppress tumor formation.

The effects of theaflavin, theaflavin gallate, and theaflavin digallate on 33-BES and 22-BES gene transferred human bronchial epithelial cell lines were studied. Theaflavin and theaflavin gallate had very little effect, indicating that gallate structure is important for growth suppression. Theaflavin digallate (25 μM) was found to cause apoptosis (possibly by cytotoxicity). It also led to hydrogen peroxide formation in 21 BES cells. Theaflavin digallate also reduced the phosphorylation of *c-Jun* protein, which lowers activator protein-1 activity and suppresses growth [24]. Aqueous black tea extract—when given as the sole source of drinking fluid to Swiss albino mice with pulmonary tumors induced by diethylnitrosamine, present in tobacco—was found to reduce the number of tumors. The inhibition of alveogenic tumors was dose-dependent. Supplementation with 4% black tea extract has significant inhibitory activity.

Black tea polyphenols were found to inhibit proliferation and increase apoptosis in Du-145 prostate carcinoma cells. Higher levels of IGF-I were found to be associated with a higher risk of the development of prostate cancer. IGF-I binding to its receptor is a part of the signal transduction pathway that causes cell proliferation. The addition of black tea polyphenol was found to block IGF-I–induced progression of cells into S phase of cell cycle. Black tea polyphenols at a dose of 10 to 20 mg/mL were required for inhibition, and complete inhibition was observed at a dose of 40 mg/mL. The effect of black tea polyphenols was found to be similar for both normal prostate cells and prostate carcinoma cells.

Black tea polyphenols were found to have an inhibitory effect on the development and progression of cancer in LNCaP prostate cancer cells. The inhibitory effect of polyphenols varied and is as follows: theoflavin digallate > theoflavin gallate-B > theoflavin gallate-A > epigallocatechin gallate > theoflavin. Androgens, although required for the normal development and functioning of the prostate, are also potential factors for the development of prostate cancer. Enzyme 5-α-reductase converts testosterone to dihydrotestosterone. 5-α-R1 and 5-α-R2 are the two isoforms of the enzyme, among these isoforms, 5-α-R1 is associated with androgen metabolism. Theaflavins were found to inhibit 5-α-R1 activity and cause an associated decrease in the growth of LNCaP prostate cancer cells, which are sensitive to androgens. A decrease in the level of androgen receptors and a decrease in the secretion of prostate-specific antigen were observed. Prostate-specific antigen is a serine protease, which acts on proteins that influence cancer initiation and progression. Its activity is androgen regulated [25]. Androgens (testosterone) play an important role in the initiation and progression of prostate cancer because they induce oxidative stress in the androgen-sensitive human prostate carcinoma cells. Antioxidants present in black tea are capable of scavenging the pro-oxidants generated by testosterone. Tea also increases the levels of antioxidant enzymes that prevent oxidative stress. Tea was found to show protection against all stages of carcinogenesis, that is, initiation promotion and progression.

Black tea was found to reduce dimethylbenz[a]anthracene-induced mammary gland tumors in female Sprague-Dawley rats fed on a high-fat diet. Rats were treated with 15 mg/kg of dimethylbenz[a]anthracene. Rats that were given water as the drinking fluid showed an increase in tumor number and weight comparable to the controls, but rats given 2% tea extract instead had a much lower tumor burden. However, no effect was observed in rats fed on the AIN-76-A diet [26]. In a cross-sectional study of postmenopausal Chinese women, the effect of tea consumption on the levels of estrogen and endostenedione in plasma were monitored. Higher levels are associated with a higher risk of breast cancer. The estrone levels in green tea drinkers were found to be 13% lower as compared with non–tea drinkers; however, in black tea drinkers, the levels were found to be 19% higher. Out of the 130 women examined, 84 were non/irregular, 27 were green tea, and 19 were black tea consumers [27]. Hence, this observation needs further study because the number of relevant participants involved was very low.

Black tea polyphenols were found to be effective inhibitors of DNA replication in HT rat hepatoma cells when black tea extract in ethyl acetate was administered at a concentration of 0.1 to 0.2 mg/mL [28]. Black tea extract was also observed to inhibit tumorigenesis in diethylnitrosamine-treated male C3H mice in a dose-dependent manner. Treatment with black tea (1.25%) decreased the number of hepatic tumors by 63% [22]. Black tea extract was also found to decrease tumor formation in the liver of male F3-44 rats, induced by diethylnitrosamine and subsequent phenobarbital treatment. Black tea extract (0.3%) considerably reduced the number of proneoplastic glutathione-S-transferase placental form–positive foci [29].

Treatment of rat ascites hepatoma cell line (AH109A) with black tea extract resulted in decreased proliferation and invasiveness. However, no inhibitory effect was seen on normal mesothelial cells. Suppression of cell proliferation was also observed for the L-929 tumor cell line. Theaflavin and theaflavin gallates were found to have maximum activity against AH109A. The higher activity against cancerous cells as compared with normal cells may be due to the higher sensitivity of tumor cells to inhibition by black tea polyphenols and the specificity of action of tea components toward cancer cells. Black tea extract was found to inhibit the invasiveness of the AH109A cell line in another *in vitro* study. The addition of ethylenediaminetetraacetic acid totally blocked the inhibitory effect of black tea, indicating that the antitumor effect of tea was due to its antioxidant activity.

Aqueous extracts of black tea were found to inhibit proliferation and increase apoptosis in malignant/nonmalignant skin tumors. In female CD-1 mice, tumor formation was induced by 7,12-dimethylbenz[a]anthracene and maintained by 12-O-tetradecanoylphorbol-13-acetate. Replacement of the drinking fluid with black tea reduced papilloma growth by 35% to 40%. Treatment with decaffeinated tea instead was found to decrease papilloma growth in one case and increase it in two others. Furthermore, black tea administration was also found to inhibit the growth (70%) of skin tumors in SKH-1 female mice induced by ultraviolet B light. There was a 58% reduction seen in the number of benign tumors and 54% reduction in squamous cell carcinomas per mouse. Tumor volume was also found to decrease by 60% and 84% for benign tumors and carcinomas, respectively. The apoptotic index was also increased by 44%, 100%, and 95% in cell papillomas, benign tumors, and carcinomas, respectively [30].

Black tea was found to protect against skin cancer induced by exposure to ultraviolet B and ultraviolet A+B in hairless mice. Consumption of tea decreased the number of skin papillomas induced by ultraviolet A+B light and the effect of black tea was found to be higher than that of green tea. The initiation of papillomas was delayed. No significant effect was observed on the number of necrotic cells or mitotic index. The number of sunburnt cells, after 24 h of ultraviolet A+B exposure in the epidermis, was lower in black tea–treated mice as compared with the controls. Neutrophil infiltration and skin redness due to increased blood flow was also slightly reduced by the tea extract. Based on the protective and healing effect of black tea on skin, a medicinal powder formulation for treating burns and scalds was developed, containing rhubarb, Fangdou tea, Dragon's bone, and premium black tea. Theaflavins present in black tea were found to inhibit ultraviolet B–induced activation of activator protein-1, which is a transcription factor, promoting cell growth and multiplication, in JB-6 mouse epidermal cell line. The inhibitory effect was found to be dose-dependent [31].

Black tea polyphenol extract in ethyl acetate was found to inhibit DNA replication in DS-19 mouse erythroleukemia cells when given at a dose of 0.1 to 0.2 mg/mL. Cell proliferation was reduced but the effect on the differentiation of cells was not significant [32]. Black tea extract was found to protect against tumor-induced apoptosis of immunocytes in Ehrlich's ascites carcinoma–bearing mice at a dose of 6 g/kg body weight per day. Ehrlich's ascites carcinoma significantly reduces the number of splenic lymphocytes, which can severely affect the host's immune system. Administration of 2.5% black tea was found to protect against Ehrlich's ascites carcinoma–induced cell death of splenic lymphocytes by downregulating the apoptotic pathway. Tumor suppressor protein p53 increases to very high levels in tumor-bearing organisms and is associated with higher apoptosis in spleen.

Treatment with black tea was found to decrease p53 levels. Black tea administration was also found to increase Bcl-2/Bax (proapoptotic/antiapoptotic protein) ratio and thus promote multiplication of splenic lymphocytes [32]. However, a recent study showed that treatment with black tea/theaflavin was found to inhibit the growth of leukemic cell lines: HL-60 and K-562. The suppression of growth was dose-dependent. The effect of black tea was found to be at par with that of green tea. The mode of action was through the promotion of apoptosis, mediated by activation of caspases 3 and 8, and a decreasing Bcl2/Bax ratio by suppressing the expression of Bcl-2 [33].

Lu et al. [30] reported that tea consumption was associated with a higher risk of bladder cancer. However, Zeegers et al. [34] reported, in their review, that tea intake was not found to be associated with higher risk of urinary tract cancer, rather tea components were found to inhibit cancer formation and proliferation in some studies. However, this effect may not be very significant. Exploiting this anticancer activity of black tea, anticancer preparations have been formulated. These can be used as anticancer tea, anticancer powder, anticancer pill, anticancer capsule, anticancer-medicated wine, anticancer beer, anticancer liquid medicine, anticancer sweets, and anticancer biscuits.

In a Japanese prospective cohort study of 8552 individuals, consumption of 10 cups of green tea per day was associated with delayed onset of cancer by 8.7 years in women and 3.0 years in men compared with those who consumed 3 cups per day [35]. In addition, a lower risk ratio was observed for lung, colon, and liver cancers. Likewise, a prospective study in postmenopausal women found that those who consumed two cups of tea (primarily black tea) had a slightly lower risk for all cancers compared with women who never or only occasionally consumed tea. On the other hand, Nagano et al. [36] found no protective relationship between tea consumption and cancer in 38,540 Japanese men and women. Thearubigens are the major fraction of black tea polyphenols and account for 20% of the solids in brewed tea. Human studies have established that these antioxidant polyphenols, in particular, epigallocatechin-3-gallate, protect against carcinogenesis. In animals, black tea significantly increased the activity of antioxidants and detoxifying enzymes, such as glutathione-S-transferase, catalase, and quinone reductase, in the lungs, liver, and small intestine. Topical administration of epigallocatechin-3-gallate, subsequent to ultraviolet radiation, significantly reduced tumor induction in mice.

20.5 Flavonoids from Fruits and Vegetables

Flavonoids are characterized by the presence of two phenyl rings that are joined together by a C3 chain and have a structure of C6-C3-C6. The generic structure of a flavonoid consists of two aromatic rings (ring A and ring B) linked by three carbons that are usually in an oxygenated heterocyclic ring or C ring. Based on differences in the generic structure of the heterocyclic C ring as well as the oxidation state and functional groups of the heterocyclic ring, they are classified as flavonols, flavones, flavonones, flavononols, anthocyanidins and isoflavones (Figure 20.5). Within each subclass, individual compounds are characterized by specific hydroxylation and conjugation patterns.

More than 4000 distinct flavonoids have been identified in fruits, vegetables, and other plant foods and have been linked to reducing the risk of cancer and other major chronic diseases (Table 20.2). Flavonoids and their polymers can alter metabolic processes and have a positive effect on health. Epidemiological studies have shown that frequent consumption

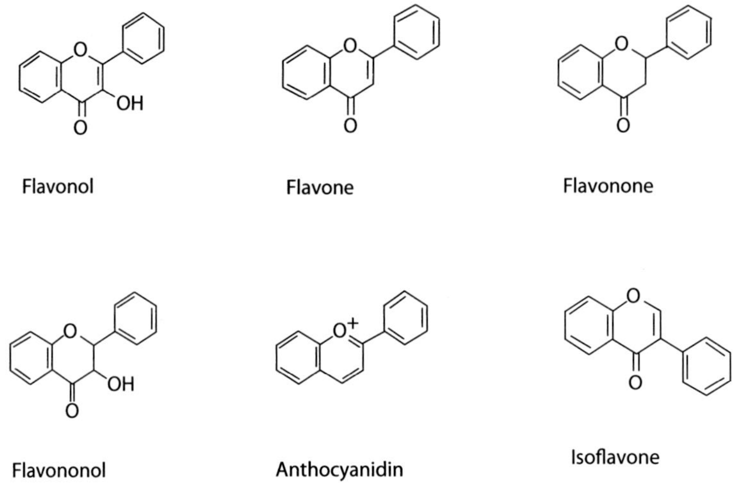

FIGURE 20.5
Structure of flavonoids (flavonol, flavone, flavonone, flavononol, anthocyanidin and isoflavone).

of fruits and vegetables is associated with low risks of various cancers. Block et al. [37] reviewed approximately 200 epidemiological studies that examined the relationship between the intake of fruits and vegetables and the incidence of cancers of the lung, colon, breast, cervix, esophagus, oral cavity, stomach, bladder, pancreas, and ovary.

The consumption of fruits and vegetables was found to have a significant protective effect in 128 of 156 dietary studies. The risk of cancer was twofold higher in persons with a low intake of fruits and vegetables than in those with a high intake. The protective effect has largely been attributed to flavonoids, which are ubiquitously present in plant-derived foods and are important constituents of the human diet, and an inverse association was shown between flavonoid intake and subsequent lung cancer incidence. In a study on 9959

TABLE 20.2

Major Group of Flavonoids and Their Food Sources

Flavonoid Subclass	Compound	Food Source
Flavonones	Hesperitin, naringin, naringenin, eriodictyol, hesperitin, pinocembrin, likvirtin	Citrus fruit, citrus peel, orange, grapefruit, lemon, peppermint, pummelo, tangelo, tangerine, tangor
Flavanols	Catechin, gallocatechin, epicatechin, epigallocatechin, epicatechin 3-gallate, epigallocatechin 3-gallate, theaflavin theaflavin 3-gallate, theaflavin 3,3′ digallate, thearubigins	Green tea, black tea, oolong tea, red wine, barley, berries, broad beans, buckwheat, grapes, rhubarb stalks
Anthocyanins	Cyanidin, delphinidin, malvidin, pelargonidin, peonidin, petunidin	Berries, cherries, elderberries, onions, beets, grapes, raspberries red/black grapes, red wine, strawberries, tea, fruit peels with dark pigments
Isoflavones	Genistein, daidzin, glycitein, formononetin	Soy, soy flour, soybeans miso, tofu, tempeh, soy milk

Source: Kale, A. et al., *Phytother. Res.* 22:567–577, 2008. With permission.

Finnish men and women aged 15 to 99 years, consumption of quercetin from onions and apples was found to be inversely associated with lung cancer risk [38].

Boyle et al. [39] showed that increased plasma levels of quercetin after a meal of onions was accompanied by increased resistance to strand breakage by lymphocyte DNA and decreased levels of some oxidative metabolites in the urine. A recent clinical trial evaluated the effect of a combination treatment of 480 mg of curcumin and 20 mg of quercetin (administered orally, three times a day) for adenomas in five familial adenomatous patients [40]. The combination of treatments was found to reduce the number and size of ileal and rectal adenomas in patients without appreciable toxicity.

The flavonoids have been reported to modulate the activities of metabolic enzymes and needs to be explored. A major hurdle in cancer treatment is the drug resistance shown by cancer cells. The drug is transported back by some efflux transporters, making the cell resistant. The cells show drug resistance because the efflux pumps maintain a subtoxic concentration of drugs within the cells. P-glycoproteins are one of the major players in drug resistance. Flavonoids have been shown to inhibit the P-glycoproteins and other efflux transporters. The stage of tumor progression or promotion, arising after the initiation has taken place, is a highly complex process and is the least understood one. Intervention at this point requires a multitargeted therapy that promotes antiproliferation and proapoptotic changes in cancer cells and should act as an antimetastatic and an antiangiogenic to suppress the spread of cancer.

There is substantial evidence from various *in vitro* studies to show that flavonoid combinations are effective at all the stages of cancer listed previously. Flavonoids are highly specific in their action on some key regulatory enzymes and receptors in our body. Flavonoids have also been reported to modulate P-glycoprotein, the multidrug-resistance protein. In addition to cytochrome P-450, flavonoids are also involved in the regulation of enzymes of phase II, which are responsible for xenobiotic biotransformations (e.g., glutathione *S*-transferase, UDP-glucuronyl transferase, *N*-acetyltransferase) and colon microflora. Because cytochromes P-450, P-glycoproteins and phase II enzymes are involved in the metabolism of drugs and in the processes of chemical carcinogenesis, interactions of flavonoids with these systems hold great promise for their therapeutic potential. These activities may be beneficial in detoxification, in chemoprevention, or in drug resistance suppression. They also inhibit many enzymes that are the targets of anticancer treatment, for example, eukaryotic DNA topoisomerase I, Cox I and II, and estrogen 2- and 4-hydroxylases.

Among the proteins that interact with flavonoids, cytochrome P-450, monooxygenases metabolizing xenobiotics (e.g., drugs and carcinogens), and endogenous substrates (e.g., steroids) play a prominent role. Flavonoids, by interacting with cytochrome P-450 enzymes, reduce the activation of procarcinogen substrates to carcinogens, which makes them putative anticancer substances [41]. An inhibitory capacity of flavonoids with respect to cytochrome P-450 activities has been extensively studied because of their potential use as agents blocking the initiation stage of carcinogenesis. *In vivo* and *in vitro* studies have shown that flavonoids can enhance or inhibit the activities of certain P-450 isozymes [42].

Cytochrome P-450 interacts with flavonoids in at least three different ways: (1) flavonoids induce biosynthesis of several cytochrome P-450s, (2) they modulate (stimulate or inhibit) enzymatic activities of cytochrome P-450, and (3) flavonoids are metabolized by several cytochrome P-450s. Synthetic and naturally occurring flavonoids are effective inhibitors of four cytochrome P-450s involved in the metabolism of xenobiotics: cytochrome P-1A1, 1A2, 1B1, and 3A4, and one steroidogenic cytochrome P (cytochrome P-19). The activation of aryl hydrocarbon receptor, a ligand-activated transcription factor, is associated with the elevation of activities of cytochrome P-1 family enzymes (cytochrome P-1A1, 1A2, and 1B1) that are responsible for the activation of carcinogens such as benzo[a]pyrene, 7,12-dimethylbenz[a]anthracene, aflatoxin B1, and meat-derived heterocyclic aromatic amines.

Increased expression of cytochrome P-1A1 in the lungs increases the risk of lung cancer [43] as well as colorectal cancer [44]. Cytochrome P-1A2 has a role in tobacco-related cancers [45]. 7-Hydroxyflavone and galangin are potent inhibitors of cytochrome P-1A1 and cytochrome P-1A2, respectively, and thereby block the process of carcinogenesis. Many flavonoids act as blocking agents of aryl hydrocarbon receptors. The inhibition of gene expression of cytochrome P-1 family enzymes through blocking aryl hydrocarbon receptors plays an important role in the cancer chemopreventive properties of flavonoids. Quercetin, one of the most abundant naturally occurring flavonoids, binds as an antagonist to aryl hydrocarbon receptors and consequently inhibits benzo[a]pyrene, dimethylbenz[a]anthracene, and aflatoxin B1 by altering the expression of cytochrome P-1A1, 1A2, and 1B1. This inhibition results in reduced benzyl[a]pyrene–DNA adduct formation. Kaempferol also prevents cytochrome P-1A1 gene transcription induced by the prototypical aryl hydrocarbon receptor ligand [46]. However, certain flavonoids (diosmin, diosmetin, and galangin) are aryl hydrocarbon receptor agonists, increasing cytochrome P-1 expression, and consequently, carcinogen activation capacity. But these compounds strongly inhibit the activities of the expressed enzymes. For instance, treatment of cells with diosmetin caused a dose-dependent expression of cytochrome P-1A1 mRNA; however, an extensive decrease in the formation of cytochrome P-1A1–mediated DNA adducts from dimethylbenz[a]anthracene was observed [46].

On the basis of available data on flavonoid–cytochrome P interactions, it can be deduced that flavonoids possessing hydroxyl groups inhibit cytochrome P–dependent monooxygenase activity, whereas those lacking hydroxyl groups can stimulate the enzyme activity. In the study by Tsyrlov et al. [41], quercetin inhibited the metabolism of aryl hydrocarbons but stimulated the activity of cDNA expressed human cytochrome P-1A2. In another study, 7,8-benzoflavone had been reported as a stimulator of cytochrome P-3A4 activity [47] and an inhibitor of human cytochrome P-1A1 and 1A2 [48], and activation of cytochrome P-3A4. This shows that a flavonoid can have different effects on different cytochrome P activities. Thus, flavonoids can either inhibit or activate human cytochromes P-450 depending on their structures, concentrations, and experimental conditions.

The beneficial properties of various flavonoids include the inhibition of cytochrome P involved in carcinogen activation and scavenging reactive species formed from carcinogens by cytochrome P–mediated reactions. Induction of cytochrome P activity by flavonoids proceeds via various mechanisms, including direct stimulation of gene expression through a specific receptor and cytochrome P protein, or mRNA stabilization [49]. UDP-glucoronyl transferase and glutathione S-transferase are two major phase-II detoxifying enzymes, which protect cells against both endogenous and exogenous carcinogens by glucuronidation and nucleophilic addition of glutathione to a variety of different substrates, respectively. Flavanones and flavones increase the activities of UDP-glucoronyl transferase and glutathione S-transferase.

Green tea catechins have been shown to activate mitogen-activated protein kinases. This activation results in the stimulation of transcription of phase II detoxifying enzymes through the antioxidant responsive element. In addition, flavonoids, being structurally similar to estrogen, show an estrogenic or antiestrogenic activity. Like natural estrogens, they can bind to the estrogen receptor and modulate its activity. They also block cytochrome P-19, a crucial enzyme involved in estrogen biosynthesis. Soy isoflavones have been studied extensively for estrogenic and antiestrogenic properties. Other flavonoids have been much less tested for steroid hormone activity. Luteolin and naringenin display the strongest estrogenicity, whereas apigenin shows a relatively strong progestational activity [50]. Flavonoids in the human diet may reduce the risk of various cancers, especially hormone-dependent breast and prostate cancers, as well as preventing menopausal symptoms.

The resistance of cancer cells to chemotherapy is a major obstacle to the success of cancer chemotherapy and has been closely associated with treatment failure. P-glycoprotein, a plasma membrane ATP-binding cassette transporter, interferes with drug bioavailability and disposition, including absorption, distribution, metabolism, and excretion, affecting the pharmacokinetics and pharmacodynamics of many herbal and synthetic drugs. An increase or overexpression of P-glycoproteins is often involved in cancer cell resistance to chemotherapy. Some isoflavones and flavones have been found to be active against P-glycoproteins [51].

Quercetin has been known to inhibit ATP-dependent drug efflux [52]. The flavone, luteolin and its 7-O-β-D-glycopyranoside, have been shown to inhibit multidrug-resistance transporters by interacting with their ATP-binding domains [53]. The prenylated flavonoids strongly inhibit drug interactions and nucleotide hydrolysis and may serve as potential modulators of multidrug resistance. The *in vitro* everted gut studies conducted by Chen et al. [54] indicated that phellamurin, a prenylated flavonoid glycoside, significantly inhibit the function of intestinal P-glycoproteins. Many isoflavone and flavone compounds have been found to be active against P-glycoproteins [51]. On the other hand, P-glycoproteins in normal tissues may serve as a cellular defense mechanism against naturally occurring xenobiotics. Due to the potential importance of P-glycoproteins in cellular defense against environmental carcinogens, the cancer-chemopreventive properties of flavonoids following the modulation of P-glycoprotein activity resulted in a flurry of research in the area. Flavonoids may upregulate the activity of P-glycoproteins.

Several commonly occurring flavonoids, especially quercetin, kaempferol, and galangin at micromolar concentrations stimulated the efflux of Adriamycin in P-glycoprotein–expressing HCT-15 colon cells [41]. Moreover, quercetin and genistein potentiate the effects of Adriamycin and aunorubicin, respectively, in a multidrug-resistant MCF-7 human breast cancer cell line [55]. Among the various cell targets of genistein, ATP-binding cassette transporters were also identified. Genistein was found to be a modulator of non–P-glycoprotein–mediated multidrug resistance, not affecting P-glycoprotein multidrug-resistant cells. In one study, quercetin, at low concentrations, stimulated the activity of P-glycoproteins, whereas at high concentrations, it inhibited P-glycoproteins [56].

Critchfield et al. [57] conducted a study to evaluate the effects of flavonols on P-glycoprotein activity in rat hepatocytes by assessing the transmembrane transport of P-glycoprotein substrates such as rhodamine-123 and doxorubicin. The results indicated that flavonols strongly upregulate the activity of P-glycoproteins in cancer cell lines. At the same time, they may differently modulate the transport of putative P-glycoprotein substrates in normal rat hepatocytes. This differential nature of flavonoids to upregulate P-glycoprotein activity in normal cells and at the same time downregulate it in cancerous cells can be of high significance in cancer therapy.

The membrane protein mediating the ATP-dependent transport of lipophilic substances conjugated to glutathione, glucuronate, or sulfate, have been identified as members of the multidrug resistance protein family. A soybean isoflavone, genistein, was found to be an inhibitor on the basis of its effect on drug accumulation in multidrug resistance protein-1–overexpressing cells. Certain members of the flavone, flavonol, flavanone, and isoflavone classes possess antiproliferative effects in different cancer cell lines [58]. The antitumor activity of several flavonoids (pinostrobin, quercetin, myricetin, and morin) is attributed to their efficiencies to inhibit topoisomerase I and II [59]. Flavonoids might slow down cell proliferation as a consequence of their binding to the estrogen receptor [60]. Certain common polyphenolic constituents of legume grains are illustrated in Table 20.3.

Complete growth retardation of androgen-independent human prostatic tumor cells was observed when they were treated with kaempferol [61]. Flavonoids can affect cancer cells by triggering the process of apoptosis [62]. Alternatively, flavonoids are also potent inhibitors of mitogen signaling processes by affecting various kinase activities. Genistein and daidzein are the two major soy isoflavones. Genistein, like other isoflavones, displays a remarkable estrogenic activity. Genistein is also a specific and potent inhibitor of tyrosine kinases and it interferes with many biochemical pathways [63].

Ahmad et al. [64] reported that epigallocatechin-3-gallate induced apoptosis and cell cycle arrest in human epidermoid carcinoma cells A-431. They also found that the apoptotic response of epigallocatechin-3-gallate was specific to cancer cells only. The study conducted by Liang et al. [65] showed that epigallocatechin-3-gallate suppresses extracellular signals and cell proliferation by binding to epidermal growth factor receptor. Their findings suggest that epigallocatechin-3-gallate blocks the cellular signal transduction pathways and might result in the inhibition of tumor formation. Epigallocatechin-3-gallate has been reported to block the induction of nitric oxide synthase by inhibiting the process of binding of nuclear factor-κB to inducible nitric oxide synthase.

TABLE 20.3

Total Phenolic Content of Various Legume Seeds

Name of the Legume Seed	Total Phenolics (mg/g DM)	References
Green bean	355	[117]
Pea	183	[117]
Lentil	1.56–60.1	[118–120]
Green pea	1.07–1.53	[119]
Yellow pea	1.13–1.67	[119]
Chickpea	1.41–1.67	[119]
Red kidney bean	1.23–5.90	[119]
Black bean	1.28–44	[119,121]
Cowpea	64.5–163.6	[122]
Red bean	93.6	[121]
Brown bean	91.4	[121]
Fava bean	55.9–80.9	[120,125]
Broad bean	23.9–60.1	[120,125]
Adzuki bean	89.7	[125]
Red bean	55.4	[125]
Pea	22.6–34.8	[120,125]
Red lentil	58.0	[125]
Green lentil	67.6	[125]
White bean	10.8	[120]
Everlasting pea	9.7	[120]
Velvet bean	49.7–50.8	[126]
Gila bean	62.3	[128]

Certain members of flavone, flavonol, flavanone, and isoflavone classes possess antiproliferative effects in different cancer cell lines. The polyphenols from tea, particularly green tea, have strong antiproliferative capacity. Valcic et al. [66] studied the antiproliferative effects of six green tea catechins [(+)-gallocatechin, (–)-epicatechin, (–)-epigallocatechin, (–)-epicatechin gallate, and (–)-epigallocatechin gallate] on four different human cancer cell lines. In their study, all catechins strongly inhibited proliferation. Epigallocatechin-3-gallate was found to be the most effective inhibitor in MCF-7 breast cancer cells, HT-29 colon cancer cells, and UACC-375 melanoma cells.

Quercetin causes cell cycle arrest in the G_0-G_1 phase in human leukemic T-cells and gastric cancer cells [67]. In human leukemic T-cells, quercetin reversibly blocked the cell cycle at a point 3 to 6 h before the start of DNA synthesis, and it suppressed DNA synthesis to 14% in gastric cancer cells. In gastric cancer cells, it specifically induced a G_1 phase arrest. It also arrests the cell cycle in the G_2-M phase [68]. Richter et al. [69], in their study on colorectal cancer cells, reported that quercetin induced growth inhibition and cell loss, and cells were preferentially retained in the S phase. It was more effective than apigenin, fisetin, robinetin, and kaempferol. This effect was attributed to the inhibition of epidermal growth factor receptor kinase by quercetin.

Current research suggests a role for quercetin and other flavonoids in cancer prevention. Epidemiologic studies have consistently demonstrated an inverse relation between flavonoid consumption and risks for certain types of cancer. Several *in vitro* and *in vivo* experiments have shown that flavonoids may interrupt various stages of the cancer process. It seems that these phytochemicals possess antioxidant activity as well as other anticarcinogenic properties. Flavonoids may act in a variety of ways beyond their antioxidant properties to interfere with carcinogenesis, such as protecting DNA from oxidative damage, deactivating carcinogens and inhibiting the expression of mutated genes and the activity of enzymes that promote carcinogenesis, as well as promoting detoxification of xenobiotics. For example, experimental studies have shown that quercetin can inhibit the initiation, promotion, and hyperproliferation of tumors in animal models.

Consumption of anthocyanins is associated with reduced risks for prostate cancer, which is a leading cause of cancer deaths among men in the United States. Extracts from four specialty potatoes and the anthocyanin fraction from genotype CO-112-F2-2 reduced cell growth and induced apoptosis in both androgen-dependent (LNCaP) and androgen-independent (PC-3) prostate cancer cell lines [70]. The anthocyanins seem to cause mitochondrial release of the proteins Endo G and AIF, which promote apoptosis. Likewise, intake of anthocyanins from purple and red potatoes might play a protective role against stomach cancer. Intake of steamed purple and red potatoes repressed the growth of mouse stomach cancer induced by benzo[*a*]pyrene [71].

Isolated anthocyanins induced apoptosis in human stomach cancer cell lines as well as suppressing benzo[*a*]pyrene-induced mouse stomach cancer proliferation indicated that anthocyanins were bioactive antitumor components. No induction of apoptosis was observed upon exposure to normal lymphocytes prepared from healthy volunteers. The exact anthocyanins involved in the antitumor properties remain to be delineated because there are approximately 10 and 8 different types of pigments present, respectively, in the purple and red potato anthocyanin fractions. The role of vegetable consumption in the prevention of cancer has not reached a consensus in the scientific community. Methanol–water extracts from four Italian short-season potato cultivars inhibited the proliferation of MCF-7 breast cancer cells [72]. Extracts from cv. Nicola were most effective, but concentrations above 1×10^{-3} μg gallic acid equivalents μL^{-1} stimulated cell growth.

20.6 Tannins from Sorghum Grains

Tannin is a collective term used to describe a variety of plant polyphenols that are used in tanning raw hides to produce leather. Depending on their structure, tannins are defined as hydrolyzable tannins or condensed tannins (proanthocyanidins; Figure 20.6). The positive effects of tannins (from sorghum or millet consumption) on cancer have been documented. Van Rensburg [73] reported that sorghum consumption consistently correlated with low incidences of esophageal cancer in various parts of the world (including several parts of Africa, Russia, India, China, Iran, etc.) whereas wheat and corn consumption correlated with increased incidences. Such regions also had deficiencies of certain minerals and vitamins in their diets. In attempting to explain this phenomenon, with considerable evidence, it is proposed that nutrient deficiencies were responsible for the high esophageal cancer incidences, and that sorghum and millet consumption promoted resistance to esophageal cancer risk.

Chen et al. [74] reported similar results in epidemiological data from Sachxi Province, China. These authors studied 21 communities within the province over a period of 6 years and found that regions that consumed the highest amounts of sorghum, and to a lesser extent, millet, had 1.4 to 3.2 times lower mortality from esophageal cancer than areas that primarily consumed wheat flour or corn. Consumption of other foods like alcohol, tea, meats, and vegetables did not contribute significantly to esophageal cancer mortality. These evidences suggest the presence of anticarcinogenic compounds in sorghum that are either lacking or are not present in significant quantities in wheat or corn. On the other hand, consumption of plants containing tannins (tea, sorghum, betel nuts, etc.) was previously implicated in incidences of cancer in the upper digestive tract. Morton [75] implicated high incidences of esophageal cancer in certain parts of South Africa, China, and Russia on the consumption of high-tannin sorghums.

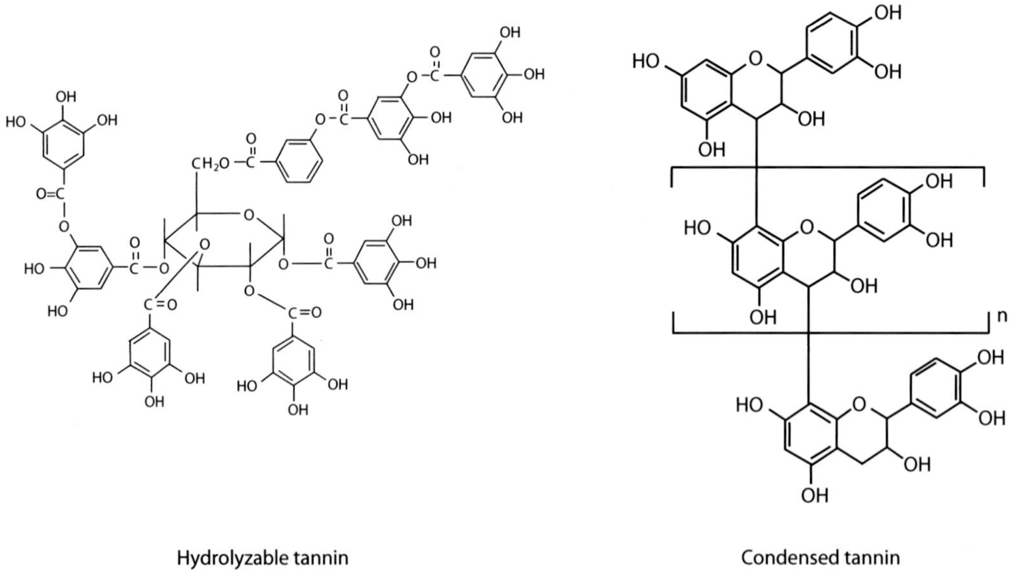

Hydrolyzable tannin   Condensed tannin

FIGURE 20.6
Structure of tannins (hydrolyzable or condensed tannins).

Oterdoom [76] theorized that the carcinogenic effect of the high-tannin sorghums reported by Morton [75] was because "tannins destroyed proteins both in the mucosa and in enzymes." However, the carcinogenic evidence suggested by Morton [75] was criticized by Yu and Swaminathan [77], and we concur, for lack of acceptable experimental design or control for confounding variables. Actually, Morton [75] did not present any data to back her claim, but merely theorized that the common denominator in regions with elevated risk of esophageal cancer was probably the consumption of plants containing tannins.

In vitro studies have also revealed anticarcinogenic properties of sorghum containing higher levels of tannins. Grimmer et al. [78] demonstrated the antimutagenicity of sorghum polyphenol extracts. They found that high molecular weight procyanidins (tannins) had the highest antimutagenic activity compared with lower molecular weight tannins. Gomez-Cordovez et al. [79] showed that sorghum tannins had anticarcinogenic activity against human melanoma cells, as well as positive melanogenic activity (melanogenesis is believed to help protect human skin against UV irradiation damage). The authors did not observe such melanogenic activity in red wine extracts. On the other hand, Parbhoo et al. [80] reported that sorghum procyanidin extracts may induce cytochrome P-450, a protein that is capable of converting certain promutagens to mutagenic derivatives in rat liver.

For sorghum tannins, available data on the anticancer effect is too limited to draw reasonable conclusions. However, the corroborative epidemiological evidence reported by Van Rensburg [73] and Chen et al. [74] against esophageal cancer warrant some follow-up. Additional *in vitro* data as well as controlled animal studies are necessary to understand how the levels and composition of polyphenols in sorghum affect cancer, and which specific components are responsible. Whole-grain consumption has long been correlated with reduced risks to other forms of digestive tract cancer, especially colon cancer. How much of these effects are contributed by the dietary fiber or the phytochemicals concentrated in the bran of the grains is still unknown.

20.7 Lycopene from Tomato

Lycopene is a bright-red carotenoid pigment mainly found in tomatoes and other fruits and vegetables such as carrots, watermelon, and papaya. Structurally, it is a tetraterpene composed of eight isoprene units (Figure 20.7). The evidence in support of lycopene in the prevention of chronic diseases comes from epidemiological studies [81] as well as tissue culture studies using human cancer cell lines [82], animal studies [83], and human clinical trials [84]. Of all the cancers, the role of lycopene in the prevention of prostate cancer has been studied the most. An inverse relationship between the consumption of tomatoes and the risk of prostate cancer was first demonstrated in a 1995 publication [85]. Lycopene was suggested as being the beneficial compound present in tomatoes (Figure 20.8). A follow-up meta-analysis of 72 different studies in 1999 showed that lycopene intake as well as serum lycopene levels were inversely related to several cancers including prostate, breast, cervical, ovarian, liver, and other organ sites [86]. Several other studies since then have demonstrated that with increased intake of lycopene and serum levels of lycopene, the risk of cancers were reduced significantly.

A study was undertaken to investigate the status of oxidative stress and antioxidants in patients with prostate cancer [87]. Results showed significant differences in the levels of serum carotenoids, biomarkers of oxidation, and prostate-specific antigen levels in these

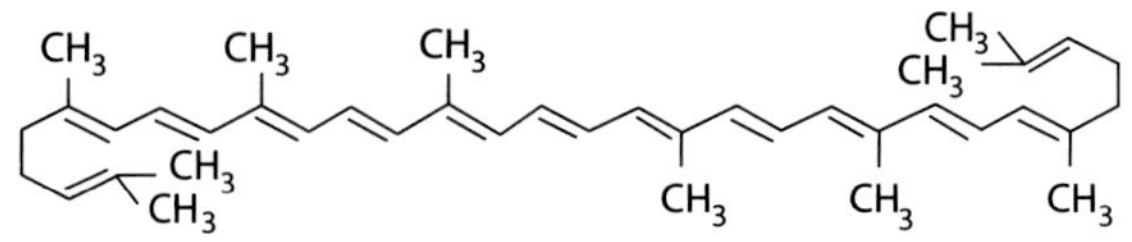

FIGURE 20.7
Structure of lycopene.

subjects. Although there were no differences in the levels of β-carotene, lutein, cryptoxanthin, and vitamins E and A between the patients with cancer and their controls, the levels of lycopene were significantly lower in the patients with cancer. As expected, the prostate-specific antigen levels were significantly elevated in the patients with cancer, who also had higher levels of lipid and protein oxidation, indicating higher levels of oxidative stress. In the same study, serum prostate-specific antigen levels were shown to be inversely related to the serum lycopene. Other carotenoids did not show a similar inverse relationship.

In more recently reported studies, lycopene was shown to decrease the levels of prostate-specific antigen as well as the growth of prostate cancer in patients with newly diagnosed prostate cancer receiving 15 mg of lycopene daily for 3 weeks prior to radical prostactomy [87]. In another study, when tomato sauce was used as a source of lycopene, providing 30 mg of lycopene/day for 3 weeks prior to prostatectomy in men diagnosed with prostate cancer, serum and prostate lycopene levels were elevated significantly [88]. Oxidative damage to DNA was reduced and serum prostate-specific antigen levels declined significantly by 20% with lycopene treatment. Although small in number, these observations raise the possibility that lycopene may be involved not only in the prevention of cancers but may play a role in the treatment of the disease.

Other than prostate cancer, there is now growing evidence in support of the protective role of lycopene in cancers in other organs including breast, lung, gastrointestinal,

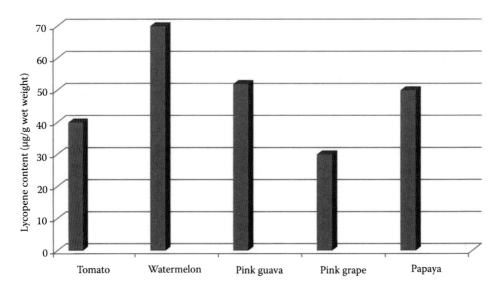

FIGURE 20.8
Lycopene content of certain common fruits and vegetables. (From Rao, A.V. and Rao, L.G., *Pharmacol. Res.* 55:207–216, 2007. With permission.)

cervical, ovarian, and pancreatic cancers [85]. Tissue culture studies using human cancer cell lines have shown that their growth is inhibited significantly in the presence of lycopene in the growth medium [89]. Similarly, several animal studies have also confirmed the inverse association between dietary lycopene and the growth of both spontaneous and transplanted tumors. Human dietary intervention studies are now beginning to be undertaken to study the role of lycopene in breast, ovarian, and cervical cancers.

Experimental studies are limited at this point, but epidemiologic studies suggest that lycopene consumption may also protect against various forms of cancer, including cancer of the prostate, cervix, pharynx, esophagus, stomach, bladder, colon, and rectum [90]. Interestingly, it seems that lycopene also may play a protective role against ultraviolet light exposure and cigarette smoke, although more research is needed. The anticarcinogenic mechanisms of lycopene remain speculative, but its antioxidant properties are believed to play a role because oxidative stress is linked to carcinogenesis. It seems that it may interfere with oxidative damage to lipids, DNA, and lipoproteins. Lycopene has been shown to be a more potent inhibitor than either α- or β-carotenes of tumor cell growth and proliferation in cell cultures and animal models.

Overall, epidemiological studies, *in vitro* tissue culture studies, animal studies, and now, some human intervention studies are showing that increased intake of lycopene will result in increased circulatory and tissue levels of lycopene. *In vivo* lycopene can act as a potent antioxidant and protect cells against oxidative damage and thereby prevent or reduce the risk of several cancers. Further studies are needed to get further proof and to gain a better understanding of the mechanisms involved.

20.8 Protease Inhibitors from Legume Grains

Protease inhibitor is a type of protein found in food sources, especially in legume grains, which can reduce/inhibit the activity of digestive protease enzymes (both trypsin and chymotrypsin). Based on their structure and function, protease inhibitors can be classified into two groups, namely, Kunitz inhibitors and Bowman–Birk inhibitors (Figure 20.9). Included among the potential protective microcomponents against cancer, which, to various extents, can be present in pulses, are protease inhibitors, saponins, phytosterols, lectins, and phytates [91]. However, according to Mathers [91], despite the fact that at least 58 studies (updated to 1997) have reported results for intakes of pulses and cancer risk, it will be very difficult, using conventional epidemiological tools, to ascertain the quantitative contribution made by pulses to cancer risk. Other conclusions are more drastic: overall, the epidemiologic data on breast cancer reduction upon soybean intake are inconclusive. Nevertheless, it is especially in this area that studies on the beneficial effects of specific and isolated legume protein components against cancer have been carried out previously.

In particular, two classes of legume proteins seem to play a role: the protease inhibitors and the lectins. Proteases are considered key players in a wide range of biological processes and the malfunction of certain proteases has been related to diseases, including cancer progression. Therefore, control of this activity by protease inhibitors seems to closely relate to the capacity of preventing or blocking certain tumoral pathologies. Studies with soybean serine protease inhibitors of the Bowman–Birk inhibitor family have provided evidence of the arrest of certain mammalian tumors [92]. More recently, the antiproliferative effects on human colon cancer cells of two recombinant wild-type Bowman–Birk

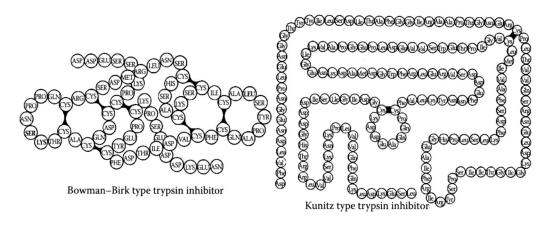

Bowman–Birk type trypsin inhibitor

Kunitz type trypsin inhibitor

FIGURE 20.9
Structure of protease inhibitor (Kunitz inhibitor and Bowman–Birk inhibitor).

inhibitors from pea seeds have been reported [93]. The measured effect was much greater than with soybean Bowman–Birk inhibitor.

According to Clemente et al. [93], the anticancer effect of Bowman–Birk inhibitors could be explained by following the identification of target proteases and by dissecting the apoptotic and mitotic processes that are affected by the inhibitor inside the test cells. Bowman–Birk inhibitors have been shown to be effective by both *in vitro* and *in vivo* approaches [92]. Other seed protease inhibitors have demonstrated their antitumoral efficacy [94]. More recently, studies on specific cancer forms have been carried out directly in humans [95]. The Bowman–Birk inhibitor is a double-headed inhibitor usually capable of inhibiting both trypsin and chymotrypsin. It has been proposed that antichymotryptic activity is more effective than antitryptic activity in the suppression of carcinogen-induced transformation [93]. Therefore, a mutagenesis program on this protein molecule aimed at converting the inhibitor to a double chymotrypsin form could be advantageous for increased biological activity. It has also been claimed that the potential beneficial effects require native protein conformation and therefore would not be attained after proper cooking of legumes. On the other hand, incomplete denaturation of the Bowman–Birk inhibitors' resistant protein structure cannot be excluded, especially in domestic preparations of legume-based dishes. This can be the basis for an effective preventive activity attributed to regular legume intake.

Besides the anticarcinogenic effects, Bowman–Birk inhibitors also showed anti-inflammatory activity by inhibiting inflammation-mediating proteases. In this context, a reduction of ulcerative colitis in mice was reported [96]. More recently, a number of patients using Bowman–Birk inhibitors for combating obesity and even degenerative and autoimmune diseases, including multiple sclerosis, Guillain–Barré syndrome and, in general, skeletal muscle atrophy have appeared. Previously, lectins were considered as promising biologically active molecules. However, their numerous observed effects, including the induction of small intestinal hyperplasia, changes in the intestinal flora, immunomodulating activity, interference with hormone secretion, and access to the systemic circulation have hampered their straightforward utilization for specific medical uses. Nevertheless, a number of reports on various animal models conclusively showed that lectins may limit tumor growth by promoting gut epithelium hyperplasia or through other effects.

20.9 Organosulfur Compounds from *Allium* Vegetables

Organosulfur compounds are organic compounds that contain sulfur. They are often associated with foul odors, but many of the sweetest compounds known are organosulfur derivatives. Allicin and ajoene (Figure 20.10) are naturally occurring organosulfur compounds in garlic, and are responsible for the characteristic flavor of garlic. Epidemiological studies continue to support the premise that dietary intake of *Allium* vegetables may be protective against various types of cancers, including stomach [97], colorectal [98], esophageal [97], and prostate cancer [99] (Table 20.4). For instance, a population-based case-control study showed that the risk of prostate cancer was significantly lower in men consuming more than 10 g/day of total *Allium* vegetables than in men with total *Allium* vegetable intakes of less than 2.2 g/day [99].

The anticarcinogenic effect of *Allium* vegetables is attributed to organosulfur compounds (e.g., diallyl sulfide, diallyl disulfide, diallyl trisulfide, S-allyl cysteine, S-allylmercaptocysteine, ajoene, etc.), which are generated upon processing (cutting or chewing) of these vegetables. *Allium* vegetable–derived organosulfurs are highly effective in affording protection against cancer in animal models induced by a variety of chemical carcinogens. For example, diallyl sulfate has been shown to inhibit aberrant crypt foci, hepatic foci, and N-nitrosomethylbenzylamine–induced esophageal tumors in rats, and polycyclic aromatic hydrocarbon–induced skin carcinogenesis in mice [100]. Likewise, diallyl disulfide has been shown to inhibit chemically induced colon carcinogenesis in rats [101], N-methyl-N-nitrosourea–induced rat mammary carcinogenesis [102], and skin tumors in mice [103].

The diallyl disulfide treatment also inhibited 2-amino-1-methyl-6-phenylimidazo(4,5-b) pyridine–induced mammary carcinogenesis in rats [104]. Previous studies have shown that diallyl disulfide suppresses the growth of H-ras oncogene–transformed tumor xenografts in nude mice without causing weight loss or any other side effects. Similarly, oral administration of 6 mmol of diallyl trisulfide (three times per week) to male nude mice significantly inhibited the growth of PC-3 human prostate cancer xenografts [105]. The protective effect of garlic oil against skin tumor promotion, garlic extract against methylcholanthrene-induced carcinogenesis of the uterine cervix in mice, and ajoene against skin tumor promotion in mice have also been documented.

A review of the literature suggests that organosulfur compounds may prevent cancer by multiple mechanisms including impairment of carcinogen activation, enhanced inactivation of carcinogenic intermediates through induction of phase 2 enzymes, inhibition of posttranslational modification of oncogenic ras, induction of apoptosis, inhibition of cell cycle progression, histone modification, inhibition of angiogenesis, and metastasis [106]. For example, diallyl sulfide has been reported to inhibit 2-amino-fluorene-DNA adduct

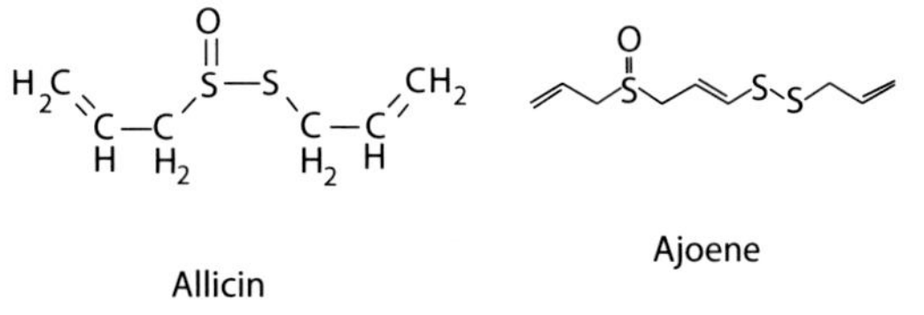

FIGURE 20.10
Organosulfur compounds in garlic (allicin and ajoene).

TABLE 20.4

Effect of Dietary Foods on the Risk of Different Types of Cancer

Mouth, Pharynx, and Larynx Cancer	
Nonstarchy vegetables	Probable decreased risk
Fruits	Probable decreased risk
Foods containing carotenoids	Probable decreased risk
Alcoholic drinks	Convincing decreased risk
Nasopharynx Cancer	
Nonstarchy vegetables	Limited-suggestive decreased risk
Fruits	Limited-suggestive decreased risk
Cantonese-style salted fish	Probable increased risk
Esophageal Cancer	
Foods containing dietary fiber	Limited suggestive decreased risk
Nonstarchy vegetables	Probable decreased risk
Fruits	Probable decreased risk
Foods containing folate	Limited-suggestive decreased risk
Foods containing β-carotene	Probable decreased risk
Foods containing vitamin C	Probable decreased risk
Foods containing pyridoxine	Limited-suggestive decreased risk
Foods containing vitamin E	Limited-suggestive decreased risk
Red meat	Limited-suggestive increased risk
Processed meat	Limited-suggestive increased risk
High temperature drinks	Limited-suggestive increased risk
Alcoholic drinks	Convincing increased risk
Lung Cancer	
Nonstarchy vegetables	Limited-suggestive decreased risk
Fruits	Probable decreased risk
Foods containing carotenoids	Probable decreased risk
Foods containing selenium	Limited-suggestive decreased risk
Foods containing quercetin	Limited suggestive decreased risk
Red meat	Limited-suggestive increased risk
Processed meat	Limited-suggestive increased risk
Total fat	Limited-suggestive increased risk
Butter	Limited-suggestive increased risk
Arsenic in drinking water	Convincing increased risk
β-Carotene (supplement)	Convincing increased risk
Selenium (supplement)	Limited-suggestive decreased risk
Retinol (Supplement)	Limited suggestive increased risk
Stomach Cancer	
Nonstarchy vegetables	Probable decreased risk
Allium vegetables	Probable decreased risk
Chili	Limited-suggestive increased risk
Fruits	Probable decreased risk
Legumes	Limited-suggestive decreased risk
Foods containing selenium	Limited-suggestive decreased risk
Processed meat	Limited-suggestive increased risk

(continued)

TABLE 20.4 (Continued)

Effect of Dietary Foods on the Risk of Different Types of Cancer

Smoked foods	Limited-suggestive increased risk
Grilled animal foods	Limited-suggestive increased risk
Salty foods	Probable increased risk

Pancreas Cancer

Fruits	Limited-suggestive decreased risk
Foods containing folate	Probable decreased risk
Red meat	Limited-suggestive increased risk
Coffee	Substantial effect on risk unlikely

Liver Cancer

Aflotoxins	Convincing increased risk
Fruits	Limited-suggestive decreased risk
Alcoholic drinks	Probable increased risk

Colorectal Cancer

Foods containing dietary fiber	Probable decreased risk
Nonstarchy vegetables	Limited-suggestive decreased risk
Garlic	Probable decreased risk
Fruits	Limited-suggestive decreased risk
Foods containing folate	Limited-suggestive decreased risk
Foods containing selenium	Limited-suggestive decreased risk
Red meat	Convincing increased risk
Processed meat	Convincing increased risk
Foods containing iron	Limited-suggestive increased risk
Fish	Limited-suggestive decreased risk
Foods containing vitamin D	Limited-suggestive decreased risk
Milk	Probable decreased risk
Cheese	Limited-suggestive increased risk
Foods containing animal fat	Limited-suggestive increased risk
Foods containing sugars	Limited-suggestive increased risk
Alcoholic drinks	Probable increased risk

Breast Cancer (Premenopause)

Alcoholic drinks	Convincing increased risk

Breast Cancer (Postmenopause)

Total fat	Limited-suggestive increased risk
Alcoholic drinks	Convincing increased risk

Ovarian Cancer

Nonstarchy vegetables	Limited-suggestive decreased risk

Endometrial Cancer

Nonstarchy vegetables	Limited-suggestive decreased risk
Red meat	Limited-suggestive increased risk

Cervix Cancer

Carrot	Limited-suggestive decreased risk

(*continued*)

TABLE 20.4 (Continued)

Effect of Dietary Foods on the Risk of Different Types of Cancer

Prostate Cancer	
Legumes	Limited-suggestive decreased risk
Foods containing β-carotene	Substantial effect on risk unlikely
Foods containing lycopene	Probable decreased risk
Foods containing selenium	Probable decreased risk
Foods containing vitamin E	Limited-suggestive decreased risk
Processed meat	Limited-suggestive increased risk
Diets high in calcium	Probable increased risk
Milk and dairy products	Limited-suggestive increased risk
β-Carotene (supplement)	Substantial effect on risk unlikely
Selenium (supplement)	Probable decreased risk
α-Tocopherol	Limited-suggestive decreased risk
Kidney Cancer	
Arsenic in drinking water	Limited-suggestive increased risk
Coffee	Substantial effect on risk unlikely
Alcoholic drinks	Substantial effect on risk unlikely
Bladder Cancer	
Milk	Limited-suggestive decreased risk
Arsenic in drinking water	Limited-suggestive increased risk
Skin Cancer	
Foods containing β-carotene	Substantial effect on risk unlikely
Arsenic in drinking water	Probable increased risk
β-Carotene (Supplement)	Substantial effect on risk unlikely
Selenium	Limited-suggestive increased risk
Retinol (Supplement)	Limited-suggestive decreased risk

Source: American Institute for Cancer Research. *A Dietician's Cancer Story.* Washington, DC: AICR. 2007. With permission.

formation in human promyelocytic leukemia cells, inhibit cyclooxygenase-2 in HEK-293-T cells and competitively inhibit the activity of cytochrome P-4502-E1 in a time-dependent and NADPH-dependent manner with pseudo–first-order kinetics.

Diallyl sulfide, diallyl disulfide, and diallyl trisulfide are potent inducers of the expression of phase 2 carcinogen-inactivating enzymes, including glutathione transferases in liver, lung, and forestomach of mice [107]. It is also becoming clear that organosulfur compounds are promiscuous because they target multiple signal transduction pathways to trigger growth arrest and programmed cell death (apoptosis). The ability of organosulfur compounds to cause cell cycle arrest and apoptosis induction was first documented by Knowles and Milner in human colon cancer cells [108]. The diallyl disulfide–mediated apoptosis in human colon cancer cells correlated with an increase in the levels of free intracellular calcium and the G_2-M phase cell cycle arrest was accompanied by a decrease in p34 cyclin–dependent kinase activity, a reduction in complex formation between cyclin-dependent kinase and cyclin B1, and a decrease in Cdc-25-C protein level [108]. Subsequently, the organosulfur-mediated cell cycle arrest and apoptosis induction was shown in human colon cancer cells, SH-SY-5Y neuroblastoma cells, and in PC-3, DU-145, and LNCaP human prostate cancer cells.

More recent studies have offered novel insights into the mechanism by which diallyl trisulfide causes cell cycle arrest and apoptosis induction in human prostate cancer cells. It was found that even a subtle change in organosulfur structure (e.g., oligosulfide chain length and the presence of terminal allyl groups) has a significant effect on its growth-inhibitory and apoptosis-inducing potency. Interestingly, a normal prostate epithelial cell line is significantly more resistant to growth arrest and apoptosis induction by diallyl trisulfide compared with prostate cancer cells. The diallyl trisulfide–treated prostate cancer cells are not only arrested in the G_2 phase of the cell cycle but also in mitosis [106].

The diallyl trisulfide–induced G_2 phase cell cycle arrest is independent of p21 but correlates with reactive oxygen species–dependent downregulation and Ser-216 phosphorylation of Cdc-25-C, a dual-specificity phosphatase partially responsible for the activation of Cdk1/cyclin B kinase complex, followed subsequently by the activation of a novel kinase-1–dependent prometaphase checkpoint in diallyl trisulfide–treated cancer cells [106]. The diallyl trisulfide–mediated prometaphase arrest was associated with the inhibition of anaphase-promoting complex/cyclosome (APC/C) as revealed by the accumulation of its substrates (cyclin A, cyclin B1, and securin) and hyperphosphorylation of core subunits (Cdc-20 and Cdh-1) of the APC/C.

Generation of the diallyl trisulfide–mediated reactive oxygen species, which also contributed to cell death by activating c-*Jun* N-terminal kinase, was caused by degradation of the iron storage protein ferritin leading to elevation of the labile iron pool. The diallyl trisulfide–mediated G_2-M phase cell cycle arrest was significantly attenuated by pretreatment with antioxidants. A schematic of the complexity of signal transduction leading to cell cycle arrest and apoptosis induction by diallyl trisulfide is shown in Figure 20.11. Despite these advances, the mechanism of diallyl trisulfide–induced cell cycle arrest is not fully understood. For instance, the mechanism by which checkpoint kinase-1 regulates diallyl trisulfide–mediated inhibition of APC/C and mitotic block is not fully understood. Likewise, it is unclear if the diallyl trisulfide–induced G_2 phase cell cycle arrest is reversible and serves to allow the cell time to repair damage.

The diallyl trisulfide–induced G_2 phase cell cycle arrest in PC-3 and DU145 human prostate cancer cells correlates with a decrease in protein levels of Cdk1 and Cdc25C, leading to the accumulation of inactive (Tyr15 phosphorylated) Cdk1 (Figure 20.11). The diallyl trisulfide treatment causes a checkpoint kinase 1–dependent prometaphase arrest in cancer cells, which correlates with the inactivation of APC/C as evidenced by the accumulation of its substrates (cyclin A, cyclin B1, and securin) and hyperphosphorylation of APC/C subunits Cdc20 and Cdh1 [106]. The mechanism by which diallyl trisulfide causes the activation of checkpoint kinase 1 remains elusive but may involve reactive oxygen species–dependent DNA double-strand breaks. The diallyl trisulfide treatment causes JNK (and to some extent, extracellular signal–regulated kinase)–dependent phosphorylation of Bcl-2 in PC-3/DU145 cells leading to reduced interaction between Bcl-2 and Bax and mitochondria-mediated caspase activation and apoptosis. The diallyl trisulfide treatment also inactivates Akt leading to reduced phosphorylation and mitochondrial translocation of proapoptotic Bad. The diallyl trisulfide inhibits angiogenic features in human umbilical vein endothelial cells in association with the inactivation of Akt and JNK activation.

The diallyl disulfides were shown to increase histone acetylation and p21 (waf1/cip1) expression. A summary of signal transduction leading to cell cycle arrest, apoptosis induction, and suppression of angiogenesis by diallyl trisulfide is shown in Figure 20.11. The diallyl trisulfide–induced G_2 phase cell cycle arrest in PC-3 and DU-145 human prostate cancer cells correlates with a decrease in protein levels of Cdk1 and Cdc-25-C, leading to an accumulation of inactive (Tyr15 phosphorylated) cysteine-dependent kinase-1. The

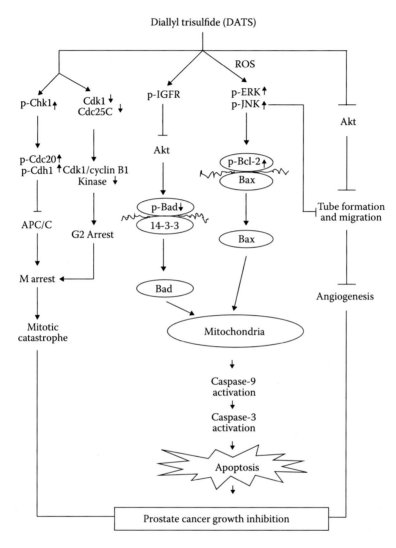

FIGURE 20.11
Summary of signal transduction leading to cell cycle arrest, apoptosis induction, and suppression of angiogenesis by diethyl trisulfide.

diallyl trisulfide treatment causes a checkpoint kinase-1–dependent prometaphase arrest in cancer cells, which correlates with the inactivation of APC/C as evidenced by the accumulation of its substrates (cyclin A, cyclin B1, and securin) and hyperphosphorylation of APC/C subunits Cdc-20 and Cdh-1 [106]. The mechanism by which diallyl trisulfide causes the activation of checkpoint kinase-1 remains elusive but may involve reactive oxygen species–dependent DNA double-strand breaks. The diallyl trisulfide treatment causes c-*Jun* N-terminal kinase–dependent phosphorylation of Bcl-2 in PC-3/DU-145 cells leading to reduced interaction between Bcl-2 and Bax and mitochondria-mediated caspase activation and apoptosis. The diallyl trisulfide treatment also inactivates Akt leading to reduced phosphorylation and mitochondrial translocation of proapoptotic Bad. The diallyl trisulfide inhibits angiogenic features in human umbilical vein endothelial cells in association with inactivation of Akt and c-*Jun* N-terminal kinase activation.

In addition, Hosono et al. [109] have documented that diallyl trisulfide can cause specific oxidative modification of cysteine residues Cys-12 and Cys-354 of β-tubulin *in vitro*. However, it remains to be seen whether diallyl trisulfide causes oxidative modification of β-tubulin cysteine *in vivo*. The diallyl trisulfide–mediated cell death in human prostate cancer cells is caspase-dependent and is regulated by c-*Jun* N-terminal kinase–mediated, and to some extent extracellular signal–regulated kinase-mediated, phosphorylation of Bcl-2 (PC-3 and DU-145 cells) leading to reduced interaction between Bcl-2 and Bax, inhibition of Akt kinase leading to reduced phosphorylation of proapoptotic Bcl-2 family member BAD and its translocation to the mitochondria, and induction of proapoptotic proteins Bax and Bak. The diallyl trisulfide–mediated suppression of PC-3 xenograft growth *in vivo* also correlates with the induction of Bax and Bak protein expression. A critical role for c-*Jun* N-terminal kinase in apoptosis induction by diallyl disulfide has also been observed in neuroblastoma cells.

Apoptosis induction by diallyl trisulfide has been shown in the BGC-823 gastric cancer cell line [110] and colon cancer cells [109]. Diallyl trisulfide treatment inhibits angiogenic features of human umbilical vein endothelial cells in association with the suppression of vascular endothelial growth factor secretion, vascular endothelial growth factor receptor-2 downregulation, and inhibition of Akt [105]. Ajoene, a lipid-soluble component of garlic, induces nuclear factor-κB activation, reactive oxygen species generation, and apoptosis and G_2-M phase cell cycle arrest in human leukemia cells. Garlic and onion oils inhibited proliferation and induced differentiation in HL-60 cells.

A fundamental question, which remains unanswered, is whether the high millimolar concentrations of organosulfur compounds needed for cancer cell growth suppression and apoptosis induction *in vitro* achievable in humans *in vivo*? Limited data exist on the bioavailability of garlic compounds. The pharmacokinetics of S-allyl cysteine was studied in rats, mice, and dogs [111]. The S-allyl cysteine was rapidly and easily absorbed in the gastrointestinal tract and distributed mainly in the plasma, liver, and kidney. The bioavailability was 98.2%, 103.0%, and 87.2% in rats, mice, and dogs, respectively. The S-allyl cysteine was found to be mainly excreted into the urine in the N-acetyl form in rats. On the other hand, mice excreted both S-allyl cysteine and the N-acetyl forms. The half-life of S-allyl cysteine was longer in dogs than in rats and mice [111].

Accumulating evidence from epidemiological studies supports the anticancer properties of garlic and its organosulfur compounds. They seem to exert their anticarcinogenic effects through multiple mechanisms [106]. A double-blind clinical trial using aged garlic extracts (for 12 months at 2.4 mL/day) has recently reported a significant suppression of both the size and number of colon adenomas in patients [112]. In a clinical trial, a combination of curcumin and quercetin reduced the number and size of ileal and rectal adenomas in patients with familial adenomatous polyposis [40].

The pharmacokinetic parameters for diallyl trisulfide in humans have not yet been measured, but the concentration of diallyl trisulfide in rat blood after treatment with 10 mg of the compound was shown to be approximately 31 mmol/L [113]. A preliminary, double-blind clinical trial using aged garlic extracts (for 12 months at 2.4 mL/day) has recently reported a significant suppression of both size and number of colon adenomas in patients [112]. Excessive consumption of garlic can cause burning sensations, diarrhea, and allergic reactions. A toxicity study in mice has reported that the 50% lethal oral dose (mg/kg body weight) for allicin is 309 mg in males and 363 mg in females; for S-allyl cysteine is 8890 mg in males and 9390 mg in females; and for diallyl disulfide is 145 mg in males and 130 mg in females [114]. Oral administration of 200 mg of synthetic diallyl trisulfide in combination with 100 mg of selenium every other day for 1 month did not cause any harmful side effects [115]. Detailed toxicology of organosulfuric compounds, especially diallyl trisulfide, due to its promising preclinical effects, followed by clinical trials are needed to determine whether constituents of *Allium* vegetables may be used to prevent cancers in humans.

The use of garlic as an effective remedy for tumors has been documented in as early as 1550 BC. More recently, animal and cell culture studies have shown garlic to be a potent inhibitor of tumorigenesis. However, epidemiologic studies have thus far not shown a strong effect of garlic intake on cancer prevention. Diallyl disulfide and diallyl sulfide seem to be the bioactive components of garlic that exert the anticarcinogenic effects. These allylic compounds stimulate glutathione S-transferase activity in the liver. This transferase binds to and detoxifies potential carcinogens.

Allicin has been found to cause a transient decrease in glutathione, which was correlated with its antiproliferative action. Organosulfur compounds derived from garlic function as antioxidants with free radical–scavenging properties to inhibit lipid peroxidation. Diallyl sulfide may function to suppress the tumor promotion phase of carcinogenesis by reducing polyamine formation inhibiting ornithine decarboxylase and possibly by stimulating DNA repair. In contrast, Fukushima et al. [116] reported that diallyl sulfide promoted rather than inhibited liver carcinogenesis. An additional mechanism by which garlic may suppress carcinogenesis is through a depression in nitrosamine formation. Although many studies report the chemopreventive effects of garlic, further studies are needed to clarify its role in cancer prevention.

20.10 Conclusion

Earlier research reports clearly evidenced the health protective effects of certain phytochemical compounds from common food sources. Among them, polyphenolics from berries, catechins from black tea, flavonoids from fruits and vegetables, tannins from sorghum grains, lycopene from tomatoes, protease inhibitors from legume seeds, and organosulfur from *Allium* vegetables are scientifically proven to be effective in preventing different types of cancers. Furthermore, the anticancer properties of various common food components are still under investigation. Hence, such food items with demonstrated anticancer effects should be included in the regular diet of human beings to prevent/eradicate the highly dangerous killer disease, cancer. Through the implementation of governmental policies, health awareness programs, various activities by food associations/NGOs and research efforts by nutritionists, it is possible to develop healthy food formulations to prevent cancer by considering the anticancer functionality of natural food products.

References

1. Bondia-Pons, I., A.M. Aura, S. Vuorela, M. Kolehmainen, H. Mykkanen, and K. Poutanen. 2009. *Journal of Cereal Science* 49:323–336.
2. McClements, D.J., E.A. Decker, Y. Park, and J. Weiss. 2009. Structural design principles for delivery of bioactive components in nutraceuticals and functional foods. *Critical Reviews in Food Science and Nutrition* 49:577–606.

3. Fontana, L. 2009. Modulating human aging and age-associated diseases. *Biochimica et Biophysica Acta* 1790:1133–1138.
4. Boukharta, M., G. Jalbert, and A. Castonguay. 1992. Biodistribution of ellagic acid and dose-related inhibition of lung tumorigenesis in A/J mice. *Nutrition and Cancer* 18:181–189.
5. Mukhtar, H., M. Das, B.J. Del Tito, and D.R. Bickers. 1984. Protection against 3-methylcholan-threne-induced skin tumorigenesis in Balb/c mice by ellagic acid. *Biochemistry and Biophysics Research Communications* 119:751–757.
6. Mandal, S., and G.D. Stoner. 1990. Inhibition of N-nitrosobenzylmethylamine-induced esophageal tumorigenesis in rats by ellagic acid. *Carcinogenesis* 11:55–61.
7. Rao, C.V., K. Tokumo, J. Rigotty, E. Zang, G. Kelloff, and B.S. Reddy. 1991. Chemoprevention of colon carcinogenesis by dietary administration of piroxicam, alpha-difluoromethylornithine, 16 alpha-fluoro-5-androsten-17-one, and ellagic acid individually and in combination. *Cancer Research* 51:4528–4534.
8. Battino, M., J. Beekwilder, B. Denoyes-Rothan, M. Laimer, G.J. McDougall, and B. Mezzetti. 2009. Bioactive compounds in berries relevant to human health. *Nutrition Reviews* 67:S145–S150.
9. Narayanan, B.A., O. Geoffroy, M.C. Willingham, G.G. Re, and D.W. Nixon. 1999. p53/p21 (WAF1/CIP1) expression and its possible role in G1 arrest and apoptosis in ellagic acid treated cancer cells. *Cancer Letters* 136:215–221.
10. Han, C., H. Ding, B. Casto, G.D. Stoner, and S. D'Ambrosio. 2005. Inhibition of the growth of premalignant and malignant human oral cell lines by extracts and components of black raspberries. *Nutrition and Cancer* 51:207–217.
11. Carlton, P.S., L.A. Kresty, J.C. Siglin, M.A. Morse, J. Lu, C. Morgan, and G.D. Stoner. 2001. Inhibition of N-nitrosomethylbenzylamine-induced tumorigenesis in the rat esophagus by dietary freeze-dried strawberries. *Carcinogenesis* 22:441–446.
12. Stoner, G.D., T. Chen, L.A. Kresty, R.M. Aziz, T. Reinemann, and R. Nines. 2006. Protection against esophageal cancer in rodents with lyophilized berries: Potential mechanism. *Nutrition and Cancer* 54:33–46.
13. Harris, G.K., A. Gupta, R.G. Nines, L.A. Kresty, S.G. Habib, W.L. Frankel et al. 2001. Effects of lyophilized black raspberries on azoxymethane-induced colon cancer and 8-hydroxy-20-deoxy-guanosine levels in the Fisher-344 rat. *Nutrition and Cancer* 40:125–133.
14. Ross, H.A., G.J. McDougall, and D. Stewart. 2007. Anti-proliferative activity is predominantly associated with ellagitannins in raspberry extracts. *Phytochemistry* 68:218–228.
15. Kresty, L.A., W. Frankel, C. Hammond, M. Baird, J.M. Mele, G.D. Stoner, and J. Fromkes. 2006. Transitioning from preclinical to clinical chemopreventive assessments of lyophilized black raspberries: Interim results show berries modulate markers of oxidative stress in Barrett's esophagus patients. *Nutrition and Cancer* 54:148–156.
16. Stoner, G.D., L.S. Wang, N. Zikri, T. Chen, S.S. Hecht, C. Huang et al. 2007. Cancer prevention with freeze-dried berries and berry components. *Seminars in Cancer Biology* 17:403–410.
17. Sun, C.L., J.M. Yuan, M.J. Lee, C.S. Young, Y.T. Gao, R.K. Ross, and M.C. Yu. 2002. Urinary tea polyphenols in relation to gastric and esophageal cancers: A prospective study of men in Shanghai, China. *Carcinogenesis* 23:1497–1503.
18. Caderni, G., C.D. Filippo, C. Luceri, M. Salvadori, A. Giannini, A. Biggeri et al. 2000. Effects of black tea, green tea and wine extracts on intestinal carcinogenesis induced by azoxymethane in F344 rats. *Carcinogenesis* 21:1965–1969.
19. Michels, K.B., W.C. Willet, C.S. Fuchs, and E. Giovannucci. 2005. Coffee, tea and caffeine consumption and incidence of colon and rectal cancer. *Journal of National Cancer Institute* 97:282–292.
20. Hong, J., T.J. Smith, C.T. Hoc, D.A. August, and C.S. Yang. 2001. Effects of purified green and black tea polyphenols on cyclooxygenase and lipoxygenase-dependent metabolism of arachidonic acid in human colon mucosa and colon tumour tissues. *Biochemical Pharmacology* 62:1175–1183.
21. IL'Yasova, D., L. Arab, A. Martinchick, A. Sdvizhkov, L. Urbanovich, and U. Weisgerber. 2003. Black tea consumption and risk of rectal cancer in Moscow population. *Annals of Epidemiology* 13:405–411.

22. Cao, J., Y. Xu, J. Chen, and J.E. Klaunig. 1996. Chemopreventive effects of green and black tea on pulmonary and hepatic carcinogenesis. *Fundamental and Applied Toxicology* 29:244–250.

23. Yang, G.Y., Z. Liu, D.N. Seril, J. Liao, W. Ding, S. Kim et al. 1997. Black tea constituents, theaflavins, inhibit 4-(methyl nitrosamino)-1-(3-pyridyl)-1-butanone (NNK)-induced lung tumorigenesis in A/J mice. *Carcinogenesis* 18:2361–2365.

24. Yang, G.Y., J. Liao, C. Li, J. Chung, E.J. Yurkow, C.T. Ho, and S. Chung. 2000. Effect of black and green tea polyphenols on C-jun phosphorylation and H_2O_2 production in transformed and non transformed human bronchial cell lines: Possible mechanisms of cell growth inhibition and apoptosis induction. *Carcinogenesis* 21:2035–2039.

25. Lee, H.H., C.T. Ho, and J.K. Lin. 2004. Theaflavin-3,3-digallate and penta-o-galloyl-*O*-D-glucose inhibit rat liver microsomal 5D-reductase activity of the expression of androgen receptor in LNCaP prostate cancer cells. *Carcinogenesis* 25:1109–1118.

26. Rogers, A.E., L.J. Hafer, Y.S. Iskander, and S. Yang. 1998. Black tea and mammary gland carcinogenesis by 7,12-dimethyl benz[a]anthracene in rats fed control or high fat diets. *Carcinogenesis* 19:1269–1273.

27. Wu, A.H., K. Arakawa, F.Z. Stanczyk, D.V.D. Berg, W.P. Koh, and M.C. Yu. 2005. Tea and circulating estrogen levels in postmenopausal Chinese women in Singapore. *Carcinogenesis* 26:976–980.

28. Lea, M.A., Q. Xiao, A.K. Sadhukhan, S. Cottle, Z.Y. Wang, and C.S. Yang. 1993. Inhibitory effects of tea extracts and epigallocatechin gallate on DANN synthesis and proliferation of hepatoma and erythroleukemia cells. *Cancer Letters* 68:231–236.

29. Matsumoto, N., K. Toshiyuki, K. Okushio, and Y. Hara. 1996. Inhibitory effects of tea catechins, black tea extract and Oolong tea extract on hepatocarcinogenesis in rat. *Japanese Journal of Cancer Research* 87:1034–1038.

30. Lu, Y.P., Y.R. Lou, J.G. Xie, P. Yen, M.T. Huang, and A.H. Conney. 1997. Inhibitory effect of black tea on the growth of established skin tumours in mice: Effects on tumour size, apoptosis, mitosis and bromodeoxyuridine incorporation into DNA. *Carcinogenesis* 18:2163–2169.

31. Nomura, M., W.Y. Ma, C. Huang, C.S. Yang, T. Bowden, K. Miyamoto, and Z. Dong. 2000. Inhibition of ultraviolet B-induced AP-1 activation by theaflavins from black tea. *Carcinogenesis* 28:148–155.

32. Bhattacharya, A., D. Mandal, L. Lahiry, G. Sa, and T. Das. 2004. Black tea protects immunocytes from tumour induced apoptosis by changing Bcl-2/Bax ratio. *Cancer Letters* 209:147–154.

33. Kundu, T., S. Dey, M. Roy, M. Siddiqui, and R.K. Bhattacharya. 2005. Induction of apoptosis in human leukemia cells by black tea and its polyphenol theaflavin. *Cancer Letters* 230:111–121.

34. Zeegers, M.P.A., F.E.S. Tan, A. Goldbohm, and P.A. Brendt. 2001. Are coffee and tea consumption associated with urinary tract cancer risk? *International Journal of Epidemiology* 30:353–362.

35. Fujiki, H., M. Suganuma, and S. Okabe. 1998. Cancer inhibition by green tea. *Mutation Research* 402:307–310.

36. Nagano, J., S. Kono, D.L. Preston, and K.A. Mabuchi. 2001. Prospective study of green tea consumption and cancer incidence, Hiroshima and Nagasaki (Japan). *Cancer Causes and Control* 112:501–508.

37. Block, G., B. Atterson, and A. Subar. 1992. Fruit, vegetables, and cancer prevention: A review of the epidemiological evidence. *Nutrition and Cancer* 18:1–29.

38. Marchand Le, L., S.P. Murphy, J.H. Hankin, L.R. Wilkens, and L.N. Kolonel. 2000. Intake of flavonoids and lung cancer. *Journal of National Cancer Institute* 92:154–160.

39. Boyle, S.P., V.L. Dobson, S.J. Duthie, J.A.M. Kyle, and A.R. Collins. 2000. Absorption and DNA protective effects of flavonoid glycosides from an onion meal. *European Journal of Nutrition* 39:213–223.

40. Cruz-Correa, M., D.A. Shoskes, and P. Sanchez. 2006. Combination treatment with curcumin and quercetin of adenomas in familial adenomatous polyposis. *Clinical Gastroenterology and Hepatology* 4:1035–1038.

41. Tsyrlov, I.B., V.M. Mikhailenko, and H.V. Gelboin. 1994. Isozyme- and species-specific susceptibility of cDNA-expressed CYP1A P450s to different flavonoids. *Biochimica et Biophysica Acta* 1205:325–335.

42. Obermeier, M.T., R.E. White, and C.S. Yang. 1995. Effects of bioflavonoids on hepatic P450 activities. *Xenobiotica* 25:575–584.

43. McLemore, T.L., S. Adelberg, M.C. Liu, and R.N. Hines. 1990. Expression of CYP1A1 gene in patients with lung cancer: Evidence for cigarette smoke-induced gene expression in normal lung tissue and for altered gene regulation in primary pulmonary carcinomas. *Journal of National Cancer Institute* 82:1333–1339.

44. Sivaraman, L., M.P. Leatham, J. Yee, L.R. Wilkens, and L.L. Marchand. 1994. CYP1A1 genetic polymorphisms and in situ colorectal cancer. *Cancer Research* 54:3692–3695.

45. Smith, T.J., Z. Guo, F.P. Guengerich, and C.S. Yang. 1996. Metabolism of 4-(methylnitrosamino)-1-(3-pyridyl)-1-butanone (NNK) by human cytochrome P450 1A2 and its inhibition by phenethyl isothiocyanate. *Carcinogenesis* 17:809–813.

46. Ciolino, H.P., P.J. Daschner, and G.C. Yeh. 1999. Dietary flavonols quercetin and kaempferol are ligands of aryl hydrocarbon receptor that affect CYP1A1 transcription differentially. *Biochemistry Journal* 340:715–722.

47. Ueng, Y.F., T. Kuwabara, Y.J. Chun, and F.P. Guengerich. 1997. Cooperativity in oxidations catalyzed by cytochrome P450 3A4. *Biochemistry* 36:370–381.

48. Tassaneeyakul, W., D.J. Birkett, and M.E. Veronese. 1993. Specificity of substrate and inhibitor probes for human cytochromes P450 1A1 and 1A2. *Journal of Pharmacology and Experimental Therapeutics* 265:401–407.

49. Shih, H., G.V. Pickwell, and L.C. Quattrochi. 2000. Differential effects of flavonoid compounds on tumor induced activation of human CYP1A2 enhancer. *Archives of Biochemistry and Biophysics* 373:287–294.

50. Zand, R.S., D.J. Jenkins, and E.P. Diamandis. 2000. Steroid hormone activity of flavonoids and related compounds. *Breast Cancer Research and Treatment* 62:35–49.

51. Ferte, J., J.M. Kuhnel, G. Chapuis, Y. Rolland, G. Lewin, and M.A. Schwaller. 1999. Flavonoid-related modulators of multidrug resistance: Synthesis, pharmacological activity and structure–activity relationships. *Journal of Medicinal Chemistry* 42:478–489.

52. Shapiro, A.B., and V. Ling. 1997. Positively cooperative sites for drug transport by P-glycoprotein with distinct drug specificities. *European Journal of Biochemistry* 250:130–137.

53. Nissler, L., R. Gebhardt, and S. Berger. 2004. Flavonoid binding to a multi-drug-resistance transporter protein: An STD-NMR study. *Analytical and Bioanalytical Chemistry* 379:1045–1049.

54. Chen, H.Y., T.S. Wu, S.F. Su, S.C. Kuo, and P.D.L. Chao. 2002. Marked decrease of cyclosporin absorption caused by phellamurin in rats. *Planta Medica* 68:138–141.

55. Chieli, E., N. Romiti, F. Cervelli, and R. Tongiani. 1995. Effects of flavonols on P-glycoprotein activity in cultured rat hepatocytes. *Life Science* 57:1741–1751.

56. Mitsunaga, Y., H. Takanaga, and H. Matsuo. 2000. Effect of bioflavonoids on vincristine transport across blood–brain barrier. *European Journal of Pharmacology* 395:193–201.

57. Critchfield, J.W., C.J. Welsh, J.M. Phang, and G.C. Yeh. 1994. Modulation of adriamycin accumulation and efflux by flavonoids in HCT-15 colon cells. *Biochemical Pharmacology* 48:1437–1445.

58. Kuntz, S., U. Wenzel, and H. Daniel. 1999. Comparative analysis of the effects of flavonoids on proliferation, cytotoxicity, and apoptosis in human colon cancer cell lines. *European Journal of Nutrition* 38:133–142.

59. Sukardiman, S., A. Darwanto, M. Tanjung, and M.O. Darmadi. 2000. Cytotoxic mechanism of flavonoid from Temu Kunci (*Kaempferia pandurata*) in cell culture of human mammary carcinoma. *Clinical Hemorheology and Microcirculation* 23:185–190.

60. Primiano, T., R. Yu, and A.N.T. Kong. 2001. Signal transduction events elicited by natural products that function as cancer chemopreventive agents. *Pharmaceutical Biology* 39:83–107.

61. Knowles, L.M., D.A. Zigrossi, R.A. Tauber, C. Hightower, and J.A. Milner. 2000. Flavonoids suppress androgen-dependent human prostatic tumor proliferation. *Nutrition and Cancer* 38:116–122.

62. Galati, G., S. Teng, M.Y. Moridani, T.S. Chan, and P.J. O'Brien. 2000. Cancer chemoprevention and apoptosis mechanisms induced by dietary polyphenolics. *Drug Metabolism and Drug Interactions* 17:311–349.

63. Polkowski, K., and A.P. Mazurek. 2000. Biological properties of genistein. A review of in vitro and in vivo data. *Acta Poloniae Pharmaceutica - Drug Research* 57:135–155.

64. Ahmad, N., D.K. Feyes, and A.L. Nieminen. 1997. Green tea constituent epigallocatechin-3-gallate and induction of apoptosis and cell cycle arrest in human carcinoma cells. *Journal of National Cancer Institute* 89:1881–1886.

65. Liang, Y.C., S.Y. Lin-shiau, C.F. Chen, and J.K. Lin. 1997. Suppression of extracellular signals and cell proliferation through EGF receptor binding by (–)-epigallocatechin gallate in human A431 epidermoid carcinoma cells. *Journal of Cellular Biochemistry* 67:55–65.

66. Valcic, S., B.N. Timmermann, and D.S. Alberts. 1996. Inhibitory effect of six green tea catechins and caffeine on the growth of four selected human tumor cell lines. *Anticancer Drugs* 7:461–468.

67. Yoshida, M., M. Yamamoto, and T. Nikaido. 1992. Quercetin arrests human leukemic T-cells in late G1 phase of the cell cycle. *Cancer Research* 2:6676–6681.

68. Choi, J.A., J.Y. Kim, and J.Y. Lee. 2001. Induction of cell cycle arrest and apoptosis in human breast cancer cells by quercetin. *International Journal of Oncology* 19:837–844.

69. Richter, M., R. Ebermann, and B. Marian. 1999. Quercetin-induced apoptosis in colorectal tumor cells: Possible role of EGF receptor signaling. *Nutrition and Cancer* 34:88–99.

70. Reddivari, L., J. Vanamala, S. Chintharlapalli, S.H. Safe, and J.C. Miller. 2007. Anthocyanin fraction from potato extracts is cytotoxic to prostate cancer cells through activation of caspase-dependent and caspase-independent pathways. *Carcinogenesis* 28:2227–2235.

71. Hayashi, K., H. Hibasami, T. Murakami, N. Terahara, M. Mori, and A. Tsukui. 2006. Induction of apoptosis in cultured human stomach cancer cells by potato anthocyanins and its inhibitory effects on growth of stomach cancer in mice. *Journal of the Japanese Society for Food Science and Technology* 12:22–26.

72. Kallio, P., M. Kolehmainen, D.E. Laaksonen, L. Pulkkinen, M. Atalay, H. Mykkanen et al. 2008. Inflammation markers are modulated by responses to diets differing in postprandial insulin responses in individuals with the metabolic syndrome. *American Journal of Clinical Nutrition* 87:1497–1503.

73. Van Rensburg, S.J. 1981. Epidemiological and dietary evidence for a specific nutritional predisposition to esophageal cancer. *Journal of the National Cancer Institute* 67:243–251.

74. Chen, F., P. Cole, Z.B. Mi, and L.Y. Xing. 1993. Corn and wheat-flour consumption and mortality from esophageal cancer in Shanxi, China. *International Journal of Cancer* 53:902–906.

75. Morton, J.F. 1970. Tentative correlations of plant usage and esophageal cancer zones. *Economic Botany* 24:217–226.

76. Oterdoom, H.J. 1985. Tannin, sorghum, and oesophageal cancer. *Lancet* 2:330–335.

77. Yu, C.L., and B. Swaminathan. 1987. Mutagenicity of proanthocyanidins. *Food and Chemical Toxicology* 25:135–139.

78. Grimmer, H.R., V. Parbhoo, and R.M. McGarth. 1992. Anti-mutagenicity of polyphenol-rich fractions from *Sorghum bicolor* grain. *Journal of Agricultural and Food Chemistry* 59:251–256.

79. Gomez-Cordovez, T., B. Bartolomez, W. Vieira, and V.M. Viradir. 2001. Effects of wine phenolics and sorghum tannins on tyrosinase activity and growth of melanoma cells. *Journal of Agricultural and Food Chemistry* 49:1620–1624.

80. Parbhoo, V., H.R. Grimmer, A. Cameron-Clarke, and R.M. McGarth. 1995. Induction of cytochrome P-450 in rat liver by a polyphenolrich extract from a bird-resistant sorghum grain. *Journal of the Science of Food and Agriculture* 69:247–252.

81. Giovannucci, E.R.E., Y. Liu, M.J. Stampfer, and W.C. Willett. 2002. A prospective study of tomato products, lycopene, and prostate cancer risk. *Journal of National Cancer Institute* 94:391–398.

82. Kim, L., A.V. Rao, and L.G. Rao. 2002. Effect of lycopene on prostate LNCaP cancer cells in culture. *Journal of Medicinal Foods* 5:181–187.

83. Guttenplan, N., C. Lee, and W.H. Frishman. 2001. Inhibition of myocardial apoptosis as a therapeutic target in cardiovascular disease prevention: Focus on caspase inhibition. *Heart Disease* 3:313–318.

84. Heath, E., S. Seren, K. Sahin, and O. Kucuk. 2006. The role of tomato lycopene in the treatment of prostate cancer. In *Tomatoes, Lycopene and Human Health*, edited by Rao, A.V., 127–140. Scotland: Caledonian Science Press.

85. Giovannocci, E., A. Ascherio, E.B. Rimm, M.J. Stampfer, G.A. Colditz, and W.C. Willet. 1995. Intake of carotenoids and retinol in relation to risk of prostate cancer. *Journal of National Cancer Institute* 87:1767–1776.

86. Giovannucci, E. 1999. Tomatoes, tomato-based products, lycopene, and cancer: Review of the epidemiologic literature. *Journal of National Cancer Institute* 91:317–331.

87. Kucuk, O., and D.P. Wood. 2002. Response of hormone rerfractory prostate cancer to lycopene. *The Journal of Urology* 167:651–658.

88. Bowen, P., L. Chen, and M. Stacewicz-Sapuntzakis. 2002. Tomato sauce supplementation and prostate cancer: Lycopene accumulation and modulation of biomarkers of carcinogenesis. *Experimental Biology and Medicine* 227:886–893.

89. Rao, A.V., and S. Agarwal. 1999. Role of lycopene as antioxidant carotenoid in the prevention of chronic diseases: A review. *Nutrition Research* 19:305–323.

90. Giovannucci, E., M. Pollak, Y. Liu, E.A. Platz, N. Majeed, E.B. Rimm, and W.C. Willett. 2003. Nutritional predictors of insulin-like growth factor I and their relationships to cancer in men. *Cancer Epidemiology, Biomarkers and Prevention* 12:84–89.

91. Mathers, J.C. 2002. Pulses and carcinogenesis: Potential for the prevention of colon, breast and other cancers. *British Journal of Nutrition* 88:273–279.

92. Kennedy, A.R. 1998. Chemopreventive agents: Protease inhibitors. *Pharmacology and Therapeutics* 78:167–209.

93. Clemente, A., D.A. McKenzie, I.T. Johnson, and C. Domoney. 2004. Investigation of legume seed protease inhibitors as potential anti-carcinogenic proteins. In *Legumes for the Benefit of Agriculture, Nutrition and the Environment*. Proceedings of 5th European Conference on Grain Legume, Dijon. 51–60, AEP.

94. Murillo, G., J.K. Choi, O. Pan, A.I. Constantinou, and R.G. Mehta. 2004. Efficacy of garbanzo and soybean flour in suppression of aberrant crypt foci in the colons of CF-1 mice. *Anticancer Research* 24:3049–3056.

95. Armstrong, W.B., A.R. Kennedy, X.S. Wang, J. Atiba, E. McLaren, and F.L. Meyskens. 2000. Single-dose administration of Bowman–Birk inhibitor concentrate in patients with oral leukoplakia. *Cancer Epidemiology and Biomarkers* 9:43–47.

96. Ware, J.H., X.S. Wan, P. Newberne, and A.R. Kennedy. 1999. Bowman–Birk inhibitor concentrate reduces colon inflammation in mice with dextran sulfate sodium–induced ulcerative colitis. *Digestive Diseases and Science* 44:986–990.

97. Gao, C.M., J.H. Takezaki, J.H. Ding, M.S. Li, and K. Tajima. 1999. Protective effect of Allium vegetables against both esophageal and stomach cancer: A simultaneous casereferent study of a high-epidemic area in Jiangsu Province, China. *Japanese Journal of Cancer Research* 90:614–621.

98. Dorant, E., P.A. Van den Brandt, and R.A. Goldbohm. 1996. A prospective cohort study on the relationship between onion and leek consumption, garlic supplement use and the risk of colorectal carcinoma in the Netherlands. *Carcinogenesis* 17:477–484.

99. Hsing, A.W., A.P. Chokkalingam, Y.T. Gao, M.P. Madigan, J. Deng, G. Gridley et al. 2002. Allium vegetables and risk of prostate cancer: A population-based study. *Journal of the National Cancer Institute* 94:1648–1651.

100. Singh, A., and Y. Shukla. 1998. Antitumor activity of diallyl sulfide on polycyclic aromatic hydrocarbon-induced mouse skin carcinogenesis. *Cancer Letters* 131:209–214.

101. Reddy, B.S., C.V. Rao, A. Rivenson, and G. Kelloff. 1993. Chemoprevention of colon carcinogenesis by organosulfur compounds. *Cancer Research* 53:3493–3496.

102. Schaffer, E.M., J.Z. Liu, J. Green, C.A. Dangler, and J.A. Milner. 1996. Garlic and associated allyl sulfur components inhibit N-methyl-N-nitrosourea-induced rat mammary carcinogenesis. *Cancer Letters* 102:199–204.

103. Dwivedi, C., S. Rohlfs, D. Jarvis, and F.N. Engineer. 1992. Chemoprevention of chemically induced skin tumor development by diallyl sulfide and diallyl disulfide. *Pharmaceutical Research* 9:1668–1670.

104. Suzui, N., S. Sugie, K.M. Rahman, M. Ohnishi, N. Yoshimi, K. Wakabayashi et al. 1997. Inhibitory effects of diallyl disulfide or aspirin on 2-amino-1-methyl-6-phenylimidazo[4,5-b]pyridine-induced mammary carcinogenesis in rats. *Japanese Journal of Cancer Research* 88:705–711.

105. Xiao, R., T.M. Badger, and F.A. Simmen. 2005. Dietary exposure to soy or whey proteins alters colonic global gene expression profiles during rat colon tumorigenesis. *Molecular Cancer* 4:1–5.

106. Herman-Antosiewicz, A., A.A. Powolny, and S.V. Singh. 2007. Molecular targets of cancer chemoprevention by garlic-derived organosulfides. *Acta Pharmacologica Sinica* 28:1355–1364.

107. Singh, S.V., S.S. Pan, S.K. Srivastava, H. Xia, X. Hu, H.A. Zaren et al. 1998. Differential induction of NAD(P)H:quinone oxidoreductase by anti-carcinogenic organosulfides from garlic. *Biochemistry and Biophysics Research Communications* 244:917–920.

108. Knowles, L.M., and J.A. Milner. 2000. Diallyl disulfide inhibits p34(cdc2) kinase activity through changes in complex formation and phosphorylation. *Carcinogenesis* 21:1129–1134.

109. Hosono, T., T. Fukao, J. Ogihara, Y. Ito, H. Shiba, T. Seki et al. 2005. Diallyl trisulfide suppresses the proliferation and induces apoptosis of human colon cancer cells through oxidative modification of beta-tubulin. *Journal of Biological Chemistry* 280:41487–41493.

110. Li, N., R. Guo, W. Li, J. Shao, S. Li, K. Zhao et al. 2006. A proteomic investigation into a human gastric cancer cell line BGC-823 treated with diallyl trisulfide. *Carcinogenesis* 27:1222–1231.

111. Nagae, S., M. Ushijima, S. Hatono, J. Imai, S. Kasuga, H. Matsuura et al. 1994. Pharmacokinetics of the garlic compound S-allyl cysteine. *Planta Medica* 60:214–217.

112. Tanaka, S., K. Haruma, and M. Yoshihara. 2006. Aged garlic extract has potential suppressive effect on colorectal adenomas in humans. *Journal of Nutrition* 136:821s–826s.

113. Sun, X., T. Guo, J. He, M. Zhao, M. Yan, F. Cui, and Y. Deng. 2006. Determination of the concentration of diallyl trisulfide in rat whole blood using gas chromatography with electroncapture detection and identification of its major metabolite with gas chromatography mass spectrometry. *Journal of the Pharmaceutical Society of Japan* 126:521–527.

114. Imada, O. 1990. Toxicity aspects of garlic. *First World Congress on the Health Significance of Garlic and Garlic Constituents, Nutrition International*, Irvine, CA 92619 – 0632, p. 47.

115. Li, H., H.Q. Li, Y. Wang, H.X. Xu, W.T. Fan, M.L. Wang et al. 2004. An intervention study to prevent gastric cancer by micro-selenium and large dose of allitridum. *Chinese Medicine Journal* 117:1155–1160.

116. Fukushima, S., N. Takada, T. Hori, and H. Wanibuchi. 1997. Cancer prevention by organosulfur compounds from garlic and onion. *Journal of Cellular Biochemistry* 27:100s–105s.

117. Turkmen, N., F. Sari, and S. Velioglu. 2005. The effect of cooking methods on total phenolics and antioxidant activity of selected green vegetables. *Food Chemistry* 93:713–718.

118. Fernandez-Orozco, R., H. Zielinski, and M.K. Piskuła. 2003. Contribution of low-molecular-weight antioxidants to the antioxidant capacity of raw and processed lentil seeds. *Nahrung/Food* 47:291–299.

119. Xu, B.J., and S.K.C. Chang. 2007. Comparative studies on phenolic profiles and antioxidant activities of legumes as affected by extraction solvents. *Journal of Food Science* 72:S159–S166.

120. Amarowicz, R., and B. Raab. 1997. Antioxidative activity of leguminous seed extracts evaluated by chemiluminescence methods. *Zeitschrift für Naturforschung* 52:709–712.

121. Madhujith, T., M. Naczk, and F. Shahidi. 2004. Antioxidant activity of common beans (*Phaseolus vulgaris* L.). *Journal of Food Lipids* 11:220–233.

122. Siddhuraju, P., and K. Becker. 2007. The antioxidant and free radical scavenging activities of processed cowpea (*Vigna unuguiculata* (L.) Walp.) seed extracts. *Food Chemistry* 101:10–19.

123. Rocha-Guzmán, N.E., A. Herzog, G.L.R. Rubén, F.J. Ibarra-Pérez, G. Zambrano-Galván, and J.A. Gallegos-Infante. 2007. Antioxidant and antimutagenic activity of phenolic compounds in three different colour groups of common bean cultivars (*Phaseolus vulgaris*). *Food Chemistry* 103:521–527.

124. Heimler, D., P. Vignolini, M.G. Dini, and A. Romani. 2005. Rapid test to assess the antioxidant activity of *Phaseolus vulgaris* L. dry bean. *Journal of Agricultural and Food Chemistry* 53:3053–3056.
125. Amarowicz, R., A. Troszynska, N. Baryłko-Pikielna, and F. Shahidi. 2004. Polyphenolics extracts from legume seeds: Correlations between total antioxidant activity, total phenolics content, tannins content and astringency. *Journal of Food Lipids* 11:278–286.
126. Vadivel, V., and M. Pugalenthi. 2008. Removal of antinutritional/toxic substances and improvement in the protein digestibility of velvet bean seeds during various processing methods. *Journal of Food Science and Technology* 45:242–246.
127. Pugalenthi, M., and V. Vadivel. 2005. Nutritional evaluation and the effect of processing methods on antinutritional factors of sword bean (*Canavalia gladiata* (Jacq.) DC). *Journal of Food Science and Technology* 42:510–516.
128. Vadivel, V., M. Pugalenthi, and M. Megha. 2008. Biological evaluation of protein quality of raw and processed seeds of gila bean (*Entada scandens* Benth.). *Tropical and Subtropical Agroecosystems* 8:125–133.
129. Kale, A., S. Gawande, and S. Kotwal. 2008. Cancer phytotherapeutics: Role for flavonoids at the cellular level. *Phytotheraphy Research* 22:567–577.
130. Crujeiras, A.B., E. Goyenechea, and J.A. Martínez. 2010. Fruit, vegetables, and legumes consumption: Role in preventing and treating obesity. In *Bioactive Foods in Promoting Health*, 359–380.
131. American Institute for Cancer Research. 2007. *A Dietician's Cancer Story*. Washington, DC: AICR.
132. Rao, A.V., and L.G. Rao. 2007. Carotenoids and human health. *Pharmacological Research* 55:207–216.

21

Genomic Resources of Agriculturally Important Animals, Insects, and Pests

Sarika, Anu Sharma, Anil Rai, Mir Asif Iquebal, and Poonam Chilana

CONTENTS

21.1 Introduction

Agriculture is changing rapidly and has been influenced by social, political, economic, environmental, and climatic factors. The need to make agriculture more sustainable is increasingly being recognized along with the fact that agriculture has many different functions in addition to food production. Advancements in the fundamental understanding of animal science, developments in molecular biology, and increasing accessibility to genomic information have provided an open challenge as well as opportunities for researchers. In fact, agriculture looks at genomics as the next "green revolution" in terms of "gene revolution." Genomics in agriculture is required to develop crops with better resistance to factors such as disease, pests, frost, drought, floods, etc.; to cultivate more nutrient-rich foods; and to breed enhanced, more resistant, and higher-quality livestock [1]. The use of genomic resources is helping to educate a new generation of scientists in cutting-edge approaches to the characterization, evaluation, management, and conservation of crop, animal, and fish genetic resources. Bioinformatics platforms and their associated databases are essential for the effective design of approaches, making the best use of genomic resources, including resource integration. Due to the importance of some animals and their products or by-products as food and nutrition for human beings, their contribution to agriculture is quite significant. Similarly, agriculturally important insects and pests play an important role in enhancing the productivity of crops, fruits, and vegetables by acting as an important medium for their pollination.

Livestock animals have been domesticated by humans since early civilizations and are an integral part of agrarian economy in terms of its contributions to the production of milk, meat, skin, hides, fertilizer, fuel, and draft power. Biologically, farm animals are placed between human and traditional model species. Understanding animal biology remains a foremost challenge in the breeding programs and production of livestock. In accordance with human genome initiatives, attempts have been made in the field of animal genomics to understand animal biology systems at the molecular level. Livestock also forms a unique resource for comparative genomics with other species, including humans. Researchers can exploit the abundant genetic variation between divergent breeds and the variation in segregating populations within breeds. The availability of a wide range of genomic tools and comparative maps help the livestock genomicists to map and identify genes at the loci for simple and complex traits [2]. Identifying functional sequences in livestock genomes helps in predicting genes and in the regulatory biological pathways of the species. Improved varieties may be developed through breeding programs with the use of precise markers for different traits related to biotic and abiotic stress [3]. The large genomic resource gives researchers unprecedented opportunities to expand knowledge of the genetic control of traits such as meat quality or disease resistance, which are difficult to measure, and develop more advanced and sustainable breeding strategies.

Beneficial insects not only help in crop pollination and crop protection but also provide useful materials like silk and honey to mankind. Entomologists know that insects are an important part of farmland biodiversity and that they provide many essential environmental services. Insects in the category of pests are responsible for crop damage, defoliation, destruction of shade, and transmission of various diseases. Fundamental entomology and applied entomology are two important braches of entomology. Fundamental entomology deals with the basic aspects of the insects, such as morphology, anatomy, physiology, and taxonomy, whereas applied entomology covers the study of insects that are either beneficial or harmful to human beings. Insects are categorized according to their economic importance to humans. Some insects are of no economic importance, as they do not benefit or harm us. Economically important insects can be divided into harmful, beneficial, and household and disease-carrying insects. Harmful insects can be severe agricultural pests, destroying up to 30% of our potential annual harvest, as well as being vectors for diseases such as yellow mosaic, wilt, and leaf curl in plants, and malaria, elephantiasis, sleeping sickness, dengue, and yellow fever in humans. Termites and cockroaches are very common household pests that have been known to wreak havoc in homes.

The whole genome sequence is an invaluable resource for the insect genomics community that allows functional genomics, comparative analysis of genomic contents and their organization, as well as functional analyses of critical parameters as insect attributes linked to their capacity to transmit disease agents, insect behavior, the ancestral relationships between major insect groups, as well as a better understanding of many individual genes and gene families. Advances in sequencing technologies have provided opportunities in bioinformatics for managing, processing, and analyzing these sequences. In this genomic era, bioinformatics is used as the bedrock of current and future biotechnologies for finding new or better alternatives such as designing potential target sites, safer insecticides, and developing transgenic insects in applied insect science.

This chapter consists of two major sections: one deals with the genomic resources of livestock and the other deals with the genomic resources of agriculturally important insects and pests. The updated status of all sequenced livestock species such as cattle, buffalo, sheep, goats, chickens, pigs, and horses as well as some important insects and pests, along with the details of their web browsers, are discussed. This chapter also contains recent releases of the respective livestock, insect, and pest assemblies as well as their collaborators, sequencing methods, coverage, and dates of release. Also, the current status of genomic data in terms of nucleotide sequence, expressed sequence tags (ESTs), unigenes, single nucleotide polymorphisms (SNPs), genes, genome survey sequence (GSS), proteins, etc., for livestock species have been compiled. We discuss comparative mapping in these farm animals followed by their genomic potentials as well as those of agriculturally important insects and pests.

21.2 Livestock Genomic Resources

The Human Genome Project contributed to the improvement of technologies and strategies in genome sequencing. It has provided the basic tools and technologies for studying other genomes including those of livestock. To understand the history of life, its genetic mechanisms, genetic disorders (pharmacogenomics), genetic control for complex phenotypes, growth, development, health, reproduction, etc., knowledge of genomes is quite important. This section deals with the current sequencing status of these livestock, their comparative mapping, and the statistics of the genomic resources.

21.2.1 The *Omics* of Livestock Species

Genomics is the scientific study of the structure, function, and interrelationship of both individual genes and genomes as a whole [4]. Genome projects aim to decode the complete genome sequence of an organism through the annotation of their protein-coding genes and other important genome-encoded features. This field commenced with a string of nucleotides in the DNA sequence and has thus far achieved whole genome sequencing, leading to comprehensive mapping and cellular function at the DNA level [5]. Today, genomics resources are organized in the form of databases for easy handling of massive amounts of biological data. Proteomics, as the science of protein structure and function, has tremendous potential for application in biomedicine. Also, advanced computational technologies in proteomics have lead to structure models that are comparable to crystal structures at low cost along with the physicochemical properties of proteins. The role of bioinformatics in the storage, retrieval, and analysis of these molecular data has led

TABLE 21.1

Genomics and Proteomics Databases of Livestock Species

Database Name	Database Type	URL	Content
GenBank	Sequence DB (for all organisms)	http://www.ncbi.nlm.nih.gov/genbank/	DNA/protein
PDB	The Protein Data Bank proteins structure	http://www.rcsb.org/pdb/home/home.do	Experimentally determined three-dimensional structure of proteins
EMBL	The EMBL nucleotide sequence database	http://www.ebi.ac.uk/embl/	Nucleotide sequences of loci
Codon usage DB	Codon usage frequency	http://www.kazusa.or.jp/codon/	Codon usage in animals and other organisms
OMIA	Online Mendelian inheritance in animals	http://www.angis.org.au/Databases/BIRX/omia/	Mendelian inheritance in animals
Swiss-Prot	Annotated protein sequence database	http://web.expasy.org/docs/swiss-prot_guideline.html	Annotated sequences of proteins
ISIS	Introns of genes	http://isis.bit.uq.edu.au/front.html	Genes of multispecies sequences introns
HomoloGene	The homologene database	http://www.ncbi.nlm.nih.gov/sites/entrez?db = homologene	Gene homology
QTLdb	Quantitative trait loci (QTL) Database	http://www.animalgenome.org/QTLdb/faq.html	QTL results from livestock species like pig, cattle, chicken, goats, horse, etc.
Interactive Bovine In Silico SNP DB	SNP and bovine mRNA database	http://www.livestockgenomics.csiro.au/IBISS3/	IBISS is searchable using human standard gene symbols, GenBank human RefSeq accession numbers, GenBank bovine accession numbers, etc.
Livestock Genome Mapping Programmes	Comparative mapping of livestock genomes	http://locus.jouy.inra.fr/cgi-bin/bovmap/livestock.pl	Houses BOVMAP, BUFFMAP, GOATMAP, RABBITMAP, HORSEMAP, comparative mapping and other outside database links
Cattle Breeds	Breeds' description	http://www.ansi.okstate.edu/breeds/cattle/	Breeds of livestock like cattle, sheep, goat, horse, sheep, etc.

to the *omics* revolution with high-end computational technologies. The availability of the genome sequences of many livestock species can bring insight into the function of conserved noncoding regions of the DNA sequence [6]. Animal genome databases contain information on proteins, biochemical, and physiological processes that occur in livestock. Table 21.1 represents the general genomics and proteomics databases of livestock species.

21.2.2 Current Sequencing Status of Livestock Species

Whole genome sequences are available for many important livestock animals such as cows, buffaloes, chickens, horses, rabbits, pigs, and sheep. Table 21.2 provides a list of web sites with genomic information for livestock species. Also, genome sequencing projects in livestock expands knowledge of basic biological process relevant to human health and

TABLE 21.2

List of Different Web Sites/Browsers Related to Animal Genomics

Web Site/Browser	URL	Remarks
Cattle (B. taurus)		
NRSP-8 Cattle Coord.—Texas	http://www.animalgenome.org/cattle/	NAGRP-U.S. Cattle Genome Research Coordination Program Has information about databases, community, genome maps, and resources
Bovine Genome Project	https://www.hgsc.bcm.edu/content/bovine-genome-project	Baylor College of Medicine
Bovine Genome Resources	http://www.ncbi.nlm.nih.gov/genome/guide/cow/	NCBI Cattle Resources
Ensembl cattle genome	http://asia.ensembl.org/Bos_taurus/Info/Index	The UMD 3.1 assembly (NCBI assembly accession GCA_000003055.3), released in December 2009, is the third release of the cow (*B. taurus*) assembly from the Center for Bioinformatics and Computational Biology at University of Maryland. The genome sequences were generated using a combination of Bacterial Artificial Chromosome (BAC)-by-BAC hierarchical (~11 million reads) and WGS (~24 million reads) sequencing methods, and assembled using the Celera Assembler version 5.2. The total length of the UMD3.1 assembly is 2.65 Gb. The N50 size is the median sequence length, i.e., 50% of the assembled genome lies in blocks of the N50 size or longer. The N50 size for contigs in the UMD3.1 assembly is 103,785. The genome assembly represented here corresponds to GenBank Assembly ID GCA_000003055.3
UCSC Cattle Genome Browser	http://genome.ucsc.edu/cgi-bin/hgGateway?org=Cow	Genome Bioinformatics group of University of California, Santa Cruz
UMD Assembly	http://www.cbcb.umd.edu/research/bos_taurus_assembly.shtml	The Center for Bioinformatics and Computational Biology at University of Maryland using the Celera Assembler. Two major releases were made public, i.e., UMD2 (April 2009) and UMD3.0 (August 2009). A minor release, UMD3.1, was made public in December 2009. The UMD3.1 assembly is identical in almost all respects to UMD3.0, except that its AGP file is deposited at GenBank. All contigs and chromosomes remain unchanged.
Bovine Genome Database (TXAM)	http://genomes.arc.georgetown.edu/drupal/bovine/	Georgetown University and the University of Adelaide. Created to support the efforts of the Bovine Genome Sequencing and Analysis Consortium by creating a model organism database that integrates bovine genomics data with structural and functional annotations of genes and the genome

(continued)

TABLE 21.2 (Continued)

List of Different Web Sites/Browsers Related to Animal Genomics

Web Site/Browser	URL	Remarks
Cattle (B. taurus)		
Ruminant Genome Biology Consortium	http://www.ruminants.org/	Formed to capture the learnings of communities who have already developed genomic tools for their own ruminant (bovine, ovine and caprine) species. The idea is to share those learnings across species which do not yet have the scientific and financial infrastructure arising from organized agriculture in the developed world with those of new ruminant species
COMRAD Cattle RH Map	—	Roslin Institute, Edinburgh Comparative radiation hybrid mapping Development of a bovine whole genome radiation hybrid map for comparative mapping across species and the identification of positional candidate genes for genetically mapped traits.
BOVMAP	http://locus.jouy.inra.fr/cgi-bin/lgbc/mapping/common/intro2.pl?BASE=cattle	Institut National de la Recherche Agronomique (INRA), France Mapping the cattle genome, loci homologies with cattle
Cattle SNP/Genome	http://www.livestockgenomics.csiro.au/cow/	CSIRO has interactive bovine in silico SNP, links to Bovine Genome Browser, Bacterial Artificial Chromosome Library Contig Project, Genome Marker Maps
Dairy Cattle QTL database	http://firefly.vetsci.usyd.edu.au/reprogen/QTL_Map/index.php?Page=QTL+Map	Studies recorded in this database include complete and partial genome scans, single chromosome scans, fine mapping studies, and contain all known reports that were published in peer-reviewed journals and readily available conference proceedings. The traits recorded in this map are milk yield, milk composition (protein yield, protein%, fat yield, fat%), and somatic cell score.
CattleQTLdb	http://www.animalgenome.org/cgi-bin/QTLdb/BT/index	Iowa State University has gathered all cattle QTL data published during the past 10+ years. The database and its peripheral tools make it possible to compare, confirm, and locate on cattle chromosomes the most feasible location for genes responsible for quantitative trait important to cattle production.
Gene Indices (GtGI)	http://compbio.dfci.harvard.edu/tgi/cgi-bin/tgi/gimain.pl?gudb=cattle	The goal of The Gene Index Project is to use the available EST and gene sequences, along with the reference genomes wherever available, to provide an inventory of likely genes and their variants and to annotate these with information regarding the functional roles played by these genes and their products. Attempts are also being made to find links between genes and pathways in different species and to provide lists of features within completed genomes that can aid in the understanding of how gene expression is regulated.

(continued)

TABLE 21.2 (Continued)

List of Different Web Sites/Browsers Related to Animal Genomics

Web Site/Browser	URL	Remarks
miROrtho: microRNA genes	http://cegg.unige.ch/mirortho	This is the catalogue of animal microRNA genes, containing predictions of precursor miRNA genes covering several animal genomes combining orthology and a support vector machine. It provides homology-extended alignments of already known miRBase families and putative miRNA families exclusively predicted by SVM and orthology pipeline.
Buffalo (B. bubalis)		
BuffaloGenome.ORG	http://buffalogenome.org/	Genome sequencing and mapping information for river buffalo
Water Buffalo Genome (India)	http://210.212.93.84/	National Bureau of Animal Genetic Resources, India The first reference-guided assembly (using the cattle genome) of Illumina short sequence reads generated from a buffalo species with robust depth and coverage. Both paired-end and mate pair data from a single Murrah female (river subspecies) with a recorded pedigree were used for this assembly. Sequence reads were mapped to the Btau 4.0 version of the cattle genome and RH-map guided buffalo pseudomolecules were constructed.
Buffalo Genome Resource (NCBI)	http://www.ncbi.nlm.nih.gov/ projects/genome/guide/buffalo/	Provides information on buffalo and buffalo-related resources from NCBI and the research community.
BUFFMAP	http://locus.jouy.inra.fr/cgi-bin/ lgbc/mapping/common/ intro2.pl?BASE=buffalo	INRA, France Mapping the buffalo genome. Loci homologies with buffalo
Sheep (O. aries)		
International Sheep Genome Consortium	http://www.sheephapmap.org/	The International Sheep Genomics Consortium is a partnership of scientists and funding agencies from a number of countries to develop public genomic resources to help researchers find genes associated with production, quality, and disease traits in sheep. A high-quality ovine BAC library available built on an existing collaboration for the International Mapping Flock that was created nearly a decade earlier.
NCBI Sheep Resources	http://www.ncbi.nlm.nih.gov/ genome/guide/sheep/	To generate an important resource for gene discovery, affecting health and biology and the growing field of comparative genomics.
UK Sheep Map	http://www.projects.roslin.ac.uk/ sheepmap/front.html	UK Sheep Genome Mapping Project, Roslin, UK The objectives were to use Suffolk, Texel and Charollais sire referencing schemes to detect and verify QTLs for growth and carcass composition traits. It also aimed at investigating candidate genes and/or chromosomal regions for associations with production traits.
Australian Sheep Gene Mapping Web Site	http:// rubens.its.unimelb.edu.au/~jillm/ jill.htm	Melbourne, Australia Has sheep markers, maps, map comparison, sheep libraries, etc.

(continued)

TABLE 21.2 (Continued)

List of Different Web Sites/Browsers Related to Animal Genomics

Web Site/Browser	URL	Remarks
Pig (Sus scrofa)		
NRSP-8 Pig Coord.	http://www.animalgenome.org/pigs/	Iowa State University Provides links to U.S. Pig Gene Mapping Coordination Program Shared Resources, Databases, Genome Maps, Swine Genetics Community Links
Genome sequencing	http://www.sanger.ac.uk/resources/downloads/othervertebrates/pig.html	Sanger Institute A physical map of the swine genome was generated by an international collaboration of four laboratories. High-throughput fingerprinting and BAC end sequencing were used to provide the template for an integrated physical map of the whole pig genome.
Genome at Pre-Ensembl	http://asia.ensembl.org/Sus_scrofa/Info/Index	The Sscrofa10.2 assembly of the pig genome was produced in August 2011 by the Swine Genome Sequencing Consortium. The genome assembly represented here corresponds to GenBank Assembly ID GCA_000003025.4
NCBI Pig Resources	http://www.ncbi.nlm.nih.gov/genome/guide/pig/	Information on porcine-related resources from NCBI and the pig research community.
ISU Pig QTLdb	http://www.animalgenome.org/cgi-bin/QTLdb/SS/index	The database and its peripheral tools make it possible to compare, confirm, and locate on pig chromosomes the most feasible location for genes responsible for quantitative trait important to pig production.
Pig Cytogenetic map—France	https://www-lgc.toulouse.inra.fr/pig/cyto/cyto.htm	Cytogenetic map of the pig. Has drawings of genes and markers mapped on pig chromosome, list of genes, and genetic markers mapped on pig cytogenetic map
PiGenome database—Korea	http://nabc.go.kr/sgd/	Animal Genomics and Bioinformatics, National Institute of Animal Science, RDA 564 Omockchun-Dong, Gwonseon-Gu Suwon 441-706, Korea 69,545 ESTs from six full-length cDNA libraries (porcine abdominal fat, porcine fat cell, porcine loin muscle, liver, and pituitary gland), 182 BAC contigs from chromosome 6 have been identified.
Chicken (G. gallus)		
NRSP-8: Chicken—Michigan	http://poultry.mph.msu.edu/	Supported by USDA-CSREES National Animal Genome research program to serve the poultry genome mapping community.
NCBI Chicken Resources	http://www.ncbi.nlm.nih.gov/genome/guide/chicken/	It provides information on chicken- and avian-related resources from NCBI and the chicken research community.
UCSC Chicken Genome Browser	http://genome.ucsc.edu/cgi-bin/hgGateway?org=Chicken	Created by Genome Bioinformatics Group of University of California, Santa Cruz It is a chicken genome browser gateway
ISU Chicken QTLdb	http://www.animalgenome.org/cgi-bin/QTLdb/GG/index	The current release of the Chicken QTLdb contains 3162 QTLs from 158 publications. Those QTLs represent 270 different traits
Gene Indices (GgGI)	http://compbio.dfci.harvard.edu/tgi/cgi-bin/tgi/gimain.pl?gudb=g_gallus	Gives functional annotation and analysis of chicken genome

(continued)

TABLE 21.2 (Continued)

List of Different Web Sites/Browsers Related to Animal Genomics

Web Site/Browser	URL	Remarks
Goat (Capra hircus)		
Goat—INRA, France	http://locus.jouy.inra.fr/cgi-bin/ lgbc/mapping/common/ intro2.pl?BASE=goat	INRA, France Has the database for goat genome
GoSh dB	http://www.itb.cnr.it/gosh/	Goat and sheep database
GoatMap V 2.0	http://locus.jouy.inra.fr/cgi-bin/ lgbc/mapping/common/ intro2.pl?BASE=goat	Goat Map Database
QTLdB	http://www.ajol.info/index.php/ sajas/article/viewFile/3967/11914	Angora QTL dB
NCBI Goat Resource	http://www.ncbi.nlm.nih.gov/ projects/genome/guide/goat/	This refers to a gateway to goat genome resources at NCBI and beyond
Horse (E. caballus)		
NRSP-8 Horse Coord.— Kentucky	http://www.uky.edu/Ag/ Horsemap/	This web site was developed with support from the USDA-NRSP8 Horse Coordinator's funds, the Dorothy Russell Havemeyer Foundation, the Morris Animal Foundation and the University of Kentucky Agricultural Experiment Station. Its aim was to understand the genetic aspects of equine physiology and disease. Genetic tools have the potential to help researchers find new therapies and treatments for diseases such as laminitis, respiratory diseases, etc.
NCBI Horse Genome	http://www.ncbi.nlm.nih.gov/ genome?term=txid9796%5Borgn%5D	Gives the assembly details and annotations of horse genome
NCBI Horse Resources	http://www.ncbi.nlm.nih.gov/ genome/guide/horse/	This site brings together information about horse data available at NCBI and other related resources from the research community, including sequence, mapping, and clone information
HorseMAP	http://locus.jouy.inra.fr/cgi-bin/ lgbc/mapping/common/ intro2.pl?BASE=horse	INRA, France Mapping the equine genome, Loci homologies with horse
Camel (Camelus dromedarius)		
Arabian Camel Genome	http://camel.kacst.edu.sa/ index.php/press-conference	A collaborative research team from King Abdulaziz City for Science and Technology, Saudi Arabia and Beijing Genomics Institute (BGI-Shenzhen), People's Republic of China, is the first in the world to successfully map the full genome of the Arabian camel (*C. dromedarius*).
Rabbit (O. cuniculus)		
Rabbit Genome	http://locus.jouy.inra.fr/cgi-bin/ lgbc/mapping/common/ intro2.pl?BASE=rabbit	INRA, France Mapping the rabbit genome. Loci homologies with rabbit
NCBI Rabbit Resources	http://www.ncbi.nlm.nih.gov/ genome?term=txid9986%5Borgn%5D	Gives the assembly details and annotations of rabbit genome

Source: http://www.animalgenome.org/. Poonam Chilana, Anu Sharma, and Anil Rai, Insect genomic resources: status, availability and future; 575, 577, 578, Vol. 102, No. 4, February 2014. Current Science Association. With permission.

provides surrogate systems for human experimentation. Knowledge of genome size is not only important for genome studies in relation to genome structure, organization, and evolution, but also for genome mapping, physical mapping, and genome sequencing of related species. Table 21.3 lists all the ongoing sequencing projects of livestock including other information such as scientific names, genome size, number of chromosomes, number of predicted genes, draft assembly, method, coverage, status, and sequencing collaborators/consortium partners of the species. Table 21.4 describes the updated lists of assemblies of livestock animals.

21.2.2.1 Cattle

Cattle (*Bos taurus*) falls under the group of eutherian mammals that are phylogenetically distant from primates [7]. It is an economically important animal and is a major source for beef and milk in the food industry. The first "genome maps" for cattle were synteny groups, genes on the same chromosome, defined by protein gene products segregating in hybrid somatic cell lines [8]. In 1988, the International Society for Animal Genetics assembled a set of families for linkage mapping. In 1990, the microsatellite markers, led to international linkage map, which was concurrent with the linkage map developed at U.S. Department of Agriculture–Meat Animal Research Center [9]. Furthermore, the next significant advancement in cattle genomics came with the development of radiation hybrid maps [10] and its use for high-resolution comparative maps [11]. The *B. taurus* genome was sequenced using a mixture of hierarchical and whole-genome shotgun (WGS) sequencing methods. Table 21.5a and b show the details of *B. taurus* genome releases, along with the descriptions, coverage, and release names in chronological order and the latest versions of the *B. taurus* assemblies.

The genome assembly version, Btau_4.0 (with accession number AAFC0000000.3.Btau_4.2), from the Bovine Genome Sequencing Project is available in GenBank, European Molecular Biology Laboratory (EMBL), and DNA Databank of Japan (DDBJ). An alternate assembly was submitted by the Center for Bioinformatics and Computational Biology at the University of Maryland (College Park, Maryland) using the Celera Assembler. UMD2 and UMD3.0, the two major releases were made public in April 2009 and August 2009, respectively. In addition, a minor release (UMD3.1) was made public in December 2009. The UMD3.1 assembly is almost identical in all respects to UMD3.0, except that its AGP file is deposited at GenBank. All contigs and chromosomes remain unchanged. UMD2 assembled 35.62 million reads into 2.85 billion base pair genomes, of which 2.61 billion (91%) base pairs were placed on chromosomes. Celera Assembler version 4.8 was used to assemble the reads from the Baylor College of Medicine (BCM). The second release, UMD3, assembled 36.82 million reads into 2.649 billion base pair genomes out of which 2.640 billion (99%) base pairs were placed on chromosomes. Celera Assembler version 5.2 was used to assemble all the genomic reads available in the National Center for Biotechnology Information (NCBI) Trace Archive with 35.62 million BCM reads along with 1.2 million reads generated by other sequencing centers. UMD3 shows significant improvement over UMD2 because it has fewer and smaller gaps, with approximately 3% more sequences placed on chromosomes. The UMD 3.1 assembly released in December 2009, with NCBI assembly accession GCA_000003055.3, is the third release of the cow assembly. The genome sequences were generated using a combination of BAC-by-BAC hierarchical (~11 million reads) and WGS (~24 million reads) sequencing methods, and assembled using the Celera Assembler version 5.2. The UMD3.1 assembly is 2.65 Gb in length, with N50 size as the median sequence length, that is, 50% of the assembled genome lies in blocks of N50 size or longer. GenBank has the updated assembly, UMD3.1 is available in GenBank with the

TABLE 21.3

Details of Genomic Projects Related to Animal Genomics

Livestock	Family	Scientific Name	Chromosomes	Method	Coverage	Status	Sequenced Animal	Collaborators/ Consortium	Release Date	Reference
Cow	Bovidae	*B. taurus*	30	WGS	7.1×	C	Hereford male L1 Domino 99375	BCM-HGSC	2009	www.hgsc.bcm.thm/projects/bovine/
Buffalo	Bovidae	*B. bubalis*	30	BAC to BAC		C		NBAGR	2010	—
Chicken	Phasianidae	*G. gallus*	33		6.6×	C		WUGSC	2004	—
Horse	Equidae	*E. caballus*	32	WGS	7×	C	Thoroughbred female Twilight	BI/MIT	2009	www.broad.mit.edu/nome/3182009
Pig	Suidae	*S. scrofa*	20	BAC to BAC,	4×	P	Duroc female	Sanger	2009	www.sanger.ac.uk/Prjects/S_scrofa/
Rabbit	Leporida	*O. cuniculus*	22	Low coverage WGS	~2×	C	—	BI/MIT	2005	—
Goat	Bovidae		—	Draft Assembly		C	—	BCM-HGSC	—	
Sheep	Bovidae	*O. aries*	27	Draft Assembly		P	—	BCM-HGSC	2010	—
Camel	Camelidae	*C. dromedarius*	—	—		P	—	—	—	

Source:　Poonam Chilana, Anu Sharma, and Anil Rai, Insect genomic resources: status, availability and future; 575, 577, 578, Vol. 102, No. 4, February 2014. Current Science Association. With permission.

Note:　BCM-HGSC, Baylor College of Medicine–Human Genome Sequence Centre; NBAGR, National Bureau of Animal Genetic Resources; WUGSC, Washington University Genomic Sequencing Center; BI/MIT, Broad Institute/Massachusetts Institute of Technology Center for Genome Research; Sanger Institute, UK; C, completed; P, in progress.

TABLE 21.4

Description of Assemblies of Livestock Animals

Organism	Cattle	Buffalo	Sheep	Rabbit	Chicken	Pig	Horse	Double-Humped Camel	Single-Humped Camel
Assembly name	Bos_taurus_UMD_3.1	ASM18099v1	Ovis_aries_1.0	OryCun2.0	Gallus_gallus-4.0	Sscrofa9.2	EquCab2.0	Not available	Not available
Last sequence update	15-Apr-2009	01-Feb-2010	25-Feb-2010	11-Jan-2010	14-Apr-2009	30-Jun-2009	14-Jul-2008	29-Jul-2009	14-Apr-2009
Highest level of assembly	Chromosome	Contigs only	Chromosome	Chromosome	Chromosome	Chromosome	Chromosome	—	—
Size (total bases)	2,670,405,961	21,675,247	2,860,496,367	2,737,445,565	1,046,915,324	2,262,484,801	2,474,912,402	16,659	16,643
Number of genes	27,142	—	—	22,971	17,516	22,491	22,887	13	13
Number of proteins	22,057	—	—	17,554	16,855	20,197	20,633	13	13
Number of chromosomes	29, X, Y	—	26, X	21, X	32, W, Z	18, X,Y	—	36, X, Y	36, X, Y
RefSeq Genome	1	1	1	2	4	3	1	1	1
Genome sequencing	5	2	2	1	2	5	2	1	1
Targeted locus (loci)	—	—	—	1	—	—	—	—	—
Transcriptome/Gene expression	166	3	35	11	165	201	17	1	1
Map	5	1	2	—	2	1	2	—	—
RefSeq Assembly	1	—	—	—	—	—	—	—	—

Source: http://www.ncbi.nlm.nih.gov/genome/assembly/.

TABLE 21.5a

B. taurus Releases

Date Released	Release Name	Coverage	Comments
April 15, 2009	Btau_4.2	7.1×	Draft assembly replaced with high-quality finished sequence where available
October 1, 2007	Btau_4.0	7.1×	Draft assembly, using WGS reads from small insert clones and BAC sequences from BACs sequenced individually or by the CAPSS clone pooling strategy. Mapped to chromosomes using refined mapping information
August 15, 2006	Btau_3.1	7.1×	Draft assembly, using WGS reads from small insert clones and BAC sequences from BACs sequenced individually or by the CAPSS clone pooling strategy
March 10, 2005	Btau_2.0	6.2×	Preliminary assembly, using WGS reads from small insert clones and BAC end sequences
September 27, 2004	Btau_1.0	3×	Preliminary assembly, using WGS reads from small insert clone

Source: http://www.ncbi.nlm.nih.gov/.

TABLE 21.5b

Latest Version Status of *B. taurus* Assemblies

Name	Submitter	Genome Representation	Assembly Level	Version Status
Bos_taurus_UMD_3.1	Center for Bioinformatics and Computational Biology, University of Maryland	Complete	Chromosome	Latest
Btau_4.6.1	Cattle Genome Sequencing International Consortium	Complete	Chromosome	Latest
Btau_ChrY_1.0	Baylor College of Medicine	Partial	Chromosome	Latest

Source: http://www.ncbi.nlm.nih.gov/.

accession no. DAAA00000000, and is also available at the University of Maryland web site (http://www.cbcb.umd.edu/research/bos_taurus_assembly.shtml).

21.2.2.2 Chicken

Chicken (*Gallus gallus*) plays a significant role as a food item in terms of eggs and meat. It comprises 41% of the meat produced in the world and is used as a model for the study of diseases and developmental biology [12]. In the evolutionary tree, chicken lies between mammals and fish. For this reason, it was believed that investigations on the chicken genome would provide significant information to understanding the evolution of the vertebrate genome. The first draft of the chicken genome was released in March 2004 and, in terms of evolution, evolved along with the boom in chicken genomic research [13]. The chicken genome was the first to be sequenced with a genome size of 1200 MB. In May 2006, the Genome Sequencing Centre at the Washington University School of Medicine (St. Louis, Missouri) submitted an improved 6.6× draft of the chicken genome assembly. In the first chicken genome assembly, chromosome 25 (i.e., LGE26C13) was missing and was later included in this improved assembly. Current chicken assembly has been used to further map other avian species. The mitochondrial genome of the white leghorn chicken sequenced at the Universite de Montreal was used for comparison to mitochondrial genomes of other vertebrates. It is believed that the sequenced chicken genome will provide the basis/framework for investigating polymorphisms of informative quantitative traits to continue their directed evolution of these species [14].

21.2.2.3 Sheep

The International Sheep Genomics Consortium took the lead in WGS assembly of the sheep (*Ovis aries*) genome with 3× coverage using 454 FLX sequencing. The contigs were assembled on the basis of alignment to the bovine genome and ordered into ovine chromosomes using BAC end sequences and the virtual sheep genome. This assembly is available in GenBank under the accession no. ACIV010000000.

21.2.2.4 Pig

Pigs (*Sus scrofa*) are evolutionarily distinct from human and rodents but have coevolved with these species. They are useful models for the study of genetic and environmental interactions with polygenic traits [15]. Sequencing of the pig genome plays a crucial role in the improvement of human health. Pig models have been used in clinical studies of infectious disease, organ transplantation, physiology, metabolic disease, pharmacology, obesity, and cardiovascular disease [16]. The Swine Genome Sequencing Consortium released Sscrofa10, a mixed BAC and WGS-based assembly of the porcine genome, which has assemblies for chromosomes 1 to 18, X and Y, majorly derived from females of the duroc breed, and some BACs from other breeds. NCBI's *Sus scrofa* 3.1 includes the Sscrofa10 assembly along with a complete mitochondrial genome from a landrace pig. This step of pig genome sequencing has generated invaluable resources in enzymology, reproduction, endocrinology, nutrition, and biochemistry research [16,17]. Further identification of gene markers for specific diseases would help breeders in raising pig stocks resistant to infectious diseases [14,18].

21.2.2.5 Rabbit

The rabbit (*Oryctolagus cuniculus*) genome was sequenced in an initiative by the Broad Institute and Massachusetts Institute of Technology (Cambridge, Massachusetts) with WGS and low coverage of approximately 2×. The Broad Institute developed OryCun2.0, genome assembly of the rabbit.

21.2.2.6 Horse

Genomics of the horse (*Equus caballus*) was initially quite slow but after 1998, three linkage maps [19–21] were produced with more than 450 markers. Broad Institute and Massachusetts Institute of Technology completed the horse genome sequencing with WGS on the thoroughbred female *Twilight* in 2009. It has a coverage of 6.8× and reported 20,436 predicted genes.

21.2.2.7 Buffalo

The buffalo (*Bubalus bubalis*) genome has been sequenced by a joint effort among the Nagoya University (Japan), National Institute of Immunology (New Delhi), National Bureau of

Animal Genetic Resources (Karnal), Central Institute for Research on Buffaloes (Hisar), and the Animal Science Division of the Indian Council of Agricultural Research (New Delhi). The buffalo (*B. bubalis*) genome was sequenced at 17× to 19× depth with 91% to 95% coverage. This is the first reference-guided assembly (using the cattle genome) of Illumina short sequence reads generated from a buffalo species with robust depth and coverage. Both paired end and mate pair data from a single Murrah female with a recorded pedigree were used for this assembly [22].

21.2.3 Genomic Resources in Livestock Species: Current Statistics

Genomic resources are ample, but the need of the hour is the exhaustive use of these resources in terms of computational approach, bioinformatics, and quantitative genetics. During the weaning process, many animals are lost, mainly because of pathogens. The use of metagenomics to reach unculturable pathogens and the use of genomics to study the relationship between animals and microbes, which may play a role in animal nutrition and to investigate the effects of abiotic stress on animals, is quite pertinent [23]. Genomics also provides opportunities to understand the molecular architecture of complex animal traits.

Comparative mapping, which is based on establishing similarities or homologies between genomics, has gained popularity in this area of research. The comparisons between these species are based on their genome maps, usually with reference to the position of the protein-coding genes or type I loci [24]. ESTs are sources of gene markers and sequence homology, genetic and physical markers, and genome maps (i.e., genetic and physical). The availability of databases is the key resource for comparative mapping. Figure 21.1 shows a graphical representation of the number of EST sequences available for livestock species. Table 21.6 shows the genomic resources, such as nucleotide sequences, EST, GSS, protein, SNP, genes, unigenes, and uniSTS in various livestock species, available

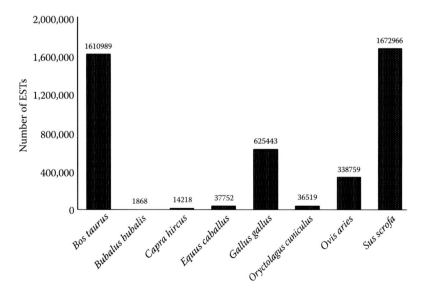

FIGURE 21.1
Graphical representation of ESTs in various livestock species.

TABLE 21.6

Genomic Resources of Livestock Species

Species	Nucleotide	EST	GSS	Protein	SNP	Gene	Unigene	UniSTS
B. taurus	210,100	1,610,989	532,064	145,563	9,587,248	83,672	52,442	18,860
B. bubalis	5,739	1,868	4,797	2,395	500	15	—	1,202
C. hircus	305,501	14,218	264	3,467	28	704	38,271	7,068
E. caballus	72,488	37,752	315,534	54,137	1,163,580	34,119	7,637	13,318
G. gallus	106,905	625,443	201,673	55,475	3,295,452	33,990	43,891	4,240
O. cuniculus	83,890	36,519	1,306	47,042	—	23,699	7,959	6,383
O. aries	18,303	338,759	435,675	10,116	2,899,215	2,124	17,424	7,151
S. scrofa	513,874	1,672,966	1,162,433	93,395	566,003	51,333	48,187	13,856

Source: http://www.ncbi.nlm.nih.gov/.

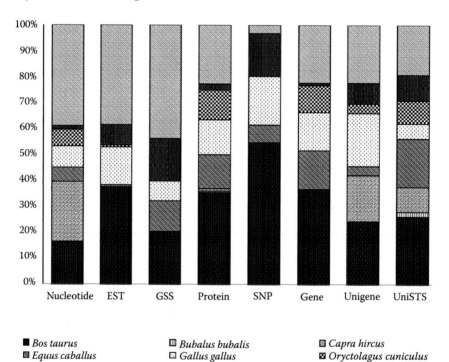

FIGURE 21.2
Graphical presentation of genomic resources of various livestock species.

in public domains until May 2012. The maximum number of nucleotides, ESTs, GSS, and unigenes have been submitted to the NCBI for the pig genome, whereas buffaloes show the least number of nucleotides, ESTs, gene, and protein sequences. Goats share the lowest number of GSS in the public domain, as well as having the least amount of SNP data among the livestock studied. Cattle genomes share the maximum number of gene, protein, SNP, and uniSTS data. No data on unigene and minimum data for uniSTS are available for the buffalo genome. Figure 21.2 shows the distribution of genomic status with respect to the availability of genomic information of various livestock.

21.2.4 Comparative Mapping in Livestock

Comparative mapping is an important tool for gene prediction and analysis of simple and complex traits in farm animals, as well as for the study of the evolution of vertebrate genomes. Comparative mapping was first applied in the prediction of the location of a specific gene in a "gene poor" map of one species from the "gene rich" map of another species [25]. Furthermore, comparative mapping by annotation and sequence similarity techniques make use of homologous DNA sequences (e.g., ESTs) to search for sequence similarity and putative orthologues in the genome of other species and then predict the chromosome location on the basis of existing comparative maps. This approach was first followed in the creation of high-resolution comparative mapping between cattle and humans [26]. The availability of EST sequences may be applied in predicting gene content of a simplified chromosomal segment to target development of new genetic or physical markers. These markers may further be used for creating high-resolution genetic maps at specific QTLs or new sequence tagged sites (STS) markers for building BAC contigs. This approach was first used to build BAC contigs and high-resolution comparative mapping between human chromosome 15 and chicken chromosome 10 [27]. Comparative mapping is also used to identify orthologous regions between farm animals and human/mouse disease loci, which are used to define candidate genes for both simple and complex traits [28].

21.3 Genomic Resources of Agriculturally Important Insects and Pests

An exponential amount of information has been generated in the field of insect science due to the latest DNA sequencing technologies. In this genomic era, bioinformatics is used for finding new or better alternatives such as designing potential target sites, safer insecticides, and developing transgenic insects in applied insect science. In this section, we gather information on the available insect genomic resources from agriculturally important insects and pests, and provide it here.

21.3.1 Fruit Fly (*Drosophila melanogaster*)—Model Insect

The fruit fly (*D. melanogaster*) is a small insect that feeds and breeds on spoiled fruit. It has been used as a valuable model organism for more than 100 years due to its biological complexity and ease of genetic manipulation. *Drosophila* research has provided insights into behavior, genetics development, and disease systems.

The *Drosophila* genome has an estimated size of 180 Mb, approximately 120 Mb of which is present as a gene-rich euchromatin. More than 95% of the total sequence now resides in scaffolds between 100 thousand bases and 1 Mb in length, and 65% in scaffolds exceeding 10 Mb in size. However, at least 1301 gaps among mapped scaffolds remain to be filled. In addition, almost all of the heterochromatic sequences, which consist mainly of repetitive sequences that cannot be stably cloned, remain inaccessible to current methods. Whole genomes of 15 other species of *Drosophila* (*D. ananassae, D. erecta, D. elegans, D. grimshawi, D. mojavensis, D. persimilis, D. pseudoobscura, D. sechellia, D. simulans, D. virilis, D. willistoni, D. takahashii, D. ficusphila, D. kikkawai,* and *D. yakuba*) have been sequenced and are accessible for comparative genomics through the Internet. The complete annotation summary of *D. melanogaster* [29] is shown in Table 21.7.

TABLE 21.7

Annotation Summary of *D. melanogaster*

Data Type	Release 2 (2000)	Release 3.1 (2002)	Release 3.2b (2004)	Release 5.1 (2006)
Annotated sequence (Mb)	3.8	12.1	14.2	24
Sequence length of repeats (Mb/%)	ND	6.3/52	6.3/75	18/77
Sequence length of exons (Mb/%)	0.15/4	0.33/2.7	0.43/3.0	1.33/5.5
Repeat nest fragments (number/Mb)	ND	ND	ND	10084/10
Full-length transposable elements	ND	ND	ND	202
Total annotations	130	447	556	11038
Protein-coding genes	130	297	472	613
Single-exon genes	43	58	195	187
Genes with finished cDNAs	48	58	92	137
Protein-coding genes with any EST/cDNA clone evidence	ND	80	142	250
Pseudogenes	0	1	7	32
ncRNAs	0	3	14	13
Recursive splice sites	ND	ND	ND	16
Miscellaneous annotations	ND	ND	ND	9
Unassembled ribosomal DNA fragments	0	6	52	67

Many genomic databases have been developed on *D. melanogaster*. This includes FlyExpress, a platform to explore the expression patterns of development-related genes in *Drosophila* embryogenesis. It contains a unique digital library of standardized images capturing the expression of thousands of genes at different developmental stages. The FlyExpress platform currently contains more than 100,000 images of expressions from more than 4000 genes derived from two high-throughput *in situ* hybridization studies [30,31], and more than 30,000 images extracted for peer-reviewed publications.

FlyBase is the primary database of integrated genetic and genomic data about the Drosophilidae, of which *D. melanogaster* is the most extensively studied species, and data types include sequence-level gene models, molecular classification of gene product functions, mutant phenotypes, mutant lesions and chromosome aberrations, gene expression patterns, transgene insertions, and anatomical images [32].

The goal of the Berkeley Drosophila Genome Project is to complete the *D. melanogaster* sequence in high quality and to generate and maintain biological annotations of this sequence. In addition to genomic sequencing, the Berkeley Drosophila Genome Project is producing gene disruptions using P-element–mediated mutagenesis on a scale unprecedented in metazoans, characterizing the sequence and expression of cDNAs and developing informatics tools that support the experimental process, identify features of the DNA sequence, and allow us to present up-to-date information about the annotated sequence to the research community. The Berkeley Drosophila Genome Project has also submitted 21,620 ESTs from the new AT adult testes cDNA library, and these clones are now available.

Since its inception in 2004, the *Drosophila* species comparative genome database (DroSpeGe) has provided genome researchers with rapid and usable access to 12 new and old *Drosophila* genomes. Scientists can use, with minimal computing expertise, the wealth of new genome information for developing new insights into insect evolution. New genome assemblies provided by several sequencing centers have been annotated with known model organism gene homologies and gene predictions to provide basic comparative data. This genome database includes homologies to *D. melanogaster* and eight other eukaryote model genomes, as well as gene predictions from several groups. BLAST searches of the newest assemblies are integrated with genome maps [33].

The *Drosophila* DNase I Footprint Database (v2.0) database was developed from 201 research articles. It provides a nonredundant set of high-quality binding site information for 87 transcription factors and 101 target genes in *Drosophila*.

FlyTF is a database of computationally predicted and experimentally verified site-specific transcription factors in the fruit fly *D. melanogaster*. These were identified from a list of candidate proteins with transcription-related gene ontology (GO) annotation as well as structural DNA-binding domain assignments.

CisTarget is a resource on predicted genomewide targets for conserved *cis*-regulatory elements of *D. melanogaster*.

The FlyNets database is a specialized database that focuses on the molecular interactions (protein–DNA, protein–RNA, and protein–protein) involved in *Drosophila* development. It is composed of the following two parts: (1) the FlyNets-base is distributed among several specific lines arranged according to a GenBank-like format and grouped into five thematic zones to improve human readability. The FlyNets database achieves a high level of integration with other databases such as FlyBase, EMBL, GenBank, and SWISS-PROT through numerous hyperlinks; and (2) FlyNets-list is a very simple and more general databank, the long-term goal of which is to report on any published molecular interaction occurring in the fly, giving direct web access to correspondents in Medline and in FlyBase. In the context of genome projects, databases describing molecular interactions and genetic networks will provide a link at the functional level between the genome, the proteome, and the transcriptome worlds of different organisms [34].

The *Drosophila melanogaster* Exon Database contains information on exons presented in a splicing graph form. The data is based on release 3.2 of the *D. melanogaster* genome annotations available at FlyBase (http://www.flybase.net). The gene structure information extracted from the annotations were checked, clustered, and transformed into a splicing graph. The splicing graph form of the gene constructs were then used for classification of the various types of alternative splicing events. In addition, the Pfam domains were mapped onto the gene structure.

FLYSNPdb provides high-resolution SNP data of *D. melanogaster*. The database currently contains 27,367 polymorphisms, including more than 3700 indels (insertions/deletions), covering all major chromosomes. These SNPs are clustered into 2238 markers, which are evenly distributed with an average density of one marker for every 50.3 kb or 6.6 genes. The database provides detailed information on the SNP data, including molecular and cytological locations (genome releases 3–5), alleles of up to five commonly used laboratory stocks, flanking sequences, SNP marker amplification primers, quality scores, and genotyping assays. Data specific for a certain region, particular stocks, or a certain genome assembly version are easily retrievable through the interface of a publicly accessible web site [35].

21.3.2 Genomic Resources on Agriculturally Beneficial Insects

Silkworm, honeybee, and nasonia wasp are beneficial insects to agriculture. Beneficial insects not only help in crop pollination and crop protection but also provide useful materials like silk and honey to mankind. This section describes the genome sequencing projects and genomic database available on these insect species.

21.3.2.1 Silkworm

The silkworm (*Bombyx mori*), domesticated for silk production for about 5000 years, is the most well-studied lepidopteran model system because of its rich repertoire of well-characterized mutations, affecting virtually every aspect of the organism's morphology, development, and behavior as well as its considerable economic importance.

B. mori has an estimated haploid nuclear genome size of 530 Mb, which is 3.6 times larger than that of the fruit fly and 1.54 times larger than that of the mosquito. The 18,510 genes in the silkworm constitute 90.09% of the total genome. Knowledge of genes allows researchers to use biomarkers to select for certain desirable traits, such as fiber quality or disease resistance, which should have the greatest potential in agricultural applications. Being a lepidopteran model, this can also give insights about the destructive larvae of related species of the most devastating polyphagous pests such as the fall armyworm and the tobacco budworm. The following are some of the genomic databases on Silkworm:

The Silkworm Knowledge Base (SilkDB) is an open-access database for genome biology of the silkworm (*B. mori*). SilkDB provides an integrated representation of the large-scale, genomewide sequence assembly, cDNAs, clusters of ESTs, transposable elements, mutants, SNPs, chromosomal mapping, microarray expression data, and functional annotations of genes with assignments to InterPro domains and GO terms [36].

The Silk Moth Microsatellite Database is a relational database of microsatellites extracted from the available ESTs and WGS sequences of the silk moth, *B. mori*. The Silk Moth Microsatellite Database also stores information on primers developed and validated in the laboratory. Users can retrieve information on the microsatellite and the protocols used, along with informative figures and polymorphism status of those microsatellites [37].

The SilkBase database presents the EST sequences of *B. mori* at various stages of growth and development, in various tissues. This database stores 35,000 ESTs from 36 cDNA libraries, which are grouped into approximately 11,000 nonredundant ESTs with an average length of 1.25 kb. A direct comparison of the silkworm EST database with FlyBase and WormBase has been added to SilkBase as an informatics tool, which is a valuable and efficient approach for revealing, at the molecular level, what makes Lepidoptera different from other insects, and providing potential candidates for targets of Lepidoptera-selective insecticides. The comparison with FlyBase suggests that the present of an EST database, SilkBase, covers more than 55% of all genes of *Bombyx*. Direct links between SilkBase and FlyBase/WormBase provide ready identification of candidate Lepidoptera-specific genes [38].

The Silkworm Genome Database (SilkDB) is a database of the integrated genome resource for the silkworm, *B. mori*. This database provides access not only to genomic data including functional annotation of genes, gene products, and chromosomal mapping, but also to extensive biological information such as microarray expression data, ESTs, and corresponding references. SilkDB is useful for the silkworm research community similar to comparative genomics.

The Silkworm Genome Research Program was established in 1994 at the National Institute of Sericultural and Entomological Science, now known as the National Institute of Agrobiological Sciences, and is being coordinated by the Insect Genome Research Team. The Silkworm Genome Research Program is currently incorporating all information from "BombMap" and "SilkBase" into a core database known as "KAIKOBase," integrating all silkworm genome data including ESTs, chromosome linkage map, and genome sequence data.

WildSilkbase is a catalogue of ESTs generated from several tissues at different developmental stages of three economically important saturniid silk moths, an Indian golden silk moth (*Antheraea assama*), an Indian tropical tasar silk moth (*Antheraea mylitta*), and the eri silk moth (*Samia cynthiaricini*). Currently, the database is provided with 57,113 ESTs, which are clustered and assembled into 4019 contigs and 10,019 singletons. WildSilkbase includes cSNP discovery, GO viewer, homologue finder, SSR finder, and links to all other related databases [39].

KAIKObase is an integrated silkworm genome database and data-mining tool with three map browsers, one gene viewer, and five independent databases. KAIKObase provides three data-mining approaches: (1) narrowing the range of data sets using PGmap or UnifiedMap of Chromosome Overview, (2) keyword and position search, and (3) sequence search using BLAST.

21.3.2.2 Honeybee and Nasonia Wasp

The honeybee, *Apis mellifera*, is a key model for social behavior and is essential to agriculture and global ecology because of its pollination activity. The honeybee genome is quite distinct from other sequenced insects due to the long evolutionary distance between them. The honeybee is a member of the insect order Hymenoptera, its genome is 50% larger than fruit flies but contains roughly the same number of genes. It has more regions rich in adenine and thymine bases, and these often contain genes. There are few transposon or retroposon families represented in the genome, another unusual characteristic compared with other insects. The honeybee (*A. mellifera*) genome was sequenced and assembled at the BCM-Human Genome Sequencing Center. The genome of the honeybee contains a total of approximately 250 million bases of DNA. Approximately 10,000 genes have been identified thus far, primarily on the basis of computer programs for gene prediction [40]. The version 4.0 assembly was released in March 2006 and published in October 2006. An upgrade, version 4.5 with additional sequence coverage, was released in 2011.

Nasonia, parasitic wasps commonly known as jewel wasps, consist of four closely related species, *Nasonia vitripennis*, *Nasonia giraulti*, *Nasonia longicornis*, and *Nasonia oneida* [41,42]. This hymenopteran wasp is considered a genetic model for parasitoids, which are important regulators of arthropod populations, including major agricultural pests and disease vectors, by laying their eggs on and killing other arthropods. Parasitoid wasps have interesting and diverse biological characteristics that make them excellent for basic studies in genetics, ecology, behavior, development, and evolution. The genome of three species, that is, *N. vitripennis*, *N. longicornis*, and *N. giraulti*, was sequenced and assembled at the BCM. These three *Nasonia* genomes will soon be joined by a fourth, that of the newly identified species, *N. oneida* [43]. Some of the available genomic databases on honeybee and nasonia wasp include:

The Hymenoptera Genome Database is an informatics resource, providing access to genome sequences and annotation for the honeybee, *A. mellifera*, the parasitoid wasp, *N. vitripennis*, and seven species of ants available through the ant genomes portal. The availability of resources across an order greatly facilitates comparative genomics and enhances our understanding of the biology of agriculturally important Hymenoptera species through genomics. The Hymenoptera Genome Database includes predicted and annotated gene sets supported with evidence tracks such as ESTs/cDNAs, small RNA sequences, and GC composition domains [44].

Honey Bee Brain EST data is a normalized unidirectional cDNA library of more than 20,000 cDNAs. Clones were partially sequenced from the normalized and subtracted libraries at the Keck Center, resulting in 15,311 vector-trimmed, high-quality sequences with an average read length of 494 bp. These sequences were assembled into 8966 putatively unique sequences, which were tested for similarity to sequences in the public databases with a variety of BLAST searches [45]. Apis ESTs were tentatively assigned molecular functions and biological processes using the GO classification system.

21.3.3 Genomic Resources on Agriculturally Important Pests

"Pest" refers to any animal or plant causing harm or damage to people or their animals, crops, or possessions, even if it only causes annoyance. Pests belong to a broad spectrum of organisms including insects, mites, ticks (and other arthropods), mice, rats (and other rodents), slugs, snails, nematodes, weeds, fungi, bacteria, and viruses (and other pathogens). This section describes the genome sequencing projects and genomic database available on flour beetle, white fly, locust, and pea aphid.

21.3.3.1 Flour Beetle

Tribolium castaneum, a powerful model organism for the study of generalized insect development, and a major devastating pest of stored agricultural products. It is the first beetle, and the first insect pest, whose genome has been sequenced. The genome was sequenced and assembled at the BCM. This genome consists of approximately 200 Mb that codes for approximately 16,000 genes (or 16,000 proteins). These sequencing efforts are extremely important for agriculture and will enable the development of new methods for the protection of food plants against beetles.

A BeetleBase genomic database is a comprehensive sequence database and important community resource for *Tribolium* genetics, genomics, and developmental biology. BeetleBase is constructed to integrate the genomic sequence data with information about genes, mutants, genetic markers, ESTs, and publications. BeetleBase uses the Chado data model and software components developed by the Generic Model Organism Database project. This strategy not only reduces the time required to develop the database query tools but also makes the data structure of BeetleBase compatible with that of other model organism databases [46].

21.3.3.2 White Fly

Bemisia tabaci is one of the most important agricultural insect pests that causes extensive damage to crops worldwide and, by feeding directly on plants, serves predominately as a vector of the plant virus genus begomoviruses. The insect damages plants by feeding on plant juices, transmitting pathogenic viruses, and promoting mold infestations. The whitefly has a relatively large genome compared with other insects, including important agricultural pests for which a genome is available [47]. It is estimated to be approximately 1020 Mb, or five times the size of the *D. melanogaster* genome. *B. tabaci* has approximately 15,000 genes, which is close to *Drosophila* and of other insects with sequenced genomes [48]. The genome sequencing is being performed by Zhejiang University.

21.3.3.3 Locust

The migratory locust (*Locusta migratoria*) is an orthopteran pest and a representative member of hemimetabolous insects for biological studies. Its transcriptomic data provides invaluable information for molecular entomology and paves a way for the comparative research of other medically, agronomically, and ecologically relevant insects. The first transcriptomic database of the locust (LocustDB), is building necessary infrastructures to integrate, organize, and retrieve data that are either currently available or to be acquired in the future. The sequencing of migratory locust is still being performed by the Beijing Research Institute.

LocustDB facilitates access and retrieves data that is either currently available or to be acquired in the future. It currently hosts 45,474 high-quality EST sequences from the locust, which were assembled into 12,161 unigenes. Migratory locust EST database, including homologous/orthologous sequences, functional annotations, pathway analysis, and codon usage are based on conserved orthologous groups, GO, protein domain (InterPro), and functional pathways (Kyoto Encyclopedia of Genes and Genomes). LocustDB also provides information from comparative analysis based on data from the migratory locust and five other invertebrate species, such as silkworm, honeybee, fruit fly, mosquito and nematode [49].

21.3.3.4 Pea Aphid

Acyrthosiphon pisum, also known as the pea aphid, is an insect responsible for hundreds of millions of dollars of crop damage every year, and is used as a model organism for studying adaptation and resistance to pesticides. It is also a powerful model system for studying bacterial endosymbiosis, developmental plasticity, and alternative modes of reproduction. The pea aphid genome was sequenced and assembled by the Human Genome Sequencing Center at the BCM with funding from the National Human Genome Research Institute. Sequencing and analysis of the pea aphid genome by the International Aphid Genomics Consortium has provided new insights into aphid development and their interactions and coevolution with obligate and facultative symbiotic bacteria. The assembled regions of the pea aphid genome have the lowest GC content of any insect genome sequenced to date; at 29.6%, pea aphid GC content is 5.2% lower than that of *A. mellifera* (34.8%). A global view of the metabolism of the pea aphid as inferred from genome sequence data is available at AcypiCyc, a dedicated BioCyc database (http://pbil.univ-lyon1.fr/software/cycads/acypicyc/home). Some of the genomic databases for pea aphid include:

AphidBase is a centralized bioinformatics resource that was developed to facilitate community annotation of the pea aphid genome by the International Aphid Genomics Consortium. The AphidBase Information System, designed to organize and distribute genomic data and annotations for a large international community, was constructed using open source software tools from the Generic Model Organism Database. The system includes Apollo and GBrowse utilities as well as a Wiki, BLAST search capabilities, and a full-text search engine. AphidBase strongly supported community cooperation and coordination in the curation of gene models during community annotation of the pea aphid genome [50].

The *A. pisum* EST database consists of the annotated clusters of all publicly available *A. pisum* ESTs. The annotation contains details of BLAST hits, domain analyses, and GO assignments. Clustering was performed with CAP3, and annotation with either BLAST or InterProScan, including applications such as Seg, Coil, Pfam, Panther, Gene3D, BlastProdom, Prints, Super Family, etc.

21.3.4 Other Genomic Databases on Insects

Some of the other important genomic databases on insects are described below:

InSatDb presents an interactive interface to query information regarding the microsatellite characteristics of five fully sequenced insect genomes (fruit fly, honeybee, malarial mosquito, red flour beetle, and silkworm). InSatDb allows users to obtain microsatellites annotated with size (in base pairs and repeat units), genomic location (exon, intron, upstream, or transposon), nature (perfect or imperfect), and sequence composition (repeat motif and GC%). One can access microsatellite cluster (compound repeats) information and a list of microsatellites with conserved flanking sequences (microsatellite family or paralogs). InSatDb is complete with the insect's information, web links for details, methodology, and a tutorial. A separate "Analysis" section illustrates the comparative genomic analysis that can be carried out using the output [51].

ButterflyBase is a unified resource for lepidopteran genomics. A total of 273,077 ESTs from more than 30 different species have been clustered to generate stable unigene sets, and robust protein translations derived from each unigene cluster. The database supports many needs of the lepidopteran research community, including molecular marker development, orthologue prediction for deep phylogenetics, and detection of rapidly evolving proteins likely involved in host–pathogen or other evolutionary processes. ButterflyBase is expanding to include additional genomic sequences as well as ecological and mapping data for key species [52].

The SPODOBASE database provides integrated access to ESTs from the lepidopteran insect, *Spodoptera frugiperda*. It is a publicly available structured database with insect pest sequences that will allow the identification of a number of genes and comprehensive cloning of gene families of interest for the scientific community. SPODOBASE contains 29,325 ESTs, which are cleaned and clustered into nonredundant sets (2294 clusters and 6103 singletons). It is constructed in such a way that other ESTs from *S. frugiperda* or other species may be added [53].

AnoBase is a database containing genomic/biological information on anopheline mosquitoes with emphasis on *Anopheles gambiae*.

VectorBase contains genome information for three mosquito species: *Aedes aegypti*, *A. gambiae*, and *Culex quinquefasciatus*; a body louse, *Pediculus humanus*; and a tick species, *Ixodes scapularis*.

The *A. gambiae* gene expression profile presents microarray experiments, functional annotation, and tools to explore gene expression in the dengue vector mosquito. Microarray analyses included at this site were based on the AffymetrixGeneChip Plasmodium/Anopheles Genome Array. The abundance of specific mRNAs represented in the array were determined for larvae (third and fourth instars), adult males (3 days postemergence), non–blood-fed females (3 days postemergence) and females at 3, 24, 48, 72, and 96 h after a blood meal, and females aged 18 days with or without a blood meal. Functional annotation integrated into the site for keyword searching combines keywords indexed in the ENSEMBL Mosquito Genome database, NCBI nonredundant databases, and conserved motifs databases (GO, Pfam, SMART). Sequence data was captured from the ENSEMBL Mosquito Genome database.

Agripestbase is a comprehensive sequence database that supports genomics research of agricultural pests. The agricultural pests database will be home to genomic information from a broad range of pests including insects, parasites, and pathogens. It will host the genomes of the Hessian fly, *Mayetiola destructor*, and tobacco hornworm, *Manduca sexta*, and in the future, the genomes of several other agriculturally important insects and pathogens.

The availability of genomic information from such a broad range of agricultural pests will result in comparative genomics and further our understanding of these species. The information available from these databases will hopefully lead to better management strategies and new methods and targets for pest control. Table 21.8 shows the insect genome databases available online.

TABLE 21.8

Insect Genome Databases Available Online

S.N	Databases	Year	Hosted By	URL
1.	Hymenoptera Genome Database	2011	Elsik Computational Genomics Laboratory, Georgetown University, Washington DC	http://hymenopteragenome.org/
2.	KAIKObase	2009	National Institute of Agrobiological Sciences, Ibarakin, Japan	http://sgp.dna.affrc.go.jp/KAIKObase/
3.	WildSilkbase	2008	Centre for DNA Fingerprinting and Diagnostics, Hyderabad, India	http://www.cdfd.org.in/WildSilkbase
4.	The Silkworm Knowledge Base (SilkDB)	2005	Beijing Genomics Institute, China	http://silkworm.genomics.org.cn
5.	Silk moth Microsatellite Database (SilkSatDb)	2005	Centre for DNA Fingerprinting and Diagnostics, Hyderabad, India	http://www.cdfd.org.in/silksatdb
6.	Silkworm Genome Database	2003	Insect Genetics and Bioscience Lab, University of Tokyo, Japan	http://papilio.ab.a.u-tokyo.ac.jp/genome/index.html
7.	SilkBase	2003	Insect Genetics and Bioscience Lab, University of Tokyo, Japan	www.ab.a.u-tokyo.ac.jp/silkbase
8.	Silkworm Genome Research Program	1994	National Institute of Agrobiological Sciences, Tsukuba, Ibaraki, Japan	http://sgp.dna.affrc.go.jp/index.html
9.	FlyExpress	2011	Arizona State University, Tempe, Arizona	http://www.flyexpress.net/
10.	FlyBase (*Drosophila*)	2009	University of Cambridge, United Kingdom	http://www.ebi.ac.uk/flybase/
11.	Berkeley *Drosophila* Genome Project	2007	Berkeley Drosophila Genome Project, Berkeley, California	http://www.fruitfly.org/
12.	DroSpeGe	2006	Genome Informatics Lab Biology, Indiana University, Bloomington, Indiana	http://insect.eugenes.org/~DroSpeGe/
13.	REDfly	2006	Center for Computational Research, State University of New York	http://www.redfly.ccr.buffalo.edu

(continued)

TABLE 21.8 (Continued)

Insect Genome Databases Available Online

S.N	Databases	Year	Hosted By	URL
14.	FlyNet	1999	IBDM, CNRS Case, Marseille, France	http://gifts.univ-mrs.fr/FlyNets/FlyNets_home_page.html
15.	FlyView—A Drosophila Image Database	1997	Institut fur Neurobiologie Badestr, Munster, Germany	http://pbio07.uni-muenster.de/
16.	Butterflybase	2008	Max Planck Institute for Chemical Ecology, Jena, Germany	http://www.butterflybase.org
17.	Aphidbase	2007	International Aphid Genomics Consortium	http://www.aphidbase.com/aphidbase/
18.	BeetleBase	2007	Bioinformatics Center, Kansas State University, Manhattan, Kansas	http://www.beetlebase.org
19.	InSatDB	2006	Centre for DNA Fingerprinting and Diagnostics, Hyderabad, India	http://www.cdfd.org.in/InSatDB
20.	AnoBase	2005	Institute of Molecular Biology and Biotechnology, Hellas, Greece	http://www.anobase.org/
21.	Spodobase	2006	Integrative Biology and Virology of Insects, INRA, France	http://bioweb.ensam.inra.fr/spodobase/
22.	LocustDB	2006	Beijing Genomics Institute, China	http://locustdb.genomics.org.cn/jsp/about.jsp
23.	Vectorbase	2009	European Bioinformatics Institute (EMBL-EBI), Wellcome Trust Genome Campus, Hinxton, Cambridgeshire, UK	http://www.vectorbase.org/
24.	FlyTF	—	—	http://www.flytf.org
25.	CisTarget	—	—	http://med.kuleuven.be/cme-mg/lng/cisTarget/
26.	Drosophila melanogaster Exon Database	—	—	http://proline.bic.nus.edu.sg/dedb/index.html
27.	FLYSNPdb	2008	University of Vienna	http://flysnp.imp.ac.at/flysnpdb.php
28.	Honey Bee Brain EST database	2002	W.M. Keck Center for Comparative and Functional Genomics, University of Illinois at Urbana Champaign	http://titan.biotec.uiuc.edu/bee/honeybee_project.htm
29.	A. pisum EST database		University of York	http://www.aphidests.org/
30.	A. gambiae gene expression profile	2006	UC Irvine	http://www.angaged.bio.uci.edu/

21.4 Future Prospects and Challenges

The development of genomic research continues to provide a foundation for future progress in examining biological phenomena in ways that would not have been possible otherwise. Genomics of livestock [54], insects, and pests have great potential in terms of biodiversity, quantitative genetics, and genetic engineering. Genomic information would help in understanding the gene structure, the nature of the mutation, and the nature of the networks of interactions within or between loci [55]. Although the causative mutation controlling a quantitative trait is complicated, combining knowledge from several lines of investigations have lead to successful identification of the gene responsible. Also, molecular genetics and genomics help improve traits related to higher production, disease resistance, and robustness for targeted selection.

The ample genomic resources available for livestock species will help researchers to identify regions within the genome that influence various traits at the molecular level with relative ease. The science of genetic engineering has paved a lot of opportunities for animal researchers. This may be applied in "pharming" (i.e., producing medicines in animals), xenotransplantation (producing tissues in animals with lower rejection risks), etc. [56]. A number of approaches exist for estimating gene expression in livestock species at both the single and whole genome levels. However, protein resources for the study are much less developed in livestock species, and hence researchers need to exploit available information from humans and biomedical animal models. Widespread use of DNA markers will have a major effect on the structure of the breeding programs and a significant effect on production systems. Selection will be based on a prediction equation derived from a reference population that has extensive phenotypic recording and genotype data. To take maximum advantage of the genomic selection, generation intervals will be shortened as much as reproductive technology will allow.

Biotechnology is providing modern improvements and a range of new tools for population control of insects in crop protection. Genetic transformation of insects is another technique that will greatly affect the future role of genomics in applied entomology. Transformations may be used for gene identification and characterization, and for creating strains with genes encoding lethality or sterility. Transgenic strains may be created to improve existing biocontrol programs such as sterile insect management technique, or potentially allow new, highly efficient control strategies. An improved understanding of the pest insect genome sequencing will stimulate the design of new classes of transgenic microorganisms to be used in pest control.

References

1. Appels, R., M. Francki, M. Cakir, and M. Bellgard. 2004. Looking through genomics: Concepts and technologies for plant and animal genomics. *Functional and Integrative Genomics* 4:7–13.
2. Phillips, T.J., and J.K. Belknap. 2002. Complex-trait genetics: Emergence of multivariate strategies. *Nature Reviews Neuroscience* 3:478–485.
3. Womack, J.E. 2005. Advances in livestock genomics: Opening the barn door. *Genome Research* 15:1699–1705.

4. Bazer, F.W., and T.E. Spencer. 2005. Reproductive biology in the era of genomics biology. *Theriogenology* 64:442–456.
5. Bilello, J.A. 2005. The agony and ecstasy of 'OMIC' technologies in drug development. *Current Molecular Medicine* 5:39–52.
6. Koyanagi, K.O., M. Hagiwara, T. Itoh, T. Gojobori, and T. Imanishi. 2005. Comparative genomics of bidirectional gene pairs and its implications for the evolution of a transcriptional regulation system. *Gene* 4:169–176.
7. Larkin, D.M., A. Everts-van der Wind, M. Rebeiz, P.A. Schweitzer, S. Bachman, C. Green et al. 2003. A cattle–human comparative map built with cattle BAC-ends and human genome sequence. *Genome Research* 13:1966–1972.
8. Gallagher, Jr., D.S., D. Threadgill, A.M. Ryan, J.E. Womack, and D.M. Irwin. 1993. Physical mapping of the lysozyme gene family in cattle. *Mammalian Genome* 4:386–373.
9. Bishop, M.D., S.M. Kappes, J.W. Keele, R.T. Stone, S.L.F. Sunden, G.A. Hawkins et al. 1994. A genetic linkage map for cattle. *Genetics* 136:619–639.
10. Williams, J.L., A. Eggen, L. Ferretti, C.J. Farr, M. Gautier, G. Amati, G. Ball, T. Caramorr, R. Critcher, S. Costa et al. 2002. A bovine whole-genome radiation hybrid panel and outline map. *Mammalian Genome* 13:469–474.
11. Itoh, T., T. Watanabe, N. Ihara, P. Mariani, C.W. Beattie, Y. Sugimoto, and A. Takasuga. 2005. A comprehensive radiation hybrid map of the bovine genome comprising 5593 loci. *Genomics* 85:413–424.
12. Dequeant, M.L., and O. Pourquie. 2005. Chicken genome: New tools and concepts. *Developmental Dynamics* 232:883–886.
13. Antin, P.B., and J.H. Konieczka. 2005. Genomic resources for chicken. *Developmental Dynamics* 232:877–882.
14. Fadiel, A., I. Anidi, and D. Kenneth. 2005. Eichenbaum Farm animal genomics and informatics: An update. *Nucleic Acids Research* 33:6308–6318.
15. Blakesley, R.W., N.F. Hansen, J.C. Mullikin, P.J. Thomas, J.C. McDowell, B. Maskeri et al. 2004. An intermediate grade of finished genomic sequence suitable for comparative analyses. *Genome Research* 14:2235–2244.
16. Rothschild, M.F. 2004. Porcine genomics delivers new tools and results: This little piggy did more than just go to market. *Genetics Research* 83:1–6.
17. Wernersson, R., M.H. Schierup, F.G. Jorgensen, J. Gorodkin, F. Panitz, H.H. Staerfeldt et al. 2005. Pigs in sequence space: A 0.66× coverage pig genome survey based on shotgun sequencing. *BMC Genomics* 6:70.
18. Klymiuk, N., and B. Aigner. 2005. Reliable classification and recombination analysis of porcine endogenous retroviruses. *Virus Genes* 3:357–362.
19. Lindgren, G., K. Sandberg, H. Persson, S. Marklund, M. Breen, B. Sandgren et al. 1998. A primary male autosomal linkage map of the horse genome. *Genome Research* 8:951–966.
20. Guérin, G., E. Bailey, D. Bernoco, I. Anderson, D.F. Antczak, K. Bell et al. 2003. The second generation of the International Equine Gene Mapping Workshop half-sibling linkage map. *Animal Genetics* 34:161–168.
21. Swinburne, J., C. Gersenberg, M. Breen, V. Aldridge, L. Lockhart, E. Marti et al. 2000. First comprehensive low-density horse linkage map based on two, three-generation, full-sibling, cross-bred horse reference families. *Genomics* 66:123–134.
22. Tantia, M.S., R.K. Vijh, V. Bhasin, Poonam, Sikka, P.K. Vij, R.S. Kataria et al. 2011. Whole-genome sequence assembly of the water buffalo (*Bubalus bubalis*). *Indian Journal of Animal Sciences* 81:38–46.
23. Andersson, L., and M. Georges. 2004. Domestic-animal genomics: Deciphering the genetics of complex traits. *Nature Reviews Genetics* 5:202–212.
24. Gellin, J., S. Brown, J.A. Marshall Graves et al. 2000. Comparative gene mapping workshop: Progress in agriculturally important animals. *Mammalian Genome* 11:140–144.
25. Andersson, L., M. Ashburner, S. Audun et al. 1996. The First International Workshop on Comparative Genome Organisation. *Mammalian Genome* 7:717–734.

26. Band, M., J.H. Larson, M. Rebeiz et al. 2000. An ordered comparative map of the cattle and human genomes. *Genome Research* 10:1359–1368.

27. Crooijmans, R.P.M.A., R.J.M. Dijkhof, T. Veenendaal et al. 2001. The gene orders on human chromosome 15 and chicken chromosome 10 reveal multiple inter- and intra-chromosomal rearrangement. *Molecular Biology and Evaluation* 18:2102–2109.

28. Andersson, L. 2001. Genetic dissection of phenotypic diversity in farm animals. *Nature Reviews Genetics* 2:130–138.

29. Smith, C.D., S.Q. Shu, C.J. Mungall, and G.H. Karpen. 2007. The release 5.1 annotation of *Drosophila melanogaster* heterochromatin. *Science* 316:1586–1591.

30. Tomancak, P. et al. 2002. Systematic determination of patterns of gene expression during *Drosophila* embryogenesis. *Genome Biology* 3:R88.

31. Lecuyer, E. et al. 2007. Global analysis of mRNA localization reveals a prominent role in organizing cellular architecture and function. *Cell* 131:174–187.

32. Drysdale, R., FlyBase Consortium. 2008. *FlyBase*: A database for the *Drosophila* research community. *Methods in Molecular Biology* 420:45–59.

33. Gilbert, D.G. 2007. DroSpeGe: Rapid access database for new *Drosophila* species genomes. *Nucleic Acids Research* 35:D480–D485.

34. Sanchez, C., C. Lachaize, F. Janody et al. 1999. Grasping at molecular interactions and genetic networks in *Drosophila melanogaster* using FlyNets, an Internet database. *Nucleic Acids Research* 27:89–94.

35. Chen, D., J. Berger, M. Fellner et al. 2009. FLYSNPdb: A high-density SNP database of *Drosophila melanogaster*. *Nucleic Acids Research* 37:D567–D570.

36. Duan, J., R. Li, D. Cheng et al. 2010. SilkDB v2.0: A platform for silkworm (*Bombyx mori*) genome biology. *Nucleic Acids Research* 38:D453-D456.

37. Prasad, M.D., M. Muthulakshmi, K.P. Arunkumar et al. 2005. SilkSatDb: A microsatellite database of the silkworm, *Bombyx mori*, *Nucleic Acids Research* 33:D403–D406.

38. Mita, K., M. Morimyo, K. Okano et al. 2003. The construction of an EST database for *Bombyx mori* and its application. *Proceedings of the National Academy of Sciences of the United States of America* 24:14121–14126.

39. Arunkumar, K.P., A. Tomar, T. Daimon, T. Shimada, and J. Nagaraju. 2008. WildSilkbase: An EST database of wild silkmoths, *BMC Genomics* 9:338.

40. Hill, M.G. 2008. *Yearbook of Science & Technology*. McGraw-Hill Companies, Inc.

41. Darling, D.C., and J.H. Werren. 1990. Biosystematics of *Nasonia* (Hymenoptera: Pteromalidae): Two new species reared from birds' nests in North America. *Annals of the Entomological Society of America* 83:352–370.

42. Raychoudhury, R., L. Baldo, D.C.S.G. Oliveira, and J.H. Werren, 2009. Modes of acquisition of *Wolbachia*: Horizontal transfer, hybrid introgression and co-divergence in the *Nasonia* species complex. *Evolution* 63:165–183.

43. Raychoudhury, R., C.A. Desjardins, J. Buellesbach et al. 2010. Behavioural and genetic characteristics of a new species of *Nasonia*. *Heredity* 104:278–288.

44. Torres, M.C., J.T. Reese, P.C. Childers et al. 2011. Hymenoptera Genome Database: Integrated community resources for insect species of the order Hymenoptera. *Nucleic Acids Research* 39:D658–D662.

45. Whitfield, C.W., M.R. Band, M.F. Bonaldo et al. 2002. Annotated expressed sequence tags and cDNA microarrays for studies of brain and behavior in the honey bee. *Genome Research* 12:555–566.

46. Wang, L., S. Wang, Y. Li, M.S. Paradesi, and S.J. Brown. 2007. BeetleBase: The model organism database for *Tribolium castaneum*. *Nucleic Acids Research* 35:D476–D479.

47. Brown, J.K., G.M. Lambert, M. Ghanim, H. Czosnek, and D.W. Galbraith. 2005. Nuclear DNA content of the whitefly *Bemisia tabaci* (Aleyrodidae: Hemiptera) estimated by flow cytometry. *Bulletin of Entomological Research* 95:309–312.

48. Stansly, P.A., and S.E. Naranjo. 2010. *Bemisia, Bionomics and Management of a Global Pest*. Dordrecht: Springer.

49. Zongyuan, M., J. Yu, and K. Le. 2006. LocustDB: A relational database for the transcriptome and biology of the migratory locust (*Locusta migratoria*), *BMC Genomics* 7:11.
50. Legeai, F., S. Shigenobu, J.P. Gauthier et al. 2010. AphidBase: A centralized bioinformatics resource for annotation of the pea aphid genome. *Insect Molecular Biology* 19:5–12.
51. Archak, S., E. Meduri, P. S. Kumar, and J. Nagaraju. 2007. InSatDb: A microsatellite database of fully sequenced insect genomes. *Nucleic Acids Research* 35:D36–D39.
52. Papanicolaou, A., S. Gebauer-Jung, M.L. Blaxter et al. 2008. ButterflyBase: A platform for lepidopteran genomics. *Nucleic Acids Research* 36:D582–D587.
53. Negre, V., T. Hotelier, A. Volkoff et al. 2006. SPODOBASE: An EST database for the lepidopteran crop pest *Spodoptera*, *BMC Bioinformatics* 7:322.
54. Howard, T.H., E.J. Homan, and R.D. Bremel. 2001. Transgenic livestock: Regulation and science in a changing environment. *Journal of Animal Science* 79:E1–E11.
55. Smith, T.J. 1994. Commercial exploitation of transgenics. *Biotechnology Advances* 12:679–686.
56. Cascalho, M., and Platt, J.L. 2001. Xenotransplantation and other means of organ replacement. *Nature Reviews Immunology* 1:154–160.

Part III

The World of Omics and Applications in Environmental Sciences

22

Applications of Molecular Spectroscopy in Environmental and Agricultural Omics

Daniel Cozzolino

CONTENTS

22.1 Introduction

Metabolomics represents a global understanding of the metabolite complement of integrated living systems and dynamic responses to the changes in both exogenous and endogenous factors [1]. In modern metabolomics, a hierarchy of "omics" terms is dividing research into areas that can be studied separately and as interacting phenomena [1–4]. From many varied definitions of metabolomics, the most generalized is the one that defines metabolomics as the systematic study of unique chemical fingerprints that specific cellular processes leave behind [5]. It is generally accepted that as an addition to the family of omics, metabolomics appeared as an integrative representation of the interactions of the genome, transcriptome, and proteome with the environment [5,6]. The combination of these four omics provides a platform to investigate the regulation of the cellular environment from the specific codes stored in the DNA through its expression by proteins (e.g., enzymes and structural proteins) to the chemical conversion of small molecules within biochemical pathways [5,6].

The development of these omics is dependent on analytical advances in spectroscopy, chromatography, and electrophoresis, sensitive and specific analytical techniques to allow the handling of large numbers of samples [1–6]. Although microchip arrays permit the testing of RNA and DNA fragments against libraries of several hundred standards at the same time, the multiparallel chromatographic, infrared (IR)/Raman spectroscopy, and mass spectrometry (MS) as well as nuclear magnetic resonance spectroscopy methods perform the separation, identification, and quantification of small molecules and fragments of proteins [2,3]. Due to the large amount of information cumulated by measurements, mathematicians have joined chemists,

TABLE 22.1

Advantages and Limitations of Molecular Spectroscopy Methods in Omics Analysis

	Advantages	Limitations
NIR		Overtones and combination bonds
	Does not require sample preparation	
		Low sensitivity and selectivity
MIR	Fundamental vibrations	
		Some sample preparation might be required
	Medium sensitivity and selectivity	

biochemists, and molecular biologists in developing algorithms that allow data analysis that would help present the results in a comprehensible manner through bioinformatics [7].

Since the start of metabolomics, classic instrumental methods such as nuclear magnetic resonance spectroscopy, ionization MS, MS, liquid chromatography, high-performance liquid chromatography, and gas chromatography have been used and applied to a vast number and type of samples in plant and biological analysis [2–7]. These technologies usually consist of several steps before or during analysis, such as separation and fractionation of the sample prior to analysis. However, less attention has been devoted to the use and development of rapid methods that are based on molecular spectroscopy, such as near-infrared (NIR) and mid-infrared (MIR) spectroscopy. The use of molecular spectroscopy techniques allows us to define the so-called metabolic fingerprint. This fingerprinting can be achieved using high-throughput methods, rapid analysis, minimal sample preparation, and nonchromatographic separation [8].

Table 22.1 summarizes some of the main advantages and limitations of the molecular spectroscopic methods commonly used in metabolomics. This chapter will introduce and discuss recent developments on the use of molecular spectroscopic methods, NIR and MIR, with special emphasis on agricultural and environmental applications.

22.2 Molecular Spectroscopy

22.2.1 IR (NIR and MIR)

Chemical bonds present in the organic matrix of most samples vibrate at specific frequencies, which are determined by the mass of the constituent atoms, the shape of the molecule, the stiffness of the bonds, and the periods of the associated vibrational coupling [9–11]. A specific vibrational bond is absorbed in the IR spectral region in which diatomic molecules have only one bond that may stretch (e.g., the distance between two atoms may increase or decrease). More complex molecules have many bonds; vibrations can also be conjugated leading to two possible modes of vibration: stretching and bending. Despite these potential problems, absorption frequencies may be used to identify specific chemical groups, and this capability has traditionally been the main role of Fourier transform-MIR (FT-MIR) spectroscopy [9–11]. The MIR region of the electromagnetic spectrum lies between 4000 and 400 cm^{-1} and can be segmented into four broad regions: the X–H stretching region (4000–2500 cm^{-1}), the triple bond region (2500–2000 cm^{-1}), the double bond region (2000–1500 cm^{-1}), and the fingerprint region (1500–400 cm^{-1}) [9–11] (Figure 22.1). Such characteristic absorption bands are associated with major components of the sample matrix. Absorptions in the fingerprint region are mainly caused by bending and skeletal vibrations, which are particularly sensitive to large wave number shifts, thereby minimizing against unambiguous identification of specific functional groups [9–11]. Analysis of such fingerprints forms the basis of many applications of MIR spectroscopy in metabolomic analysis. In recent years, dedicated FT-MIR instruments have become readily available and have been used extensively for routine analysis. The application of FT-MIR in metabolomics is of special interest due to the presence of sharp and specific absorption bands for constituents.

More recently, the development of sampling accessories attached to a wide range of IR spectrophotometers, such as attenuated total reflectance cells, has led to major improvements in routine IR analysis by simplifying sample handling and avoiding measurement problems often found using transmission cells. In conventional FT-MIR spectrophotometers commonly used in analysis, samples are analyzed through a short–path length transmission cell [12,13]. The advantage of using transmission cells is that it provides very accurate and reproducible spectroscopic measurements and is easily temperature-controlled. On the other hand, the use of transmission cells has several drawbacks, such as filling and cleaning the cell, and variation of sample path length due to window wear and turbidity of the sample, among others factors. Therefore, the use of attenuated total reflectance cells make this an attractive sample presentation to use combined with MIR for metabolomic analysis.

Other types of sensors based on molecular spectroscopy in the IR region include base sensors in NIR. In the last 40 years, NIR spectroscopy has become one of the most attractive and used methods of analysis, providing simultaneous, rapid, and nondestructive quantification of major components in many agriculture-related products and plant materials. The use of NIR for the analysis of beverages is characterized by low molar absorptivities and scattering, which leaves a nearly effortless evaluation of pure materials [14–16].

Spectral "signatures" in the MIR result from the fundamental stretching, bending, and rotating vibrations of the sample molecules, whereas NIR spectra result from complex overtones and high-frequency combinations at shorter wavelengths. Spectral peaks in the MIR frequencies are often sharper and better resolved than in the NIR domain, the higher overtones of the O–H (oxygen–hydrogen), N–H (nitrogen–hydrogen), C–H (carbon–hydrogen),

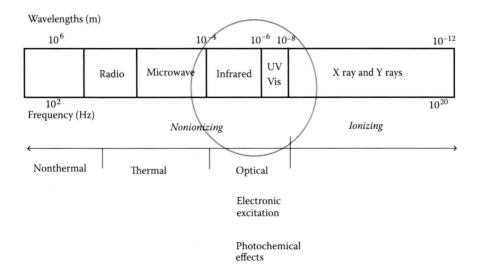

FIGURE 22.1
The electromagnetic spectrum and the location of the IR region.

and S–H (sulfur–hydrogen) bands from the MIR wavelengths are still observed in the NIR region, although much weaker than the fundamental frequencies in the MIR. In addition to the existence of combination bands (i.e., CO stretch and NH bend in protein), gives rise to a crowded NIR spectrum with strongly overlapping bands [14–17]. A major disadvantage of this characteristic overlap and complexity in the NIR spectra has been the difficulty of quantification and interpretation of data from NIR spectra. On the other hand, the broad overlapping bands can diminish the need to use a large number of wavelengths during the analysis.

In recent years, new instrumentation and computer algorithms have taken advantage of this complexity and have made the technique much more powerful and simple to use. In addition, the advent of inexpensive and powerful computers has contributed to the surge of new NIR applications [16–18].

22.2.2 Raman Spectroscopy

Raman spectroscopy is based on fundamental vibration modes that can be assigned to specific chemical functional groups within a sample molecule and therefore can potentially provide useful information for fingerprinting or qualitative identification [12]. Raman spectroscopy has also been explored as an emerging methodology in metabolomics studies [8,12]. Raman spectroscopy, similar to NIR and MIR, is a constantly developing technique that has also allowed for rapid, nondestructive, reagent-less, and high-throughput analysis for a diverse range of sample types. Very recently, Raman spectroscopy was also introduced as a metabolic fingerprinting technique within the plant sciences [2,19–20]. Raman spectroscopy coupled with microscopy has recently been used with great success, making the identification and quantification of photochemical distribution directly from the plant tissues possible [21–23]. Despite the advantages of Raman spectroscopy, when compared with MIR or NIR, the development of this technique has been less extensive and slower because of several issues such as instrumentation and high cost, among others [12].

22.3 Multivariate Data Analysis

Chemical information contained in such instrumental methods resides in the occurrence of peaks, band positions, intensities, and shapes [9–11]. The most successful approach to extracting quantitative, qualitative, or structural information from such spectra is to use multivariate mathematical analysis or chemometrics [24–27]. In modern chemical measurements, we are often confronted with so much data that the essential information may not be readily evident. Certainly, that can be the case with chromatographic or spectral data, for which many different observations (peaks or wavelengths) have been collected. Each different measurement can be thought of as a different dimension. Traditionally, as analysts, we strive to eliminate matrix interference in our methods by isolating or extracting the analyte to be measured, making the measurement apparently simple and certain. However, this ignores the possible effects of chemical and physical interactions between the large amounts of constituents present in the sample. Univariate models do not consider the contributions of more than one variable source and can result in models that could be an oversimplification. Therefore, we need to look at the sample in its entirety and not just at a single component if we wish to untangle all the complicated interactions between the constituents and understand their combined effects on the whole matrix [24–27].

Multivariate methods provide the means to move beyond the one-dimensional (univariate) world. In many cases, multivariate analysis can reveal constituents that are important through the various interferences and interactions [24–27].

Today, many modern instrumental measurement techniques are multivariate and based on indirect measurements of the chemical and physical properties [24–27]. A typical characteristic of many of the most useful of these instrumental techniques is that, paradoxically, the measurement variable might not have a direct relationship with the property of interest, for instance, the concentration of a particular chemical in the sample. Molecular spectroscopy techniques provide the possibility of obtaining more information from a single measurement because these techniques can record responses at many wavelengths simultaneously, and it then becomes essential to use multivariate analysis to extract the information. These powerful methods and the computer technology necessary to use them have only become readily available in recent years, but their use has become a significant feature in the development of instrumental applications.

A broad range of techniques is now available including data reduction tools, regression techniques, and classification methods. Figure 22.2 summarizes some of the uses and applications of multivariate data methods. Principal component analysis (PCA) is a commonly used data compression and visualization tool, reducing a spectral data set into a small (generally <20) number of new, orthogonal (e.g., noncorrelated) variables on each of which a score (or value) for each sample is calculated [25,26–28]. Graphical display of these scores can often reveal patterns or clustering within a data set because similar samples are expected to locate close to each other; unexpected sample locations in this hyperspace may alert the analyst to unusual or outlying samples, which may be reanalyzed or, as a final resort, deleted from the data set before further data processing. Principal component scores may be used in further mathematical operations to classify samples into different, naturally occurring groups [26,27]. A number of procedures are available for sample classification or discrimination: soft independent

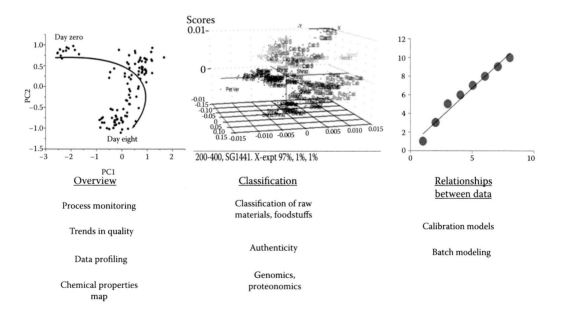

FIGURE 22.2
Overview of the multivariate methods applied in omics analysis of instrumental methods.

modeling of class analogy is an example of a popular class-modeling method, whereas linear discriminant analysis, hierarchical cluster analysis, factorial discriminant analysis, artificial neural networks, and discriminant partial least-squares are examples of much-used discriminant methods [9,28]. Class-modeling methods focus on characterizing each of the classes of sample being analyzed and involve the calculation of a model and boundaries within which samples of each particular type may be expected to be found. Discriminant methods focus on characterizing the boundaries between samples of different classes and do not involve the calculation of statistically robust confidence limits for each class [9,28]. The application of artificial neural networks is a more recent technique for data and knowledge processing, which is characterized by its analogy with a biological neuron [9,28]. When the firing frequency of a neuron is compared with that of a computer, then for a neuron, this frequency is rather low [26–28]. In the biological neuron, the input signal from the dendrites travel through the axons to the synapse. There, the information is transformed and sent across the synapse to the dendrites of the next neuron [26–28]. For more comprehensive coverage of chemometric tools and procedures, the interested reader is referred to other sources [24,26–28].

Figure 22.3 shows the NIR spectra of a set of corn samples from different genotypes. The application of PCA analysis allows us to interpret hidden information in the NIR spectra

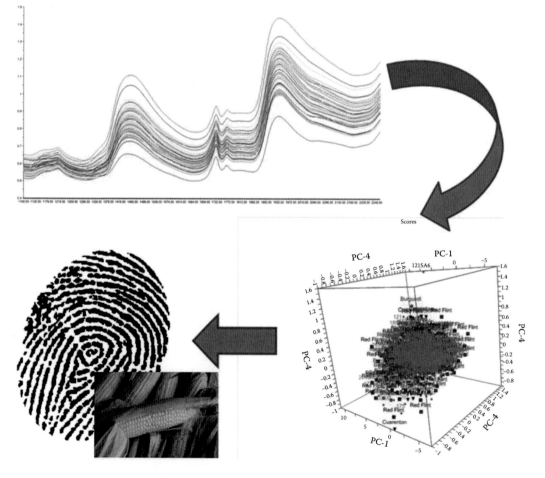

FIGURE 22.3
(See color insert.) Example of combining NIR analysis and PCA to fingerprint corn samples for high oil content.

of the samples, which it is difficult to visualize. The end point of the process is the finger-print of the sample that can be used to classify or discriminate unknown samples.

22.4 Applications

It is well known that IR spectroscopy (NIR and MIR) is a well-established analytical tech-nique that enables the rapid and nondestructive, reagent-less, and high-throughput analysis of a diverse range of sample types [8]. Due to its holistic nature, IR spectroscopy is a valu-able metabolic fingerprinting tool because of its ability to analyze carbohydrates, amino acids, lipids, and fatty acids as well as proteins and polysaccharides simultaneously [8].

To date, the vast majority of metabolomic studies undertaken using molecular spectros-copy have been focused on the application of MIR spectroscopy. Few reports, however, can be found on the literature that relates with the application of both NIR and FT-NIR spectroscopy to assess its suitability as tools for metabolomic analysis such as yeast identi-fication [29] and to detect and identify bacterial strains [30]. However, the use of NIR spec-troscopy as a metabolomic technique has not been extensively applied because this region is predominantly dominated by overtones and combination vibrations, and is considered to less rich in information compared with MIR [8].

One of the main advantages of using NIR spectroscopy in metabolomic studies is its ability to provide comprehensive spectral data without the need for preliminary separa-tion or preprocessing of the sample. However, the NIR spectrum is a complex multivariate data set and requires the use of chemometrics to interpret the data generated to use its information for high-throughput metabolomic fingerprinting [31]. Although sample pre-sentation is better compared with other molecular spectroscopic methods (e.g., MIR), the main drawback of NIR from the analytical point of view is its inherent low sensitivity (0.1% w/w) compared with MIR [32]. Table 22.2 summarizes some examples using IR spec-troscopy for omic analysis in agricultural and environmental applications.

22.4.1 Agricultural and Food Applications

IR technologies have been introduced as metabolic fingerprinting techniques within the agricultural and food industries. Such studies have been included into plant-to-plant

TABLE 22.2

Examples of Use of IR Spectroscopy Methods for the "Omic" Analysis in Agriculture and Environmental Applications

Applications	Method	Reference
Phenotyping and genotype characterization of forages during harvest	NIR	[34]
Omic analysis of plants	NIR and IR	[35]
Study gene expression in barley	NR	[38]
Effect of mutations on barley endosperm	NR	[39]
Effect of mutations on corn endosperm	NIR	[43]
Transgenic corn	NIR	[45]
Bacterial contamination in foods	MIR	[55]
Bioactive compounds in several plant species and crops	NIR, MIR	[61]

interactions, in which global metabolite changes were associated with abiotic and biotic perturbations and interactions [33]. Other applications included the metabolite finger-printing of salt-stressed tomatoes, in which functional groups of importance have been related to tomato salinity [34].

To unravel the genetic basis of complex traits, it is necessary to associate genotypic information with the corresponding phenotypic data. Recent progress in DNA marker assays and sequencing technologies have enabled high-throughput genotyping of many individual plants at relatively low cost [35]. By comparison, the phenotyping of large map-ping populations for several traits in field trials is still laborious and expensive. Research and developments in NIR spectroscopy on agricultural harvesters [36,37] and spectral reflectance of plant canopy [38] present new opportunities to develop novel phenotyping platforms that enable large-scale screenings of genotypes for several traits in multiloca-tion field trials. These phenotyping techniques could bring remarkable progress in plant genetic research to unravel the genetic basis of dynamic traits, such as biomass accumula-tion and drought or frost stress tolerance. Currently, these traits are treated as static (i.e., are measured only once), but ignoring their dynamic nature entails a tremendous loss of information regarding the analysis of genes and gene networks that are active at differ-ent phases of plant development and in reactions to environmental stresses [35]. The use of NIR spectroscopy on agricultural harvesters reduces the manpower and expenditure required for the determination of relevant traits. In contrast to conventional sample-based methods, NIR spectroscopy on agricultural harvesters secures a good distribution of mea-surements within plots and covers substantially larger amounts of plot material [35,36]. Consequently, agricultural harvesters equipped with NIR spectroscopy reduce the sam-pling error and yield more representative measurements of the plot material, and can also be used successfully to determine dry matter, starch, and crude protein contents in maize grain [35]. In silage maize, the potential of this technology has also been reported for dry matter, starch, and soluble sugars [35,36]. Measurements of crude protein, digestibility, fiber contents, and energy-related traits can be used successfully to classify genotypes; however, for the precise quantitative trait evaluation, further technical improvements for an optimal sample presentation are crucial [35,36]. The use of NIR spectroscopy on agri-cultural harvesters represents a high-throughput phenotyping technique with substan-tially reduced sampling error, whereas spectral reflectance of plant canopies facilitates the determination of dynamic traits in a noninvasive mode. Combining these novel pheno-typing methods with high-throughput genotyping based on different omic technologies and analyzing the data with sophisticated mathematical models promises a major break-through in the elucidation of the underlying genetic basis of complex plant traits, such as drought or frost tolerance [35–38].

Munck and collaborators [39] assessed the discriminatory power of NIR spectroscopy combined with PCA to evaluate different lysine mutants in barley (*Hordeum vulgare*). These authors were able to identify different mutants based on the NIR fingerprint of the sample. It has been shown that it is possible to select improved malting barley genotypes based on its NIR spectrum as a total physical–chemical spectral fingerprint of the sample and interpreted using PCA [39–41]. NIR spectroscopy was also evaluated to discriminate bar-ley flour containing high levels of lysine amino acids [39–41]. It was found that differences between these two barley flours might be observed through the interpretation of the PCA scores [42]. Thereby, these two variants of barley flour were classified correctly in two groups according to their different NIR spectra. The use of NIR spectroscopy was also evaluated to assess the effects of different environments on transgenic and nontransgenic plant materials [39–43]. For example, four genotypes consisting of common genotypes and

LYS3 (including four alleles of a, b, c, and m) were used [39–43]. The main phenotypic changes caused by manipulating the alleles were recognized by combining NIR spectroscopy with PCA [39–43]. Several other reports can be found on the use of NIR spectroscopy to discriminate between transgenic and nontransgenic grains [44–49]. All these authors reported that classification methods such as PCA yield good performances to separate completely transgenic from nontransgenic grains, concluding that the utilization of NIR spectroscopy as a nondestructive method had a suitable efficiency to detect transgenic grains [44–49].

Molecular spectroscopy was also used to evaluate the full-length cDNA overexpressing (FOX) gene hunting system's usefulness for genomewide gain-of-function analysis [48]. The screening of FOX lines requires a high-throughput metabolomic method that can detect a wide range of metabolites. The use of FT-NIR spectroscopy combined with chemometric techniques has been reported to analyze 3000 rice crosses with *Arabidopsis* FOX lines [48]. The candidate lines exhibiting alteration in their metabolite fingerprints were also analyzed by gas chromatography/time-of-flight/MS. Orthogonal projection to latent structures was used to develop calibration models by integration of the data sets obtained from FT-NIR and gas chromatography/time-of-flight/MS analyses. The authors concluded that FT-NIR enables untargeted analysis and high-throughput screening focusing on the alteration of metabolite composition. This method is time-saving in that it can be used to detect the metabolite fingerprints of seed material without pretreatment [48]. The advantage of FT-NIR spectroscopy is that it can be used to detect the composition of a variety of metabolites. This technique, combined with chemometrics, can be used for large-scale screening of gain-of-function mutant resources. Moreover, this technology will be effective in the analysis of useful plant gene functions using loss-of-function mutants such as knockout (TILLING or insertion mutation) or knockdown (RNAi) genes [48].

In the last few years, both microbial and plant metabolite analysis have shifted from specific assays toward methods offering both high accuracy and sensitivity in highly complex mixtures of compounds [50,51]. Large-scale metabolome analysis is based on the use of gas chromatography/MS and liquid chromatography/MS [50,51]. Both NIR and FT-NIR spectroscopy have been examined as tools for yeast identification [29]. The potential of combining NIR spectroscopy and multivariate analysis as a rapid screening technique to discriminate different yeast strains with particular metabolic profiles was reported. Samples used were fermentation supernatants of different *Saccharomyces cerevisiae* deletion strains, which were scanned using a VIS-NIR spectrophotometer [49]. Multivariate classification models were developed using the spectra of the yeast strains. PCA models for each yeast deletion strain were performed and used to develop classification models based on soft independent model class analogy [49]. These yeast models were based on the wild-type laboratory strain. The models showed that the deletion strains were correctly classified as different from the wild-type laboratory strain [49].

There is a continuing requirement for rapid, accurate, and preferably automated methods for the identification of microorganisms [52–60]. Ideal techniques for rapid microbial characterization would include those that require minimal sample preparation, permit the automatic analysis of many samples with negligible reagent costs, allow rapid characterization against a stable database, are easy to use, and can be operated under the control of a computer [52–60]. With recent developments in analytical instrumentation, these requirements are being fulfilled by physicochemical spectroscopic methods, often referred to as "whole-organism fingerprinting" and, more recently, as "metabolic fingerprinting" [52–60].

Several studies reported the use of MIR spectroscopy to interpret the basis of the separation obtained between different microorganisms as a metabolomic approach. Many cellular components have absorption bands between 1500 and 1200 cm^{-1}, for example, nucleic acids and phospholipids [52–61]. Spectral intervals between 1200 and 900 cm^{-1} provide information about the structure of polysaccharides [52–61]. The MIR spectra of organic acids are characterized by several common absorption bands: C=O stretching of acids, C–O stretching of acids and alcohols, O–H bending in acids and alcohols, and C–H stretching modes in aliphatic carboxylic acids [52–61]. Similarly, regions in the NIR such as O–H absorption bands (second overtone), O–H stretch and C=O second overtone combinations, O–H first overtone, and C–H and C=C tones, can be associated with several organic compounds such as aromatic groups, amino acids, and sugars [61].

An exciting prospective use of IR spectroscopy was recently demonstrated by screening of malolactic bacterial strains for metabolic activity in red and white wines resulting in different aroma and flavor profiles [60]. Natural bacterial strains have vast genetic diversity, and pinpointing those strains that have the desirable metabolic diversity with IR (NIR and MIR) spectroscopy would be an efficient and effective strategy for identifying potentially useful commercial wine strains [60]. The study has demonstrated that both MIR and NIR spectroscopic technologies can be used to rapidly differentiate red and white wines produced with different bacteria strains which are the first step in achieving this goal. Overall IR spectroscopy can be applied directly to the surface of the food and produce biochemically interpretable IR "fingerprint" which represents the overall changes or patterns in the data set or sample [60,62,63].

22.4.2 Environmental Applications

Advances in molecular spectroscopy have provided with new methods and techniques in environmental applications. Although MIR is used in the vast majority of these applications, other classes of molecular spectroscopy such as NIR or Raman spectroscopy are making progress in these applications [64–66]. Some work has been carried out using NIR spectroscopy, having significant potential for metabolomics studies sharing many of the same advantages of the FT-MIR. Both MIR and NIR spectroscopy have been used to monitor changes in soils related with agronomic practices, waste management as well as other applications related with the environment (e.g., biofuels) [64–66].

NIR spectroscopy has reportedly been used to monitor anaerobic digestion processes online or inline to control and monitor biogas production [67]. Variables such as volatile fatty acids, ammonia, total solids, total nitrogen, and carbon can be monitored. NIR spectroscopy has also been reportedly used to monitor methanogenesis during the biogas production process [68]. Using the same principles, the combination of NIR spectroscopy and multivariate data analysis can be used to monitor municipal waste, in turn monitoring the entire process [66–68]. Other applications can be found in the bioethanol and biodiesel industries.

The emission of organic pollutants into the environment, either man-made or through natural processes, is a topic that generates considerable interest and public concern [66]. A number of methods have been proposed to tackle these controversial issues. Various methods are available for screening and assessing the effect of hydrocarbon pollutants on the biosphere. By exploring certain energies in the electromagnetic spectrum, a great deal of information can be obtained about the structure and identity of some hydrocarbon molecules [66].

It is well-established that every hydrocarbon molecule generates a characteristic absorption pattern or fingerprint and the ability to distinguish between different compounds based on the IR spectrum. Growth rates, photosystem II photosynthesis, and the levels of chlorophyll A and secondary metabolites of *Chlorella ovalis* were estimated to determine if they were enhanced by the addition of swine urine (BM) or cow compost water (EP) that had been fermented by soil bacteria to deep seawater (DSW) in an attempt to develop a medium that could enable batch mass culture at lower costs. The use of FT-MIR spectroscopy revealed that *C. ovalis* grown in DSW + EP60% had consistent peaks and various biochemical pool shifts compared with those grown in other types of media. Together, the results of this study indicate that the use of DSW + EP at 60% to culture *C. ovalis* could reduce maintenance expenses and promote higher yields in bacterial growth.

Volatile organic compounds (VOCs) are harmful gaseous pollutants in the ambient air [70–77]. A simulation of the radiance profiles acquired from airborne passive multispectral IR imaging measurements of ground sources of VOCs was developed by Sulub and Small [72]. The simulation model allows the superposition of pure-component laboratory spectra of VOCs onto spectral backgrounds, which simulate those acquired during field measurements. The instrument used in the research was a downward-looking IR line scanner mounted on an aircraft that flew at an altitude of 2000 to 3000 m [72]. Wavelength selectivity was done through the use of a multichannel Hg:Cd:Te detector with up to 16 integrated optical filters. Those filters allowed the discrimination of absorption and emission signatures of VOCs superimposed on the upwelling IR background radiance within the instrument's field of view. By combining simulated radiance profiles containing analyte signatures with field-collected background signatures, supervised pattern recognition methods can be employed to train automated classifiers for use in detecting the signatures of VOCs during field measurements [72]. The targeted application for this methodology is for use in imaging systems to detect the release of VOCs during emergency response scenarios. In the work described here, the simulation model was combined with Stepwise linear discriminant analysis (SLDA) to build automated classifiers for detecting ethanol and methanol. Field data collected during controlled releases of ethanol, and an accidental methanol release from an industrial facility, were used to evaluate the methodology [71–77].

FT-IR remote sensing measurements were applied to the detection of methanol vapor fumes released from a chemical manufacturing facility [77]. The spectrometer was mounted in a downward-looking mode on a fixed-wing aircraft, flights over the facility are made during the methanol release. Signal processing and pattern recognition methods were applied to the acquired data to develop an automated classification algorithm for methanol detection [71–77]. The analysis was based on the use of short, digitally filtered segments of the raw interferogram data collected by the spectrometer. The classifiers were trained with data collected on the ground through an experimental protocol designed to simulate background conditions observed from the air. Optimization of digital filtering and interferogram segment parameters lead to successful classification based on 100 or 120 interferogram points. The optimal interferogram segment location is found to be 95 points displaced from the center burst. The best-performing digital filters were centered on the methanol C–O stretching band at 1036 cm^{-1}, and had a pass band full-width at half-maximum of 100 to 160 cm^{-1}. The best classification models achieved classification errors of less than 1% and were resistant to possible interference effects from species such as ethanol and ozone. This work demonstrates the utility of airborne FT-MIR remote sensing measurements of VOCs under complex background conditions such as those encountered while monitoring an operating industrial facility [73–77].

Continuous and online monitoring of VOCs based on near infrared reflectance spectroscopy (NIRS) was explored. Du and collaborators [73–77] analyzed the characteristic spectra of two types of VOCs, propane and isobutene using FT-MIR. A prediction model was established using partial least-squares in the NIR wavelength range of 5600 to 6200 cm^{-1}. The concentration of the two gases from the validation set showed that the calibration models and the relative error was less than 5% [73–77]. Monitoring multiple gas components from any gas *in situ* by NIRS is of great significance in the field. The spectral characteristics of the three types of the VOCs, propane, propylene, and methylbenzene, were analyzed. For propylene, the NIR spectra from 1620 to 1750 nm, which includes the absorption bands of the three VOCs, were acquired. Partial least-squares calibration models for propane and isobutene concentrations were developed and validated using an independent validation set. The results of the experiment indicated that NIRS could easily determine the content of the multiple components from gas and can be used for monitoring VOCs *in situ* [73–78].

Quantitative FT-MIR analysis was used for the determination of the adsorption capacity of VOC under dynamic conditions. The analytical method also showed the possibility of distinguishing between reversible and irreversible adsorption and further detection of the adsorbed VOC transformation. The adsorbed amounts were used for the determination of the heat of adsorption and the activation energy of desorption using isosteric and temperature-programmed desorption methods, respectively. The approach was applied to explore the potential use of local clay as an adsorbent material for VOC pollutants [73–78].

Remote sensing FTIR is one of the most important technologies in atmospheric pollutant monitoring. It has at least four prevailing advantages: (1) high resolution and high selectivity, (2) it does not require sampling and sample preprocessing, (3) capability of detecting several compounds simultaneously, and (4) real-time, long-distance, and automatic monitoring. This makes this technology appropriate for remote, real-time, and dynamic monitoring of air contaminants, especially toxic VOCs [73–78].

Classic bacteria culture media, as well as domestic or industrial wastewater treated by biological processes, have a complex composition. The online or *in situ* determination of some substances is possible, but expensive, because sample collection and pretreatment in still conditions are often necessary. More global methods can be used to rapidly detect "accidental events" such as the appearance of undesirable by-products in a fermentation broths or substances in wastewater. Among potential candidates, spectroscopy can be the basis for noninvasive and nondestructive measuring systems. Some of them have already been tested *in situ*: UV-VIS, IR, fluorescence (mono-dimensional, two-dimensional, or synchronous), dielectric methods, MS, coupled or not to pyrolysis, nuclear magnetic resonance, and Raman spectroscopy. All these methods provide spectra or large data sets, from which meaningful information should be extracted for further analysis. The use of data-mining techniques such as PCA, or projection on latent structures, is a necessary step for compressing large data sets. A review of spectroscopic techniques with examples from the bioengineering field is reported elsewhere [78].

A deployable fiber-optic sensor system for the continuous determination of a range of environmentally relevant VOCs in seawater was developed by Kraft and collaborators [78]. The prototype of a robust, miniaturized FT-MIR spectrometer for *in situ* underwater pollutant monitoring was designed, developed, and built. The assembled instrument was enclosed in a sealed aluminum pressure vessel and was capable of maintenance-free operation in an oceanic environment down to depths of at least 300 m. The whole system could be incorporated either in a tow frame or as remotely operated vehicle. A fiber-optic sensor head was developed, optimized in terms of sensitivity and hydrodynamics, and connected to the underwater FT-MIR spectrometer. Due to a modular system design, several sensor head configurations could be realized and tested, ensuring the versatility of the instrument. The sensor system was characterized in a series of laboratory and simulated field tests. The sensor proved to be capable of quantitatively detecting a range of chlorinated hydrocarbons and monocyclic aromatic hydrocarbons in seawater down to the low parts per billion (μg/L) concentration range, in mixtures of up to six components. It was demonstrated that varying amounts of salinity, turbidity, or humic acids, as well as interfering seawater pollutants, such as aliphatic hydrocarbons or phenols, did not significantly influence the sensor characteristics. In addition, the sensor exhibited sufficient long-time stability and a low susceptibility to sensor fouling [79].

22.5 Final Considerations

These applications and examples have demonstrated the potential of combining molecular spectroscopy (NIR and MIR) and multivariate techniques as a rapid tool in research and development for fingerprint sampling in food and environmental studies. Although these techniques and methods are based on molecular spectroscopy and generally cannot measure molecules with low concentrations, the indirect effects of such differences can be observed in the IR spectrum of a given sample (fingerprint). Molecular spectroscopy has been successfully applied for natural product composition analysis, product quality assessment, and in production control, and it has been shown that molecular spectroscopy has a place in metabolomic studies. The global signature of composition (fingerprint) which, with the application of chemometric techniques (e.g., PCA or discriminant analysis), can be used to elucidate particular compositional characteristics not easily detected by traditional targeted chemical analysis, is one of the main advantages of this method.

The other important advantages of this technique over the traditional chemical and chromatographic methods are the rapidity and ease of use in routine operation. Moreover, IR is a nondestructive technique which requires minimal or zero sample preparation. Therefore, molecular spectroscopy can be suggested as the first line of tools to be used in metabolomic studies.

The instrumental cost and complexity of IR spectroscopy methods are reduced compared with classic chromatographic methods. However, the main drawback of molecular spectroscopy, in particular, NIR spectroscopy, is that the overtone and combination bands tend to be very broad and overlapping relative to the sharp, well-separated fundamental features found in the MIR region. As a result, multivariate algorithms are required for analysis of NIR spectra to extract chemically relevant information. Therefore, NIR analysis is not typically used for target metabolite identification, but is used for overall spectral profile comparisons. However, due to the low absorbance and correspondingly longer path lengths in NIR, multivariate models built from NIR data tend to be more accurate for quantitative purposes than those derived from MIR data.

The potential savings, reduction of analysis time and cost, and the environmentally friendly nature of the technology has positioned molecular spectroscopy as a very attractive technique with a bright future in metabolomics. It is clear that the breadth of applications based on molecular spectroscopy, either in routine use or under development, is showing no sign of diminishing. The combination of different analytical techniques with multivariate methods could be used as a tool for fingerprinting samples on a large scale. However, the chemical basis of this separation is not addressed using this type of methodology, and other analytical techniques (i.e., gas chromatography-MS) need to be used or combined to reveal the fundamental causes of the separation/classification.

The development and implementation of other techniques such as hyperspectral imaging, microspectroscopy, and other new algorithms will soon place these techniques as one of the most useful tools in metabolomic studies. However, the lack of formal education in both molecular spectroscopy and chemometrics are still a barrier for the widespread acceptance of this technology as a tool in metabolomic studies.

References

1. Zhang, A., H. Sun, Z. Wang, W. Sun, P. Wang, and X. Wang. 2010. Metabolomics: Towards understanding traditional Chinese medicine. *Planta Medica* 76:2026–2035.
2. Sumner, L.W., P. Mendes, and R.A. Dixon. 2003. Plant metabolomics: Large-scale phytochemistry in the functional genomics era. *Phytochemistry* 62:817–836.
3. Brown, S.C., G. Kruppa, and J.L. Dasseux. 2005. Metabolomics applications of FT-ICR mass spectrometry. *Mass Spectrometry Reviews* 24:223–231.
4. Nicholson, J.K., and I.D. Wilson. 2003. Understanding 'global' systems biology: Metabonomics and the continuum of metabolism. *Nature Reviews Drug Discovery* 2:668–676.
5. Fiehn, O. 2002. Metabolomics—the link between genotypes and phenotypes. *Plant Molecular Biology* 48:155–171.
6. Fiehn, O. 2001. Combining genomics, metabolomics analysis and biochemical modelling to understand metabolomic networks. *Comparative and Functional Genomics* 2:155–168.
7. Taylor, J., R.D. King, T. Altmann, and O. Fiehn. 2002. Application of metabolomics to plant genotype discrimination using statistics and machine learning. *Bioinformatics* 18:S241–248.
8. Dunn, W.B., and D.I. Ellis. 2005. Metabolomics: Current analytical platforms and methodologies. *Trends Analytical Chemistry* 24:285–294.
9. Woodcock, T., G. Downey, and C.P. O'Donnell, 2008. Better quality food and beverages: The role of near infrared spectroscopy. *Journal of Near Infrared Spectroscopy* 16:1–29.
10. Karoui, R., G. Downey, and Ch. Blecker. 2010. Mid-infrared spectroscopy coupled with chemometrics: A tool for the analysis of intact food systems and the exploration of their molecular structure–quality relationships—A review. *Chemical Reviews* 110:6144–6168.
11. Subramanian, A., and L. Rodrigez-Saona. 2009. Fourier transform infrared (FTIR) spectroscopy. In *Infrared Spectroscopy for Food Quality Analysis and Control*, edited by Da Wen Sun, 146–174. Oxford, UK: Academic Press.
12. Li-Chan, E.C.Y. 2010. Introduction to vibrational spectroscopy. In *Applications of Vibrational Spectroscopy in Food Science*, edited by Li-Chan, E., P.R. Griffiths, and J.M. Chalmers. Wiley and Sons.
13. McClure, W.F. 2003. 204 years of near infrared technology: 1800–2003. *Journal of Near Infrared Spectroscopy* 11:487–518.
14. Cozzolino, D. 2009. Near infrared spectroscopy in natural products analysis. *Planta Medica* 75:746–757.
15. Smyth, H.E., and D. Cozzolino. 2011. Applications of infrared spectroscopy for quantitative analysis of volatile and secondary metabolites in plant materials. *Current Bioactive Compounds* 7:66–74.
16. Nicolai, B.M., K. Beullens, E. Bobelyn, A. Peirs, W. Saeys, K.I. Theron et al. 2007. Non-destructive measurement of fruit and vegetable quality by means of NIR spectroscopy: A review. *Postharvest Biology and Technology* 46:99–118.
17. Ellis, D.I., and R. Goodacre. 2002. Rapid and quantitative detection of the microbial spoilage of muscle foods: Current status and future trends. *Trends in Food Science & Technology* 12:413–423.
18. Baranska, M., and H. Schultz. 2006. Application of infrared and Raman spectroscopy for analysis of selected medicinal and spice plants. *Journal of Medicinal & Spice Plants (Zeitschrift für Arznei- & Gewürzpflanzen)* 2:72–80.

19. Baranska, M., H. Schulz, P. Rösch, M.A. Strehle, and J. Popp. 2004. Identification of secondary metabolites in medicinal and spice plants by NIR-FT-Raman microspectroscopic mapping. *The Analyst* 129:926–930.
20. Quilitzsch, R., M. Baranska, H. Schulz, and E. Hoberg. 2005. Fast determination of carrot quality by spectroscopy methods in the UV-VIS, NIR and IR range, *Journal of Applied Botany and Food Quality* 79:163–167.
21. Roessner, U., A. Luedemann, D. Brust, O. Fiehn, T. Linke, L. Willmitzer et al. 2001. Metabolic profiling allows comprehensive phenotyping of genetically or environmentally modified plant systems, *The Plant Cell* 13:11–29.
22. Schulz, H., M. Baranska, and R. Baranski. 2005. Potential of NIR-FT-Raman spectroscopy in natural carotenoid analysis. *Biopolymers* 77:212–221.
23. Wold, S. 1995. Chemometrics; What do we mean with it, and what do we want from it? *Chemometrics and Intelligent Laboratory Systems* 30:109–115.
24. Naes, T., T. Isaksson, T. Fearn, and T. Davies. 2002. *A User-friendly Guide to Multivariate Calibration and Classification*. Chichester, UK: NIR Publications.
25. Otto, M. 1999. Chemometrics: Statistics and computer application in analytical chemistry. Chichester, UK: Wiley-VCH.
26. Brereton, R.G. 2003. *Chemometrics: Data Analysis for the Laboratory and Chemical Plant*. Chichester, UK: John Wiley & Sons, Ltd.
27. Brereton, R.G. 2007. *Applied Chemometrics for Scientists*. Chichester, UK: John Wiley & Sons, Ltd.
28. Halasz, A., A. Hassan, A. Toth, and M. Varadi. 1997. NIR techniques in yeast identification. *Zeitschrift für Lebensmittel-Untersuchung und -Forschung* 204:72–74.
29. Rodriguez-Saona, L.E., F.M. Khambaty, F.S. Fry, and E.M. Calvey. 2001. Rapid detection and identification of bacterial strains by Fourier transform near infrared spectroscopy. *Journal of Agricultural and Food Chemistry* 49:574–579.
30. Ryan, D., and K. Robach. 2006. Metabolomics: The greatest omics of them all? *Analytical Chemistry* 78:7954–7958.
31. Gottlieb, D.M., J. Schultz, S.W. Bruun, S. Jacobsen, and Ib. Søndergaard. 2004. Multivariate approaches in plant science. *Phytochemistry* 65:1531–1548.
32. Gidman, E., R. Goodacre, B. Emmett, A.R. Smith, and D. Gwynn-Jones. 2003. Investigating plant–plant interference by metabolic fingerprinting. *Phytochemistry* 63:705–710.
33. Johnson, H.E., D. Broadhurst, R. Goodacre, and A.R. Smith. 2003. Metabolic fingerprinting of salt-stressed tomatoes. *Phytochemistry* 62:919–928.
34. Montes, J.M., A.E. Melchinger, and J.C. Reif. 2007. Novel throughput phenotyping platforms in plant genetic studies. *Trends in Plant Science* 12:433–436.
35. Peleman, J.D., and J.R. van der Voort. 2003. Breeding by design. *Trends in Plant Science* 8:330–334.
36. Welle, R., W. Greten, T. Müller, G. Weber, and H. Wehrmanna. 2005. Application of near infrared spectroscopy on combine in corn grain breeding. *Journal of Near Infrared Spectroscopy* 13:69–75.
37. Montes, J.M., H.F. Utz, W. Schipprack, B. Kusterer, J. Muminovic, C. Paul et al. 2006. Near-infrared spectroscopy on combine harvesters to measure maize grain dry matter content and quality parameters. *Plant Breeding* 125:591–595.
38. Munck, L., J. Pram Nielsen, B. Møller, S. Jacobsen, I. Søndergaard, S.B. Engelsen et al. 2001. Exploring the phenotypic expression of a regulatory proteome-altering gene by spectroscopy and chemometrics. *Analytica Chimica Acta* 446:171–186.
39. Munck, L., B. Møller, S. Jacobsen, and S. Søndergaard. 2004. Near infrared spectra indicate specific mutant endosperm genes and reveal a new mechanism for substituting starch with (1/3, 1/4)-b-glucan in barley. *Journal of Cereal Science* 40:213–222.
40. Munck, L. 2007. A new holistic exploratory approach to systems biology by near infrared spectroscopy evaluated by chemometrics and data inspection. *Journal of Chemometrics* 21:406–426.
41. Munck, L., and B. Møller. 2005. Principal component analysis of near infrared spectra as a tool of endosperm mutant characterisation and in barley breeding for quality. *Czech Journal of Genetics and Plant Breeding* 41:89–95.

42. Munck, L. 2006. Conceptual validation of self-organisation studied by spectroscopy in an endosperm gene model as a data driven logistic strategy in chemometrics. *Chemometrics and Intelligent Laboratory Systems* 84:26–32.

43. Campbell, M.R., J. Sykes, and D.V. Glover. 2000. Classification of single and double-mutant corn endosperm genotypes by near-infrared transmittance spectroscopy. *Cereal Chemistry* 77:774–778.

44. Rossel, S.A., C.L. Hardy, C.R. Hurburgh, and G.R. Rippke. 2001. Application of near-infrared diffuse reflectance spectroscopy to the detection and identification of transgenic corn. *Applied Spectroscopy* 55:1425–1432.

45. Rui, Y., Y. Luo, K. Haung, W. Wang, and L. Zhang. 2005. Discrimination of transgenic corns using NIR diffuse reflectance spectroscopy and back propagation (BP). *Spectroscopy and Spectral Analysis* 25:1581–1592.

46. Alishahi, H., H. Farahmand, N. Prieto, and D. Cozzolino. 2010. Identification of transgenic foods using NIR spectroscopy: A review. *Spectrochimica Acta Part A: Molecular and Biomolecular Spectroscopy* 75:1–7.

47. Suzuki, M., M. Kusano, H. Takahashi, Y. Nakamura, N. Hayashi, M. Kobayashi, et al. 2010. Rice-Arabidopsis FOX line screening with FT-NIR based fingerprinting for GC-TOF/MS based metabolomic profiling. *Metabolomics* 6:137–145.

48. Cozzolino, D., L. Flood, J. Bellon, M. Gishen, and M. De Barros Lopes. 2007. Combining near infrared spectroscopy and multivariate analysis: A tool to differentiate different strains of *Saccharomyces cerevisiae*: a metabolomic study. *Yeast* 23:1089–1096.

49. Sweetlove, L.J., R.L. Last, and A.R. Fernie. 2004. Predictive metabolic engineering: A goal for systems biology. *Plant Physiology* 132:420–425.

50. Ellis, D.I., and R. Goodacre. 2006. Metabolic fingerprinting in disease diagnosis: Biomedical applications of infrared and Raman spectroscopy. *The Analyst* 131:875–885.

51. Goodacre, R. 2003. Explanatory analysis of spectroscopic data using machine learning of simple, interpretable rules. *Vibrational Spectroscopy* 32:33–45.

52. Goodacre, R., E.M. Timmins, P.J. Rooney, J.J. Rowland, and D.B. Kell. 1996. Rapid identification of *Streptococcus* and *Enterococcus* species using diffuse reflectance spectroscopy and artificial neural networks. *FEMS Microbiology Letters* 140:233–239.

53. Goodacre, R., S. Vaidyanathan, W.B. Dunn, G.G. Harrigan, and D.B. Kell. 2004. Metabolomics by numbers: Acquiring and understanding global metabolite data. *Trends in Biotechnology* 22:245–252.

54. Al-Qadiri, H.M., N.I. Al-Alami, M.A. Al-Holy, and B.A. Rasco. 2008. Using Fourier transform infrared (FT-IR) absorbance spectroscopy and multivariate analysis to study the effect of chlorine-induced bacterial injury in water. *Journal of Agricultural and Food Chemistry* 56:8992–8997.

55. Burgula, Y., D. Khali, S. Kim, S.S. Krishnan, M.A. Cousin, and J.P. Gore. 2007. Review of mid-infrared Fourier transform–infrared spectroscopy applications for bacterial detection. *Journal of Rapid Methods and Automation in Microbiology* 15:146–175.

56. Irudayaraj, J., H. Yang, and S. Sakhamuri. 2002. Differentiation and detection of microorganism using Fourier transform infrared photoacoustic spectroscopy. *Journal of Molecular Structure* 606:181–188.

57. Naumann, D., D. Helm, and H. Labischinski. 1991. Microbiological characterizations by FT-IR spectroscopy. *Nature* 351:81–82.

58. Beekes, M., P. Lasch, and D. Naumann. 2007. Analytical applications of Fourier transform-infrared (FT-IR) spectroscopy in microbiology and prion research. *Veterinary Microbiology* 123:305–319.

59. Cozzolino, D., J. McCarthy, and E. Bartowsky. 2012. Comparison of near infrared and mid infrared spectroscopy to discriminate between wines produced by different *Oenococcus oeni* strains after malolactic fermentation: A feasibility study. *Food Control* 26:81–87.

60. Workman, J., and L. Weyer. 2008. *Practical Guide to Interpretive Near-Infrared Spectroscopy*. Boca Raton: CRC Press/Taylor & Francis Group.

61. McGoverin, C.M., J. Weeranantanaphan, G. Downey, and M. Manley. 2010. The application of near infrared spectroscopy to the measurement of bioactive compounds in food commodities. *Journal of Near Infrared Spectroscopy* 18:87–111.

62. Nielsen, J., and S. Oliver. 2005. The next wave in metabolome analysis. *Trends in Biotechnology* 23:544–546.

63. Hounsome, N., B. Hounsome, D. Tomos, and G. Edward-Jones. 2008. Plant metabolites and nutritional quality of vegetables. *Journal of Food Science* 73:R48–R65.

64. Martson, A., and K. Hostettmann. 2009. Natural product analyses over the last decades. *Planta Medica* 75:672–683.

65. Hines, A., G.S. Oladiran, J.P. Bignell, G.D. Stentiford, and M.R. Viant. 2007. Direct sampling of organisms from the field and knowledge of their phenotype: Key recommendations for environmental metabolomics. *Environmental Science & Technology* 41:3375–3381.

66. Holm-Nielsen, J.B., H. Andree, H. Lindorfer, and K.H. Esbensen. 2007. Transflexive embedded near infrared monitoring for key process intermediates in anaerobic digestion/biogas production. *Journal of Near Infrared Spectroscopy* 15:125–135.

67. Zhang, Y., Z. Zhang, N. Sugiura, and T. Maekawa. 2002. Monitoring of methanogen density using near-infrared spectroscopy. *Biomass and Bioenergy* 22:489–495.

68. Hashimoto, A., and T. Kameoka. 2008. Applications of infrared spectroscopy to biochemical, food, and agricultural processes. *Applied Spectroscopy Review* 43:416–451.

69. Kim, M.K., and K.H. Jeune. 2009. Use of FT-IR to identify enhanced biomass production and biochemical pool shifts in the marine microalgae, Chlorella ovalis, cultured in media composed of different ratios of deep seawater and fermented animal wastewater. *Journal of Microbiology and Biotechnology* 19(10):1206–1212.

70. Du, Z.H., Y.Q. Zhai, J.Y. Li, and B. Hu. 2009. Techniques of on-line monitoring volatile organic compounds in ambient air with optical spectroscopy. *Spectroscopy and Spectral Analysis* 29:3199–3203.

71. Sulub, Y., and G.W. Small. 2008. Simulated radiance profiles for automating the interpretation of airborne passive multi-spectral infrared images. *Applied Spectroscopy* 62:1049–1059.

72. Wan, B.Y., and G.W. Small. 2008. Airborne passive Fourier transform infrared remote sensing of methanol vapor from industrial emissions. *The Analyst* 133:1776–1784.

73. Du, Z.-H., R.-B. Qi, H.-M. Zhang, X. Yin, and K.-X. Xu. 2008. Quantitative detection of propane and isobutene based on NIR spectroscopy. *Journal of Tianjin University* 589–592.

74. Qi, R.-B., X. Yin, L. Yang, Z.-H. Du, J. Liu, and K.-X. Xu. 2008. Application of NIR spectroscopy to multiple gas components, identification. *Spectroscopy and Spectral Analysis* 2008:2855–2858.

75. Chafik, T., H. Zaitan, S. Harti, A. Darir, and O. Achak. 2007. Determination of the heat of adsorption and desorption of a volatile organic compound under dynamic conditions using Fourier-transform infrared spectroscopy. *Spectroscopy Letters* 40:763–775.

76. Hu, L.P., Y. Li, L. Zhang, L.M. Zhang, and J.D. Wang. 2006. Advanced development of remote sensing FTIR in air environment monitoring. *Spectroscopy and Spectral Analysis* 26:1863–1867.

77. Pons, M.N., S. Le Bonte, and O. Potier. 2004. Spectral analysis and fingerprinting for biomedia characterisation. *Journal of Biotechnology* 113:211–230.

78. Kraft, M., M. Jakusch, M. Karlowatz, A. Katzir, and B. Mizaikoff. 2003. New frontiers for mid-infrared sensors: Towards deep sea monitoring with a submarine FT-IR sensor system. *Applied Spectroscopy* 57:591–599.

23

Metagenomics: Techniques, Applications, and Challenges

Jyoti Vakhlu and Puja Gupta

CONTENTS

23.1 Introduction

Metagenomics combines the power of genomics, bioinformatics, and systems biology. It involves the simultaneous study of collective genomes of many microorganisms and is also advantageous for accessing microbial diversity. Only 1% of the microbial diversity can be accessed by traditional cultivation techniques because 99% of microbes resist cultivation in the laboratory, and thus, their potential also remains unexplored. The term *metagenomics* was coined by Jo Handelsman [1] in 1998 and it implies direct cloning of environmental DNA into large clone libraries to facilitate the analysis of genes and the sequences within the libraries. The discipline of metagenomics was born when Torsvik and Goksoyr [2] and Pace et al. [3] independently gave the idea that the genomes of a microorganism can be characterized without cultivating them. Initially, metagenomics was used to study the diversity of uncultivable microbiota, but with the advent of efficient cloning vectors like bacterial artificial chromosomes, cosmids, and many advanced screening technologies, it is now possible to express large fragments of DNA, and screen large clone libraries. Many novel antibiotics, enzymes, and drugs have been explored through functional screening. Functional screening has also been supplemented with homology-based screening and resulted in the cloning of genes like polyketide synthases [4]. High-throughput, next-generation sequencing methods have been introduced, increasing the number and size of metagenomic projects and microbial diversity accessed thereby. In this chapter, an effort has been made to discuss the various techniques used in metagenomics and thus the benefits of this study. The field is in its infancy (20 years), and because of some initial hiccups, the full potential genetic and biochemical diversity of yet-to-be-cultivated microbes has not yet been harvested. An attempt has been also done to throw some light on these bottlenecks.

23.1.1 Habitat Selection

Habitat is the place or an environment where an organism normally lives and grows [5]. Habitat selection and sampling from that particular environment is the most important step of any microbiological study and the same is true of metagenomic research. Habitat selection for metagenomic analysis will depend on the question being addressed and sampling strategies depend on the type of microbial community, its environment, the population density, and the presence of contaminating substances. It is very important that the collected sample must be representative of the habitat to account for heterogeneity. Therefore, multiple samples are taken, pooled, and then homogenized to produce a final sample, called the composite sample, which is thought to be representative of the entire mixed sample. The conclusions drawn will also depend on the size, scale, number, and timing of sampling [6,7].

Complex or unusual habitats are usually selected to find the existence of life, microbial diversity, and novel genes and genomes. As previously mentioned, habitat and sampling varies with the aim; for instance, if the aim is to look for microbial diversity of nitrogen-fixation genes in various microbes, the habitat will be selected on the basis of utilization of nitrogen, as was the case in metagenomic analysis of acid mine drainage biofilm [8]. Likewise, the human gut was selected as a habitat to study gut microbial diversity, and it was discovered that their contribution to our metabolism is more than that of own cells [9]. In addition, comparative metagenomics is also done to correlate between different habitats and the microbial communities associated therein. Comparative metagenomics is done to discover the effect of environment on microbial diversity and also on the diversity of genes [10–12].

Microbial habitats, in general, can be divided into two types: biotic and abiotic habitats (Figure 23.1). Biotic habitats can further be divided into animal, insect, and plant bodies. The animal body, as a habitat, includes the gut [13], skin [14], oral cavity [15], vaginal tract [16], and (in cattle) rumen [17]. Insects' gut can serve as a habitat for many symbiotic microbes [18]. The plant body, as a habitat, includes the endosphere [19], rhizosphere [20], and phyllosphere [21]. In the past, microbes were associated with human disease, but as knowledge about their beneficial role came to light, we were in for huge surprises.

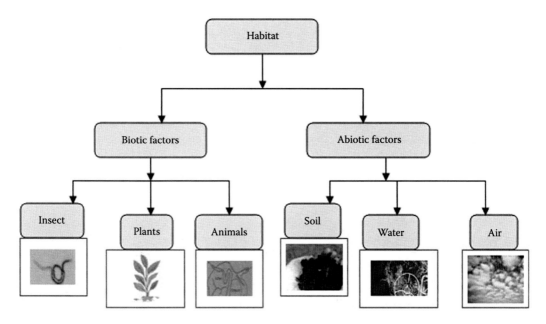

FIGURE 23.1
(See color insert.) Habitats for metagenomic analysis.

A metagenomic study of the human body to unravel the microflora, as well as their contributions, was one of the initial metagenomic projects. The first effort to study the human gut metagenome was undertaken in 2005 by members of the Relman laboratory at the Department of Medicine, Stanford University and at the Institute for Genomic Research [22]. A majority of the bacterial sequences corresponded to uncultivated species and novel microorganisms. Knowledge about this immense diversity would be very beneficial in understanding the role of these bacteria in gut metabolism.

The plant body, as a biotic habitat for microbes, is divided into three main categories for metagenomic studies: rhizosphere, endosphere, phyllosphere. Among all three categories, a huge amount of metagenomic work has been done on rhizospheric soil samples [23,24].

Among insects, termites are widely studied owing to their economic importance as wood-degrading organisms with essential environmental roles in the turnover of carbon. In addition, they are also prospective sources of biochemical catalysts. The hindgut and midgut of the wood-feeding "higher" termite and the gypsy moth (*Lymantrisexia dispar*) have been selected as habitats for isolating microbes/genes for cellulose degradation [18,25].

Abiotic habitats include the soil, water, and air. Soil has been of great interest to microbiologists and soil metagenome, as expected, tops the priority list of metagenomists. The microbial diversity in soils exceeds that of other environments and is far greater than that of eukaryotic organisms. It has been already reported that 1 g of soil can contain up to 10 billion microorganisms of possibly thousands of different species [26,27]. Different components of soil (sand, silt, clay, and organic matter) provide a wide variety of microbial niches and thus lead to high microbial diversity. These microorganisms are localized in close association with soil particles, such as complexes of clay–organic matter [17,28].

Various water bodies have been selected as habitats to study microbial load and diversity, and using metagenomic analysis, it was found that water harbors a greater diversity and load of microbes than that thought of earlier. These conclusions were drawn mainly by researchers from the J. Craig Venter Institute [29]. They reported the discovery of millions of new genes, thousands of new protein families, and specifically, the characterization of thousands of new protein kinases from ocean microbes by using whole environment shot-gun sequencing and new computational tools from the Sorcerer II Global Ocean Sampling Expedition.

Air is another source of metagenomes as airborne microbes are often attached to dust particles or water droplets from sneezes and coughs. On evaporation of water in aerosols, the microbes become droplet nuclei and clumps and thus can stay airborne and drift with air flows [30].

23.2 Metagenomic Techniques

23.2.1 Isolation and Purification of Metagenomic DNA

Metagenomic analysis is initiated by the isolation of DNA from environmental samples. There are two ways of extracting DNA: (1) direct, *in situ*, extraction in which the cells are lysed within the matrix/sample and then the DNA is recovered; and (2) indirect extraction techniques, in which the cells are first recovered from the sample and then lysed for DNA recovery [31–34]. In both techniques, the common step is to break open the cell. The proce-dure to break open the cell is either to shear it mechanically or chemically. Both methods have their own advantages and limitations. Chemical treatment of the sample is a gentle method and results in the recovery of higher molecular weight DNA, but chemical lysis can select certain species only by exploiting their biochemical characteristics. Mechanical shearing, on the other hand, does not show such bias and is known to recover nucleic acid from more diverse cells as compared with the chemical method. However, the quality of the DNA is not so good. Therefore, microbial (metagenomic) community DNA isolation is a compromise between vigorous extractions required for the representation of all micro-bial genomes and minimization of the DNA shearing.

Environmental genomic DNA often contains contaminants such as polyphenols, humic acids, and polysaccharides that reduce template purity and are usually difficult to remove. So, purification, in the case of the metagenomic DNA isolation, is a more complicated and important step because the source of contaminants is not only biological (as in case of culture-based isolation) but also abiotic. Apart from the traditional methods of purifi-cation by using phenol-chloroform, ethanol precipitation or a cesium chloride gradient, various modified methods are used that include using PVPP before lysis, treatment with formamide within agarose plug [35], SephadexG-200 column purification after extraction [36], Q-Sepharose ion exchange chromatography [37], and silica-based methods [38,39]. Multiple displacement amplification is yet another technique that is used to remove the interfering contaminants and at the same time to increase the concentration of the meta-genomic DNA [40,41].

23.2.2 Harvesting the Metagenome and Its Analysis

Once the source of the metagenome is identified and the DNA isolated and purified, the next step is the analysis of the metagenome. Metagenome can be analyzed by either cloning the isolated metagenome or by direct analysis based on mass sequencing or microarray.

23.2.2.1 Cloning-Dependent Approach

This approach involves the cloning of DNA/complementary DNA (cDNA) fragments into suitable vectors that are further analyzed. Libraries constructed using DNA and cDNA are known as metagenomic and metatranscriptomic libraries, respectively. These libraries can be screened for phylogenetic anchors to assess microbial diversity, or function-based screening is done to isolate novel genes. In addition to genomic libraries, an amplicon library is made wherein specific primers are used to amplify either a specific gene or specific phylogenetic marker. The advantage of this technique is that the search is narrowed down, but at the same time, sequence-based bias will be more prevalent and novel genes and genomes may escape detection.

23.2.2.1.1 Metagenomic Library-Based Metagenome Analysis

In the last 10 years, quite a few genetic and biochemical potential of yet-to-be-cultivated microbes have been assessed by direct construction and analysis of metagenomic libraries [42–46]. Whole metagenome shotgun sequencing approaches have been mostly employed for the cloning and sequencing of microbial DNA from various environments. This involves the generation of small-insert DNA clone libraries, and their subsequent analysis using Sanger dideoxy sequencing methods, which generate sequences that can then be used to query the known databases for function or phylogenetic relationship. This approach can give read lengths ranging from 600 to 900 bp, which can be extended through the entire fosmid clone [46–48].

Another example of metagenome analysis is that of the microbial community associated with the biofilm of acid mine drainage [8]. Metagenomic DNA was sheared into small fragments to construct a small insert plasmid library (average insert size, 3.2 kb for random shotgun sequencing). A total of 76.2 million base pairs of DNA sequence were generated from 103,462 high-quality reads (averaging 737 bp per read) and the shotgun data set was assembled with JAZZ, a whole-genome shotgun assembler. This sequencing effort resulted in the construction of the complete genome of three bacteria and two archaea that dominate the acid mine drainage (AMD).

The biochemical diversity of the metagenome can be assessed by the construction of a metagenomic expression library. The choice vectors are usually cosmids or bacterial artificial chromosomes because they can accommodate large-sized DNA fragments and allow entire functional operons to express together. Operons, for several multigenic pathways, have been successfully cloned and isolated by using such high-capacity insert vectors [50]. Beja [51] reported the construction of fosmid libraries with inserts of 40 kb of foreign DNA.

Functional screening requires faithful transcription and translation of the gene(s) of interest, production of gene product, and an efficient assay to identify the product. *Escherichia coli* are still the preferred hosts for the cloning and expression of any metagenome-derived genes. Other hosts, such as *Streptomyces lividans* and *Pseudomonas putida*, have been employed to identify the genes involved in the synthesis of novel antibiotics [52]. Screening can be a simple activity-based approach in which the detection of recombinant *E. coli* clones exhibiting specific activity like protease activity on indicator agar containing skimmed milk as a protease substrate [53,54] or lipolytic activity by employing indicator agar containing tributyrin or tricaprylin as lipase substrates [55–57]. Clones with proteolytic or lipolytic activity are identified by halo formation on solidified indicator medium. Yun and Ryu [58] introduced a high-throughput screening strategy termed substrate gene expression screening, which is based on induced gene expression. It can be induced by a target substrate and display catabolic gene expression. Upon induction by the substrate, the gfp gene is coexpressed, and positive clones harboring the catabolic genes can rapidly be separated from other clones by fluorescence-activated cell sorting [45,59]. This method was applied successfully to isolate aromatic hydrocarbon-induced genes from a metagenomic library. However, this method has certain limitations. First, it is sensitive to the structure and orientation of the genes with the desired traits. Second, it misses the active clones that have transcription terminators between catabolic genes and the following gfp. Particularly for the second reason, substrate gene expression screening is not suitable for application to metagenomic libraries harboring large-insert DNA due to the abundance of transcription terminators. Williamson et al. [60] reported a similar screening strategy termed metabolite-regulated expression, in which metagenomic clones producing small molecules are identified by a biosensor that detects small diffusible signal molecules. These signals induce quorum sensing. When a threshold concentration of the signal molecule is exceeded, GFP is produced. Thus, the positive clones can rapidly be separated from other clones by fluorescent microscopy.

23.2.2.1.2 *Metatranscriptomic Library-Based Metagenome Analysis*

Metatranscriptomic libraries involve the analysis of community transcripts isolated directly from the environment or from microcosms [61]. Metatranscriptomics is advantageous for gene discovery, especially in targeted environments (specialized communities or enriched communities), because it requires a much smaller sequence space compared with metagenomics and it focuses on the expressed subset of genes [62]. It connects the taxonomic makeup of the community directly to its *in situ* activity (function) via profiling of most abundant transcripts and correlating them with specific environmental conditions.

This approach involves the extraction of total RNA from an environmental sample. However, constructing libraries from metagenomic environmental messenger RNA (mRNA) is relatively tedious due to the difficulty associated with RNA isolation, separation of mRNA from other RNA species, and instability of mRNA. To overcome the abovementioned difficulty, cDNA libraries have been constructed from mRNA that has been isolated from environmental samples [63].

The construction and analysis of cDNA libraries from diverse environments has revealed several unique sequences and the potential to uncover a high degree of novelty within microbial communities [64]. Frias-Lopez et al. [65] constructed cDNA libraries from mRNA derived from microbial metagenome of ocean surface water. These cDNA libraries were further pyrosequenced and the generated data set was compared with available databases. It was found that approximately 50% of all detected transcripts were unique, indicating that a large unknown metabolic diversity is present in the ocean.

23.2.2.1.3 *Amplicon Library*

Environmental DNA can be directly amplified by using universal primers that specifically amplify the 16S ribosomal RNA (rRNA) region. 16S rRNA sequences are relatively short, often conserved within a species, and generally different between species, thus they can be used to explore phylogenetic diversity. Metagenomic clone libraries of environmental DNA are also screened for 16S rRNA genes to identify clones that can be used to explore the metabolic potential of a particular bacterial group [66].

Phylogenetic analysis of metagenomic clones bearing 16S rRNA genes from the surface waters of the Pacific Ocean identified several groups of archaea and bacteria that are common in marine water [67]. E-DNA can also be amplified by using gene-specific primers. These have been extensively used to probe communities for microorganisms with specific metabolic or biodegradable capabilities. For example, targeting methane monooxygenase, methanol dehydrogenase, and ammonia monooxygenase genes were used to identify methanotrophic [68] and chemolithotrophic ammonium oxidizing bacteria [69,70]. The biodegradative potential of indigenous microbial populations has been assessed by screening the metagenome for catechol 2,3-dioxygenase, chlorocatechol dioxygenase, and phenol hydroxylase genes [71–73]. However, as a tool for discovering biocatalysts, gene-specific polymerase chain reactions (PCR) have two major drawbacks. First, the design of primers is dependent on existing sequence information, and therefore, it skews the search in favor of known sequences. Moreover, functionally similar genes resulting from convergent evolution are not likely to be detected by a single gene family–specific set of PCR primers. Second, only a fragment of a structural gene will be amplified by gene-specific PCR, necessitating additional steps to access the full length of genes. Conventionally, amplicons can be labeled as probes to identify the putative full-length gene from metagenomic libraries [50].

23.2.2.2 Cloning Independent Approach

Cloning independent analysis of metagenomes relies either on high-throughput sequencing technologies or hybridization techniques, for example, fluorescent *in situ* hybridization (FISH) and microarrays.

23.2.2.2.1 Environmental Shotgun Sequencing

In this technique, metagenomic DNA isolated directly from an environmental sample is broken into small fragments, and then portions of these fragments were sequenced and assembled into larger pieces by looking for overlaps in the sequence. The generated sequences were then compared with the known databases for function or phylogenetic relationship [62]. This includes high-throughput sequencing technologies that are intended to produce thousands or millions of sequences at once and also lower the cost of DNA sequencing [74,75]. Pyrosequencing, Illumina/Solexa, and SOLiD technologies are some of the latest high-throughput sequencing technologies currently available.

Pyrosequencing technique has been used to study marine samples and the studies revealed that oceans harbor several orders of magnitude more species than previously imagined [76], and are complemented by major groups that were not thought to occur in the ocean, such as single-stranded viral sequences [77]. It is the most feasible sequencing approach to apply to deep sediment because of the extremely low DNA yields from this environment [78,79]. Pyrosequencing is also used in the determination of the taxonomic diversity of various environments, including an acid mine biofilm [8], the Soudan mine [80], honey bee colonies [81], and the Peru Margin subsea floor [79]. This technique has led to the generation of the largest metagenomic data set within the framework of the Global Ocean Sampling expedition [82,83]. The Global Ocean Sampling data set is an extension of the Sargasso Sea data set [47]. The phylogenetic diversity stored in this data set comprises 7.7 million sequences (6.3 billion bp) that were assessed by analysis of the 16S rRNA gene sequences present in the metagenomic libraries [82,84]. Random insert libraries were constructed from DNA isolated from bacterioplankton derived from 41 marine surface environments and a few non–marine aquatic samples. In general, the alphaproteobacteria were the dominant phylogenetic group in ocean surface waters, whereas the abundance of other phyla differed depending on the type of environment [84]. These pyrosequencing-based platforms preceded other high-throughput platforms such as the Illumina/Solexa and SOLiD technologies.

23.2.2.2.2 Fluorescent In Situ Hybridization

This technique uses direct analysis of microbial communities through metagenomics. FISH allows the direct observation and estimation of microorganisms of specific species, genera, families, or phyla in a given environmental sample. This technique is mostly used to compare the genomes of two biological species to deduce their evolutionary relationships. FISH probes are often primers for the 16S rRNA region. This method has been used successfully to study the spatial distribution of bacteria in biofilms [85]. Biofilms are composed of complex (often) multispecies bacterial organizations. In this technique, DNA probes for one species are prepared and used to visualize the distribution of this specific species within the biofilm. Two different probes (in two different colors) corresponding to two different species allows us to visualize/study the colocalization of these two species in the biofilm, and can be useful in determining the fine architecture of the biofilm.

Pernthaler and colleagues [86] used this technique in combination with immunomagnetic cell capture to purify syntrophic anaerobic methane oxidizing ANME-2c archaea and physically associated microorganisms directly from deep-sea marine sediments. This technique was also used by Singh and his coworkers [87], along with microautoradiography for the phylogenetic identification of uncultured bacteria in natural environments using fluorescent group-specific phylogenetic probes (targeting rRNA) and fluorescence microscopy. The combination of FISH and microautoradiography has been used successfully in several studies, such as in substrate utilization by ammonia-oxidizing, inorganic phosphate–accumulating bacteria, in other members of the proteobacteria present in an activated sludge system, and in various natural environments.

FISH or hybridization of nucleic acids extracted directly from environmental samples suffers from one limitation, that is, the lack of sensitivity. The sequences are detected only if present in high copy number such as from dominant species, otherwise, the signal is hard to detect.

23.2.2.2.3 Microarray

Representing powerful high-throughput systems for the analysis of genes, microarrays are primarily based on DNA–DNA hybridization. Different types of arrays have been used for looking at environmental DNA that differ on the kind of DNA being investigated.

23.2.2.2.3.1 Phylogenetic Genome Arrays Made with oligonucleotides from rRNA genes, phylogenetic genome arrays have great potential for the detection, identification, and characterization of microorganisms in their natural habitats [88]. Loy and his group [89] constructed a microarray with 132 oligonucleotide probes consisting of 18 nucleotides targeting 16S rRNA. They used this array to determine the diversity of sulfate-reducing prokaryotes in periodontal tooth pockets and a hypersaline cyanobacterial mat. This microarray could be used to distinguish most of the reference strains as it represented all recognized groups of sulfate-reducing prokaryotes.

Microarrays of immobilized oligonucleotide gene targets have also been used to select appropriate biotope samples for screening the metagenomic library [90]. Such arrays can be used for the affinity capture of targets as a means of enrichment before the construction of the metagenomic library. This technology can be used before shotgun sequencing for the preselection of genes from metagenomic libraries, thereby reducing the sequencing burden and reducing the proportion of sequences unassigned by the database sequence similarity searches [91].

23.2.2.2.3.2 Community Genome Arrays Made with genomic DNA isolated from environmental samples or pure culture, community genome array is also called genome-typing, and can be conducted both within species and between species in various lineages to discover gene gain and loss events, and to correlate the gain/loss patterns to parameters such as gene function, expression level, involvement in protein complexes, and timing of gene duplication. These types of studies can also generate empirical estimates of gene loss rates [92–94].

Microarrays are also used in functional analysis of nitrogen and carbon cycling genes across many ecosystems to understand the forces driving important processes in nutrient cycling. Yergeau et al. [95] used microarrays to examine soil-borne microbial communities via a functional gene microarray approach across an Antarctic latitudinal transect. The oligonucleotide probes were designed to target genes involved in nitrogen, carbon, sulfur and phosphorus cycling, metal reduction, and resistance and organic contaminant degradation [96]. It was detected that N- and C-cycle genes were significantly different across different sampling locations and vegetation types. With respect to the N-cycle, denitrification genes were linked to higher soil temperatures, and the presence of N_2-fixation genes linked to sites that were mainly vegetated by lichens.

Despite the abovementioned successes, microarrays have significant challenges remaining with regard to specificity, sensitivity, and quantification of microbes in their natural habitats because microbial communities contain highly heterogeneous groups of organisms with undefined/unknown genomic relationships. The distribution of microbial species, the potential of cross-hybridization between closely related species, and the genetic variation among and within strains influence the interpretation of the data.

23.3 Metagenomics Applications

23.3.1 Bioprospecting

Metagenomics techniques have been used to identify a significant number of novel genes encoding for biocatalysts and secondary metabolites with potential use in industries.

23.3.1.1 Bioprospecting for Enzymes

Metagenomic literature is replete with information on the isolation and identification of genes encoding enzymes directly from the selected environmental metagenome. Enzyme encoding genes have been the most targeted genes as far as metagenomic retrieval of genes is concerned. Mainly because of environmental concerns, industrial processes have to become "green," and so there is a great requirement for resilient enzymes.

Metagenome-derived DNA libraries have mainly focused on many classes of enzymes, among these, lipases and esterases are prominent [97–100]. Three esterases and one lipase were identified from metagenomic libraries constructed from marine samples, that is, the deep-sea sediment, the intertidal flat sediment, and the Arctic seashore sediment [101]. Oxidoreductases are another example of metagenomic-derived enzymes useful for the synthesis of carbonyl compounds, hydroxy acids, amino acids. and chiral alcohols [102]. Several reports are also available on the isolation of novel amylolytic enzymes from metagenomic DNA libraries [99,103,104]. The list of cellulases isolated and identified by

TABLE 23.1

Products of Metagenomics

Target Gene/Enzyme	Source	Screening Method	Reference No.
Esterases	Soil metagenome	Functional	[115]
	Deep-sea sediment	Functional	[116]
	Surface seawater	Functional	[117]
	Arctic sediment	Functional	[101]
	Plant rhizosphere soil	Functional	[118]
	Cow gut	Functional	[119]
Lipases	Sediment	Functional	[120,121]
	Soil, compost	Functional	[122]
	Deep-sea sediment	Functional	[101]
	Soil	Functional	[123]
	Lake water samples	Functional	[105]
Cellulase	Sargasso Sea	Sequencing	[67]
	Rabbit cecum contents	Functional	[106]
		Sequencing	[124]
		Functional	[125]
Cellulase	Soda lake	Functional	[126]
	Soil	Functional	[108]
Oxidoreductase	Soil	Functional	[102]
Amylase	Soil	Functional	[99]
	Soil from the junction of the groundwater table	Functional	[103]
Vitamin biosynthesis	—	—	[107,109]
Turbomycin	Soil	—	[127]
Indirubin	Soil	—	[111]
β-Lactamases	Soil	Functional	[114]

metagenomic techniques is quite impressive [67,105,106]. The list of clutch enzymes isolated from metagenomes of various selected environments is shown in Table 23.1 for easy comprehension.

23.3.1.2 Bioprospecting for Secondary Metabolites and Antibiotics

Due to the rediscovery of similar antibiotics, it was believed that perhaps all the antibiotics available in nature have been isolated. This belief changed once it was realized that we are not able to isolate any new microbial compounds, specifically antibiotics, because of the fact that not more than 1% of the microbes have been cultivated thus far. Some searches have also focused on the isolation of genes involved in vitamin biosynthesis, for example, genes related to the synthesis of 2,5-diketo-D-gluconic acid, using glucose as a substrate, were isolated by a metagenomic approach because 2,5-diketo-D-gluconic acid is a precursor to vitamin C (ascorbic acid) production [107]. Similarly, biotin biosynthesis genes are linked to the construction of biotin-overproducing bacteria for large-scale fermentation of this vitamin [108]. Entcheva and his coworkers [109] reported that biotin production can be increased by avidin enrichment of various environmental samples. A range of antibodies have been detected in metagenomic libraries, namely, indirubin [110], turbomycin [111], deoxyviolacein, and the broad spectrum antibiotic, violacein [112]. Brady and

Clardy constructed and screened seven different soil metagenomic libraries from multiple geographically distinct environmental samples, for example, Ithaca, New York; Boston, Massachusetts; and Costa Rica. They found 11 clones producing long-chain N-acyltyrosine antibiotics, and further analysis revealed that 10 of them were novel [112]. These libraries have also been used for isolating natural antibiotic resistance genes, as well as the results of that search. Riesenfeld and his labmates [113] identified nine aminoglycoside and one tetracycline antibiotic resistance genes from soil metagenomes. They also identified nine clones from two previously reported libraries. These clones expressed resistance to aminoglycoside antibiotics and only one expressed tetracycline resistance. Amino acid sequencing showed that, except for one clone, all the sequences were considerably different from previously reported sequences. Allen et al. [114] reported β-lactamases in remote Alaskan soil. The Alaskan β-lactamases confer resistance on *E. coli* without manipulating its gene expression machinery, demonstrating the potential for soil resistance genes to compromise human health, if transferred to pathogens.

23.3.2 Human Health

"Despite our monumental achievements in philosophy, technology, and arts, to bacteria, humans are no more than an organic mass to be utilized for growth and reproduction" [128]. As far as humans are concerned, the so-called highly evolved organism, neither they nor the planet they survive on, can do without microbes.

Human bodies are like ecosystems with multiple ecological niches and habitats in which a variety of cellular species collaborate and compete. The human body is considered as a superorganism that incorporates multiple symbiotic cell species into a single individual with blurry boundaries. The microorganisms that live on and inside humans are estimated to outnumber human cells by a factor of 10 [129]. The National Institutes of Health initiated the "Human Microbiome Project" in this context. In this project, metagenomic techniques are used to determine the complexity of microbial diversity at five main sites, namely, the skin, oral cavity, gastrointestinal tracts, and vagina. This preliminary study will provide knowledge of the core and variegated microbiome at each site. Metagenomic techniques were also used to find if variation in the microbiome at each site could be related to human phenotype or differences between healthy and diseased states. The skin acts as the first line of defense and is surrounded by commensal and pathogenic bacteria that contribute to both human health and disease. Culture-based techniques previously used showed the presence of only *Staphylococcus aureus* and *Staphylococcus epidermis*. Grice et al. [14] analyzed the complexity and identity of the microbes inhabiting the skin by 16S ribotyping and found that *S. aureus* and *S. epidermis* represent only 5% of the microflora. The other 95% was represented by actinobacteria (51.8%), firmicutes (24.4%), proteobacteria (16.5%), and bacteroides (6.3%). This study of human skin microbiota will serve to direct future research addressing the role of skin microbiota in health and disease, and metagenomic projects addressing the complex physiological interactions between the skin and the microbes that inhabit it. Understanding such interactions between skin and microbes is also essential for new treatments for skin conditions and diseases, such as acne and atopic dermatitis (eczema). It can also be helpful in identifying compounds released or produced by such microbes that can be beneficial for human health.

Another area under investigation for the Human Microbiome Project is the oral cavity. Dental biofilms are complex and multispecies ecosystems in which oral bacteria interact cooperatively or competitively with other members [130]. More than 100 million bacteria are found in every milliliter of saliva and more than 600 different species are found in the mouth. A metagenomic approach has been used to study the microbial diversity present in the biofilm community that sticks to the teeth and gums. The National Institute of Dental and Craniofacial research, which is part of the National Institutes of Health, has sponsored a project to find and examine all the microbes found in the oral cavity of humans. Understanding the diversity and composition of microbial communities residing in the biofilm will lead to the cure of many oral diseases. It has been reported that the accumulation of dental biofilms leads to a change in bacterial composition. Dental biofilms mainly consist of Gram-positive bacteria in healthy individuals but it is dominated by an increased number of Gram-negative anaerobic rods in diseased individuals (as usually observed in periodontitis) [131].

A vast diversity and load of microbes is found in the gastrointestinal tract, where they perform various functions such as extracting nutrients and calories from the diet and processing essential vitamins, amino acids, and detoxifying potentially harmful chemicals contained in our diet. The human body contains both harmful and beneficial microbes. Understanding such microbes can help researchers learn about various disorders. Many diseases originating from microbial disorders, such as autism, which may be related to the overgrowth of neurotoxin-producing bacteria living in the intestines. Recently, metagenomics has been used to compare the genes in the microbial communities in the gut of obese mice with leaner relatives. The results suggested that obesity is associated with a shift in the proportions of the two major groups of gut bacteria. Researchers transplanted gut microbial communities harvested from both obese and lean animals into lean germ-free mice and the result was that the lean mice that received microbial communities transplanted from obese mice gained more fat. Therefore, the calorific value of the food we eat may vary depending on the composition of the person's gut microbes [129].

The vaginal microbiota affects the health of women as well as the success of pregnancy. The vagina hosts a unique consortia of microbes suggesting selection for these key organisms. Zhou and coworkers [132] first characterized the vaginal microbiota by 16S clone library sequencing and found that *Lactobacilli* and *Atopobium* were the predominant organisms. They also reported the first identification of a *Megasphaera* species in the vagina. Sundquist and coworkers [133] studied the human vagina during pregnancy by a full 16S pyrosequencing strategy and identified *Lactobacillus* as the dominant genus as well. There was a significant presence of other genera such as *Psychrobacter*, *Magnetobacterium*, *Prevotella*, *Bifidobacterium*, and *Veillonella*. However, *Lactobacillus* can be missing in healthy vaginal microbiota and replaced by other predominant genera such as *Gardnerella*, *Pseudomonas*, or *Streptococcus* [134].

23.3.3 Plant Health

Plants play a central role in providing nutrient input into the soil, both through microbial-mediated decomposition of plant matter and through direct provision of photosynthate-derived root exudates. These nutrients support large and diverse microbial communities. These microbes are inhabitants of the rhizosphere (directly surrounding the root) and the phyllosphere (on the surface of above-ground leaves and shoots). Interactions between plants and microbes involve complex chemical signaling, many of which provide direct benefit to the plant.

The rhizosphere is an area where there is close association between plant roots and adhered soil. This association affects the biological, chemical, and physical characteristics of soil. Microbial populations associated with the rhizosphere produce biologically active substances that affect plant growth, for example, biotin and pantothenic acid. *Agrobacterium tumefaciens* releases agrobactin, which increases iron uptake by the root. Plant roots support the growth and activities of a wide variety of microorganisms that may have a profound effect on the growth and health of plants. These beneficial bacteria are termed plant growth–promoting rhizobacteria. Metagenomics have made major contributions toward plant growth–promoting gene, gene products, and characterization of yet-to-be-cultured plant growth–promoting rhizobacteria [135]. The combination of shotgun sequencing and enrichment strategy allows the metagenomic analysis of rhizobacteria with a particular function of interest. Erkel and colleagues [136] used metagenomic techniques to reconstruct the 3.18 Mbp genome of rice cluster I (RC-I) archaea originating from the rice rhizosphere. The aerial habitat colonized by microbes is termed the phyllosphere, and the inhabitants are called epiphytes. Culture-independent molecular techniques were utilized for exploring bacterial diversity and selection of putative UV exposure marker sequences of bacterial assemblages on *Zea mays* (maize) leaves. Few sequences corresponded with previously characterized phyllosphere bacteria. Denaturing gel gradient electrophoresis was done to observe overall diversity. A significant difference in community structure was found between the bacterial samples from control and solar UV-B exposed plants [137]. A cultivation-independent technique has also been used to see the difference in microbial diversity of unaged and aging flue-cured tobacco K326. Flue-cured tobacco leaves (FCTL) contain abundant bacteria, and these bacteria play very important roles in the tobacco aging process. A comparison of the number of operational taxonomic units between the cloned libraries from the unaged and aging FCTL were made by restriction fragment length polymorphism analysis. Twenty-three bacterial species were identified from the unaged FCTL only, whereas 15 species were identified from the aging FCTL only. Interestingly, more uncultured bacteria species were found in aging FCTL than in the unaged FCTL [138].

23.3.4 Planet Health

Microbes have dominated the history of the Earth, being its first inhabitants, playing a major role in shaping its chemical and physical properties. These tiny living structures are the most widely distributed organisms on the planet, accounting for half of the world's biomass. Microbes play a major role in maintaining the habitability of the entire planet by regulating various biogeochemical cycles like the carbon cycle, oxygen cycle, nitrogen cycle, sulfur cycle, and cycling of metals [139].

A great diversity of such microbes is harbored in the oceans. Different functional groups of bacteria, archaea, and protists arise from this diversity and drive globally important biogeochemical cycles. The collective metabolism of marine microbial communities has global effects on the fluxes of energy and matter in the sea, on the composition of the Earth's atmosphere, and on global climate. Falkowski and colleagues [140], in a review of the global biogeochemical cycles that keep our planet habitable, wrote "In essence, microbes can be viewed as vessels that ferry metabolic machines through strong environmental perturbations into vast stretches of relatively mundane geological landscapes. The individual taxonomic units evolve and go extinct, yet the core machines survive surprisingly unperturbed." Metagenomics enable us to know the identity, abundance, and physiology of such marine microbes that carry out carbon fixation or degradation processes.

Metagenomic sampling of microbial communities in the open ocean has revealed a surprising new pathway for energy conservation by marine heterotrophs. Proteorhodopsin, a protein functioning as a light-driven proton pump in cell membranes, has been detected in a wide range of ocean habitats. Previously, it was thought that proteorhodopsin exist only in archaeal extremophiles living in salt ponds. Genes encoding these proteins are ubiquitous in marine bacterioplankton, such as the SAR cluster, which was originally isolated from samples taken in the Sargasso Sea (more than 782 rhodopsin-like photoreceptors were identified from those samples) [47]. The common occurrence of bacterioplankton harboring this protein in surface waters worldwide suggests a potential mechanism for widespread mixotrophic energy conservation in marine environments. Mixotrophy is a form of growth in which two methods of energy generation are used simultaneously. In this form, bacteria would augment energy derived from the consumption of organic substrates and conserve carbon resources by creating an additional proton gradient using light energy to drive synthesis of ATP, a multifunctional nucleotide responsible for cellular energy transfer and storage [44].

23.3.5 Fossil Discovery and Analyses by Metagenomics

It is very difficult to gather information from fossil remains due to the poor preservation of their DNA and the limited ability to retrieve nuclear DNA. Most of the extracted DNA is either a mixture of bacterial, fungal, or often human contaminants, complicating the isolation of endogenous DNA. Moreover, the isolated DNA is truncated into fragments of very short length, that is, less than 300 bp due to hydrolysis of the DNA backbone, and oxidation of pyrimidines [141–143]. The truncated fragment hinders PCR by preventing the extension by Taq DNA polymerase. Previously, these problems were overcome indirectly by concentrating on the small number of genes present on the maternally inherited mitochondrial genome, which is present in high copy number in animal cells. But this approach gives incomplete access to the storehouses of genetic information.

Metagenomics is used for the analysis of extinct organisms because it provides a method for examining the nuclear genomes of ancient organisms. This approach has been applied to cave bears. The Cave Bear Project was initiated in 2005. In this project, DNA was extracted from the two Austrian cave bear bones that were more than 40,000 years old and cloned into a metagenomic library, consisting of bacterial DNA, fungal DNA, human DNA, and some cave bear DNA. The cloned fragments were then directly sequenced and comparison of each sequence was done with the genome of the dog, a modern relative of the bear, to make sure that each portion of DNA was really from the bears rather than a contaminating source. Nearly 6% of the sequences analyzed from one of their animal samples belonged to ancient bears. The rest of the DNA probably came from soil microbes or the paleontologists handling the bones [144]. The PCR step was skipped as it is very important in limiting contamination (because modern DNA amplifies much more readily than ancient DNA), and because the cave bear became extinct tens of thousands of years ago. The results provided insights into the evolutionary relationship of the cave bear to its modern relatives, the polar, brown, and black bears [142].

The Neanderthal Project (2006) was initiated to sequence the genome of the closest hominid relative to humans whose remains are roughly as old as those of the cave bear. In 2010, the results of this project were published in the journal *Science* [145], and it determined that some mixture of genes occurred between Neanderthals and anatomically modern humans and also presented evidence that elements of their genome remain in that of non-African modern humans. On the other hand, the Mammoth Project (2006) was initiated to sequence the genome of mammoths. A mammoth is any species of the extinct genus *Mammuthus* and is a close relative of modern elephants. They lived from the Pliocene epoch from approximately 4.8 million to 4500 years ago. Deciphering the woolly mammoth genome might help solve the mystery of what killed off these animals as the last ice age waned approximately 12,000 years ago, and provide hints about the demise of their fellow megafauna [142,146].

23.3.6 Forensic Sciences

Microorganisms have been used as weapons by bioterrorists in criminal acts [147]. The solution to such bioterrorism events would rely heavily on forensic science. With the help of forensic science, many such events could be readily perpetrated without the availability of eyewitnesses. Microbial forensics is a scientific discipline dedicated to analyzing evidence from a bioterrorist attack, biocrime, or inadvertent microorganism/toxin release for attribution purposes (who was responsible for the crime) [148]. Since the anthrax letter attacks of 2001, when letters containing anthrax spores were mailed to several news media offices and two Democratic senators in the United States, killing five people and infecting 17 others, funding agencies in the United States and other countries have prioritized research projects on organisms that might potentially challenge our security and economy. The *Bacillus anthracis* spores found in the mailed envelopes were related to the Ames strain, commonly used in research in more than 20 laboratories. Since the Ames strain was created, unique point mutations arose separately in distinct populations grown in separate laboratories. Many of the spores produced by *B. anthracis* harbored mutations, posing a great difficulty in finding the origin of mutations. Elucidating the origins of this anthrax strain required a large amount of genomewide sequencing and analyses to generate sufficient data. Metagenomics techniques were employed to obtain the diversity of mutations within the envelope samples.

A number of criminal investigations have involved forensic analyses of HIV. A forensic investigation in Florida looking into the source and transmission of HIV led to the discovery that the dreaded virus originated from a dentist who passed it on to his patients. The case was solved by sequencing the amplified viral genes (obtained by PCR) of both the doctor and the patients [149].

Detection of novel human viruses is greatly needed as physicians and epidemiologists increasingly deal with infectious diseases caused by new or previously unrecognized pathogens. Viruses are suspected to play a role in many clinical syndromes. In addition, new viruses continue to challenge the human population. Metagenomics-based tools, such as microarrays and high-throughput sequencing, are ideal for responding to these challenges. Pan-viral microarrays, containing representative sequences from all known viruses, have been used to detect novel and distantly related variants of known viruses [150]. Sequencing-based methods have also been successfully employed to detect novel viruses and have the potential to detect the full spectrum of viruses, including those present in low numbers.

23.3.7 Microbial Diversity

Microbial diversity is a general term that includes genetic diversity and ecological diversity. Genetic diversity is the distribution of genetic information within microbial species, whereas ecological diversity includes variation in community structure, complexity of interactions, and number of trophic levels. Knowing the extent of diversity in a particular habitat will aid in estimating the sample sizes required to draw conclusions and may determine the choice of habitat for particular types of study. For example, searches for antibiotics might focus on environments that contain a high diversity of actinobacteria, the phylum that has yielded the most antibiotic-producing cultured organisms.

The most common approach to exploring microbial diversity is usually by analysis of conserved ribosomal RNA gene sequences [151]. Traditional surveys of environmental prokaryotic communities are based on the amplification and cloning of 16S rRNA gene before sequence analysis. Many reference databases have been generated due to sequencing of ribosomal gene such as ribosomal data base project II, Greengenes or SILVA [44]. These databases allow the classification of environmental 16S rRNA gene sequences. In addition to the huge wealth of microbial diversity hidden in databases, DNA-based stable-isotope probing has also aided in exploring microbial diversity by identifying the microorganisms in environmental samples that use a particular growth substrate. In DNA-based stable-isotope probing, substrate containing a stable-isotope (such as C13) is incorporated into the cells in a community that allows their specific separation (by density) and aids in identification (with microarrays) of their genes [152].

Quaiser [66] provided the first information beyond 16S rRNA gene sequences about the uncultured Acidobacterium subgroups based on partial and full-length sequencing. Many metagenomic clones belonging to the phylum Acidobacterium have been fully sequenced, revealing many genes with homology to housekeeping genes involved in DNA repair, transport, cell division, translation, and purine biosynthesis. Other gene sequences include those with homology to genes encoding cyclic β1-2 glucan synthetase, polyhydroxybutyrate depolymerase, *Bacteroides fragilis* aerotolerance functions, and an operon distantly related to the lincomycin biosynthesis pathway of *Streptomyces lincolnensis*, which provides hints about the ecological roles of Acidobacteria [152].

The microbial diversity of extreme environments have always fascinated researchers due to their novelty. Metagenomic sampling of microbial communities that survive in extreme environments of temperature and pH can be helpful to understand the adaptations or mechanisms that enable them to tolerate such conditions that, in turn, have an effect on the geochemistry of the environment [8,153–156]. One such example is acid drainage of the Richmond mine, one of the most extreme environments on Earth. The microbial community residing in the acid drainage of the Richmond mine survives in extremely low pH (0–1) with high levels of Fe, Zn, Cu, As, and no source of carbon or nitrogen other than the gaseous forms present in the air. The mine is rich in sulfide minerals, including pyrite (FeS_2) dissolved as the result of oxidation catalyzed by microbial activity [156]. Tyson et al. [8] cloned total DNA and sequenced it. It was found that the GC content of the genomes of the dominant taxa in the mine differ substantially. Further sequence alignment of 16S rRNA and tRNA synthetase genes was performed, and it confirmed that the community is dominated by three bacterial genera: Leptospirillum, Sulfobacillus, and Acidomicrobium along with two archaeal species (*Ferroplasma acidarmanus*, Thermoplasmatales). All the genomes in the acid mine drainage are rich in genes associated with removing toxic elements from the cell. Proton efflux systems are likely responsible for maintaining the nearly neutral intracellular pH and metal resistance determinants pump metals out of the cells, maintaining nontoxic levels in the interior of the cells.

23.4 Challenges

23.4.1 Metagenomic DNA Isolation and Purification Techniques

No single DNA extraction method can provide the true picture of microbial diversity and its potential [34,157]. Certain biases could be introduced by the DNA extraction method due to different lysis efficiencies of different species [158]. The association of microbes with soil and other matrix in the soil also affects the yield and quality of metagenomic DNA isolated. Moreover, with soil DNA, further manipulation is required as humic acids are coextracted with DNA and hinders further enzymatic reactions. A number of methods are available that can be used to remove humic acids, but none are 100% efficient. Thus, improvement of DNA extraction methods, as well their purification, would make metagenomic applications more feasible.

23.4.2 Heterologous Gene Expression for Discovery of Novel Genes and Gene Functions

Functional analysis of the metagenome requires the expression of metagenomic DNA in heterologous hosts but the major problem with heterologous gene expression is that genes either are not expressed or, if expressed, the level of expression is too low to be detected. The heterologous hosts used in metagenomics are *E. coli, P. putida, S. lividans, Sinorhizobium meliloti*, and *Rhizobium leguminosarum, Ralstonia metallidurans* [159–163]. Although, *E. coli* is often used as a host for heterologous gene expression, it is estimated to express only 40% of the genes derived from diverse microbial origins and it is strongly biased against the expression of genes from certain groups of distantly related organisms [164].

Wang [162] reported that different positive clones are detected when different hosts are used. For example, *S. meliloti* and *E. coli bdhA* mutants were used as hosts to screen for D-3-hydroxybutyrate dehydrogenase. Only 1 of 25 clones detected using *S. meliloti* was able to complement *E. coli* and none of the clones found using *E. coli* could be used to complement *S. meliloti*). Thus, screening of metagenomic libraries can be improved when a range of hosts are used. We suggest that the hit rate can be increased by (a) increasing the amount of environmental sample investigated, (b) using different DNA isolation protocols, or (c) using different hosts. Much effort is still required to explore host systems with different expression capabilities that could increase the probability of recovering exploitable activities by metagenome cloning.

23.4.3 Bioinformatics Tools to Understand Metadata

Although metagenomics is a recent addition to the life sciences, it has made remarkable progress in the last few years with the aid of high-throughput sequencing technologies. These technologies have added plenty of data that would not have been possible with traditional Sanger methods. Many databases and bioinformatics tools have been developed to analyze this data, for example, the taxonomical content of metagenome is usually estimated by homology-based approaches. BLAST and MEGAN are bioinformatics tools used for homology-based approaches. Venter et al. [47] shotgun-sequenced fragments from Sargasso Sea and BLAST these sequences against the comprehensive GenBank database [165]. But classifying genomic fragments on the basis of hits using BLAST can be reliable only if close relatives are available for comparison. MEGAN ("MEtaGenome ANalyzer") is another program that relies on BLAST for analysis. This software allows the analysis of large metagenomic data sets in which sets of DNA reads (or contigs) are compared against databases of known sequences using BLAST or another comparison tool. These bioinformatics tools are mostly used to identify short reads. Short reads tend to miss distantly related sequences and miss a significant amount of homologues found in long reads.

Apart from taxonomic classification, bioinformatics tools are also used to study comparative metagenomics: the effect of habitat on the function of microbial communities. Tringe et al. [30] compared samples from different habitats, for example, agricultural, deep-sea whale-fall carcasses, the Sargasso Sea, and the acid mine drainage environments using a clustering-based approach. They showed that profiles of the microbial communities from each environment clustered with those of others in the same community, and concluded that the "functional profile of a community is influenced by its environment."

These days, bioinformatics tools are also used to find the functional profile of the community. For this, the important step is to find the regions of DNA that encode for proteins. The National Center for Biotechnology Information Open Reading Frame Finder tool focuses on finding open reading frames in the DNA sequence. An open reading frame is generally defined as a sequence of DNA that begins with a start codon and ends with one of the stop codons [166]. This method has a limitation in that it gives us an idea of the open reading frames but does not indicate which regions actually encode proteins. Other methods such as GENIE [167], GENSCAN [168], GENEMARK [169], and GLIMMER [170] are efficient at predicting the regions encoding proteins in prokaryotic organisms.

Despite the development of many bioinformatics tools, there is no single database that can provide complete information about a particular metagenomic sample. The biggest reason behind this is that most bacteria and their genes have not yet been sequenced and thus are not available for cross-referencing. This limitation needs to be worked out in the future.

23.4.4 Removal of Bias toward Metagenome Analysis of Selective Microbes

The rRNA approach is the most widely used tool to assess microbial diversity [171]. The rRNA molecules and their genes consist of highly conserved domains interspersed with variable regions, so that comparative analysis of these sequences is a powerful means to infer phylogenetic relatedness among organisms. However, PCR amplification primers targeted against these conserved sequences and DNA extraction methods possess a great degree of bias. A single combination of PCR primers/DNA extraction techniques recovers only 50% diversity. Sipos et al. [172] reported that PCR primers are discriminated against certain templates or show inhibition toward certain templates by self-annealing. Even different sets of universal primers amplify different portions of ribosomal RNA. Hong et al. [173] use different sets of primers and multiple DNA extraction techniques to minimize the associated biases. By this modification, more species were recovered compared with the classic methods. But even this modification does not represent true sample diversity.

23.5 Conclusion

Metagenomics has the ability to directly examine the genomic content of microbial communities of a particular environment. The diversity and biotechnological potentials of the metagenome of various biotic and abiotic habitats have been analyzed thus far using this technique. Success has been achieved in discovering novel genes, novel genomes, and novel functions performed by these microbes in a particular environment. Recent cloning and high-throughput sequencing techniques have speeded up the efficiency of analyzing simple and complex communities. Despite the impressive success stories from this emerging technique, it suffers from a few limitations and, to realize the full potential of metagenomics, these obstacles need to be overcome. The focus should be to improve DNA extraction techniques, followed by the development of efficient heterologous gene expression systems. Development of efficient bioinformatics tools and databases for reference is most pressing because an abundance of metadata has been generated by recent high-throughput sequencing technologies, which need to be explored.

References

1. Handelsman, J. 2004. Metagenomics: Application of genomics to uncultured microorganisms. *Microbiology and Molecular Biology Reviews* 68(4):669–685.
2. Torsvik, K.V., and J. Goksoyr. 1978. Determination of bacterial DNA in soil. *Soil Biology and Biochemistry* 10:7–12.
3. Pace, N.R., D.A. Stahl, D.J. Lane, and G.J. Olsen. 1986. The analysis of natural microbial populations by ribosomal RNA. *Advances in Microbial Ecology* 9:1–55.
4. Seow, K.T., G. Meurer, M. Gerlitz, P.E. Wendt, C.R. Hutchinson, and J. Davies. 1997. A study of iterative type II polyketide synthases, using bacterial genes cloned from soil DNA a means to access and use genes from uncultured microorganisms. *Journal of Bacteriology* 179:7360–7368.
5. Morrison, N., G. Cochrane, N. Faruque, T. Tatusova, Y. Tateno, D. Hancock et al. 2006. Concept of sample in omics technology OMICS. *A Journal of Integrative Biology* 10(2):127–137.
6. Ranjard, L., D.P.H. Lejon, C. Mougel, L. Scheher, D. Merdinoglu, and R. Chaussod. 2003. Sampling strategy in molecular microbial ecology: Influence of soil sample size on DNA Fingerprinting of fungal and bacterial communities. *Environmental Microbiology* 5:1111–1120.
7. Kang, S., and A. Mills. 2006. The effect of sample size in studies of soil microbial community. *Journal of Microbiological Methods* 66:242–250.
8. Tyson, G.W., J. Chapman, P. Hugenholtz, E.E. Allen, R.J. Ram, P.M. Richardson et al. 2004. Community structure and metabolism through reconstruction of microbial genomes from the environment. *Nature* 428(4):37–43.
9. Woyke, T., H. Teeling, N.N. Ivanova, M. Huntemann, M. Richter, F.O. Gloeckner et al. 2006. Symbiosis insights through metagenomic analysis of a microbial consortium. *Nature* 443(7114):950–955.
10. Fierer, N., M. Breitbart, J. Nulton, P. Salamon, C. Lozupone, R. Jones et al. 2007. Metagenomic and small subunit rRNA analyses reveal the genetic diversity of bacteria, archaea, fungi, and viruses in soil. *Applied and Environmental Microbiology* 73(21):7059–7066.
11. Raes, J., K.U. Foerstner, and P. Bork. 2007. Get the most out of your metagenome: Computational analysis of environmental sequence data. *Current Opinion in Microbiology* 10(5):490–498.
12. Allison, S.D., and J.B.H. Martiny. 2008. Resistance, resilience, and redundancy in microbial communities. *Proceedings of the National Academy of Sciences of the United States of America* 105:11512–11519.
13. Jones, B.V., M. Begley, C. Hill, C.G. Gahan, and J.R. Marchesi. 2008. Functional and comparative metagenomic analysis of bile salt hydrolase activity in the human gut microbiome. *Proceedings of the National Academy of Sciences of the United States of America* 105(36):13580–13585.
14. Grice, E.A., H.H. Kong, G. Renaud, A.C. Young, G.G. Bouffard, R.W. Blakesley et al. 2008. A diversity profile of the human skin microbiota. *Genome Research* 18:1043–1050.
15. Hojo, K., S. Nagaoka, T. Ohshima, and N. Maeda. 2009. Bacterial interactions in dental biofilm development. *Journal of Dental Research* 88:982.
16. Hyman, R.W., M. Fukushima, L. Diamond, J. Kumm, L.C. Guidice, and R.W. Davis. 2005. Microbes on the human vaginal epithelium. *Proceedings of the National Academy of Sciences of the United States of America* 102:7952–7957.
17. Ghazanfar, S., and A. Azim. 2009. Metagenomics and its application in rumen ecosystem: Potential biotechnological prospects. *Pakistan Journal of Nutrition* 8(8):1309–1315.

18. Sugimoto, A., D.E. Bigneil, and J.A. MacDonald. 2000. Global impact of termites on the carbon cycle and atmospheric trace gases. In *Termites: Evolution, Sociality, Symbiosis, Ecology*, edited by Abe, T., D.E. Bignell, and M. Higashi, 409–435. Dordrecht, Boston, London: Kluwer Academic Publishers.

19. Compant, S., C. Clément, and A. Sessitsch. 2010. Plant growth–promoting bacteria in the rhizo and endosphere of plants: Their role, colonization, mechanisms involved and prospects for utilization. *Soil Biology and Biochemistry* 42:669–678.

20. Gray, E.J., and D.L. Smith. 2005. Intracellular and extracellular PGPR: Commonalities and distinctions in the plant–bacterium signaling processes. *Soil Biology and Biochemistry* 37:395–412.

21. Lindow, S.E., and M.T. Brandl. 2003. Microbiology of the phyllosphere. *Applied and Environmental Microbiology* 69(4):1875–1883.

22. Eckburg, P.B., E.M. Bik, C.N. Bernstein, E. Purdom, L. Dethlefsen, M. Sargent et al. 2005. Diversity of the human intestinal microbial flora. *Science* 308:1635–1638.

23. Ryan, P.R., Y. Dessaux, L.S. Thomashow, and D.M. Weller. 2009. Rhizosphere engineering and management. *Plant Soil* 321:363–383.

24. Teixeira, L.C.R.S., R.S. Peixoto, J.C. Cury, W.J. Sul, V.H. Pellizari, J. Tiedje et al. 2010. Bacterial diversity in rhizosphere soil from Antarctic vascular plants of Admiralty Bay, maritime Antarctica. *The ISME Journal* 4:989–1001.

25. Warnecke, F., and M. Hess. 2009. A perspective: Metatranscriptomics as a tool for the discovery of novel biocatalysts. *Journal of Biotechnology* 142:91–95.

26. Rosello, M.R., and R. Amann. 2001. The species concept for prokaryotes. *FEMS Microbiology Reviews* 25:39–67.

27. Curtis, T.P., W.T. Sloan, and J.W. Scannell. 2002. Estimating prokaryotic diversity and its limits. *Proceedings of the National Academy of Sciences of the United States of America* 99:10494–10499.

28. Garbeva, P., V.J.A. Van, and V.J.D. Elsas. 2004. Microbial diversity in soil: Selection of the microbial populations by plant and soil type and implementations for disease suppressivenss. *Annual Review Phytopathology* 42:243–270.

29. J. Craig Venter Institute. 2007. Millions of new genes, thousands of new protein families found in ocean sampling expedition. *Science Daily*. Retrieved March 20, 2011, from http://www.sciencedaily.com.

30. Tringe S.G., T. Zhang, X. Liu, Y. Yu, W.H. Lee, J. Yap et al. 2008. The airborne metagenome in an indoor urban environment. *PLoS One* 3(4):1862.

31. Brady, S.F. 2007. Construction of soil environmental DNA cosmid libraries and screening for clones that produce biologically active small molecules *Nature Protocols* 2:1297–1305.

32. Schmeisser, C., C. Stockigt, C. Raasch, J. Wingender, K.N. Timmis, D.F. Wenderothet et al. 2003. Metagenome survey of biofilms in drinking-water networks. *Applied and Environmental Microbiology* 69:7298–7309.

33. Pang, M.F., N. Abdullah, C.W. Lee, and C.C. Ng. 2008. Isolation of high molecular weight DNA from forest topsoil for metagenomic analysis. *Asia Pacific Journal of Molecular Biology and Biotechnology* 16(2):35–41.

34. Inceoglu, O., E.F. Hoogwout, P. Hill, and J.D.V. Elsas. 2010. Effect of DNA extraction method on the apparent microbial diversity of soil. *Applied and Environmental Microbiology* 76(10):3378–3382.

35. Liles, M.R., L.L. Williamson, J. Rodbumrer, V. Torsvik, S.C. Parseley, R.M. Goodman et al. 2009. Isolation and cloning of high-molecular-weight metagenomic DNA from soil microorganisms. *Cold Spring Harbor Protocols*: prot5271R.

36. Amorim, J.H., T.N.S. Macena, J.G.V. Lacerda, R.P. Rezende, J.C.T. Dias, M. Brendel et al. 2008. An improved extraction protocol for metagenomic DNA from a soil of the Brazilian Atlantic rainforest. *Genetics and Molecular Research* 7(4):1226–1232.

37. Sharma, P.K., N. Capalash, and J. Kaur. 2007. An improved method for single step purification of metagenomic DNA. *Molecular Biotechnology* 36:61–63.

38. Herrera, R.R., J.N. Zapata, M.Z. Maya, and M.E.M. Martınez. 2008. A simple silica-based method for metagenomic DNA extraction from soil and sediments. *Biotechnology* 40:13–17.

39. Purohit, M.K., and S.P. Singh. 2009. Assessment of various methods for extraction of metagenomic DNA from saline habitats of coastal Gujarat (India) to explore molecular diversity. *Letters in Applied Microbiology* 49(3):293–415.

40. Hosono, S., A.F. Faruqi, F.B. Dean, Y. Du, Z. Sun, X. Wu et al. 2003. Unbiased whole-genome amplification directly from clinical sample. *Genome Research* 13:954–964.

41. Shoaib, M., S. Baconnais, U. Mechold, E.L. Cam, M. Lipinski, and V. Ogryzko. 2008. Multiple displacement amplification for complex mixtures of DNA fragments. *BMC Genomics* 9:415.

42. Streit, W.R., and R.A. Schmitz. 2004. Metagenomics—the key to the uncultured microbes. *Current Opinion in Microbiology* 7:492–498.

43. Schmeisser, C., H. Steele, and W.R. Streit. 2007. Metagenomics, biotechnology with non-culturable microbes. *Applied Microbiology and Biotechnology* 75:955.

44. Simon, C., and D. Daniel. 2009. Achievements and new knowledge unraveled by metagenomic approaches. *Applied Microbiology and Biotechnology* 85:265–276.

45. Uchiyama, T., T. Abe, T. Ikemura, and K. Watanabe. 2005. Substrate-induced gene expression screening of environmental metagenome libraries for isolation of catabolic genes. *Nature Biotechnology* 23:88–93.

46. Craig, J.W., F.Y. Chang, J.H. Kim, S.C. Obiajulu, and S.F. Brady. 2010. Expanding small-molecule functional metagenomics through parallel screening of broad-host-range cosmid environmental DNA libraries in diverse *Proteobacteria*. *Applied and Environmental Microbiology* 76(5):1633–1641.

47. Venter J.C., K. Remington, J.F. Heidelberg, A.L. Halpern, D. Rusch, J.A. Eisen et al. 2004. Environmental genome shotgun sequencing of the Sargasso Sea. *Science* 304:66–74.

48. Eisen, J.A. 2007. Environmental shotgun sequencing: Its potential and challenges for studying the hidden world of microbes. *PLoS Biology* 5(3):382–388.

49. Kennedy, J., B. Flemer, S. Jackson, D.P.H. Lejon, P.M. John, F.O. Gara et al. 2010. Marine metagenomics: New tools for the study and exploitation of marine microbial metabolism. *Marine Drugs* 8:608–628.

50. Cowan, D., Q. Meyer, W. Stafford, S. Muyanga, R. Cameron, and P. Wittwer. 2005. Metagenomic gene discovery: Past, present and future. *Trends in Biotechnology* 23:321–329.

51. Beja, O. 2004. To BAC or not to BAC: Marine ecogenomics. *Current Opinion in Biotechnology* 15:187–190.

52. Courtois, S. 2003. Recombinant environmental libraries provide access to microbial diversity for drug discovery from natural products. *Applied and Environmental Microbiology* 69:49–55.

53. Lee, D.G., J.H. Jeon, M.K. Jang, N.Y. Kim, J.H. Lee, S.J. Kim et al. 2007. Screening and characterization of a novel fibrinolytic metalloprotease from a metagenomic library. *Biotechnology Letters* 29:465–472.

54. Waschkowitz, T., S. Rockstroh, and R. Daniel. 2009. Isolation and characterization of metalloproteases with a novel domain structure by construction and screening of metagenomic libraries. *Applied and Environmental Microbiology* 75:2506–2516.

55. Lee, M.H., C.H. Lee, T.K. Oh, J.K. Song, and J.H. Yoon. 2006. Isolation and characterization of a novel lipase from a metagenomic library of tidal flat sediments: Evidence for a new family of bacterial lipases. *Applied and Environmental Microbiology* 72:7406–7409.

56. Hårdeman, F., and S. Sjöling. 2007. Metagenomic approach for the isolation of a novel low-temperature-active lipase from uncultured bacteria of marine sediment. *FEMS Microbiology Ecology* 59:524–534.

57. Heath. C., X.P. Hu, S.C. Cary, and D. Cowan. 2009. Isolation and characterization of a novel, low-temperature-active alkaliphilic esterase from an Antarctic desert soil metagenome. *Applied and Environmental Microbiology* 75:4657–4659.

58. Yun. J., and S. Ryu. 2005. Screening for novel enzymes from metagenome and SIGEX, as a way to improve it. *Microbial Cell Factories* 4:8.

59. Handelsman, J. 2005. Sorting out metagenomes. *Nature Biotechnology* 23:38–39.

60. Williamson, L.L., B.R. Borlee, P.D. Schloss, C. Guan, H.k. Allen, and J. Handelsman. 2005. Intracellular screen to identify metagenomic clones that induce or inhibit a quorum-sensing biosensor. *Applied and Environmental Microbiology* 71:6335–6344.

61. Chistoserdova, L. 2010. Recent progress and new challenges in metagenomics for biotechnology. *Biotechnology Letters* 32:1351.
62. Warnecke, F., and P. Hugenholtz. 2007. Building on basic metagenomics with complementary technologies *Genome Biology* 8:231.
63. Gilbert, J.A., S. Thomas, N.A. Cooley, A. Kulakova, D. Field, T. Booth et al. 2009. Potential for phosphonoacetate utilization by marine bacteria in temperate coastal waters. *Environmental Microbiology* 11:111–125.
64. McGrath, K.C., R. Skye, H. Thomas, C.T. Cheng, L. Leo, A. Alexa et al. 2008. Isolation and analysis of mRNA from environmental microbial communities. *Journal of Microbiological Methods* 75(2):172–176.
65. Frias-Lopez, J., Y. Shi, G.W. Tyson, M.L. Coleman, S.C. Schuster, S.W. Chisholm et al. 2008. Microbial community gene expression in ocean surface waters. *Proceedings of the National Academy of Sciences of the United States of America* 105:3805–3810.
66. Quaiser, A., T. Ochsenreiter, C. Lanz, S.C. Schuster, A.H. Treusch, J. Eck et al. 2003. Acidobacteria form a coherent but highly diverse group within the bacterial domain: Evidence from environmental genomics. *Molecular Microbiology* 50:563–575.
67. Cottrell, M.T., L. Yu, and D.L. Kirchman. 2005. Sequence and expression analyses of cytophaga-like hydrolases in a western arctic metagenomic library and the Sargasso Sea. *Applied and Environmental Microbiology* 71:8506–8513.
68. Henckel, T., M. Friedrich, and R. Conrad. 1999. Molecular analyses of the methane oxidizing microbial community in rice field soil by targeting the genes of the 16S rRNA particulate methane monooxygenase, and methanol dehydrogenase. *Applied and Environmental Microbiology* 65:1980–1990.
69. MacDonald, I.R., E.M. Kenna, and J.C. Murrel. 1995. Detection of methanotrophic bacteria in environmental samples with the PCR. *Applied and Environmental Microbiology* 61:116–121.
70. Henckel, T., U. Jackel, S. Schnell, and R. Conrad. 2000. Molecular analyses of novel methanotrophic communities in forest soil that oxidize atmospheric methane. *Applied and Environmental Microbiology* 66:1801–1808.
71. Watanabe, K., M. Teramoto, H. Futamata, and S. Harayama. 1998. Molecular detection, isolation, and physiological characterization of functionally dominant phenol degrading bacteria in activated sludge. *Applied and Environmental Microbiology* 64:4396–4402.
72. Mesarch, M.B., C.H. Nakabsu, and L. Nies. 2000. Development of catechol 2,3 dioxygenase specific primers for monitoring bioremediation by competitive quantitative PCR. *Applied and Environmental Microbiology* 66:678–683.
73. Futamata, H., S. Harayama, and W. Kazuya. 2001. Group specific monitoring of phenol hydroxylase genes for a functional assessment of phenol stimulated trichloroethylene bioremediation. *Applied and Environmental Microbiology* 67:4671–4677.
74. Church, G.M. 2006. Genomes for all. *Scientific American* 294(1):46–54.
75. Hall, N. 2007. Advanced sequencing technologies and their wider impact in microbiology. *Journal of Experimental Biology* 210(9):1518–1525.
76. Sogin, M.L., H.G. Morrison, J.A. Huber, and D.M. Welch. 2006. Microbial diversity in the deep sea and the underexplored rare biosphere. *Proceedings of the National Academy of Sciences of the United States of America* 103:12115–12120.
77. Angly, F.E., B.M. Felts, P. Breitbart, R.A. Salamon, R.A. Edwards, C. Carlson et al. 2006. The marine viromes of four oceanic regions. *PLoS Biology* 4(11):2121–2131.
78. Webster, G., C.J. Newberry, J.C. Fry, and A.J. Weightman. 2003. Assessment of bacterial community structure in the deep sub-seafloor biosphere by 16S rDNA-based techniques: A cautionary tale. *Journal of Microbiological Methods* 55:155–164.
79. Biddle, J.F., G.S. Fitz, S.C. Schuster, J.E. Brenchley, and C.H. House. 2008. Metagenomic signatures of the Peru Margin subseafloor biosphere show a genetically distinct environment. *Proceedings of the National Academy of Sciences of the United States of America* 105:10583–10588.
80. Edwards, R.A., B.B. Rodriguez, L. Wegley, M. Haynes, M. Breitbart, D.M. Peterson et al. 2006. Using pyrosequencing to shed light on deep mine microbial ecology. *BMC Genomics* 7:57.

81. Cox-Foster, D.L., S. Conlan, E.C. Holmes, G. Palacios, J.D. Evans, N.A. Moran et al. 2007. A metagenomic survey of microbes in honey bee colony collapse disorder. *Science* 318:283–287.

82. Rusch, D.B., A.L. Halpern, G. Sutton, K.B. Heidelberg, S. Williamson, S. Yooseph et al. 2007. The Sorcerer II Global Ocean Sampling Expedition: Northwest Atlantic through Eastern Tropical Pacific. *PLoS Biology* 5:77.

83. Yooseph, S., G. Sutton, D.B. Rusch, A.L. Halpern, S.J. Williamson, K. Remington et al. 2007. The Sorcerer II Global Ocean Sampling expedition: Expanding the universe of protein families. *PLoS Biology* 5(3):e16.

84. Biers, E.J., S. Sun, and E.C. Howard. 2009. Prokaryotic genomes and diversity in surface ocean waters: Interrogating the global ocean sampling metagenome. *Applied and Environmental Microbiology* 75:2221–2229.

85. Schramm, A., L.H. Larsen, N.P. Revsbech, N.B. Ramsing, R. Amann, and K.H. Schleifer. 1996. Structure and function of a nitrifying biofilm as determined by *in situ* hybridization and the use of microelectrodes. *Applied and Environmental Microbiology* 62:4641–4647.

86. Pernthaler, A., A.E. Dekas, C.T. Brown, S.K.G. Tsegereda Embaye, and V.J. Orphan. 2008. Diverse syntrophic partnerships from deep-sea methane vents revealed by direct cell capture and metagenomics *Proceedings of the National Academy of Sciences of the United States of America* 105(19):7052–7057.

87. Singh, B.K., P. Millard, A.S. Whiteley, and J.C. Murrell. 2004. Unravelling rhizosphere microbial interactions: Opportunities and limitations. *Trends in Microbiology* 12(8):386–394.

88. Wu, L., D.K. Thompson, and X. Liu. 2004. Development and evaluation of microarray-based whole-genome hybridization for detection of microorganisms within the context of environmental applications. *Environmental Science and Technology* 38:6775–6782.

89. Loy, A., J. Lehner, N. Lee, J. Adamczyk, H. Meier, J. Ernst et al. 2002. Oligonucleotide microarray for 16S rRNA gene based detection of all recognised lineages of sulfate-reducing prokaryotes in the environment. *Applied and Environmental Microbiology* 68:5064–5081.

90. Kim, C.C., E.A. Joyce, K. Chan, and S. Falkow. 2002. Improved analytical methods for microarray based genome composition analysis. *Genome Biology* 3(11): Research 0065.1–0065.17.

91. Sebat, J.L., F.S. Colwell, and R.L. Crawford. 2009. Metagenomic profiling: Microarray analysis of an environmental genomic library. *Applied and Environmental Microbiology* 69:4927–4934.

92. Doolittle, W.F. 1999. Lateral genomics. *Trends in Cell Biology* 9:M5–M8.

93. Jain, R., M.C. Rivera, and J.A. Lake. 1999. Horizontal gene transfer among genomes: The complexity hypothesis. *Proceedings of the National Academy of Sciences of the United States of America* 96:3801–3806.

94. Shiu, S.H., and J.O. Borevitz. 2008. The next generation of microarray research: Applications in evolutionary and ecological genomics *Heredity* 100:141–149.

95. Yergeau, E., S. Kang, Z. He, J. Zhou, and G.A. Kowalchuk. 2007. Functional microarray analysis of nitrogen and carbon cycling genes across an Antarctic latitudinal transect. *The ISME Journal* 1:163–179.

96. He, Z.L., T.J. Gentry, C.W. Schadt, L. Wu, J. Liebich, S.C. Chong et al. 2007. GeoChip: A comprehensive microarray for investigating biogeochemical, ecological and environmental processes. *The ISME Journal* 1:67–77.

97. Henne, A. R.A. Schmitz, M. Bomeke, G. Gottschalk, and R. Daniel. 2000. Screening of environmental DNA libraries for the presence of genes conferring lipolytic activity on *Escherichia coli*. *Applied and Environmental Microbiology* 66:3113–3116.

98. Bell, P.J., A. Sunna, M.D. Gibbs, N.C. Curach, H. Nevalainen, and P.L. Bergquist. 2002. Prospecting for novel lipase genes using PCR. *Microbiology* 148:2283–2291.

99. Voget, S., C. Leggewie, A. Uesbeck, C. Raasch, K.E. Jaeger, and W.R. Streit. 2003. Prospecting for novel biocatalysts in a soil metagenome. *Applied and Environmental Microbiology* 69:6235–6242.

100. Schmeisser, C., C. Stockigt, C. Raasch, and J. Wingender. 2003. Metagenome survey of biofilms in drinking-water networks. *Applied and Environmental Microbiology* 69:7298–7309.

101. Jeon, J.H., J.T. Kim, Y.J. Kim, H.K. Kim, H.S. Lee, S.G. Kang et al. 2009. Cloning and characterization of a new cold-active lipase from a deep-sea sediment metagenome *Applied and Environmental Microbiology* 81:865–874.

102. Knietsch, A., T. Waschkowitz, S. Bowien, A. Henne, and R. Daniel. 2003. Construction and screening of metagenomic libraries derived from enrichment cultures: Generation of a gene bank for genes conferring alcohol oxidoreductase activity on *Escherichia coli*. *Applied and Environmental Microbiology* 69:1408–1416.

103. Yun, J., S. Kang, S. Park, H. Yoon, M.J. Kim, S. Heu, and S. Ryu. 2004. Characterization of a novel amylolytic enzyme encoded by a gene from a soil-derived metagenomic library. *Applied and Environmental Microbiology* 70:7229–7235.

104. Ferrer, M., O.V. Golyshina, T.N. Chernikova, A.N. Khachane, D. Reyes-Duarte, V.A. Santos et al. 2005. Novel hydrolase diversity retrieved from a metagenome library of bovine rumen microflora. *Environmental Microbiology* 7:1996–2010.

105. Rees, H.C., S. Grant, B. Jones, W.D. Grant, and S. Heaphy. 2003. Detecting cellulase and esterase enzyme activities encoded by novel genes present in environmental DNA libraries. *Extremophiles* 7:415–421.

106. Feng Y.D., C.J. Duan, H. Pang, X.C. Mo, C.F. Wu, Y. Yu et al. 2007. Cloning and identification of novel cellulase genes from uncultured microorganisms in rabbit caecum and characterization of the expressed cellulases. *Applied and Environmental Microbiology* 75:319–328.

107. Eschenfeldt W.H., L. Stols, H. Rosenbaum, Z.S. Khambatta, E. Quaite-Randall, S. Wu et al. 2001. DNA from uncultured organisms as a source of 2, 5-diketo-D-gluconic acid reductases. *Applied and Environmental Microbiology* 67:4206–4214.

108. Streit, W.R., and P. Entcheva. 2003. Biotin in microbes, the genes involved in its biosynthesis, its biochemical role and perspectives for biotechnological production. *Applied and Environmental Microbiology* 61:21–31.

109. Entcheva, P., W. Liebl, A. Johann, T. Hartsch, and W.R. Streit. 2001. Direct cloning from enrichment cultures, a reliable strategy for isolation of complete operons and genes from microbial consortia. *Applied and Environmental Microbiology* 67:89–99.

110. MacNeil I.A., C.L. Tiong, C. Minor, P.R. August, T.H. Grossman, K.A. Loiacono et al. 2001. Expression and isolation of antimicrobial small molecules from soil DNA libraries. *Journal of Molecular and Microbial Biotechnology* 3(2):301–308.

111. Lim, H.K., E.J. Chung, J.C. Kim, G.J. Choi, K.S. Jang, Y.R. Chung et al. 2005. Characterisation of a forest soil metagenome clone that confers indirubin and indigo production on *Escherichia coli*. *Applied and Environmental Microbiology* 71:7768–7777.

112. Brady, S.F., and J. Clardy. 2004. Palmitoylputrescine, an antibiotic isolated from the heterologous expression of DNA extracted from bromeliad tank water. *Journal of Natural Products* 67:1283–1286.

113. Riesenfeld, C.S., R.M. Goodman, and J. Handelsman. 2004. Uncultured soil bacteria are a reservoir of new antibiotic resistance genes. *Environmental Microbiology* 6:981–989.

114. Allen H.K., L.A. Moe, J. Rodbumrer, and A. Gaarder. 2008. Functional metagenomics reveals diverse b-lactamases in a remote Alaskan soil. *ISME Journal* 3(2):243–251.

115. Elend, C., C. Schmeisser, C. Leggewie, P. Babiak, J.D. Carballeira, H.L. Steele et al. 2006. Isolation and biochemical characterization of two novel metagenome-derived esterases. *Applied and Environmental Microbiology* 72:3637–3645.

116. Park, H.J., J.H. Jeon, S.G. Kang, and J.H. Lee. 2007. Functional expression and refolding of new alkaline esterase, EM2L8 from deep-sea sediment metagenome. *Protein Expression and Purification* 52:340–347.

117. Chu, X., H. He, C. Guo, and B. Sun. 2008. Identification of two novel esterases from a marine metagenomic library derived from South China Sea. *Applied Microbiology and Biotechnology* 80:615–625.

118. Lee, M.H., K.S. Hong, S. Malhotra, H. Park, E.C. Hwang, H.K. Choi et al. 2010. A new esterase EstD2 isolated from plant rhizosphere soil metagenome. *Applied Microbiology and Biotechnology* 88:1125–1134.

119. Ferrer, M., O.V. Golyshina, T.N. Chernikova, A.N. Khachane, D.R. Duarte, V.A.P. Santos, et al. 2005. Novel hydrolase diversity retrieved from a metagenome library of bovine rumen microflora. *Environmental Microbiology* 7:1996–2010.

120. Lee, M.H., C.H. Lee, T.K. Oh, J.K. Song, and J.H. Yoon. 2006. Isolation and characterization of a novel lipase from a metagenomic library of tidal flat sediments: Evidence for a new family of bacterial lipases. *Applied and Environmental Microbiology* 72:7406–7409.

121. Hårdeman, F., and S. Sjöling. 2007. Metagenomic approach for the isolation of a novel low-temperature-active lipase from uncultured bacteria of marine sediment. *FEMS Microbial Ecology* 59:524–534.

122. Lämmle, K., H. Zipper, M. Breuer, B. Hauer, C. Buta, H. Brunner et al. 2007. Identification of novel enzymes with different hydrolytic activities by metagenome expression cloning. *Journal of Biotechnology* 127:575–592.

123. Nacke, H., C. Will, S. Herzog, B. Nowka, M. Engelhaupt, and R. Daniel. 2011. Identification of novel lipolytic genes and gene families by screening of metagenomic libraries derived from soil samples of the German Biodiversity Exploratories. *FEMS Microbiology Ecology* (doi: 10.1111/j.1574-6941.2011.01088.x).

124. Mewis, K., M. Taupp, and S.J. Hallam. 2011. A high throughput screen for biomining cellulase activity from metagenomic libraries. *Journal of Visual Expresssion* 48:2461.

125. Wang, Y., M. Radosevich, D. Hayes, and N. Labbé. 2011. Compatible ionic liquid-cellulases system for hydrolysis of lignocellulosic biomass. *Biotechnology and Bioengineering* 108(5):1042–1048.

126. Grant, S., W.D. Grant, D.A. Cowan, B.E. Jones, Y. Ma, A. Ventosa et al. 2006. Identification of eukaryotic open reading frames in metagenomic cDNA libraries made from environmental samples. *Applied and Environmental Microbiology* 72:135–143.

127. Gillespie, D.E., S.F. Brady, A.D. Bettermann, N.P. Cianciotto, M.R. Liles, M.R. Rondon et al. 2002. Isolation of antibiotics turbomycin A and B from a metagenomic library of soil microbial DNA. *Applied and Environmental Microbiology* 68(9):4301–4306.

128. Sokurenko, E.V., D.L. Hasty, and D.E. Dykhuzien. 1999. Pathoadaptive mutations: Gene loss and variation in bacterial pathogens. *Trends in Microbiology* 5:191–195.

129. Turnbaugh, P.J., M. Hamady, T. Yatsunenko, B.L. Cantarel, A. Duncan, R.E. Ley et al. 2009. A core gut microbiome in obese and lean twins. *Nature* 457:480–484.

130. Baehni, P.C., and Y. Takeuchi. 2003. Anti-plaque agents in the prevention of biofilm-associated oral diseases. *Oral Disease* 9:23–29.

131. Hojo, K., S. Nagaoka, T. Ohshima, and N. Maeda. 2009. Bacterial interactions in dental biofilm development. *Journal of Dental Research* 88:982.

132. Zhou, X., S.J. Bent, M.G. Schneider, C.C. Davis, M.R. Islam, and L.J. Forney. 2004. Characterization of vaginal microbial communities in adult healthy women using cultivation-independent methods. *Microbiology* 150:2565–2573.

133. Sundquist, A., S. Bigdeli, R. Jalili, M.L. Druzin, S. Waller, K.M. Pullen et al. 2007. Bacterial flora-typing with targeted, Chip-based pyrosequencing. *BMC Microbiology* 7:108.

134. Hyman, R.W., M. Fukushima, L. Diamond, J. Kumm, L.C. Giudice, and R.W. Davis. 2005. Microbes on the human vaginal epithelium. *Proceedings of the National Academy of Sciences of the United States of America* 102:7952–7957.

135. Johan, H., and J. Leveau. 2007. The magic and menace of metagenomics: Prospects for the study of plant growth-promoting rhizobacteria. *European Journal of Plant Pathology* 119:279–300.

136. Erkel, C., M. Kube, R. Reinhardt, and W. Liesack. 2006. Genome of Rice Cluster I archaea—the key methane producers in the rice rhizosphere. *Science* 313:370–372.

137. Kadivar, H., and A.E. Stapleton. 2003. Ultraviolet radiation alters maize phyllosphere bacterial diversity. *Microbial Ecology* 45(4):353–361.

138. Huang, J., J. Yang, Y. Duan, W. Gu, X. Gong, W. Zhe et al. 2010. Bacterial diversities on unaged and aging flue-cured tobacco leaves estimated by 16S rRNA sequence analysis. *Applied Microbial and Cell Physiology* 88(2):553–562.

139. Pedros, A.C. 2006. Genomics and marine microbial ecology. *International Microbiology* 9(3): 191–197.

140. Falkowski, P.G., T. Fenchel, and E.F. Delong. 2008. The microbial engines that drive earth's biogeochemical cycles. *Science* 1034–1039.
141. Paabo, S., H. Poinar, D. Serre, V. Jaenicke, J. Hebler, N. Rohland et al. 2004. Genetic analyses from ancient DNA. *Annual Review of Genetics* 38:645–679.
142. Poinar, H.N., C. Schwarz, J. Qi, B. Shapiro, R.D. Macphee, B. Buigues et al. 2005. Metagenomics to paleogenomics: Large-scale sequencing of mammoth DNA. *Science* 311(5759):392–394.
143. Hoss, M., A. Dilling, A. Currant, and S. Paabo. 1996. Molecular phylogeny of the extinct ground sloth *Mylodon darwinii*. *Proceedings of the National Academy of Sciences of the United States of America* 93:181–185.
144. Noonan, J.P., G. Coop, S. Kudaravalli, D. Smith, J. Krause, J. Alessi et al. 2006. Sequencing and analysis of Neanderthal genomic DNA. *Science* 314(5802):1113–1118.
145. Green, R.E., J. Krause, A.W. Briggs, T. Maricic, U. Stenzel, M. Kircher et al. 2010. A draft sequence of the Neanderthal genome. *Science* 328:710–722.
146. Stiller, M., R.E. Green, M. Ronan, J.F. Simons, L. Du, W. He et al. 2006. Patterns of nucleotide misincorporations during enzymatic amplification and direct large-scale sequencing of ancient DNA. *Proceedings of the National Academy of Sciences of the United States of America* 103(37):14977.
147. Lederberg, J. 2000. Infectious history. *Science* 288:287–293.
148. Budowle, B. 2004. Genetics and attribution issues that confront the microbial forensics field. *Forensic Science International* Suppl. 146:185–188.
149. Anonymous. 1999. Epidemiologic notes and reports update. Transmission of HIV infection during an invasive dental procedures—Florida, *Morbidity and Mortality Weekly Report* 40:22–27.
150. Tang, P., and C. Charles. 2010. Metagenomics for the discovery of novel human viruses. *Future Microbiology* 5(2):177–189.
151. Woese, C.R. 1987. Bacterial evolution. *Microbiological Reviews* 51(2):221–271.
152. Dumont, M.G., and J.C. Murrell. 2005. Stable isotope probing-linking microbial identity to function. *Nature Reviews in Microbiology* 3:499–504.
153. Allen, E.E., and J.F. Banfield. 2005. Community genomics in microbial ecology and evolution. *Nature Reviews in Microbiology* 3(6):489–498.
154. Tyson, G.W., and J.F. Banfield. 2005. Cultivating the uncultivated: A community genomics perspective. *Trends in Microbiology* 13(9):411–415.
155. Ram, R.J., N.C. Ver Berkmoes, M.P. Thelen, G.W. Tyson, B.J. Baker, R.C. II Blake et al. 2005. Community proteomics of a natural microbial biofilm. *Science* 308:1915–1920.
156. Baker, B.J., and J.F. Banfield. 2003. Microbial communities in acid mine drainage. *FEMS Microbiology and Ecology* 44:139–152.
157. Kunin, V., A. Copeland, A. Lapidus, K. Mavromatis, and P. Hugenholtz. 2008. A bioinformaticians guide to metagenomics. *Microbiology and Molecular Biology Reviews* 72(4):557–578.
158. Kauffmann, I.M., J. Schmitt, and R.D. Schmid. 2004. DNA isolation from soil samples for cloning in different hosts. *Applied Microbiology and Biotechnology* 64:665–670.
159. Martinez, A., S.J. Kolvek, C.L.T. Yip, J. Hopke, K.A. Brown, I.A. MacNeil et al. 2004. Genetically modified bacterial strains and novel bacterial artificial chromosome shuttle vectors for constructing environmental libraries and detecting heterologous natural products in multiple expression hosts. *Applied and Environmental Microbiology* 70:2452–2463.
160. Li, Y., M. Wexler, D.J. Richardson, P.L. Bond, and A.W.B. Johnston. 2005. Screening a wide host-range, waste-water metagenomic library in tryptophan auxotrophs of *Rhizobium leguminosarum* and of *Escherichia coli* reveals different classes of cloned *trp* genes. *Environmental Microbiology* 7:1927–1936.
161. Wexler, M., P.L. Bond, D.J. Richardson, and A.W. Johnston. 2005. A wide host range metagenomic library from a waste water treatment plant yields a novel alcohol/aldehyde dehydrogenase. *Environmental Microbiology* 7(12):1917–1926.
162. Wang, G. 2006. Diversity and biotechnological potential of the sponge-associated microbial consortia. *Journal of Industrial Microbiology and Biotechnology* 33:545–551.

163. Craig, J.W., F.Y. Chang, J.H. Kim, S.C. Obiajulu, and S.F. Brady. 2010. Expanding small-molecule functional metagenomics through parallel screening of broad-host-range cosmid environmental DNA libraries in diverse *Proteobacteria. Applied and Environmental Microbiology* 76(5):1633–1641.

164. Gabor, E.M., W.B. Alkema, and D.B. Janssen. 2004. Quantifying the accessibility of the metagenome by random expression cloning techniques *Environmental Microbiology* 6:879–886.

165. Tress, M.L., D. Cozzetto, A. Tramontano, and A. Valencia. 2006. An analysis of the Sargasso Sea resource and the consequences for database composition. *BMC Bioinformatics* 7:213.

166. National Center for Biotechnology Information (US). Available from: http://www.ncbi.nlm.nih.gov/projects/gorf.

167. Kulp, D., D. Haussler, M.G. Reese, and F.H. Eeckman. A generalized hidden Markov model for the recognition of human genes in DNA. In *Proceedings Conference on Intelligent Systems in Molecular Biology*, edited by States, D., P. Agarwal, T. Gaasterland, L., Hunter, and R. Smith, 134–142. Menlo Park, CA: AAAI Press.

168. Burge, C., and S. Karlin. 1997. Prediction of complete gene structures in human genomic DNA. *Journal of Molecular Biology* 268(1):78–94.

169. Lukashin, A.V., and M. Borodovsky. 1997. GeneMark.hmm: New solutions for gene finding. *Nucleic Acids Research* 26(4):1107–1115.

170. Salzberg, S.L., A.L. Delcher, S. Kasif, and O. White. 1998. Microbial gene identification using interpolated Markov models. *Nucleic Acids Research* 26(2):544–548.

171. Woese, C.R., O. Kandler, and M.L. Wheelis. 1990. Towards a natural system of organisms: Proposal for the domains Archaea, Bacteria, and Eucarya. *Proceedings of the National Academy of Sciences of the United States of America* 87:4576–4579.

172. Sipos, R., A.J. Szekely, M. Palatinszky, S. Revesz, K. Marialigeti, and M. Nikolausz. 2007. Effect of primer mismatch, annealing temperature and PCR cycle number on 16S rRNA gene-targeting bacterial community analysis. *FEMS Microbiology Ecology* 60:341–350.

173. Hong, S.H., J. Bunge, C. Leslin, S. Jeon, and S.S. Epstein. 2009. Polymerase chain reaction primers miss half of rRNA microbial diversity. *ISME Journal* 3(12):1365–1373.

24

Toxicogenomics in the Assessment of Environmental Pollutants

Luciana B. Crotti and Luis A. Espinoza

CONTENTS

24.1 Introduction

Modern toxicology has evolved significantly in the last decade with the application of genomewide expression profiling technologies that have created a hot new field called *toxicogenomics*. Making this field applicable for molecular biology analysis is the emergence of microarray technology, an approach that is now being widely applied in toxicological studies to measure the simultaneous expression of a large number of genes in a given sample in response to toxicity. This new discipline advances the information provided by conventional pharmacological and toxicological approaches and has the potential to contribute to in-depth investigation of the molecular mechanisms that may explain a toxicant's mode of action. An array containing many DNA samples utilizes the ability of a given messenger RNA (mRNA) molecule to bind specifically to, or hybridize to, the DNA template from which it originated, thereby scientists could determine, in a single experiment, the expression levels of thousands of genes within a single sample by measuring the amount of mRNA bound to each site on the array.

The intention behind microarray data analysis is to identify patterns of expression across multiple gene families; hence, the analysis, under these conditions, assumes that genes interact and that their expression is thus correlated with one another. This data can generate a profile of gene expression within a single sample that can be compared with the signatures of other samples. Consequently, changes in expression profiles may provide clues for the identification of putative biomarkers that may predict risk of clinical outcome or guide for appropriate therapy. In this chapter, we describe how toxicogenomics allows us to observe how a given chemical targets gene function within a living cell and the repercussions in a specific tissue. By inducing gene expression patterns, in terms of cell repair mechanisms and survivability, it may become possible to predict the outcome of toxicity before the appearance of histological or clinical pathologic changes.

24.2 Application of Toxicogenomics

The application of toxicogenomic technologies to RNA extracted from cultured human cells or tissues can provide a complete picture of the biological effects of the test substance and especially envisage the potential damage to the human organism (Figure 24.1). In this respect, DNA microarrays may contribute to understanding how molecular events are altered inside of cells in response to a specific agent, which can then be used to determine mechanistic insights into immunotoxicity as well as the identification of the hazardous effects of existing and novel toxic agents [1].

The parameters to consider during the interpretation of toxicogenomic data are composition, doses, and time of exposure of toxicants. Careful evaluation of precision and accuracy of the parameters mentioned above to assure the validity of results is therefore of great importance. For example, *in vitro* studies have demonstrated that several toxicants induce a variety of complex molecular perturbations in multiple signaling pathways, provoking differential gene expression at the transcript and functional protein level, leading to pathological outcomes [2,3]. This valuable information can also be used for the development of *in vitro* screening assays and for the prevention of or treatment for exposure to certain environmental agents. Some studies have shown that the toxic effects of environmental pollutants in the immune system evoke a series of events leading to the activation of genes that may trigger irreversible cell death by apoptosis, ultimately resulting in decreased immune function in related cell and organ types [4–7]. These detectable early intracellular changes in gene profiles in response to even lower doses of toxicants have been considered as potential indicators of toxicity, which in turn may identify the molecular pathways affected by the toxicant. Such an understanding may outperform clinical parameters in predicting disease outcome [8]. In this respect, it is imperative to evaluate

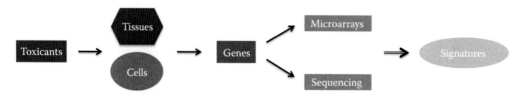

FIGURE 24.1
A flowchart describing the steps in the utilization of microarray analysis to identify gene expression changes in samples exposed to environmental toxicants.

dose–response relationships as well as the effect(s) of several components of a toxicant. For example, fuels are mixtures of a variety of chemicals that may act differently if some of the components are tested [9,10]. Another variable is the duration of the exposure, which may be a determinant to the type of evaluation for unknown toxicants. Although it is well known that the acute effects of air pollution result in pulmonary malfunction or diseases, there is little evidence associating the long-term effects of air pollution to increased risk of developing diseases [11,12].

Laboratory animals have been used to simulate *in vivo* situations of exposure to environmental toxicants. The evaluation of toxicogenomics in samples derived from these animal models has allowed the identification of patterns of toxicity, which are represented in the altered expression of a group of genes [1]. These data can be useful to compare *in vitro* results with intact animal expression changes and the different signatures that may eventually be predictors of the nocive effects of exposure to toxicants. Gene signatures can also be very informative for building a toxicity classification that may be dependent on the concentration of the toxicant.

Although it is expected that different concentrations of specific transcripts reflect the concentrations and activities of the protein products of the genes, this relationship is not always the case. Recent technological advances allow testing the effect of toxicants at lower doses or at earlier time points. Thus, for example, the threshold dose–response effect (hormesis) may provide different types of effects, either beneficial or harmful, depending of the situation [13]. These approaches can also facilitate the follow-up of transcriptional and translational events of genes and even the entire genome in response to specific toxic agents.

24.2.1 Microarray Technology

Microarray technology has significantly facilitated monitoring on the expression of thousands of genes in a single experiment; this has dramatically accelerated many types of investigation, such as those on diseases and conditions resulting from environmental pollution. Before microarray technology was available, methods in molecular biology had only allowed scientists to focus on the expression of a single gene per experiment. The concept of arrays was introduced by Fodor et al. [14], who defined it as microscopic groups of thousands of molecules (DNA) of known sequences attached to a solid surface such as a nylon membrane or a simple glass microscope slide. In a cell, genes may or may not be active and, because of that, the mechanism that regulates for gene expression is often compared with that of an on and off switch, which refers to whether a gene is active or inactive. When a gene is activated, the cellular machinery begins to copy certain segments of that gene. The resulting product is known as the mRNA, which is the body's template for creating proteins. The mRNA produced by the cell is complementary, and therefore will bind to the original portion of the DNA strand from which it was copied (Figure 24.2).

Because gene expression is a complex process characterized by a high degree of regulation, studying which genes are active and which are inactive in response to a toxicant helps to understand both how these cells function normally and how they are affected when various genes do not perform properly. In addition, gene expression profiles can provide key information in the identification of a signature that can be used to assess corresponding biological responses in other potential candidate(s) that are indicative of toxicant effects and pathological end points. Methods of microarrays have evolved over the last several years, starting with membrane microarrays, and progressing through microarrays on glass, and more recently, the oligonucleotide-based microarray system (Table 24.1).

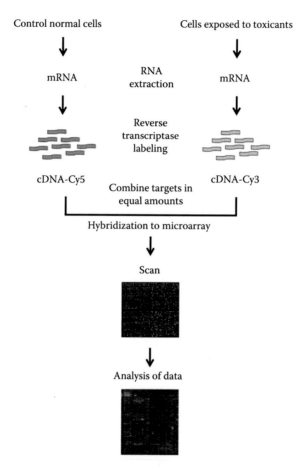

FIGURE 24.2
The flow of information during the course of a microarray experiment and the analysis procedure.

24.2.1.1 Microarrays Spotted onto Glass

To determine which genes are turned on or turned off in a given cell exposed to a toxicant, first, mRNA molecules present in that cell have to be isolated. Then, each mRNA molecule has to be labeled by using a reverse transcriptase enzyme that generates a complementary DNA (cDNA) to the mRNA. During that process, fluorescent nucleotides of the cyanine dye family Cy3 (green) and Cy5 (red) are attached to the cDNA. Treated and normal samples are labeled with different fluorescent dyes; generally, samples exposed to a toxicant are labeled with a red dye and normal samples are labeled with a green dye. Next, the labeled cDNAs are placed onto a DNA microarray slide. The labeled cDNAs that represent mRNAs in the cell will then hybridize to their synthetic cDNAs attached on the microarray slide, leaving their fluorescent tag(s). The cohybridizing of the probes (e.g., labeled cDNAs) will lead to a competition for the synthetic cDNAs on the microarray slide. If the spot is red, that specific gene is more expressed in treated than in normal sample (upregulated); if a spot is green, that gene is more expressed in the normal sample (downregulated). Furthermore, if a spot is yellow, that specific gene is equally expressed in both normal and treated samples; then, a special scanner is utilized to measure the fluorescent intensity to obtain the fold change in expression levels for each spot/area on the microarray slide.

TABLE 24.1

Microarray Platforms Used to Identify Changes in Gene Profiles of Cell Lines and Tissues in Response to Environmental Toxicants

Macroarray		Microarray									
Clontech	Intergri-Derm	Affymetrix	Affymetrix	Affymetrix	Applied Microarrays	Applied Microarrays	Agilent	Agilent	Illumina	Illumina	Illumina
Atlas human apoptosis and stress cDNA expression arrays	Human derm array GeneFilters DNA microarray	GeneChip human genome U133 array	GeneChip rat genome U34 array	Mouse genome 430 array	CodeLink human whole genome arrays	CodeLink rat whole genome	SurePrint G3 human gene expression	Rat gene expression microarray	Human HT-12 v4 expression BeadChip	Mouse WG-6 v2 expression BeadChip	RatRef 12 expression BeadChip
439 probes	5,171 probes	45,000 probes	8000 probes	45,000 probes	54,841 probes	33,849 probes	60,000 probes	44,000 probes	48,804 probes	45,281 probes	22,523 probes
Radioactive-labeling	Radioactive-labeling	Biotin-labeling	Biotin-labeling	Biotin-labeling	Cy5-Streptavidin	Cy5-Streptavidin	Cy3 and/or CY5-CTP	Cy3 and/or CY5-CTP	Cy3-Streptavidin	Cy3-Streptavidin	Cy3-Streptavidin

24.2.1.2 Macroarrays/Microarrays Spotted onto Membranes

Clontech (Mountain View, CA-Atlas Arrays), and Integriderm/MediQuest Therapeutics (Bothell, WA-DermArray Gene Filters) developed nylon membrane–based macroarrays. This format was very popular in the last decade because it did not demand the expensive equipment required to use glass chip microarrays. In these arrays, each cDNA is printed in duplicate, side-by-side. The labeling of the Atlas Array was optimized for the phosphorous isotopes ^{32}P, whereas the Integriderm system used ^{33}P to label probes as approaches to radiation detection for DNA microarray analysis.

24.2.1.3 Oligonucleotide Microarrays

The probes in oligonucleotide microarrays are short sequences designed to match parts of the sequence of known or predicted open reading frames. Rather than depositing intact sequences on a specific platform, oligonucleotide arrays are produced by printing short oligonucleotide sequences designed to represent a single gene or family of gene splice-variants directly onto the array surface. Companies have developed several methods for creating such arrays. Recently, there has been an increased utilization of the silicon chip, which is generally known as a Gen Chip when an Affymetrix (Santa Clara, CA) chip is used [15]. Other microarray platforms, such as Illumina (San Diego, CA), use microscopic beads (Bead Chip), instead of a large solid support, such as nylon membranes [16]. Another system provides single-channel microarrays from Agilent (Santa Clara, CA) [17]. These DNA arrays are different from other types of microarrays only in that they either measure DNA or use DNA as part of the detection system. Sequences may be longer (60-mer probes) such as the Agilent system or shorter (25-mer probes) such as the Illumina platform, depending on the desired purpose; longer probes are more specific to individual target genes, and shorter probes may be spotted across the array in higher densities and are cheaper to manufacture. One technique used to produce oligonucleotide arrays involves photolithographic synthesis on a silica substrate in which light and light-sensitive masking agents are used to build a sequence of one nucleotide at a time across the entire array. Each applicable probe is selectively unmasked before bathing the array in a solution containing just one nucleotide. Then, a masking reaction takes place and the next set of probes is unmasked in preparation for a different nucleotide exposure.

24.2.2 Analysis of Microarrays

A series of multiple steps is necessary for the experimental design approach and analysis of microarray data, which includes data processing and normalization, quality control metrics and thresholds for quality assessment, handling flags and low-intensity transcripts, identification of outlier hybridizations, identification of lists of differentially expressed genes, models for classification and prediction of biological end points, and biological interpretation of microarray results [18]. These steps identify the biological markers generated by the microarray database and are indicated in Table 24.2.

Furthermore, the appearance of multiple platforms, different protocols for the analysis of signatures, and the lack of interlaboratory and intralaboratory assay reproducibility on the same type of sample have undermined the acceptance of potential prognostic biomarkers based on microarray expression data. Therefore, the MicroArray Quality Control project developed by the U.S. Food and Drug Association–National Center for Toxicological Research, focuses on the identification of quality standards for evaluating gene expression measurement platforms, and on factors determining performance assessments and

TABLE 24.2

Flowchart for the Identification of Toxicant-Sensitive Genes by Using Microarray Technology

1. Statistical experimental design
2. Processing and storage of samples. Extraction of RNA from samples
3. Evaluation of RNA quantity and quality
4. Microarray sample preparation and platform selection
5. Hybridization and raw data management
6. Establish the appropriate bioinformatic tools. Convert images in pixels to raw expression evels and examine images for quality control
7. Filter, normalize, and standardize expression level of data
8. Statistical analysis of gene expression microarray data
9. Validation of potential candidate genes by RT-PCR, Western blot, or immunohistochemistry
10. Utilization of publicly available data to predict functional significance of each potential gene candidate
11. Identification and comparative analysis of gene clusters in terms of known signaling pathways
12. Identification of candidate genes in terms of their chromosomal locations with regions of susceptibility to toxicant effects
13. Determine overlap of candidate gene list with genes identified in microarray studies of animals
14. Consistence in Gene Ontology terms and pathways in intersite and cross-platform comparisons
15. Utilization of biomarkers for the development of therapeutic strategies

limitations of the different data analysis methods [19]. This initiative is a powerful tool to build a consensus on the utilization of microarray data for clinical purposes.

The creation of specialized bioinformatics tools, such as the development of new algorithms and the creation of new software, is critical for researchers relying on the use of these platforms to apply the best analytical approaches to microarray data; such platforms allow high-throughput quantitative comparison of the transcriptional activity of a potentially large number of individual genes that distinguish untreated and treated groups in a single experiment. New analytical strategies are now available to process the massive data sets generated by microarray studies and to define the significance of implicated genes. One important parameter is the reliability of the statistical significance of the gene profile data. Unless appropriate statistical methods are utilized, microarray data can be a source of false findings because of the elevated number of genes available for analysis. Therefore, choosing a statistical method is a critical step to generate representative differential gene profile data. All these tools make it possible to identify enormous amounts of data, such as the variation in expression levels of genes including those involved in stress response and cellular detoxification, which defines the molecular response to toxicant exposure [4,20–22]. Large-scale analysis of gene expression can also generate global patterns that hopefully will allow us to better understand the complex responses to toxicants by individuals with and without occupational exposure [21]. Accordingly, several studies are reporting that gene expression data allows building carcinogenicity prediction models, which may be used in determining toxicant responses and prevention approaches [23]. In this respect, the long list of genes identified in several toxicogenomic studies [2,3,24–26] is a valuable source for designing potential therapeutic targets.

Microarray results must be validated by methods such as *in situ* hybridization, real-time polymerase chain reaction (RT-PCR), immunohistochemistry, or immunoblotting. *In situ* hybridization histochemistry provides one important confirmation approach of the expression of genes and can be used at multiple levels of resolution to confirm expression alteration of relevant mRNA in specific tissues, its localization to subnuclei or layers, and the type of cells in which it is expressed. One limitation of this technique is the availability of tissue samples in a sufficiently well-preserved state to permit immunocytochemistry. Besides the commonly used immunohistochemistry and immunoblotting techniques, microarray results can be further analyzed by radioactive *in situ* hybridization histochemistry. This technique can be applied in a quantitative manner to confirm the expression levels demonstrated by array studies, and it can also be used to quantify the number of cells in a specific tissue in cases in which the concentration of signal over cells is sufficiently dense.

RT-PCR is another relatively high-throughput technique that can be used for the quantification of steady-state mRNA levels. This technique provides high sensitivity so that weakly expressed transcripts can be detected. It may also be used to independently detect messages (signatures) from a reduced number of cells or small sections of tissue. RT-PCR involves PCR amplification of a segment of the gene from mRNA that is turned into cDNA, and measurement by fluorescence of the PCR product formed by interactions of a green dye with the double-stranded DNA product. Other RT-PCR methods, such as the TaqMan probe assay manufactured by Applied Biosystems with florescent-labeled probes, have increased the specificity of RT-PCR detections by way of mechanisms that activate the fluorescent signal only when the fluorescent-labeled probe is specifically bound to a target sequence. The following sections will discuss the effect of fuels on environmental pollution and the application of toxicogenomics to identify the effects of this type of pollutant on gene expression using cultured cell lines, primary cells, and animal models.

24.3 Environmental Pollutants

The World Health Organization estimates that about a quarter of the diseases facing mankind today are due to prolonged exposure to environmental pollution [27]. Most of these environment-related diseases are, however, not easily detected and may be acquired during childhood and manifested later in adulthood. In addition, more than half of the world's population lives in cities, a trend that is rapidly accelerating, especially in developing countries. In all big cities, areas are merging into huge megapolitan areas, especially along major roadways. Cities require and use large quantities of energy derived from fuels, processing them and generating large quantities of waste products and pollutants, resulting in unsustainable environments that adversely affect human health [28,29]. In this regard, the main source of environmental pollutants is the combustion of fuels, which are very frequent in transportation and in stationary sources, including residential, commercial, and industrial heating. Individuals are exposed to fuels by breathing the air contaminated with fuel components when they drive or ride in a vehicle, jog or bike along roads or park in a public garage, or when they fill up their vehicle's fuel tank. People who work in or live near freeways, refineries, chemical plants, loading and storage facilities or other places that handle crude oil and petroleum products are also exposed to even higher levels of fuel components than the general public and face higher health risks [29,30].

The harmful substances in fuels can disperse into the environment throughout the entire cycle of fuel production: manufacturing, transportation, storage, distribution, and usage. Most commonly, the harmful substances come out the tailpipes of vehicles as exhaust or unburned fuel. Fuel vapors escaping directly from automobile engines and gas tanks, especially on a hot day, are also real sources of environmental contamination. The vapors can also spill into the air during refueling, or when liquid fuel evaporates from a spill. Fuels can contaminate lakes and reservoirs through accidental spills or from motorized boats and personal watercraft. Fuels spilled on the ground or leaking from fuel storage tanks can potentially contaminate groundwater. Also, fuel components are released into the environment during oil drilling, refining and transportation. Indeed, vehicles using diesel are responsible for the largest amount of diesel emissions (~85%), with areas varying widely in the relative proportion contributed by on-road and nonroad sources. Area sources of diesel exhaust (DE) include shipyards, warehouses, heavy equipment repair yards, and oil and gas production operations. The primary point sources that have reported emissions of DE are heavy construction, electrical services, and crude petroleum and natural gas extraction [31,32].

Furthermore, jet airplane traffic and of course airports are being constructed at a heady rate worldwide [33]. Chemical compounds in airborne engine exhaust settle directly onto water, soil, and vegetation, or they can be washed down onto these surfaces when it rains. Even groundwater can be contaminated in this way. Because this traffic is releasing large quantities of greenhouse gases directly into the stratosphere around the globe, this is perhaps a major contributor to the observed global warming. Unfortunately, the very serious environmental consequences and increased risk of diseases from increasing airplane traffic have not thus far been recognized by the scientific community or grabbed the attention of the general public [34,35].

24.3.1 Jet Propulsion Fuel-8

Jet propulsion fuel-8 (JP-8), a kerosene-based fuel used by the U.S. and NATO forces, is a complex mixture of both aliphatic and aromatic hydrocarbons, the latter of which is associated with DNA damage and carcinogenesis. JP-8 is toxic if absorbed into the body in sufficient amounts. Its low volatility characteristics allow it to be a potential toxic and irritant substance on the respiratory system in aerosol, vapor, or liquid forms. JP-8 jet fuel is essentially the same as commercial jet fuel (Jet A and A-1, ASTM D1655 and DEFSTAN 91-91, respectively) except for the inclusion of several additives that inhibit icing, reduce corrosion, and dissipate static electricity [36]. JP-8 was developed as a replacement fuel for JP-4 and, with safety in mind, was designed to reduce the inherent risks of fire and toxicity associated with JP-4. In contrast to the gasoline-like nature of JP-4, JP-8 is more kerosene-like. That is, relative to JP-4, JP-8 is formulated to contain more long-chain aliphatic hydrocarbons and fewer low–molecular weight constituents that tend to increase the flash point and lower the vapor pressure of the fuel allowing JP-8 to remain in its liquid phase much longer than JP-4 [37]. Because of the multipurpose nature of the JP-8 fuel, any varied types of exposure to JP-8, including occupational exposure for those who work on jet engines, occur during cold aircraft engine starts, aircraft fueling and defueling, engine and fuel cell maintenance, maintenance of related equipment and machinery, use of tent heaters, and cleaning or degreasing with fuel, fuel transportation, and accidental spills. This means that if JP-8 is unintentionally released into the environment, then fuel exposure times could increase along with health risks [38].

One of the first studies on JP-8 was carried out on Swedish jet-motor workers who displayed acute central nervous system effects, including dizziness, headache, nausea, and fatigue [39]. The preliminary findings indicated that there were acute effects on cognitive and motor processes associated with jet fuel exposure and significant chronic carryover effects on conditioned response, balance, cognitive, and motor responses associated with fuel maintenance work in the military personnel who worked with the fuel. Exposure to JP-8 has also been associated with symptoms such as fatigue, headache, and skin irritation [40,41]. Several studies with laboratory animals have shown that exposure to JP-8 in vapor, aerosol, or liquid form affects immunologic [42,43], neurological, and pulmonary [44,45] functions. Low concentrations of aerosolized JP-8 in mice also resulted in a marked decrease in immune cell numbers and organ weight as well as suppression of immune function [43]. Continued exposure to JP-8 targeted morphological alterations in the bronchiolar epithelium, inducing interstitial edema, Clara cell vacuolization, and necrosis. Alveolar changes included sporadic pulmonary edema, intra-alveolar hemorrhage, alterations in alveolar epithelial type II cells, and accumulation of inflammatory cells, demonstrating histological (i.e., morphological) damage in the pulmonary system as a direct effect of chronic JP-8 exposure [46]. Even at lower doses that mimic occupational exposure in humans, lungs of rats exposed to an aerosol of JP-8 showed deterioration of the alveolar–capillary barrier and sporadic areas of red blood cell accumulation within alveolar spaces, and occasional formation of perivascular edema was also apparent [2].

Ultrastructural evaluation of lung tissue focused on alveolar type II epithelial cells and the terminal bronchial airway epithelium indicated that the number and size of surfactant-producing lamellar bodies seemed to be increased in alveolar type II epithelial cells of rats exposed to JP-8 [2]. Other studies on JP-8 pulmotoxicology have revealed that jet fuel exposure provoked significant physiological [45,47,48], cellular [4,48,49], proteomic [50,51], and genomic changes [2]. Pathological damage in airways was characterized by loss of epithelial barrier integrity and alterations of ventilatory function in bronchial and bronchiolar airways [45,47,48,52]. These parameters were suggested as important indicators for lung injury promoted by the fuel.

24.3.2 Diesel

Diesel fuel is widely used throughout our society, powering the trucks that deliver products to our communities, the buses that carry us to school and work, the agricultural equipment that plants and harvests our food, and the backup generators that can provide electricity during emergencies. Diesel engines have historically been more versatile and cheaper to run than gasoline engines or other sources of power. Unfortunately, the exhaust from these engines contains substances that can pose a risk to human health. Diesel exhaust is a complex mixture of thousands of gases, fine particles, and vapors released by diesel-fueled compression-ignition engines. The health effects of diesel exposure include eye, throat, and bronchial irritation, coughing, phlegm, and neurophysiological symptoms [53]. Pulmonary exposure to diesel may result from aspiration of liquid during manual siphoning or inhaling the aerosol of micrometer-sized diesel particles liberated by leaking of diesel onto hot engine manifolds [54]. Diesel also includes many known or suspected cancer-causing substances, such as benzene, arsenic, and formaldehyde. Accordingly, evidence from occupational studies has suggested that chronic DE exposure contributes to lung cancer mortality [55–57]. Diesel fuel also contains other harmful pollutants, including nitrogen oxides (a component of urban smog). Occupational exposure to diesel may potentially occur during manual filling or discharge operations within the petrochemical industry, repair or service of diesel engines, or from practices where diesel is used as a cleaning agent or solvent. Large-scale environmental contamination has occurred following the release of diesel from storage tanks and sea tankers, and some concern has been expressed over health effects of vapor arising from contaminated soil [53].

Diesel exhaust particles (DEP) are composed of a carbon core coated in organics, hydro-carbons, sulfates, and metals. Diesel ultrafine particles or nanoparticles are so small they can enter the bloodstream from the lungs. Several studies have found that DEP at lower levels promote the release of cytokines, chemokines, immunoglobulins, and oxi-dants in the upper and lower airways, whereas at high levels it can act as a nonspecific airway irritant. DEP has also been associated with worsening respiratory symptoms in individuals with preexisting allergies or asthma [58,59]. The induction of interleukin-8, granulocyte–macrophage colony-stimulating factor, and interleukin-1β release by DEP in a time-dependent manner suggested that DEP itself might potentially promote asthmatic symptoms at high exposure levels [60–62]. DEP inducing apoptosis in pulmonary alveolar macrophages (AM) has been associated with reactive oxygen species (ROS) production and impairment of mitochondrial function before and at the onset of apoptosis. Mitochondrial perturbation leading to an orderly sequence of events started with a decrease in mitochon-drial membrane potential, followed by cytochrome *c* release and development of mem-brane asymmetry [63].

The worldwide multipurpose use of diesel and JP-8 jet fuels provides opportunities for many varied types of exposures to the unburned fuel. Repeated exposure to hydrocarbon fuel is a very real risk for both civilian and military populations. In many commercial or industrial workplaces, occupational assignment to enclosed areas routinely containing one of more types of hydrocarbon fuel in the form of vapor or aerosol originating from leakage, spillage, and normal venting is very common [64,65]. Moreover, the exposure to residual fuel on skin or clothing also increases the risk of developing pathological condi-tions in family members of aircraft workers [66]. Therefore, the utilization of toxicogenom-ics is important to determine the effect on individuals with short and extensive exposure to fuels. Studies using toxicogenomics can assess health outcomes and provide recom-mendations and guidance for the prevention of harmful effects, or provide the basis for the development of a therapeutic strategy to counteract exposure to fuels.

24.4 Toxicogenomics in Fuels

24.4.1 Toxicogenomic Analysis of the Effects of JP-8

Treatment of several human cell lines with JP-8 induced cell death that exhibits various biochemical and morphological characteristics of apoptosis, such as caspase-3 activation and the classic apoptotic biochemical and morphological alteration events before apoptotic cell death [4]. Macroarray analysis identified 26 genes out of a total of 439 apoptosis-related or stress response–related genes analyzed whose expressions were either upregulated or downregulated, indicating that the toxicity of this fuel is mediated by complex signaling pathways with prominent expressions of genes with functions mainly associated with termination of proliferation or the induction of apoptosis [22]. Genes whose expressions were increased by JP-8 were p38 MAPK kinase, growth arrest–specific 1 (GAS1), tran-scription factor Dp-1 (TFDP1), E2F transcription factor (E2F1), E2F transcription factor 5, p130-binding (E2F-5), caspase-3, apoptosis-related cysteine peptidase (CASP3), caspase 9, apoptosis-related cysteine peptidase (CASP9), heat shock 70 kDa protein 8 (HSC70), heat shock protein 90 kDa α (HSP90), and glutathione S-transferase ω1 (GSTO1). Furthermore, vimentin (VIM) gene expression was consistently downregulated in Jurkat cells treated with JP-8. Consistent with these findings, another microarray study showed that exposure of rats to the same toxicant resulted in changes in the expression of genes that mainly express stress response and apoptosis induction in brain tissue. Interestingly, downregulation of VIM was also observed in the brain tissue of rats exposed to the fuel [67]. Changes in these genes of rat brains exposed to JP-8 may explain the neurological and neurobehavioral deficits in airport workers experiencing long-term, multiroute jet fuel exposure [39,40].

Constant exposure to pollutants may induce altered cell growth, which may reflect more general phenomena in carcinogenesis. This postulate is supported by several studies revealing that dermal exposure to JP-8 induces disruption of the skin barrier function, skin irritation, and alteration of the skin structure [68,69]. Furthermore, chronic application of a jet fuel after treatment with dimethylbenzanthracene has been shown to act as a tumor promoter in a mouse model, possibly resulting from frequent irritation [70]. Chronic exposure to DE or intratracheal ingestion promoted lung cancer in rats [71,72]. Because the occurrence of oxidative stress and changes in cellular redox is believed to contribute to the process of carcinogenesis, it is important to define the protective mechanisms that may be activated in cells exposed to a specific toxicant.

On cold days, it is especially troublesome for military flight-line personnel, whose necessary proximity to the aircraft during engine start-up virtually ensures some skin and clothing exposure to JP-8, even when proper protective gear is donned. Because there is a high risk for a large number of military and civilian personnel exposed to jet fuel JP-8 to suffer percutaneous absorption of this fuel, one microarray analysis in normal human epidermal keratinocytes was intended to identify the target genes responsive to the JP-8 jet fuel toxicity [21]. Increased transcription of genes involved in stress response, such as transmembrane protease, serine 2 (TMPRSS2), plasminogen activator, urokinase (PLAU), and its receptor PLAUR mRNAs were observed. Transcripts involved in metabolism, such as acyl-CoA thioesterase 12 (ACOT12) and pyruvate dehydrogenase α1 (lipoamide; PDHA1), were upregulated whereas succinate dehydrogenase complex, subunit C, integral membrane (SDHC), and aldolase B, fructose-biphosphate (ALDOB) showed reduced expression. No apoptotic genes were overexpressed, which was consistent with the cell viability data, in which the death of keratinocytes was rarely detected. Additionally, upregulated genes encoding for the structural protein include keratin 18 (K18) and VIM, and downregulation of myosin light-chain kinase (MYCK), a serine–threonine kinase, was also detected. This signature seemed to be relevant to the increased resistance of keratinocytes against JP-8 toxicity [21].

Another approach used microarrays to measure the changes in mRNA levels in the epidermis of Fisher rats exposed to JP-8 [73]. Genes associated with the biological processes of metabolism, development, and death were upregulated at 1 h in the epidermis exposed to the fuel. Such changes affecting genes whose products serve as triggers or initiators of the JP-8–induced inflammatory process. ERK/MAPK pathway, interleukin-6, platelet-derived growth factor, and p38 MAPK–signaling pathways had the most significant changes in gene expression due to JP-8 exposure. Several transcription factors related to the changed signaling pathways, especially FBJ murine osteosarcoma viral oncogene homologue (FOS), Jun oncogene (JUN), and phosphoinositide-3-kinase regulatory subunit 2β (PIK3R2) were changed at the first skin sample collection time [74]. Four hours after exposure, the chemokine and platelet-derived growth factor–signaling pathways were still altered. By this time, T-cell signaling and G-coupled receptor–signaling pathways also were significantly changed. In addition, metabolic pathways for sterol biosynthesis, synthesis, and degradation of ketone bodies, along with valine, leucine, and isoleucine degradation were altered for the first time at 4 h after the beginning of the JP-8 exposure. By 8 h, thirteen additional metabolic pathways were significantly altered in addition to the three metabolic pathways altered at 4 h. The majority (64%) of the gene transcripts related to metabolism that were changed at 8 h were decreased; the exceptions were pathways for sterol biosynthesis, as well as synthesis and degradation of ketone bodies. Although none of the metabolic pathways that were significantly changed in the later time points were changed at 1 h, there were two metabolism-related genes in those pathways that were increased at all three skin-sampling points; cytochrome P450, subfamily 1, polypeptide 1 isoform (CYP1B1), and ornithine decarboxylase 1 (ODC1) [74].

Because JP-8 jet fuel is composed of hundreds of hydrocarbons, it has been of enormous interest to determine the correlations between toxic responses to JP-8 and global changes in the overall genetic expression profiles of skin cells exposed to this complex yet important toxicant. In this regard, a study tried to correlate if aromatic or aliphatic components such as undecane, tetradecane, trimethylbenzene, or dimethylnaphtalene could mimic the JP-8–induced gene expression response in rat skin [73]. Although significant gene alterations were induced with each of these chemicals, the signature promoted by JP-8 definitely could not be mimicked by any of these components alone. Altogether, the signatures observed in skin cells clearly indicate that JP-8 induced marked changes in the expression of various genes with diverse functions, which are probably responsible for inducing skin irritation in occupational exposure [73,74].

To identify genes whose expression is substantially altered in the lungs of rats subjected to direct inhalation of JP-8 under controlled conditions, lung tissue isolated from rats exposed to JP-8 at 171 or 352 mg/m^3 for 1 h per day for 7 days were compared with that of lung tissue derived from control animals exposed to air alone were analyzed using the GeneChip Rat Genome U34 Array A (Affymetrix) [2]. The higher dose (352 mg/m^3) is estimated to mimic the level of occupational exposure in humans. The rationale for the low dose was to assess whether the gene expression patterns, which are affected at the higher dose, are maintained or negatively altered at the much lower exposure dose. The expression of most (48/56, 86%) of the genes affected by JP-8 at 171 mg/m^3 was downregulated, whereas that of 42% (28/66) of the genes affected by JP-8 at 352 mg/m^3 was upregulated. The lungs of rats exposed to JP-8 at the occupationally relevant dose of 352 mg/m^3 manifested a 5.5-fold increase in expression of the gene for γ-synuclein, a centrosomal protein that plays an important role in the regulation of cell growth and which is present at a relatively low level in normal lung tissue. The expression of genes whose products contribute to the cellular response to oxidative stress or to toxicants, including glutathione S-transferases (Gsta1, GstaYc2) and cytochrome P450 (CYP2C13, Cyp2e1), was also prominently increased in lung tissue from rats exposed to JP-8 at the higher dose. In contrast, the expression of none of the apoptosis-related genes represented on the microarray was affected by JP-8 at either dose, which was consistent with the histological analysis showing minimal cell damage in the lungs of the JP-8–treated rats. The abundance of mRNAs for various structural proteins, including myosin heavy chain 7 (Myh7), α-actin, and β-actin, was increased by exposure to JP-8 at 352 mg/m^3, whereas that of the mRNA for high–molecular weight microtubule-associated protein 2 (HMW-MAP2) was decreased. Expression of the gene for the inositol 1,4,5-trisphosphate receptor (InsP3R1), a ligand-gated Ca^{2+} channel that mediates the release of Ca^{2+} from intracellular stores, was reduced by a factor of 10 in the lungs of rats exposed to JP-8 at 352 mg/m^3. Expression of the genes for aquaporin 1 (Aqp1) and aquaporin 4 (Aqp4), proteins that mediate water transport in various tissues including the lungs, was increased and decreased, respectively, in the lungs of rats exposed to JP-8 at 352 mg/m^3 [2].

As mentioned in the Introduction, the purpose behind microarray data analysis is to discover patterns of expression across multiple gene families; hence, the analysis, under these conditions, assumes that genes can interact and that their expression is thus correlated with one another. Therefore, a large-scale analysis of gene expression using different types of samples (human cell lines and rat lung tissues) exposed to JP-8 have generated patterns that allow a better understanding about the complex responses to JP-8 by occupational exposure. Initially, it was proposed that the molecular signatures of JP-8 in cells or tissues may be caused in part by only a few of the components of JP-8 when they were analyzed separately [21]. However, different experiments using aromatic or aliphatic components of JP-8 proved that components in JP-8 produced different profile changes when compared with the signatures promoted by the mixture [21,73–75].

There is no doubt that it is necessary to integrate proteomic analysis with gene expression profiles for the prediction of cellular perturbations that are regulated by the JP-8 toxicant in simulated occupational models. Accordingly, the comparison of proteomic experiments, albeit in different rodents using approximately the same dose of JP-8, have already revealed a substantial agreement in genes associated with extracellular matrices such as those encoding to actin and myosin, which were upregulated as a consequence of JP-8 treatment [50,76]. Increased expression of these contractile filament genes may imply a protective response of cells to the oxidative stress induced by the jet fuel. Furthermore, proteomic data obtained from lungs of animals exposed at higher doses of JP-8 (1000 and 2500 mg/m^3) [51] compared with that used for gene expression analysis [2] also demonstrated a remarkable correspondence in the upregulation of certain detoxification genes such as GSTs, which are known to play an important role in the defense mechanisms against the toxic effects of JP-8 components [49].

It is logical to propose that the modulation of several genes encoding proteins that are part of the signaling pathways in response to the fuel indicates that JP-8 causes toxicity. Although the value of these studies is the identification of changes in the genes involved in various intracellular signaling cascades, genes that have significance to toxicity are of particular interest for monitoring occupational exposure to JP-8, which can cause clinically adverse outcomes. The information provided by toxicogenomic studies may also provide important leads for further investigations into the molecular basis of JP-8 cytotoxicity and design mechanisms of protection against it.

24.4.2 Toxicogenomic Analysis of the Effects of Diesel

DEP consist of carbon cores that adsorb many organic compounds, which may be dissolved in the lungs after deposition, including polycyclic aromatic hydrocarbons, heterocyclic organic compounds, quinines, aldehydes, and aliphatic hydrocarbons. Several studies have suggested that adsorbed organic compounds affect cells by producing ROS, which can promote inflammation and even induce tumor formation [63]. It is known that DEP and organic compounds induce oxidative and inflammatory effects in lungs, AMs, and endothelial cells. Inhaled DEP generates ROS in the lungs of mice and are probably responsible for protein oxidation, lipid oxidation, and DNA damage in target cells such as macrophages and epithelial cells in the lung [77]. Other adverse effects of particulate matter (PM), including DEP, are dysfunctions of the respiratory and cardiovascular systems. In the lung, DEP can also exacerbate allergic diseases and promote thrombosis and carcinogenesis [78].

Most of these effects may be a consequence of changes in the expression of various genes such as the overexpression of the v-raf murine sarcoma 3611 viral oncogene homologue

(A-raf) and the proliferating cell nuclear antigen [79]. These upregulations of the proto-oncogene A-raf in lung tissue of rats exposed to 2.5 and 7.5 mg/m^3 of DE for 4 continuous weeks were associated with transformation activity and stimulation of the Raf pathway, whereas augmented expression of proliferating cell nuclear antigen may repair DNA damage induced by DE. These events were identified as early indicators of lung carcinogenesis in rats chronically exposed to DE.

As an interesting approach to a comprehensive analysis of the alterations of gene expression on the effects of DEP on pulmonary cells, a gene expression analysis was carried out on AM exposed for a short period to DEP [80]. This study envisaged analyzing AMs and lung epithelial cells, which are important cellular targets for environmental pollutants such as DEP in the lung. AM ingests inhaled foreign substances such as particles and microorganisms. During ingestion, AM transmits stimulatory signals indicating the presence of a foreign substance to other cells by releasing chemical mediators and pro-inflammatory cytokines. This mechanism is thought to be involved in the induction of aggravation of pulmonary disease, including asthma, by DEP [81]. Thus, the early response of these lung cells to the toxicological effects of the extract was investigated using an Atlas Rat Toxicology Array II, which includes 450 rat cDNAs. The gene expression signature indicated that mRNA levels of heme oxygenase 1 (decycling; Hmox1), heme oxygenase 2 (decycling; Hmox2), peroxiredoxin (Prdx2), glutathione S-transferase π1 (Gstp1), NAD(P) H dehydrogenase quinone 1 (Nqo1), and proliferating cell nuclear antigen were elevated when compared with the control values. Particularly relevant were the expression levels of Hmox1, an inducible protein that accumulates in response to oxidative stress, which was significantly augmented in AM exposed to DEP in a dose-dependent manner. These results suggested that AM may play a crucial role in pulmonary defense via acute inflammation by DEP and against other pollutants through the induction of antioxidative enzymes [82].

A complimentary study by Kobayashi's group [83] focused on addressing the mechanisms underlying the effects of DEP on lung diseases. For this purpose, they examined the change of expression in the rat alveolar type II epithelial cell line SV40T2 after exposure to an organic extract of DEP. The gene expression analysis revealed pronounced changes in genes encoding to drug metabolism, antioxidation, cell cycle, and coagulation. Noteworthy was the upregulation of Hmox1 and the downregulation of transglutaminase 2 and C polypeptide (Tgm2) mRNA, a regulator of coagulation. These alterations were consistently reduced by the thiol antioxidant *N*-acetyl-cysteine, suggesting that augmented expression of Hmox1 *in vivo* [80] and *in vitro* [83] microarray data as well as in other toxicological studies [84,85] is an oxidative stress response from DEP-induced pulmonary defense that is regulated by the activation of the redox-sensitive transcription factors nuclear factor (erythroid-derived 2)-like 2 (Nrf2), nuclear factor κ-light-chain enhancer of activated B cells (NFκB), and activator protein 1 (AP-1) [83]. The treatment with thiol antioxidants inhibiting the adjuvant effect of DEP in the induction of ROS production clearly shows that oxidative stress is a key component causing the biological effects of ambient PM and may explain the exacerbation of allergic inflammation by DEP [83].

An epidemiological study determined that PM with a diameter of less than 10 μm (PM10) are pollutants accounting for 20% to 80% of the mass of airborne particles from vehicular activities, promoting increased cardiorespiratory problems, morbidity, and mortality rates [86]. Another study addressed the bioreactivity effect of exhaust particles in rat lungs. PM10 is composed of a mixture of particles including minerals, metals oxides, sea salt, biological components, soluble ionic species, and organic micropollutants (polycyclic aromatic hydrocarbons and nitro-polyaromatic hydrocarbons). Nitro-polyaromatic hydrocarbons induce mutations in mammalian and bacterial cells and also induce sister chromatid

exchange and chromosomal aberrations (numerical and structural) in cultured mammalian cells [87,88]. This mixture can vary from city to city, from day to day, and from hour
to hour. To identify signatures from rat lung in response to a small instilled mass of DEP,
NaCl (control) and 1.25 mg of DEP were instilled into rat lung and responses were determined by alterations in lung permeability and the presence of inflammation and epithelial
cell markers in lavage fluid [89]. DEP caused a slight pulmonary edema and approximately
5% of the rat stress genes were altered in response to DEP instillation. Other genes that
showed consistent change in expression were the N-terminal EF-hand calcium binding
protein 2 (Necab2), cyclin D1 (Ccnd1), mitogen-activated protein kinase kinase 5 (Map2k5),
and cell division cycle 25 homologue B (Cdc25b). The authors also observed a strong con-
cordance between the macroarray data and conventional toxicology and also showed that
DEP induced a low bioreactive response in a healthy rat lung [89]. This work indicated that
the macroarray is a powerful tool that may provide reliable data about the effects of both
soluble and nonsoluble components of the PM10 mixture.

To gain insight into the acute mechanisms of action of PM10 and to identify more specific biomarkers of PM10-induced damage/repair in the lung, another group made comparisons of gene profiles from the lung tissue of male rats instilled with whole urban PM10
or the soluble fraction alongside sham-treated animals (vehicle instillation only) [90]. A
mild but significant change in lung permeability was observed after the instillation of
a high (10 mg) dose of the whole PM10 as adjudged by increases in lung-to-body weight
ratio and total acellular lavage fraction. Such effects were less marked after instillation of
a water-soluble fraction (80% of the total mass) but histological examination showed that
lung capillaries were swollen in size with this treatment. The toxicogenomic assessment
evidenced significant change in a group of nine lung genes whose abundance was significantly increased (≥2.0-fold) or decreased (≤2.0-fold). Seven of the transcripts, which had
functional classification in G protein–coupled receptors, growth factors, hormones, and
intracellular kinase were downregulated and only two genes with hormone and interleukin functions were upregulated. Of interest was the upregulation of interleukin 2, a stimulatory factor that modulates the proliferation of pulmonary epithelial type II cells, which
has been indicated as a nonspecific cellular repair response following epithelial injury [90].

Many epidemiologic studies have shown that exposure to ambient PM leads to increases
in morbidity and daily mortality caused by respiratory diseases. Because DEP are the main
constituents of PM 2.5 (which are PM with a <2.5 μm diameter), their access to the body may
promote a variety of respiratory diseases, including asthma, pulmonary edema, and lung
cancer. In this regard, studies have shown that intratracheal inoculation of DEP enhances
antigen-specific IgG$_1$ production, eosinophilic airway inflammation, and the expression of
cytokines in the murine lung [81,91]. In addition, DEP synergistically enhances acute lung
injury related to lipopolysaccharide (LPS), a component of the cell wall of Gram-negative
bacteria that induces or exacerbates a variety of lung conditions such as asthma [92] and
acute lung injury [93]. A similar rodent study showed an increase in the secretion of inter-
leukin-1 β in peripheral blood mononuclear cells when compared with rodents exposed to
LPS alone, indicating a synergistic enhancement of these two compounds in neutrophilic
lung inflammation [82,94]. This synergistic enhancement has also been concomitant with
the increased gene expression of proinflammatory molecules such as interleukin-1β and
macrophage inflammatory protein 1a [81,91]. In this respect, a cDNA microarray study was
conducted to elucidate the effects of DEP on the global pattern of gene expression related to
LPS in the murine lung 4 h after intratracheal instillation in each group. Accordingly, 1073
genes were found upregulated in the LPS group, although only 38 and 204 were upregulated in the DEP and DEP + LPS groups, respectively. In the set of genes whose expression

was greater than or equal to 6, twenty-six genes were upregulated after combined instillation of DEP and LPS as compared with eighteen genes upregulated after LPS instillation and three upregulated by DEP administration. The concomitant administration of DEP and LPS induced a dramatic upregulation of several genes, including chemokine (C-X-C motif) ligand 10 (Cxcl10), lipocalin 2 (Lcn2), metallothionein 1 (MT-1), metallothionein 2 (MT-2), and S100 calcium binding protein A9 (S100a9), all of whose gene expression levels were increased by 20-fold in the DEP + LPS group but not in the LPS group [95]. These findings confirmed that DEP affects the respiratory and immune systems *in vivo* and *in vitro* in the presence or absence of LPS with a broad effect on gene expression changes that seems to reflect the synergistic aggravation of acute lung injury by LPS and DEP when compared with the signature of tissues exposed to DEP alone.

Experiments in mice have shown that short-term exposures to diluted DE enhanced the development of allergic eosinophilia. The objectives of these studies were to investigate the adjuvant effects, postantigen challenge (ovoalbumin), of a short-term inhalation exposure to diesel DE diluted to yield 500 and 2000 µg/m^3 [96]. Exposures were conducted for 4 h/day over 5 consecutive days. On days 0, 1, and 2, mice were intranasally instilled with ovoalbumin or saline. On day 18, mice were either challenged with ovoalbumin or saline, and all mice were challenged with ovoalbumin on day 28. Effects were assessed after the second challenge to confirm that mild adjuvancy was accomplished. Gene expression analysis was carried out in lung tissues taken 4 h after the last DE exposure to identify sets that were altered by the treatments and may be associated with later development of clinical disease. There was a good consistency in the gene expression data with a group of genes such as CD177 antigen (Cd177), CD14 antigen (Cd14,) chemokine (C-X-C motif) ligand 1 (Cxcl1), chemokine (C-X-C motif) ligand 5 (Cxcl5), interferon-induced transmembrane protein 1 (Ifitm1), lipocalin 2 (Lcn2), polymeric immunoglobulin receptor (Pigr), prominin 1 (Prom1), regenerating islet-derived 3γ (Reg3g), and resistin-like α (Retnla). These results showed changed alteration in animals exposed to two different DE dilutions and demonstrated that relatively short exposures to DE at concentrations that mimic occupational exposure induced immunological sensitization to allergens [96].

Because DEP are deposited mainly in the alveolar region of the lungs where they may affect alveolar epithelial cells adversely, another study used rat type II alveolar epithelial (SV40T2) cells with the purpose of identifying and characterizing genes whose expression is altered by exposure to fractions of DEP [97]. To determine the relationship among the properties of chemicals in DEP extracts and the gene expression that may be regulated by exposure to fractions of DEP extracts, a dichloromethane-soluble fraction of DEP was further fractionated into *n*-hexane soluble fraction and *n*-hexane insoluble fraction. This study revealed that DEP extracts upregulated the expression of genes related to drug metabolism, antioxidant enzymes, cell cycle/apoptosis and coagulation, and in addition, that the *n*-hexane soluble fraction of DEP extracts contains aliphatic and polycyclic aromatic hydrocarbons, and the *n*-hexane insoluble fraction contains oxygenated compounds and has strong oxidative properties. Rat lung epithelial cells SV40T2 were exposed to these fractions (30 µg/mL) for 6 h. The dichloromethane-soluble fraction predominantly upregulated genes associated with drug metabolism: cytochrome P450, family 1, subfamily a, polypeptide 1 (Cyp1a1), and glutathione S-transferase α3 (Gsta3); oxidative stress response: heme oxygenase (decycling) 1 (Hmox-1) and sulfiredoxin 1 homologue (Srxn1); and cell cycle/apoptosis: oxidative stress–induced growth inhibitor 1 (Osgin1). The genes upregulated by the *n*-hexane soluble fraction were mainly associated with drug metabolism (Cyp1a1 and Gsta3). The genes upregulated by *n*-hexane insoluble fraction included antioxidant enzymes (Hmox-1 and Srxn1), genes responsive to cell damage, such as those functioning in cell cycle regulation or apoptosis (Osgin1), and genes that modulate coagulation pathways. These findings suggested that the *n*-hexane soluble and insoluble fractions regulated specific genes that respond differently to the chemical properties of each fraction. Similar findings were also detected in the adenocarcinomic human alveolar basal epithelial cell line (A549) exposed to high concentrations of PM, which induced the expression of cytochrome P450, family 1, subfamily A, polypeptide 1 (CYP1A1), and heme oxygenase (decycling) 1 (HMOX-1) mRNAs [98]. Although in this study, several cancer-related genes were also overexpressed, it is important to note that although A549 cells are frequently used as an *in vitro* model for a type II pulmonary epithelial cell model for drug metabolism, this cell line is probably not the most appropriate to determine the toxicological effects of DEP due to its high capacity for tumorigenicity.

To test the effect of DEP in human cells, Verheyen et al. [99] used *in vitro* cultures of the human acute monocytic leukemia cell line (THP-1), focusing on determining the biochemical pathways related to immunotoxicology that are influenced by exposure to DEP; this approach used a microarray of 13,000 clones to evaluate the changes in gene expression of THP-1 cells that was differentiated to macrophages and exposed to DEP at two different time points (6 and 24 h). From the 6-h time course experiment, 50 genes were upregulated and 39 genes were downregulated, whereas for the 24-h period, 54 transcripts were upregulated and 60 transcripts were downregulated. It is evident that a variety of biological processes can be influenced by DEP, and among them, nine genes were upregulated at both time points: BTG family, member 2 (BTG2), chromosome 1 open reading frame 66 (C1orf66), cytochrome P450, family 1, subfamily B, polypeptide 1 (CYP1B1), exocyst complex component 6B (EXOC6B), interleukin 1β (IL-1β), integrin-β7 (ITGB7), leupaxin (LPXN) thrombomodulin (THBD), tumor necrosis factor receptor superfamily, member 1B (TNFRS1B); and seven were consistently downregulated: butyrobetaine (γ), 2-oxoglutarate dioxygenase (γ-butyrobetaine hydroxylase) 1 (BBOX1), CD36 molecule thrombospondin receptor (CD36), CDC-like kinase 1 (CLK1), high-mobility group box 2 (HMGB2), peroxiredoxin 1 (PRDX1), protein kinase, cAMP-dependent, catalytic, β (PRKACB), and stathmin 1 (STMN1). Using this strategy, the results obtained were quite modest (<3-fold induction/repression) on gene expression variation, with less than 1% of the genes showing significant regulation by DEP, although a broad range of pathways were affected. These evidences strongly indicate that in human AM the toxic effects of DEP seem to affect several pathways [99].

An exploratory study assessed the global gene expression profile in peripheral blood mononuclear cells of healthy human volunteers that were exposed to realistic ambient of DE exhaust [100]. This study identified altered expression of genes involved in oxidative stress, inflammation, leukocyte activation, cell adhesion, cell migration, and vascular homeostasis, which have been previously associated with pathophysiological mechanisms associated with PM health effects. Evidence showing that DE promotes skewed immune response of T helper type 2 (Th2) and T cells were the basis for clarifying the molecular mechanism related with the occurrence of this phenomena in immune cells exposed to DEP [101]. Accordingly, the gene expression profiling of stimulated human T cells pretreated with DE identified pronounced upregulation of ferritin heavy polypeptide 1 (FTH1), heme oxygenase (decycling 1; HMOX1), glutamate-cysteine ligase (GCL), and thioredoxin reductase 1 (TXNRD1) transcripts. These genes were previously demonstrated to be induced by ROS accumulation and to depend on the Nrf2–Keap-1 pathway activation. Using the same approach previously described by Koike et al. [83], pretreatment with the thiol antioxidant *N*-acetyl-cysteine caused a significant suppression on gene changes. This study strongly supported the notion that ROS production is the key driver of DEP-mediated oxidative stress and is also responsible for transcriptional deregulation [101].

Several reports have associated DEP in aggravating pulmonary diseases such as asthma and chronic bronchitis. Therefore, the identification of genes specifically altered by DEP exposure would provide not only a better understanding of mechanisms responsible for the biological effects of DEP but also may identify molecular markers that are sensitive to the toxic exposure of ambient PM. Microarray technology proved to be an excellent tool for monitoring the simultaneous expression of thousands of genes, and it provided rapid and immediate information regarding the genes that are affected by diesel components. In this respect, gene expression arrays have already been successfully used to examine the gene expression changes associated with DEP instillation *in vivo* and *in vitro*, and they showed that consistent and reproducible gene changes occur in the lung in response to the fuel. The incorporation of microarray analysis along with conventional techniques to assess pulmonary damage resulting from the instillation of well-characterized urban PM10 samples have also proved that this technology is capable of providing lists of candidate biomarkers for additional specific studies that may be useful in the identification of biological end points in lung toxicology such as chronic respiratory disorders.

24.5 Conclusions

The genomic examination of cells or tissues exposed to environmental toxicants can provide enormous information about changes in gene expression and associated diseases. The sensitivity and specificity reported in the different toxicogenomics studies suggest that microarray technology has the potential to confirm and predict the potential adverse effects of environmental pollutants. The correlation of microarrays with proteomics should improve our understanding of the molecular basis of exposure to environmental pollutants as well as to the toxicological effects that may efficiently monitor complex molecular responses and the prediction of potentially harmful toxicants. Of course, important issues that have to be considered are standardization and validation for the assessment of toxicity. Most importantly, the standardization of these parameters can clearly identify the molecular events that are affected by environmental toxicants. The altered expression of specific genes related to inflammation, immune response, and oxidative stress are major characteristics of increased oxidative stress in samples exposed to pollutants and, for that reason, transcripts that are modulated in response to environmental toxicants exposure. These events may represent a unique molecular signature of the toxicant effect that potentially would serve as a diagnostic or predictive signature for levels of toxicity. In this respect, specific expression profiles may lead us to hypothesize which signature is preferentially targeted by a specific environmental toxicant. Finally, the wide list of genes that are identified could be a source of potential therapeutic targets because most of these genes may be involved in key mechanisms that overcome diseases.

References

1. Baken, K.A., R.J. Vandebriel, J.L. Pennings, J.C. Kleinjans, and H. van Loveren. 2007. Toxicogenomics in the assessment of immunotoxicity. *Methods* 41:132–141.

2. Espinoza, L.A., M. Valikhani, M.J. Cossio, T. Carr, M. Jung, J. Hyde et al. 2005. Altered expression of gamma-synuclein and detoxification-related genes in lungs of rats exposed to JP-8. *American Journal of Respiratory Cell and Molecular Biology* 32:192–200.

3. Kaposi-Novak, P., J.S. Lee, L. Gomez-Quiroz, C. Coulouarn, V.M. Factor, and S.S. Thorgeirsson. 2006. Met-regulated expression signature defines a subset of human hepatocellular carcinomas with poor prognosis and aggressive phenotype. *Journal of Clinical Investigation* 116:1582–1595.

4. Boulares, A.H., F.J. Contreras, L.A. Espinoza, and M.E. Smulson. 2002. Roles of oxidative stress and glutathione depletion in JP-8 jet fuel–induced apoptosis in rat lung epithelial cells. *Toxicology and Applied Pharmacology* 180:92–99.

5. Espinoza, L.A., F. Tenzin, A.O. Cecchi, Z. Chen, M.L. Witten, and M.E. Smulson. 2006. Expression of JP-8–induced inflammatory genes in AEII cells is mediated by NF-kappaB and PARP-1. *American Journal of Respiratory Cell and Molecular Biology* 35:479–487.

6. Harris, D.T., D. Sakiestewa, R.F. Robledo, and M. Witten. 1997. Immunotoxicological effects of JP-8 jet fuel exposure. *Toxicology and Industrial Health* 13:43–55.

7. Harris, D.T., D. Sakiestewa, D. Titone, X. He, J. Hyde, and M. Witten. 2008. JP-8 jet fuel exposure suppresses the immune response to viral infections. *Toxicology and Industrial Health* 24:209–216.

8. Gatzidou, E.T., A.N. Zira, and S.E. Theocharis. 2007. Toxicogenomics: A pivotal piece in the puzzle of toxicological research. *Journal of Applied Toxicology* 27:302–309.

9. Inman, A.O., N.A. Monteiro-Riviere, and J.E. Riviere. 2008. Inhibition of jet fuel aliphatic hydrocarbon induced toxicity in human epidermal keratinocytes. *Journal of Applied Toxicology* 28:543–553.

10. Muhammad, F., N.A. Monteiro-Riviere, R.E. Baynes, and J.E. Riviere. 2005. Effect of in vivo jet fuel exposure on subsequent in vitro dermal absorption of individual aromatic and aliphatic hydrocarbon fuel constituents. *Journal of Toxicology and Environmental Health. Part A* 68:719–737.

11. Forbes, L.J., V. Kapetanakis, A.R. Rudnicka, D.G. Cook, T. Bush, J.R. Stedman et al. 2009. Chronic exposure to outdoor air pollution and lung function in adults. *Thorax* 64:657–663.

12. Forbes, L.J., M.D. Patel, A.R. Rudnicka, D.G. Cook, T. Bush, J.R. Stedman et al. 2009. Chronic exposure to outdoor air pollution and markers of systemic inflammation. *Epidemiology* 20:245–253.

13. Calabrese, E.J. 2010. Hormesis is central to toxicology, pharmacology and risk assessment. *Human & Experimental Toxicology* 29:249–261.

14. Fodor, S.P., J.L. Read, M.C. Pirrung, L. Stryer, A.T. Lu, and D. Solas. 1991. Light-directed, spatially addressable parallel chemical synthesis. *Science* 251:767–773.

15. Dalma-Weiszhausz, D.D., J. Warrington, E.Y. Tanimoto, and C.G. Miyada. 2006. The Affymetrix GeneChip platform: An overview. *Methods in Enzymology* 410:3–28.

16. Kreil, D.P., R.R. Russell, and S. Russell. 2006. Microarray oligonucleotide probes. *Methods in Enzymology* 410:73–98.

17. Wolber, P.K., P.J. Collins, A.B. Lucas, A. De Witte, and K.W. Shannon. 2006. The Agilent in situ-synthesized microarray platform. *Methods in Enzymology* 410:28–57.

18. Shi, L., R.G. Perkins, H. Fang, and W. Tong. 2008. Reproducible and reliable microarray results through quality control: Good laboratory proficiency and appropriate data analysis practices are essential. *Current Opinion in Biotechnology* 19:10–18.

19. Shi, L., L.H. Reid, W.D. Jones, R. Shippy, J.A. Warrington, S.C. Baker et al. 2006. The MicroArray Quality Control (MAQC) project shows inter- and intraplatform reproducibility of gene expression measurements. *Nature Biotechnology* 24:1151–1161.

20. Auerbach, S.R., C. Manlhiot, S. Reddy, C. Kinnear, M.E. Richmond, D. Gruber et al. 2009. Recipient genotype is a predictor of allograft cytokine expression and outcomes after pediatric cardiac transplantation. *Journal of the American College of Cardiology* 53:1909–1917.

21. Espinoza, L.A., P. Li, R.Y. Lee, Y. Wang, A.H. Boulares, R. Clarke et al. 2004. Evaluation of gene expression profile of keratinocytes in response to JP-8 jet fuel. *Toxicology and Applied Pharmacology* 200:93–102.

22. Espinoza, L.A., and M.E. Smulson. 2003. Macroarray analysis of the effects of JP-8 jet fuel on gene expression in Jurkat cells. *Toxicology* 189:181–190.

23. Ellinger-Ziegelbauer, H., J. Aubrecht, J.C. Kleinjans, and H.J. Ahr. 2009. Application of toxicogenomics to study mechanisms of genotoxity and carcinogenicity. *Toxicology Letters* 186:36–44.

24. Auerbach, S.S., R.R. Shah, D. Mav, C.S. Smith, N.J. Walker, M.K. Vallant et al. 2010. Predicting the hepatocarcinogenic potential of alkenylbenzene flavoring agents using toxicogenomics and machine learning. *Toxicology and Applied Pharmacology* 243:300–314.

25. Hu, T., D.P. Gibson, G.J. Carr, S.M. Torontali, J.P. Tiesman, J.G. Chaney et al. 2004. Identification of a gene expression profile that discriminates indirect-acting genotoxins from direct-acting genotoxins. *Mutation Research* 549:5–27.

26. Jayaraman, A., M.L. Yarmush, and C.M. Roth. 2005. Evaluation of an in vitro model of hepatic inflammatory response by gene expression profiling. *Tissue Engineering* 11:50–63.

27. World Health Organization. 2007. Threats to Public Health Security. Geneva: World Health Organization.

28. Balmes, J.R., G. Earnest, P.P. Katz, E.H. Yelin, M.D. Eisner, H. Chen et al. 2009. Exposure to traffic: Lung function and health status in adults with asthma. *The Journal of Allergy and Clinical Immunology* 123:626–631.

29. Meng, Y.Y., M. Wilhelm, R.P. Rull, P. English, and B. Ritz. 2007. Traffic and outdoor air pollution levels near residences and poorly controlled asthma in adults. *Annals of Allergy, Asthma & Immunology* 98:455–463.

30. Balmes, J.R. 2009. Can traffic-related air pollution cause asthma? *Thorax* 64:646–647.

31. Lloyd, A.C., and T.A. Cackette. 2001. Diesel engines: Environmental impact and control. *Journal of the Air & Waste Management Association* 51:809–847.

32. Woodcock, J., D. Banister, P. Edwards, A.M. Prentice, and I. Roberts. 2007. Energy and transport. *Lancet* 370:1078–1088.

33. Ritchie, G., K. Still, J. Rossi, 3rd, M. Bekkedal, A. Bobb, and D. Arfsten. 2003. Biological and health effects of exposure to kerosene-based jet fuels and performance additives. *Journal of Toxicology and Environmental Health. Part B, Critical Reviews* 6:357–451.

34. Cavallo, D., C.L. Ursini, G. Carelli, I. Iavicoli, A. Ciervo, B. Perniconi et al. 2006. Occupational exposure in airport personnel: Characterization and evaluation of genotoxic and oxidative effects. *Toxicology* 223:26–35.

35. Tunnicliffe, W.S., S.P. O'Hickey, T.J. Fletcher, J.F. Miles, P.S. Burge, and J.G. Ayres. 1999. Pulmonary function and respiratory symptoms in a population of airport workers. *Occupational and Environmental Medicine* 56:118–123.

36. Shafer, L.M., R.C. Striebich, J. Gomach, and T. Edwards. 2003. *Chemical Class Composition of Commercial Jet Fuels and Other Specialty Kerosene Fuels.* AIAA2006–7972.

37. Andersen, M.E., and B.J. Walker. 2003. *Toxicologic Assessment of Jet-Propulsion Fuel 8.* Washington, DC: National Academy Press.

38. Martel, C.R. 1944. *Military Jet Fuels.* 1944–1987.

39. Struwe, G., B. Knave, and P. Mindus. 1983. Neuropsychiatric symptoms in workers occupationally exposed to jet fuel— a combined epidemiological and casuistic study. *Acta Psychiatrica Scandinavica Supplementum* 303:55–67.

40. Smith, L.B., A. Bhattacharya, G. Lemasters, P. Succop, E. Puhala, 2nd, M. Medvedovic et al. 1997. Effect of chronic low-level exposure to jet fuel on postural balance of US Air Force personnel. *Journal of Occupational and Environmental Medicine* 39:623–632.

41. Zeiger, E., and L. Smith. 1998. The first international conference on the environmental health and safety of jet fuel. *Environmental Health Perspectives* 106:763–764.

42. Harris, D.T., D. Sakiestewa, R.F. Robledo, and M. Witten. 1997. Protection from JP-8 jet fuel induced immunotoxicity by administration of aerosolized substance P. *Toxicology and Industrial Health* 13:571–588.

43. Harris, D.T., D. Sakiestewa, R.F. Robledo, and M. Witten. 1997. Short-term exposure to JP-8 jet fuel results in long-term immunotoxicity. *Toxicology and Industrial Health* 13:559–570.

44. Pfaff, J., K. Parton, R.C. Lantz, H. Chen, A.M. Hays, and M.L. Witten. 1995. Inhalation exposure to JP-8 jet fuel alters pulmonary function and substance P levels in Fischer 344 rats. *Journal of Applied Toxicology* 15:249–256.

45. Robledo, R.F., R.S. Young, R.C. Lantz, and M.L. Witten. 2000. Short-term pulmonary response to inhaled JP-8 jet fuel aerosol in mice. *Toxicologic Pathology* 28:656–663.

46. Wang, S., R.S. Young, and M.L. Witten. 2001. Age-related differences in pulmonary inflammatory responses to JP-8 jet fuel aerosol inhalation. *Toxicology and Industrial Health* 17:23–29.

47. Hays, A.M., G. Parliman, J.K. Pfaff, R.C. Lantz, J. Tinajero, B. Tollinger et al. 1995. Changes in lung permeability correlate with lung histology in a chronic exposure model. *Toxicology and Industrial Health* 11:325–336.

48. Wong, S.S., J. Hyde, N.N. Sun, R.C. Lantz, and M.L. Witten. 2004. Inflammatory responses in mice sequentially exposed to JP-8 jet fuel and influenza virus. *Toxicology* 197:139–147.

49. Espinoza, L.A., M.E. Smulson, and Z. Chen. 2007. Prolonged poly(ADP-ribose) polymerase-1 activity regulates JP-8–induced sustained cytokine expression in alveolar macrophages. *Free Radical Biology & Medicine* 42:1430–1440.

50. Drake, M.G., F.A. Witzmann, J. Hyde, and M.L. Witten. 2003. JP-8 jet fuel exposure alters protein expression in the lung. *Toxicology* 191:199–210.

51. Witzmann, F.A., M.D. Bauer, A.M. Fieno, R.A. Grant, T.W. Keough, S.E. Kornguth et al. 1999. Proteomic analysis of simulated occupational jet fuel exposure in the lung. *Electrophoresis* 20:3659–3669.

52. Wong, S.S., J. Vargas, A. Thomas, C. Fastje, M. McLaughlin, R. Camponovo et al. 2008. In vivo comparison of epithelial responses for S-8 versus JP-8 jet fuels below permissible exposure limit. *Toxicology* 254:106–111.

53. Ris, C. 2007. U.S. EPA health assessment for diesel engine exhaust: A review. *Inhalation Toxicology* 19 Suppl 1:229–239.

54. Heinrich, U., R. Fuhst, S. Rittinghausen, O. Creutzenberg, B. Bellmann, W. Koch et al. 1995. Chronic inhalation exposure of Wistar rats and two different strains of mice to diesel engine exhaust, carbon black, and titanium dioxide. *Inhalation Toxicology* 7:533–556.

55. Bhatia, R., P. Lopipero, and A.H. Smith. 1998. Diesel exhaust exposure and lung cancer. *Epidemiology* 9:84–91.

56. Garshick, E., F. Laden, J.E. Hart, B. Rosner, T.J. Smith, D.W. Dockery et al. 2004. Lung cancer in railroad workers exposed to diesel exhaust. *Environmental Health Perspectives* 112:1539–1543.

57. Garshick, E., M.B. Schenker, A. Munoz, M. Segal, T.J. Smith, S.R. Woskie et al. 1988. A retrospective cohort study of lung cancer and diesel exhaust exposure in railroad workers. *The American Review of Respiratory Disease* 137:820–825.

58. Takahashi, G., H. Tanaka, K. Wakahara, R. Nasu, M. Hashimoto, K. Miyoshi et al. 2010. Effect of diesel exhaust particles on house dust mite-induced airway eosinophilic inflammation and remodeling in mice. *Journal of Pharmacological Sciences* 112:192–202.

59. Zhang, J.J., J.E. McCreanor, P. Cullinan, K.F. Chung, P. Ohman-Strickland, I.K. Han et al. 2009. Health effects of real-world exposure to diesel exhaust in persons with asthma. *Research report (Health Effects Institute)* 5-109. discussion 11–23.

60. Hirota, R., K. Akimaru, and H. Nakamura. 2008. In vitro toxicity evaluation of diesel exhaust particles on human eosinophilic cell. *Toxicology In Vitro* 22:988–994.

61. Porter, M., M. Karp, S. Killedar, S.M. Bauer, J. Guo, D. Williams et al. 2007. Diesel-enriched particulate matter functionally activates human dendritic cells. *American Journal of Respiratory Cell and Molecular Biology* 37:706–719.

62. Boland, S., A. Baeza-Squiban, T. Fournier, O. Houcine, M.C. Gendron, M. Chevrier et al. 1999. Diesel exhaust particles are taken up by human airway epithelial cells in vitro and alter cytokine production. *American Journal of Physiology* 276:L604–613.

63. Hiura, T.S., N. Li, R. Kaplan, M. Horwitz, J.C. Seagrave, and A.E. Nel. 2000. The role of a mitochondrial pathway in the induction of apoptosis by chemicals extracted from diesel exhaust particles. *The Journal of Immunology* 165:2703–2711.

64. Ritchie, G.D., J. Rossi, 3rd, A.F. Nordholm, K.R. Still, R.L. Carpenter, G.R., Wenger et al. 2001. Effects of repeated exposure to JP-8 jet fuel vapor on learning of simple and difficult operant tasks by rats. *Journal of Toxicology and Environmental Health. Part A* 64:385–415.

65. Ritchie, G.D., K.R. Still, W.K. Alexander, A.F. Nordholm, C.L. Wilson, J. Rossi, 3rd et al. 2001. A review of the neurotoxicity risk of selected hydrocarbon fuels. *Journal of Toxicology and Environmental Health. Part B, Critical Reviews* 4:223–312.

66. Pleil, J.D., L.B., Smith, and S.D. Zelnick. 2000. Personal exposure to JP-8 jet fuel vapors and exhaust at air force bases. *Environmental Health Perspectives* 108:183–192.

67. Lin, B., G.D. Ritchie, J. Rossi, 3rd, and J.J. Pancrazio. 2004. Gene expression profiles in the rat central nervous system induced by JP-8 jet fuel vapor exposure. *Neuroscience Letters* 363(3):233–238.

68. Kabbur, M.B., J.V. Rogers, P.G. Gunasekar, C.M. Garrett, K.T. Geiss, W.W. Brinkley et al. 2001. Effect of JP-8 jet fuel on molecular and histological parameters related to acute skin irritation. *Toxicology and Applied Pharmacology* 175:83–88.

69. Kanikkannan, N., B.R. Locke, and M. Singh. 2002. Effect of jet fuels on the skin morphology and irritation in hairless rats. *Toxicology* 175:35–47.

70. Nessel, C.S., J.J. Freeman, R.C. Forgash, and R.H. McKee. 1999. The role of dermal irritation in the skin tumor promoting activity of petroleum middle distillates. *Toxicological Sciences* 49:48–55.

71. Iwai, K., T. Udagawa, M. Yamagishi, and H. Yamada. 1986. Long-term inhalation studies of diesel exhaust on F344 SPF rats. Incidence of lung cancer and lymphoma. *Developments in Toxicology and Environmental Science* 13:349–360.

72. Nagashima, M., H. Kasai, J. Yokota, Y. Nagamachi, T. Ichinose, and M. Sagai. 1995. Formation of an oxidative DNA damage, 8-hydroxydeoxyguanosine, in mouse lung DNA after intratracheal instillation of diesel exhaust particles and effects of high dietary fat and beta-carotene on this process. *Carcinogenesis* 16:1441–1445.

73. McDougal, J.N., and C.M. Garrett. 2007. Gene expression and target tissue dose in the rat epidermis after brief JP-8 and JP-8 aromatic and aliphatic component exposures. *Toxicological Sciences* 97:569–581.

74. McDougal, J.N., C.M. Garrett, C.M. Amato, and S.J. Berberich. 2007. Effects of brief cutaneous JP-8 jet fuel exposures on time course of gene expression in the epidermis. *Toxicological Sciences* 95:495–510.

75. Chou, C.C., J.H. Yang, S.D. Chen, N.A. Monteiro-Riviere, H.N. Li, and J.J. Chen. 2006. Expression profiling of human epidermal keratinocyte response following 1-minute JP-8 exposure. *Cutaneous and Ocular Toxicology* 25:141–153.

76. Witzmann, F.A., N.A. Monteiro-Riviere, A.O. Inman, M.A. Kimpel, N.M. Pedrick, H.N. Ringham et al. 2005. Effect of JP-8 jet fuel exposure on protein expression in human keratinocyte cells in culture. *Toxicology Letters* 160:8–21.

77. Danielsen, P.H., S. Loft, and P. Moller. 2008. DNA damage and cytotoxicity in type II lung epithelial (A549) cell cultures after exposure to diesel exhaust and urban street particles. *Particle and Fibre Toxicology* 5:6.

78. Landvik, N.E., M. Gorria, V.M. Arlt, N. Asare, A. Solhaug, D. Lagadic-Gossmann et al. 2007. Effects of nitrated-polycyclic aromatic hydrocarbons and diesel exhaust particle extracts on cell signalling related to apoptosis: Possible implications for their mutagenic and carcinogenic effects. *Toxicology* 231:159–174.

79. Sato, H., M. Sagai, K.T. Suzuki, and Y. Aoki. 1999. Identification, by cDNA microarray, of A-raf and proliferating cell nuclear antigen as genes induced in rat lung by exposure to diesel exhaust. *Research Communications in Molecular Pathology and Pharmacology* 105:77–86.

80. Koike, E., S. Hirano, N. Shimojo, and T. Kobayashi. 2002. cDNA microarray analysis of gene expression in rat alveolar macrophages in response to organic extract of diesel exhaust particles. *Toxicological Sciences* 67:241–246.

81. Takano, H., T. Yoshikawa, T. Ichinose, Y. Miyabara, K. Imaoka, and M. Sagai. 1997. Diesel exhaust particles enhance antigen-induced airway inflammation and local cytokine expression in mice. *American Journal of Respiratory and Critical Care Medicine* 156:36–42.

82. Takano, H., R. Yanagisawa, T. Ichinose, K. Sadakane, S. Yoshino, T. Yoshikawa et al. 2002. Diesel exhaust particles enhance lung injury related to bacterial endotoxin through expression of pro-inflammatory cytokines, chemokines, and intercellular adhesion molecule-1. *American Journal of Respiratory and Critical Care Medicine* 165:1329–1335.

83. Koike, E., S. Hirano, A. Furuyama, and T. Kobayashi. 2004. cDNA microarray analysis of rat alveolar epithelial cells following exposure to organic extract of diesel exhaust particles. *Toxicology and Applied Pharmacology* 201:178–185.

84. Camhi, S.L., J. Alam, L. Otterbein, S.L. Sylvester, and A.M. Choi. 1995. Induction of heme oxygenase-1 gene expression by lipopolysaccharide is mediated by AP-1 activation. *American Journal of Respiratory Cell and Molecular Biology* 13:387–398.

85. Itoh, K., T. Chiba, S. Takahashi, T. Ishii, K. Igarashi, Y. Katoh et al. 1997. 2An Nrf2/small Maf heterodimer mediates the induction of phase II detoxifying enzyme genes through antioxidant response elements. *Biochemical and Biophysical Research Communications* 36:313–322.

86. Anderson, H.R., E.S., Limb, J.M. Bland, A. Ponce de Leon, D.P. Strachan, and J.S. Bower. 1995. Health effects of an air pollution episode in London, December 1991. *Thorax* 50:1188–1193.

87. Hayakawa, K., A. Nakamura, N. Terai, R. Kizu, and K. Ando. 1997. Nitroarene concentrations and direct-acting mutagenicity of diesel exhaust particulates fractionated by silica-gel column chromatography. *Chemical & Pharmaceutical Bulletin* 45:1820–1822.

88. Salmeen, I., A.M. Durisin, T.J. Prater, T. Riley, and D. Schuetzle. 1982. Contribution of 1-nitropyrene to direct-acting Ames assay mutagenicities of diesel particulate extracts. *Mutation Research* 104:17–23.

89. Reynolds, L.J., and R.J. Richards. 2001. Can toxicogenomics provide information on the bioreactivity of diesel exhaust particles? *Toxicology* 165:145–152.

90. Wise, H., D. Balharry, L.J. Reynolds, K. Sexton, and R.J. Richards. 2006. Conventional and toxicogenomic assessment of the acute pulmonary damage induced by the instillation of Cardiff PM10 into the rat lung. *The Science of the Total Environment* 360:60–67.

91. Ichinose, T., H. Takano, Y. Miyabara, R. Yanagisawa, and M. Sagai. 1997. Murine strain differences in allergic airway inflammation and immunoglobulin production by a combination of antigen and diesel exhaust particles. *Toxicology* 22:183–192.

92. Tulic, M.K., J.L. Wale, P.G. Holt, and P.D. Sly. 2000. Modification of the inflammatory response to allergen challenge after exposure to bacterial lipopolysaccharide. *American Journal of Respiratory Cell and Molecular Biology* 22:604–612.

93. Dahlem, P., A.P. Bos, J.J. Haitsma, M.J. Schultz, E.K. Wolthuis, J.C. Meijers et al. 2006. Mechanical ventilation affects alveolar fibrinolysis in LPS-induced lung injury. *The European Respiratory Journal* 28:992–998.

94. Inoue, K., H. Takano, R. Yanagisawa, T. Ichinose, K. Sadakane, S. Yoshino et al. 2004. Components of diesel exhaust particles differentially affect lung expression of cyclooxygenase-2 related to bacterial endotoxin. *Journal of Applied Toxicology* 24:415–418.

95. Yanagisawa, R., H. Takano, K. Inoue, T. Ichinose, S. Yoshida, K. Sadakane et al. 2004. Complementary DNA microarray analysis in acute lung injury induced by lipopolysaccharide and diesel exhaust particles. *Experimental Biology and Medicine (Maywood)* 229:1081–1087.

96. Stevens, T., Q.T. Krantz, W.P. Linak, S. Hester, and M.I. Gilmour. 2008. Increased transcription of immune and metabolic pathways in naive and allergic mice exposed to diesel exhaust. *Toxicological Sciences* 102:359–370.

97. Omura, S., E. Koike, and T. Kobayashi. 2009. Microarray analysis of gene expression in rat alveolar epithelial cells exposed to fractionated organic extracts of diesel exhaust particles. *Toxicology* 262:65–72.

98. Tsukue, N., H. Okumura, T. Ito, G. Sugiyama, and T. Nakajima. 2010. Toxicological evaluation of diesel emissions on A549 cells. *Toxicology In Vitro* 24:363–369.

99. Verheyen, G.R., J.M. Nuijten, P. Van Hummelen, and G.R. Schoeters. 2004. Microarray analysis of the effect of diesel exhaust particles on in vitro cultured macrophages. *Toxicology In Vitro* 18:377–391.

100. Peretz, A., E.C. Peck, T.K. Bammler, R.P. Beyer, J.H. Sullivan, C.A. Trenga et al. 2007. Diesel exhaust inhalation and assessment of peripheral blood mononuclear cell gene transcription effects: An exploratory study of healthy human volunteers. *Inhalation Toxicology* 19:1107–1119.
101. Sasaki, Y., T. Ohtani, Y. Ito, M. Mizuashi, S. Nakagawa, T. Furukawa et al. 2009. Molecular events in human T cells treated with diesel exhaust particles or formaldehyde that underlie their diminished interferon-gamma and interleukin-10 production. *International Archives of Allergy and Immunology* 148:239–250.

25

Omics Approaches in Biofuel Production for a Green Environment

Atul Grover, Patade Vikas Yadav, Maya Kumari,
Sanjay Mohan Gupta, Mohommad Arif, and Zakwan Ahmed

CONTENTS

25.1 Introduction

In light of the ever-diminishing resources of fossil fuels and the perilous rates at which climate change is becoming visible, there is an increasing worldwide economic interest and scientific focus on developing biofuel crops [1]. Worldwide energy consumption has increased 13-fold in the 20th century alone, tripling since 1960, which is more rapid than the population explosion [2–3]. Obviously, it is difficult to keep up with the high energy demands from the finite resources alone. Biofuels, in fact, are the only alternative source of liquid transportation fuels compatible with our existing fleet of automobiles.

The term "biofuel" refers to the liquid fuel that is obtained upon the conversion of plant biomass through fermentation [4], or some other enzymatic, chemical, or thermochemical treatment (Table 25.1). Biofuels are sources of renewable, green, and clean energy. They aim at carbon credit generation by cutting down on the emission of greenhouse gases, increasing community self-reliance, and providing a spur for local job creation and growth. They also cut dependence on fuel wood, which is often scarce and causes immense health problems through indoor air pollution. In principle, biofuels can be obtained from any biomass; however, nonedible oilseeds, animal fats, plant biomass, and other waste biomatter such as dairy waste, agricultural waste, forestal waste, kitchen waste, livestock waste, etc., have been recognized as major sources of biofuels. Existing crop plants, which can directly or indirectly be used for biofuel production, have been called "biofuel crops." The term also includes upcoming woody and nonwoody crops as well. Biofuel crops in arid regions

TABLE 25.1

Important Terms and Definitions

Association mapping	Determination of linkage disequilibrium between given loci in a genome
Biodiesel	Vegetable oil– or animal fat–based liquid fuel consisting of long-chain alkyl esters
Bioenergy	A renewable form of energy originating from biological sources
Bioenergy crop	A crop whose harvest is targeted for the power or transportation (or both) sectors
Biofuel	A fuel whose stored energy is derived from natural process of biological fixation
Biomass	Sum total of matter in living organisms
Biomass crop	A crop whose economically important component is its biomass, to be utilized for the power or transportation sectors
Biopower crop	A crop whose harvest is targeted for the power sector
Carbon sequestering	The process of removing carbon from the atmosphere and depositing it in a biotic reservoir
Climate change	A significant and lasting change in the distribution of weather patterns of the earth over a period of time
First-generation biofuel	Biofuels obtained from edible feedstocks like sugars and vegetable oils
Genomics	A branch of study that deals with the structure, function, and mapping of genomes
Linkage disequilibrium	Occurrence of linked alleles at a frequency higher than expected
Metagenome	Entirety of DNA material in an ecological community
Metagenomics	A branch of study that deals with the analysis of genomic material of a given ecological community
Omics	Group of molecular biology–based advanced studies dealing with the structure, function, annotation, evolution, and mapping of different biomolecules
Proteomics	A branch of study that deals with the characterization of structure and function of all the proteins encoded by the genome of a cell
Second-generation biofuel	Biofuels obtained from sustainable biological feedstocks. Examples include cellulosic ethanol, algal fuel, biohydrogen, biomethanol, Fischer–Tropsch diesel, mixed alcohols (bioethanol + biopropanol + biobutanol) and wood diesel

can act as windbreakers and stabilize soil from erosion, which may lead to its conservation. Plants fulfilling these requirements should also meet agronomic, environmental, and societal parameters for successful deployment as a source of energy. The efficient growth strategies of these plants rely on newly assimilated and recycled carbon and remobilized nitrogen in a continually shifting balance between sources and sinks [1]. The success of bioenergy crops also depends on soil quality, invasiveness, landscape diversity, displacement of carbon sequestering native vegetation, and proper water use [5–8].

For present-day farmers, harvesting and selecting crops for bioenergy and associated traits is a novel activity, and one which has not been carried out in previous centuries. The improvement of bioenergy traits can be achieved through two basic routes—crossing and selection programs aided by knowledge of the genetic basis of the traits and the identification of molecular markers for marker-assisted selection [9]. The alternative approach utilizes transgenic or genetic modification technologies to introduce new genes, modify existing genes, or interfere with gene expression [10]. For both routes, advances in molecular mapping, whole-genome sequencing, "omics" (transcriptomics, proteomics, metabolomics, etc.), whole-genome scans, and bioinformatics provide powerful approaches for gene discovery. However, identifying whether traits are determined by major genes or by quantitative trait loci is of foremost importance [1]. Realizing that the proper agricultural management of bioenergy crops and the use of biofuels are already the realities of 21st century, in the following sections, we provide an insight into the current omic developments in bioenergy crops for their accelerated domestication and proper exploitation for industrial use, contributing to a green environment.

25.2 Bioenergy Crops and Omics Advancements

Advances in sequencing technologies, feasibility of high-throughput expression profiling, and genome mapping technologies have encouraged the rapid development of bioenergy crops.

Omics studies aim to understand (at the molecular level) the mechanisms of all the activities of an organism that we see at the macro level. In the context of the current century, plant improvement strategies are required for food, feed, fiber, and fuel. The list of current bioenergy crops that can also be used as models owing to the availability of their whole-genome information in public domains and the long history of genetic improvement, is increasing day by day (Table 25.2). Examples include sorghum and maize among monocots, and *Ricinus*, *Populus*, and *Glycine max* among dicots. Furthermore, plants like *Brachypodium* and *Arabidopsis*, for which whole-genome information is available, can also be used as genomic models for plants with bioenergy potentials belonging to families Poaceae and Brassicaceae, respectively.

Bioenergy grasses are ideal for planning the future biofuel scenario. They have shorter life cycles enabling their fast replenishment, offer flexibility, are well-suited to existing farm equipment and agro-infrastructure, and are amenable to current genetic improvement methods. From the pyrolysis point of view, due to the absence of lignin in their cell walls, it is easier to process them. Although whole-genome sequencing efforts have enhanced our understanding of the basic biology of plants, these have not translated into any significant crop improvement efforts in grasses with bioenergy potentials such as maize, wheat, and barley. The size and complexity of the Triticaceae genomes (Table 25.2) have been the main obstacles for the efficient development of genome sequencing projects for these species. Advancements in sequencing technology, however, have led the international community to now consider whole-genome sequencing in wheat [11].

Maize (*Zea mays*), on the other hand, has its genome well characterized and mapped. Despite a smaller amount of land being cultivated with maize, greater yields are obtained due to the utilization of more efficient C_4 photosynthetic pathways by maize. C_4 plants facilitate CO_2 concentration in the bundle sheath, owing to several biochemical, physiological, and morphological adaptations. As a result, these plants have greater N-use and water-use efficiencies, and higher rates of photosynthesis with stoichiometrically lower number of Rubisco molecules. Thus, more efficient conversion of energy to biomass is carried out by these plants [12]. In fact, C_4 grasses are among the most productive plants on the planet, and thus, the most promising biofuel feedstocks. Successful implementation of these grasses in biofuel planning depends on the improvement of the critical crop characteristics. that is, their biomass yields as well as their biotic and abiotic stress tolerances [13]. Marker-assisted selection and transgenic approaches are the current solutions for the improvement of these traits. In addition to biomass, C_4 grasses are also attractive sources of ethanol, which is an option for replacing gasoline and reducing CO emissions from fossil oil. The overall economy of ethanol production will rely on how well these crops are engineered [14].

TABLE 25.2

Major Present-Day Bioenergy Crops and the Online Sources for Their Genomic Information

Crop	Family	Potential (C/LC/O)	Genome Size (Mbp)	Chromosome Number (2n)	Life Cycle	Institute/Consortium and Web Link for Whole Genome Information
Triticum aestivum (wheat)	Poaceae	C/LC	16,000	42	Annual	International Wheat Genome Sequencing Consortium (http://www.wheatgenome.org)
Zea mays (maize)	Poaceae	C/LC	2500	20	Annual	MaizeGDB Sequencing Information Portal (http://www.maizegdb.org/genome/)
Sorghum bicolor (sorghum)	Poaceae	C/LC	760	20	Annual	Sorghum bicolor Genome Project (http://www.plantgdb.org/SbGDB/)
Saccharum officinarum (sugarcane)	Poaceae	C/LC	~930 (monoploid)	80	Perennial	Sugarcane Genome Sequencing Initiative (http://sugarcanegenome.org/)
Eucalyptus spp. (Eucalyptus)	Myrtaceae	LC	0.16–655	22	Perennial	US Department of Energy Joint Genome Institute (http://web.up.ac.za/eucagen/; http://www.phytozome.net/)
Populus spp. (poplar)	Salicaceae	LC	450–550	38	Perennial	International Populus Genome Consortium (http://www.ornl.gov/sci/ipgc/)
Glycine max (soybean)	Leguminoseae	O	1115	40	Annual	The Soybean (*Glycine max*) Genome Project (http://www.phytozome.net/soybean)
Brassica rapa (canola)	Brassicaceae	O	529	20	Annual	Brassica rapa Genome Sequencing Project Consortium (www.ivfcaas.ac.cn)
Phoenix dactylifera (date palm)	Palmaceae	O	~658	26–36	Perennial	Weill Cornell Medical College in Qatar, Doha, Qatar (http://qatar-weill.cornell.edu/research/datepalmGenome/)
Ricinus communis (castor bean)	Euphorbiaceae	O	323	20	Perennial	TIGR Castor Bean Genome Database (http://www.tigr.org/msc)
Jatropha curcas (physic nut)	Euphorbiaceae	O	416	22	Perennial	Jatropha Genome Database, Kazusa DNA Research Institute (http://www.kazusa.or.jp/jatropha/)
Helianthus annuus (sunflower)	Compositae	O	3500	34	Annual	Compositae Genome Project (http://cgb.indiana.edu/research/projects/9)

Note: C, cellulose; LC, lignocellulose; O, oilseed.

Maize has a long history of genetic improvements, leading to a 50% to 60% overall increase in yield [15] and significant improvements in associated traits such as pest and disease resistance. However, the use of maize as a bioenergy crop is controversial because of the projected competition between food and fuels [16]. Furthermore, high input requirements for water and agrochemicals also defeat the purpose owing to smaller environmental benefits relative to the use of fossil fuels [7]. However, corn stover, which is abundantly available, can be used as a feedstock for pyrolysis and for the effective utilization of maize as a biomass crop. Gene expression analysis from developing ovary tissue revealed an increase in expression levels even up to 100-fold [17,18]. As it undergoes changes involving both cell wall and sugar metabolism, this data is highly relevant for future manipulation of maize as a bioenergy crop. Optimizing water and nutrient uptake will improve the sustainability of maize production. Transgenic technology can also be applied relatively safely in maize [19].

Unfortunately, most C_4 biofuel species, such as *Miscanthus*, switchgrass, sugarcane, energy cane, miscane, Bermuda grass, Napier grass, etc., are polyploid–diploid to octoploid, and in some cases, their chromosome numbers even exceed 100, thus complicating studies on a genomic level, that is, mapping and mutagenesis. *Sorghum bicolor* is an appropriate model for this group of bioenergy plants. It is diploid with a genome spanning 730 Mb that has been completely sequenced [20]. Sorghum is drought-tolerant, has a long history of genetic improvement, and produces lignocellulose, sugar, and starch.

Sugarcane is one of the most prominent bioenergy crops today due to the success of sugar-based fuel ethanols in Brazil [21]. In addition, it is an attractive lignocellulosic feedstock as well. Unfortunately, genetic studies of sugarcane are challenging, primarily because of complicated ploidy. Genetic maps of sugarcane based on molecular markers have been developed [22,23]. Sugarcane, being a glycophyte, exhibits salt toxicity symptoms leading to reduced growth, yield, and quality of produce. Patade et al. [24] suggested set priming with salt as a simple and effective approach for enhanced salt tolerance in sugarcane. Furthermore, molecular mechanisms in terms of transcript regulation of stress-responsive genes underlying the priming-mediated stress tolerance in sugarcane are investigated [25]. Additionally, PCR-based suppression subtractive hybridization has been successfully employed for the identification of novel salt stress–expressed genes with high-, middle-, low-, or rare-abundance transcripts, which can be cloned with equal probability [26]. Besides, as most agronomic traits are polygenic and quantitatively inherited, rapid progress has been achieved through the characterization of the transcriptome of sugarcane using gene expression analysis [27–30]. Developing transgenic sugarcane is an attractive approach for genetic enhancement of yield potential [31]. Furthermore, there is a limited risk of propagation via pollen or seed, as sugarcane is mostly propagated by stem cuttings. The Sugarcane Genome Project is under way (Table 25.2) for sequencing of genome of sugarcane. However, large genome size (~10 Gbp) of sugarcane is currently a major challenge. Understanding the genetic basis of sugar accumulation would have a significant effect on the classic and marker-assisted selection in sugarcane. Microarray analysis of sugarcane genotypes varying in sugar content (from the stem) has led to the identification of a number of transcription factors and protein kinases [32].

Purpose-grown trees will become an integral part of bioenergy planning and management, mainly because end-use markets of tree biomass already exists along with the associated infrastructure [33]. Furthermore, trees require little maintenance, they are rich in biomass, and are easy to store. *Populus* is a model system for tree species. It is also a well-suited model for the biomass of woody crops, considering its small haploid genome size (480 Mbp; $2n = 38$; paleopolyploid), rapid juvenile growth, ease of clonal propagation, high-throughput transformation, and availability of extensive genetic maps [34]. The small genome size of poplar, coupled with high-density physical maps, facilitate effective gene isolation, gene mapping, and gene tagging efforts [35]. A number of quantitative trait loci for many traits have been identified in poplar [36], and an increasing number of candidate genes are being identified. A number of advancements through transgenic technology have already been achieved using poplar [37].

Trees can benefit both from traditional breeding coupled with selection as well as by the introduction of genes. Traditional breeding of poplars for wood production in short-rotation forestry has been extremely successful. Hybrid poplars are faster growing and more productive [1]. Trees, currently identified for their bioenergy potential mostly undergo C_3 photosynthesis and have high CO_2 exchange rates, light-use efficiencies, and photosynthetic capacities [38].

Edible and nonedible oilseeds are other attractive sources of bioenergy, particularly biodiesel (Table 25.2), and many countries in the world are concentrating their efforts on the development of such plants as the main bioenergy crops. Although soybean, canola, and rapeseed are leading sources of vegetable oils that are converted to biodiesel, most of the third-world countries have concentrated their efforts on the development of nonedible oilseeds as main bioenergy crops. In India, Jatropha has been identified as the main bioenergy crop, with 30% to 35% extractable oil in the seeds. Other plants like neem, mahua, cheura, camelina, and pongamia are also lucrative sources of biodiesel. However, it has to be recognized that current oil yields from these plants are insufficient to meet the bioenergy demand. With regard to Jatropha, more than 30,000 expressed sequence tags are currently available, and efforts are under way to sequence the entire genome of the plant [39]. Planting through cuttings is generally advocated in Jatropha. Furthermore, nonavailability of superior clones/varieties is the major factor limiting large-scale plantations. Tissue culturing techniques for Jatropha have long been standardized, but no economically important somaclonal variants or transgenics have thus far been reported. We are involved in Jatropha improvement through transgenic and conventional methods. It is speculated that by using genetic transformation of these crops with candidate genes like glycerol-3-phosphate acyltransferase (GPAT), lysophosphatidic acid acyltransferase (LPAT) and diacylglycerol acyltransferase (DGAT) would enhance the lipid content in seeds.

Interestingly, photosynthetic microorganisms (e.g., algae and cyanobacteria) have the ability to produce 8 to 24 times more lipids per unit area for biofuel production than the most productive land plants [40]. Aside from feasibility studies, the production of algae to harvest oil for biodiesel has not yet been undertaken on a commercial scale. In addition to its projected high yield, algal culture, unlike crop-based biofuels, does not entail a decrease in food production because it requires neither farmland nor fresh water. A number of cyanobacterial and eukaryotic algal genome sequences are currently available in the public domain, including *Chlamydomonas reinhardtii*, *Volvox carteri*, *Cyanidioschizon merolae*, etc., which have the abilities to produce high amounts of storage lipid contents. For *C. reinhardtii*, extensive genomic, biological, and physiological data is already available in the public domain [41,42] qualifying it as a suitable model eukaryotic algal species with biofuel potential.

25.3 Role of Comparative Genomics

Comparative genomics relies on the colinearity of the genomes of closely related species. Widespread colinearity of genomes has been well reported in Poaceae [43], Brassicaceae

[44], and in poplar and willow [45]. Interestingly, good number of bioenergy crops have been identified in each of the abovementioned pairs. Elucidation of gene function is a direct outcome of comparative genomics of such pairs of genomes. For example, the cellulose-synthase–like genes (*csl*), that is, *cslF* and *cslH*, were found to be unique to grasses [46]. Similarly, Penning et al. [18] compared the GT47 gene family encoding glycosyl transferases in maize, rice, and Arabidopsis and identified a grass-specific clade. Genomic information earned from model plants can be applied to other members in the same family. Genetic improvements for bioenergy need higher yields (biomass, cellulose, or lignocellulose) and other related traits like plant size, cell wall composition, distribution of vascular tissue, flowering time, and stover yield and composition.

A better understanding of oil accumulation in seeds has emerged from comparative genomics of *Brassica* and *Arabidopsis*. Disruption in the homeobox protein encoding gene *GLABRA2* in *Arabidopsis* has reportedly resulted in a mutant accumulating 8% more seed oil. Genomic data of *Arabidopsis* has been exploited a great deal to build a reference data set on fatty acid composition. This data can be used for the mining and prediction of genes in other oilseed crops. As most metabolic pathways are conserved across plant species, information from model plants may easily be adapted for improvement of oil content. This cause has further been facilitated by the online databases, which make genomic, proteomic, metabolomic, and other relevant information available for research use.

25.4 Omics for Accelerated Domestication of Upcoming Bioenergy Crops

For accelerated domestication of present-day underdeveloped plant species into futuristic bioenergy crops, omics has a great role to play. In the last decade, tens of thousands of genes have been characterized through major genome sequencing efforts (Table 25.2), and many of these genes have been found relevant to accelerated domestication efforts (Table 25.3). With the wealth of genomic information available in semantic space, targeted gene expression in the species of interest can be carried out using comparative genomics. A candidate gene approach for the identification of genes affecting biofuel traits (improved productivity, feedstock uniformity, stress tolerance, branching, response to competition, etc.) in a novel crop should be an effective strategy [47]. Limited success in this direction has been achieved in a few of the upcoming bioenergy crops. In addition to the genomic parameters, invasiveness is a desired character for novel biofuel crops. It can be contributed by adaptive characters [48].

As a matter of fact, the domestication process has best been characterized based on genetic elements in various crops [49], namely, maize, sorghum, wheat, and sunflower, which are valued as potential biofuels. Thus, the sequences of the homologous domestication genes can be easily identified and cloned in upcoming biofuel crops. The gene constructs can be prepared for ectopic expression, dominance, upregulation, gain-of-function, etc., as the case may be under the control of a constitutive promoter. Similarly, silencing constructs of candidate genes can also be prepared. The identified gene scan can be upregulated or downregulated within the same species as well, which may eliminate the need for transformation using foreign species.

When plants are intended to be used as a source of energy, the aim is to maximize the net energy content per unit of production area. An ideal bioenergy crop is one that requires the least energy extensive inputs (irrigation, fertilizer, and pesticides) to produce

TABLE 25.3

Tools/Parameters to Affect Rapid Domestication of New Energy Crops

High efficiency (next-generation) sequencing
Efficient transfer of developed traits
C/N productivity
Agronomic productivity
Water use efficiency
Nutrient use efficiency
Drought tolerance
Biotic stress tolerance

large amounts of energy-dense agronomic character with efficient processing. Thus, the overall energy balance of an ideal bioenergy crop should be on the positive side. All current and future domestication efforts should be directed toward improving these traits.

25.5 Lignocellulose Manipulation

Secondary plant cell walls are primary components of plant biomass that have found wide applications as raw feedstock for liquid biofuel production. The matrix of hemicellulose and lignin surrounds the cellulose in the plant's cell wall. This complexity of the cell wall presents operating challenges to the conversion process of lignocellulosic biomass to liquid fuel [50]. Interestingly, lignin content in the cell walls also inversely determines sugar release and thus the ethanol extraction process [51,52]. Hence, biomass composition or cell wall composition is a major trait for improvement of biofuel crops. Modern genetic engineering tools provide an option of pathway-specific manipulation for affecting chemical and physical properties of biomass. This can be done either by the discovery of regulatory genes involved in key pathways or by discovery of promoters that can drive the expression of the introduced genes. Even though major enzymatic steps in lignin, cellulose, and hemicellulose metabolisms have already been identified [51,53,54], little information is thus far available on lignin manipulation. Unfortunately, the brown midrib, is probably the most widely studied trait, originally in corn, and subsequently in sorghum as well as in millet. At the present state of scientific knowledge and expertise, any feedstock with reduced or modified lignin, or else increased cellulose and hemicelluloses, will be a significant enhancement, as this will improve the efficiency of feedstock conversion [33] as the heat of combustion of lignin is greater than that of polysaccharides.

25.6 Role of Metagenomics

Nowadays, there is an increasing interest in the development of more efficient and less time-consuming methods to assess the presence of microorganisms, as well as their viability for bioprocess control and improvement [55]. Novel enzyme discovery is central to the conversion of biomass, sucrose, or vegetable oil to biofuel. Many such enzymes are produced in nature, for example, in the guts of termites or ruminants. Metagenome sequencing efforts and classic metagenomics help a great deal in such discoveries (Table 25.1). The discovery of genes from communities or organisms like *Thermoanaerobacter, Ethanolicus, Pichia stipitis, Phanerochaete chrysosporium, Clostridium thermocellum, Saccharophagus degradans* and *Acidothermus cellulolyticus* will be sources of novel enzymes in the coming years [56].

25.7 Molecular Markers and Their Exploitation for Improvement of Bioenergy Crops

Whole-genome sequences usually represent a single genotype. Thus, to assess the entire variation in the gene pool from an industrial point of view, high-throughput expression profiling or molecular mapping has to be utilized. The genome sequence data nevertheless constitutes an ideal source to identify molecular markers [57–59]. Molecular markers help in the mapping of traits that can otherwise not be scored easily. This may include the flowering-related or polygenic traits. Development of high-throughput genotyping methods has been a revolutionary leap in this area of genomics. Map-based cloning is now practical [60] and perennials will be best affected by the use of such technologies. For example, this may theoretically reduce the breeding cycle of oil palm from 19 years to 13 years [61]. Nevertheless, mapping depends on the successful identification of recombination events and therefore it is most easily implemented in diploids. Polyploid bioenergy crops like sugarcane are therefore still at a disadvantage. However, some success has been achieved by the use of F_1 progeny derived from two cultivated parents [62,63], an interspecific *Saccharum officinarum* × *Saccharum spontaneum* cross [64], or a cultivated parent and a wild species.

Molecular markers also contribute in the estimation of gene pool diversity through DNA fingerprinting, which can be implemented in parent selections [65]. Monsanto's successful marker-assisted breeding programs [66,67] in soybean and sunflower were based on high-throughput genotyping and the resulting lines were reported to show higher grain yields and oil content. Higher genetic gains were reported in early populations in each of the above programs. Present-day transcriptomic markers like expressed sequence tag–simple sequence repeats, diversity arrays technology, and single-nucleotide polymorphism can efficiently be applied for high-throughput germplasm characterizations because they are robust and often cross-species transferable [68,69]. Furthermore, they have a greater ability to be developed as trait-linked markers [68]. Diversity array technology markers do not require sequence information and can easily be implemented in crop improvement programs [70]. Several hundred polymorphic genomics as well as transcriptomic microsatellite markers are already available for a variety of crop plants, including the bioenergy crops [57,71], which can easily be transferred to closely related species [72,73].

Continued advancements in marker technology are likely to accelerate the application of marker-assisted selection for quantitatively inherited traits as well. Recombinant inbred lines have been extensively employed for mapping quantitative trait loci in energy crops [74]. Association mapping may be employed for those species for which no known population structures exist, or when individuals are distantly related [75]. It also allows breeders to focus on the most valuable genetic resources. However, association mapping may suffer due to the low linkage disequilibrium in some species, as may be expected in currently nondomesticated potential biofuel crops in which thousands of markers may be required for a full genome scan in a natural population. In such cases, following a candidate gene association mapping approach may be helpful. In recent years, the advent of nested association mapping has combined the benefits of both recombinant inbred line and association mapping [76]. Nested association mapping has thus far been implemented in genetically highly characterized maize and the success can also be replicated in other crops like sorghum.

25.8 Planning and Management Issues

Among all the alternatives to fossil fuels proposed to meet the current energy crisis, the development of cellulosic and lignocellulosic biomass seems to be the most viable option because the cell wall is the most abundant renewable source of energy on the planet. Current land-use requirements and process inefficiencies are primary hurdles in the large-scale deployment of biomass-to-liquid technologies [77]. Little information is available on the likely supply and spatial yield data of these crops, leading to uncertainties in future land requirements.

Transgenic bioenergy crops can be designed to yield more produce per hectare. Most transgenic crops, at present, comprise either herbicide or insect resistance. Interestingly, current bioenergy crops, such as soybean, corn, populus, sugarbeet, maize, and canola, have all been transformed with foreign genes, and the genetically modified crops thus developed are all under cultivation in different parts of world. Miscanthus, switchgrass, and other novel crops are outcrossers—creating dangers of gene flow if transgenics are developed for these crops.

Importantly, diverse sources of bioenergy species and the technologies will be important to the biofuel industry, as any single crop by itself will not be able to sustain the global fuel demands. It is generally agreed that in the years to come, a combination of current bioenergy crops and new biofuel crops will be required. Furthermore, crop resources must be diverse and agricultural practices should at least be based on crop rotation of biotypes. Presently, Europe is relying on *Miscanthus* × *giganteus*, China on corn and cassava, Brazil on sugarcane, the United States on maize and soybean, and India on Jatropha for their bioenergy needs. Thus, under the monoculture model(s) the world is following presently, a devastating crop loss would lead to serious energy shortages. Interestingly, in the production of biodiesel, the use of crop residues offer lower production costs compared with the use of a dedicated biomass crop [78], and thus, any agricultural waste can be incorporated into the bioconversion process in a cost-effective manner.

25.9 Summary

Recent advancements in the field of genomics have opened the gates to systematic improvements of underutilized crops with biofuel potential. It is also now possible to identify the genes responsible for biofuel traits in the current food crops and, if sustainable, divert their use to biofuel production. Current agricultural practices, however, are energy-intensive. An ideal biofuel crop would have fewer demands of irrigation and agrochemicals, thereby minimizing the adverse effects on the environment. An omics-powered search is also on for the discovery of suitable enzymes and processes that can make postharvest practices for biofuel production environment friendly. The replacement of fossil fuels by biofuels has its own advantages, as these are less polluting, renewable, and carbon neutral.

References

1. Karp, A., and I. Shield. 2008. Bioenergy from plants and the sustainable yield challenge. *New Phytologist* 179:15–32.
2. Aghamiri, S.F., K. Kabiri, and G. Emtiazi. 2011. A novel approach for optimization of crude oil bioremediation in soil by the Taguchi Method. *Journal of Petroleum & Environmental Biotechnology* 2:108.
3. Ghoodjani, E., R. Kharrat, M. Vossoughi, and S.H. Bolouri. 2011. A review on thermal enhanced heavy oil recovery from fractured carbonate reservoirs. *Journal of Petroleum & Environmental Biotechnology* 2:109.
4. Lawrence, C.J., and V. Walbot. 2007. Traditional genomics for bioenergy production from fuel-stock grasses: Maize as the model species. *Plant Cell* 19:2091–2094.
5. Wilhelm, W.W., J.M.F. Johnson, D.L. Karlen, and D.T. Lightle. 2007. Corn stover to sustain soil organic carbon further constrains biomass supply. *Agronomy Journal* 99:1665–1667.
6. Firbank, L. 2008. Assessing the ecological impacts of bioenergy. *BioEnergy Research* 1:12–19.
7. Fargione, J., J. Hill, D. Tilman, S. Polasky, and P. Hawthorne. 2008. Land clearing and the biofuel carbon debt. *Science* 319:1235–1238.
8. Vermerris, W. 2011. Survey of genomics approaches to improve bioenergy traits in maize, sorghum and sugarcane. *Journal of Integrative Plant Biology* 53:106–119.
9. Price, A.H. 2006. Believe it or not, QTLs are accurate. *Trends in Plant Science* 11:213–216.
10. Torney, F., L. Moeller, A. Scarpa, and K. Wang. 2007. Genetic engineering approaches to improve bioethanol production from maize. *Current Opinion in Biotechnology* 18:193–199.
11. Gupta, P.K., R.R. Mir, A. Mohan, and J. Kumar. 2008. Wheat genomics: Present status and future prospects. *International Journal of Plant Genomics* 2008:896451.
12. Heaton, E.A., P. Mascia, R. Flavell, S. Thomas, P.S. Long, and F.G. Dohleman. 2008. Energy crop development: Current progress and future prospects. *Current Opinion in Biotechnology* 19:202–209.
13. Jakob, K., F. Zhou, and A.H. Paterson. 2009. Genetic improvement of C4 grasses as cellulosic biofuel feedstocks. *In Vitro Cellular & Developmental Biology. Plant* 45:291–305.
14. Wyman, C.E. 2007. What is (and is not) vital to advancing cellulosic ethanol. *Trends in Biotechnology* 25:153–157.
15. Duvick, D.N., and K.G. Cassman. 1999. Post-green revolution trends in yield potential of temperate maize in the north-central United States. *Crop Science* 39:1622–1630.
16. Tenenbaum, D.J. 2008. Food vs fuel: Diversion of crops could cause more hunger. *Environmental Health Perspectives* 116:A254–A257.
17. Eveland, A.L., D.R. McCarty, and K.E. Koch. 2008. Transcript profiling by 3′-untranslated region sequencing resolves expression of gene families. *Plant Physiology* 146:31–44.
18. Penning, B., R. Tayengawa, C.T. Hunter, III et al. 2009. Genetic resources for functional genomics of cell wall biology. *Plant Physiology* 151:1703–1728.
19. Frame, B.R., H. Shou, R.K. Chikwamba et al. 2002. *Agrobacterium tumefaciens* mediated transformation of maize embryos using a standard binary vector system. *Plant Physiology* 129:13–22.
20. Paterson, A.H., J.E. Bowers, R. Bruggmann et al. 2009. The *Sorghum bicolor* genome and the diversification of grasses. *Nature* 457:551–556.

21. Matsuka, S., J. Ferro, and P. Arruda. 2009. The Brazilian experience of sugarcane ethanol industry. *In Vitro Cellular & Developmental Biology Plant* 45:372–381.
22. Garcia, A.A., E.A. Kido, A.N. Meza et al. 2006. Development of an integrated genetic map of sugarcane (*Saccharum* spp.) commercial cross, based on a maximum-likelihood approach for estimation of linkage and linkage phases. *Theoretical and Applied Genetics* 112:298–314.
23. Raboin, L.M., K.M. Oliveira, L. Lecunff et al. 2006. Genetic mapping in sugarcane, a high polyploid, using bi-parental progeny: Identification of a gene controlling stalk colour and a new rust resistance gene. *Theoretical and Applied Genetics* 112:1382–1391.
24. Patade, V.Y., S. Bhargava, and P. Suprasanna. 2009. Halopriming imparts tolerance to salt and PEG induced drought stress in sugarcane. *Agriculture, Ecosystems & Environment* 134:24–28.
25. Patade, V.Y., S. Bhargava, and P. Suprasanna. 2012. Halopriming mediated salt and iso-osmotic PEG stress tolerance and, gene expression profiling in sugarcane (*Saccharum officinarum* L.). *Molecular Biology Reports* 39:9563–9572.
26. Patade, V.Y., A.N. Rai, and P. Suprasanna. 2011. Expression analysis of sugarcane *shaggy-like kinase (SuSK)* gene identified through cDNA subtractive hybridization in sugarcane (*Saccharum officinarum* L.). *Protoplasma* 248:613–621.
27. Menossi, M.C., M. Siva-Filho, M.A. Vincentz, and G.M. Van-Sluys. 2008. Sugarcane functional genomics: Gene discovery for agronomic trait development. *International Journal of Plant Genomics* 2008:1–11.
28. Rodrigues, F.A., M.L. de Laia, and S.M. Zingaretti. 2009. Analysis of gene expression profiles under water stress in tolerant and sensitive sugarcane plants. *Plant Science* 17:286–302.
29. Patade, V.Y., and P. Suprasanna. 2010. Short-term salt and PEG stresses regulate expression of microRNA, miR159 in sugarcane leaves. *Journal of Crop Science and Biotechnology* 13:177–182.
30. Patade, V.Y., S. Bhargava, and P. Suprasanna. 2012. Transcript expression profiling of stress responsive genes in response to short-term salt or PEG stress in sugarcane leaves. *Molecular Biology Reports* 39:3311–3318.
31. Suprasanna, P., V.Y. Patade, and V.A. Bapat. 2008. Sugarcane biotechnology. A perspective on recent developments and emerging opportunities. In *Advances in Plant Biotechnology*, edited by Rao, G.P. 303–331. Studium Publication, LLC, Texas, USA.
32. Papini-Terzi, F., F.R. Rocha, R.Z.N. Vêncio et al. 2009. Sugarcane genes associated with sucrose content. *BMC Genomics* 10:120–141.
33. Hinchee, M., W. Rottmann, L. Mullinax et al. 2011. Short rotation woody crops for bioenergy and biofuel applications. *In Vitro Cellular & Developmental Biology. Plant* 45:619–629.
34. Tuskan, G.A., S. DiFazio, S. Jansson et al. 2006. The genome of black cottonwood, *Populus trichocarpa* (Torr. & Gray). *Science* 313:1596–1604.
35. Jansson, S., and C.J. Douglas. 2007. *Populus*: A model system for plant biology. *Annual Review of Plant Biology* 58:435–458.
36. Fabbrini, F., M. Gaudet, C. Bastien et al. 2012. Phenotypic plasticity, QTL mapping and genomic characterization of bud set in black poplar. *BMC Plant Biology* 12:47.
37. Du, N., X. Li, Y. Liu et al. 2012. Genetic transformation of *Populus tomentosa* to improve salt tolerance. *Plant Cell, Tissue and Organ Culture* 108:181–189.
38. Ceulemans, R., A.J.S. McDonald, and J.S. Pereira. 1996. A comparison among eucalypt, poplar and willow characteristics with particular reference to a coppice, growth-modelling approach. *Biomass & Bioenergy* 11:215–231.
39. Costa, G.G.L., K.C. Cardoso, L.E.V. Del Bem et al. 2010. Transcriptome analysis of the oil-rich seed of the bioenergy crop *Jatropha curcas* L. *BMC Genomics* 11:462.
40. Arumugam, M., A. Agarwal, M.C. Arya, and Z. Ahmed. 2011. Microalgae: A renewable source for second generation biofuels. *Current Science* 100:1141–1142.
41. Grossman, A. 2005. Paths toward algal genomics. *Plant Physiology* 137:410–427.
42. Mus, F., A. Dubini, M. Seibert, M.C. Posewitz, and A.R. Grossman. 2007. Anaerobic adaptation in *Chlamydomonas reinhardtii*: Anoxic gene expression, hydrogenase induction and metabolic pathways. *Journal of Biological Chemistry* 282:25475–25486.
43. Devos, K.M., and M.D. Gale. 2000. The grass model in current research. *Plant Cell* 12:637–646.

44. Schranz, M.E., B.H. Song, A.J. Windson, and T. Michell-Olds. 2007. Comparative genomics in the Brassicaceae: A family-wide perspective. *Current Opinion in Plant Biology* 10:168–175.
45. Hanley, S.J., M.D. Mallott, and A.T. Karp. 2006. Alignment of a *Salix* linkage map to the *Populus* genomic sequence reveals macrosynteny between willow and poplar genomes. *Tree Genetics & Genomes* 3:35–48.
46. Hazen, S.P., J.S. Scott-Craig, and J.D. Walton. 2002. Cellulose synthase-like genes of rice. *Plant Physiology* 128:336–340.
47. McIntyre, C.L., R.E. Casu, J. Drenth et al. 2005. Resistance gene analogues in sugarcane and sorghum and their association with quantitative trait loci for rust resistance. *Genome* 48:391–400.
48. Jang, C.S., T.L. Kamps, H. Tang et al. 2008. Evolutionary fate of rhizome-specific genes in a non-rhizomatous *Sorghum* genotype. *Heredity* 102:266–273.
49. Doebley, J.F., B.S. Gaut, and B.D. Smith. 2006. The molecular genetics of crop domestication. *Cell* 127:1309–1321.
50. Gupta, S.M., S. Gupta, and A. Kumar. 2009. Development of bed reactor using brick dust immobilized CM-cellulase from seeds of cowpea (*Vigna sinensis* L). *Journal of Plant Biochemistry and Biotechnology* 18:113–116.
51. Chen, F., and R.A. Dixon. 2007. Lignin modification improves fermentable sugar yields for biofuel production. *Nature Biotechnology* 25:759–761.
52. Vermerris, W., A. Saballos, G. Ejeta et al. 2007. Molecular breeding to enhance ethanol production from corn and sorghum stover. *Crop Science* 47:S145–S153.
53. Molhoj, M., S. Pagant, and H. Hofte. 2002. Towards understanding the role of membrane bound endo beta1,4-glucanases in cellulose biosynthesis. *Plant & Cell Physiology* 43:1399–1406.
54. Hisano, H., R. Nandakumar, and Z.Y. Wang. 2009. Genetic modification of lignin biosynthesis for improved biofuel production. *In Vitro Cellular & Developmental Biology. Plant* 45:306–313.
55. Lopes, F., F. Motta, C.C.P. Andrade, M.I. Rodrigues, and F. Maugeri-Filho. 2011. Thermo-stable xylanases from nonconventional yeasts. *Journal of Microbial & Biochemical Technology* 3:36–42.
56. Graber, J.R., J.R. Leadbetter, and J.A. Breznak. 2004. Description of *Treponema azotonutricium* sp. nov. and *Treponema primitia* sp. nov., the first Spirochetes isolated from termite guts. *Applied and Environmental Microbiology* 70:1315–1320.
57. Aishwarya, V., A. Grover, and P.C. Sharma. 2007. EuMicrosatdb: A database for microsatellites in the sequenced genomes of eukaryotes. *BMC Genomics* 8:225.
58. Sharma, P.C., A. Grover, and G. Kahl. 2007. Mining microsatellites in eukaryotic genomes. *Trends in Biotechnology* 25:490–498.
59. Grover, A., V. Aishwarya, and P.C. Sharma. 2012. Searching microsatellites in DNA sequences: Approaches used and tools developed. *Physiology and Molecular Biology of Plants* 18:11–19.
60. Salvi, S., G. Sponza, M. Morgante et al. 2007. Conserved noncoding genomic sequences associated with a flowering-time quantitative trait locus in maize. *Proceedings of the National Academy of Sciences of the United States of America* 104:11376–11381.
61. Wong, C.K., and R. Bernardo. 2008. Genome wide selection in oil palm: increasing selection gain per unit time and cost with small populations. *Theoretical and Applied Genetics* 116:815–824.
62. Ming, R., S. Liu, P.H. Moore, J.E. Irvine, and A.H. Paterson. 2001. QTL analysis in a complex autopolyploid: Genetic control of sugar content in sugarcane. *Genome Research* 11:2075–2084.
63. Pinto, L.R., A.A.F. Garcia, M.M. Pastina et al. 2010. Analysis of genomic and functional RFLP derived markers associated with sucrose content, fiber and yield QTLs in a sugarcane (*Saccharum* spp.) commercial cross. *Euphytica* 172:313–327.
64. Alwala, S., C.A. Kimbeng, J.C. Veremis, and K.A. Gravois. 2009. Identification of molecular markers associated with sugar-related traits in a *Saccharum* interspecific cross. *Euphytica* 167:127–142.
65. Anderson, J.A., S. Chao, and S. Liu. 2007. Molecular breeding using a major QTL for *Fusarium* head blight resistance in wheat. *Crop Science* 47:S112–S119.
66. Eathington, S.R., T.M. Crosbie, M.D. Edwards, R.S. Reiter, and J.K. Bull. 2007. Molecular markers in a commercial breeding program. *Crop Science* 47:S154–S163.

67. Edgerton, M.D. 2009. Increasing crop productivity to meet global needs for feed, food, and fuel. *Plant Physiology* 149:7–13.
68. Varshney, R.K., A. Graner, and M.E. Sorrells. 2005. Genetic microsatellites in plants: Features and applications. *Trends in Biotechnology* 23:48–55.
69. Gupta, P.K., S. Rustgi, and R.R. Mir. 2008. Array-based high-throughput DNA markers for crop improvement. *Heredity* 101:5–18.
70. Mace, E.S., L. Xia, D.R. Jordan et al. 2008. DArT markers: diversity analyses and mapping in *Sorghum bicolor*. *BMC Genomics* 9:26.
71. Aishwarya, V., and P.C. Sharma. 2008. UgMicroSatdb: Database for mining microsatellites from unigenes. *Nucleic Acids Research* 36:D53–D56.
72. Grover, A., B. Ramesh, and P.C. Sharma. 2009. Development of microsatellite markers in potato and their transferability in some members of Solanaceae. *Physiology and Molecular Biology of Plants* 15:343–358.
73. Jain, A., R. Ghangal, A. Grover, S. Raghuvanshi, and P.C. Sharma. 2010. Development of EST-based new SSR markers in seabuckthorn. *Physiology and Molecular Biology of Plants* 16:375–378.
74. Haddadi, P., A. Langlade, N.B. Ebrahimi et al. 2012. Genetic dissection of tocopherol and phytosterol in recombinant inbred lines of sunflower through quantitative trait locus analysis and the candidate gene approach. *Molecular Breeding* 29:717–729.
75. Rafalski, J.A. 2010. Association genetics in crop improvement. *Current Opinion in Plant Biology* 13:174–180.
76. McMullen, M.D., S. Kresovich, H.S. Villeda et al. 2009. Genetic properties of the maize nested association mapping population. *Science* 325:737–740.
77. Rubin, E.M. 2008. Genomics of cellulosic biofuels. *Nature* 454:841–845.
78. Simon, D., W.E. Tyner, and F. Jacquet. 2010. Economic analysis of the potential of cellulosic biomass available in France from agricultural residue and energy crops. *Bioenergy Research* 3:183–192.

26

Environomics: Omics for the Environment

Dinesh K. Yadav, Neelam Yadav, and Satyendra Mohan Paul Khurana

CONTENTS

26.1 Overview of Environmental Omics: Opportunities and Challenges

Etymological analysis suggests that the suffix "ome" is derived from the Sanskrit *om* "the completeness and fullness" [1]. By combining "gene" and "ome," Hans Winkler created the term genom(e), referring to "the haploid chromosome set, which, together with the pertinent protoplasm, specifies the material foundations of the species" [1,2]. Alternatively, the *Oxford English Dictionary* suggests that Winkler used genom(e) as a portmanteau of gene and chromosome [3]. Victor McKusick and Frank Ruddle added "genomics" to the scientific lexicon as the title for a new journal they cofounded in 1987, with emphasis on linear gene mapping, DNA sequencing, and comparison of genomes from different species [4]. Thus, the neologism "omics" is used to refer to the study of large sets of biological molecules [5], such as genomics, proteomics, metabolomics, or ionomics. The related suffix *ome* is used in molecular biology to address the *totality* of the objects of study. The omics technologies allow the generation of copious amounts of data at multiple levels of biology from gene sequence and expression to protein and metabolite patterns underlying variability in cellular networks and function of whole organ systems [6,7].

The recent development of omics technologies have provided the ability to explore critical details and characterize them at the molecular level regarding the interaction of organisms and communities and their response to and interaction with the environment. The idea that the field of molecular biology needed to move from studying isolated biological molecules to a broad analysis of large sets of biological molecules was underscored by the completion of the Human Genome Project [8,9]. The Human Genome Project demonstrated that a relatively limited number of genes could be identified in the human genome, which substantiated the theory that complex biological processes were regulated on levels other than DNA sequence alone. This realization triggered the rapid development of several fields in molecular biology that together are described by the term *omics* (Table 26.1) [10].

Omic technologies include genomics (annotation of the functions of as many genes as possible of a given species), transcriptomics (gene expression profiling), proteomics (annotation of the entire complement of proteins, their structures and functions including the modifications made to a particular set of proteins produced by an organism or system), ionomics (the study of the ionome, involving quantitative and simultaneous measurement of the elemental composition of living organisms and changes in this composition in response to physiological stimuli, developmental state, and genetic modifications), and metabolomics (quantitative measurement of the dynamic multiparametric metabolic response of living systems to pathophysiological stimuli or genetic modification) [11]. We are only beginning to appreciate the full complexity and the multidimensional nature of the biochemical networks operating in all living organisms (Figure 26.1) [12]. Studies of metabolic networks, gene regulatory networks, and protein–protein interaction networks in microbial organisms have significantly contributed to this and indeed to the identity of systems biology.

An integrative approach and global measurements have amplified the throughput, which has altered the process of fundamental research in "omic science." The omics approach has reversed the "first hypothesize-then-experiment" tradition into a "first experiment-then-hypothesize" mode of operation, and promises to discover unprecedented pathophysiological mechanisms of disease as well as response and toxicity to drugs and nutrition. The omics science and technologies suggest a marked improvement in the simplistic and reductionist experimental models that offer a merely temporal view of the much more complex, longitudinal, and dynamic nature of biological networks (and their fluctuations in response to social/environmental exposures) that fundamentally govern human health and disease.

Biomonitoring of environmental issues requires integrating approaches to sum up the multiple variables and factors that contribute to ecosystem behavior. Living organisms are usually the most unambiguous bioindicators of environmental concern because they can reflect the effect of contaminants on cellular metabolisms and global homeostasis. Generally, environmental stress situations can be assessed at the molecular level using different parameters (biomarkers). They include hundreds of cytochromes P450, which add polar groups into an organic substrate, for example, oxidation and subsequent

TABLE 26.1

Overview of the Different Omics Technologies

Omics	Molecules of Interest	Definition	Temporal Variance	Influence by Disease Status
Genomics	DNA	Assessment of variability in DNA sequence in the genome	None	No
Transcriptomics	RNA	Assessment of variability in composition and abundance of the transcriptome	High	Yes
Proteomics	Proteins	Assessment of variability in composition and abundance of the proteome	High	Yes
Metabolomics	Small molecules	Assessment of variability in composition and abundance of the metabolome	High	Yes
Ionomics	Ions	Assessment of variability in composition and abundance of the ionome	High	Yes

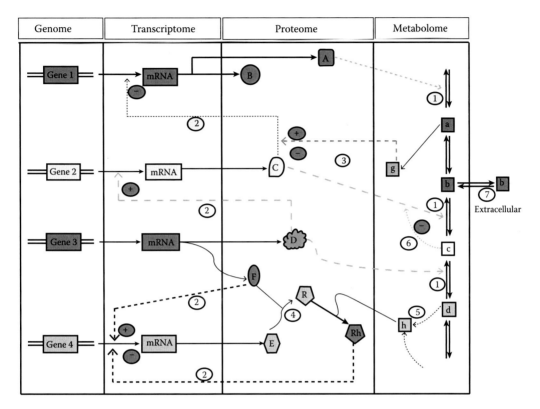

FIGURE 26.1
Biochemical network showing complexity and mutual interdependence of genome, transcriptome, proteome, and metabolome. In biological systems, a large number of structurally and functionally diverse components (genes, proteins, and metabolites) are involved in dynamic, nonlinear interactions, which in turn involve a range of timescales and interaction strengths. Solid lines show direct conversions of species, whereas dashed lines represent some possible interactions (not necessarily one step). Biochemical interactions shown include (1) enzyme catalysis, (2) posttranscriptional control of gene expression by proteins/protein complexes, including mechanisms that act on mRNAs (deadenylation, storage granulation) and mechanisms that act either directly or indirectly on DNA (histone modification, methylation), (3) effect of metabolite on gene transcription mediated by a protein, (4) protein–protein interaction, (5) effect of a downstream ("reporter") metabolite on transcription through binding to a protein, (6) feedback inhibition/activation of an enzyme by a downstream metabolite, and (7) exchange of a metabolite with extracellular system.

hydroxylation (RH → ROH), or number of glutathione-*S*-transferases, which conjugate electrophiles to reduced glutathione. In addition, heat shock proteins refold denatured proteins whereas metallothionines protect organisms from toxic metals [13].

Aside from primary antioxidant enzymes, such as superoxide dismutases, glutathione peroxidases, and peroxiredoxin, secondary antioxidant enzymes, including glucose-6-phosphate dehydrogenase and glutathione reductase, and antioxidant proteins, such as thioredoxins and glutaredoxins are also induced by exposure to many pollutants. Finally, low–molecular weight antioxidants, including reduced glutathione and vitamins C or E, are also greatly enhanced in exposed organisms [14,15]. Other biomarkers arise from the damage of key biomolecules by toxic chemicals when they exceed pollutant-elicited defenses. This includes malondialdehyde and 4-hydroxy-2-nonenal, products of lipid peroxidation and oxidized bases, abasic sites, and chain breaks generated after nucleic acid damage. In proteins, disulfide bridges are formed and Met or His are oxidized to methionine sulfoxide and 2-oxohistidine, respectively. Reduced glutathione (GSH) is oxidized (GS-) and two oxidized gluthatione molecules (GS-SG) forms mixed disulfides with protein thiols. These molecular damages trigger repair enzymes, such as phospholipases, glutathione peroxidase (GSHPx) and peroxiredoxins in lipids injury, glycosidases for nucleic acids, methionine sulfoxide reductase and thioredoxins for proteins, and glutaredoxins and glutathione reductase for GGSG and mixed disulfides [14,15]. These numerous conventional biomarkers are good tools in pollution assessment, but require a deep knowledge of their mechanisms of toxicity and assess a limited number of well-known proteins, excluding others that were also altered but whose relationship with pollution were unknown.

Omic technologies were proposed as an alternative to conventional biomarkers because these techniques quantitatively monitor many biological molecules in a high-throughput manner and thus provide a general appraisal of biological responses altered by exposure to contaminants. Therefore, massive identification of proteins by proteomic approaches can provide a general appraisal of proteins altered under pollutant exposure [13]. Environmental proteomics provides a more comprehensive assessment of the toxic and defense mechanisms triggered by pollutants without requiring any previous knowledge. However, this approach has problems in environmental studies that generally use nonmodel sentinel organisms, whose genetic sequences are not included in databases. This fact makes the identification of proteins difficult with high-throughput proteomic methods based on mass spectroscopic analyses of tryptic digests of two-dimensional electrophoresis spots. It is due to the requirement of expensive and cumbersome *de novo* sequencing of limited applicability to the vast number of proteins involved in such studies [13]. Although genes typically exert their functions at the protein level, genetic responses to stress conditions are often regulated at the transcription level. Nowadays, microarray technology is used to generate genomewide transcriptional profiles. Certainly, this methodology is making a huge contribution to functional genomics as a whole, but it also has a number of shortcomings. One critical issue in microarray-based transcript quantification is sensitivity. The most robust methodology for investigations aimed at quantitative analysis is reverse transcription followed by polymerase chain reaction (PCR). In fact, changes detected by microarrays have to be confirmed by quantitative PCR in a relative or absolute manner. It provides the fold variation in the level of a particular messenger RNA (mRNA) increase or decrease in a problem relative to a reference sample, or it provides the actual number of mRNA molecules. Omic technologies were proposed as an alternative to conventional biomarkers because these techniques quantitatively monitor many biological molecules in a high-throughput manner and thus provide a general appraisal of biological responses altered by exposure to various environmental contaminants. As the number of studies using omic technologies increases, it is becoming clear that any single omic approach may not be sufficient to characterize the complexity of ecosystems; hence, the integration of omic technologies with analytical chemistry, biochemical, and molecular biological approaches will improve our understanding of the biological effect of emerging pollutants to the environment and on human health.

26.2 Environmental Genomics: Techniques and Applications

The genome constrains the life but does not dictate the features of an organism. The phenotype is the result of a complex interaction of genes with the niche of an organism. An evolving environment leads organisms to explore and adjust to different states consistent with their genome. Environmental pressures force a species to adapt in changing ecological conditions. It leads to reprogramming of the genome in response to the environment, that is, phenotypic plasticity. A trait is plastic when the same genotype results in the development of a different phenotype depending on the environment the organism faces [16,17]. Plastic responses range from morphological modifications to drastic changes in physiology, life history, and behavior, and often involve changes in a suite of traits. In many cases, it has been shown that not only is a plastic response adaptive and allows individuals to meet the challenges brought about by a variable environment, but that it can also be costly to the organism as compared with the constitutive expression of the trait [16,18]. The presence of genetic variation for a trait, along with the force of selection, determines whether selection acting on that trait will result in an evolutionary response. Therefore, if different genotypes in a population produce different reaction norms, and if the slope of the reaction norm is positively correlated with fitness, increased plasticity should evolve in that population for that particular trait [19]. The amount of plasticity can thus evolve differently in distinct populations and species, depending on the selection pressures they each face. Furthermore, if certain genotypes are plastic within a population and others are not, it is possible that genetic variation for the trait will be measured as null in one environment (all genotypes develop the same phenotype) and high in another in which plasticity is expressed and genotypes differ in phenotypes [20]. Therefore, the response to selection on the basis of the same trait can vary among environments for the same set of genotypes [18]. It has been proposed that plasticity can accelerate evolution by allowing individuals to exploit a novel environment if they possess the capacity to develop a new phenotype when facing that environment [21,22].

Traditional microbiology and its genome sequencing and genomics rely on cultivated clonal cultures. Early environmental gene sequencing cloned specific genes, often the 16S rRNA, to produce phylogenetic diversity in a natural sample. Such work revealed that the vast majority of microbial biodiversity had been missed by cultivation-based methods [23]. A more complete picture of life on and even within the Earth has recently become possible by extracting and sequencing DNA from an environmental sample, a process called environmental genomics or metagenomics [24–31]. This approach allowed us to identify members of microbial communities and to characterize the abilities of the dominant members even when the isolation of those organisms had proven intractable. However, with a few exceptions [28,30], assembling complete or even near-complete genomes for a substantial portion of the member species is usually hampered by the complexity of natural microbial communities. In addition to elevated temperatures and a lack of oxygen, conditions within the Earth's crust at depths of more than 1 km are fundamentally different from those of the surface and deep ocean environments. Severe nutrient limitation is believed to result in cell doubling time ranging from hundreds to thousands of years [32–34], and as a result, subsurface microorganisms might be expected to reduce their reproductive burden and exhibit the streamlined genomes of specialists or spend most of their time in a state of semisenescence, waiting for the return of favorable conditions. Such microorganisms are of particular interest because they permit insight into a mode of life independent of the photosphere.

Technology will play an essential role in enabling solutions to the most pressing environmental challenges. Developing technologies required for innovative environmental genomics involve the recovery of longer genomic DNA sequences from environmental systems without contaminating the DNA of the organism from unrelated ecological systems. International drilling projects and their technologies for the study of microbial communities in the hot, deep-subsurface biosphere are quite useful for culture-independent community analysis of hyperthermophilic archaea in core samples recovered from deep-subsurface geothermal unicellular ecosystems. Environmental genomics requires an enrichment of the genomic database of members from each trophic level of diverse ecological systems. It requires an advanced molecular biological technique enabling the construction of large genomic libraries and high-throughput DNA sequencing methodologies. Proliferation in bioinformatics and advancements in DNA sequencing have greatly aided the analyses of DNA sequences recovered from metagenomes. The shotgun sequencing approach provides qualitative information about organisms and possible metabolic processes present in that metagenomic community. This can be helpful in understanding the environmental phylogeny in that community. Shotgun metagenomics is also capable of sequencing nearly complete microbial genomes directly from the environment (Figure 26.2) [28,35].

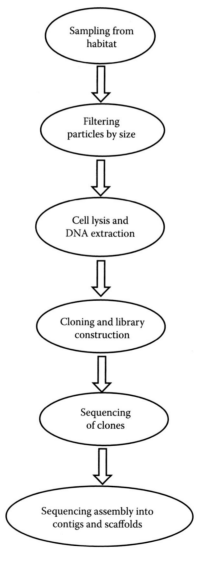

FIGURE 26.2
Flowchart showing environmental shotgun sequencing.

High-throughput sequencing using pyrosequencing of metagenomes generates megabase/gigabase sequences with an additional advantage that this technique does not require DNA cloning before sequencing, removing one of the main biases in environmental sampling. Metagenomic DNA sequence data are usually less redundant and error-prone short read lengths. With the development of various DNA assembly programs, such as Phrap or Celera Assembler [35], information from paired-end tags can be used to improve the accuracy of DNA assemblies. After assembly, gene annotations are provided to connect community composition and function in metagenomes by a similarity-based association of a particular sequence with an organism. Comparative analyses between metagenomes can provide additional insights into the function of complex microbial communities and their role in that particular ecosystem [36]. Pairwise or multiple comparisons between metagenomes can be made at the level of sequence composition (comparing guanine/cytosine (GC) or genome size), taxonomic diversity, or functional complement. Comparison of population structure and phylogenetic diversity can be made on the basis of 16S rRNA and other phylogenetic marker genes. Metagenomics has the potential to advance knowledge in a wide variety of fields including health, biofuel, environmental remediation, and agriculture. It can also be applied to solving practical challenges in medicine, environmental engineering, and sustainability.

26.3 Environmental Transcriptomics: Techniques and Applications

Omics science and technologies offer the promise of improving the simplistic and reductionist experimental models that hitherto merely presented a temporal snapshot of the much more complex, longitudinal and dynamic nature of biological networks and their fluctuations in response to socioenvironmental exposures. Although metagenomics allows researchers to access the functional and metabolic diversity of microbial communities, still it cannot show the metabolically active biological processes of the system [37]. Thus, high-throughput analyses of differential metagenomic mRNA expression (metatranscriptome) is a powerful tool for discovering novel genes or for gaining additional information about the regulation and expression profiles of biological processes in complex communities at a genomic scale.

Changes in mRNA steady state levels are mostly accomplished by changing the transcriptional rates of genes. Such fluctuations in relative mRNA abundance are indicative of changes in the environment and developmental program or reflect responses to all kinds of stimuli. Properly understanding a gene's function requires critical knowledge about when, where, and up to what extent a gene is expressed, it also essentially requires the discovery of other related genes which are coregulating with the gene of interest. Hence, monitoring the transcriptome by measuring mRNA concentrations of defined genes in a multiparallel and quantitative manner allows us to assign function to a multitude of unknown genes. The short half-life of mRNAs limits the collection and profiling of an environmental transcriptome; hence, there are relatively fewer *in situ* metatranscriptomic studies available thus far. Transcriptional profiling using microarrays has developed into the most prominent tool for functional genomics and has convincingly demonstrated how information from raw sequence data can be converted into a broader understanding of gene function.

The advantage of microarray-based expression analysis is that a large number of genes, from anonymous or defined sequences, can be monitored at the same time using a small amount of biological sample. Originally restricted to microarray technology, metatranscriptomic studies can now use direct high-throughput cDNA sequencing to provide whole-genome expression and quantification of a microbial community.

26.4 Environmental Proteomics: Techniques and Applications

"Environmental proteomics" or "community proteogenomics" is the holistic study of all protein samples expressed in and recovered directly from environmental sources. It deals with all the genes and proteins identified from complex communities, in which individuals cannot be binned into species or organism types. Analysis of the collective proteome of microbial communities is known as metaproteomics. Here, the community is viewed as a "metaorganism," in which population and metaproteome shifts are forms of functional responses. Prokaryotic and eukaryotic microorganisms make a vital contribution to biogeochemical cycles by decomposing virtually all natural compounds and thereby exert a lasting effect on the biosphere and climate. Despite a growing knowledge of the range of microbial diversity, most of the microorganisms seen in natural environments are uncultivated and their functional roles and interactions are unknown. The rapidly growing number of metagenomic sequences, together with revolutionary advances in bioinformatics and protein analyses, has opened completely new horizons to investigate the molecular basis of response to complex processes like starvation, desiccation, or heat shock cycles.

Proteomics has contributed substantially to our understanding of individual organisms at the cellular level as it offers excellent possibilities to probe many protein functions, interactions, and responses simultaneously. Proteomics provides a global view of the protein complement of biological systems and, in combination with other omic technologies, has a crucial role in uncovering the mechanisms of varied cellular processes and thereby advance the development of environomics. However, it has not yet been widely applied in microbial ecology, although most proteins have an intrinsic metabolic function that can be used to relate microbial activities to the identity of defined organisms in multispecies communities.

Environmental proteomics is much less developed than other proteomics applications but enables simple protein cataloging, comparative and semiquantitative proteomics, analyses of protein localization, discovery and effect of posttranslational modifications, and genotyping by strain-resolved proteogenomics (Table 26.2) [38].

An important recent advancement in environmental proteomics is the ability to identify proteins from unsequenced organisms with the use of modern bioinformatic techniques. Cross-species protein identification [39,40] and protein sequence similarity searches [41] are the most common strategies used to identify proteins when the genomic sequence is not available. These techniques have shown great potential in the evaluation of biological processes in a complex community without isolating organisms. It also allows a view of organism interactions, which are impossible to determine using pure cultures.

Environmental proteomics, including metaproteomics, yield better results in combination with other omics approaches such as metabolomics and transcriptomics. In addition, proteomics allows us to confirm the existence of gene products predicted from genomics, and has proven to be an effective complement to nucleic acid–based methods as a

TABLE 26.2

Relationship between Feasibility of Proteomic Studies and Sample Complexity

Experimental System	Research Objective	Complexity	Feasibility
Isolation in laboratory	Metabolic pathways, stress responses, and adaptation		
Community in laboratory	Community function under controlled conditions, model communities to study their interactions	↓	↑
Community in environment	Community interactomics under complex ecologically relevant conditions		

problem-solving tool in molecular biology [42]. In addition, proteomics can be used for the phylogenetic classification of microbial communities, by using either two-dimensional maps [43] or peptide sequences obtained from mass spectroscopy [44]. Proteomics has the advantage of not being limited to organisms for which the genomic sequence is available. In addition, the proteome represents the actual enzyme content in a system, going beyond potential gene expression as determined by microarrays, and providing information about the role of posttranslational modifications.

Ecological studies focus on naturally occurring bacterial adaptations to their environments. Proteomic studies provide insights into the mechanisms of several phenomena like adaptation, especially to extremes of temperature. Proteins of hyperthermophilic organisms are of particular importance because they have an enhanced conformational stability, allowing them to be active at high temperatures. This property can be used to investigate the molecular basis of protein folding and conformational stability. Monitoring environmental pollution using biomarkers requires a comprehensive knowledge about the markers, and many only allow a partial assessment of pollution. Environmental proteomics can identify and validate new alternative pollution marker proteins. In model organisms (yeast, *Arabidopsis*, rat, cells, or mice) exposed to model contaminants, environmental proteomics has developed the concept of protein expression signatures. Samples from each environment had typical species compositions and their protein profiles varied according to their environments. Hence, the development of a sustainable remediation strategy requires a complexity profile of the affected microbial community of that ecosystem. Environmental proteomics can provide this profile to adopt a sustainable wastewater management/treatment program.

Proteomic studies of the effect of metabolic engineering are essential for the identification and quantification of the changes in host cell physiology reflected by protein production or other cellular processes. Proteomics approaches are used to gain insights into the physiological responses of microorganisms to temperature, chemical, and other stresses. The choice of proteomics as the primary experimental tool is a reflection of the ability to obtain system-wide information for non–model organisms to obtain protein identifications. Even environmental proteomics has a high potential and prospects; however, there are many challenges to overcome in the field of current environmental proteomic methods. For example, it is not yet possible to acquire proteome data present in complex communities, mainly due to a wide range of protein concentrations, the lack of an appropriate procedure for the enrichment of low-abundance proteins from uncultured or unsequenced organisms. Thus, it is often an obvious advantage to complement proteomics with other omics tools. Some of the main challenges in metaproteomics are the difficulties related to the evaluation of such a large number of gene products as well as the lack of genome sequences for the large majority of environmental bacteria.

The primary challenge lies in the extraction of complex proteomes from soil, high-solid matrices, and other hyperthermophilic ecosystems, in the laboratory. In these samples, the presence of the proteins' adsorbent surfaces and high concentrations of interfering compounds with properties similar to proteins interfere with proteome isolation and urges new strategies for protein extraction, purification and separation. Improved bioinformatic tools are also required to aid the identification of proteins from unsequenced microbial communities. Another challenge of environmental proteomics is that the environment of interest is uncontrolled and is difficult to simulate in the laboratory.

26.5 Environmental Metabolomics: Techniques and Applications

Metabolomics is the study of the naturally occurring low–molecular weight organic metabolites (~1000 Da) within a cell, tissue, or biofluid using chemometric techniques for data interpretation [45–51]. In essence, environmental metabolomics is the application of metabolomics to characterize the interactions of living organisms with their environment. Metabolomics can provide considerable insight encompassing the comprehensive and simultaneous systematic profiling of metabolite levels, patterns, and the systematic and temporal alterations in response to environment, disease, toxicity, genetic manipulations, and pharmaceuticals within an organism.

Environmental sciences in general have been viewed as a science that uses crude and often outdated methodologies. However, the scenario is changing, and today, the most advanced instrumentation available is being used. Nuclear magnetic resonance spectroscopy is one of the most widely utilized methods for identification and quantification of metabolites within complex biological samples. But the limited sensitivity of this approach hampers the detection of a large fraction of the metabolome. *Daphnia magna* is a Cladoceran freshwater flea that has been internationally recognized as model organism for toxicity testing and hence the ability to acquire spectra from the biofluid of such a small organism opens up the potential for metabolomics-based chemical risk assessments in this and similar ecologically important test species.

In recent years, mass spectroscopic approaches have become more widespread in environmental metabolomics. This includes liquid chromatography-mass spectrometry (for example to characterize steroid profiles in the gonads of roach, a freshwater fish, after exposure to environmental estrogens) [52] and gas chromatography-mass spectrometry (to identify endogenous metabolites in the microbe *Pseudomonas putida* that could discriminate between exposure to six pharmaceuticals associated with wastewater treatment plants) [53]. A novel direct-infusion Fourier transform ion cyclotron resonance mass spectrometry method, which detected more than 3000 mass features in liver extracts of a marine flatfish, *Limanda limanda*, has been reported [54]. This method is widely accepted as one of the most powerful tools for complex mixture analyses because it offers the highest mass accuracy and resolution of all mass spectrometers [55–57].

In the past few years, metabolomic studies of mammalian urine have yielded a wealth of information related to toxicology [58]. Urine that is ideally suited for such studies as it is an indicative of an organism's systemic metabolic condition, resulting from the integration of one or several organ responses. Furthermore, urine can be collected noninvasively, not only allowing an organism to serve as its own control but also facilitating time course investigations from predose to postdose, and into a recovery phase [59]. Another important aspect of metabolomic studies is metabolite extraction, which, due to considerable diversity in the matrices of environmental samples, can require considerable method development. Several studies have been reported recently, including an optimized protocol for extracting metabolites from the earthworm *Eisenia fetida*, which is frequently used in ecotoxicology studies [60]. Several solvent systems have been evaluated for extracting metabolites from fish, including muscle from the Chinook salmon (*Onchorhyncus tshawytscha*) and liver from chub (*Leuciscus cephalus*) [61].

Metabolomics affords several advantages for studying organism–environment interactions and for assessing organism function and health at the molecular level. Measurements of metabolomic data on the actual functional status of the organism (cell, tissue, or biofluid) can be mechanistically related to organism phenotype. Another advantage of metabolomics is that it can discover unexpected relationships and metabolite responses, which in itself can lead to hypothesis generation. Metabolomics offers a wide array of applications in the environmental sciences, ranging from understanding organisms' responses to abiotic stresses, including both natural factors such as temperature and anthropogenic factors such as pollution, to investigate biotic–biotic interactions, biomarker development, and risk assessment of toxicant exposure, and disease diagnosis and monitoring [34]. This approach can be used to study aquatic organisms, including the challenges of measuring metabolites and their variability as well as the importance of genotypic and phenotypic anchoring to facilitate the interpretation of multivariate metabolomic data [62]. Characterizing the metabolic responses of the organisms to anthropogenic stressors highlights another important application of environmental metabolomics. This approach would use sentinel organisms of a particular ecosystem to reveal the condition of the environment. These interactions can be studied in individuals or in populations of individuals, and can be related to the fields of ecophysiology and ecology as well as to the effects of these interactions over evolutionary timescales, thus enabling the studies of genetic adaptation. Ecophysiology includes the metabolic responses to temperature, water, food availability, light and circadian rhythms, atmospheric gases, and season. Metabolomics-based studies of organism–environment interactions, such as seasonal cycles associated with reproduction and hibernation, and the effects of feeding behaviors on metabolism, help determine the molecular basis to the range of organisms and adaptation to natural and anthropogenic environmental stressors. Such studies are of great value particularly for non–model organisms from remote areas of our planet, for which known metabolic pathway maps have no or limited relevance.

Although much progress has been made in environmental metabolomics in the past few years, researchers have only scratched the surface in terms of potential applications. This is partly because this approach is still technically complicated, limiting its widespread introduction into environmental laboratories. Indeed, considerable work still remains in developing the chemical and computational technologies that underpin metabolomics. As the technology advances, the scientific world will better realize and exploit the advantages of metabolomics for studying disease and toxicity in wildlife.

26.6 Omics Approaches for Genetoxicity and Environmental Diseases

The genotoxicity of chemical agents is an intrinsic chemical character, based on the agent's electrophilic potential to bind with such nucleophilic sites in the cellular macromolecules as DNA. Thus, genotoxicity is the toxicity manifested in the genetic material of the cells.

The broader definition of genotoxicity includes both direct and indirect effects in DNA: (i) the induction of mutations (gene, chromosomal, genomic, and recombinational) at the molecular level are similar to events involved in carcinogenesis, (ii) indirect surrogate events associated with mutagenesis (e.g., unscheduled DNA synthesis and sister chromatid exchange), or (iii) DNA damage (e.g., the formation of adducts), which may eventually lead to mutations.

Genotoxicity can be due to many physicochemical agents that result in a wide variety of possible damage to the genetic material, ranging from various DNA adducts to single-strand and double-strand breakages, DNA–DNA and DNA–protein cross-links, or even chromosomal breakage (Figure 26.3) [63–65]. The major challenge in genotoxicity testing resides in developing methods that can sensibly detect a wide array of damage or a general cellular response to genotoxic insult. It is known that no single test can detect every genotoxin; therefore, the concept of test batteries has been implemented in many regulatory guidelines [66]. Numerous complex and efficient damage signaling and repair mechanisms have been unraveled in prokaryote, eukaryote, and mammalian systems [67,68]. Despite DNA alterations in the nucleotide sequences and in the arrangement of DNA strands arising from mistakes in the repair process, agents interfering with damage signaling and repair mechanisms are generally not considered in safety testing. They should, however, be detected as they could impair indirect genotoxicity by facilitating the activity of genotoxic agents such as direct genotoxins, reactive oxygen species, and radiation (UV and gamma) [69–71].

Toxicogenomics, aims to study the interaction between the structure and activity of the genome and the adverse biological effects of exogenous agents [72,73]. This discipline is based on the concept that the toxic effects of xenobiotics on biological systems are generally reflected at the cellular level by their effect on transcriptomics, proteomics, and metabolomics (Figure 26.4) [72,74–78].

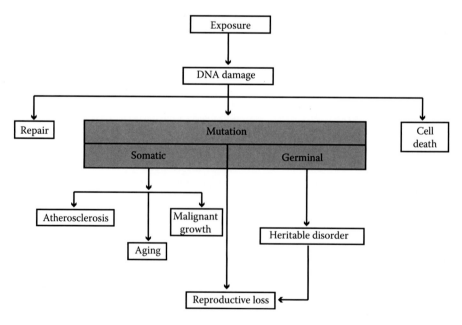

FIGURE 26.3
Schematic view of the scientific paradigm in genotoxicology and human health effects.

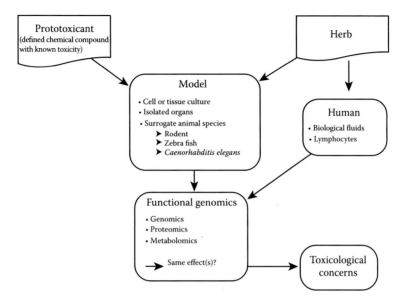

FIGURE 26.4
Testing strategy for omics methods.

Human biological monitoring uses body fluids or other easily obtainable biological samples and their metabolites to measure the biological effects of exposure to specific substances. Biological monitoring allows the estimation of total individual exposure through different exposure pathways (lungs, skin, and gastrointestinal tract) and different sources of exposure (air, diet, lifestyle, or occupation). In a complex exposure situation, different exposing agents may interact with one another to either enhance or inhibit the effects of the individual compounds. Individuals with different genetic constitutions exhibit variability in their response to chemical exposures. Thus, it may be more reasonable to look for early effects directly in the exposed individuals or groups than to predict potential hazards of the complex exposure patterns from data pertaining to single compounds. This is an advantage of genetic biomonitoring for early effects, an approach employing techniques that focus on cytogenetic damage, point mutations, or DNA adducts in surrogate human tissue.

Transcriptomics analyzes the expression level of genes by measuring the transcriptome, the genomewide mRNA expression [79,80]. Transcriptomics uses high-density and high-throughput methods, such as cDNA microarrays and quantitative PCR, for the assessment of mRNA expression. DNA microarrays or biochips are the most common approach used for gene expression profiling [79], generated by immobilizing a large number of oligonucleotides on an extremely small surface (up to 200,000 spots/cm^2). Based on the target sequences, significant changes in the mRNA can be estimated for thousands of genes [81–83]. Specialized subsets of gene expression profiling and quantitative PCR-based approaches that focus on specific genes have also been developed. Quantitative PCR is an authentic tool for quantifying gene expression with improved sensitivity and high specificity [80].

Other technologies such as serial analysis of gene expression, massively parallel signature sequencing, and total gene expression analysis are used to detect changes in transcriptomes [84–86]. Unlike microarray technology (which measures transcript abundance with preselected, known probe sequences), these approaches are "open systems" for gene discovery and offer linear gene expression quantification over a wide range [87]. The microarray genomic studies have been quite inconclusive for genotoxicity prediction but are specific to certain target genes like, *GADD45a, p53R2, Ephx1, Btg2, Cbr3*, and *Perp*, among which a robust induction of expression was noted for a series of genotoxins with apparently high sensitivity and specificity [88–90]. There is a considerable interest in the genes involved in tissue development, cell death, cell-to-cell signaling, cell cycle and cellular growth, proliferation, DNA damage signaling, and DNA repair [91,92].

The tumor suppressor protein, p53, plays an important role by regulating the expression of a series of genes that promote genomic stability, DNA repair, cell cycle arrest, and induction of apoptosis in response to DNA damage [93,94]. These include *p53R2* (subunit of ribonucleotide reductase), *CDKN1A* (cyclin-dependent kinase inhibitor 1A), and *GADD45a* (growth arrest and DNA damage) [95,96]. p53R2 is activated by gamma rays, UV irradiation, and several genotoxic compounds [97–99]. A luciferase reporter plasmid, dependent on three tandem repeat sequences of the p53-binding site derived from the p53R2 gene, has been developed for genotoxicity testing [89]. This high-throughput assay is available for wild-type p53 human cell lines, requires only a small number of test samples, and gives fewer false-positive results [100].

GADD45a plays an important role in cell cycle control, DNA repair mechanisms, and signal transduction [101]. Other cellular signaling pathways, including BRCA1, c-MYC, and NF-κB are implicated in GADD45a induction [102–104]. GADD45a-dependent reporter assays, based on green fluorescent protein or luciferase, are commercially available for testing genotoxicity response [88,105]. The genes involved in cell cycle (CDKN1A, GADD45, cyclin E), apoptosis (*bax, bcl-xl*), DNA repair (XPC, DDB2, GADD45), and various physiological processes (FOS/JUN, MDM2, FRA-1 IL-8, HSP70), which are found to be altered by ionizing radiation, could be predictive biomarkers of genotoxicity [106,107]. Transcriptomics has a major limitation in which changes in gene expression levels may predict changes in the protein profiles of cells, tissues, or organisms but there are cases in which a functional protein is not produced despite gene expression; and so, changes in the transcriptome do not necessarily reflect a change in the profile of the end-products [79]. The availability of a large database of proteomic "fingerprints" for a series of known carcinogens will make it possible to identify changes in biochemical pathways and assessment of the toxicity of unknown compounds. The products of the genes identified thus far in transcriptomic studies are probably promising candidates as proteomics markers.

Toxicometabolomics concerns the analyses, either in organs, blood, or urine, of metabolites and metabolic pathway modifications after a toxic insult [83]. This implies the quantitative and qualitative study of a wide range of low–molecular weight molecules produced as the net result of cellular functions [108,109] using nuclear magnetic resonance spectroscopy, Fourier transform infrared, and near-infrared spectroscopy or mass spectroscopy. Metabolomics have been rarely applied to genotoxicity studies. Its major applications concern the urinary profiling of damaged bases excreted upon DNA excision repair and the search for activated metabolites of precarcinogens [110].

26.7 Environmental Omics: Applications in Bioactivity and Ecosystem

Omics technologies are increasingly being used by researchers to answer fundamental queries about changes in the natural environment over time and the effects of anthropogenic activities on environmentally relevant organisms, populations, and complex communities. Omics technologies can be utilized as relatively new biomarker discovery tools that can be applied to study large sets of biological molecules. The ability of organisms to rapidly identify and respond to changes in a particular niche represents the foundation for application of omics-based assays to identify the signature profiles of ecological species. These signature motifs of ecological species may reside within the genome, transcriptome, proteome, or metabolome. It might be indicative of a particular environmental perturbation. Their application in observational studies is now possible due to increases in the sensitivity, resolution, and throughput of omics-based assays. The effect of anthropogenic activities and different pollutants such as oil spills, benzene, arsenic, etc., can be viewed in marine teleosts, mammals, invertebrates, and plants as well as in human health. The biological information obtained from omics can provide a framework in the detection and evaluation of the effects of environmental chemical contaminants on the health of flora and fauna. Repeated evaluations across different seasons and local environs will lead to discrimination/identification between signature profiles representing normal variation within the complex milieu of environmental factors that trigger biological response in a given sentinel species. It may provide a greater understanding of normal *versus* anthropogenic-associated modulation of biological pathways, which prove detrimental to the flora and fauna of that ecosystem. It is anticipated that the incorporation of contaminant-specific molecular signatures into current risk assessment paradigms will lead to enhanced environmental health management strategies. A significant benefit of adding support for the capture of environmental information about bioactivity and ecosystem used to describe data sets is the increased probability that data sets of environmental relevance can be integrated. Hence, by the inclusion of geographical coordinates, omics data can be mapped onto a standard reference grid with other variables, for example, to explore the distribution of specific organism or phenotypic traits within different environments, inclusion of information on the geographical location helps significantly for the global viewing of all isolates collected from a specific location.

These analyses would be complemented by the identification of characteristic fingerprints of bioactive molecules (transcriptomic, proteomic, or metabolomic profiles) detected within key sentinel organisms, which reveal predictive markers indicative of the presence of pollutants in general or specific classes of toxicants [111,112]. These markers may be composed of holistic profiles or more likely selected groups of functionally linked molecules for the early detection of pollutants. Omics tools as applied to microbial systems are important for risk assessment of chemicals before environmental release and may possibly aid the reduction, refinement, and replacement of current vertebrate-based testing regimes.

26.8 Environmental Integrative Omics: Applications in Environment Monitoring

Omics made a major and direct contribution to environmental monitoring. The use of a community-based omics approach, that is, metagenomics, metatranscriptomics, metaproteomics, or metametabolomics, to diversity profiling of either complex taxa (diatoms, phytoplankton, or collembolan) or whole bacterial or mesofaunal communities will provide future objective and automatable tools to assess the health of the ecosystem. These analyses would be complemented by the identification of characteristic fingerprints of molecules (transcriptomic, proteomic, or metabolomic profiles) detected within key sentinel organisms, which reveal predictive markers indicative of the presence of pollutants in general or specific classes of toxicants [111,112]. These markers may be composed of holistic profiles or more likely selected groups of functionally linked molecules for the early detection of pollutants. Ultimately, this could lead to a fundamental shift from chemical-based monitoring to a more cost-effective, information-rich biological effects–based environmental assessment strategy.

Omics tools, as applied to microbial systems, are important for risk assessment of chemicals before environmental release and may help in the reduction, refinement, and replacement of current mammalian-based testing regimes. The importance of microbes, invertebrates, and fish species (and cell lines derived from them) remains a useful area for investigation. Omics technology brings the potential to evaluate the true natural variations within populations and communities and precise assessment of the effects of global or local environmental change. This ability can be used in proactive environmental management to mitigate the future effects associated with global warming. Omics also plays a vital role in understanding the response of organisms as individuals, as a community, as a population, or as an ecosystem to long-term environmental changes. It arms us with novel tools to address the capacity and resilience of species and ecosystems to respond to environmental changes. It also provides knowledge about the key biological feedbacks and their regulation. This valuable information, obtained through omics approaches, allows us to develop substantial biogeoengineering methods to deal with ecosystem health and environmental decline.

26.9 Omics for Environmental Assessment of Wildlife, Marine Ecology, and Oceans

Monitoring environmental issues requires the integration of omics methodologies to sum up the multiple variables and factors that contribute to ecosystem behavior. Living organisms are usually the most unambiguous bioindicators of environmental concern because they can reflect the effect of contaminants on cellular metabolism and global homeostasis. Environomics provides a more comprehensive assessment of the toxic and defense mechanisms triggered by pollutants. An overall evaluation of changes that contaminants induce in cells is only possible by the integration of omics technologies. It is because the transcripts induced by pollutants (transcriptomics) encode proteins with altered expression profiles, which undergo posttranslational modifications (proteomics). For example, transcriptomic studies indicate considerable gene sequence similarities between the wildlife species *Mus spretus* and, as a reference, the gene/protein sequence databases from the model species *Mus musculus*. It has been demonstrated that the high-throughput proteomic methods and its database are applicable in wild, free-living *M. spretus* [113].

The marine environment and its wealthy bioresources remained uncharted and untapped for a long time as compared with terrestrial resources. However, advancements in science and technology, particularly omics technologies, are expanding our knowledge of the wildlife and marine environment, and is providing new tools to study the marine ecosystems. The ocean provides the basis for great biodiversity and contains some of the most challenging environmental extremes found on earth. The marine environment has enabled the evolution of organisms with unique structures, metabolic pathways, reproductive systems, and sensory and defense mechanisms. Within the oceans, marine bioresources are providing a number of important ecosystem services for the planet and its inhabitants. Marine organisms: microalgae, fish, and invertebrates, are a rich source of food for billions of people and livestock. The oceans are well known as regulators of global temperatures and filters of pollution. They are also sinks for carbon and nitrogen, and a source of oxygen and food. Living in the ocean's surface water, in easy reach of light, phytoplankton are estimated to produce half of the oxygen which humans and animals breathe. In addition to this important function, phytoplankton have an important position in the food web and play an important role in carbon cycling: locking away carbon dioxide and nitrogen, which is eventually deposited on the ocean bottom—slowing the effects of global warming. These ecosystem services are a function of biodiversity. They are vital for proper functioning of the biome and need to be maintained.

Metagenomic sequencing is now being used to study microbial ecosystems and for the identification of new genes with novel functions. Marine microbes seem to hold particular promise for marine biotechnology and are revealing new limits of biodiversity.

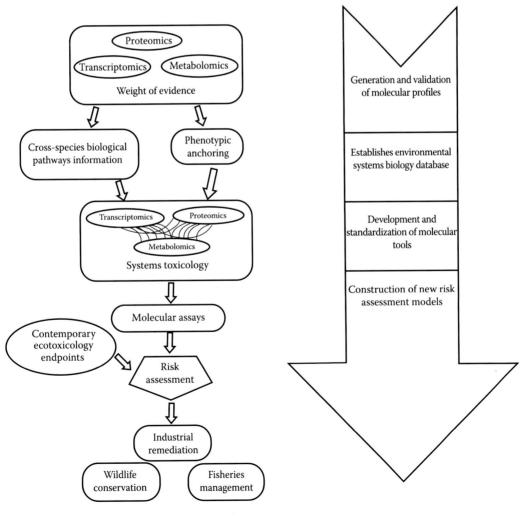

FIGURE 26.5
Present challenges and future prospects for omics-derived biological information in marine ecosystems.

Use of omics technology for ocean studies remains in its nascent phase and, at this point, the majority of effort can be considered proof-of-concept in nature. Longitudinal assessment combined with extensive evaluation of phenotypically anchored molecular data are required to establish a foundation of molecular profile signatures across marine sentinels as they relate to changing environmental factors. Subsequent determination of omics cross talk and the creation of biology-based integrative systems response signatures would require the incorporation of cross-species molecular pathways information. The resultant knowledge can establish focused molecular assays that combine maximal performance in the characterization of sentinel–environment interactions with minimal run costs. The utmost challenge is how the ensuing molecular profiling information will be employed to help augment current risk assessments performed in a regulatory framework and provide direction in the variety of issues affecting animal species in the marine environment (Figure 26.5) [114].

26.10 Bioinformatics and Mathematical Modeling for the Environment

Modeling and simulation is a discipline for developing an understanding of the interaction of the parts of a system, and of the system as a whole. The level of understanding developed by modeling and simulation is seldom achievable by any other discipline. The fitness of a model depends on the extent to which it promotes understanding. A model should include details of relevant interactions and promote understanding. At the same time, it should not contain every minute detail that could make it complicated and actually preclude the development of understanding. A simulation refers to a computerized version of the model which is run over time to study the implications of the defined interactions. Thus, the goal of omic approaches is to understand complex biological systems by modeling the relationship between multiple measured attributes of biomolecular organization and the phenotype of the organism. Mathematical modeling is the process of describing phenomena in terms of mathematical equations. It can be used in context with the goal of promoting applications of calculator-based models and simulations as a tool for planning, implementing, and monitoring a sustainable future for environmental resources.

Modeling is an indispensable skill and tool, when decisions for positive human–environment relations are based on integrated multidisciplinary data and knowledge. The development of high-throughput methods related to genomics, transcriptomics, proteomics, and metabolomics have generated large amounts of information. This data can be gathered at distinct levels of biological organization, allowing genotype, mRNA expression, protein, metabolic, and physiological data to be gathered for a particular organism under a specified set of conditions. These large volumes of complex data necessitate computational management. A unison analysis offers us the most comprehensive understanding of the fundamental complexities, such as climate change, developing cleaner energy sources, combating environmental pollution, preservation of biodiversity, improving nutrition, and formulating directed heath care, of complex communities. To address these challenges using omic technologies, it is important to be able to place the data in proper context. Modern scientific research depends on computer technology to organize and analyze large data sets. Evolutionary bioinformatics is devoted to leveraging the power of nature's experiment of evolution to extract key findings from sequence and experimental data.

Genomic data sets are accumulating at an exponential rate and have changed the landscape of data collection with the inherent challenges of new methods of statistical analyses and modeling. This challenge is addressed by adopting mathematical and statistical software, computer modeling, and other computational and engineering methods. As a result, bioinformatics has become the latest engineering discipline. As computers provide the ability to process the complex models, high-performance computer languages have become a necessity for implementing state-of-the-art algorithms and methods. Reconstructed networks from environmental genomics data can be analyzed using various methods of mathematical modeling [115–118], which can assess and quantify their dynamic properties and generate hypotheses on community and ecosystem functioning. Hypothesis testing can be verified by experimental and environmental approaches, with the subsequent possibility of iterations between the different steps of the process (Figure 26.6) [115].

The decreasing costs of DNA sequencing has triggered a large number of metagenome projects. However, there is an absence of easy, accessible, and user-friendly tools, and the complex data generated from each project, and its analyses, are obscured. As a result, we are witnessing an emerging field within computational biology aiming not only at the development of those tools but also at an understanding of the ecosystem imprinted in the sequence data. Recently, several computational methods have been developed and applied to analyze the functional and phylogenetic composition of environments and to derive various properties of the inhabiting microbial communities. In addition, the results of comparative metagenomics allow us to draw more general conclusions about the relationship between metagenome properties and the habitat of their origin.

The flow of information of a typical metagenomics data analysis starts from raw data assembly to increase fragment length and gain insights into the population structure. After performing

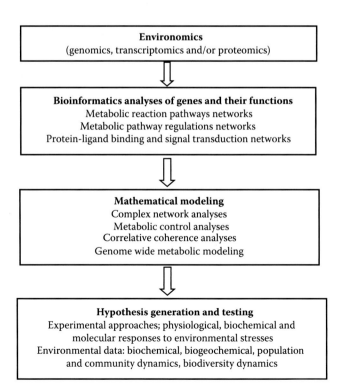

FIGURE 26.6
Mathematical modeling in environmental genomics analyses.

gene calling, the most functional analyses are done at the protein level. Furthermore, the higher-level metagenome descriptors such as sequence composition, species composition, functional composition, and population properties are derived. These descriptors provide first insights in the communities but become even more powerful when compared with environments using comparative metagenomics approaches. Computer-based data analyses generally include computational experiments, mathematical calculations, or scientific analyses of substances [120]. Computer models help predict functions such as (*i*) estimation of a DNA-reactive moiety in a molecule [121], (*ii*) quantitative structure–activity relationship models based on "electro-topological" descriptors [122–125], and (*iii*) three-dimensional computational DNA-docking models to identify molecules capable of noncovalent DNA interaction [126].

These *in silico* prediction systems are cheaper, more rapid, have higher reproducibility, have lower compound synthesis requirements, can undergo constant optimization, and have the potential to reduce or replace the use of animals [120,127]. The major limitations are the lack of available toxicity data, inappropriate modeling of some end points, and poor domain applicability of models [128]. The microbial communities that are major contributors (~50% of biomass) to the biosphere play crucial roles in agriculture. Microbial ecosystems are essential for the growth of plants, animals, and humans and are involved in geochemical cycles, bioremediation, and sustainable energy production. The complete genome sequences [129–131], along with gene predictions [132], have developed the data that is helping to analyze the transcriptome and the proteome of model organisms and humans. The metaomics and its recent *in silico* application in genomics and proteomics include protein–protein, protein–DNA, or other "component–component" interaction mapping (interactome mapping), systematic phenotypic analyses (phenome mapping), and transcript or protein localization mapping. Computational methods can then be used to model biological processes based on integrated data. The resulting models can be tested either by "synthetic biology" (*de novo* design and generation of biological modules based on suspected network properties) or by systematic perturbations, or both. Systems biology strategies can be viewed as a combination of omic approaches, data integration, modeling, and synthetic biology (Figure 26.7) [133].

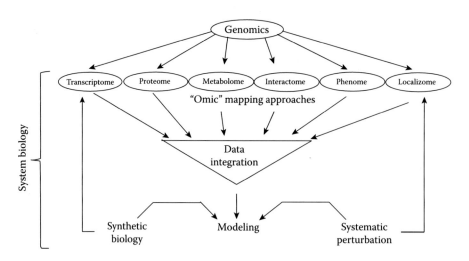

FIGURE 26.7
Integrating omic information.

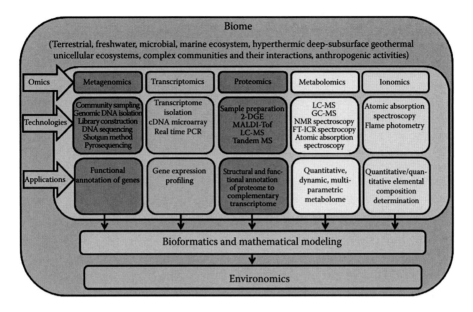

FIGURE 26.8
Omics technologies and their applications. LC-MS, liquid chromatography-mass spectrometry; NMR, nuclear magnetic resonance; FT-ICR, Fourier transform ion cyclotron resonance.

Thus, metaomics, along with recent technological breakthroughs and their applications (Figure 26.8), are helping to investigate and understand these ecosystems. Omics approaches are applied to many biological processes, leading to large lists of genes actively involved in the corresponding modules [131–139], and to clusters of genes or proteins behaving identically under various experimental conditions [140–144].

26.11 Omics for Environmental Engineering and Protection

Environmental engineering is the application of science and engineering principles to improve the natural environment for humans and for other organisms. The major objective of environmental engineering is to address three great present-day challenges: to ensure an adequate and safe food supply; to protect and remediate the world's natural resources, including water, soil, air, biodiversity, and energy; and to provide innovative and effective solutions for environmentally sustainable systems that monitor, replace, or intervene in the mechanisms of living organisms. This involves wastewater management by the use of omic data about appropriate microorganisms that can help in the remediation of air pollution control, recycling, waste disposal, radiation protection, industrial hygiene, environmental sustainability, and public health issues. Omics can help environmental engineers conduct hazardous waste management studies to evaluate the significance of such hazards, provide appropriate advice on treatment and containment, and develop regulations to prevent mishaps and for the protection of ecosystems and the environment.

26.12 Conclusions

Integrated technological advancements in different fields of omics have considerably enriched our knowledge of physiological processes and their diversity. A higher level of integrated metaomic approaches armed with validated high-throughput technologies, appropriate study designs, better sampling and handling methods, and advanced statistical and computational methods for data interpretation have shown great promise in investigating complex microbial communities as a whole and addressing the important aspect of uncultured flora and fauna growing in diverse ecosystems from polar ice, the human intestine, hyperthermic oceans, to polluted industrial sites and will provide a closer functional measurement of the phenotype (i.e., proteomics, metabolomics, and ionomics). Combined with classic methods of physicochemistry and microbiology, these endeavors will provide us with an integrated view of the adaptation of microorganisms to particular ecological niches and participate in the dynamics of ecosystems.

References

1. Lederberg, J., and A.T. McCray. 2001. "Ome sweet omics"—a genealogical treasury of words. *Scientist* 15, 2.
2. Winkler, H. 1920. Verbreitung und Ursache der Parthenogenesis im Pflanzen- und Tierreiche. Jena: Verlag von Gustav Fischer.
3. Gregory, T.R. 2005. Genome size evolution in animals. In *The Evolution of the Genome*, edited by Gregory, T.R., 3–87. San Diego, CA: Elsevier.
4. McKusick, V.A., and F.H. Ruddle. 1987. Toward a complete map of the human genome. *Genomics* 1:103–106.
5. Smith, M.T., R. Vermeulen, G. Li, L. Zhang, Q. Lan, A.E. Hubbard et al. 2005. Use of "omic" technologies to study humans exposed to benzene. *Chemico-Biological Interactions* 153–154, 123–127.
6. Suarez-Kurtz, G., and I. Cascorbi. 2008. São Paulo Research Conference on Molecular Medicine and Pharmacogenetics, A Joint Meeting with the IUPHAR Sub-Committee on Pharmacogenetics and the Brazilian Pharmacogenetics Network, São Paulo, Brazil, September 18–20, 2008. *Current Pharmacogenomics and Personalized Medicine* 6, 234–238.
7. Wilke, R.A., RK. Mareedu, and J.H. Moore. 2008. The pathway less traveled: Moving from candidate genes to candidate pathways in the analysis of genome-wide data from large scale pharmacogenetic association studies. *Current Pharmacogenomics and Personalized Medicine* 6, 150–159.
8. Sachidanandam, R., D. Weissman, S.C. Schmidt, J.M. Kakol, L.D. Stein, G. Marth et al. 2001. A map of human genome sequence variation containing 1.42 million single nucleotide polymorphisms. *Nature* 409(6822):928–933.
9. Venter, J.C., M.D. Adams, E.W. Myers, P.W. Li, R.J. Mural, G.G. Sutton et al. 2001. The sequence of the human genome. *Science* 291(5507):1304–1351.
10. Vlaanderen, J., L.E. Moore, M.T. Smith, Q. Lan, L. Zhang, C.F. Skibola et al. 2010. Application of omics technologies in occupational and environmental health research; Current status and projections. *Occupational and Environmental Medicine* 67(2):136–143, doi:10.1136/oem.2008.042788.
11. Holtorf, H., M.C. Guitton, and R. Reski. 2002. Plant functional genomics. *Naturwissenschaften* 89:235–249, doi:10.1007/s00114-002-0321-3.

12. Likic, V.A., M.J. McConville, T. Lithgow, and A. Bacic. 2010. Systems biology: The next frontier for bioinformatics. *Advances in Bioinformatics* 1–10, doi:10.1155/2010/268925.

13. López-Barea, J., and J.L. Gómez-Ariza. 2006. Environmental proteomics and metallomics. *Proteomics* 6, Suppl. 1:S51.

14. López-Barea, J. 1995. Biomarkers in ecotoxicology: An overview. *Archives of Toxicology. Supplement* 17, 57.

15. López-Barea, J., and C. Pueyo. 1998. Mutagen content and metabolic activation of promutagens by molluscs as biomarkers of marine pollution. *Mutation Research* 399:3–15.

16. Pigliucci, M. 2001. *Phenotypic Plasticity: Beyond Nature and Nurture.* Baltimore, MD: Johns Hopkins University Press.

17. West-Eberhard, M.J. 2003. *Developmental Plasticity and Evolution.* Oxford: Oxford University Press.

18. Pigliucci, M. 2005. Evolution of phenotypic plasticity: Where are we going now? *Trends in Ecology & Evolution* 20:481–486.

19. Schlichting, C.D., and M. Pigliucci. 1998. *Phenotypic Evolution: A Reaction Norm Perspective.* Sunderland, MA: Sinauer Associates, Inc.

20. Landry, C.R., J. Oh, D.L. Hartl, and D. Cavalieri. 2006. Genome-wide scan reveals that genetic variation for transcriptional plasticity in yeast is biased towards multi-copy and dispensable genes. *Genes* 366(2):343–351.

21. Price, T.D., A. Qvarnström, and D.E. Irwin. 2003. The role of phenotypic plasticity in driving genetic evolution. *Proceedings of the Royal Society of London* 70:1433–1440.

22. Yeh, P.J., and T.D. Price. 2004. Adaptive phenotypic plasticity and the successful colonization of a novel environment. *The American Naturalist* 164(5):569–580.

23. Hugenholz, P., B.M. Goebel, and N.R. Pace. 1998. Impact of culture-independent studies on the emerging phylogenetic view of bacterial diversity. *Journal of Bacteriology* 180(18):4765–4774.

24. Deutschbauer, A.M., D. Chivian, and A.P. Arkin. 2006. Genomics for environmental microbiology. *Current Opinion in Biotechnology* 17:229–235.

25. Beja, O., L. Aravind, E.V. Koonin, M.T. Suzuki, A. Hadd, L.P. Nguyen et al. 2000. Bacterial rhodopsin: Evidence for a new type of phototrophy in the sea. *Science* 289(5486):1902–1906.

26. Rondon, M.R., P.R. August, A.D. Bettermann, S.F. Brady, T.H. Grossman, M.R. Liles et al. 2000. Cloning the soil metagenome: A strategy for accessing the genetic and functional diversity of uncultured microorganisms. *Applied and Environmental Microbiology* 66:2541–2547.

27. Venter, J.C., K. Remington, J.F. Heidelberg, A.L. Halpern, D. Rusch, J.A. Eisen et al. 2004. Environmental genome shotgun sequencing of the Sargasso Sea. *Science* 304(5667):66–74.

28. Tyson, G.W., J. Chapman, P. Hugenholtz, E.E. Allen, R.J. Ram, P.M. Richardson et al. 2004. Insights into community structure and metabolism by reconstruction of microbial genomes from the environment. *Nature* 428(6978):37–43. doi:10.1038/nature02340.

29. Tringe, S.G., C. von Mering, A. Kobayashi, A.A. Salamov, K. Chen, H.W. Chang et al. 2005. Comparative metagenomics of microbial communities. *Science* 308:554–557.

30. Strous, M., E. Pelletier, S. Mangenot, T. Rattel, A. Lehner, M.W. Taylor et al. 2006. Deciphering the evolution and metabolism of an anammox bacterium from a community genome. *Nature* 440:790–794.

31. Rusch, D.B., A.L. Halpern, G. Sutton, K.B. Heidelberg, S. Williamson, S. Yooseph et al. 2007. The *Sorcerer II* global ocean sampling expedition: Northwest Atlantic through Eastern Tropical Pacific. *PLoS Biology* 5, e77. doi:10.1371/journal.pbio.0050077.

32. Phelps, T.J., E.M. Murphy, S.M. Pfiffner, and D.C. White. 1994. Comparison between geochemical and biological estimates of subsurface microbial activities. *Microbial Ecology* 28:335–349.

33. Jørgensen, B.B., and S.D. Hondt. 2006. A starving majority deep beneath the seafloor. *Science* 314(5801):932–934.

34. Lin, C.Y., M.R. Viant, and R.S. Tjeerdema. 2006. Metabolomics: Methodologies and applications in the environmental sciences. *Journal of Pesticide Science* 31:245–251. doi:10.1584/jpestics.31.245.

35. Wooley, J.C., A. Godzik, and I. Friedberg. 2010. A primer on metagenomics. *PLoS Computational Biology* 6(2):e1000667. doi:10.1371/journal.pcbi.1000667.
36. Kurokawa, K., I. Takehiko, K. Tomomi, O. Kenshiro, T. Hidehiro, T. Atsushi et al. 2007. Comparative metagenomics revealed commonly enriched gene sets in human gut microbiomes. *DNA Research* 14(4):169–181. doi:10.1093/dnares/dsm018.
37. Simon, C., and R. Daniel. 2010. Metagenomic analyses: Past and future trends. *Applied and Environmental Microbiology* 77(4):1153–1161. doi:10.1128/AEM.02345-10.
38. Lacerda, C.M., and F.R. Kenneth. 2009. Environmental proteomics: Applications of proteome profiling in environmental microbiology and biotechnology. *Briefings in Functional Genomics and Proteomics* 8(1):75–87.
39. Cordwell, S.J., M.R. Wilkins, A. Cerpapoljak, A.A. Gooley, M. Duncan, K.L. Williams et al. 1995. Cross species identification of proteins separated by 2-dimensional gel-electrophoresis using matrix-assisted laser-desorption ionization time-of-flight mass-spectrometry and amino acid-composition. *Electrophoresis* 16:438–443.
40. Cordwell, S.J., D.J. Basseal, and S.I. Humphery. 1997. Proteome analysis of *Spiroplasma melliferum* (A56) and protein characterisation across species boundaries. *Electrophoresis* 18:1335–1346.
41. Shevchenko, A., S. Sunyaev, A. Loboda, A. Shevchenko, P. Bork, W. Ens et al. 2001. Charting the proteomes of organisms with unsequenced genomes by MALDI-quadrupole time of flight mass spectrometry and BLAST homology searching. *Analytical Chemistry* 73:1917–1926.
42. Humphery-Smith, I., M.J. Cordwell, and W.P. Blackstock. 1997. Proteome research: Complementarity and limitations with respect to the RNA and DNA worlds. *Electrophoresis* 18:1217–1242.
43. Dopson, M., C. Baker-Austin, and P.L. Bond. 2004. First use of two-dimensional polyacrylamide gel electrophoresis to determine phylogenetic relationships. *Journal of Microbiological Methods* 58:297–302.
44. Dworzanski, J.P., and A.P. Snyder. 2005. Classification and identification of bacteria using mass spectrometry-based proteomics. *Expert Review of Proteomics* 2:863–878.
45. Lindon, J.C., J.K. Nicholson, and E. Holmes. 2006. *The Handbook of Metabonomics and Metabolomics*. London: Elsevier Science.
46. Lindon, J.C., E. Holmes, and J.K. Nicholson. 2007. Metabonomics in pharmaceutical R&D. *FEBS Journal* 274:140–151.
47. Kell, D.B. 2004. Metabolomics and systems biology: Making sense of the soup. *Current Opinion in Microbiology* 7(3):296–307.
48. Griffiths, W. 2007. *Metabolomics, Metabonomics and Metabolite Profiling*. Cambridge: Royal Society of Chemistry.
49. Harrigan, G.G., and R. Goodacre. 2003. *Metabolic Profiling: Its Role in Biomarker Discovery and Gene Function Analysis*. Boston: Springer.
50. Nicholson, J.K., J.C. Lindon, and E. Holmes. 1999. Metabonomics: Understanding the metabolic responses of living systems to pathophysiological stimuli via multivariate statistical analysis of biological NMR spectroscopic data. *Xenobiotica* 29:1181–1189.
51. Nicholson, J.K., J. Connelly, J.C. Lindon, and E. Holmes. 2002. Metabonomics: A platform for studying drug toxicity and gene function. *Nature Reviews. Drug Discovery* 1:153–161.
52. Flores, A., and E. Hill. 2007. Proceedings of the Metabolomics Society's 3rd Annual International Conference, Manchester, UK.
53. Currie, F., D. Broadhurst, W. Dunn, and R. Goodacre. 2007. Proceedings of the Metabolomics Society's 3rd Annual International Conference, Manchester, UK.
54. Southam, A.D., T.G. Payne, H.J. Cooper, T.N. Arvanitis, and M.R. Viant. 2007. Spectral stitching method increases the dynamic range and mass accuracy of wide-scan direct infusion nanoelectrospray fourier transform ion cyclotron resonance mass spectrometry-based metabolomics. *Analytical Chemistry* 79:4595–4602.
55. Brown, S.C., G. Kruppa, and J.L. Dasseux. 2005. Metabolomics applications of FT-ICR mass spectrometry. *Mass Spectrometry Reviews* 24:223–231.

56. Breitling, R., A.R. Pitt, and M.P. Barrett. 2006. Precision mapping of the metabolome. *Trends in Biotechnology* 24(12):543–548.
57. Breitling, R., D. Vitkup, and M.P. Barrett. 2008. New surveyor tools for charting microbial metabolic maps. *Nature Reviews. Microbiology* 6(2):156–161.
58. Robertson, D.G. 2005. Metabonomics in toxicology: A review. *Toxicological Sciences* 85:809–822.
59. Clayton, T.A., J.C. Lindon, H. Antti, C. Charuel, G. Hanton, J.P. Provost et al. 2006. Pharmaco-metabonomic phenotyping and personalised drug treatment. *Nature* 440(7087):1073–1077.
60. Brown, S.A.E., A.J. Simpson, and M.J. Simpson. 2008. Evaluation of sample preparation methods for nuclear magnetic resonance metabolic profiling studies with *Eisenia fetida*. *Environmental Toxicology and Chemistry* 27:828–836. doi:10.1897/07-412.1.
61. Lin, C.Y., H.F. Wu, R.S. Tjeerdema, and M.R. Viant. 2007. Evaluation of metabolite extraction strategies from tissue samples using NMR metabolomics. *Metabolomics* 3(1):55–67.
62. Viant, M.R. 2007. Metabolomics of aquatic organisms: The new "omics" on the block. *Marine Ecology Progress Series* 332:301–306. doi:10.3354/meps332301.
63. Ogura, R., N. Ikeda, K. Yuki, O. Morita, K. Saigo, C. Blackstock, N. Nishiyama, and T. Kasamatsu. 2008. Genotoxicity studies on green tea catechin. *Food and Chemical Toxicology* 46:2190–2200.
64. Cavalcanti, B.C., JRO, Ferreiraa DJ, Moura RM, Rosa GV, Furtado RR, Burbano et al. 2010. Structure–mutagenicity relationship of kaurenoic acid from *Xylopia sericeae* (Annonaceae). *Mutation Research* 701:153–163.
65. Wang, J., S. Yu, S. Jiao, X. Lv, M. Ma, B.Z. Zhu et al. 2012. Characterization of TCHQ-induced genotoxicity and mutagenesis using the pSP189 shuttle vector in mammalian cells. *Mutation Research* 729(1–2):16–23.
66. Billinton, N., P.W. Hastwell, D. Beerens, L. Birrell, P. Ellis, S. Maskell et al. 2008. Interlaboratory assessment of the GreenScreen HC GADD45a-GFP genotoxicity screening assay: An enabling study for independent validation as an alternative method. *Mutation Research* 653:23–33.
67. Moller, P., and H. Wallin. 1998. Adduct formation, mutagenesis and nucleotide excision repair of DNA damage produced by reactive oxygen species and lipid peroxidation product. *Mutation Research* 410:271–290.
68. Bootsma, D., K.H. Kraemer, J.E. Cleaver, and J.H. Hoeijmakers. 2001. Nucleotide excision repair syndromes: *Xeroderma pigmentosum*, Cockayne syndrome and trichothiodystrophy. In *The Metabolic and Molecular Bases of Inherited Disease*, edited by Scriver, C.R., A.L. Beaudet, W.S. Sly, and D. Valle, 677–703. New York: McGraw-Hill.
69. Johnson, M.K., and G. Loo. 2000. Effects of epigallocatechin gallate and quercetin on oxidative damage to cellular DNA. *Mutation Research* 459:211–218.
70. Kelly, M.R., J. Xu, K.E. Alexander, and G. Loo. 2001. Disparate effects of similar phenolic phytochemicals as inhibitors of oxidative damage to cellular DNA. *Mutation Research* 485:309–318.
71. Azqueta, A., Y. Lorenzo, and A.R. Collins. 2009. In vitro comet assay for DNA repair: A warning concerning application to cultured cells. *Mutagenesis* 24:379–381.
72. Borner, F.U., H. Schutz, and P. Wiedemann. 2011. The fragility of omics risk and benefit perceptions. *Toxicology Letters* 201:249–257.
73. Bishop, W.E., D.P. Clarke, and C.C. Travis. 2001. The genomic revolution: What does it mean for risk assessment? *Risk Analysis* 21:983–987.
74. Aardema, M.J. and J.T. MacGregor. 2002. Toxicology and genetic toxicology in the new era of "toxicogenomics": Impact of "-omics" technologies. *Mutation Research* 499:13–25.
75. Marchant, G.E. 2002. Toxicogenomics and toxic torts. *Trends in Biotechnology* 20:329–332.
76. Heijne, W.H., A.S. Kienhuis, B. van Ommen, R.H. Stierum, and J.P. Groten. 2005. Systems toxicology: Applications of toxicogenomics, transcriptomics, proteomics and metabolomics in toxicology. *Expert Review of Proteomics* 2:767–780.

77. Marques, A., H.M. Lourenco, M.L. Nunes, C. Roseiro, C. Santos, A. Barranco et al. 2011. New tools to assess toxicity, bioaccessibility and uptake of chemical contaminants in meat and sea-food. *Food Research International* 44:510–522.

78. Ouedraogo, M., T. Baudoux, C. Stévigny, J. Nortier, J.M. Colet, T. Efferth et al. 2012. Review of current and "omics" methods for assessing the toxicity (genotoxicity, teratogenicity and nephrotoxicity) of herbal medicines and mushrooms. *Journal of Ethnopharmacology* 140:492–512.

79. Davies, H. 2010. A role for "omics" technologies in food safety assessment. *Food Control* 21:1601–1610.

80. Wilson, V.S., N. Keshava, S. Hester, D. Segal, W. Chiu, C.M. Thompson et al. 2011. Utilizing toxicogenomic data to understand chemical mechanism of action in risk assessment. *Toxicology and Applied Pharmacology* doi:10.1016/j.taap.2011.01.017.

81. Eisenbrand, G., B. Pool-Zobel, V. Baker, M. Balls, B.J. Blaauboer, A. Boobis et al. 2002. Methods of in vitro toxicology. *Food and Chemical Toxicology* 40:193–236.

82. Oberemm, A., L. Onyon, and U. Gundert-Remy. 2005. How can toxicogenomics inform risk assessment? *Toxicology and Applied Pharmacology* 207:592–598.

83. Ulrich-Merzenich, G., H. Zeitler, D. Jobst, D. Panek, H. Vetter, and H. Wagner. 2007 Application of the "-omic-" technologies in phytomedicine. *Phytomedicine* 14:70–82.

84. Velculescu, V.E., L. Zhang, B. Vogelstein, and K.W. Kinzler. 1995. Serial analysis of gene expression. *Science* 270:484–487.

85. Brenner, S., M. Johnson, J. Bridgham, G. Golda, D.H. Lloyd, D. Johnson et al. 2000. Gene expression analysis by massively parallel signature sequencing (MPSS) on microbead arrays. *Nature Biotechnology* 18:630–634.

86. Sutcliffe, J.G., P.E. Foye, M.G. Erlander, B.S. Hilbush, L.J. Bodzin, J.T. Durham et al. 2000. TOGA: An automated parsing technology for analyzing expression of nearly all genes. *Proceedings of the National Academy of Sciences of the United States of America* 97:1976–1981.

87. Kussmann, M., F. Raymond, and M. Affolter. 2006. OMICS-driven biomarker discovery in nutrition and health. *Journal of Biotechnology* 124:758–787.

88. Hastwell, P.W., L.L. Chai, K.J. Roberts, T.W. Webster, J.S. Harvey, R.W. Rees et al. 2006. High-specificity and high-sensitivity genotoxicity assessment in human cell line: Validation of the GreenScreen HC GADD45a-GFP genotoxicity assay. *Mutation Research* 607:160–175.

89. Ohno, K., Y. Tanaka-Azuma, Y. Yoneda, and T. Yamada. 2005. Genotoxicity test system based on p53R2 gene expression in human cells: Examination with 80 chemicals. *Mutation Research* 588:47–57.

90. Hendriks, G., M. Atallah, M. Raamsman, B. Morolli, H. van der Putten, H. Jaadar et al. 2011. Sensitive DsRed fluorescence-based reporter cell systems for genotoxicity and oxidative stress assessment. *Mutation Research* 709–710:49–59.

91. Ellinger-Ziegelbauer, H., J. Aubrecht, J.C. Kleinjans, and H.J. Ahr. 2009. Application of toxicogenomics to study mechanisms of genotoxicity and carcinogenicity. *Toxicology Letters* 186:36–44.

92. Jordan, S.A., D.G. Cunningham, and R.J. Marles. 2010. Assessment of herbal medicinal products: Challenges, and opportunities to increase the knowledge base for safety assessment. *Toxicology and Applied Pharmacology* 243:198–216.

93. Levine, A.J. 1997. p53, the cellular gatekeeper for growth and division. *Cell* 88:323–331.

94. Nakamura, Y. 2004. Isolation of p53-target genes and their functional analysis. *Cancer Science* 95:7–11.

95. Corn, P.G., and W.S. El-Deiry. 2007. Microarray analysis of p53-dependent gene expression in response to hypoxia and DNA damage. *Cancer Biology & Therapy* 6:1858–1866.

96. Lu, X., J. Shao, H. Li, and Y. Yu. 2009. Early whole-genome transcriptional response induced by benzo(a)pyrene diol epoxide in a normal human cell line. *Genomics* 93:332–342.

97. Tanaka, H., H. Arakawa, T. Yamaguchi, K. Shiraishi, K. Fukuda, K. Matsui et al. 2000. A ribonucleotide reductase gene involved in a p53-dependent cell-cycle checkpoint for DNA damage. *Nature* 404:42–49.

98. Guittet, O., P. Hakansson, N. Voevodskaya, S. Fridd, A. Graslund, H. Arakawa et al. 2001. Mammalian p53R2 protein forms an active ribonucleotide reductase in vitro with the R1 protein, which is expressed both in resting cells in response to DNA damage and in proliferating cells. *The Journal of Biological Chemistry* 276:40647–40651.

99. Xue, L., B. Zhou, X. Liu, W. Qiu, Z. Jin, and Y. Yen. 2003. Wild-type p53 regulates human ribonucleotide reductase by protein–protein interaction with p53R2 as well as hRRM2 subunits. *Cancer Research* 63:980–986.

100. Ohno, K., K. Ishihata, Y. Tanaka-Azuma, and T. Yamada. 2008. A genotoxicity test system based on p53R2 gene expression in human cells: Assessment of its reactivity to various classes of genotoxic chemicals. *Mutation Research* 656:27–35.

101. Zhan, Q. 2005. Gadd45a, a p53- and BRCA1-regulated stress protein, in cellular response to DNA damage. *Mutation Research* 569:133–143.

102. Harkin, D.P., J.M. Bean, D. Miklos, Y.H. Song, V.B. Truong, C. Englert et al. 1999. Induction of GADD45 and JNK/SAPK-dependent apoptosis following inducible expression of BRCA1. *Cell* 97:575–586.

103. Barsyte-Lovejoy, D., D.Y. Mao, and L.Z. Penn. 2004. c-Myc represses the proximal promoters of GADD45a and GADD153 by a post-RNA polymerase II recruitment mechanism. *Oncogene* 23:3481–3486.

104. Zheng, X., Y. Zhang, Y.Q. Chen, V. Castranova, X. Shi, and F. Chen. 2005. Inhibition of NFkappaB stabilizes gadd45alpha mRNA. *Biochemical and Biophysical Research Communications* 329:95–99.

105. Adler, S., D. Basketter, S. Creton, O. Pelkonen, J. van Benthem, V. Zuang et al. 2011. Alternative (non-animal) methods for cosmetics testing: Current status and future prospects-2010. *Archives of Toxicology* 85:367–485.

106. Amundson, S.A., M. Bittner, and A.J. Fornace, Jr. 2003. Functional genomics as a window on radiation stress signaling. *Oncogene* 22:5828–5833.

107. Snyder, A.R., and W.F. Morgan. 2004. Gene expression profiling after irradiation: Clues to understanding acute and persistent responses? *Cancer Metastasis Reviews* 23:259–268.

108. Lindon, J.C., E. Holmes, and J.K. Nicholson. 2004. Toxicological applications of magnetic resonance. *Progress in Nuclear Magnetic Resonance Spectroscopy* 45:109–143.

109. van Ravenzwaay, B., G.C.P. Cunha, E. Leibold, R. Looser, W. Mellert, A. Prokoudine et al. 2007. The use of metabolomics for the discovery of new biomarkers of effect. *Toxicology Letters* 172:21–28.

110. Kirkland, D., M, Aardema L. Henderson, and L. Muller. 2005. Evaluation of the ability of a battery of three in vitro genotoxicity tests to discriminate rodent carcinogens and non-carcinogens I. Sensitivity, specificity and relative predictivity. *Mutation Research* 584:1–256.

111. Ankley, G.T., and A.N. Miracle. 2006. *Genomics in Regulatory Ecotoxicology*. SETAC Press.

112. Van Aggelen, G., G.T. Ankley, W.S. Baldwin, D.W. Bearden, W.H. Benson, J.K. Chipman et al. 2010. Integrating omic technologies into aquatic ecological risk assessment and environmental monitoring: hurdles, achievements, and future outlook. *Environmental Health Perspectives* 118:1–5.

113. González-Fernández, M., T. García-Barrera, J. Jurado, M.J. Prieto-Álamo, C. Pueyo, J. López-Barea et al. 2008. Integrated application of transcriptomics, proteomics, and metallomics in environmental studies. *Pure and Applied Chemistry* 80(12):2609–2626.

114. Veldhoen, N., G. Michael, and C.C. Helbing. 2012. Molecular profiling of marine fauna: Integration of omics with environmental assessment of the world's oceans. *Ecotoxicology and Environment Safety* 76(1):23–38.

115. Fuhrman, J.A. 2009. Microbial community structure and its functional implications. *Nature* 459:193–199.

116. Getz, W.M. 2003. Correlative coherence analysis: Variation from intrinsic and extrinsic sources in competing populations. *Theoretical Population Biology* 64:89–99.

117. Feist, A.M., M.J. Herrgard, I. Thiele, J.L. Reed, and B.O. Palsson. 2008. Reconstruction of biochemical networks in microorganisms. *Nature Reviews Microbiology* 7:129–143.

118. Westerhoff, H.V., and B.O. Palsson. 2004. The evolution of molecular biology into systems biology. *Nature Biotechnology* 22:1249–1252.

119. Palsson, B.O. 2006. *Systems Biology, Properties of Reconstructed Networks*, 1st ed. Cambridge, New York: Cambridge University Press.

120. Valerio, Jr., L.G. 2009. In silico toxicology for the pharmaceutical sciences. *Toxicology and Applied Pharmacology* 241:356–370.

121. Greene, N. 2002. Computer systems for the prediction of toxicity: An update. *Advanced Drug Delivery Reviews* 54:417–431.

122. Johnson, D.E., and G.H.I. Wolfgang. 2000. Predicting human safety: Screening and computational approaches. *Drug Discovery Today* 5:445–454.

123. Mattioni, B.E., G.W. Kauffman, and P.C. Jurs. 2003. Predicting the genotoxicity of secondary and aromatic amines using data subsetting to generate a model ensemble. *Journal of Chemical Information and Computer Sciences* 43:949–963.

124. Serra, J.R., E.D. Thompson, and P.C. Jurs. 2003. Development of binary classification of structural chromosome aberrations for a diverse set of organic compounds from molecular structure. *Chemical Research in Toxicology* 16:153–163.

125. Votano, J.R., M. Parham, L.H. Hall, L.B. Kier, S. Oloff, A. Tropsha et al. 2004. Three new consensus QSAR models for the prediction of Ames genotoxicity. *Mutagenesis* 19:365–377.

126. Snyder, R.D., D.E. Ewing, and L.B. Hendry. 2004. Evaluation of DNA intercalation potential of pharmaceuticals and other chemicals by cell-based and three-dimensional computational approaches. *Environmental and Molecular Mutagenesis* 44:163–173.

127. Hofer, T., I. Gerner, U. Gundert-Remy, M. Liebsch, A. Schulte, H. Spielmann et al. 2004. Animal testing and alternative approaches for the human health risk assessment under the proposed new European chemicals regulation. *Archives of Toxicology* 78:549–564.

128. Cronin, M.T.D. 2002. The current status and future applicability of quantitative structure–activity relationships (QSARs) in predicting toxicity. *Alternatives to Laboratory Animals: ATLA* 30:81–84.

129. Blattner, F.R., G. Plunkett, C.A. Bloch, N.T. Perna, V. Burland, M. Riley et al. 1997. The complete genome sequence of *Escherichia coli* K-12. *Science* 277:1453–1474.

130. Goffeau, A., R. Aert, M.L. Agostini-Carbone, A. Ahmed, M. Aigle, L. Alberghina et al. 1997. The yeast genome directory. *Nature* 387(6632 suppl.):5.

131. Adams, M.D., S.E. Celniker, R.A. Holt, C.A. Evans, J.D. Gocayne, P.G. Amanatides et al. 2000. The genome sequence of *Drosophila melanogaster*. *Science* 287:2185–2195.

132. Mathe, C., M.F. Sagot, T. Schie, and P. Rouze. 2002. Current methods of gene prediction, their strengths and weaknesses. *Nucleic Acids Research* 30:4103–4117.

133. Ge, H., A.J.M. Walhout, and M. Vidal. 2003. Integrating "omic" information: A bridge between genomics and systems biology. *Trends in Genetics* 19(10):551–560.

134. Lockhart, D.J., and E.A. Winzeler. 2000. Genomics, gene expression and DNA arrays. *Nature* 405:827–835.

135. Pandey, A., and M. Mann. 2000. Proteomics to study genes and genomes. *Nature* 405:837–846.

136. Walhout, A.J.M., and M. Vidal. 2001. Protein interaction maps for model organisms. *Nature Reviews. Molecular Cell Biology* 2:55–62.

137. Sternberg, P.W. 2001. Working in the post-genomic *C. elegans* world. *Cell* 105:173–176.

138. Hope, I.A. 2001. Broadcast interference-functional genomics. *Trends in Genetics* 17:297–299.

139. Zhu, H., and M. Snyder. 2002. "Omic" approaches for unraveling signaling networks. *Current Opinion in Cell Biology* 14:173–179.

140. Eisen, M.B., P.T. Spellman, P.O. Brown, and D. Botstein. 1998. Cluster analysis and display of genome-wide expression patterns. *Proceedings of the National Academy of Sciences of the United States of America* 95:14863–14868.

141. Tamayo, P., D. Slonim, J. Mesirov, Q. Zhu, S. Kitareewan, E. Dmitrovsky et al. 1999. Interpreting patterns of gene expression with self-organizing maps: Methods and application to hematopoietic differentiation. *Proceedings of the National Academy of Sciences of the United States of America* 96:2907–2912.

142. Brown, M.P., W.N. Grundy, D. Lin, N. Cristianini, C.W. Sugnet, T.S. Furey et al. 2000. Knowledge-based analysis of microarray gene expression data by using support vector. *Proceedings of the National Academy of Sciences of the United States of America* 97:262–267.

143. Boulton, S.J., A. Gartner, J. Reboul, P. Vaglio, N. Dyson, D.E. Hill et al. 2002. Combined functional genomic maps of the *C. elegans* DNA damage response. *Science* 295:127–131.

144. Piano, F., A.J. Schetter, D.G. Morton, K.C. Gunsalus, V. Reinke, S.K. Kim et al. 2002. Gene clustering based on RNAi phenotypes of ovary-enriched genes in *C. elegans*. *Current Biology* 12:1959–1964.

Index

Page numbers followed by *f* and *t* indicate figures and tables, respectively.